Image and Video Processing and Recognition Based on Artificial Intelligence

Image and Video Processing and Recognition Based on Artificial Intelligence

Editors

Kang Ryoung Park
Sangyoun Lee
Euntai Kim

MDPI • Basel • Beijing • Wuhan • Barcelona • Belgrade • Manchester • Tokyo • Cluj • Tianjin

Editors

Kang Ryoung Park
Division of Electronics and
Electrical Engineering
Dongguk University
Seoul
Korea, South

Sangyoun Lee
School of Electrical and
Electronic Engineering
Yonsei University
Seoul
Korea, South

Euntai Kim
School of Electrical and
Electronic Engineering
Yonsei University
Seoul
Korea, South

Editorial Office
MDPI
St. Alban-Anlage 66
4052 Basel, Switzerland

This is a reprint of articles from the Special Issue published online in the open access journal *Sensors* (ISSN 1424-8220) (available at: www.mdpi.com/journal/sensors/special_issues/IVPR_artificial_intelligence).

For citation purposes, cite each article independently as indicated on the article page online and as indicated below:

LastName, A.A.; LastName, B.B.; LastName, C.C. Article Title. *Journal Name* **Year**, *Volume Number*, Page Range.

ISBN 978-3-0365-1592-2 (Hbk)
ISBN 978-3-0365-1591-5 (PDF)

Contents

About the Editors

Kang Ryoung Park

Kang Ryoung Park received his B.S. and M.S. degrees in electronic engineering from Yonsei University, Seoul, South Korea, in 1994 and 1996, respectively. He received his Ph.D. degree in electrical and computer engineering from Yonsei University in 2000. He has been a professor in the division of electronics and electrical engineering at Dongguk University since March 2013. His research interests include image processing and deep learning.

Sangyoun Lee

Sangyoun Lee received the B.S. and M.S. degrees in electrical and electronic engineering from Yonsei University, Seoul, South Korea, in 1987 and 1989, respectively, and the Ph.D. degree in electrical and computer engineering from the Georgia Institute of Technology, Atlanta, GA, USA, in 1999. He is currently a Professor and the Head of Electrical and Electronic Engineering with the Graduate School, and the Head of the Image and Video Pattern Recognition Laboratory, Yonsei University. His research interests include all aspects of computer vision, with a special focus on pattern recognition for face detection and recognition, advanced driver-assistance systems, and video codecs.

Euntai Kim

Euntai Kim received B.S., M.S., and Ph.D. degrees in Electronic Engineering, all from Yonsei University, Seoul, Korea, in 1992, 1994, and 1999, respectively. From 1999 to 2002, he was a Full-Time Lecturer in the Department of Control and Instrumentation Engineering, Hankyong National University, Kyonggi-do, Korea. Since 2002, he has been with the faculty of the School of Electrical and Electronic Engineering, Yonsei University, where he is currently a Professor. He was a Visiting Researcher with the Berkeley Initiative in Soft Computing, University of California, Berkeley, CA, USA, in 2008. He was also a Visiting Researcher with Korea Institute of Science and Technology (KIST), Korea, in 2018. His current research interests include computational intelligence, statistical machine learning and deep learning and their application to intelligent robotics, autonomous vehicles, and robot vision.

Preface to "Image and Video Processing and Recognition Based on Artificial Intelligence"

Recent developments have led to the powerful application of artificial intelligence (AI) techniques to image and video processing and recognition. While the state-of-the-art technology has matured, its performance is still affected by various environmental conditions and heterogeneous databases. The purpose of this Special Issue was to invite high-quality and state-of-the-art academic papers on challenging issues in the field of AI-based image and video processing and recognition. We solicited original papers of unpublished and completed research that were not under review by any other conference, magazine, or journal. Topics of interest included but were not limited to the following:

- AI-based image processing, understanding, recognition, compression, and reconstruction;
- AI-based video processing, understanding, recognition, compression, and reconstruction;
- Computer vision based on AI;
- AI-based biometrics;
- AI-based object detection and tracking;
- Approaches that combine AI techniques and conventional methods for image and video processing and recognition;
- Explainable AI (XAI) for image and video processing and recognition;
- Generative adversarial network (GAN)-based image and video processing and recognition;
- Approaches that combine AI techniques and blockchain methods for image and video processing and recognition.

Kang Ryoung Park, Sangyoun Lee, Euntai Kim
Editors

Article

A Robust Handwritten Numeral Recognition Using Hybrid Orthogonal Polynomials and Moments

Sadiq H. Abdulhussain [1], **Basheera M. Mahmmod** [1], **Marwah Abdulrazzaq Naser** [2], **Muntadher Qasim Alsabah** [3], **Roslizah Ali** [4] and **S. A. R. Al-Haddad** [4,*]

[1] Department of Computer Engineering, University of Baghdad, Al-Jadriya 10071, Iraq; sadiqhabeeb@coeng.uobaghdad.edu.iq (S.H.A.); basheera.m@coeng.uobaghdad.edu.iq (B.M.M.)
[2] Continuous Education Center, University of Baghdad, Baghdad 10001, Iraq; marwah@dcec.uobaghdad.edu.iq
[3] Department of Electronic and Electrical Engineering, University of Sheffield, Sheffield S1 4ET, UK; mqalsabah1@sheffield.ac.uk
[4] Department of Computer and Communication Systems Engineering, Faculty of Engineering, Universiti Putra Malaysia, Serdang 43400, Malaysia; roslizah@upm.edu.my
[*] Correspondence: sar@upm.edu.my

Abstract: Numeral recognition is considered an essential preliminary step for optical character recognition, document understanding, and others. Although several handwritten numeral recognition algorithms have been proposed so far, achieving adequate recognition accuracy and execution time remain challenging to date. In particular, recognition accuracy depends on the features extraction mechanism. As such, a fast and robust numeral recognition method is essential, which meets the desired accuracy by extracting the features efficiently while maintaining fast implementation time. Furthermore, to date most of the existing studies are focused on evaluating their methods based on clean environments, thus limiting understanding of their potential application in more realistic noise environments. Therefore, finding a feasible and accurate handwritten numeral recognition method that is accurate in the more practical noisy environment is crucial. To this end, this paper proposes a new scheme for handwritten numeral recognition using Hybrid orthogonal polynomials. Gradient and smoothed features are extracted using the hybrid orthogonal polynomial. To reduce the complexity of feature extraction, the embedded image kernel technique has been adopted. In addition, support vector machine is used to classify the extracted features for the different numerals. The proposed scheme is evaluated under three different numeral recognition datasets: Roman, Arabic, and Devanagari. We compare the accuracy of the proposed numeral recognition method with the accuracy achieved by the state-of-the-art recognition methods. In addition, we compare the proposed method with the most updated method of a convolutional neural network. The results show that the proposed method achieves almost the highest recognition accuracy in comparison with the existing recognition methods in all the scenarios considered. Importantly, the results demonstrate that the proposed method is robust against the noise distortion and outperforms the convolutional neural network considerably, which signifies the feasibility and the effectiveness of the proposed approach in comparison to the state-of-the-art recognition methods under both clean noise and more realistic noise environments.

Keywords: character recognition; orthogonal polynomials; orthogonal moments; Krawtchouk polynomials; Tchebichef polynomials; support vector machine

Citation: Abdulhussain, S.H.; Mahmmod, B.M.; Naser, M.A.; Alsabah, M.; Ali, R.; Al-Haddad, S.A.R. A Robust Handwritten Numeral Recognition Using Hybrid Orthogonal Polynomials and Moments. *Sensors* **2021**, *21*, 1999. https://doi.org/10.3390/s21061999

Academic Editor: Kang Ryoung Park

Received: 25 December 2020
Accepted: 24 January 2021
Published: 12 March 2021

Publisher's Note: MDPI stays neutral with regard to jurisdictional claims in published maps and institutional affiliations.

1. Introduction

Due to the advancement of computational capabilities of computers, the automatic processing of commercial applications that deals with handwriting has gained substantial attention around the world. In particular, handwriting digit recognition plays a vital role in the area of pattern recognition and computer vision, which are considered as an essential subfield of optical character recognition (OCR) [1]. Both handwritten and printed numerals recognition have gained considerable attention in different practical applications. Specifically, handwritten numeral recognition can be used for several purposes: (a) processing cheques in the bank sector, (b) scanning zip code in the post offices, (c) recognition of vehicle number, (d) processing medical data, and (e) recognition of the street number [1–4]. However, automatic recognition of handwritten numeral is to date considered to be a challenging issue. This is due to the huge variations in handwriting styles, which imply that the same numeral can be written in various ways depending on the font size, font orientation, and the usage of different writing materials. Therefore, to address these challenges, developing an accurate handwritten numeral recognition scheme is crucial. In addition, the trade-off between the process of extracting the most informative features, which have the ability to enhance the classification accuracy, and the complexity reduction is to date one of the most critical issue in this area of research. Furthermore, handwritten numeral recognition is defined as the process of recognizing and classifying numerals from 0 to 9 without any human involvement [5]. While extensive research works have been carried out for the recognition of handwritten numerals and several different techniques have already been proposed, improving the recognition accuracy rates are still required [6].

2. Related Works

In [7], different feature extraction classifiers have been developed using cooperation of four support vector machine (SVM) for Roman handwritten digit recognition. To this end, four different types of feature extractions, namely, projection histograms, ring-zones, contour profiles, and Kirsch features, have been investigated. Chen et al. [8] examined a max-min pseudo-probabilities approach for Roman handwritten numerals recognition. In this study, 256 different dimensions features have been extracted from an input image through the use of principal component analysis (PCA). Note that the research works in [7,8] have been carried out using the National Institute of Standards and Technology (NIST) digital database in [9]. Shi et al. [10] proposed an algorithm using gradient and curvature features. These features are fused to compose a feature vector to classify Roman numerals. The work in [11] examined a feature extraction for handwritten numeral recognition using a sparse coding strategy with a local maximum operation. This method has been evaluated using a modified NIST (MNIST) digital database in [12]. In [13], a handwritten numeral recognition approach, which utilizes multiple feature extraction methods and classifier ensembles, has been proposed. In this proposed scheme, six different features of extraction methods have been evaluated. In [6], a moment-based approach has been proposed for handwritten digit recognition. In particular, the proposed scheme has been evaluated under several scripts: (a) Indo-Arabic, (b) Bangla, (c) Devanagari, (d) Roman, and (e) Telugu. The proposed method in [6] has been examined using the databases from the Center for Microprocessor Applications for Training Education and Research (CMATER) and MNIST.

Besides the research studies that used feature extraction approaches for handwritten digit recognition, another line of research studies has focused on using the neural network (NN) method for numeral recognition. To this end, several NN-based architectures have been proposed for numeral recognition. For example, in [14,15] a trainable feature extraction has been introduced based on convolutional neural network architecture (CNN) with SVM based classifier. The proposed scheme has been applied on the MNIST database. In [16], a new limited receptive area (LiRA) features has been investigated for the handwritten Roman recognition based on MNIST database, which has been developed in [17]. Subsequently, two different NN classifiers have been considered: (a) a modified 3-layer perceptron LiRA and (b) a modular assembly neural network. A new method based on Boltzmann machine (RBM) and CNN deep learning method has been proposed in [18] for Arabic handwritten digit recognition. The proposed method has been applied in the CMATERDB 3.3.1 Arabic handwritten digit datasets. In [19], a CNN algorithm has been proposed for the Arabic numeral recognition, which uses several convolutional layers along with ReLU activation. In [20], an offline handwritten recognition based on DNN has been developed for both digits and letters. The proposed method has been examined using the MNIST and EMNIST databases. Several studies have investigated the enhancement of the digits misrecognitions by adopting the deep learning-based deep belief network (DBN), see, e.g., in [21,22]. However, the DBN methods suffer from a poor recognition accuracy and high running time, which make such methods unfeasible for practical implementation. To overcome these issues, a combined method based on feature extraction and decision making of reinforcement learning has been proposed, see, e.g., in [23,24]. In particular, an adaptive deep Q-learning method has been proposed in [3] aiming to enhance the recognition accuracy and reducing the running time for handwritten digit recognition. In [1], two CNN-based approaches have been investigated for handwritten Arabic digit recognition. In the research, a new development in the size of the Arabic numerals database from 3,000 to 72,000 has also been introduced. Both the CNN models have achieved a recognition accuracy that is close to the state-of-the-art methods for Arabic numerals recognition.

Two shape-based feature descriptors, which are termed as a Point-Light Source-based Shadow (PLSS) and Histogram of Oriented Pixel Positions (HOPP), have been proposed in [25]. The proposed framework has been implemented using ten of the available datasets of handwritten digits, which are written in eight different languages and one numeric string recognition dataset. The nine offline handwritten digits datasets include Bangla, Arabic, Telugu, Nepali, Assamese, two versions of Gurumukhi Latin and two versions of Devanagari, and one online Assamese numeral dataset.

In [26], a hybrid of PCA and modular PCA (MPCA) recognition method with quadtree-based hierarchically derived longest-run (QTLR) features has been proposed for optical character recognition of handwritten digits using SVM classifier. In this study, five popular Indian sub-continent scripts have been evaluated, which include Arabic, Bangla, Devanagari, Latin, and Telugu. The authors of [27] have proposed the 5-layer CNN method with SVM classifier for efficient handwritten recognition. In this work, 50-class Bangla datasets samples have been used in the training. In particular, this method has extracted the features of five different 10-class problems of Indian scripts, which are English, Devanagari, Bangla, Telugu, and Oriya. In [28], a novel 196-element Regional Weighted Run Length (RWRL) feature has been proposed for handwritten Devanagari numerals recognition. The proposed scheme has been evaluated by using SVM classifier. In this work, the authors generated their own samples, which are given by 6000 handwritten Devanagari digit. Recently, an improved handwritten recognition method using a CNN (IHRS-CNN) method has been proposed in [29] for improving the performance of handwritten digit recognition while maintaining the computational complexity. In particular, the IHRS-CNN method makes use of a pure CNN architecture only without the requirement of using any ensemble architecture, which could increase both the cost and computational complexity. The results show that a better performance can be achieved using pure CNN architecture while reducing the computational cost and operational complexity.

Although many deep learning-based classification algorithms have been studied for handwritten digits recognition, the recognition accuracy and the running time remain major issues that need to be addressed. In particular, the current techniques do not provide results that meet the desired accuracy with fast execution time. As such, a careful investigation of a new fast and robust technique is essential to meet the desired accuracy. Furthermore, to date most of the existing research works do not consider the effect of the noise, limiting understanding of their potential application in noise environments. We believe that investigating the robustness of the proposed solution against the noise is crucial to characterize the effectiveness of the features extraction.

To address the aforementioned research challenges, the present paper employs a combination of two orthogonal polynomials—Krawtchouk polynomials (KPs) and Tchebichef polynomials (TPs)—for numeral recognition system. These two well-known orthogonal polynomials are widely used and frequently encountered in the open literature of image representation and signal compression [30,31]. This is due to their powerful capabilities in analyzing the signals' components and retaining the significant features of the signals quickly and efficiently. In this paper, we adopt these powerful polynomials, more accurately their combination to produce a hybrid orthogonal polynomials, which we call squared Krawtchouk–Tchebichef polynomial (SKTP). In addition, to expedite the process of extracting different types of features (gradient and smooth), the embedded kernel technique is employed. The performance of the proposed approach is evaluated and compared under different numeral recognition datasets. In particular, the proposed method is evaluated under the noise-free environment, which we call it as clean noise. In addition, we investigate the effectiveness of the proposed approach for numeral recognition in the presence of noise distortion. To the best of our knowledge, no previous research works have investigated the combination of two hybrid orthogonal polynomials for handwritten numeral recognition. To this end, we provide the mathematical formulation of the utilized orthogonal polynomials and their moments. Furthermore, we compare the proposed character recognition system with the state-of-the-art recognition methods. The results illustrate that the proposed method achieves almost the highest recognition accuracy in comparison with the state-of-the-art recognition methods in all the scenarios considered. Importantly, the results show that the proposed method is robust against the noise distortion and outperforms the CNN remarkably, which signifies the feasibility and the effectiveness of our proposed method in comparison to the state-of-the-art recognition methods under both clean noise and more practical noise environments

The paper is organized as follows. In Section 3, the system models of the orthogonal polynomials and orthogonal moment along with their mathematical formulations are introduced. In Section 4, we explain the methodology of the feature extraction process together with the classification process. In Section 5, numerical results are provided in order to characterize the performance of the proposed method and also to validate the effectiveness of the proposed approach in the presence of noise distortion. Finally, the paper is concluded in Section 6.

Notation: In this paper, an upper boldface symbol stands for a matrix whereas a lower boldface symbol stands for a vector. The operator transpose is denoted by $(\cdot)^{\mathrm{T}}$.

3. Mathematical Models of Orthogonal Polynomials and Moments

This section provides the mathematical analysis of the employed orthogonal polynomials and the computation of their moments for two-dimensional signals.

3.1. Orthogonal Polynomials

The basic principle of the orthogonal polynomials is to project a signal, which could be a speech or image, in the orthogonal polynomials basis. Orthogonal polynomials have the potential to represent the features of signal in an enhanced, efficient, and non-redundant way. Orthogonal polynomials consist of a square matrix with two axes, where these axes represent the signal index (x) and polynomial order (n). The elements of the generated matrix are referred to the orthogonal polynomial coefficients. Motivated by the properties of orthogonal polynomials, which allow a combination of any two orthogonal polynomials matrices to generate orthogonal polynomials matrix, this paper explores Tchebichef polynomials and Krawtchouk polynomials and their moments to produce a hybrid form of orthogonal polynomials. This combination results in a squared matrix with orthogonal polynomials termed as SKTP [32], which has unique features combined from both Tchebichef and Krawtchouk polynomials. The hybrid polynomials show a remarkable performance improvement in terms of energy compaction and localization properties in comparison to any sorts of orthogonal polynomials, i.e., Krawtchouk–Tchebichef polynomials only [33]. Such polynomial combinations could also be efficiently applied in communication signal processing to reduce the complexity of RZF and RZFBF precoders [34,35] or to minimize the feedback overhead [36,37]. To this end, this paper proposes an approach based on SKTP aiming to provide a fast handwritten numeral recognition with high accuracy. Accordingly, the n-th order of the SKTP, $\mathcal{R}_n$, can be written as [32]

$$\mathcal{R}_n(x; p) = \sum_{i=0}^{N-1} \sum_{j=0}^{N-1} \sum_{l=0}^{N-1} \mathcal{K}_j(i; p) \, \mathcal{T}_j(x) \, \mathcal{K}_l(n; p) \, \mathcal{T}_l(i), \tag{1}$$

$$n, x = 0, 1, \cdots, N-1,$$

where parameters $\mathcal{T}$ and $\mathcal{K}$ are the TPs and KPs coefficients, respectively. Furthermore, parameter p is the polynomial degree of Krawtchouk polynomial. The n-th order normalized Krawtchouk polynomial can be written as [38]

$$\mathcal{K}_n(n; p) = \sqrt{\frac{\omega_{\mathcal{K}}(x)}{\rho_{\mathcal{K}}(n)}} \, _2F_1\left(-n, -x; -N+1; \frac{1}{p}\right), \tag{2}$$

where parameters $\omega_{\mathcal{K}}, \rho_{\mathcal{K}}$ represent the weight and norm of the Krawtchouk polynomial, respectively. On the other hand, the n-th order normalized Tchebichef polynomial can be expressed as [39]

$$\mathcal{T}_n(x) = \sqrt{\frac{\omega_{\mathcal{T}}(x)}{\rho_{\mathcal{T}}(n)}} \, (1-N)_n \, _3F_2(-n, -x, 1+n; 1-N; 1), \tag{3}$$

where parameters $\omega_{\mathcal{K}} \, \rho_{\mathcal{K}}$ are defined as the weight and norm functions of the Tchebichef polynomial, respectively. The employed SKTP matrix in (1) can be computed by using matrix–matrix multiplication, which preserves the orthogonality due to the unique feature of the orthogonal polynomials. As such, the employed SKTP matrix $\mathbf{R}$ can be expressed using a matrix multiplication as

$$\mathbf{R} = (\mathbf{R}_k \times \mathbf{R}_t)^2, \tag{4}$$

where matrices $\mathbf{R}_k$ and $\mathbf{R}_t$ denote Krawtchouk polynomial and Tchebichef polynomial, respectively. It is worth noting that the computation of the Tchebichef polynomial and Krawtchouk polynomial is conducted here using a three-term recurrence algorithm. This is due to the fact that the hypergeometric series of these polynomials ($_2F_1$ and $_3F_2$) and the gamma functions are computationally cost, and thus could be infeasible for practical implementation [40]. As such, the three terms recurrence algorithm allows a computational effect solution, which justifies its use here.

3.2. Orthogonal Moments

The orthogonal moments (OMs) have recently received considerable attention as key fundamental tools in different digital processing systems [41]. In particular, OMs can be effectively utilized to reduce the noise distortion effect and improve the features representation. The OMs are computationally efficient and have the capability to reduce the numerical error in comparison to the continuous orthogonal moments (COMs). In addition, OMs are scalar quantities that can discover any small changes or distortions that could be introduced or occurred to the signals [42]. Owing to these beneficial capabilities, OMs have been widely used in various signal processing applications. The basic principle of OMs are to project the signal/image on the polynomial basis functions, which could result in scalar quantities that are used in retaining the significant features of the signal/image [40]. This implies that OMs can be considered as shape descriptors (features). OMs can be divided into two types, which are low-order moments (LOMs) and high-order moments (HOMs). The LOMs preserve most of the signal energy, which, in principle, represents the most effective features/information about the signal. On the other hand, the HOMs contain the other details of the signal [43]. In this paper, the moments are computed using the hybrid SKTP. As such, we call the SKTP with the moments combination as a squared

Krawtchouk–Tchebichef moments (SKTMs). To this end, for any two-dimensional signal, the SKTM ($\mathcal{M}_{nm}$) can be computed as

$$
\begin{aligned}
\mathcal{M}_{nm} &= \sum_{x=0}^{N_1-1} \sum_{y=0}^{N_2-1} \mathcal{R}_n(x; p, N_1) \mathcal{R}_m(y; p, N_2) f(x,y), \\
n &= \frac{N_1}{2} - 1, \frac{N_1}{2}, \ldots, \frac{N_1 - O_n}{2}, \frac{N_1 + O_n}{2} - 1, \\
m &= \frac{N_2}{2} - 1, \frac{N_2}{2}, \ldots, \frac{N_2 - O_m}{2}, \frac{N_2 + O_m}{2} - 1,
\end{aligned}
\tag{5}
$$

where parameters O_n and O_m represent the maximum order of moments used to represent the two-dimensional signal, and $f(x,y)$ represents the two-dimensional signal/image. Specifically, to achieve a fast computation time and efficient implementation of SKTMs, matrix–matrix multiplication is utilized. Accordingly, the resulting matrix of the moments, **M**, with nth and m-th elements, $\mathcal{M}_{nm}$, can be expressed as

$$
\mathbf{M} = \mathbf{R}_1 \times \mathbf{F} \times \mathbf{R}_2^{\mathrm{T}},
\tag{6}
$$

where **F** denotes the matrix form of the image $f(x,y)$, and $\mathbf{R}_1$ and $\mathbf{R}_2$ represent the matrix form of orthogonal polynomials with nth and m-th elements, which are given by $\mathcal{R}_n$ and $\mathcal{R}_m$. It is noteworthy that the basis functions of orthogonal polynomials can be utilized as an approximate solution for differential equations as discussed in [44].

4. The Proposed Methodology for Handwritten Numeral Recognition

In this section, the proposed methodology for handwritten numeral recognition is provided. In particular, to make it simple and easy to clarify, this section is divided into two subsections: the feature extraction process and classification process.

4.1. Feature Extraction Process

The feature extraction process is considered as key fundamental part in the recognition system, which is beneficial for signal representation. As such, to enable an efficiently numeral recognition system, a global feature extraction is utilized instead of the local feature extraction. Although an approach based on SKTMs has shown reasonable performance in signal representation [45], this was based on clean noise environment. In particular, the vast majority of research studies on numeral recognition have focused on clean noise environment, thus preventing the characterization of efficient numeral recognition methods in more practical environments. However, when the numeral characters exhibit a noisy environment, numeral recognition would be severely affected, and thus the accuracy of signal/image recognition will be significantly reduced. This paper addresses the aforementioned challenge by taking into account the effect of the noise, which underpins the main contribution of this work. Specifically, this paper applies smoothed and gradient operator [46] to the input image in order to extract the most effective features. To this end, a smoothed kernel is used in this paper to minimize the effect of noise. In addition, this paper exploits the gradient kernel to compute the gradient of the input image for efficient numeral recognition outcome. It is worth pointing out that applying the aforementioned kernels directly is computationally and cost inefficient. Therefore, to increase the computational time, an orthogonal polynomial embedded image kernel technique [46] is explored here. In order to compute the smoothed kernel moments $\mathbf{Y}_s$, the following formula is used,

$$
\mathbf{Y}_s = \mathbf{S}_y \times \mathbf{F} \times \mathbf{S}_x,
\tag{7}
$$

where parameters $\mathbf{S}_x$ and $\mathbf{S}_y$ are the SKTP embedded with smoothed operator in the x and y directions, respectively. To this end, parameters $\mathbf{S}_x$ and $\mathbf{S}_y$ can be expressed as [46]

$$\mathbf{S}_x = \mathbf{R} \times \mathbf{H}_{sx}, \tag{8}$$

$$\mathbf{S}_y = \mathbf{R} \times \mathbf{H}_{sy}, \tag{9}$$

where matrix $\mathbf{R}$ is given in (4), which represents the combination of Krawtchouk and Tchebichef polynomials, and $\mathbf{H}_{sx}$ and $\mathbf{H}_{sy}$ represent the Toeplitz matrix (An $N \times N$ matrix $\mathbf{A}$ is a Toeplitz matrix if the i, j elements of $\mathbf{A}$, i.e., $[\mathbf{A}]_{i,j}$, satisfies the following $[\mathbf{A}]_{i,j} = \mathbf{A}_{i+1,j+1} = \mathbf{a}_{i-j}$.) matrices of the smoothed operators of vectors $\mathbf{h}_{sx}$ and $\mathbf{h}_{sy}$, respectively. This paper uses the circularly symmetric complex Gaussian distribution with mean $\mathbf{0}$ as a smoothing kernel. Accordingly, the smoothing kernel can be expressed as

$$h_{sx} = \frac{1}{2\pi\sigma_x^2} e^{-\frac{x^2}{2\sigma_x^2}}, \tag{10}$$

$$h_{sy} = \frac{1}{2\pi\sigma_y^2} e^{-\frac{y^2}{2\sigma_y^2}}, \tag{11}$$

where σ^2 represents the standard deviation of the Gaussian distribution. The expression in (8) and (9) indicates that no preprocessing operation is required before extracting features.

To compute the moments of gradient image, the SKTMs of the gradient image, which are given in $\mathbf{\Psi}_x$ and $\mathbf{\Psi}_y$, can be directly computed from the smoothed moments. Accordingly, the moments of gradient image are obtained as

$$\mathbf{\Psi}_x = \mathbf{\Psi}_s \times \mathbf{P}_x, \tag{12}$$

$$\mathbf{\Psi}_y = \mathbf{P}_y \times \mathbf{\Psi}_s, \tag{13}$$

where matrices $\mathbf{P}_x$ and $\mathbf{P}_y$ can be determined as

$$\mathbf{P}_x = \mathbf{S}_x \times \mathbf{G}_x^T \tag{14}$$

$$\mathbf{P}_y = \mathbf{G}_y \times \mathbf{S}_y^T \tag{15}$$

where matrices $\mathbf{G}_x$ and $\mathbf{G}_y$ represent the SKTP embedded with gradient operator in the x and y directions, respectively. Matrices $\mathbf{G}_x$ and $\mathbf{G}_y$ can be computed as [46]

$$\mathbf{G}_x = \mathbf{R} \times \mathbf{H}_{gx}, \tag{16}$$

$$\mathbf{G}_y = \mathbf{R} \times \mathbf{H}_{gy}, \tag{17}$$

where matrices $\mathbf{H}_{gx}$ and $\mathbf{H}_{gy}$ represent the Toeplitz matrices of the smoothed operators $\mathbf{h}_{gx}$ and $\mathbf{h}_{gy}$, respectively. In this paper, a simple gradient operator is exploited as a gradient kernel, which is given as

$$\mathbf{h}_{gx} = [-1 \ 1], \tag{18}$$

$$\mathbf{h}_{gy} = [-1 \ 1]^{\mathrm{T}}. \tag{19}$$

After the moments of the smoothed and gradient are numerically obtained, they can be concatenated in order to form a unique feature vector (FV). The obtained feature vector is then normalized to ensure similar dynamic range [38,47]. The normalization process is used because the values of features lie within wide ranges; thus, the effect of large features values dominate small features values [40].

Figure 1a shows samples of feature vectors without normalization and Figure 1b shows samples of features with normalization using zero mean and standard deviation of 1. Unlike the normalize feature vectors, the feature vectors without normalization show values with a wide range $[-400, 600]$.

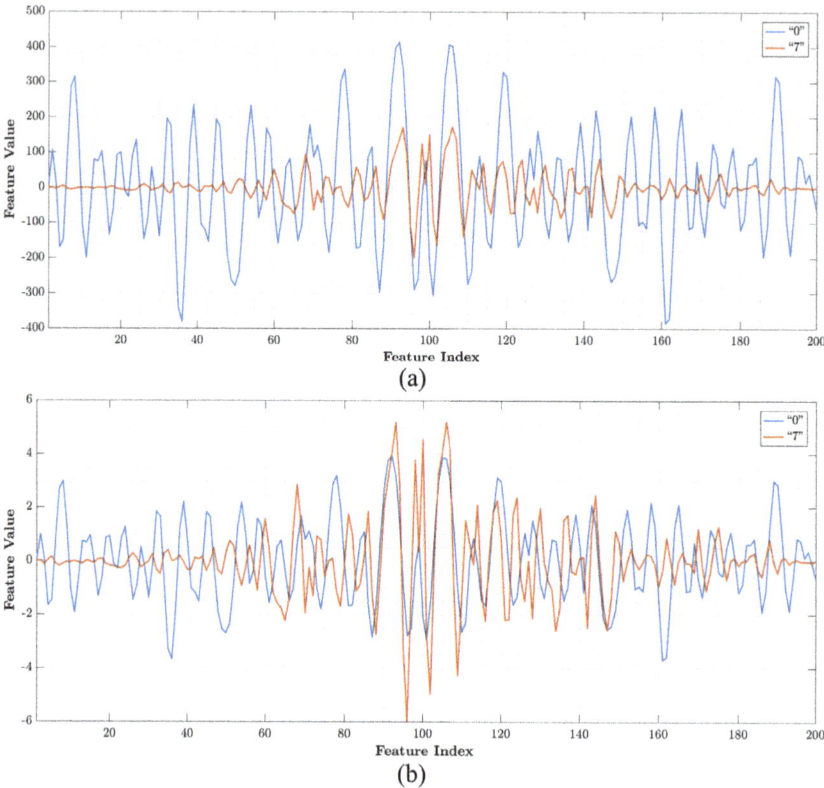

Figure 1. Plot of feature vector samples with indices from 1 to 200 (**a**) without normalization and (**b**) with normalization.

In order to make the proposed methodology of feature extraction more clearly, the block diagram of the feature extraction process is provided in Figure 2.

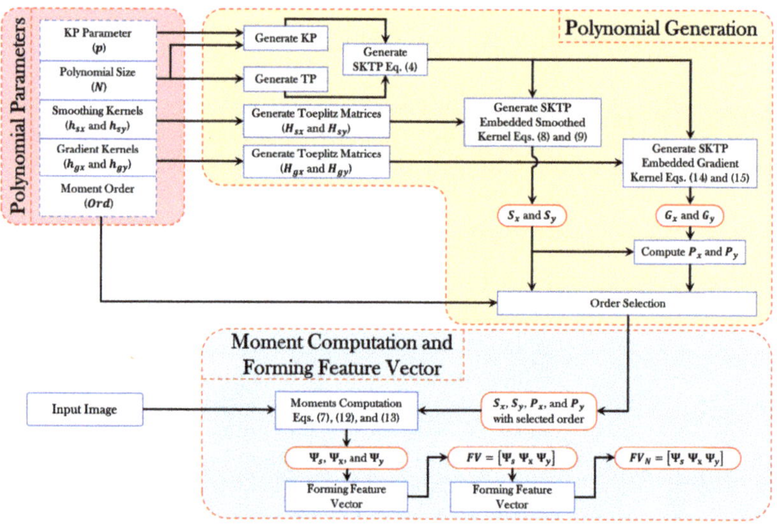

Figure 2. A flow chart of the feature extraction process.

4.2. Classification Process

After obtaining the normalized feature vector, an identified number value (ID) for each input image is obtained based on a classifier. Note that the feature vector is considered as an input to the classifier. A method using SVM is utilized in this paper to accomplish the classification process. The reason behind selecting the SVM method is due to its capability to maximize the margin between separation classes of the hyperplane and data, which is achieved by generating a hyperplane [48–50]. This SVM can also minimize the structural risk by controlling the out-of-sample error [48,49]. In addition, SVM is more adequate for recognition as it is more resistant to the noisy environment [49]. By generating a hyperplane, SVM can separate the positive and negative images [50]. The LIB-SVM with kernel function is used for SVM implementation [47,51].

Figure 3 shows a schematic diagram of the classification process using the SVM method, which includes training and testing process. The SVM kernels utilize polynomial and radial basis function (RBF). These choices are considered as an effective classification mechanism since these kernels show nonlinear separation between classes [47]. To ensure high prediction accuracy, the cross-validation process is carried out. This allows the best kernel parameters to be obtained. Note that the cost and gamma are essential parameters that are required to be tuned. To this end, five-fold cross-validation is applied. The ranges of the parameters for cost and gamma are considered to be $(2^0, 2^1, \ldots, 2^5)$ and $(2^{-10}, 2^{-9}, \ldots, 2^0)$. The cost and gamma parameters show high accuracy on the testing set.

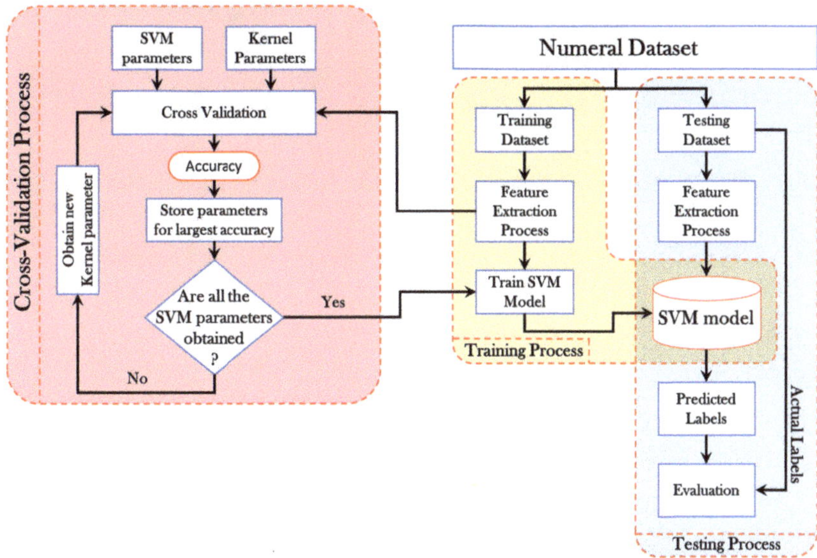

Figure 3. A schematic diagram for support vector machine (SVM) training and testing process.

5. Experimental Results and Discussion

This section presents several simulation results, which characterize the performance of the proposed numeral recognition in different datasets. We first evaluate different choices of kernels in order to select the best SVM kernel. Furthermore, we compare the proposed numeral recognition method with the state-of-the-art recognition methods. The following subsection describes the datasets used in this paper.

5.1. Database Description

In this paper, several benchmarks handwritten numeral datasets are used in the experiments. These datasets include Roman, Devanagari, and Arabic scripts. Among the diverse set of datasets, Roman, which originally came from the Greek alphabet [6], seems to be the most popular handwritten numeral. This is mainly due to the fact that Roman is considered as a second language in the vast majority of countries around the world. Arabic handwritten numerals are widely used in the Middle East countries as well as part of Asia such as in India. On the other hand, Devanagari script is also considered in this paper, which is not only the most popular script used for the Hindi language, but also used by more than 120 different languages around the world [6]. This can justify our focus on these particularly essential handwritten numerals in our experiments. To this end, the MINST numeral dataset provided in [12] is used in this paper. The MINST includes 10 numerals, i.e., 10 classes with 1000 images for each class. The Arabic numeral dataset (CMATERdb 3.3.1) is obtained from the work in [52], wherein over 10,000 images are explored with over 1000 images for each class. Finally, we obtain the dataset of Devanagari (CMATERdb 3.2.1) from the work in [52]. In the Devanagari case, over 20,000 images are used with about 2000 images for each numeral. Figure 4 elaborates different samples for different datasets, which include Roman, Arabic, and Devanagari. The following subsection evaluates the performance of the proposed numeral recognition method based on the above discussed datasets.

Class ID	Value
0	Zero
1	One
2	Two
3	Three
4	Four
5	Five
6	Six
7	Seven
8	Eight
9	Nine

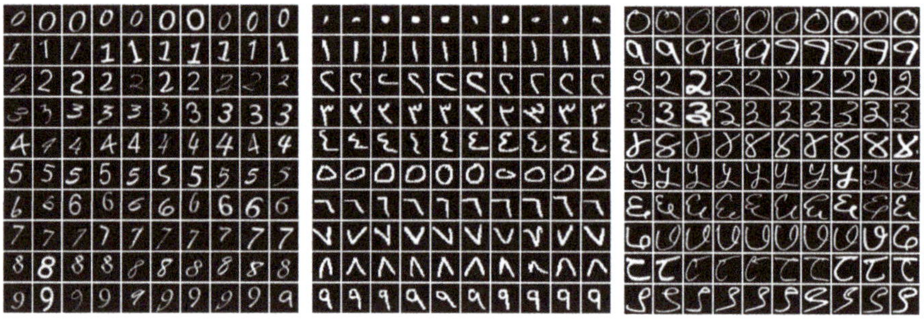

(a)   (b)   (c)

Figure 4. Samples from dataset images (**a**) Roman, (**b**) Arabic, and (**c**) Devanagari.

5.2. Characterizing the Performance of the Proposed Numeral Recognition Using Different Kernels Methods

The main objective of this paper is to identify the best handwritten numeral recognition that achieves the best accuracy. To this end, for a given input image, we identity the best recognition from different types of numerals. The feature extraction process and the training and testing process provided in Figures 2 and 3, respectively, and discussed in detail in Section 4, are exploited here for numeral recognition. In particular, for each dataset (e.g., Roman, Arabic, and Devanagari), the samples are divided equally into two parts, i.e., one for the training phase and the other part for testing phase (see Table 1 for details). For each phase, the feature extraction process is shown in Figure 2, where this process is utilized to represent each image by its related features.

Table 1. Size of the sample for each dataset, and the size of the training and testing sets.

Dataset	Sample Size	The Size of the Training Set	The Size of the Testing Set
Roman	10,000	5000 (50%)	5000 (50%)
Arabic	10,000	5000 (50%)	5000 (50%)
Devanagari	20,000	10,000 (50%)	10,000 (50%)

The proposed numeral recognition method is implemented based on two SVM kernels, which are defined as the polynomial and radial basis function (RBF) kernels. For the radial basis function kernel, the tuned parameters are given by the cost (C) and gamma (γ). On the other hand, for the polynomial kernel, the parameters are given by the cost (C), gamma (γ), and the kernel coefficient (*coef*0). More details about the polynomial as well as radial basis function kernels can be found in [51]. The obtained kernels coefficients are shown in Table 2.

Table 2. Comparing the polynomial and RBF kernels coefficients based on the SVM method for Roman, Arabic, and Devanagari datasets.

Dataset	Kernel	C	γ	*coef*0	Degree	Accuracy
Roman	RBF	2^6	$2^{-9.6}$	-	-	99.80
	Polynomial	2^6	$2^{-7.2}$	0	4	100
Arabic	RBF	2^6	$2^{-9.6}$	-	-	98.94
	Polynomial	2^6	$2^{-9.6}$	-0.15	4	99.32
Devanagari	RBF	2^6	$2^{-9.6}$	-	-	99.12
	Polynomial	2^6	$2^{-9.6}$	-0.16	4	99.28

Figure 2 depicts that the polynomial kernel using the SVM method achieves better overall preference in terms of accuracy than the RBF kernel. As such, the polynomial kernel can be considered as a more robust recognition technique in comparison to the RBF kernel. For more clarification about the differences between polynomial and RBF kernels, we have provided Figure 5. In particular, Figure 5 shows the accuracy comparison between polynomial and RBF based on Roman, Arabic, and Devanagari datasets considered. From Figure 5a, we observe that for classes 3, 5, 7, and 8, the RBF kernel is less accurate than the polynomial kernel when the Roman dataset is used. For the Arabic dataset in Figure 5b, however, the class accuracy for polynomial kernel is slightly degraded in comparison to Figure 5a. However, the results in Figure 5b illustrate that the RBF kernel is considerably degraded when Arabic dataset is used, especially for classes 3, 4, and 9. Overall, the results show that the polynomial kernel is more accurate than the RBF kernel for almost all the classes considered when Arabic dataset is used. On the other hand, Figure 5c investigates the accuracy of polynomial and RBF kernel based on the Devanagari dataset. The results show that for Devanagari dataset, again, the RBF kernel is less robust in recognition than the polynomial kernel where the accuracy has been considerably dropped, especially when we considered classes 2, 5, 6, and 9. For more elucidation, the confusion matrices polynomial and RBF kernels are evaluated based on all the datasets. To achieve this purpose, Figure 6 conducts a comparison of the accuracy between the polynomial and RBF kernels.

Specifically, Figure 6a–c provides a confusion matrix of the numeral recognition using polynomial kernel based on Roman, Arabica, and Devanagari, respectively. Figure 6d–f illustrates the results of the confusion matrix of the numeral recognition using RBF kernel based on Roman, Arabica, and Devanagari, respectively. Figure 6 demonstrates clearly that, based on the confusion matrix of the numeral recognition, the polynomial kernel outperforms RBF kernel in all datasets considered. Based on the experiments above, and due to the accuracy that polynomial kernel provides over a wide range of comparisons, we consider the polynomial kernel with SVM method for feature extraction in the rest of the paper. In particular, polynomial kernel is used here with SVM in the classification process for numeral recognition system and through the comparison with the state-of-the-art methods.

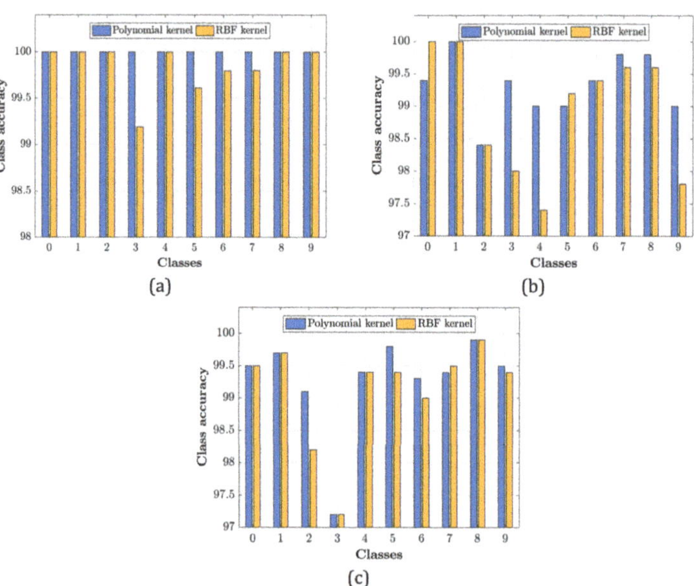

Figure 5. Comparison of classes accuracy using polynomial and RBF kernels for (**a**) Roman, (**b**) Arabic, and (**c**) Devanagari.

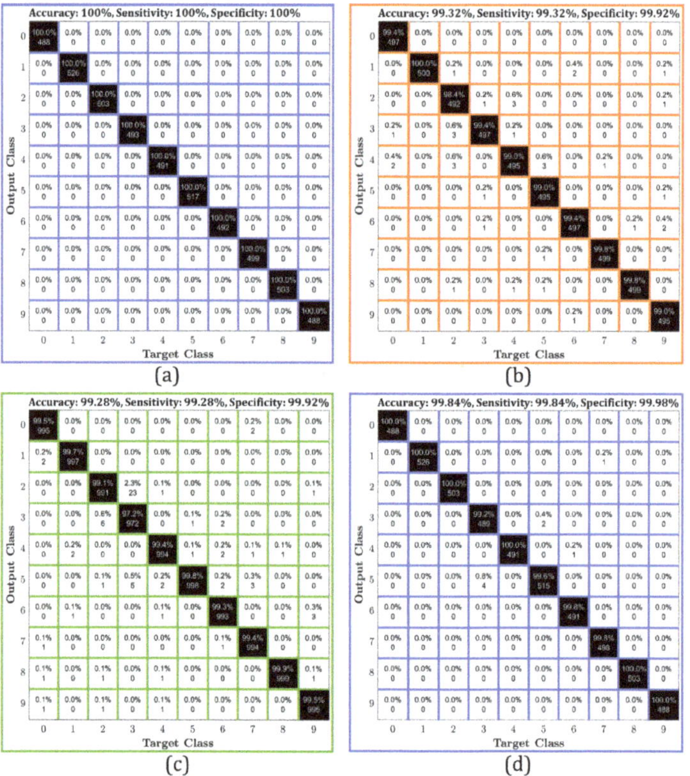

Figure 6. *Cont.*

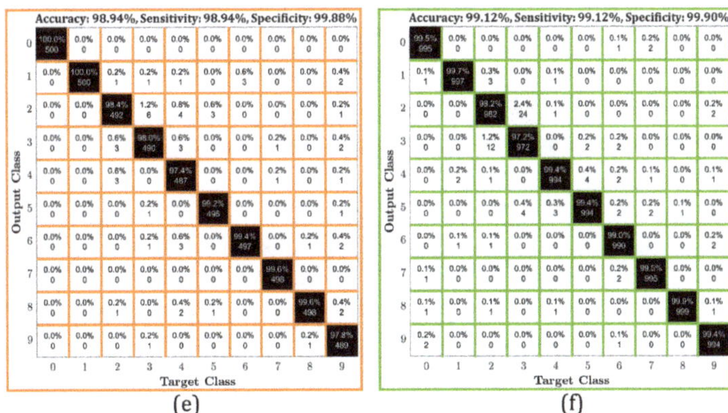

Figure 6. Confusion matrix of the numeral recognition system using (**a**) polynomial kernel for Roman dataset, (**b**) polynomial kernel for Arabic dataset, (**c**) polynomial kernel for Devanagari dataset, (**d**) RBF kernel for Roman dataset, (**e**) RBF kernel for Arabic dataset, and (**f**) RBF kernel for Devanagari dataset.

5.3. Comparing Performances of the Proposed Method and the Existing State-of-the-Art Methods for Numeral Recognition

To evaluate the performance of the proposed algorithm, different comparisons with existing recognition methods are carried out in this subsection. Unless otherwise specified, the comparisons are conducted based on clean noise environment, i.e., without introducing the noise. Our proposed method use SVM as a classifier with polynomial kernel. Table 3 provides a comparison based on the Roman numeral recognition. Different recognition methods, which use different classifier types, are compared with each other and with our proposed method. To this end, a handwritten recognition based on moments fusion (MF) [6] is used. MF recognition method uses two different classifier types, which are defined as a multilayer perceptron (MLP) and SVM. Conventional neural network CNN, which uses 5-layers CNN, has been proposed in [27]. We consider this 5-layers CNN method in the comparison and we term it CNN-5 for brevity. We also compare our method with the statistical–topological feature combination (STFC) proposed in [26]. The fourth recognition method that we consider in the comparison is IHRS-CNN. As such, the classifier type for the CNN-5 and IHRS-CNN methods is CNN.

The results in Table 3 show that the STFC and MF methods that use SVM classifier achieve the lowest accuracy recognition levels, which are below 99%. On the other hand, CNN-5-based CNN classifier achieves 99.10% recognition accuracy, while the MF-based MLP classifier and IHRS-based CNN classifier achieve the same recognition accuracy of 99.77%. Interestingly, the proposed method achieves the highest accuracy level, i.e., 100%, among all the methods considered. This signifies the feasibly of our proposed method in comparison with other methods used for handwritten Roman numerals recognition. Particularly, the proposed method provides an accuracy improvement of 0.23% in comparison with the MF-based MLP and IHRS methods, and 1.25% improvement in comparison with MF-based SVM.

Having demonstrated the feasibility of the proposed method with the Roman numeral recognition, it is pertinent to compare the proposed algorithm performance recognition with that obtained from the conventional recognition methods based on Arabic numeral recognition. To this end, the proposed algorithm is compared with several different algorithms. For example, shape features (SF) [25] using random forest and MLP for classification are considered in the comparison based on Arabic numeral recognition. In addition, convolutional neural network (CNN), which is proposed in [19] for Arabic handwritten numerals, is also used here in the comparison. The rest of the methods that are used in the comparison are similar to those already considered with the Roman numeral recognition, which are MF-based MLP and SVM classifiers [6] and STFC with SVM classifier [26]. Table 4 provides an accuracy comparison of the aforementioned methods with our proposed method. The results show that CNN [19] method based on CNN classifier achieves the lowest accuracy recognition level, which is below 97.40%. The results demonstrate that the proposed numeral recognition algorithm outperform the existing algorithms for handwritten Arabic numeral recognition, which achieves 99.32% recognition accuracy.

Table 3. Comparison between the proposed method and the state-of-the-art methods based on Roman numeral recognition in clean environment.

Method	Classifier Type	Dataset	Accuracy %
MF [6]	MLP	MNIST	99.77
MF [6]	SVM	MNIST	98.75
STFC [26]	SVM	MNIST	98.90
CNN-5 [27]	CNN	MNIST	99.10
IHRS-CNN [29]	CNN	MNIST	99.77
Proposed	SVM	MNIST	100.00

Table 4. Comparison between the proposed method and the state-of-the-art methods for Arabic numeral recognition in clean environment.

Method	Classifier Type	Dataset	Accuracy %
SF [25]	random forest	CMATERDB 3.3.1	98.40
SF [25]	MLP	CMATERDB 3.3.1	98.20
MF [6]	MLP	CMATERDB 3.3.1	98.92
MF [6]	SVM	CMATERDB 3.3.1	97.95
STFC [26]	SVM	CMATERDB 3.3.1	98.40
IHRS-CNN [29]	CNN	CMATERDB 3.2.1	98.42
CNN [19]	CNN	CMATERDB 3.3.1	97.40
Proposed	SVM	CMATERDB 3.3.1	99.32

So far, the results presented are based on the Roman and Arabic dataset, in what follows we compare the proposed method with the state-of-the-art methods based on Devanagari handwritten numeral. To this end, a regional weighted run length feature (RWRLF) [28] algorithm, wherein the SVM and MLP classifier are employed for classification, is used here for comparison. The rest of the methods that are compared here are similar to those already used with the Roman and Arabic numeral recognition, which are MF-based MLP and SVM classifiers [6], SF [25] with random forest and MLP classifier, and STFC with SVM classifier [26]. Table 5 presents the accuracy comparison of different recognition with their corresponding classifiers. The results show that the MF based on MLP classifier method achieves the best accuracy performance of 99.30%. However, the proposed algorithm still achieves a comparable results to that obtained based on the MF method with only 0.02% accuracy difference that is considered negligible. Nevertheless, our proposed method outperforms the MF method in both Roman and Arabic handwritten numeral recognition. Furthermore, the proposed method considerably outperforms all the rest of the methods considered in the comparison for the Devanagari handwritten numeral. This again signifies the feasibility of our method in comparison to the other methods considered.

Table 5. Comparison between the proposed method and the state-of-the-art methods for Devanagari numeral recognition in clean environment.

Method	Classifier Type	Dataset	Accuracy %
SF [25]	Random Forest	CMATERDB 3.2.1	98.01
SF [25]	MLP	CMATERDB 3.2.1	97.40
MF [6]	MLP	CMATERDB 3.2.1	99.30
MF [6]	SVM	CMATERDB 3.2.1	97.98
RWRLF [28]	SVM	CMATERDB 3.2.1	95.03
RWRLF [28]	MLP	CMATERDB 3.2.1	94.47
STFC [26]	SVM	CMATERDB 3.2.1	98.70
IHRS-CNN [29]	CNN	CMATERDB 3.2.1	97.60
Proposed	SVM	CMATERDB 3.2.1	99.28

From the previous comparison, the proposed method seems to be almost outperforming all the existing algorithms based on the clean environment. To illustrate the effectiveness and the robustness of our proposed algorithm, different noisy environments are also considered. We select the updated state-of-the-art algorithm, which is IHRS-CNN [29]. The implementation of IHRS-CNN algorithm is also straightforward. This can justify its use in the comparison with different noisy environments. The noisy environments are chosen to be Gaussian noise with two different variance values, i.e., ($\sigma^2 = 0.01$ and $\sigma^2 = 0.05$). Moreover, the Salt and Pepper noise with two different density values, i.e., $d = 0.01$ and $d = 0.05$ are also considered. Finally, we use the blur noise in the comparison, which is implemented using an averaging filter with size of 3×3. In addition, as in previously presented results, the comparison is conducted based on the Roman, Arabic, and Devanagari numerals. The results of the comparison are provided in Table 6.

Table 6. Comparison between the proposed method and IHRS-CNN [29] method for Roman, Arabic, and Devanagari numeral recognition in both clean and noisy environments.

Roman			Arabic			Devanagari		
Environment	Proposed	IHRS-CNN [29]	Environment	Proposed	IHRS-CNN [29]	Environment	Proposed	IHRS-CNN [29]
Clean	100.00	99.77	Clean	99.32	98.42	Clean	99.23	97.6
Gaussian noise $\sigma^2 = 0.01$	100.00	97.08	Gaussian noise $\sigma^2 = 0.01$	99.12	97.78	Gaussian noise $\sigma^2 = 0.01$	99.17	97.32
Gaussian noise $\sigma^2 = 0.05$	96.68	90.00	Gaussian noise $\sigma^2 = 0.05$	99.00	94.76	Gaussian noise $\sigma^2 = 0.05$	98.66	95.69
Salt & Pepper noise $(d = 0.01)$	100.00	93.68	Salt & Pepper noise $(d = 0.01)$	99.08	97.68	Salt & Pepper noise $(d = 0.01)$	99.22	97.4
Salt & Pepper noise $(d = 0.05)$	99.90	90.76	Salt & Pepper noise $(d = 0.05)$	99.12	93.84	Salt & Pepper noise $(d = 0.05)$	99.14	96.69
Blur (filter size = 3×3)	100.00	96.88	Blur (filter size = 3×3)	99.02	98.08	Blur (filter size = 3×3)	99.23	97.27

As illustrated in Table 6, for the Roman numerals, the proposed algorithm shows a stable recognition accuracy in clean and noisy environments. However, there is a slight degradation in the recognition accuracy when the standard deviation of the Gaussian noise is increased, i.e., $\sigma^2 = 0.05$. On the other hand, the recognition accuracy performance of the IHRS-CNN method degrades in all the noisy environments considered. For example, the IHRS-CNN shows 2.7% and 9.8% for the Gaussian noise with $\sigma^2 = 0.01$ and $\sigma^2 = 0.05$, respectively. For the Arabic numerals, the proposed method achieves almost the same accuracy of about 99% in all the noisy environments considered. Clearly, the results show that the proposed method again is more accurate than the state-of-the-art method in all cases. In particular, the proposed method achieves a maximum gain in the accuracy of 5.28% in the Salt and Pepper noisy environment with density $d = 0.05$. For the Devanagari numeral recognition, the proposed method again considerably outperforms the IHRS-CNN method. Here, the proposed method achieves a maximum gain of 2.97 over the IHRS-CNN method in the Gaussian noise with standard deviation of $\sigma^2 = 0.05$.

In terms of execution time, the proposed method with Roman and Arabic numerals are executed with 24.2 and 15.1 s, respectively, while the CNN spends 26.7 and 17.9 s, respectively, to get executed. Note that the time is measured for the entire test set for each type of numerals. This clearly implies that the proposed method has nearly comparable results with the existing methods in terms of execution time. Overall, the results presented show that the proposed algorithm is still robust in the noisy environments in comparison to the state-of-the-art IHRS-CNN method.

6. Conclusions

This paper presented a new codesign of orthogonal polynomials with their associated moments to improve handwritten numeral recognition accuracy. The proposed algorithm have been evaluated for three different numeral recognitions: Roman, Arabic, and Devanagari. The results demonstrated that the proposed approach achieves the highest recognition accuracy in comparison to the state-of-the-art numeral recognition methods. Importantly, the numerical results showed that the proposed approach is robust against the noise distortion, which signifies the effectiveness of the proposed approach under realistic environments.

Author Contributions: Conceptualization, S.H.A. and B.M.M.; methodology, S.H.A., B.M.M. and M.A.N.; software, S.H.A.; validation, S.H.A. and B.M.M.; investigation, M.A.N., B.M.M., S.H.A. and M.Q.A.; resources, S.A.R.A.-H. and R.A.; data curation, M.Q.A. and S.A.R.A.-H.; writing–original draft preparation, B.M.M., M.Q.A. and M.A.N.; writing–review and editing, M.A.N., M.Q.A. and R.A.; visualization, S.H.A., B.M.M. and R.A.; project administration, S.A.R.A.-H.; funding acquisition, S.A.R.A.-H. and R.A. All authors have read and agreed to the published version of the manuscript.

Funding: This research was funded by Research Management Center (RMC), Universiti Putra Malaysia (UPM), UPM Journal Publication Fund (9001103).

Institutional Review Board Statement: Not applicable.

Informed Consent Statement: Not applicable.

Data Availability Statement: The data presented in this study are available on request from the corresponding author.

Conflicts of Interest: The authors declare no conflicts of interest.

Abbreviations

The following abbreviations are used in this manuscript:

OCR	optical character recognition
SVM	support vector machine
NIST	national institute of standards and technology
MNIST	modified NIST
CMATER	Center for Microprocessor Applications for Training Education and Research
CNN	convolutional neural network
HOPP	Histogram of Oriented Pixel Positions
PLSS	Point-Light Source-based Shadow
KP	Krawtchouk polynomials
TP	Tchebichef polynomials
SKTP	squared Krawtchouk–Tchebichef polynomial
COM	continuous orthogonal moment
OM	orthogonal moment
LOM	low-order moments
HOM	high-order moments
SKTM	squared Krawtchouk–Tchebichef moment
FV	feature vector

References

1. Ahamed, P.; Kundu, S.; Khan, T.; Bhateja, V.; Sarkar, R.; Mollah, A.F. Handwritten Arabic numerals recognition using convolutional neural network. *J. Ambient Intell. Humaniz. Comput.* **2020**, *11*, 5445–5457. [CrossRef]
2. Tuba, E.; Tuba, M.; Simian, D. *Handwritten Digit Recognition by Support Vector Machine Optimized by Bat Algorithm*; Václav Skala-UNION Agency: Plzen, Czech Republic, 2016.
3. Qiao, J.; Wang, G.; Li, W.; Chen, M. An adaptive deep Q-learning strategy for handwritten digit recognition. *Neural Netw.* **2018**, *107*, 61–71. [CrossRef]
4. Aradhya, V.M.; Kumar, G.H.; Noushath, S. Robust Unconstrained Handwritten Digit Recognition using Radon Transform. In Proceedings of the 2007 International Conference on Signal Processing, Communications and Networking, Chennai, India, 22–24 February 2007; pp. 626–629. [CrossRef]
5. Bag, S.; Harit, G. A survey on optical character recognition for Bangla and Devanagari scripts. *Sadhana* **2013**, *38*, 133–168. [CrossRef]
6. Singh, P.K.; Sarkar, R.; Nasipuri, M. A study of moment based features on handwritten digit recognition. *Appl. Comput. Intell. Soft Comput.* **2016**, *2016*, 2796863. [CrossRef]
7. Gorgevik, D.; Cakmakov, D. Handwritten digit recognition by combining SVM classifiers. In Proceedings of the EUROCON 2005—The International Conference on "Computer as a Tool", Belgrade, Serbia, 21–24 November 2005; Volume 2, pp. 1393–1396.
8. Chen, X.; Liu, X.; Jia, Y. Learning handwritten digit recognition by the max-min posterior pseudo-probabilities method. In Proceedings of the Ninth International Conference on Document Analysis and Recognition (ICDAR 2007), Parana, Brazil, 23–26 September 2007; Volume 1, pp. 342–346.
9. Garris, M.D.; Blue, J.L.; Candela, G.T.; Grother, P.J.; Janet, S.; Wilson, C.L. *NIST Form-Based Handprint Recognition System*; US Department of Commerce, Technology Administration, National Institute of Standards and Technology: Gaithersburg, MD, USA, 1997.
10. Shi, M.; Fujisawa, Y.; Wakabayashi, T.; Kimura, F. Handwritten numeral recognition using gradient and curvature of gray scale image. *Pattern Recognit.* **2002**, *35*, 2051–2059. [CrossRef]
11. Labusch, K.; Barth, E.; Martinetz, T. Simple method for high-performance digit recognition based on sparse coding. *IEEE Trans. Neural Netw.* **2008**, *19*, 1985–1989. [CrossRef] [PubMed]
12. LeCun, Y.; Bottou, L.; Bengio, Y.; Haffner, P. Gradient-based learning applied to document recognition. *Proc. IEEE* **1998**, *86*, 2278–2324. [CrossRef]
13. Cruz, R.M.; Cavalcanti, G.D.; Ren, T.I. Handwritten digit recognition using multiple feature extraction techniques and classifier ensemble. In Proceedings of the 17th International Conference on Systems, Signals and Image Processing, Rio de Janeiro, Brazil, 17–19 June 2010; pp. 215–218.
14. Lauer, F.; Suen, C.Y.; Bloch, G. A trainable feature extractor for handwritten digit recognition. *Pattern Recognit.* **2007**, *40*, 1816–1824. [CrossRef]
15. Niu, X.X.; Suen, C.Y. A novel hybrid CNN–SVM classifier for recognizing handwritten digits. *Pattern Recognit.* **2012**, *45*, 1318–1325. [CrossRef]
16. Goltsev, A.; Gritsenko, V. Investigation of efficient features for image recognition by neural networks. *Neural Netw.* **2012**, *28*, 15–23. [CrossRef]
17. LeCun, Y. The MNIST Database of Handwritten Digits. 1998. Available online: http://yann.lecun.com/exdb/mnist/ (accessed on 24 January 2021).
18. Alani, A.A. Arabic handwritten digit recognition based on restricted Boltzmann machine and convolutional neural networks. *Information* **2017**, *8*, 142. [CrossRef]
19. Ashiquzzaman, A.; Tushar, A.K. Handwritten Arabic numeral recognition using deep learning neural networks. In Proceedings of the 2017 IEEE International Conference on Imaging, Vision & Pattern Recognition (icIVPR), Dhaka, Bangladesh, 13–14 February 2017; pp. 1–4.
20. Gunawan, T.S.; Noor, A.F.R.M.; Kartiwi, M. Development of english handwritten recognition using deep neural network. *Indones. J. Electr. Eng. Comput.* **2018**, *10*, 562–568. [CrossRef]
21. Hinton, G.E.; Osindero, S.; Teh, Y.W. A fast learning algorithm for deep belief nets. *Neural Comput.* **2006**, *18*, 1527–1554. [CrossRef] [PubMed]
22. Papa, J.P.; Scheirer, W.; Cox, D.D. Fine-tuning deep belief networks using harmony search. *Appl. Soft Comput.* **2016**, *46*, 875–885. [CrossRef]
23. Mnih, V.; Kavukcuoglu, K.; Silver, D.; Rusu, A.A.; Veness, J.; Bellemare, M.G.; Graves, A.; Riedmiller, M.; Fidjeland, A.K.; Ostrovski, G.; et al. Human-level control through deep reinforcement learning. *Nature* **2015**, *518*, 529–533. [CrossRef]
24. Deng, Y.; Bao, F.; Kong, Y.; Ren, Z.; Dai, Q. Deep direct reinforcement learning for financial signal representation and trading. *IEEE Trans. Neural Netw. Learn. Syst.* **2016**, *28*, 653–664. [CrossRef]
25. Ghosh, S.; Chatterjee, A.; Singh, P.K.; Bhowmik, S.; Sarkar, R. Language-invariant novel feature descriptors for handwritten numeral recognition. *Vis. Comput.* **2020**. [CrossRef]
26. Das, N.; Reddy, J.M.; Sarkar, R.; Basu, S.; Kundu, M.; Nasipuri, M.; Basu, D.K. A statistical–topological feature combination for recognition of handwritten numerals. *Appl. Soft Comput.* **2012**, *12*, 2486–2495. [CrossRef]

27. Maitra, D.S.; Bhattacharya, U.; Parui, S.K. CNN based common approach to handwritten character recognition of multiple scripts. In Proceedings of the 2015 13th International Conference on Document Analysis and Recognition (ICDAR), Tunis, Tunisia, 23–26 August 2015; pp. 1021–1025. [CrossRef]

28. Singh, P.K.; Das, S.; Sarkar, R.; Nasipuri, M. Recognition of offline handwriten Devanagari numerals using regional weighted run length features. In Proceedings of the 2016 International Conference on Computer, Electrical & Communication Engineering (ICCECE), Kolkata, India, 16–17 December 2016; Volume 110, pp. 1–6. [CrossRef]

29. Ahlawat, S.; Choudhary, A.; Nayyar, A.; Singh, S.; Yoon, B. Improved Handwritten Digit Recognition Using Convolutional Neural Networks (CNN). *Sensors* **2020**, *20*, 3344. [CrossRef]

30. Radeaf, H.S.; Mahmmod, B.M.; Abdulhussain, S.H.; Al-Jumaeily, D. A steganography based on orthogonal moments. In Proceedings of the International Conference on Information and Communication Technology—ICICT'19, Baghdad, Iraq, 15–16 April 2019; ACM Press: New York, NY, USA, 2019; pp. 147–153. [CrossRef]

31. Mahmmod, B.M.; Ramli, A.R.; Baker, T.; Al-Obeidat, F.; Abdulhussain, S.H.; Jassim, W.A. Speech Enhancement Algorithm Based on Super-Gaussian Modeling and Orthogonal Polynomials. *IEEE Access* **2019**, *7*, 103485–103504. [CrossRef]

32. Abdulhussain, S.H.; Ramli, A.R.; Mahmmod, B.M.; Saripan, M.I.; Al-Haddad, S.; Jassim, W.A. A New Hybrid form of Krawtchouk and Tchebichef Polynomials: Design and Application. *J. Math. Imaging Vis.* **2019**, *61*, 555–570. [CrossRef]

33. Mahmmod, B.M.; bin Ramli, A.R.; Abdulhussain, S.H.; Al-Haddad, S.A.R.; Jassim, W.A. Signal compression and enhancement using a new orthogonal-polynomial-based discrete transform. *IET Signal Process.* **2018**, *12*, 129–142. [CrossRef]

34. Alsabah, M.; Vehkapera, M.; O'Farrell, T. Non-Iterative Downlink Training Sequence Design Based on Sum Rate Maximization in FDD Massive MIMO Systems. *IEEE Access* **2020**, *8*, 108731–108747. [CrossRef]

35. Naser, M.A.; Alsabah, M.; Mahmmod, B.M.; Noordin, N.K.; Abdulhussain, S.H.; Baker, T. Downlink Training Design for FDD Massive MIMO Systems in the Presence of Colored Noise. *Electronics* **2020**, *9*, 2155. [CrossRef]

36. Abdulhasan, M.Q.; Salman, M.I.; Ng, C.K.; Noordin, N.K.; Hashim, S.J.; Hashim, F.B. Approximate linear minimum mean square error estimation based on channel quality indicator feedback in LTE systems. In Proceedings of the 2013 IEEE 11th Malaysia International Conference on Communications (MICC), Kuala Lumpur, Malaysia, 26–28 November 2013; pp. 446–451.

37. Abdulhasan, M.Q.; Salman, M.I.; Ng, C.K.; Noordin, N.K.; Hashim, S.J.; Hashim, F. An adaptive threshold feedback compression scheme based on channel quality indicator (CQI) in long term evolution (LTE) system. *Wirel. Pers. Commun.* **2015**, *82*, 2323–2349. [CrossRef]

38. Abdulhussain, S.H.; Ramli, A.R.; Al-Haddad, S.A.R.; Mahmmod, B.M.; Jassim, W.A. Fast Recursive Computation of Krawtchouk Polynomials. *J. Math. Imaging Vis.* **2018**, *60*, 285–303. [CrossRef]

39. Abdulhussain, S.H.; Ramli, A.R.; Al-Haddad, S.A.R.; Mahmmod, B.M.; Jassim, W.A. On Computational Aspects of Tchebichef Polynomials for Higher Polynomial Order. *IEEE Access* **2017**, *5*, 2470–2478. [CrossRef]

40. Abdulhussain, S.H.; Al-Haddad, S.A.R.; Saripan, M.I.; Mahmmod, B.M.; Hussien, A. Fast Temporal Video Segmentation Based on Krawtchouk-Tchebichef Moments. *IEEE Access* **2020**, *8*, 72347–72359. [CrossRef]

41. Mukundan, R.; Raveendran, P.; Jassim, W. New orthogonal polynomials for speech signal and image processing. *IET Signal Process.* **2012**, *6*, 713–723. [CrossRef]

42. Thung, K.H.; Paramesran, R.; Lim, C.L. Content-based image quality metric using similarity measure of moment vectors. *Pattern Recognit.* **2012**, *45*, 2193–2204. [CrossRef]

43. Jassim, W.; Paramesran, R.; Zilany, M. Enhancing noisy speech signals using orthogonal moments. *IET Signal Process.* **2014**, *8*, 891–905. [CrossRef]

44. Mizel, A.K.E. Orthogonal Functions Solving Linear functional Differential EquationsUsing Chebyshev Polynomial. *Baghdad Sci. J.* **2008**, *5*, 143–148.

45. Abdulhussain, S.H.; Ramli, A.R.; Mahmmod, B.M.; Saripan, M.I.; Al-Haddad, S.A.R.; Jassim, W.A. Shot boundary detection based on orthogonal polynomial. *Multimed. Tools Appl.* **2019**, *78*, 20361–20382. [CrossRef]

46. Abdulhussain, S.H.; Ramli, A.R.; Hussain, A.J.; Mahmmod, B.M.; Jassim, W.A. Orthogonal polynomial embedded image kernel. In Proceedings of the International Conference on Information and Communication Technology—ICICT'19, Baghdad, Iraq, 15–16 April 2019; ACM Press: New York, NY, USA, 2019; pp. 215–221. [CrossRef]

47. Hsu, C.W.; Chang, C.C.; Lin, C.J. *A Practical Guide to Support Vector Classification*; Department of Computer Science and Information Engineering: Taipei, Taiwan, 2003.

48. Byun, H.; Lee, S.W. A survey on pattern recognition applications of support vector machines. *Int. J. Pattern Recognit. Artif.* **2003**, *17*, 459–486. [CrossRef]

49. Awad, M.; Motai, Y. Dynamic classification for video stream using support vector machine. *Appl. Soft Comput.* **2008**, *8*, 1314–1325. [CrossRef]

50. Nigam, S.; Deb, K.; Khare, A. Moment invariants based object recognition for different pose and appearances in real scenes. In Proceedings of the 2013 International Conference on Informatics, Electronics and Vision (ICIEV), Dhaka, Bangladesh, 17–18 May 2013; pp. 1–5. [CrossRef]

51. Chang, C.C.; Lin, C.J. LIBSVM: A library for support vector machines. *ACM Trans. Intell. Syst. Technol.* **2011**, *2*, 27:1–27:27. [CrossRef]

52. CMATERdb: The Pattern Recognition Database Repository. 2020. Available online: https://code.google.com/archive/p/cmaterdb/ (accessed on 27 January 2021).

Article

Fast Approximation for Sparse Coding with Applications to Object Recognition

Zhenzhen Sun and Yuanlong Yu *

The College of Mathematics and Computer Science, Fuzhou University, Fuzhou 350116, China; zhenzhen_sun@foxmail.com
* Correspondence: yu.yuanlong@fzu.edu.cn

Abstract: Sparse Coding (SC) has been widely studied and shown its superiority in the fields of signal processing, statistics, and machine learning. However, due to the high computational cost of the optimization algorithms required to compute the sparse feature, the applicability of SC to real-time object recognition tasks is limited. Many deep neural networks have been constructed to low fast estimate the sparse feature with the help of a large number of training samples, which is not suitable for small-scale datasets. Therefore, this work presents a simple and efficient fast approximation method for SC, in which a special single-hidden-layer neural network (SLNNs) is constructed to perform the approximation task, and the optimal sparse features of training samples exactly computed by sparse coding algorithm are used as ground truth to train the SLNNs. After training, the proposed SLNNs can quickly estimate sparse features for testing samples. Ten benchmark data sets taken from UCI databases and two face image datasets are used for experiment, and the low root mean square error (RMSE) results between the approximated sparse features and the optimal ones have verified the approximation performance of this proposed method. Furthermore, the recognition results demonstrate that the proposed method can effectively reduce the computational time of testing process while maintaining the recognition performance, and outperforms several state-of-the-art fast approximation sparse coding methods, as well as the exact sparse coding algorithms.

Keywords: sparse coding; fast approximation; homotopy iterative hard thresholding; object recognition

Citation: Sun, Z.; Yu, Y. Fast Approximation for Sparse Coding with Applications to Object Recognition. *Sensors* **2021**, *21*, 1442. https://doi.org/10.3390/s21041442

Academic Editor: Antonio Guerrieri
Received: 19 December 2020
Accepted: 17 February 2021
Published: 19 February 2021

Publisher's Note: MDPI stays neutral with regard to jurisdictional claims in published maps and institutional affiliations.

1. Introduction

Object recognition is a fundamental problem in machine learning, and has been widely researched for many years. The performance of object recognition methods largely relies on feature representation. Traditional methods used handcrafted features to represent objects, i.e., scale-invariant feature transform (SIFT) [1], histograms of oriented gradients (HOG) [2], etc. Inspired by biological finding [3,4], learning sparse representation is more beneficial for object recognition, because mapping features from low-dimensional space to a high-dimensional space makes the features more likely to be linearly separable. Therefore, many sparse coding (SC) algorithms have been proposed to learn a good sparse representation for natural signals [5–7].

In general, SC is the problem of reconstructing input signal using a linear combination of an over-complete dictionary with sparse coefficients, i.e., for an observed signal $x \in R^p$ an over-complete dictionary $D \in R^{p \times K} (p \ll K)$, SC aims to find a representation $\alpha \in R^K$ to reconstruct x by using only a small number of atoms chosen from D. The problem of SC is formulated as

$$\min_{\alpha} : ||x - D\alpha||_2^2 + \lambda ||\alpha||_0, \tag{1}$$

where the l_0-norm is defined as the number of non-zero elements of α, and λ is the regularization factor. Several optimization algorithms have been proposed for the numerical solution of (1). However, the high computational cost induced by these optimization algorithms is a major drawback for real-time applications, especially when a large-sized dictionary is used.

To get rid of this problem, many works focusing on fast approximation for sparse coding have been proposed. Kavukcuoglu et al. [8] proposed a method named Predictive Sparse Decomposition (PSD) that used a non-linear regressor to approximate the sparse feature, and applied this method to objection recognition. However, the predictor is simplistic and produces crude approximation, and the regressor training procedure is somewhat time-consuming because of the gradient descent training method. Recently, deep learning showed its widespread success on many inference problems, which provides another way to design fast approximation methods for sparse coding algorithms. The idea is first proposed by Gegor et al. [9] who constructed two deep learning networks to approximate the iterative soft thresholding and coordinate descent algorithms, leading to the so-called LISTA and LCoD methods, respectively. LISTA showed its superiority on calculation and approximation, and many recent variants of LISTA have been proposed for miscellaneous applications, see [10,11] for some examples. Inspired by [9], many fast approximation sparse coding methods based on deep learning have been proposed and shown their effectiveness on unfolding the corresponding sparse coding algorithms, i.e., LAMP [12], LVAMP [12], etc.

Though these methods perform well in large-scale datasets, there are three defects. First, they are not suitable for small-scale datasets, in which the number of training samples is far less then ten thousand. The performance of deep neural network is sensitive to the scale of training data, when the number of training samples is small, the deep network model is over-parameterized and may result in over-fitting. Second, deep networks involve lots of hyper-parameters, whose training requires large computational and storage resources because of the gradient-based back-propagation method, and is easy to get stuck in a local optimal solution. Last but not least, each deep network architecture is designed only for the corresponding sparse coding algorithm that cannot be generalized to other algorithms. Therefore, the extendibility of these methods are limited.

To solve the problems mentioned above, a simple and effective fast approximation sparse coding method is proposed for small-scale datasets object recognition task in this paper. Differing from the deep learning-based methods, a special single-hidden-layer neural network (SLNNs) is constructed to perform the approximation task, and the training process of this SLNNs can be easily implemented by the least squared method. The proposed method includes two steps. In the first step, the optimal sparse features of training samples are exactly computed by sparse coding algorithm (in this paper, the homotopy iterative hard thresholding (HIHT) algorithm [13] is used), and in the second step the optimal sparse features are used as ground truth to train the especially constructed SLNNs. After training, the input layer and hidden layer of this SLNNs can be used to implement the nonlinear feature mapping from the input space to sparse feature space, which only involves simple inner product calculation with a non-linear activation function. Therefore, the sparse features of new samples can be estimated quickly. Ten benchmark datasets taken from UCI databases and two face image datasets are used to validate the proposed method, and the root mean square error (RMSE) results on testing data have verified the approximation performance of this proposed method. Furthermore, the approximated sparse features have been applied to object recognition task, and the recognition results demonstrate that this proposed approximation sparse coding method is beneficial for object recognition in terms of recognition accuracy and testing time.

The main contributions of this paper can be concluded as

1. A fast approximation sparse coding method is proposed for small-scale datasets object recognition task, which can quickly estimate the sparse features for testing samples.
2. A special SLNNs architecture has been constructed to perform the approximation task, whose parameters can be optimized easily by the least squared method, avoiding the multifarious procedure induced by the gradient-based back-propagation training.

3. Experiment results on ten benchmark UCI datasets and two face image datasets show that our approach is more effective than current state-of-the-art deep learning-based fast approximation sparse coding methods both in RMSE, recognition accuracy and testing time.

The remainder of this paper is organized as follows. Section 2 briefly reviews the sparse coding algorithms and fast approximation sparse coding methods. Section 3 details the proposed method. Section 4 describes implementation details and presents experimental results. Finally, conclusions are given in Section 5.

2. Related Work

2.1. Sparse Coding Algorithms

As described in Section 1, the problem of SC can be formulated as problem (1). However, problem (1) is NP-hard, which is difficult to be solved. There are three common methods for approximations/relaxations of this problem: (1) iterative greedy algorithms [14–16]; (2) l_1-norm convex relaxation methods (which are called basis pursuit) [17]; (3) l_p-norm $(0 < p < 1)$ relaxation methods [18–22]. Among these methods, BP has been studied more widely, in which the l_0 norm is replaced by l_1 norm to make a convex relaxation for the problem (1), i.e.,

$$\min_{\boldsymbol{\alpha}} : ||\boldsymbol{x} - \boldsymbol{D}\boldsymbol{\alpha}||_2^2 + \lambda ||\boldsymbol{\alpha}||_1, \tag{2}$$

where the l_1-norm is defined as the sum of absolute values of all elements of $\boldsymbol{\alpha}$.

BP methods were proven to give the same solutions to (1) when the dictionary satisfies the Restricted Isometry Property (RIP) condition [23,24]. Many research works focusing on efficiently solving problem (2) have been proposed, [25] provides a comprehensive review of five representative algorithms, namely *Gradient Projection* (GP) [26,27], *Homotopy* [28,29], *Iterative Soft Shrinkage-Thresholding Algorithm* (ISTA) [30–32], *Proximal Gradient* (PG) [33,34], and *Augmented Lagrange Multiplier* (ALM) [35]. Among these algorithms, ISTA is the most popular algorithm, and lots of heuristic strategies have been proposed to reduce the computational time of ISTA, i.e., TwIST [36], FISTA [33], etc. Recently, a kind of pathwise coordinate optimization method called PICASSO [37–39] has been proposed to solve the l_p $(0 < p \le 1)$ least squared problem, which showed superior empirical performance compared with other state-of-the-art sparse coding algorithms mentioned above.

Although satisfactory results can be achieved by using the approximation/relaxation methods, the l_0-norm is more desirable from the sparsity perspective. In recent years, researchers have attempted to solve problem (1) directly, with iterative hard thresholding (IHT) [13,40,41] being the most popular method. The IHT methods have strong theoretical guarantees, and the extensive experimental results show that the IHT methods can improve the sparse representation reconstruction results.

2.2. Fast Approximation for Sparse Coding

The sparse coding algorithms mentioned in Section 2.1 involve a lot of iterative operations, which induces high computational cost and prohibits them from real-time applications. To get rid of this problem, some research focusing on fast approximation for sparse coding was proposed. Kavukcuoglu et al. [8] proposed the PSD method to approximate sparse coding algorithms using a non-linear regressor. In inspired by this, Chalasani et al. [42] extended PSD to estimate convolutional sparse features. However, the approximation performance of non-linear regressor is limited. As the development of deep learning, some researchers have constructed deep networks to solve the fast approximation sparse coding problem. Given a large set of training examples $\{(\boldsymbol{x}_i, \boldsymbol{\alpha}_i)\}_{i=1}^N$, a many-layer neural network is optimized to minimize the reconstruction mean squared error between network outputs and $\{\boldsymbol{\alpha}_i\}_{i=1}^N$. After training, the approximation of sparse representation for a new signal $\boldsymbol{x}_{new}$ can be quickly predicted by the deep network. The idea is first proposed by Gregor et al. [9] who constructed two deep learning networks to approximate the iterative soft thresholding and coordinate descent algorithms, leading to the so-called LISTA and LCoD methods, respectively. Inspired by [9], Xin et al. [43] translated the iterative hard thresholding algorithm into a deep learning framework. Borgerding et al. [12] proposed two deep neural-network architectures to unfold the approximate message passing (AMP) algorithm [44] and "vector AMP" (VAMP) algorithm [45] respectively, namely LAMP and LVAMP. In [46], the authors proposed a deep learning framework for the approximation of sparse representation of a signal with the aid of a correlated signal, the so-called side information. The learned deep networks perform steps similar to those implemented by corresponding sparse coding algorithms; however, the trained network can reduce the computational cost when calculating the sparse representation of new samples effectively, which is critical in large-scale data settings and real-time applications.

3. Materials and Methods

3.1. Homotopy Iterative Hard Thresholding Algorithm

The homotopy iterative hard thresholding (HIHT) [13] is an extension of IHT for the l_0-norm regularized problem

$$\min_{\alpha} : \theta_\lambda(\alpha) = f(\alpha) + \lambda||\alpha||_0, \tag{3}$$

where $f(\alpha) = \frac{1}{2}||x - D\alpha||_2^2$ is a differentiable convex function, whose gradient $\nabla f(\alpha)$ satisfies the Lipschitz continuous condition with parameter $L_f > 0$. Therefore, $f(\alpha)$ can be approximately iteratively updated by the projected gradient method

$$\alpha^{k+1} = argmin \, f(\alpha^k) + \nabla f(\alpha^k)^T(\alpha - \alpha^k) + \frac{L}{2}||\alpha - \alpha^k||_2^2, \tag{4}$$

where $L \geq 0$ is a constant, which should satisfies the condition of $L \geq L_f$.

Adding $\lambda||\alpha||_0$ into both side of (4), the solution of (3) can be obtained by iteratively solving the subproblem

$$\begin{aligned} \alpha^{k+1} &= arg \min \, p_{L,\lambda}(\alpha^k, \alpha) \\ &= f(\alpha^k) + \nabla f(\alpha^k)^T(\alpha - \alpha^k) + \frac{L}{2}||\alpha - \alpha^k||_2^2 + \lambda||\alpha||_0. \end{aligned} \tag{5}$$

The optimization of (5) is the same as follows (by removing or adding some constant items which are independent on α):

$$\min_{\alpha} \frac{L}{2}\left[||\alpha - (\alpha^k - \frac{1}{L}\nabla f(\alpha^k))||_2^2 + \frac{2\lambda}{L}||\alpha||_0\right]. \tag{6}$$

If denote

$$T_L(\alpha^{k+1}) = arg \min_{\alpha} \, p_{L,\lambda}(\alpha^k, \alpha), \tag{7}$$

then the closed form solution of $T_L(\alpha^{k+1})$ is given by the following lemma.

Lemma 1. *[32,41] The solution $T_L(\alpha^{k+1})$ of (7) is give by*

$$[T_L(\alpha^{k+1})]_i = \begin{cases} [s_L(\alpha^k)]_i, & if \, [s_L(\alpha^k)]_i^2 > \frac{2\lambda}{L}; \\ 0, & if \, [s_L(\alpha^k)]_i^2 \leq \frac{2\lambda}{L}. \end{cases} \tag{8}$$

where $s_L(\alpha) = \alpha - \frac{1}{L}\nabla f(\alpha)$, and $[.]_i$ refers to the i-th element of a vector.

In (8), the parameter L needs to be tuned. The upper bound on Lipschitz constant L_f is unknown or may not be easily calculated, thus we use the line search method to search L as suggested in [41] until the objective value descends.

Homotopy Strategy: many works [13,26,39] have verified that the sparse coding approaches benefit from a good starting point. Therefore, we use a recursive process automatically tunes regularization factor λ. This process begins from a large initial value λ^0. At the end of each λ-tuning iterations indexed by k, an optimal solution α^k is obtained given λ^k. Then λ is updated as $\lambda^{k+1} = \rho\lambda^k$, where $\rho \in [0,1]$, and α^k is used as the initial solution for the next iteration $k+1$. The process stops once λ is small enough (given a positive lower-bound target, the stop condition is $\lambda_k \leq \lambda_{target}$). An outline of HIHT algorithm is described as Algorithm 1.

Algorithm 1 $\{\alpha^\star, L^\star\} \leftarrow HIHT(L_0, \lambda_0, \alpha^0)$

(Input:) $L_0, \lambda_0, \alpha^0, D, L_{min}, L_{max}; // L_0 \in [L_{min}, L_{max}]$
(Output:) $\alpha^\star, L^\star$;
initialize $\rho \in (0,1), \eta > 0, \gamma > 1, \epsilon > 0, k \leftarrow 0$;
repeat
  $i \leftarrow 0$;
  $\alpha^{k,0} = \alpha^k$;
  $L_{k,0} \leftarrow L_k$;
  repeat
    An L-tuning iteration indexed by i
    $\alpha^{k,i+1} \leftarrow T_{L_{k,i}}(\alpha^{k,i})$;
    while $\theta_{\lambda_k}(\alpha^{k,i}) - \theta_{\lambda_k}(\alpha^{k,i+1}) < \frac{\eta}{2}||\alpha^{k,i} - \alpha^{k,i+1}||^2$ **do**
      $L_{k,i} \leftarrow min\{\gamma L_{k,i}, L_{max}\}$;
      $\alpha^{k,i+1} \leftarrow T_{L_{k,i}}(\alpha^{k,i})$;
    end while
    $L_{k,i+1} \leftarrow L_{k,i}$;
    $i \leftarrow i+1$;
  until$||\alpha^{k,i} - \alpha^{k,i+1}||_2^2 \leq \epsilon$
  $\alpha^{k+1} \leftarrow \alpha^{k,i}$;
  $L_{k+1} \leftarrow L_{k,i}$.
  $\lambda_{k+1} \leftarrow \rho\lambda_k$;
  $k \leftarrow k+1$;
until $\lambda_{k+1} \leq \lambda_{target}$
$\alpha^\star \leftarrow \alpha^k$;
$L^\star \leftarrow L_k$.

3.2. Proposed Method

Figure 1 illustrates the schematic diagram of this proposed method. As it can be seen, for the given training dataset $X = \{x_1, x_2, ..., x_N\} \in R^{p*N}$ and the over-complete dictionary D, the HIHT algorithm described in Section 3.1 is used to calculate the optimal sparse features $A = \{\alpha_1, \alpha_2, ..., \alpha_N\} \in R^{K*N}$ of training data in the first step. After that, these optimal sparse features are used to train the SLNNs in second step.

As Figure 1 shows, the architecture of the neural network consists of an input layer, a feature layer and an output layer. The number of hidden neurons is the same as that of output neurons, which is set as the dimension of the sparse feature. Each hidden neuron is only connected to its corresponding output neuron with weight 1. Our goal is to obtain a optimal input weights $\hat{W}$ to make the outputs of hidden layer as equal to A as possible, that is

$$||g(\hat{W}^T X) - A||_F \rightarrow 0, \tag{9}$$

where $g(.)$ refers to a non-linear activation function.

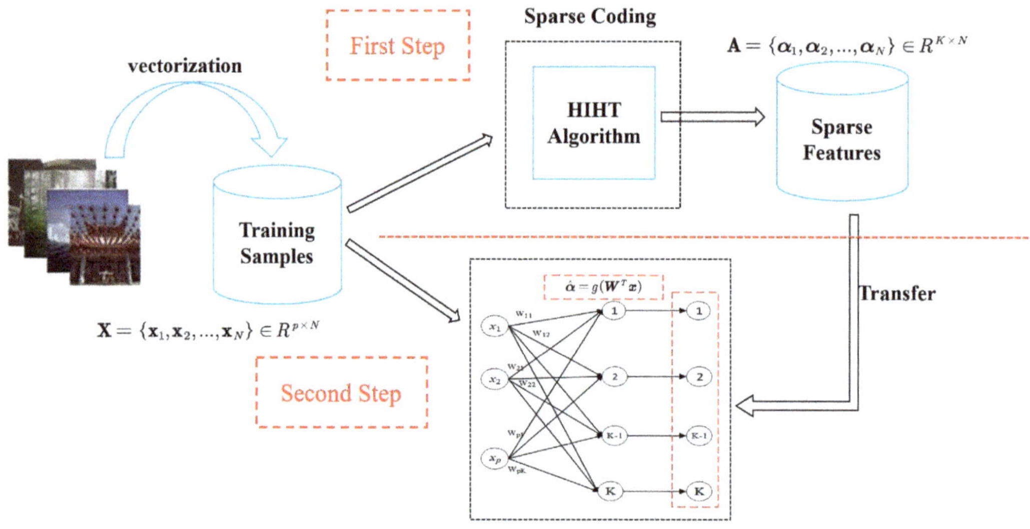

Figure 1. The schematic diagram of this proposed method.

There are two strategies to optimize the input weights $\boldsymbol{W}$:

(1) If the activation function is known, we chose *tanh* function as the activation function, where $g(x) = tanh(x)$. We firstly calculate $arctanh(\boldsymbol{A})$, and denote the result as $\boldsymbol{Z}$, that is

$$\boldsymbol{Z} = arctanh(\boldsymbol{A}), \tag{10}$$

then we formulate the objective function of the SLNNs as

$$\text{Minimize:} \quad \frac{1}{2}\|\boldsymbol{Z} - \boldsymbol{W}^T\boldsymbol{X}\|_F^2 + \frac{C_1}{2}\|\boldsymbol{W}\|_F^2, \tag{11}$$

where constant C_1 refers to the regularization factor used to control the trade-off between the smoothness of the mapping function and the closeness to $\boldsymbol{Z}$.

By setting the derivative of (11) with respect to $\boldsymbol{W}$ to zero and solve this equality, then the optimal solution of $\boldsymbol{W}$ is obtained as follows:

$$\hat{\boldsymbol{W}} = (\frac{\boldsymbol{I}}{C_1} + \boldsymbol{X}\boldsymbol{X}^T)^{-1}\boldsymbol{X}\boldsymbol{Z}^T. \tag{12}$$

In addition, the sparse feature of a testing sample x_{test} can be quickly estimated as

$$\hat{\boldsymbol{\alpha}}_{test} = tanh(((\frac{\boldsymbol{I}}{C_1} + \boldsymbol{X}\boldsymbol{X}^T)^{-1}\boldsymbol{X}\boldsymbol{Z}^T)^T x_{test}). \tag{13}$$

(2) If the activation function is unknown, a kernel trick based on Mercer's condition can be used to calculate the approximated sparse feature of testing data x_{test} directly instead of training the weights $\boldsymbol{W}$,

$$\hat{\boldsymbol{\alpha}}_{test} = g(\boldsymbol{W}^T * x_{test}) = Ker(x_{test}, \boldsymbol{X})(\frac{\boldsymbol{I}}{C_1} + \Omega_{train})^{-1}\boldsymbol{A}, \tag{14}$$

where $\Omega_{train} == Ker(\boldsymbol{X}, \boldsymbol{X})$ and *Ker* stands for the kernel function.

In this proposed method, Gaussian function is used as the kernel function *Ker*:

$$Ker(x_1, x_2) = exp(-\frac{||x_1 - x_2||^2}{\sigma^2}), \tag{15}$$

where σ denotes the standard deviation of the Gaussian function.

4. Results and Discussion

4.1. Data Sets Description

Ten benchmark datasets taken from *UCI Machine Learning Repository* [47] and two image datasets: the Extended YaleB [48] and the AR dataset [49], are used to validate the proposed method. The ten UCI datasets include 5 binary-classification cases and 5 multi-classification cases. The details of these datasets are shown in Table 1. In this table, column "Random Perm" shows whether the training and testing data are randomly assigned or not. In the experiments, $\frac{2}{3}$ of samples per class are randomly selected for training, and the rest samples are responsible for testing if "Random Perm" is Yes.

Table 1. UCI Data sets Used in Our Experiments.

Datasets	Training	Testing	Features	Classes	Random Perm
Australian Credit	459	231	14	2	Yes
Diabetes	511	257	8	2	Yes
Glass	140	74	9	6	Yes
Image segmentation	1540	770	19	7	Yes
LiverDisorders	229	116	6	2	Yes
Madelon	2000	600	500	2	No
Satimage	4435	2000	36	6	No
Vehicle	562	284	18	4	Yes
Wine	118	60	13	3	Yes
Wisconsin Breast Cancer	379	190	30	2	Yes

The extended YaleB dataset [48] contains 38 different people with 2414 frontal face images, and each class has about 64 samples. This dataset is challenging from varying expressions and illumination conditions, see Figure 2 for some examples. The random face feature descriptor generated in [7] is used as raw feature, in which a cropped image with 192 × 168 pixels was projected onto a 504-dimensional vector by a random normal distributed matrix. In the experiment, 50% of samples per class are randomly selected for training and the rest are responsible for testing.

Figure 2. Extended YaleB.

The AR face dataset contains over 126 people with more than 4000 face images. There are 26 images per person taken during two different sessions. The images have large variations in terms of disguise, facial expressions, and illumination conditions. A few samples from the AR dataset are shown in Figure 3 for illustration. A subset of 2600 images pertaining to 50 males and 50 females objects are used for experiment. For each object, 20 samples are randomly chosen for training and the rest for testing. The images with 165 × 120 pixels were projected onto a 540-dimensional vector by using a random projection matrix.

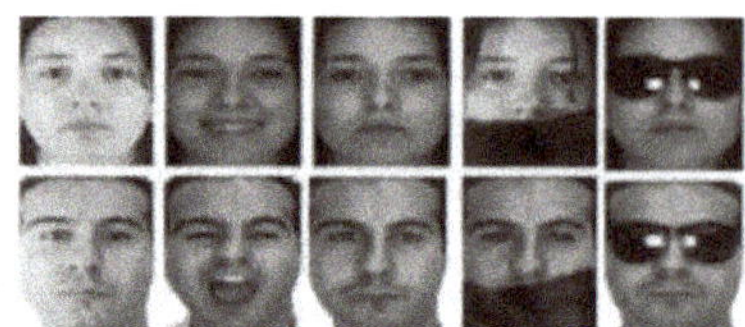

Figure 3. AR Face.

4.2. Implementation Details

The experiments are mainly divided into two parts: (1) The RMSE between the approximated sparse features and the optimal features of testing data is calculated to verify the approximation performance of this proposed method, and the results of several state-of-the-art fast approximation sparse coding methods are also reported for comparison. (2) Classification experiments are implemented to validate the recognition performance of the approximated sparse features estimated by the proposed SLNNs. The compared methods can be categorized as follows: (a) Different representation learning methods: ELM [50] with random feature mapping, and ScELM [51] with optimal sparse features computed by HIHT; (b) Different fast approximation sparse coding methods: PSD [8], LISTA [9], LAMP [12], and LVAMP [12], detailed descriptions to these methods are provided in Section 2.

Implementations of ELM, ScELM, PSD, and this proposed method are based on Matlab codes and others are based on Python. A random normal distributed matrix is used as the dictionary in each sparse coding algorithm, and the number of atoms or hidden nodes K is set to 100 if the dimension of dataset is less than 100, otherwise 1000. The parameter C_1 is searched for in the grid of $\{2^{-25}, 2^{-20}, ..., 2^{25}\}$, and the σ is searched for in $\{100, 200, ..., 1000\}$. The number of hidden layers of LISTA, LAMP, and LVAMP are set as 6, 5, and 4, respectively, if not stated otherwise. Other parameters are default as the authors suggested. For the randomly training-testing assigned datasets, ten repeated trials are carried out in the following experiments, and the average result and standard deviation are recorded.

In object recognition experiments, the trained network of each method is used to compute the approximated sparse features for training and testing samples, and the approximated sparse features are used as the input of the classifier. The ridge regression model is used as the classifier in our experiments, whose objective function is

$$\text{Minimize:} \quad \tfrac{1}{2}\|Y - \beta^T \hat{A}\|_F^2 + \tfrac{C_2}{2}\|\beta\|_F^2, \tag{16}$$

where Y is the label matrix of training data X, and β is the weights of the classifier model. For a testing sample x_{test}, the predicted label for it is calculated as

$$identity(x_{test}) = argmax_i(\beta^T \hat{\alpha}_{test}). \tag{17}$$

The hyper-parameter of the classifier C_2 is searched for in the grid of $\{2^{-25}, 2^{-20}, ..., 2^{25}\}$, and a value with best validation accuracy is selected. We compare our method with others in terms of recognition accuracy and testing time, where the recognition accuracy is defined as the ratio of the number of correctly classified testing samples to that of all testing samples, and the testing time refers to the total spending time of testing samples' feature calculation and classification.

A standard PC is used in our experiments and its hardware configuration as follows:

1. CPU: Intel(R) Pentium(R) CPU G2030 @3.40GHz;
2. Memory: 32.00GB;
3. Graphics Processing Unit (GPU): None.

4.3. Root Mean Square Error Results

For testing data X_{test}, whose optimal sparse features computing by sparse coding algorithm is denoted as A_{test}, and the approximated sparse features computing by the fast approximation method is denoted as $\hat{A}_{test}$, the RMSE between A_{test} and $\hat{A}_{test}$ is defined as

$$RMSE(A_{test}, \hat{A}_{test}) = \sqrt{\frac{1}{N_{test}K}||A_{test} - \hat{A}_{test}||_F^2}, \tag{18}$$

where N_{test} denotes the amount of testing samples.

Some UCI datasets are used in this experiment, and we reported the results of our method, LISTA, LAMP and LVAMP to compare their approximation performance, Table 2 shows the results. As it can be seen from this table, our approach can achieve a lower RMSE result than other methods on the most datasets, which indicates that the approximated sparse features estimated by our approach are more closer to the optimal ones than that estimated by the compared methods. For the Glass dataset, our method has achieved a significant improvement, and for LiverDisorders, though the result of our approach is not the best one, it is very close to the best one.

Table 2. The root mean square error of compared methods on UCI datasets (bold one represents the best result).

Datasets	LISTA	LAMP	LVAMP	Ours	
				Tanh	Kernel
Australian Credit	0.0418	0.0614	0.0497	0.0591	**0.0391**
Diabetes	0.0460	0.0356	0.0446	0.0494	**0.0350**
Glass	0.0527	0.0436	0.0550	**0.0111**	0.0144
Image segmentation	0.0324	0.0406	0.0462	0.0375	**0.0317**
LiverDisorders	0.0440	**0.0403**	0.0435	0.0476	0.0597
Vehicle	0.0267	0.0442	0.0496	0.0141	**0.0122**
Wine	0.0415	0.0394	0.0462	0.0149	**0.0094**
Wisconsin Breast Cancer	0.0295	0.0551	0.0427	0.0222	**0.0179**

4.4. Objection Recognition Results

4.4.1. The Evaluation of HIHT

The existing literature on sparse coding only compared different sparse coding algorithms in terms of reconstruction error and convergence speed, but did not compare their classification performance when applying these algorithms in object recognition. To show why this paper uses the HIHT algorithm to compute the optimal sparse features, we implemented some experiments to validate the superiority of HIHT compared with several state-of-the-art sparse coding algorithms when used in object recognition. The compared methods include IHT, homotopy GPSR (HGPSR) [26], PGH [34], and PICASSO [39].

(1) *Effectiveness on Object Recognition*: the binary-classification datasets listed in Table 1 are used in this experiment. Firstly, the sparse coding algorithms are used to compute sparse features for the experimental datasets using the same dictionary, and the measure of *cross entropy* is used to show how different the sparse features are between class 1 and class 2. A higher value means that the sparse features computed by corresponding algorithm are more discriminative and more beneficial for object recognition. The measure of *cross entropy* is estimated as follows: we accumulate a histogram $h(\alpha_k|v)$ along feature dimensions over all sparse features α_k that belongs to the same class v ($v \in \{1,2\}$), then normalize the histogram as the probability $p(v)$ of class v, the *cross entropy* between class 1 and class 2 is estimated as

$$cross\ entropy(p(1), p(2)) = -\sum_{k=1}^{K} p_k(1) \log \frac{1}{p_k(2)}, \tag{19}$$

where $p_k(v)$ is the p-th element of the probability $p(v)$.

Table 3 shows the *cross entropy* results. It can be seen that the HIHT algorithm can achieve the best result on the most datasets than the other four algorithms. It indicated that the sparse features computed by HIHT can distinguish different classes more effectively, which is more useful for classification, especially when a simple linear classifier is used.

Table 3. Cross entropy of sparse features between different classes.

Datasets	HGPSR	IHT	PGH	PICASSO	HIHT
Austrain Credit	6.840	6.785	6.933	6.299	**7.268**
Diabetes	4.381	6.164	3.899	4.118	**6.410**
LiverDisorders	3.087	5.461	3.267	5.894	**6.346**
Madelon	8.040	9.335	8.076	9.280	**9.680**
Wisconsin Breast Cancer	4.652	**6.256**	4.643	5.632	5.878

Subsequently, we use these sparse coding algorithms to compute the optimal sparse features of training data to train the proposed SLNNs, and compare the final recognition results, which is shown in Table 4. From this table it can be seen that these sparse coding algorithms can achieve similar classification performance on most datasets when used in the proposed method, while HIHT outperforms the other three algorithms in some datasets (i.e., Glass and Vehicle) significantly. From the view of standard deviation, the results show that the optimal sparse features computed by HIHT are more robust to classification than other algorithms.

Table 4. Recognition Results of the Proposed Method Using different sparse coding algorithms.

Datasets	HGPSR	IHT	PICASSO	HIHT
Australian Credit	85.59 ± 1.83	85.87 ± 1.87	85.67 ± 1.87	86.08 ± 1.73
Diabetes	76.06 ± 1.96	76.73 ± 1.68	74.96 ± 2.28	76.99 ± 1.66
Glass	60.97 ± 4.83	61.64 ± 3.38	64.59 ± 4.63	66.60 ± 3.20
Image segmentation	90.08 ± 1.02	90.81 ± 0.97	90.91 ± 0.89	91.90 ± 0.73
LiverDisorders	68.59 ± 3.62	72.34 ± 4.27	69.57 ± 3.96	73.52 ± 2.56
Madelon	59.67	60.88	58.33	60.67
Satimage	82.12	83.37	83.00	84.30
Vehicle	78.46 ± 2.47	78.68 ± 2.07	76.20 ± 1.81	81.00 ± 1.51
Wine	96.80 ± 1.92	98.53 ± 1.47	97.33 ± 1.81	98.56 ± 1.07
Wisconsin Breast Cancer	93.87 ± 1.68	95.31 ± 1.09	96.00 ± 2.89	95.92 ± 1.81

(2) *Parameter Sensitivity*: In HIHT algorithm, different values of the regularization factor λ_{target} and dictionary D will product different sparse features, which will cause the proposed method to estimate different approximated sparse features and influence final recognition result. In this experiment, the sensitivities of λ_{target} and D in final recognition performance are verified, and the two face image datasets are used for testing.

Firstly, the influence of λ_{target} is investigated. By fixing other parameters (i.e., dictionary, parameters of the classifier), λ_{target} is searched for in the grid of $\{10^{-10}, 10^{-8}, ..., 10^{2}\}$, and the corresponding recognition accuracy is recorded. From the results in Figure 4, we can conclude that the final recognition result is not very sensitive to the λ_{target}, so it is no need to spend much time turning λ_{target} when uses HIHT to compute the optimal sparse features in this proposed method.

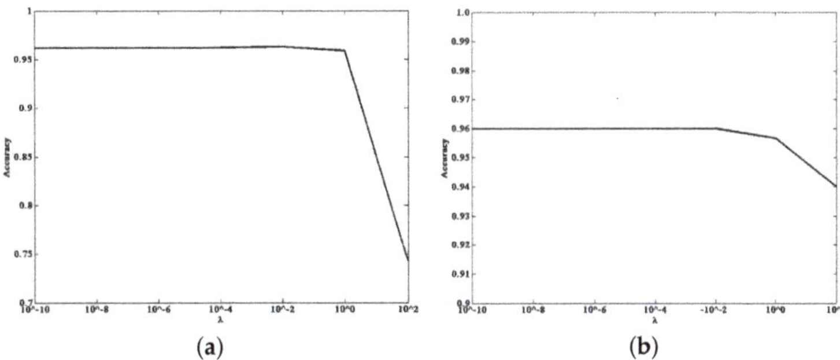

Figure 4. The influence of the target value λ_{target} of HIHT on the recognition performance. (**a**) Extended YaleB. (**b**) AR Face.

Subsequently, we investigate the influence of D. An unsupervised learned dictionary by the Lagrangian dual method [52] is used to compare with a random dictionary generated by normal distribution. The number of iterations in dictionary learning is set as 5, and 10 times with random selection of training and testing data are repeated, the average accuracy is recorded for comparison. As Table 5 shows, the final recognition accuracy achieved by using learned dictionary are close to that by using random dictionary. However, the computational time of optimal sparse features calculation with dictionary learning is five times (equal to the number of iterations) that with the random dictionary. Thus, in the following experiments we use random dictionary to compute optimal sparse features in HIHT algorithm.

Table 5. Comparison of final recognition performance with or without dictionary learning in HIHT algorithm.

Datasets	Learned Dictionary	Random Dictionary
YaleB	95.95 ± 0.65	96.34 ± 0.69
AR	96.63 ± 0.77	97.37 ± 0.39

4.4.2. Evaluation on UCI Datasets

The average recognition accuracies on UCI datasets are listed in Tables 6 and 7 presents the testing time. From these two tables we can conclude that the proposed approach outperforms other methods in terms of accuracy and testing time simultaneously. For most datasets, the approximated sparse features estimated by our approach can obtain the highest accuracy, and is approximately 100 times faster than ScELM (exact sparse coding algorithm), especially in high-dimensional datasets. Compared with other approximation sparse coding methods, our approach can achieve higher recognition accuracy with simpler network training, and the testing time of the proposed method and PSD are much less than LISTA, LAMP and LVAMP. It is worth noting that the performances of activation function *tanh* and kernel function of this approach are similar, but kernel function outperforms *tanh* when the dataset is a litter complex, (i.e., Satimage, Madelon), which will be confirmed in next experiments.

Table 6. The average accuracy of compared methods on UCI datasets (red is the best result and blue is the second one).

Datasets	ELM	ScELM	PSD	LISTA	LAMP	LVAMP	Ours	
							Tanh	Kernel
Australian Credit	86.13	83.52	86.22	81.97	80.26	-	86.08	86.15
Diabetes	65.32	69.26	65.01	73.98	68.19	74.59	76.99	75.14
Glass	57.30	62.38	63.46	67.50	60.50	61.75	66.60	67.23
Image segmentation	77.32	88.77	89.30	88.13	87.39	90.63	91.90	91.86
LiverDisorders	67.45	65.52	69.17	70.34	73.39	66.10	73.52	74.25
Madelon	58.17	59.17	59.35	51.67	50.92	56.93	60.67	64.83
Satimage	71.70	80.25	78.55	78.10	77.60	74.25	84.30	89.15
Vehicle	73.99	75.30	78.85	74.22	78.61	72.83	81.00	80.56
Wine	95.13	94.00	95.00	93.25	96.83	94.29	98.56	98.00
Wisconsin Breast Cancer	87.54	94.40	94.80	93.23	91.56	93.75	95.92	96.48

Table 7. The testing time of compared methods on UCI datasets.

Datasets	ELM	ScELM	PSD	LISTA	LAMP	LVAMP	Ours	
							Tanh	Kernel
Australian Credit	0.0079	0.6696	2.006×10^{-4}	0.0728	0.0811	0.0728	3.181×10^{-4}	0.0033
Diabetes	0.0136	1.7844	6.704×10^{-5}	0.0756	0.1332	0.0927	3.536×10^{-4}	0.0057
Glass	0.0013	0.9932	5.282×10^{-5}	0.0527	0.0714	0.0505	1.276×10^{-4}	6.061×10^{-4}
Image segmentation	0.0029	22.226	2.258×10^{-4}	0.1833	0.4511	0.4998	9.128×10^{-4}	0.0365
LiverDisorders	0.0011	0.7055	6.198×10^{-5}	0.0797	0.2499	0.5070	2.052×10^{-4}	9.227×10^{-4}
Madelon	0.0494	148.59	0.0062	1.2150	1.2560	0.9656	0.0293	0.073
Satimage	0.0134	39.218	7.130×10^{-4}	0.4272	0.3601	0.3287	0.0028	0.3018
Vehicle	0.0015	1.2062	1.013×10^{-4}	0.1249	0.3214	0.0991	4.392×10^{-4}	0.0047
Wine	7.232×10^{-7}	0.2457	5.687×10^{-5}	0.1042	0.0783	0.0556	4.392×10^{-4}	4.291×10^{-4}
Wisconsin Breast Cancer	0.0021	2.7363	$2.086e \times 10^{-4}$	0.0625	0.0929	0.0731	3.099×10^{-4}	0.0134

Figure 5 (The Tanh and Kernel mean the *tanh* version and kernel version of our method, respectively.) shows the confusion matrices obtained by this proposed method, PSD, LISTA, LAMP and LVAMP on Satimage dataset, in which the kernel version of this proposed method achieved a much better recognition result than others. It can be seen from this figure that all methods almost fail to correctly classify the test samples of class 4 except the kernel version of our method. It indicates that the features computed by this proposed method are more discriminative than that of other approximation methods. Figure 6 shows two examples of the receiver operating characteristic (ROC) curves of the approximation methods, where the red lines report the performance of our approach. It is clear that the Areas Under ROC curves (AUC) of our approach is much higher than others.

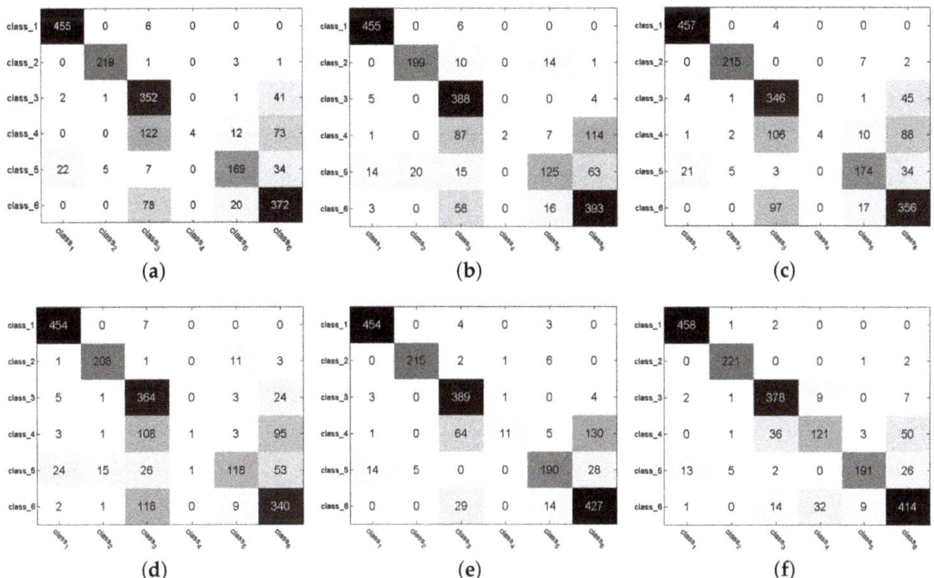

Figure 5. Confusion matrices on Satimage dataset. (**a**) PSD. (**b**) LISTA. (**c**) LAMP. (**d**) LVAMP. (**e**) Tanh. (**f**) Kernel.

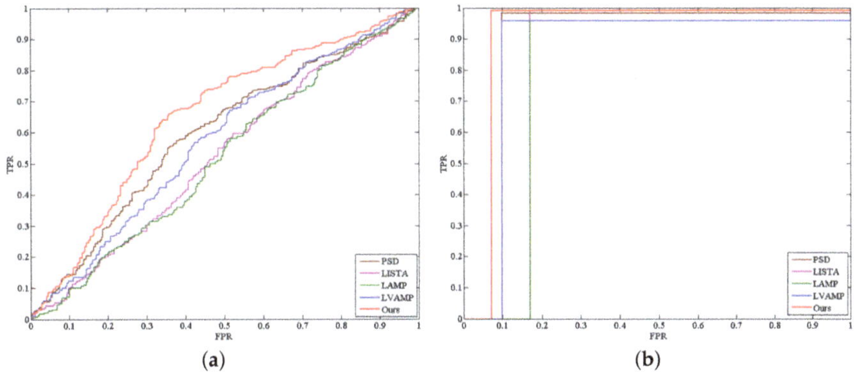

Figure 6. The receiver operating characteristic (ROC) curves of the approximate methods. (**a**) Madelon. (**b**) Breast.

4.4.3. Evaluation on Extended YaleB Dataset

Table 8 lists the recognition accuracies and testing time on Extended YaleB dataset, in which the famous sparse representation-based face recognition algorithm SRC [53] and collaborative representation-based classification (CRC) [54] are also used for comparison. Furthermore, a result obtained by raw features is set as the error bar (denoted as Baseline), and we set different number of hidden layers (denoted as T) for LISTA to show its influence on object recognition performance. As Table 8 shows, all methods beat the Baseline, indicating the benefit of feature learning. The kernel version of this proposed method obtains the best result with the value of 98.33%, and is 1.89% higher than the second one. In testing process, the proposed method is approximately 21 times faster than the deep learning-based approximation methods, and 182 times faster than SRC, also much faster than CRC. For LISTA, if the number of layers is small ($T = 2$), the recognition performance will degrade much, and as the number of layers increases, the recognition results tend to be stable. Thus, the recognition performance is somewhat sensitive to the number of layers of deep network.

Figure 7 shows the patterns of confusion across classes obtained by this proposed method, in which coordinates in X-axis and Y-axis represent 38 face classes. Color at coordinates (x, y) represents the number of test samples whose ground truth are x while machine's output labels are y. From this figure it can be seen that our approach shows fewer points in the non-diagonal region (i.e., fewer false positives and false negatives), indicting that the proposed method can classify most testing samples correctly.

Table 8. Average Recognition Accuracy with Random-Face Features on the Extended YaleB Database.

Method	Accuracy (%)	Time (s)
Baseline	91.54 ± 1.14	0.0009
SRC	96.27 ± 0.64	16.1676
CRC	96.82 ± 0.58	1.5006
ELM	96.44 ± 0.60	0.0256
ScELM	92.43 ± 0.93	168.3158
PSD	93.01 ± 0.69	0.057
LISTA (T = 2)	92.19 ± 1.21	2.0129
LISTA (T = 6)	95.16 ± 1.18	2.2535
LISTA (T = 10)	95.34 ± 0.82	4.6385
LAMP	94.07 ± 0.72	2.0119
LVAMP	94.62 ± 1.12	1.900
Tanh	96.34 ± 0.69	0.0256
kernel	**98.33 ± 0.40**	0.0888

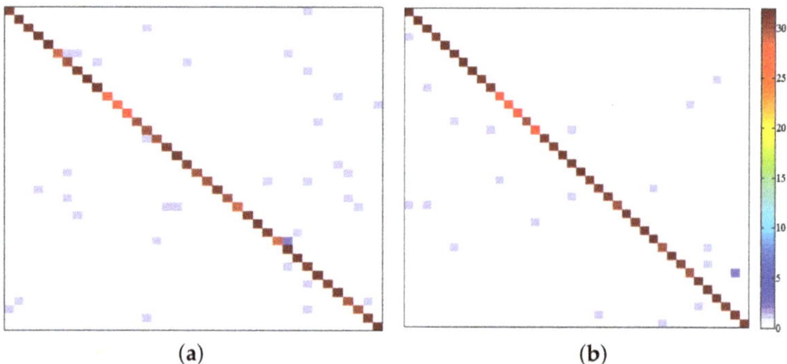

(a)   (b)

Figure 7. Patterns of confusion on Extended YaleB. (a) Tanh. (b) Kernel.

4.4.4. Evaluation on AR Dataset

For the AR dataset, a protocol (e.g., only five training samples per class or all training samples are used) is established in our experiments, and the corresponding results are list in Table 9. As we can see, the kernel version of this proposed method achieves the best result in both cases. In addition, the *tanh* version of this method gets comparable result with LAMP and ELM, but still better than SRC when all training samples were used. In terms of testing time, the proposed method is approximately 12 times faster than the deep learning-based approximation methods, 300 times faster than SRC, and 24 times faster than CRC. It is worth noting that the computational speed of kernel version is a little slower than that of *tanh* version, since it needs to compute the kernel matrix between testing samples and training samples, while it is still much faster than the deep learning-based approximation methods.

We use a confusion matrix to give the detailed evaluation at the class-level. Figure 8 shows the results, in which coordinates in x- and y-axis denote 100 face classes. Red point with coordinates (x, y) represents the misclassified test samples. It can be seen from this figure that this proposed method shows rare points in the non-diagonal region than other methods, indicating that this proposed method performs better than other methods in object recognition.

Table 9. Average Recognition Accuracy with Random-Face Features on the AR Face Database. The four column is the result when only 5 training samples per class are used.

Method	Accuracy (%)	Time (s)	Accuracy (%)
Baseline	93.00 ± 0.84	0.0011	78.67
SRC	95.42 ± 0.59	27.1421	84.00
CRC	96.92 ± 0.76	1.9682	84.83
ELM	97.05 ± 0.78	0.0178	83.67
ScELM	94.50 ± 1.09	62.1872	73.50
PSD	96.80 ± 0.66	0.0305	77.17
LISTA (T=3)	95.57 ± 0.85	0.8888	-
LISTA (T=6)	96.47 ± 0.61	1.2335	80.33
LISTA (T=8)	96.80 ± 0.73	3.1065	-
LAMP	97.37 ± 0.59	1.0505	81.67
LVAMP	95.30 ± 0.97	1.0314	84.50
Tanh	97.37 ± 0.39	0.0189	85.33
Kernel	**98.40 ± 0.46**	0.0815	**86.67**

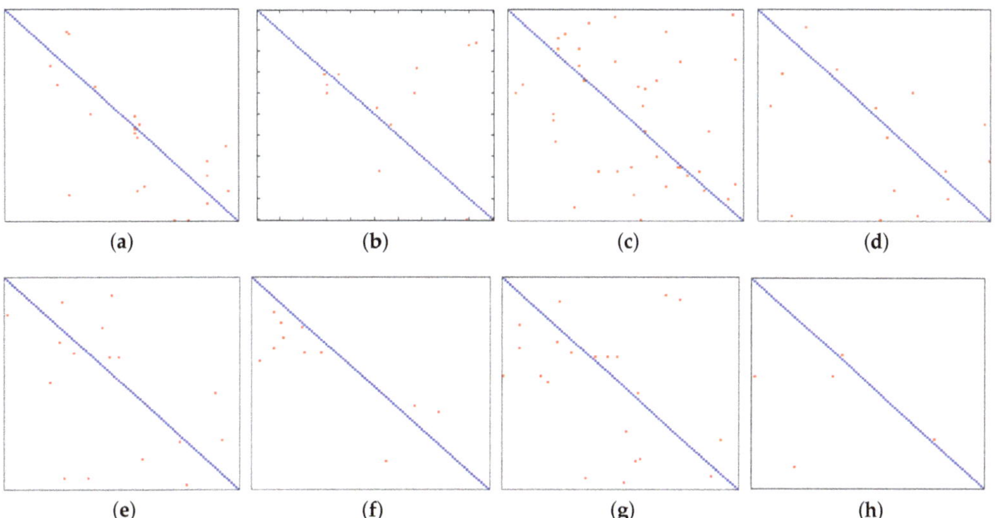

Figure 8. Patterns of confusion on AR dataset. (**a**) SRC. (**b**) CRC. (**c**) ScELM. (**d**) PSD. (**e**) LISTA. (**f**) LAMP. (**g**) LVAMP. (**h**) Kernel.

To give an intuitive illustration, Figure 9 shows all misclassified images obtained by LAMP method (which achieves the second best result) and this proposed method (kernel version). It can be seen that images with exaggerated facial expressions is the main reason causing misclassification for both methods. Another interesting point can be seen that most images with facial "disguises" are misclassified by LAMP method while they are correctly recognized by this proposed method. It indicates that the approximate sparse features estimated by this proposed method is robustness to facial occlusion or corruption than LAMP.

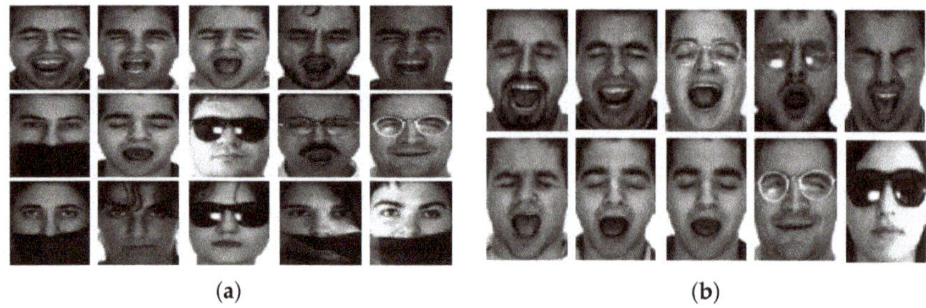

(a)   (b)

Figure 9. All misclassified images produced by (**a**) LAMP and (**b**) Kernel.

5. Conclusions

This paper proposes a simple fast approximation sparse coding method for small-scale datasets object recognition task, in which the optimal sparse features of training data computed by HIHT algorithm are used as ground truth to train a succinct and special SLNNs, thus make the representation learning in object recognition task more practical and efficient. Extensive experimental results on publicly available datasets show that this approach outperforms the compared approximation methods in terms of approximation performance, recognition accuracy and computational time. The high recognition and computational efficiency makes the proposed method very promising for real-time applications. Moreover, experimental results have demonstrated that this proposed method is robust to parameters on recognition performance, that make it more practical. Future work includes supervised sparse coding algorithms and autonomously finding an over-complete dictionary.

Author Contributions: Conceptualization, Z.S. and Y.Y.; methodology, Z.S.and Y.Y.; validation, Z.S.; data curation, Z.S.; writing—original draft preparation, Z.S.; writing—review and editing, Z.S. and Y.Y.; funding acquisition, Y.Y. All authors have read and agreed to the published version of the manuscript.

Funding: This research was funded by the National Natural Science Foundation of China (NSFC) under Grant No. 61873067.

Institutional Review Board Statement: Not applicable.

Informed Consent Statement: Not applicable.

Data Availability Statement: Publicly available datasets were analyzed in this study. These datasets can be found here: http://archive.ics.uci.edu/ml/index.php.

Conflicts of Interest: The authors declare no conflict of interest.

References

1. Lowe, D.G. Distinctive Image Features from Scale-Invariant Key-points. *Int. J. Comput. Vis.* **2004**, *60*, 91–110. [CrossRef]
2. Dalal, N.; Triggs, B. Histograms of Oriented Gradients for Human Detection. In Proceedings of the IEEE Computer Society Conference on Computer Vision and Pattern Recognition, San Diegol, CA, USA, 20–26 June 2005. [CrossRef]
3. Hubel, D.; Wiesel, T. Receptive fields of signal neurons in the cat's striate cortex. *J. Physiol.* **1959**, *148*, 574–591. [CrossRef] [PubMed]
4. Roll, E.; Tovee, M. Sparseness of the neuronal representation of stmuli in the primate temporal visual cortex. *J. Neurophysiol.* **1992**, *173*, 713–726. [CrossRef]
5. Mairal, J.; Bach, F.; Ponce, J.; Sapiro, G.; Zisserman, A. Discriminative learned dictionaries for local image analysis. In Proceedings of the IEEE Conference on Computer Vision and Pattern Recognition, Anchorage, AK, USA, 23–28 June 2008. [CrossRef]

6. Wang, J.; Yang, J.; Yu, K.; Lv, F.; Huang, T.; Gong, Y. Locality-constrained Linear Coding for image classification. In Proceedings of the 2010 IEEE Computer Society Conference on Computer Vision and Pattern Recognition, San Francisco, CA, USA, 13–18 June 2010; pp. 3360–3367. [CrossRef]

7. Jiang, Z.; Lin, Z.; Davis, L. Label consistent K-SVD: Learning a discriminative dictionary for recognition. *IEEE Trans. Pattern Anal. Mach. Intell.* **2013**, *35*, 2651–2663. [CrossRef]

8. Kavukcuoglu, K.; Ranzato, M.; LeCun, Y. Fast Inference in Sparse Coding Algorithms with Applications to Object Recognition. In *Technical Report CBLL-TR-2008-12-01, Computational and Biological Learning Lab, Courant Institude, NYU*; New York University: New York, NY, USA, 2008.

9. Gregor, K.; LeCun, Y. Learning fast approximations of sparse coding. In Proceedings of the International Conference on Machine Learning, Haifa, Israel, 21–24 June 2010.

10. Deng, X.; Dragotti, P.L. Deep Coupled ISTA Network for Multi-Modal Image Super-Resolution. *IEEE Trans. Image Process.* **2020**, *29*, 1683–1698. [CrossRef]

11. Qian, Y.; Xiong, F.; Qian, Q.; Zhou, J. Spectral Mixture Model Inspired Network Architectures for Hyperspectral Unmixing. *IEEE Trans. Geosci. Remote Sens.* **2020**, *58*, 7418–7434. [CrossRef]

12. Borgerding, M.; Schniter, P.; Rangan, S. Amp-inspired deep networks for sparse linear inverse problem. *IEEE Trans. Signal Process.* **2017**, *65*, 4293–4348. [CrossRef]

13. Dong, Z.; Zhu, W. Homotopy methods based on l_0-norm for compressed sensing. *IEEE Trans. Neural Networks Learn. Syst.* **2018**, *29*, 1132–1146. [CrossRef]

14. Mallat, S.; Zhang, Z. Matching pursuits with time-frequency dictionaries. *IEEE Trans. Signal Process.* **1993**, *41*, 3397–3415. [CrossRef]

15. Pati, Y.; Rezaiifar, R.; Krishnaprasad, P. Orthogonal matching pursuit: recursive function approximation with applications to wavelet decomposition. In Proceedings of the IEEE International Conference on Signal, Systems and Computers, Pacific Grove, CA, USA, 1–3 November 1993; pp. 40–44. [CrossRef]

16. Loza, C. RobOMP: Robust variants of Orthogonal Matching Pursuit for sparse representations. *PeerJ Comput. Sci.* **2019**, *5*, e192. [CrossRef]

17. Chen, S.; Donoho, D.; Saunders, M. Atomic decomposition by basis pursuit. *SIAM Rev.* **2001**, *43*, 129–159. [CrossRef]

18. Chartrand, R. Exact Reconstruction of Sparse Signals via Nonconvex Minimization. *IEEE Signal Process. Lett.* **2007**, *14*, 707–710. [CrossRef]

19. Xu, Z.; Chang, X.; Xu, F.; Zhangm, H. $L_{1/2}$ regularization: A thresholding representation theory and a fast solver. *IEEE Trans. Neural Networks Learn. Syst.* **2012**, *23*, 1013–1027. [CrossRef]

20. Chen, X.; Ng, M.K.; Zhang, C. Non-Lipschitz l_p-Regularization and Box Constrained Model for Image Restoration. *IEEE Trans. Image Process.* **2012**, *21*, 4709–4721. [CrossRef] [PubMed]

21. Qin, L.; Lin, Z.C.; She, Y.; Chao, Z. A comparison of typical L_p minimization algorithms. *Neurocomputing* **2013**, *119*, 413–424. [CrossRef]

22. Qiu, Y.; Jiang, H.; Ching, W.; Ng, M.K. On predicting epithelial mesenchymal transition by integrating RNA-binding proteins and correlation data via $L_{1/2}$-regularization method. *Artif. Intell. Med.* **2019**, *95*, 96–103. [CrossRef]

23. Donoho, D.; Elad, M. Optimally sparse representation in general (nonorthogonal) dictionaries via l^1 minimization. *Proc. Natl. Acad. Sci. USA* **2003**, *100*, 2197–2202. [CrossRef]

24. Candes, E.J.; Tao, T. Near-Optimal Signal Recovery From Random Projections: Universal Encoding Strategies? *IEEE Trans. Inf. Theory* **2006**, *52*, 5406–5425. [CrossRef]

25. Yang, A.; Ganesh, A.; Zhou, Z.; Sastry, S.; Ma, Y. A Review of Fast l_1-Minimization Algorithm for Robust Face Recognition. In Proceedings of the International Conference on Computer Vision and Pattern Recognition, Las Vegas, NV, USA, 12–15 July 2010; pp. 1–36.

26. Figueiredo, M.; Nowak, R.; Wright, S. Gradient Projection for Sparse Reconstruction: Application to Compressed Sensing and other Inverse Problems. *IEEE J. Sel. Top. Signal Process.* **2007**, *1*, 586–597. [CrossRef]

27. Kim, S.J.; Koh, K.; Boyd, S. An Interior-Point Method for Large-Scale l_1-Regularized Least Squares. *IEEE J. Sel. Top. Signal Process.* **2007**, *1*, 606–617. [CrossRef]

28. Osborne, M.R.; Presnell, B.; Turlach, B.A. A new approach to variable selection in least squares problems. *IMA J. Numer. Anal.* **2000**, *20*, 389–404. [CrossRef]

29. Malioutov, D.M.; Cetin, M.; Willsky, A.S. Homotopy continuation for sparse signal representation. In Proceedings of the IEEE International Conference on Acoustics, Speech, and Signal Processing, Philadelphia, PA, USA, 18–23 March 2005. [CrossRef]

30. Combettes, P.; Wajs, V. Signal recovery by proximal forward-backward splitting. *SIAM Multiscale Model. Simul.* **2005**, *4*, 1168–1200. [CrossRef]

31. Hale, E.; Yin, W.; Zhang, Y. *A Fixed-Point Continuation Method for l_1-Regularized Minimization with Applications to Compressed Sensing*; CAAM Tech Report TR07-07; Rice University: Houston, TX, USA, 7 July 2007; pp. 1–45. [CrossRef]

32. Wright, S.J.; Nowak, R.D.; Figueiredo, M.A. Sparse reconstruction by separable approximation. *IEEE Trans. Signal Process.* **2009**, *57*, 2479–2493. [CrossRef]

33. Beck, A.; Teboulle, M. A fast iterative shrinkage-thresholding algorithm for linear inverse problems. *SIAM J. Imaging Sci.* **2009**, *2*, 183–202. [CrossRef]

34. Lin, X.; Tong, Z. A proximal-gradient homotopy method for the sparse least-squares problem. *SIAM J. Optim.* **2013**, *23*, 1062–1091. [CrossRef]

35. Yang, J.; Zhang, Y. Alternating direction algorithms for l_1-problems inbcompressive sensing. *SIAM J. Sci. Comput.* **2011**, *31*, 250–278. [CrossRef]

36. Bioucas-Dias, J.M.; Figueiredo, M.A.T. A New TwIST: Two-Step Iterative Shrinkage/Thresholding Algorithms for Image Restoration. *IEEE Trans. Image Process.* **2007**, *16*, 2992–3004. [CrossRef]

37. Li, X.G.; Ge, J.; Jiang, H.M.; Wang, M.D.; Hong, M.Y.; Zhao, T. *Boosting Pathwise Coordinate Optimization in High Dimensions: Sequential Screening and Proximal Sub-Sampled Newton Algorithm*; Technical Report; Georgia Tech: Atlanta, GA, USA, 2017.

38. Zhao, T.; Liu, H.; Zhang, T. Pathwise coordinate optimization for nonconvex sparse learning: Algorithm and theory. *Ann. Stat.* **2018**, *46*, 180–218. [CrossRef]

39. Ge, J.; Li, X.; Jiang, H.; Liu, H.; Zhang, T.; Wang, M.; Zhao, T. Picasso: A sparse learning library for high dimensional data analysis in R and Python. *J. Mach. Learn. Res.* **2019**, *20*, 1–5.

40. Blumensath, T.; Davies, M. Iterative thresholding for sparse approximations. *Fourier Anal. Appl.* **2008**, *14*, 629–654. [CrossRef]

41. Lu, Z. Iterative hard thresholding methods for l_0 regularized convex cone programming. *Math. Program.* **2014**, *147*, 125–154. [CrossRef]

42. Chalasani, R.; Principe, J.C.; Ramakrishnan, N. A fast proximal method for convolutional sparse coding. In Proceedings of the 2013 International Joint Conference on Neural Networks (IJCNN), Dallas, TX, USA, 4–9 August 2013; pp. 1–5. [CrossRef]

43. Xin, B.; Wang, Y.; Gao, W.; Wipf, D.; Wang, B. Maximal sparsity with deep networks? In *Advances in Neural Information Processing Systems*; Curran Associates Inc.: Red Hook, NY, USA, 2016; pp. 4340–4348.

44. Donoho, D.L.; Maleki, A.; Montanari, A. Message passing algorithms for compressed sensing. *Proc. Natl. Acad. Sci. USA* **2009**, *106*, 18914–18919. [CrossRef]

45. Rangan, S.; Schniter, P.; Fletche, A.K. Vector approximate message passing. *arXiv* **2016**, arXiv:1610.03082.

46. Tsiligianni, E.; Deligiannis, N. Deep coupled-representation learning for sparse linear inverse problems with side information. *IEEE Signal Process. Lett.* **2019**, *26*, 1768–1772. [CrossRef]

47. Dua, D. and Graff, C. *UCI Machine Learning Repository*; University of California: Oakland, CA, USA, 2017.

48. Georghiades, A.; Belhumeur, P.; Kriegman, D. From few to many: Illumination cone models for face recognition under variable lighting and pose. *IEEE Trans. Pattern Anal. Mach. Intell.* **2001**, *23*, 643–660. [CrossRef]

49. Martinez, A.; Benavente, R. *The AR Face Database*; Tech. Rep; Comput. Vis. Center, Purdue University: West Lafayette, IN, USA, 1998.

50. Huang, G.; Zhou, H.; Ding, X.; Zhang, R. Extreme learning machine for regression and multiclass classification. *IEEE Trans. Syst. Man, Cybern. Part Cybern.* **2012**, *42*, 513–529. [CrossRef]

51. Yu, Y.; Sun, Z. Sparse coding extreme learning machine for classification. *Neurocomputing* **2017**, *261*, 50–56. [CrossRef]

52. Lee, H.; Battle, A.; Raina, R.; Ng, A. Efficient sparse coding algorithm. In *Advances in Neural Information Processing Systems*; Curran Associates Inc.: Vancouver, BC, Canada, 2006; pp. 801–808.

53. Wright, J.; Yang, A.; Ganesh, A.; Sastry, S.; Ma, Y. Robust Face Recognition via Sparse Representation. *IEEE Trans. Pattern Anal. Mach. Intell.* **2009**, *31*, 210–227. [CrossRef] [PubMed]

54. Zhang, L.; Yang, M.; Feng, X. Sparse representation or collaborative representation: Which helps face recognition? In Proceedings of the 2011 International Conference on Computer Vision, Barcelona, Spain, 6–13 November 2011; pp. 471–478. [CrossRef]

sensors

Article

Multi-Block Color-Binarized Statistical Images for Single-Sample Face Recognition

Insaf Adjabi [1], Abdeldjalil Ouahabi [1,2,*], Amir Benzaoui [3] and Sébastien Jacques [4]

[1] Department of Computer Science, LIMPAF, University of Bouira, Bouira 10000, Algeria; i.adjabi@univ-bouira.dz

[2] Polytech Tours, Imaging and Brain, INSERM U930, University of Tours, 37200 Tours, France

[3] Department of Electrical Engineering, University of Bouira, Bouira 10000, Algeria; a.benzaoui@univ-bouira.dz

[4] GREMAN UMR 7347, University of Tours, CNRS, INSA Centre Val-de-Loire, 37200 Tours, France; sebastien.jacques@univ-tours.fr

* Correspondence: abdeldjalil.ouahabi@univ-tours.fr; Tel.: +33-2-4736-1323

Abstract: Single-Sample Face Recognition (SSFR) is a computer vision challenge. In this scenario, there is only one example from each individual on which to train the system, making it difficult to identify persons in unconstrained environments, mainly when dealing with changes in facial expression, posture, lighting, and occlusion. This paper discusses the relevance of an original method for SSFR, called Multi-Block Color-Binarized Statistical Image Features (MB-C-BSIF), which exploits several kinds of features, namely, local, regional, global, and textured-color characteristics. First, the MB-C-BSIF method decomposes a facial image into three channels (e.g., red, green, and blue), then it divides each channel into equal non-overlapping blocks to select the local facial characteristics that are consequently employed in the classification phase. Finally, the identity is determined by calculating the similarities among the characteristic vectors adopting a distance measurement of the K-nearest neighbors (K-NN) classifier. Extensive experiments on several subsets of the unconstrained Alex and Robert (AR) and Labeled Faces in the Wild (LFW) databases show that the MB-C-BSIF achieves superior and competitive results in unconstrained situations when compared to current state-of-the-art methods, especially when dealing with changes in facial expression, lighting, and occlusion. The average classification accuracies are 96.17% and 99% for the AR database with two specific protocols (i.e., Protocols I and II, respectively), and 38.01% for the challenging LFW database. These performances are clearly superior to those obtained by state-of-the-art methods. Furthermore, the proposed method uses algorithms based only on simple and elementary image processing operations that do not imply higher computational costs as in holistic, sparse or deep learning methods, making it ideal for real-time identification.

Keywords: biometrics; face recognition; single-sample face recognition; binarized statistical image features; K-nearest neighbors

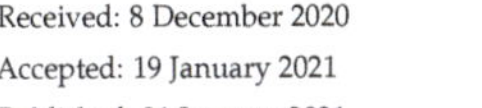

Citation: Adjabi, I.; Ouahabi, A.; Benzaoui, A.; Jacques, S. Multi-Block Color-Binarized Statistical Images for Single-Sample Face Recognition. *Sensors* **2021**, *21*, 728. https://doi.org/10.3390/s21030728

Academic Editor: Kang Ryoung Park

Received: 8 December 2020

Accepted: 19 January 2021

Published: 21 January 2021

Publisher's Note: MDPI stays neutral with regard to jurisdictional claims in published maps and institutional affiliations.

1. Introduction

Generally speaking, biometrics aims to identify or verify an individual's identity according to some physical or behavioral characteristics [1]. Biometric practices replace conventional knowledge-based solutions, such as passwords or PINs, and possession-based strategies, such as ID cards or badges [2]. Several biometric methods have been developed to varying degrees and are being implemented and used in numerous commercial applications [3].

Fingerprints are the biometric features most commonly used to identify criminals [4]. The first automated fingerprint authentication device was commercialized in the early 1960s. Multiple studies have shown that the iris of the eye is the most accurate modality since its texture remains stable throughout a person's life [5]. However, those techniques have the significant drawback of being invasive, which significantly restricts their applications.

37

Besides, iris recognition remains problematic for users who do not wish to put their eyes in front of a sensor. On the contrary, biometric recognition based on facial analysis does not pose any such user constraints. In contrast to other biometric modalities, face recognition is a modality that can be employed without any user–sensor co-operation and can be applied discreetly in surveillance applications. Face recognition has many advantages: the sensor device (i.e., the camera) is simple to mount; it is not costly; it does not require subject co-operation; there are no hygiene issues; and, being passive, people much prefer this modality [6].

Two-dimensional face recognition with Single-Sample Face Recognition (SSFR) (i.e., using a Single- Sample Per Person (SSPP) in the training set) has already matured as a technology. Although the latest studies on the Face Recognition Grand Challenge (FRGC) [7] project have shown that computer vision systems [8] offer better performance than human visual systems in controlled conditions [9], research into face recognition, however, needs to be geared towards more realistic uncontrolled conditions. In an uncontrolled scenario, human visual systems are more robust when dealing with the numerous possibilities that can impact the recognition process [10], such as variations in lighting, facial orientation, facial expression, and facial appearance due to the presence of sunglasses, a scarf, a beard, or makeup. Solving these challenges will make 2D face recognition techniques a much more important technology for identification or identity verification.

Several methods and algorithms have been suggested in the face recognition literature. They can be subdivided into four fundamental approaches depending on the method used for feature extraction and classification: holistic, local, hybrid, and deep learning approaches [11]. The deep learning class [12], which applies consecutive layers of information processing arranged hierarchically for representation, learning, and classification, has dramatically increased state-of-the-art performance, especially with unconstrained large-scale databases, and encouraged real-world applications [13,14].

Most current methods in the literature use several facial images (samples) per person in the training set. Nevertheless, in real-world systems (e.g., in fugitive tracking, identity cards, immigration management, or passports), only SSFR systems are used (due to the limited storage and privacy policy), which employ a single sample per person in the training stage (generally neutral images acquired in controlled conditions), i.e., just one example of the person to be recognized is recorded in the database and accessible for the recognition task [15]. Since there are insufficient data (i.e., we do not have several samples per person) to perform supervised learning, many well-known algorithms may not work particularly well. For instance, Deep Neural Networks (DNNs) [13] can be used in powerful face recognition techniques. Nonetheless, they necessitate a considerable volume of training data to work well. Vapnik and Chervonenkis [16] showed that vast training data must ensure learning systems' generalization in their statistical learning theorem. In addition, the use of three-dimensional (3D) imaging instead of two-dimensional representation (2D) has made it possible to cover several issues related to image acquisition conditions, in particular pose, lighting and make-up variations. While 3D models offer a better representation of the face shape for a clear distinction between persons [17,18], they are often not suitable for real-time applications because they require expensive and sophisticated calculations and specific sensors. We infer that SSFR remains an unsolved issue in academic and business circles, particularly with respect to the major efforts and growth in face recognition.

In this paper, we tackle the SSFR issue in unconstrained conditions by proposing an efficient method based on a variant of the local texture operator Binarized Statistical Image Features (BSIF) [19] called Multi-Block Color-binarized Statistical Image Features (MB-C-BSIF). It employs local color texture information to obtain honest and precise representation. The BSIF descriptor has been widely used in texture analysis [20,21] and has proven its utility in many computer vision tasks. In the first step, the proposed method uses preprocessing to enhance the quality of facial photographs and remove noise [22–24]. The color image is then decomposed into three channels (e.g., red, green, and blue for the RGB color-space). Next, to find the optimum configuration, several multi-block decompositions are checked and examined under various color-spaces (i.e., we tested RGB, Hue Saturation Value (HSV), in addition to the YCbCr color-spaces, where Y is the luma component; Cb and Cr are the blue-difference and red-difference chroma components, respectively). Finally, classification is undertaken using the distance measurement of the K-nearest neighbors (K-NN) classifier. Compared to several related works, the advantage of our method lies in exploiting several kinds of information: local, regional, global, and color-texture. Besides, the algorithm of our method is simple and does not require greater complexity, which makes it suitable for real-time applications (e.g., surveillance systems or real-time identification). Our system is based on only basic and simple image processing operations (e.g., median filtering, a simple convolution, or histogram calculation), involving a much lower computational cost than existing systems. For example, (1) Subspace or sparse representation-based methods involve many calculations and higher time in dimensionality reduction, or (2) Deep learning methods involve very high complexity cost and require many computations. For such systems, GPUs' need clearly shows that many calculations must be done in parallel; GPUs are designed to run concurrently with thousands of processor cores, making for extensive parallelism where each core is concentrated on making accurate calculations. With a standard CPU, a considerable amount of time for training and testing will be needed for deep learning systems.

The rest of the paper is structured as follows. We discuss relevant research about SSFR in Section 2. Section 3 describes our suggested method. In Section 4, the experimental study, key findings, and comparisons are performed and presented to show our method's superiority. Section 5 of the paper presents key findings and discusses research perspectives.

2. Related Work

Current methods designed to resolve the SSFR issue can be categorized into four fundamental classes [25], namely: virtual sample generating, generic learning, image partitioning, and deep learning methods.

2.1. Virtual Sample Generating Methods

The methods in this category produce some additional virtual training samples for each individual to augment the gallery (i.e., data augmentation), so that discriminative sub-space learning can be employed to extract features. For example, Vetter (1998) [26] proposed a robust SSFR algorithm by generating 3D facial models through the recovery of high-fidelity reflectance and geometry. Zhang et al. (2005) [27] and Gao et al. (2008) [28] developed two techniques to tackle the issue of SSFR based on the singular value decomposition (SVD). Hu et al. (2015) [29] suggested a different SSFR system based on the lower-upper (LU) algorithm. In their approach, each single subject was decomposed and transposed employing the LU procedure and each raw image was rearranged according to its energy. Dong et al. (2018) [30] proposed an effective method for the completion of SSFR tasks called K-Nearest Neighbors virtual image set-based Multi-manifold Discriminant Learning (KNNMMDL). They also suggested an algorithm named K-Nearest Neighbor-based Virtual Sample Generating (KNNVSG) to augment the information of intra-class variation in the training samples. They also proposed the Image Set-based Multi-manifold Discriminant Learning algorithm (ISMMDL) to exploit intra-class variation information. While these methods can somewhat alleviate the SSFR problem, their main disadvantage lies in the strong correlation between the virtual images, which cannot be regarded as independent examples for the selection of features.

2.2. Generic Learning Methods

The methods in this category first extract discriminant characteristics from a supplementary generic training set that includes several examples per individual and then use those characteristics for SSFR tasks. Deng et al. (2012) [31] developed the Extended Sparse Representation Classifier (ESRC) technique in which the intra-class variant dictionary is created from generic persons not incorporated in the gallery set to increase the efficiency of the identification process. In a method called Sparse Variation Dictionary Learning (SVDL), Yang et al. (2013) [32] trained a sparse variation dictionary by considering the relation between the training set and the outside generic set, disregarding the distinctive features of various organs of the human face. Zhu et al. (2014) [33] suggested a system for SSFR based on Local Generic Representation (LGR), which leverages the benefits of both image partitioning and generic learning and takes into account the fact that the intra-class face variation can be spread among various subjects.

2.3. Image Partitioning Methods

The methods in this category divide each person's images into local blocks, extract the discriminant characteristics, and, finally, perform classifications based on the selected discriminant characteristics. Zhu et al. (2012) [34] developed a Patch-based CRC (PCRC) algorithm that applies the original method proposed by Zhang et al. (2011) [35], named Collaborative Representation-based Classification (CRC), to each block. Lu et al. (2012) [36] suggested a technique called Discriminant Multi-manifold Analysis (DMMA) that divides any registered image into multiple non-overlapping blocks and then learns several feature spaces to optimize the various margins of different individuals. Zhang et al. (2018) [37] developed local histogram-based face image operators. They decomposed each image into different non-overlapping blocks. Next, they tried to derive a matrix to project the blocks into an optimal subspace to maximize the different margins of different individuals. Each column was then redesigned to an image filter to treat facial images and the filter responses were binarized using a fixed threshold. Gu et al. (2018) [38] proposed a method called Local Robust Sparse Representation (LRSR). The main idea of this technique is to merge a local sparse representation model with a block-based generic variation dictionary learning model to determine the possible facial intra-class variations of the test images. Zhang et al. (2020) [39] introduced a novel Nearest Neighbor Classifier (NNC) distance measurement to resolve SSFR problems. The suggested technique, entitled Dissimilarity-based Nearest Neighbor Classifier (DNNC), divides all images into equal non-overlapping blocks and produces an organized image block-set. The dissimilarities among the given query image block-set and the training image block-sets are calculated and considered by the NNC distance metric.

2.4. Deep Learning Methods

The methods in this category employ consecutive hidden layers of information-processing arranged hierarchically for representation, learning, and classification. They can automatically determine complex non-linear data structures [40]. Zeng et al. (2017) [41] proposed a method that uses Deep Convolutional Neural Networks (DCNNs). Firstly, they propose using an expanding sample technique to augment the training sample set, and then a trained DCNN model is implemented and fine-tuned by those expanding samples to be used in the classification process. Ding et al. (2017) [42] developed a deep learning technique centered on a Kernel Principal Component Analysis Network (KPCANet) and a novel weighted voting technique. First, the aligned facial image is segmented into multiple non-overlapping blocks to create the training set. Then, a KPCANet is employed to get filters and banks of features. Lastly, recognition of the unlabeled probe is achieved by applying the weighted voting form. Zhang and Peng (2018) [43] introduced a different method to generate intra-class variances using a deep auto-encoder. They then used these intra-class variations to expand the new examples. First, a generalized deep auto-encoder is used to train facial images in the gallery. Second, a Class-specific Deep Auto-encoder (CDA) is fine-tuned with a single example. Finally, the corresponding CDA is employed to expand the new samples. Du and Da (2020) [44] proposed a method entitled Block Dictionary Learning (BDL) that fuses Sparse Representation (SR) with CNNs. SR is implemented to augment CNN efficiency by improving the inter-class feature variations and creating a global-to-local dictionary learning process to increase the method's robustness.

It is clear that the deep learning approach for face recognition has gained particular attention in recent years, but it suffers considerably with SSFR systems as they still require a significant amount of information in the training set.

Motivated by the successes of the third approach, "image partitioning", and the reliability of the local texture descriptor BSIF, in this paper, we propose an image partitioning method to address the problems of SSFR. The proposed method, called MB-C-BSIF, decomposes each image into several color channels, divides each color component into various equal non-overlapping blocks, and applies the BSIF descriptor to each block-component to extract the discriminative features. In the following section, the framework of the MB-C-BSIF is explained in detail.

3. Proposed Method

This section details the MB-C-BSIF method (see Figure 1) proposed in this article to solve the SSFR problem. MB-C-BSIF is an approach based on image partitioning and consists of three key steps: image pre-processing, feature extraction based on MB-C-BSIF, and classification. In the following subsections, we present these three phases in detail.

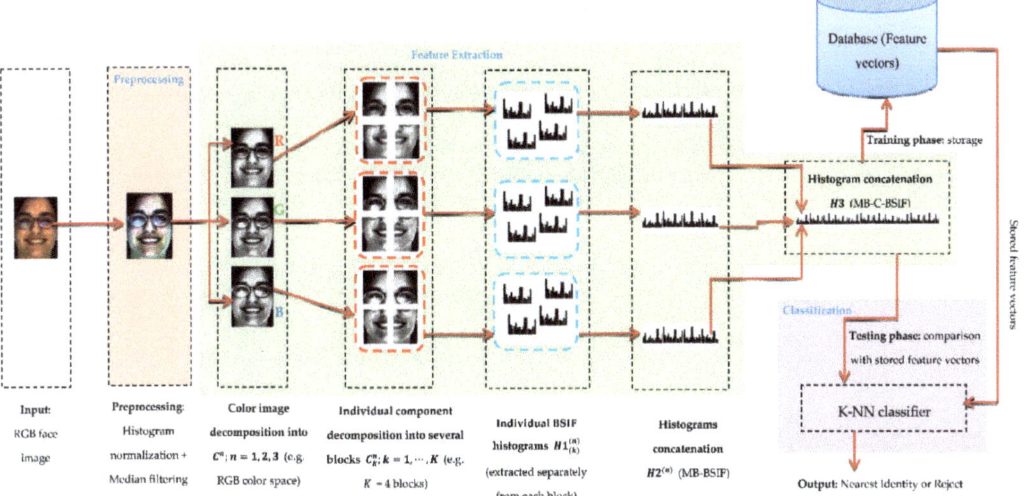

Figure 1. Schematic of the proposed Single-Sample Face Recognition (SSFR) system based on the Multi-Block Color-Binarized Statistical Image Features (MB-C-BSIF) descriptor.

3.1. Preprocessing

The suggested feature extraction and classification rules compose the essential steps in our proposed SSFR. However, before driving these two steps, pre-processing is necessary to improve the visual quality of the captured image. The facial image is enhanced by applying histogram normalization and then filtered with a non-linear filter. The median filter [45] was adopted to minimize noise while preserving the facial appearance and enhancing the operational outcomes [46].

3.2. MB-C-BSIF-Based Feature Extraction

Our advanced feature extraction technique is based on the multi-block color representation of the BSIF descriptor, entitled Multi-Block Color BSIF (MB-C-BSIF). The BSIF operator proposed by Kannala and Rahtu [16] is an efficient and robust descriptor for texture analysis [47,48]. BSIF focuses on creating local image descriptors that powerfully encode texture information and are appropriate for describing image regions in the form of histograms. The method calculates a binary code for all pixels by linearly projecting local image blocks onto a subspace whose basis vectors are learned from natural pictures through Independent Component Analysis (ICA) [45] and by binarizing the coordinates through thresholding. The number of basis vectors defines the length of the binary code string. Image regions can be conveniently represented with histograms of the pixels' binary codes. Other descriptors that generate binary codes, such as the Local Binary Pattern (LBP) [49] and the Local Phase Quantization (LPQ) [50], have inspired the BSIF process. However, the BSIF is based on natural image statistics rather than heuristic or handcrafted code constructions, enhancing its modeling capabilities.

Technically speaking, the s_i filter response is calculated, for a given picture patch X of size $l \times l$ pixels and a linear filter W_i of the same size, by:

$$s_i = \sum_{u,v} W_i(u,v) X(u,v) \tag{1}$$

where the index i in W_i indicates the ith filter.

The binarized b_i feature is calculated as follows:

$$b_i = \begin{cases} 1 & \text{if } s_i > 0 \\ 0 & \text{otherwise} \end{cases} \tag{2}$$

The BSIF descriptor has two key parameters: the filter size $l \times l$ and the bit string length n. Using ICA, W_i filters are trained by optimizing s_i's statistical independence. The training of W_i filters is based on different choices of parameter values. In particular, each filter set was trained using 50,000 image patches. Figure 2 displays some examples of the filters obtained with $l \times l = 7 \times 7$ and $n = 8$. Figure 3 provides some examples of facial images and their respective BSIF representations (with $l \times l = 7 \times 7$ and $n = 8$).

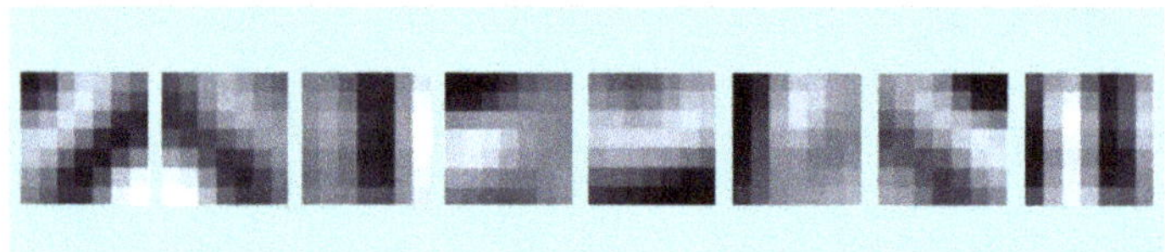

Figure 2. Examples of 7×7 BSIF filter banks learned from natural pictures.

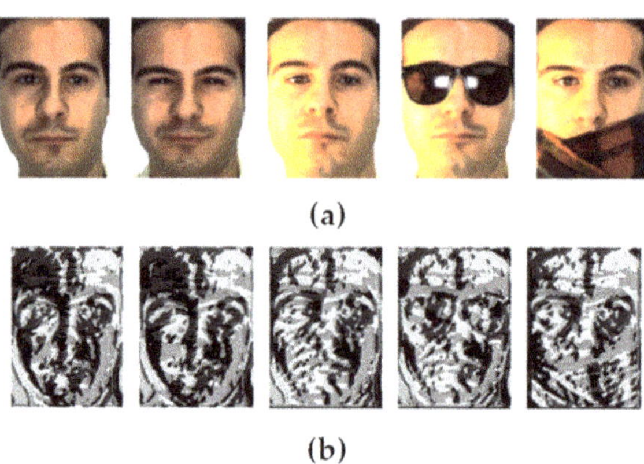

(a)

(b)

Figure 3. (a) Examples of facial images, and (b) their parallel BSIF representations.

Like LBP and LPQ methodologies, the BSIF codes' co-occurrences are collected in a histogram $H1$, which is employed as a feature vector.

However, the simple BSIF operator based on a single block does not possess information that dominates the texture characteristics, which is forceful for the image's occlusion and rotation. To address those limitations, an extension of the basic BSIF, the Multi-Block BSIF (MB-BSIF), is used. The concept is based on partitioning the original image into non-overlapping blocks. An undefined facial image may be split equally along the horizontal and vertical directions. As an illustration, we can derive 1, 4, or 16 blocks by segmenting the image into grids of 1×1, 2×2, or 4×4, as shown in Figure 4. Each block possesses details about its composition, such as the nose, eyes, or eyebrows. Overall, these blocks provide information about position relationships, such as nose to mouth and eye to eye. The blocks and the data between them are thus essential for SSFR tasks.

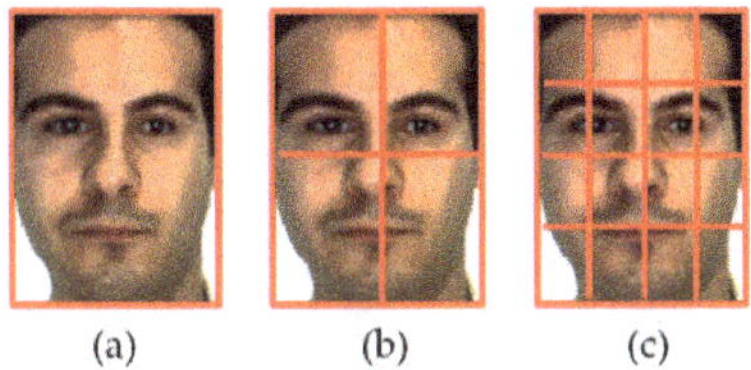

(a)   (b)   (c)

Figure 4. Examples of multi-block (MB) image decomposition: (**a**) 1×1, (**b**) 2×2, and (**c**) 4×4.

Our idea was to segment the image into equal non-overlapping blocks and calculate the BSIF operator's histograms related to the different blocks. The histogram $H2$ represents the fusion of the regular histograms calculated for the different blocks, as shown in Figure 5.

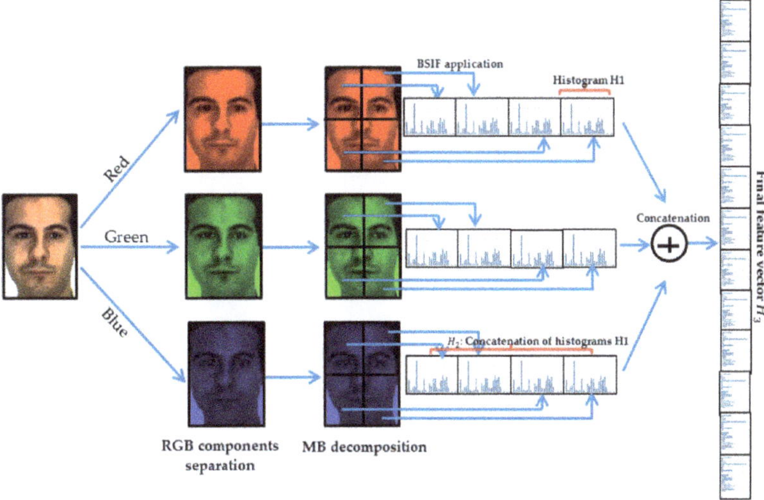

Figure 5. Structure of the proposed feature extraction approach: MB-C-BSIF.

In the face recognition literature, some works have concentrated solely on analyzing the luminance details of facial images (i.e., grayscale). This paper suggests a different and exciting technique that exploits color texture information and shows that analysis of chrominance can be beneficial to SSFR systems. To prove this idea, we can separate the RGB facial image into three channels (i.e., red, green, and blue) and then compute the MB-BSIF separately for each channel. The final feature vector is the concatenation of their histograms in a global histogram $H3$. This approach is called Multi-Block Color BSIF (MB-C-BSIF). Figure 5 provides a schematic illustration of the proposed MB-C-BSIF framework.

We note that the RGB is the most commonly employed color-space for detecting, modeling, and displaying color images. Nevertheless, its use in image interpretation is restricted due to the broad connection between the three color channels (i.e., red, green, and blue) and the inadequate separation of details in terms of luminance and chrominance. To identify captured objects, the various color channels can be highly discriminative and offer excellent contrast for several visual indicators from natural skin tones. In addition to the RGB, we studied and tested two additional color-spaces—HSV and YCbCr—to exploit color texture details. These color-spaces are based on separating components of the chrominance and luminance. For the HSV color-space, the dimensions of hue and saturation determine the image's chrominance while the dimension of brightness (v) matches the luminance. The YCbCr color-space divides the components of the RGB into luminance (Y), chrominance blue (Cb), and chrominance red (Cr). We should note that the representation of chrominance components in the HSV and YCbCr domains is dissimilar, and consequently, they can offer additional color texture descriptions for SSFR systems.

3.3. K-Nearest Neighbors (K-NN) Classifier

During the classification process, each tested facial image is compared with those saved in the dataset. To assign the corresponding label (i.e., identity) to the tested image, we used the K-NN classifier associated with a distance metric. In scenarios of general usage, K-NNs show excellent flexibility and usability in substantial applications.

Technically speaking, for a presented training set $\{(x_i, y_i)\ i = 1, 2, \ldots, s\}$, where $x_i \in R^D$ denotes the ith person's feature vector, y_i denotes this person's label, D is the dimension of the characteristic vector, and s represents the number of persons. For a test person $x' \in R^D$ that is expected to be classified, the K-NN is used to determine a training person x^* resembling to x' based on the distance rate and then attribute the label of x^* to x'.

K-NN can be implemented with various distance measurements. We evaluated and compared three widely used distance metrics in this work: Hamming, Euclidean, and city block (also called Manhattan distance).

The Hamming distance between x' and x_i is calculated as follows:

$$d(x', x_i) = \sum_{j=1}^{D} \left(x_j' - x_{ij} \right)^2 \tag{3}$$

The Euclidean distance between x' and x_i is formulated as follows:

$$d(x', x_i) = \sqrt{\sum_{j=1}^{D} \left(x_j' - x_{ij} \right)^2} \tag{4}$$

The city block distance between x' and x_i is measured as follows:

$$d(x', x_i) = \sum_{j=1}^{D} \left(x_j' - x_{ij} \right) \tag{5}$$

where x' and x_i are two vectors of dimension D, while x_{ij} is the jth feature of x_i, and x_j' is the jth feature of x'.

The corresponding label of x' can be determined by:

$$y' = y_{i*} \tag{6}$$

where

$$i^* = arg_{i=1,\ldots,s} \left(min \left(d(x', x_i) \right) \right) \tag{7}$$

The distance metric in SSFR corresponds to calculating the similarities between the test example and the training examples.

The Algorithm 1 sums up our proposed method of SSFR recognition.

Algorithm 1 SSFR based on MB-C-BSIF and K-NN

Input: Facial image X
1.   Apply histogram normalization on X
2.   Apply median filtering on X
3.   Divide X into three components (red, green, blue): $C^n; n = 1, 2, 3$
4.   **for** $n = 1$ to 3
5.   Divide C^n into K equivalent blocks: $C_k^n; k = 1, \ldots, K$
6.   **for** $k = 1$ to K
7.   Compute BSIF on the block-component C_k^n: $H1_{(k)}^{(n)}$
8.   **end for**
9.   Concatenate the computed MB-BSIF features of the component C_n:
10. $H2^{(n)} = H1_{(1)}^{(n)} + H1_{(2)}^{(n)} + \cdots + H1_{(K)}^{(n)}$
11.   **end for**
12.   Concatenate the computed MB-C-BSIF features: $H3 = H2^{(1)} + H2^{(2)} + H2^{(3)}$
13.   Apply K-NN associated with a metric distance
Output: Identification decision

---

4. Experimental Analysis

The proposed SSFR was evaluated using the unconstrained Alex and Robert (AR) [51] and Labeled Faces in the Wild (LFW) [52] databases. In this section, we present the specifications of each utilized database and their experimental setups. Furthermore, we analyze the findings obtained from our proposed SSFR method and compare the accuracy of recognition with other current state-of-the-art approaches.

4.1. Experiments on the AR Database

4.1.1. Database Description

The Alex and Robert (AR) face database [51] includes more than 4000 colored facial photographs of 126 individuals (56 females and 70 males); each individual has 26 different images with a frontal face taken with several facial expressions, lighting conditions, and occlusions. These photographs were acquired at an interval of two-weeks and their analysis was in two sessions (shots 1 and 2). Each session comprised 13 facial photographs per subject. A subset of facial photographs of 100 distinct individuals (50 males and 50 females) was selected in the subsequent experiments. Figure 6 displays the 26 facial images of the first individual from the AR database, along with detailed descriptions of them.

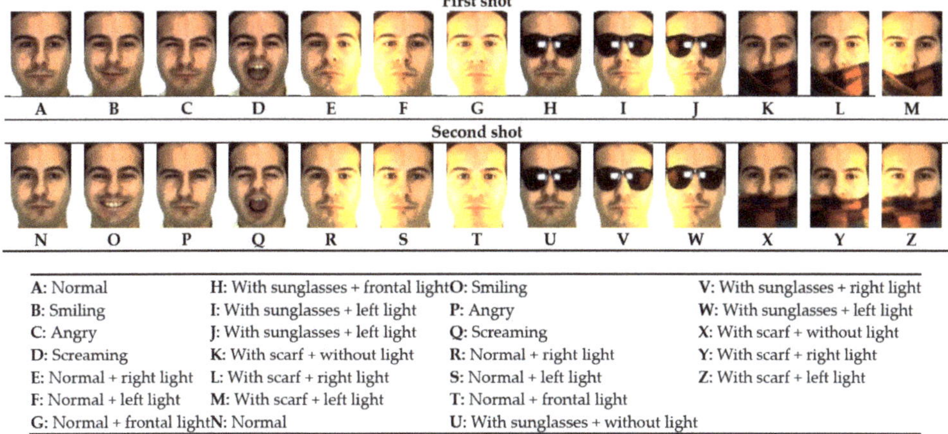

A: Normal	H: With sunglasses + frontal light	O: Smiling	V: With sunglasses + right light
B: Smiling	I: With sunglasses + left light	P: Angry	W: With sunglasses + left light
C: Angry	J: With sunglasses + left light	Q: Screaming	X: With scarf + without light
D: Screaming	K: With scarf + without light	R: Normal + right light	Y: With scarf + right light
E: Normal + right light	L: With scarf + right light	S: Normal + left light	Z: With scarf + left light
F: Normal + left light	M: With scarf + left light	T: Normal + frontal light	
G: Normal + frontal light	N: Normal	U: With sunglasses + without light	

Figure 6. The 26 facial images of the first individual from the AR database and their detailed descriptions.

4.1.2. Setups

To determine the efficiency of the proposed MB-C-BSIF in dealing with changes in facial expression, subset A (normal-1) was used as the training set and subsets B (smiling-1), C (angry-1), D (screaming-1), N (normal-2), O (smiling-2), P (angry-2), and Q (screaming-2) were employed for the test set. The facial images from the eight subsets displayed different facial expressions and were used in two different sessions. For the training set, we employed 100 images of the normal-1 type (100 images for 100 persons, i.e., one image per person). Moreover, we employed 700 images in the test set (smiling-1, angry-1, screaming-1, normal-2, smiling-2, angry-2, and screaming-2). These 700 images were divided into seven subsets for testing, with each subset containing 100 images.

As shown in Figure 6, two forms of occlusion are found in 12 subsets. The first is occlusion by sunglasses, as seen in subsets H, I, J, U, V, and W, while the second is occlusion by a scarf in subsets K, L, M, X, Y, and Z. In these 12 subsets, each individual's photographs have various illumination conditions and were acquired in two distinct stages. There are 100 different items in each subset and the total number of facial photographs used in the test set was 1200. To examine the performance of the suggested MB-C-BSIF under conditions of object occlusion, we considered subset A as the training set and the 12 occlusion subjects as the test set, which was similar to the initial setup.

4.1.3. Experiment #1 (Effects of BSIF Parameters)

As stated in Section 3.2, the BSIF operator is based on two parameters: filter kernel size $l \times l$ and bit string length n. In this test, we assessed the proposed method by testing various BSIF parameters to obtain the best configuration, i.e., the one that yielded the best recognition accuracy. We transformed the image into a grayscale level, we did not segment the image into non-overlapping blocks (i.e., 1×1 block), and we used the city block distance associated with K-NN. Tables 1–3 show comprehensive details and comparisons of results obtained using some (key) BSIF configurations for facial expression variation subsets, occlusion subsets for sunglasses, and occlusion subsets for scarfs, respectively. The best results are in bold.

We note that using the parameters $l \times l = 17 \times 17$ and $n = 12$ for the BSIF operator achieves the best performance in identification compared to other configurations considered in this experiment. Furthermore, an increase in the identification rate appears when we augment the values of l or n. The implemented configuration can achieve better accuracy for changes in facial expression with all seven subsets. However, for subset Q, which is characterized by considerable variation in facial expression, the accuracy of recognition was very low (71%). Lastly, the performance of this implemented configuration under conditions of occlusion by an object is unsatisfactory, especially with occlusion by a scarf, and needs further improvement.

Table 1. Comparison of the results obtained using six BSIF configurations with changes in facial expression.

$l \times l$ (Pixels)	n (Bits)	Accuracy (%)							Average Accuracy (%)
		B	**C**	**D**	**N**	**O**	**P**	**Q**	
3×3	5	70	72	38	36	20	24	14	39.14
5×5	9	94	97	59	75	60	66	30	68.71
9×9	12	100	100	91	95	90	92	53	88.71
11×11	8	97	99	74	85	70	75	43	77.57
15×15	12	100	100	96	97	96	96	73	94.00
17×17	12	100	100	98	97	96	97	71	94.14

Table 2. Comparison of the results obtained using six BSIF configurations with occlusion by sunglasses.

$l \times l$ (Pixels)	n (Bits)	Accuracy (%)						Average Accuracy (%)
		H	I	J	U	V	W	
3×3	5	29	8	4	12	4	3	10.00
5×5	9	70	24	14	28	14	8	26.50
9×9	12	98	80	61	80	38	30	61.50
11×11	8	78	34	23	48	26	15	37.33
15×15	12	100	84	85	87	50	46	75.33
17×17	12	100	91	87	89	58	46	78.50

Table 3. Comparison of the results obtained using six BSIF configurations with occlusion by scarf.

$l \times l$ (Pixels)	n (Bits)	Accuracy (%)						Average Accuracy (%)
		K	L	M	X	Y	Z	
3×3	5	7	4	2	3	2	2	3.33
5×5	9	22	9	6	12	6	2	9.50
9×9	12	88	54	34	52	31	15	45.67
11×11	8	52	12	90	22	9	7	32.00
15×15	12	97	69	64	79	48	37	65.67
17×17	12	98	80	63	90	48	31	68.33

4.1.4. Experiment #2 (Effects of Distance)

In this experiment, we evaluated the last configuration (i.e., grayscale level image, 1×1 block $l \times l = 17 \times 17$, and $n = 12$) by checking various distances associated with K-NN for classification. Tables 4–6 compare the results achieved by adopting the city block distance and other well-known distances with facial expression variation subsets, occlusion subsets for sunglasses, and occlusion subsets for scarfs, respectively. The best results are in bold.

Table 4. Comparison of the results obtained using different distances with changes in facial expression.

Distance	Accuracy (%)							Average Accuracy (%)
	B	C	D	N	O	P	Q	
Hamming	63	79	9	69	23	40	6	41.29
Euclidean	99	100	80	90	83	82	43	82.43
City block	100	100	98	97	96	97	71	94.14

Table 5. Comparison of the results obtained using different distances with occlusion by sunglasses.

Distance	Accuracy (%)						Average Accuracy (%)
	H	I	J	U	V	W	
Hamming	37	5	6	11	4	2	10.83
Euclidean	96	68	42	68	31	17	53.67
City block	100	91	87	89	58	46	78.50

Table 6. Comparison of the results obtained using different distances with occlusion by scarf.

Distance	Accuracy (%)						Average Accuracy (%)
	K	**L**	**M**	**X**	**Y**	**Z**	
Hamming	34	5	8	20	4	4	12.50
Euclidean	79	32	16	41	22	5	32.50
City block	98	80	63	90	48	31	68.33

We note that the city block distance produced the most reliable recognition performance compared to the other distances analyzed in this test, such as the Hamming and Euclidean distances. As such, we can say that the city block distance is the most suitable for our method.

4.1.5. Experiment #3 (Effects of Image Segmentation)

To improve recognition accuracy, especially under conditions of occlusion, we proposed decomposing the image into several non-overlapping blocks, as discussed in Section 3.2. The objective of this test was to estimate identification performance when MB-BSIF features are used instead of their global computation over an entire image. In this paper, three methods for image segmentation are considered and compared. Each original image was divided into 1×1 (i.e., global information), 2×2, and 4×4 blocks (i.e., local information). In other terms, an image was divided into 1 block (i.e., the original image), 4 blocks, and 16 blocks. For the last two cases, the feature vectors (i.e., histograms H1) derived from each block were fused to create the entire image extracted feature vector (Histogram H2). Tables 7–9 present and compare the recognition accuracy of the tested MB-BSIF for various blocks with subsets of facial expression variation, occlusion subsets for sunglasses, and occlusion subsets for scarfs, respectively (with grayscale images, city block distance, $l \times l = 17 \times 17$, and $n = 12$). The best results are in bold.

Table 7. Comparison of the results obtained using different divided blocks with changes in facial expression.

Segmentation	Accuracy (%)							Average Accuracy (%)
	B	**C**	**D**	**N**	**O**	**P**	**Q**	
(1×1)	100	100	98	97	96	97	71	94.14
(2×2)	100	100	95	98	92	91	60	90.86
(4×4)	100	100	99	98	92	97	76	94.57

Table 8. Comparison of the results obtained using different divided blocks with occlusion by sunglasses.

Segmentation	Accuracy (%)						Average Accuracy (%)
	H	**I**	**J**	**U**	**V**	**W**	
(1×1)	100	91	87	89	58	46	78.50
(2×2)	100	99	98	91	83	71	90.33
(4×4)	100	99	99	93	81	79	91.83

Table 9. Comparison of the results obtained using different divided blocks with occlusion by scarf.

Segmentation	Accuracy (%)						Average Accuracy (%)
	K	L	M	X	Y	Z	
(1×1)	98	80	63	90	48	31	68.33
(2×2)	98	95	92	92	79	72	88.00
(4×4)	99	98	95	93	84	77	91.00

From the resulting outputs, we can observe that:

- For subsets of facial expression variation, a small change arises because the results of the previous experiment were already reasonable (e.g., subsets A, B, D, N, and P). However, the accuracy rises from 71% to 76% for subset Q, which is characterized by significant changes in facial expression.
- For occluded subsets, there was a significant increase in recognition accuracy when the number of blocks was augmented. As an illustration, when we applied 1 to 16 patches, the accuracy grew from 31% to 71% for subset Z, from 46% to 79% for subset W, and from 48% to 84% for subset Y.
- As such, in the case of partial occlusion, we may claim that local information is essential. It helps to go deeper in extracting relevant information from the face like details about the facial structure, such as the nose, eyes, or mouth, and information about position relationships, such as nose to mouth, eye to eye, and so on.
- Finally, we note that the 4×4 blocks provided the optimum configuration with the best accuracy for subsets of facial expression, occlusion by sunglasses, and scarf occlusion.

4.1.6. Experiment #4 (Effects of Color Texture Information)

For this analysis, we evaluated the performance of the last configuration (i.e., segmentation of the image into 4×4 blocks, K-NN associated with city block distance, $l \times l = 17 \times 17$, and $n = 12$) by testing three color-spaces, namely, RGB, HSV, and YCbCr, instead of transforming the image into grayscale. This feature extraction method is called MB-C-BSIF, as described in Section 3.2. The AR database images are already in RGB and so do not need a transformation of the first color-space. However, the images must be converted from RGB to HSV and RGB to YCbCr for the other color-spaces. Tables 10–12 display and compare the recognition accuracy of the MB-C-BSIF using several color-spaces with subsets of facial expression variations, occlusion by sunglasses, and occlusion by a scarf, respectively. The best results are in bold.

Table 10. Comparison of the results obtained using different color-spaces with changes in facial expression.

Color-Space	Accuracy (%)							Average Accuracy (%)
	B	C	D	N	O	P	Q	
Gray Scale	100	100	99	98	92	97	76	94.57
RGB	100	100	95	97	92	93	67	92.00
HSV	100	100	99	97	96	95	77	94.86
YCbCr	100	100	96	98	93	93	73	93.29

Table 11. Comparison of the results obtained using different color-spaces with occlusion by sunglasses.

Color-Space	Accuracy (%)						Average Accuracy (%)
	H	I	J	U	V	W	
Gray Scale	100	99	99	93	81	79	91.83
RGB	100	99	100	93	85	84	93.50
HSV	100	97	99	96	82	80	92.33
YCbCr	100	99	98	93	81	80	91.83

Table 12. Comparison of the results obtained using different color-spaces with occlusion by scarf.

Color-Space	Accuracy (%)						Average Accuracy (%)
	K	L	M	X	Y	Z	
Gray Scale	99	98	95	93	84	77	91.00
RGB	99	97	97	94	88	81	92.67
HSV	99	96	90	95	75	74	88.17
YCbCr	98	98	96	93	87	78	91.67

From the resulting outputs, we can see that:

- The results are almost identical for subsets of facial expression variation with all checked color-spaces. In fact, with the HSV color-space, a slight improvement is reported, although slight degradations are observed with both RGB and YCbCr color-spaces.
- All color-spaces see enhanced recognition accuracy compared to the grayscale standard for sunglasses occlusion subsets. RGB is the color-space with the highest output, seeing an increase from 91.83% to 93.50% in terms of average accuracy.
- HSV shows some regression for scarf occlusion subsets, but both the RGB and YCbCr color-spaces display some progress compared to the grayscale norm. Additionally, RGB remains the color-space with the highest output.
- The most significant observation is that the RGB color-space saw significantly improved performance in the V, W, Y, and Z subsets (from 81% to 85% with V; 79% to 84% with W; 84% to 88% with Y; and 77% to 87% with Z). Note that images of these occluded subsets are characterized by light degradation (either to the right or left, as shown in Figure 6).
- Finally, we note that the optimum color-space, providing a perfect balance between lighting restoration and improvement in identification, was the RGB.

4.1.7. Comparison #1 (Protocol I)

To confirm that our suggested method produces superior recognition performance with variations in facial expression, we compared the collected results with several state-of-the-art methods recently employed to tackle the SSFR issue. Table 13 presents the highest accuracies obtained using the same subsets and the same assessment protocol with Subset A as the training set and subsets of facial expression variations B, C, D, N, O, and P constituting the test set. The results presented in Table 13 are taken from several references [36,39,53,54]. "- -" signifies that the considered method has no experimental results. The best results are in bold.

Table 13. Comparison of 18 methods of facial expression variation subsets.

Authors	Year	Method	Accuracy						Average Accuracy (%)
			B	C	D	N	O	P	
Turk, Pentland [55]	1991	PCA	97.00	87.00	60.00	77.00	76.00	67.00	77.33
Wu and Zhou [56]	2002	$(PC)^2A$	97.00	87.00	62.00	77.00	74.00	67.00	77.33
Chen et al. [57]	2004	$E(PC)^2A$	97.00	87.00	63.00	77.00	75.00	68.00	77.83
Yang et al. [58]	2004	2DPCA	97.00	87.00	60.00	76.00	76.00	67.00	77.17
Gottumukkal and Asari [59]	2004	Block-PCA	97.00	87.00	60.00	77.00	76.00	67.00	77.33
Chen et al. [60]	2004	Block-LDA	85.00	79.00	29.00	73.00	59.00	59.00	64.00
Zhang and Zhou [61]	2005	$(2D)^2PCA$	98.00	89.00	60.00	71.00	76.00	66.00	76.70
Tan et al. [62]	2005	SOM	98.00	88.00	64.00	73.00	77.00	70.00	78.30
He et al. [63]	2005	LPP	94.00	87.00	36.00	86.00	74.00	78.00	75.83
Zhang et al. [27]	2005	SVD-LDA	73.00	75.00	29.00	75.00	56.00	58.00	61.00
Deng et al. [64]	2010	UP	98.00	88.00	59.00	77.00	74.00	66.00	77.00
Lu et al. [36]	2012	DMMA	99.00	93.00	69.00	88.00	85.00	85.50	79.00
Mehrasa et al. [53]	2017	SLPMM	99.00	94.00	65.00	- -	- -	- -	- -
Ji et al. [54]	2017	CPL	92.22	88.06	83.61	83.59	77.95	72.82	83.04
Zhang et al. [37]	2018	DMF	100.00	99.00	66.00	- -	- -	- -	- -
Chu et al. [65]	2019	MFSA+	100.00	100.00	74.00	93.00	85.00	86.00	89.66
Pang et al. [66]	2019	RHDA	97.08	97.00	96.25	- -	- -	- -	- -
Zhang et al. [39]	2020	DNNC	100.00	98.00	69.00	92.00	76.00	85.00	86.67
Our method	**2021**	**MB-C-BSIF**	**100.00**	**100.00**	**95.00**	**97.00**	**92.00**	**93.00**	**96.17**

The outcomes obtained validate the robustness and reliability of our proposed SSFR system compared to state-of-the-art methods when assessed with identical subsets. We suggest a competitive technique that has achieved a desirable level of identification accuracy with the six subsets of up to: 100.00% for B and C; 95.00% for D; 97.00% for N; 92.00% for O; and 93.00% for P.

For all subsets, our suggested technique surpasses the state-of-the-art methods analyzed in this paper, i.e., the proposed MB-C-BSIF can achieve excellent identification performance under the condition of variation in facial expression.

4.1.8. Comparison #2 (Protocol II)

To further demonstrate the efficacy of our proposed SSFR system, we also compared the best configuration of the MB-C-BSIF (i.e., RGB color-space, segmentation of the image into 4×4 blocks, city block distance, $l \times l = 17 \times 17$, and $n = 12$) with recently published work under unconstrained conditions. We followed the same experimental protocol described in [33,39]. Table 14 displays the accuracies of the works compared on the tested subsets H + K (i.e., occlusion by sunglasses and scarf) and subsets J + M (i.e., occlusion by sunglasses and scarf with variations in lighting). The best results are in bold.

In Table 14, we can observe that the work presented by Zhu et al. [33], called LGR, shows a comparable level, but the identification accuracy of our MB-C-BSIF procedure is much higher than all the methods considered for both test sessions.

Compared to related SSFRs, which can be categorized as either generic learning methods (e.g., ESRC [31], SVDL [32], and LGR [33], image partitioning methods (e.g., CRC [35], PCRC [34], and DNNC [39]) or deep learning methods (e.g., DCNN [41] and BDL [44]), the capabilities of our method can be explained in terms of its exploitation of different forms of information. This can be summarized as follows:

- The BSIF descriptor scans the image pixel by pixel, i.e., we consider the benefits of local information.

- The image is decomposed into several blocks, i.e., we exploit regional information.
- BSIF descriptor occurrences are accumulated in a global histogram, i.e., we manipulate global information.
- The MB-BSIF is applied to all RGB image components, i.e., color texture information is exploited.

Table 14. Comparison of 12 methods on occlusion and lighting-occlusion sessions.

Authors	Year	Method	Occlusion (H + K) (%)	Lighting + Occlusion (J + M) (%)	Average Accuracy (%)
Zhang et al. [35]	2011	CRC	58.10	23.80	40.95
Deng et al. [31]	2012	ESRC	83.10	68.60	75.85
Zhu et al. [34]	2012	PCRC	95.60	81.30	88.45
Yang et al. [32]	2013	SVDL	86.30	79.40	82.85
Lu et al. [36]	2012	DMMA	46.90	30.90	38.90
Zhu et al. [33]	2014	LGR	98.80	96.30	97.55
Ref. [67]	2016	SeetaFace	63.13	55.63	59.39
Zeng et al. [41]	2017	DCNN	96.5	88.3	92.20
Chu et al. [65]	2019	MFSA+	91.3	79.00	85.20
Cuculo et al. [68]	2019	SSLD	90.18	82.02	86.10
Zhang et al. [39]	2020	DNNC	92.50	79.50	86.00
Du and Da [44]	2020	BDL	93.03	91.55	92.29
Our method	**2021**	**MB-C-BSIF**	**99.5**	**98.5**	**99.00**

To summarize this first experiment, the performance of the proposed approach was evaluated using the AR database. In this experiment, the issues studied were changes in facial expression, lighting and occlusion by sunglasses and headscarf, which are the most common cases in real-world applications. As presented in Tables 13 and 14, our system obtained very good results (i.e., 96.17% with Protocol I and 99% with Protocol II) that surpass all the approaches compared (including the handcrafted and deep-learning-based approaches), i.e., that the approach we propose is appropriate and effective in the presence of the problems mentioned above.

4.2. Experiments on the LFW Database

4.2.1. Database Description

The Labeled Faces in the Wild (LFW) database [52] comprises more than 13,000 photos collected from the World Wide Web of 5749 diverse subjects in challenging situations, of which 1680 subjects possess two or more shots per individual. Our tests employed the LFW-a, a variant of the standard LFW where the facial images are aligned with a commercial normalization tool. It can be observed that the intra-class differences in this database are very high compared to the well-known constrained databases and face normalization has been carried out. The size of each image is 250×250 pixels and uses the jpeg extension. LFW is a very challenging database: it aims to investigate the unconstrained issues of face recognition, such as changes in lighting, age, clothing, focus, facial expression, color saturation, posture, race, hairstyle, background, camera quality, gender, ethnicity, and other factors, as presented in Figure 7.

4.2.2. Experimental Protocol

This study followed the experimental protocol presented in [30,32–34]. From the LFW-a database, we selected only those subjects possessing more than 10 images to obtain a subset containing the facial images of 158 individuals. We cropped each image to a size of 120×120 pixels and then resized it to 80×80 pixels. We considered the first 50 subjects' facial photographs to create the training set and the test set. We randomly selected one shot from each subject for the training set, while the remaining images were employed in the test set. This process was repeated for five permutations and the average result for each was taken into consideration.

Figure 7. Examples of two different subjects from the Labeled Faces in the Wild (LFW)-a database.

4.2.3. Limitations of SSFR Systems

In this section, the SSFR systems, and particularly the method we propose, will be voluntarily tested in a situation that is not adapted to their application: they are applicable in the case where only one sample is available and, very often, this sample is captured in very poor conditions.

We are particularly interested in cases where hundreds of samples are available, as in the LFW database, or when the training stage is based on millions of samples. In such a situation, deep learning approaches must be obviously chosen.

Therefore, the objective of this section is to assess the limitations of our approach.

Table 15 summarizes the performance of several rival approaches in terms of identification accuracy. Our best result was obtained by adopting the following configuration:

- BSIF descriptor with filter size $l \times l = 17 \times 17$ and bit string length $n = 12$.
- K-NN classifier associated with city block distance.
- Segmentation of the image into blocks of 40×40 and 20×20 pixels.
- RGB color-space.

Table 15. Identification accuracies using the LFW database.

Authors	Year	Method	Accuracy (%)
Chen et al. [60]	2004	Block LDA	16.40
Zhang et al. [27]	2005	SVD-FLDA	15.50
Wright et al. [69]	2009	SRC	20.40
Su et al. [70]	2010	AGL	19.20
Zhang et al. [35]	2011	CRC	19.80
Deng et al. [31]	2012	ESRC	27.30
Zhu et al. [34]	2012	PCRC	24.20
Yang et al. [32]	2013	SVDL	28.60
Lu et al. [36]	2012	DMMA	17.80
Zhu et al. [33]	2014	LGR	30.40
Ji et al. [54]	2017	CPL	25.20
Dong et al. [30]	2018	KNNMMDL	32.30
Chu et al. [65]	2019	MFSA+	26.23
Pang et al. [66]	2019	RHDA	32.89
Zhou et al. [71]	2019	DpLSA	37.55
Our method	2021	MB-C-BSIF	38.01
Parkhi et al. [12]	2015	Deep-Face	62.63
Zeng et al. [72]	2018	TDL	74.00

We can observe that the traditional approaches did not achieve particularly good identification accuracies. This is primarily because the photographs in the LFW database have been taken in unregulated conditions, which generates facial images with rich intra-class differences and increases face recognition complexity. As a consequence, the efficiency of the SSFR procedure is reduced. However, our recommended solution is better than the other competing traditional approaches. The superiority of our method can be explained by its exploitation of different forms of information, namely: local, regional, global, and color texture information. SVDL [32] and LGR [33] also achieved success in SSFR because the intra-class variance information obtained from other subjects in the standardized training set (i.e., augmenting the training-data) helped boost the performance of the system. Additionally, KNNMMDL [30] achieved good performance because it uses the Weber-face algorithm in the preprocessing step, which handles the illumination variation issue and employs data augmentation to enrich the intra-class variation in the training set.

In another experiment, we implemented and tested the successful DeepFace algorithm [12], whose weights were trained on millions of images from the ImageNet database that are close to real-life situations. As presented in Table 15, the DeepFace algorithm shows significant superiority to the compared methods. This success is down to the profound and specific training of the weights in addition to the significant number of images employed in its operation.

In a recent work by Zeng et al. [72], the authors combined traditional (handcrafted) and deep learning (TDL) characteristics to overcome the limitation of each class. They reached an identification accuracy of near 74%, which is something of a quantum leap in this challenging topic.

In the comparative study presented in [73], we can see that current face recognition systems employing several examples in the training set achieve very high accuracy with the LFW database, especially with deep-learning-based methods. However, SSFR systems suffer considerably when using the challenging LFW database and further research is required to improve their reliability.

In the situation where the learning stage is based on millions of images, the proposed SSFR technique cannot be used. In such a situation, References [12,72], which use deep learning techniques with data augmentation [12] or deep learning features combined with handcrafted features [72], allow one to obtain better accuracy.

Finally, the proposed SSFR method is reserved for the case where only one sample per person is available, which is the most common case in the real world through remote surveillance or unmanned aerial vehicles' shots. In these applications, faces are most often captured under harsh conditions, such as changing lighting, posture, or if the person is wearing accessories such as glasses, masks, or disguises. In these cases, the method proposed here is by far the most accurate. Finally, it would be interesting to explore and test some proven approaches that have shown good performance in solving real-world problems, in order to evaluate their performance using the same protocol and database, such as multi-scale principal component analysis (MSPCA) [74], signal decomposition methods [75,76], generative adversarial neural networks (GAN) [77], and centroid-displacement-based-K-NN [78].

5. Conclusions and Perspectives

In this paper, we have presented an original method for Single-Sample Face Recognition (SSFR) based on the Multi-Block Color-binarized Statistical Image Features (MB-C-BSIF) descriptor. It allows for the extraction of features for classification by the K-nearest neighbors (K-NN) method. The proposed method exploits various kinds of information, including local, regional, global, and color texture information. In our experiments, the MB-C-BSIF has been evaluated on several subsets of images from the unconstrained AR and LFW databases. Experiments conducted on the AR database have shown that our method significantly improves the performance of SSFR classification when dealing with several variations of facial recognition. The proposed feature extraction strategy achieves a high accuracy, with an average value of 96.17% and 99% for the AR database with Protocols I and II, respectively. These significant results validate the effectiveness of the proposed method compared to state-of-the-art methods. The potential applications of the method are oriented towards a computer-aided technology that can be used for real-time identification.

In the future, we aim to explore the effectiveness of combining both deep learning and traditional methods in addressing the SSFR issue. Hybrid features combine handcrafted features with deep characteristics to collect richer information than those obtained by a single feature extraction method, thus improving the level of recognition. Besides, we plan to develop a deep learning method based on semantic information, such as age, gender, and ethnicity, to solve the problem of SSFR, which is an area that deserves further study. We also aim to investigate and analyze the SSFR issue in unconstrained environments using large-scale databases that hold millions of facial images.

Author Contributions: Investigation, software, writing original draft, I.A.; project administration, supervision, validation, writing, review and editing, A.O.; methodology, validation, writing, review and editing, A.B.; validation, writing, review and editing, S.J. All authors have read and agreed to the published version of the manuscript.

Funding: This research received no external funding.

Institutional Review Board Statement: Not applicable.

Informed Consent Statement: Not applicable.

Data Availability Statement: Data sharing not applicable.

Conflicts of Interest: The authors declare no conflict of interest.

References

1. Alay, N.; Al-Baity, H.H. Deep Learning Approach for Multimodal Biometric Recognition System Based on Fusion of Iris, Face, and Finger Vein Traits. *Sensors* **2020**, *20*, 5523. [CrossRef] [PubMed]
2. Pagnin, E.; Mitrokotsa, A. Privacy-Preserving Biometric Authentication: Challenges and Directions. *Secur. Commun. Netw.* **2017**, *2017*, 1–9. [CrossRef]
3. Mahfouz, A.; Mahmoud, T.M.; Sharaf Eldin, A. A Survey on Behavioral Biometric Authentication on Smartphones. *J. Inf. Secur. Appl.* **2017**, *37*, 28–37. [CrossRef]
4. Ferrara, M.; Cappelli, R.; Maltoni, D. On the Feasibility of Creating Double-Identity Fingerprints. *IEEE Trans. Inf. Forensics Secur.* **2017**, *12*, 892–900. [CrossRef]
5. Thompson, J.; Flynn, P.; Boehnen, C.; Santos-Villalobos, H. Assessing the Impact of Corneal Refraction and Iris Tissue Non-Planarity on Iris Recognition. *IEEE Trans. Inf. Forensics Secur.* **2019**, *14*, 2102–2112. [CrossRef]
6. Benzaoui, A.; Bourouba, H.; Boukrouche, A. System for Automatic Faces Detection. In Proceedings of the 3rd International Conference on Image Processing, Theory, Tools, and Applications (IPTA), Istanbul, Turkey, 15–18 October 2012; pp. 354–358.
7. Phillips, P.J.; Flynn, P.J.; Scruggs, T.; Bowyer, K.W.; Chang, J.; Hoffman, K.; Marques, J.; Min, J.; Worek, W. Overview of the Face Recognition Grand Challenge. In Proceedings of the IEEE Computer Society Conference on Computer Vision and Pattern Recognition (CVPR), San Diego, CA, USA, 20–26 June 2005; pp. 947–954.
8. Femmam, S.; M'Sirdi, N.K.; Ouahabi, A. Perception and Characterization of Materials Using Signal Processing Techniques. *IEEE Trans. Instrum. Meas.* **2001**, *50*, 1203–1211. [CrossRef]
9. Ring, T. Humans vs Machines: The Future of Facial Recognition. *Biom. Technol. Today* **2016**, *4*, 5–8. [CrossRef]
10. Phillips, P.J.; Yates, A.N.; Hu, Y.; Hahn, A.C.; Noyes, E.; Jackson, K.; Cavazos, J.G.; Jeckeln, G.; Ranjan, R.; Sankaranarayanan, S.; et al. Face Recognition Accuracy of Forensic Examiners, Superrecognizers, and Face Recognition Algorithms. *Proc. Natl. Acad. Sci. USA* **2018**, *115*, 6171–6176. [CrossRef]
11. Kortli, Y.; Jridi, M.; Al Falou, A.; Atri, M. Face Recognition Systems: A Survey. *Sensors* **2020**, *20*, 342. [CrossRef]
12. Ouahabi, A.; Taleb-Ahmed, A. Deep learning for real-time semantic segmentation: Application in ultrasound imaging. *Pattern Recognition Letters* **2021**, *144*, 27–34.
13. Rahman, J.U.; Chen, Q.; Yang, Z. Additive Parameter for Deep Face Recognition. *Commun. Math. Stat.* **2019**, *8*, 203–217. [CrossRef]
14. Fan, Z.; Jamil, M.; Sadiq, M.T.; Huang, X.; Yu, X. Exploiting Multiple Optimizers with Transfer Learning Techniques for the Identification of COVID-19 Patients. *J. Healthc. Eng.* **2020**, *2020*, 1–13.
15. Benzaoui, A.; Boukrouche, A. Ear Recognition Using Local Color Texture Descriptors from One Sample Image Per Person. In Proceedings of the 4th International Conference on Control, Decision and Information Technologies (CoDIT), Barcelona, Spain, 5–7 April 2017; pp. 827–832.
16. Vapnik, V.N.; Chervonenkis, A. Learning Theory and Its Applications. *IEEE Trans. Neural Netw.* **1999**, *10*, 985–987.

17. Vezzetti, E.; Marcolin, F.; Tornincasa, S.; Ulrich, L.; Dagnes, N. 3D Geometry-Based Automatic Landmark Localization in Presence of Facial Occlusions. *Multimed. Tools Appl.* **2017**, *77*, 14177–14205. [CrossRef]

18. Echeagaray-Patron, B.A.; Miramontes-Jaramillo, D.; Kober, V. Conformal Parameterization and Curvature Analysis for 3D Facial Recognition. In Proceedings of the 2015 International Conference on Computational Science and Computational Intelligence (CSCI), Las Vegas, NV, USA, 7–9 December 2015; pp. 843–844.

19. Kannala, J.; Rahtu, E. BSIF: Binarized Statistical Image Features. In Proceedings of the 21th International Conference on Pattern Recognition (ICPR), Tsukuba, Japan, 11–15 November 2012; pp. 1363–1366.

20. Djeddi, M.; Ouahabi, A.; Batatia, H.; Basarab, A.; Kouamé, D. Discrete Wavelet for Multifractal Texture Classification: Application to Medical Ultrasound Imaging. In Proceedings of the 2010 IEEE International Conference on Image Processing, Hong Kong, China, 26–29 September 2010; pp. 637–640.

21. Ouahabi, A. Multifractal Analysis for Texture Characterization: A New Approach Based on DWT. In Proceedings of the 10th International Conference on Information Science, Signal Processing and their Applications (ISSPA 2010), Kuala Lumpur, Malaysia, 10–13 May 2010; pp. 698–703.

22. Ouahabi, A. *Signal and Image Multiresolution Analysis*, 1st ed.; ISTE-Wiley: London, UK, 2012.

23. Ouahabi, A. A Review of Wavelet Denoising in Medical Imaging. In Proceedings of the 8th International Workshop on Systems, Signal Processing and their Applications (WoSSPA), Tipaza, Algeria, 12–15 May 2013; pp. 19–26.

24. Sidahmed, S.; Messali, Z.; Ouahabi, A.; Trépout, S.; Messaoudi, C.; Marco, S. Nonparametric Denoising Methods Based on Contourlet Transform with Sharp Frequency Localization: Application to Low Exposure Time Electron Microscopy Images. *Entropy* **2015**, *17*, 3461–3478.

25. Kumar, N.; Garg, V. Single Sample Face Recognition in the Last Decade: A Survey. *Int. J. Pattern Recognit. Artif. Intell.* **2019**, *33*, 1956009. [CrossRef]

26. Vetter, T. Synthesis of Novel Views from a Single Face Image. *Int. J. Comput. Vis.* **1998**, *28*, 103–116. [CrossRef]

27. Zhang, D.; Chen, S.; Zhou, Z.H. A New Face Recognition Method Based on SVD Perturbation for Single Example Image per Person. *Appl. Math. Comput.* **2005**, *163*, 895–907. [CrossRef]

28. Gao, Q.X.; Zhang, L.; Zhang, D. Face Recognition Using FLDA with Single Training Image per Person. *Appl. Math. Comput.* **2008**, *205*, 726–734. [CrossRef]

29. Hu, C.; Ye, M.; Ji, S.; Zeng, W.; Lu, X. A New Face Recognition Method Based on Image Decomposition for Single Sample per Person Problem. *Neurocomputing* **2015**, *160*, 287–299. [CrossRef]

30. Dong, X.; Wu, F.; Jing, X.Y. Generic Training Set Based Multimanifold Discriminant Learning for Single Sample Face Recognition. *KSII Trans. Internet Inf. Syst.* **2018**, *12*, 368–391.

31. Deng, W.; Hu, J.; Guo, J. Extended SRC: Undersampled Face Recognition via Intraclass Variant Dictionary. *IEEE Trans. Pattern Anal. Mach. Intell.* **2012**, *34*, 1864–1870. [CrossRef] [PubMed]

32. Yang, M.; Van, L.V.; Zhang, L. Sparse Variation Dictionary Learning for Face Recognition with a Single Training Sample per Person. In Proceedings of the IEEE International Conference on Computer Vision (ICCV), Sydney, Australia, 1–8 December 2013; pp. 689–696.

33. Zhu, P.; Yang, M.; Zhang, L.; Lee, L. Local Generic Representation for Face Recognition with Single Sample per Person. In Proceedings of the Asian Conference on Computer Vision (ACCV), Singapore, 1–5 November 2014; pp. 34–50.

34. Zhu, P.; Zhang, L.; Hu, Q.; Shiu, S.C.K. Multi-Scale Patch Based Collaborative Representation for Face Recognition with Margin Distribution Optimization. In Proceedings of the European Conference on Computer Vision (ECCV), Florence, Italy, 7–13 October 2012; pp. 822–835.

35. Zhang, L.; Yang, M.; Feng, X. Sparse Representation or Collaborative Representation: Which Helps Face Recognition? In Proceedings of the International Conference on Computer Vision (ICCV), Barcelona, Spain, 6–13 November 2011; pp. 471–478.

36. Lu, J.; Tan, Y.P.; Wang, G. Discriminative Multimanifold Analysis for Face Recognition from a Single Training Sample per Person. *IEEE Trans. Pattern Anal. Mach. Intell.* **2012**, *35*, 39–51. [CrossRef] [PubMed]

37. Zhang, W.; Xu, Z.; Wang, Y.; Lu, Z.; Li, W.; Liao, Q. Binarized Features with Discriminant Manifold Filters for Robust Single-Sample Face Recognition. *Signal Process. Image Commun.* **2018**, *65*, 1–10. [CrossRef]

38. Gu, J.; Hu, H.; Li, H. Local Robust Sparse Representation for Face Recognition with Single Sample per Person. *IEEE/CAA J. Autom. Sin.* **2018**, *5*, 547–554. [CrossRef]

39. Zhang, Z.; Zhang, L.; Zhang, M. Dissimilarity-Based Nearest Neighbor Classifier for Single-Sample Face Recognition. *Vis. Comput.* **2020**, 1–12. [CrossRef]

40. Mimouna, A.; Alouani, I.; Ben Khalifa, A.; El Hillali, Y.; Taleb-Ahmed, A.; Menhaj, A.; Ouahabi, A.; Ben Amara, N.E. OLIMP: A Heterogeneous Multimodal Dataset for Advanced Environment Perception. *Electronics* **2020**, *9*, 560. [CrossRef]

41. Zeng, J.; Zhao, X.; Qin, C.; Lin, Z. Single Sample per Person Face Recognition Based on Deep Convolutional Neural Network. In Proceedings of the 3rd IEEE International Conference on Computer and Communications (ICCC), Chengdu, China, 13–16 December 2017; pp. 1647–1651.

42. Ding, C.; Bao, T.; Karmoshi, S.; Zhu, M. Single Sample per Person Face Recognition with KPCANet and a Weighted Voting Scheme. *Signal Image Video Process.* **2017**, *11*, 1213–1220. [CrossRef]

43. Zhang, Y.; Peng, H. Sample Reconstruction with Deep Autoencoder for One Sample per Person Face Recognition. *IET Comput. Vis.* **2018**, *11*, 471–478. [CrossRef]

44. Du, Q.; Da, F. Block Dictionary Learning-Driven Convolutional Neural Networks for Few-Shot Face Recognition. *Vis. Comput.* **2020**, 1–10.

45. Stone, J.V. Independent Component Analysis: An Introduction. *Trends Cogn. Sci.* **2002**, *6*, 59–64. [CrossRef]

46. Ataman, E.; Aatre, V.; Wong, K. A Fast Method for Real-Time Median Filtering. *IEEE Trans. Acoust. Speech Signal Process.* **1980**, *28*, 415–421. [CrossRef]

47. Benzaoui, A.; Hadid, A.; Boukrouche, A. Ear Biometric Recognition Using Local Texture Descriptors. *J. Electron. Imaging* **2014**, *23*, 053008. [CrossRef]

48. Zehani, S.; Ouahabi, A.; Oussalah, M.; Mimi, M.; Taleb-Ahmed, A. Bone Microarchitecture Characterization Based on Fractal Analysis in Spatial Frequency Domain Imaging. *Int. J. Imaging Syst. Technol.* **2020**, 1–19.

49. Ojala, T.; Pietikainen, M.; Maenpaa, T. Multiresolution Gray-Scale and Rotation Invariant Texture Classification with Local Binary Patterns. *IEEE Trans. Pattern Anal. Mach. Intell.* **2002**, *24*, 971–987. [CrossRef]

50. Ojansivu, V.; Heikkil, J. Blur Insensitive Texture Classification Using Local Phase Quantization. In Proceedings of the 3rd International Conference on Image and Signal Processing (ICSIP), Paris, France, 7–8 July 2012; pp. 236–243.

51. Martinez, A.M.; Benavente, R. The AR Face Database. *CVC Tech. Rep.* **1998**, *24*, 1–10.

52. Huang, G.B.; Mattar, M.; Berg, T.; Learned-Miller, E. *Labeled Faces in the Wild: A Database for Studying Face Recognition in Unconstrained Environments*; Technical Report 07-49; University of Massachusetts: Amherst, MA, USA, 2007; pp. 7–49.

53. Mehrasa, N.; Ali, A.; Homayun, M. A Supervised Multimanifold Method with Locality Preserving for Face Recognition Using Single Sample per Person. *J. Cent. South Univ.* **2017**, *24*, 2853–2861. [CrossRef]

54. Ji, H.K.; Sun, Q.S.; Ji, Z.X.; Yuan, Y.H.; Zhang, G.Q. Collaborative Probabilistic Labels for Face Recognition from Single Sample per Person. *Pattern Recognit.* **2017**, *62*, 125–134. [CrossRef]

55. Turk, M.; Pentland, A. Eigenfaces for Recognition. *J. Cogn. Neurosci.* **1991**, *3*, 71–86. [CrossRef]

56. Wu, J.; Zhou, Z.H. Face Recognition with One Training Image per Person. *Pattern Recognit. Lett.* **2002**, *23*, 1711–1719. [CrossRef]

57. Chen, S.; Zhang, D.; Zhou, Z.H. Enhanced (PC)2A for Face Recognition with One Training Image per Person. *Pattern Recognit. Lett.* **2004**, *25*, 1173–1181. [CrossRef]

58. Yang, J.; Zhang, D.; Frangi, A.F.; Yang, J.Y. Two-Dimensional PCA: A New Approach to Appearance-Based Face Representation and Recognition. *IEEE Trans. Pattern Anal. Mach. Intell.* **2004**, *26*, 131–137. [CrossRef] [PubMed]

59. Gottumukkal, R.; Asari, V.K. An Improved Face Recognition Technique Based on Modular PCA Approach. *Pattern Recognit. Lett.* **2004**, *25*, 429–436. [CrossRef]

60. Chen, S.; Liu, J.; Zhou, Z.H. Making FLDA Applicable to Face Recognition with One Sample per Person. *Pattern Recognit.* **2004**, *37*, 1553–1555. [CrossRef]

61. Zhang, D.; Zhou, Z.H. (2D)2PCA: Two-Directional Two-Dimensional PCA for Efficient Face Representation and Recognition. *Neurocomputing* **2005**, *69*, 224–231. [CrossRef]

62. Tan, X.; Chen, S.; Zhou, Z.H.; Zhang, F. Recognizing Partially Occluded, Expression Variant Faces from Single Training Image per Person with SOM and Soft K-NN Ensemble. *IEEE Trans. Neural Netw.* **2005**, *16*, 875–886. [CrossRef]

63. He, X.; Yan, S.; Hu, Y.; Niyogi, P.; Zhang, H.J. Face Recognition Using Laplacian Faces. *IEEE Trans. Pattern Anal. Mach. Intell.* **2005**, *27*, 328–340.

64. Deng, W.; Hu, J.; Guo, J.; Cai, W.; Fenf, D. Robust, Accurate and Efficient Face Recognition from a Single Training Image: A Uniform Pursuit Approach. *Pattern Recognit.* **2010**, *43*, 1748–1762. [CrossRef]

65. Chu, Y.; Zhao, L.; Ahmad, T. Multiple Feature Subspaces Analysis for Single Sample per Person Face Recognition. *Vis. Comput.* **2019**, *35*, 239–256. [CrossRef]

66. Pang, M.; Cheung, Y.; Wang, B.; Liu, R. Robust Heterogeneous Discriminative Analysis for Face Recognition with Single Sample per Person. *Pattern Recognit.* **2019**, *89*, 91–107. [CrossRef]

67. Seetafaceengine. 2016. Available online: https://github.com/seetaface/SeetaFaceEngine (accessed on 1 September 2020).

68. Cuculo, V.; D'Amelio, A.; Grossi, G.; Lanzarotti, R.; Lin, J. Robust Single-Sample Face Recognition by Sparsity-Driven Sub-Dictionary Learning Using Deep Features. *Sensors* **2019**, *19*, 146. [CrossRef] [PubMed]

69. Wright, J.; Yang, A.Y.; Ganesh, A.; Sastry, S.S.; Ma, Y. Robust Face Recognition via Sparse Representation. *IEEE Trans. Pattern Anal. Mach. Intell.* **2009**, *31*, 210–227. [CrossRef] [PubMed]

70. Su, Y.; Shan, S.; Chen, X.; Gao, W. Adaptive Generic Learning for Face Recognition from a Single Sample per Person. In Proceedings of the 2010 IEEE Computer Society Conference on Computer Vision and Pattern Recognition, San Francisco, CA, USA, 13–18 June 2010; pp. 2699–2706.

71. Zhou, D.; Yang, D.; Zhang, X.; Huang, S.; Feng, S. Discriminative Probabilistic Latent Semantic Analysis with Application to Single Sample Face Recognition. *Neural Process. Lett.* **2019**, *49*, 1273–1298. [CrossRef]

72. Zeng, J.; Zhao, X.; Gan, J.; Mai, C.; Zhai, Y.; Wang, F. Deep Convolutional Neural Network Used in Single Sample per Person Face Recognition. *Comput. Intell. Neurosci.* **2018**, *2018*, 1–11. [CrossRef]

73. Adjabi, I.; Ouahabi, A.; Benzaoui, A.; Taleb-Ahmed, A. Past, Present, and Future of Face Recognition: A Review. *Electronics* **2020**, *9*, 1188. [CrossRef]

74. Sadiq, M.T.; Yu, X.; Yuan, Z.; Aziz, M.Z. Motor Imagery BCI Classification Based on Novel Two-Dimensional Modelling in Empirical Wavelet Transform. *Electron. Lett.* **2020**, *56*, 1367–1369. [CrossRef]

75. Sadiq, M.T.; Yu, X.; Yuan, Z.; Fan, Z.; Rehman, A.U.; Li, G.; Xiao, G. Motor Imagery EEG Signals Classification Based on Mode Amplitude and Frequency Components Using Empirical Wavelet Transform. *IEEE Access* **2019**, *7*, 127678–127692. [CrossRef]
76. Sadiq, M.T.; Yu, X.; Yuan, Z. Exploiting Dimensionality Reduction and Neural Network Techniques for the Development of Expert Brain—Computer Interfaces. *Expert Syst. Appl.* **2021**, *164*, 114031. [CrossRef]
77. Khaldi, Y.; Benzaoui, A. A New Framework for Grayscale Ear Images Recognition Using Generative Adversarial Networks under Unconstrained Conditions. *Evol. Syst.* **2020**. [CrossRef]
78. Nguyen, B.P.; Tay, W.L.; Chui, C.K. Robust Biometric Recognition from Palm Depth Images for Gloved Hands. *IEEE Trans. Hum. Mach. Syst.* **2015**, *45*, 799–804. [CrossRef]

Article

Finger-Vein Recognition Using Heterogeneous Databases by Domain Adaption Based on a Cycle-Consistent Adversarial Network

Kyoung Jun Noh, Jiho Choi, Jin Seong Hong and Kang Ryoung Park *

Division of Electronics and Electrical Engineering, Dongguk University, 30 Pildong-ro, 1-gil, Jung-gu, Seoul 04620, Korea; nohkyungjun@dongguk.edu (K.J.N.); choijh1027@dongguk.edu (J.C.); turtle1990@dgu.ac.kr (J.S.H.)
* Correspondence: parkgr@dongguk.edu; Tel.: +82-10-3111-7022; Fax: +82-2-2277-8735

Abstract: The conventional finger-vein recognition system is trained using one type of database and entails the serious problem of performance degradation when tested with different types of databases. This degradation is caused by changes in image characteristics due to variable factors such as position of camera, finger, and lighting. Therefore, each database has varying characteristics despite the same finger-vein modality. However, previous researches on improving the recognition accuracy of unobserved or heterogeneous databases is lacking. To overcome this problem, we propose a method to improve the finger-vein recognition accuracy using domain adaptation between heterogeneous databases using cycle-consistent adversarial networks (CycleGAN), which enhances the recognition accuracy of unobserved data. The experiments were performed with two open databases—Shandong University homologous multi-modal traits finger-vein database (SDUMLA-HMT-DB) and Hong Kong Polytech University finger-image database (HKPolyU-DB). They showed that the equal error rate (EER) of finger-vein recognition was 0.85% in case of training with SDUMLA-HMT-DB and testing with HKPolyU-DB, which had an improvement of 33.1% compared to the second best method. The EER was 3.4% in case of training with HKPolyU-DB and testing with SDUMLA-HMT-DB, which also had an improvement of 4.8% compared to the second best method.

Keywords: finger-vein recognition; camera position; finger position; lighting; unobserved database; heterogeneous database; domain adaptation; cycle-consistent adversarial networks; SDUMLA-HMT-DB; HKPolyU-DB

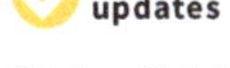

Citation: Noh, K.J.; Choi, J.; Hong, J.S.; Park, K.R. Finger-Vein Recognition Using Heterogeneous Databases by Domain Adaption Based on a Cycle-Consistent Adversarial Network. *Sensors* **2021**, *21*, 524. https://doi.org/10.3390/s21020524

Received: 24 December 2020
Accepted: 10 January 2021
Published: 13 January 2021

Publisher's Note: MDPI stays neutral with regard to jurisdictional claims in published maps and institutional affiliations.

1. Introduction

Finger-vein images are difficult to forge and easy to obtain, but the image qualities are easily affected by the shades inevitably generated by other biological tissues (e.g., bone and fingernail) [1,2]. A finger-vein recognition system employs a small amount of feature for recognition because of this fundamental characteristic of data [3]. Therefore, models trained using such a dataset are ineffective for unobserved data.

To consider this issue, non-training-based finger-vein recognition methods have been studied extensively to overcome this drawback. However, they exhibit significantly poorer performance than training-based methods because a large amount of information is removed by noise, thus making the classifier incapable of making an accurate decision [1–3]. Moreover, variations in the environment when acquiring images such as the camera position, lighting position, and lighting intensity create a large discrepancy between each dataset domain. This also deteriorates the performance of non-training-based methods.

The existing non-training-based finger-vein recognition method extracts specific features using local binary patterns for recognition [1]; however, these features are significantly affected by misalignment or image quality, making them unsuitable for finger-vein recognition. Subsequently, local directional patterns (LDPs) [2] and optimal filter-based finger-vein recognition [3] have been proposed, which can solve the misalignment problem but cannot solve the fundamental problem of image quality or removed information.

Hence, training-based finger-vein recognition methods have been extensively researched [4,5]. In [4], authors increased the number of training images by five times based on the data augmentation of image translation and cropping. In [5], they also increased the number of training images by 121 times based on the data augmentation of image translation and cropping. Although the similarity among augmented images increased by simple image translation and cropping, the training of proposed models was successfully performed with the augmented images, and the consequent accuracies of recognition were enhanced in their methods [4,5]. These methods exhibit good performance for finger-vein images with low quality by extracting features using a filter optimized for the distribution of input data rather than extracting features of a fixed form.

Although the training-based methods exhibit better recognition performance than non-training-based methods, their recognition rate in the cross-domain environment is significantly lower. The training-based methods are trained for optimizing the distribution of the training data used as input; thus, they exhibit poorer generality in the cross-domain than non-training-based methods, which extract features of a fixed form regardless of the training data. Moreover, the distances between domains are inevitably increased as training is repeated with a small amount of information of finger-vein data. In general, a specific dataset used for training refers to one domain, and the model trained using this dataset is optimized for this specific domain. However, if the dataset of a different domain is used for testing, the performance is significantly deteriorated because the data encountered by the model are different from those used to train the model (the problem of heterogeneity).

For mitigating the trade-off between recognition performance and generality, this study proposes a method for improving the finger-vein recognition rate of cross-domain databases through finger-vein domain adaptation using cycle-consistent adversarial networks (Cycle-GAN).

This paper is organized as follows. Section 2 presents previous studies related to the finger-vein recognition method, domain transfer, and domain adaptation, and Section 3 presents the contributions of this study. Section 4 provides the details of the proposed method, and Sections 5 and 6 present the experimental results of this study and discussions, respectively. Lastly, Section 7 concludes this study.

2. Related Work

Research on finger-vein recognition in which domain adaptation is considered is lacking. In this section, the scope of previous studies is expanded to include hand-based biometrics; whether domain adaptation was performed, was analyzed by dividing the studies into non-training-based and training-based methods.

2.1. Non-Training-Based Methods

Lu et al. performed domain adaptation to some extent by reducing the difference in brightness present in each finger-vein dataset using a peak-value-based method (PVM) [6]. The difference in brightness occurs when different sensors are used during acquisition of the dataset; this study focused on the difference between domains from this perspective. Jia et al. attempted to solve the cross-sensor problem using various dimension reduction algorithms and orientation coding methods [7].

Wang et al. performed a simple normalization to reduce the heterogeneity between domains for a dorsal hand-vein database obtained from various sensors and then performed segmentation to remove unimportant information which could increase heterogeneity [8]. In this study, matching was based on the scale-invariant feature transform (SIFT). The generality was high because matching was performed using a non-training-based algorithm; however, the performance was not suitable for biometric systems which require a high level of security. Wang et al. then performed soft domain adaptation using the same normalization algorithm followed by matching using an improved SIFT algorithm. This model was a more general and robust dorsal hand-vein recognition system [9].

Alshehri et al. used various handcrafted features to solve the problem of heterogeneity generated by different sensors when acquiring a fingerprint dataset, and in particular, ridge pattern, orientation, and minutiae points present in fingerprint images were used [10]. Binary gradient pattern (BGP) and Gabor-histogram of oriented gradients (Gabor-HoG) were used as descriptors, and the Sobel operator was used to compute the gradient. A robust fingerprint recognition system was proposed by performing score level fusion of the scores obtained from each descriptor. Ghiani et al. confirmed the problem with the accuracy of a fingerprint spoof attack detection system being abruptly reduced in the cross-sensor environment [11]. A least squares-based domain transformation function was adopted to reduce the extent of changes in the distribution caused by cross-sensors.

2.2. Training-Based Methods

Kute et al. used the Bregman divergence regularization method to reduce the distribution gap between domains; the researchers used Fisher linear discriminant analysis (FLDA) subspace learning algorithm to find a subspace through a projection matrix between fully heterogeneous data and then used the subspace to perform recognition using a support vector machine and K-nearest neighbor classifier [12]. Gajawada et al. performed domain adaptation between spoof attack databases to perform augmentation to improve the generality of a fingerprint spoof attack detector [13]. Here, a synthetic spoof attack patch was created using a universal material translator wrapper.

Anand et al. customized the DeepDomainPore network, which is a pore detection network trained with high-resolution images to enable the pore information observed only in high-resolution fingerprint images to be used in low-resolution images [14]. Domain adaptation was performed for inserting pore information in the low-resolution image. Using this method, pores, which are a level 3 feature, can be exploited even when low-resolution images are input in a fingerprint recognition system. Shao et al. proposed PalmGAN, which generates synthetic data using a palmprint dataset with labels [15]. Fake labeled data were generated using the palmprint dataset without labels as the target and the palmprint dataset with labels as the source. The fake labeled data were then used as new data with a newly inserted label while maintaining the identity information of the target domain, i.e., domain adapted data. The data were input to a deep hash network to perform palmprint recognition.

Moreover, the researchers attempted to solve the cross-domain problem by performing domain adaptation using an auto-encoder structured model [16]. Malhotra et al. highlighted the need to reinforce the touch-based biometric recognition system as the coronavirus disease (COVID-19) is increasingly becoming a serious issue across the globe [17]. Accordingly, the system was reinforced so that the fingerprint authentication system implements matching using a finger-selfie image. The finger-selfie image is segmented primarily using a handcrafted method to reduce the difference between the enrolled finger-scan image and finger-selfie domain. The segmented finger-selfie image and enrolled image undergo feature extraction through a deep ScatNet to allow matching with the trained random decision forest (RDF) model.

Jalilian et al. performed finger-vein segmentation using a fully convolutional network (FCN) [18]. The recognition performance was assessed in the cross-domain environment using the segmented image. However, the performance was not satisfactory in the cross-domain environment even when recognition was performed using only compact information. Dabouei et al. verified the performance in the cross-sensor environment using a conditional generative adversarial network (CGAN) for fingerprint ridge map reconstruction [19].

Nogueira et al. performed fingerprint spoof attack detection using visual geometry group (VGG)-16 and a convolutional neural network (CNN) and verified that a deep learning-based method is not effective in the cross-data, cross-sensor environment, even though this study was not related to recognition [20]. Chugh et al. confirmed that fingerprint spoof detection based on the minutiae-based local patch approach and MobileNet did not exhibit good performance in the cross-sensor environment [21]. Thus, training the distribution of the training data in the cross-domain, cross-sensor environment without using specific domain adaptation methods is ineffective for unobserved databases.

Although it is not the hand-based biometrics, Chui et al. proposed a CGAN and improved fuzzy c-means clustering (IFCM) algorithm called CGAN-IFCM for the multi-class voice disorder detection of three common types of voice disorders for smart healthcare applications [22].

To overcome the drawbacks of previous studies, we propose a method to improve the finger-vein recognition rate in cross-domain databases through finger-vein domain adaptation based on a CycleGAN. The reason for using CycleGAN in our method is that there is no paired data of input and target images in our experiments. That is, two finger-vein images from two different open databases (Shandong University homologous multi-modal traits finger-vein database (SDUMLA-HMT-DB) and Hong Kong Polytech University finger-image database (HKPolyU-DB)) are respectively used in our experiments. Because they are not from same class, there is no target image about input image in our case, and one of them can be used as input and the other can only be used as the reference image for the unpaired cases. Due to this reason, we use CycleGAN, which can use this kind of unpaired images. It is different from other types of GAN such as conditional GAN, which requires the paired data of input and target images [23].

CycleGAN can perform a task where the information of the source domain data is retained to some extent while reflecting the target domain information, instead of carrying out a task for simply making the source and target identical [24]. It is confirmed that our CycleGAN-based method showed better performances than other types of GAN.

3. Contributions

Our research is novel in the following five ways compared to previous works:

- This is the first study to examine GAN-based domain adaptation to solve the problem of performance deterioration of the finger-vein recognition system in a heterogeneous cross dataset.
- Domain adaptation was performed through a CycleGAN so that the existing training-based finger-vein recognition method can handle unobserved data. Each finger-vein dataset has different numbers of classes. Therefore, we used CycleGAN, which can deal with unpaired datasets.
- The proposed finger-vein recognition system does not have to be trained again when unobserved data are input into the system.
- The experiments with two open databases of SDUMLA-HMT-DB and HKPolyU-DB showed that the equal error rate (EER) of finger-vein recognition was 0.85% in case of training with SDUMLA-HMT-DB and testing with HKPolyU-DB, which is the improvement of 33.1% compared to the second best method. The EER was 3.4% in case of training with HKPolyU-DB and testing with SDUMLA-HMT-DB, which is also the improvement of 14.1% compared to the second best method.
- CycleGAN-based domain adaptation models and finger-vein recognition models trained with our domain adapted dataset proposed in this study are disclosed for a fair assessment of performance [25] by other researchers. On the website (http://dm.dgu.edu/link.html) explained in [25], we include the instructions of how other researchers can obtain our CycleGAN-based domain adaptation models and finger-vein recognition models.

4. Proposed Method

In this section, we would explain the overview of the proposed method in Section 4.1, our preprocessing method in Section 4.2, and proposed data adaption method based on CycleGAN in Section 4.3. In addition, we would explain the method of generating composite image for the input to CNN in Section 4.4, and finger-vein recognition method by DenseNet and shift matching in Section 4.5.

4.1. Overview of the Proposed Method

Figure 1 shows the overall procedure of the proposed finger-vein recognition method. The method involves preprocessing to remove unnecessary information generated by near-infrared light (NIR) used while acquiring images of finger veins, other biological tissues (e.g., bone or fingernail), or parts where information has been removed by shades [26] (Step 2 of Figure 1).

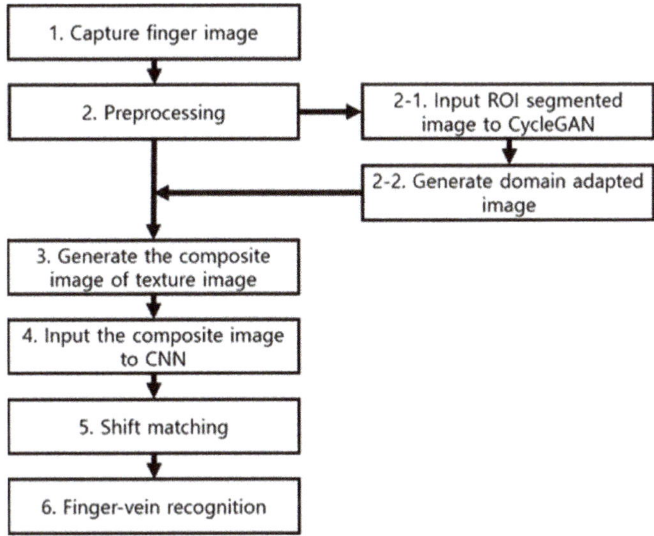

Figure 1. Overall procedure for the proposed finger-vein recognition method.

First, binary thresholding is performed to distinguish the finger region from the background region. The image that has undergone binary thresholding is used as a mask of the original finger-image and then undergoes linear stretching to fit the input size of a CNN subsequently. The finger region is not stretched uniformly if burrs are present in the mask during linear stretching. Thus, boundary smoothing enables the finger region to be stretched uniformly, thus minimizing information loss.

In addition, misalignment may occur when the user's finger trembles or is not fixed properly when acquiring finger-vein images. Misalignment is a major factor that reduces the finger-vein recognition performance. Hence, in-plane rotation compensation is performed to eliminate the misalignment problem. During in-plane rotation compensation, second-order moments of the entire image are found with respect to the finger-shape, and then, rotation is performed accordingly. In general, both edges of the finger image are thick and thus are affected more by biological tissues than other regions, or shades are generated by fingernails. Therefore, it is difficult to obtain the essential information of the finger vein. To overcome this problem, the parts are removed in the preprocessing step. Only the regions with the best finger-vein representation are segmented using the final mask obtained to be used as an input for finger-vein recognition.

The existing finger-vein recognition system has improved the performance of finger-vein recognition while being biased to the training dataset. The proposed method, in contrast, adds a domain adaptation stage to the acquired finger-vein images using a CycleGAN to better handle unobserved data, thus improving the generality of the finger-vein recognition system. After inputting the actual finger-vein images obtained in the preprocessing stage to the CycleGAN, the mapping function needed for domain adaptation is found during training. The mapping function converts the source domain into the target domain. Owing to the unpaired trait of the CycleGAN, a completely one-to-one mapping function is not observed; instead, training is continued to identify style information of the target domain. Therefore, the main structure of the data of the source domain is fairly maintained to create a new image to which the distribution characteristics of the target domain are transferred (Steps 2-1 and 2-2 of Figure 1). This process mitigates the heterogeneity between datasets.

Subsequently, a composite image is generated using the new image obtained with a CycleGAN (Step 3 of Figure 1), and it is then input to a densely connected network (DenseNet)-161 (Step 4 of Figure 1). Then, finger-vein recognition is finally performed via shift matching (Steps 5 and 6 of Figure 1).

4.2. Preprocessing

The obtained finger image has both a background and finger region; therefore, the finger region and the background region need to be primarily segmented to obtain only the finger region in the preprocessing step. Figure 2 shows each preprocessing stage. Binary thresholding and segmentation are performed using the Sobel edge detector and Otsu thresholding method [27]. The image for which binary thresholding has been performed becomes a masked image filled with 255 in the finger region and with 0 for other regions. If the background region and both edges of the finger region have areas with a small pixel value, areas can be mis-classified as the finger region. To remove such areas, both edges are removed and the image is corrected again with component labeling. The boundary of this mask has numerous burrs; thus, a smoothing process would be required to perform accurate linear stretching. Then, in-plane rotation compensation is performed to ensure that the angles of all data are identical. Misalignment in the input image is a major factor that causes false rejection in particular and thus needs to be removed. In-plane rotation compensation involves calculating the second-order angle moments of the binarized mask as shown in Equations (1)–(4), thereby performing misalignment compensation so that all images can have the same angle with respect to the central axis [28].

$$\varnothing_{11} = \frac{\sum_{(a,b)\in M}(b - m_b)^2 I(a,b)}{\sum_{(a,b)\in M} I(a,b)}, \tag{1}$$

$$\varnothing_{22} = \frac{\sum_{(a,b)\in M}(a - m_a)^2 I(a,b)}{\sum_{(a,b)\in M} I(a,b)}, \tag{2}$$

$$\varnothing_{12} = \frac{\sum_{(a,b)\in M}(b - m_b)(a - m_a) I(a,b)}{\sum_{(a,b)\in M} I(a,b)}, \tag{3}$$

$$\tau = \begin{pmatrix} arctan\left(\frac{\varnothing_{11}-\varnothing_{22}+\sqrt{(\varnothing_{11}-\varnothing_{22})^2+4\varnothing_{12}^2}}{-2\varnothing_{12}}\right) \ (if \ \varnothing_{11} > \varnothing_{22}) \\ arctan\left(\frac{-2\varnothing_{12}}{\varnothing_{22}-\varnothing_{11}+\sqrt{(\varnothing_{22}-\varnothing_{11})^2+4\varnothing_{12}^2}}\right) \ (otherwise) \end{pmatrix}, \tag{4}$$

where $I(a,b)$ and (m_a, m_b) represent the pixel value and center index in the (a,b) index of the input, $M(a,b)$ represents the pixel value of the mask obtained through binary segmentation; its value should be 255 for the actual finger region and 0 for all other regions. $\varnothing$ is the second-order moments for each axis based on which the rotation compensation angle τ is calculated. In detail, $\varnothing_{11}$ and $\varnothing_{22}$ represent the correlation values in the vertical and horizontal directions, respectively. In addition, $\varnothing_{12}$ shows that in the diagonal direction. For example, if $\varnothing_{11}$ is larger than $\varnothing_{22}$, the correlation value of input ($I(a,b)$) with mask ($M(a,b)$) in the vertical direction is larger than that in the horizontal direction, which indicates that the input ($I(a,b)$) with mask ($M(a,b)$) has the elliptical shape, which is longer in the vertical direction than the horizontal direction. If $\varnothing_{12}$ is larger than $\varnothing_{11}$ and $\varnothing_{22}$, the correlation value of input ($I(a,b)$) with mask ($M(a,b)$) in the diagonal direction is larger than those in the vertical and horizontal directions, which indicates that the input ($I(a,b)$) with mask ($M(a,b)$) has the elliptical shape, which is longer in the diagonal direction than the vertical and horizontal directions. Based on this information, the rotation compensation angle τ is calculated by Equation (4) [28]. With respect to the central axis, in-plane rotation is performed for the initial finger image and binary mask based on this rotation compensation angle; then, the final finger-vein region is obtained by taking the mask as a condition. In the obtained finger region, the areas in which a finger-vein region cannot be observed easily due to the thickness of the finger or areas in which finger-vein information has been removed due to shades created by the fingernail or bone need to be removed. Therefore, removing a certain portion in the left and right sides of the mask used for acquiring the finger-vein region presents confident finger-vein information.

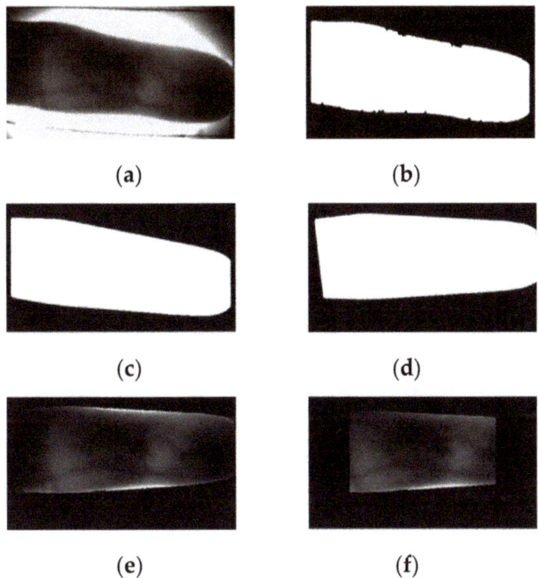

Figure 2. Sample images of each preprocessing stage: (**a**) original image, (**b**) image obtained after Sobel edge detection and thresholding, (**c**) image after edge smoothing, (**d**) image after in-plane rotation compensation, (**e**) finger-vein image obtained by region of interest (ROI) mask, and (**f**) finally cropped finger-vein ROI image.

Certain areas in the mask region, such as the background region represented as a dark area, may be mis-segmented as the finger region during binary thresholding; such areas need to be removed by component labeling [27]. Moreover, if there are areas eroded by additional noise in the finger-shape area, the final ROI mask is obtained through compensation during the smoothing process for removing such areas [26]. The finger region obtained thus undergoes linear interpolation to a size of 256×256 to be used as an input of the CycleGAN, which is detailed in the next section.

4.3. Domain Adaptation

The existing finger-vein recognition systems are specialized for training data to simply improve performance. However, a finger-vein recognition system is generally used for security purposes; therefore, performance improvement for unobserved data needs to be prioritized. If the image characteristics including brightness, shape, and texture between datasets are different, the network trained with a specific dataset experiences serious performance deterioration when tested with a different dataset. This problem implies that the model lacks generality, and its performance will fluctuate when it is applied in the real world, thus inhibiting the construction of a stable security system. In this study, therefore, both performance and generality are guaranteed by improving the generality in the distribution of the fundamental data through domain adaptation. The network used for domain adaptation in the proposed finger-vein system for this purpose is a CycleGAN.

4.3.1. CycleGAN Architecture

When performing domain adaptation for finger-vein images, there is a high possibility that the features generated in a latent space cannot encompass all the data distribution of each domain if the shape information of the finger-vein is transformed to a high extent. Thus, the image should be generated in a form such that texture information can be inserted while maintaining a shape information of specific domain.

A generative adversarial network which exploits unpaired data is most appropriate for this study because finger-vein image datasets have a different number of classes and thus require unpaired data to be utilized. The purpose is to find the latent space of a new domain between each domain. A CycleGAN uses unpaired data as the source and target; therefore, it can perform a task where the information of the source domain data is retained to some extent while reflecting the target domain information, instead of carrying out a task for simply making the source and target identical [24]. Therefore, a CycleGAN is most appropriate considering these circumstances. A CycleGAN is a network consisting of two discriminators and two generators.

A 70×70 PatchGAN [23] was used as the discriminator. Unlike a general discriminator, PatchGAN is a classifier that discriminates images at a patch unit. The prediction made by a discriminator of a typical GAN is output in an image unit, whereas the prediction made by a discriminator of a PatchGAN is output in a specific patch unit. In other words, the chronic problem of a GAN where blurry output is generated occurs less frequently by determining whether a specific patch region is fake or real, and the process is faster. When the finger-vein shape information used for recognition becomes blurry, the gradient between the finger-vein boundary and skin region is reduced, which implies that it cannot be used effectively. Accordingly, a CycleGAN was selected for domain adaptation in this study.

Table 1 shows the architecture of a 70×70 PatchGAN based discriminator. The fake image and original image created in the generator are concatenated to be input. Because it uses a 70×70 PatchGAN based method, it is parameter efficient and the relationship between adjacent pixels can be clearly identified based on a local-level discrimination rather than by determining real or fake data in the entire image.

Table 1. Architecture of the discriminator used in CycleGAN.

Layer	Filter (Number/Size/Stride)	Input Size	Output Size
Input layer		$256 \times 256 \times 3\ (\times 2)$	$256 \times 256 \times 6$
Conv1 *	$64/4 \times 4 \times 3/2$	$256 \times 256 \times 6$	$128 \times 128 \times 64$
Conv2 *	$128/4 \times 4 \times 64/2$	$128 \times 128 \times 64$	$64 \times 64 \times 128$
Conv3 *	$256/4 \times 4 \times 128/2$	$64 \times 64 \times 128$	$32 \times 32 \times 256$
Conv4 *	$512/4 \times 4 \times 256/1$	$32 \times 32 \times 256$	$31 \times 31 \times 512$
Conv5	$1/4 \times 4 \times 512/1$	$31 \times 31 \times 512$	$30 \times 30 \times 1$

* denotes that the convolutional layer is followed by instance normalization and a leaky rectified linear unit (ReLU) with a slope parameter of 0.2.

For the generator, a residual network (ResNet) based on an encoder-decoder structural network was used. Figure 3 shows the overall structure of the CycleGAN. Table 2 presents the detailed network architecture of the generator. We use the same settings of parameters and number of layers to those of [24] in Tables 1 and 2.

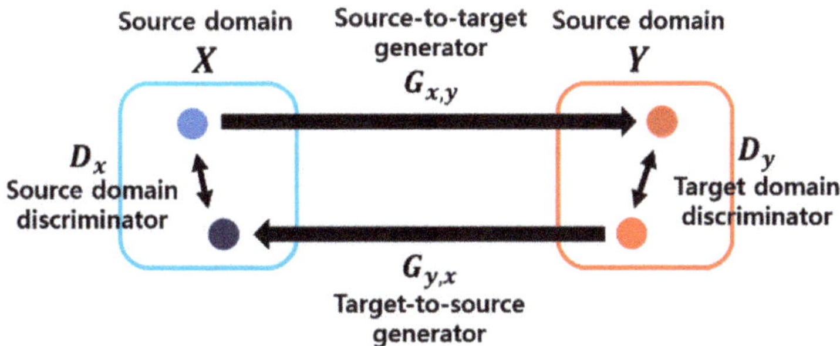

Figure 3. Summary of the CycleGAN structure.

Table 2. Architecture of the generator used in CycleGAN.

Layer	Filter (Number/Size/Stride)	Input Size	Output Size
Input layer		$256 \times 256 \times 3$	$256 \times 256 \times 3$
Conv1	$64/7 \times 7 \times 3/1$	$256 \times 256 \times 3$	$256 \times 256 \times 64$
Conv2 *	$128/3 \times 3 \times 64/2$	$256 \times 256 \times 64$	$128 \times 128 \times 128$
Conv3 *	$256/3 \times 3 \times 128/2$	$128 \times 128 \times 128$	$64 \times 64 \times 256$
Res1	$(256/3 \times 3 \times 256/1) \times 3$ **	$64 \times 64 \times 256$	$64 \times 64 \times 256$
Res2	$(256/3 \times 3 \times 256/1) \times 3$ **	$64 \times 64 \times 256$	$64 \times 64 \times 256$
Res3	$(256/3 \times 3 \times 256/1) \times 3$ **	$64 \times 64 \times 256$	$64 \times 64 \times 256$
Res4	$(256/3 \times 3 \times 256/1) \times 3$ **	$64 \times 64 \times 256$	$64 \times 64 \times 256$
Res5	$(256/3 \times 3 \times 256/1) \times 3$ **	$64 \times 64 \times 256$	$64 \times 64 \times 256$
Res6	$(256/3 \times 3 \times 256/1) \times 3$ **	$64 \times 64 \times 256$	$64 \times 64 \times 256$
Res7	$(256/3 \times 3 \times 256/1) \times 3$ **	$64 \times 64 \times 256$	$64 \times 64 \times 256$
Res8	$(256/3 \times 3 \times 256/1) \times 3$ **	$64 \times 64 \times 256$	$64 \times 64 \times 256$
Res9	$(256/3 \times 3 \times 256/1) \times 3$ **	$64 \times 64 \times 256$	$64 \times 64 \times 256$
Up-conv1	$128/3 \times 3 \times 256/2$	$64 \times 64 \times 256$	$128 \times 128 \times 128$
Up-conv2	$64/3 \times 3 \times 256/2$	$128 \times 128 \times 128$	$256 \times 256 \times 64$
Conv4	$3/7 \times 7 \times 3/1$	$256 \times 256 \times 64$	$256 \times 256 \times 3$

* denotes that the convolutional layer is followed by instance normalization and ReLU. ** denotes that the Res(k) is a residual block where an input feature map is added to the output of each residual block, and each residual block includes three convolutional layers.

4.3.2. Generating a Domain Adapted Finger-Vein Image

The data of each domain are used as a source and a target of the CycleGAN to generate an image for which domain adaptation has been applied. Figure 4 shows an example of the domain adapted image. It resembles the shape of an image used as a source and shows the shape in which the distribution of lighting intensity or contrast of the target domain is reflected. Hence, an image of a new domain is obtained for which information is composited.

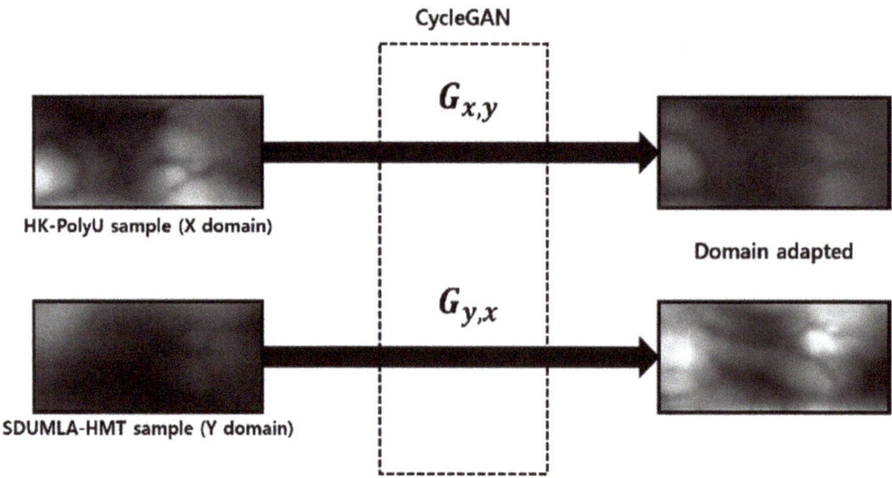

Figure 4. Examples of the domain translated image.

The loss function of a CycleGAN is the weighted sum of adversarial loss and cycle-consistency loss (see Equations (7)–(9)). The purpose of a generator is to deceive a discriminator by generating fake data that resemble the real data as much as possible, whereas a discriminator is trained to distinguish fake data from real data. Comparing the real data and simply generated data generates adversarial loss, as shown in Equations (5)–(7), while cycle-consistency loss helps in building a robust model through reconstruction by comparing the real data with source data, as shown in Equation (8). Ultimately, the loss function in which both adversarial loss and cycle-consistency loss are considered, as shown in Equation (9), is used. We use the same loss functions of Equations (5)–(9) to those in traditional CycleGAN [24].

$$Loss_{adv}\left(G_{x,y}, D_y,\ X\right) = \frac{1}{m}\sum_{i=1}^{m}\left(1 - D_y\left(G_{x,y}(x_i)\right)\right)^2, \tag{5}$$

$$Loss_{adv}\left(G_{y,x}, D_x,\ Y\right) = \frac{1}{m}\sum_{i=1}^{m}\left(1 - D_x\left(G_{y,x}(y_i)\right)\right)^2, \tag{6}$$

$$Loss_{adv} = Loss_{adv}\left(G_{x,y}, D_y,\ X\right) + Loss_{adv}\left(G_{y,x}, D_x,\ Y\right), \tag{7}$$

$$Loss_{cyc}\left(G_{x,y}, G_{y,x}, X,\ Y\right) = \frac{1}{m}\sum_{i=1}^{m}\left(\left(G_{x,y}\left(G_{y,x}(y_i)\right)\right) - y_i\right) - \left(\left(G_{y,x}\left(G_{x,y}(x_i)\right)\right) - x_i\right), \tag{8}$$

$$Loss_{total} = Loss_{adv} + \lambda Loss_{cyc}, \tag{9}$$

where G and D represent the generator and discriminator, respectively, x_i and y_i are the source image and target image selected in the X, Y domain, respectively, and m is the total number of data of each domain. λ is a cycle-consistency coefficient; a value of 10 was used in this study. Processing heterogeneous data through domain adaptation, as proposed in this study, enables us to retain the shape information of a specific domain while generating new domain data through adaptation of the texture information of a different domain. Thus, for a proper mixture of shape information and texture information, cycle-consistency loss value and adversarial loss value were adjusted using λ.

4.4. Generating Composite Image

A composite image is generated using the domain adapted image [26]. It is generated for a matching case, and it maximizes the network utilization rate more than the feature-based Euclidean distance matching method used in conventional finger-vein recognition systems. For the feature-based Euclidean distance matching method, matching is performed using the features extracted before the fully connected layer in a trained CNN model for the finger-vein recognition system. Thus, a trained fully connected layer cannot be used. In contrast, when generating authentic and imposter matching images as composite images, all layers in the trained CNN model for finger-vein recognition including the fully connected layer can be used. Furthermore, a data augmentation effect is observed during training because composite images are generated for the number of matching cases, and it is more robust for noise than difference image-based matching [5]. As shown in Figure 5, a composite image is an image generated by having an enrolled image, a matched image, and a concatenated image in each channel. The concatenated image is created by resizing the enrolled image and the matched image into 1/2 size images and then concatenating vertically. As a result, a three-channel shape image is generated and input in the CNN classifier. The composite image-based method does not involve Euclidean distance calculation by a n-dimensional feature vector, thus requiring a shorter time during inference compared to feature distance-based matching.

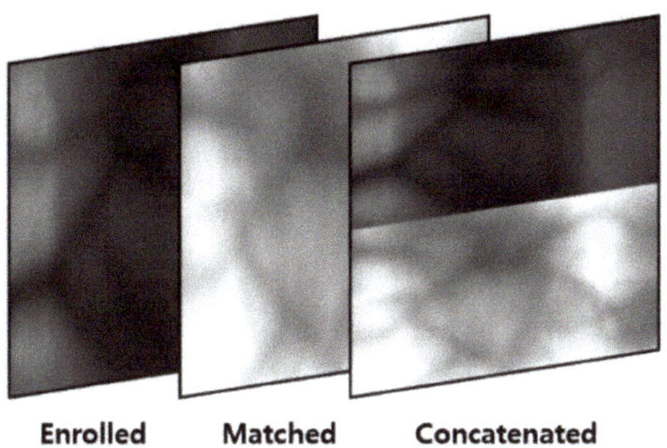

Figure 5. Example of composite image.

4.5. Finger-Vein Recognition Based on Deep Densenet and Shift Matching

In this study, DenseNet-161 was used as the model for finger-vein recognition [26,29]. Table 3 represents architecture of DenseNet-161 that used in this study. We use the same settings of parameters and number of layers to those of [29] in Table 3. In the DenseNet-161 used for proposed method, the growth rate was set to 48. The original structure of DenseNet was designed for ImageNet classification [29]. The output of the fully connected layer was a 1000-dimensional vector. As only two types of output—authentic matching score and imposter matching score—are used in this study, the existing fully connected layer was removed and fine tuning was performed after replacing it with a fully connected layer that outputs a two-dimensional score vector. DenseNet can effectively convey low level features to deeper layers through a dense connection.

Table 3. Architecture of DenseNet-161.

Layer	Filter (Number/Size/Stride)	Input Size	Output Size
Input layer		$224 \times 224 \times 3$	$224 \times 224 \times 3$
Conv	$(96/7 \times 7 \times 96/2)$	$224 \times 224 \times 3$	$112 \times 112 \times 96$
Max pool	$(96/2 \times 2 \times 1/2)$	$112 \times 112 \times 96$	$57 \times 57 \times 96$
Dense block	$(6/(1 \times 1 \times 192, 3 \times 3 \times 48)/1)$	$57 \times 57 \times 96$	$57 \times 57 \times 384$
Transition block	$(1/(1 \times 1 \times 192, 2 \times 2 \times 192)\ ^*/1)$	$57 \times 57 \times 384$	$29 \times 29 \times 192$
Dense block	$(12/(1 \times 1 \times 192, 3 \times 3 \times 48)/1)$	$29 \times 29 \times 192$	$29 \times 29 \times 768$
Transition block	$(1/(1 \times 1 \times 384, 2 \times 2 \times 384)\ ^*/1)$	$29 \times 29 \times 768$	$15 \times 15 \times 384$
Dense block	$(36/(1 \times 1 \times 192, 3 \times 3 \times 48)/1)$	$15 \times 15 \times 384$	$15 \times 15 \times 2112$
Transition block	$(1/(1 \times 1 \times 1056, 2 \times 2 \times 1056)\ ^*/1)$	$15 \times 15 \times 2112$	$8 \times 8 \times 1056$
Dense block	$(24/(1 \times 1 \times 192, 3 \times 3 \times 48)/1)$	$8 \times 8 \times 1056$	$8 \times 8 \times 2208$
Global average pool	$(2208/8 \times 8 \times 1/1)$	$8 \times 8 \times 2208$	$1 \times 1 \times 2208$
Fully connected layer		$1 \times 1 \times 2208$	$1 \times 1 \times 2$

* denotes the shape of the convolutional filter and average pooling filter, respectively.

Therefore, DenseNet was determined to be a very suitable classifier because low level features such as a ridge are the core components of the vein shape information present in the finger-vein data used in this study. For the composite image generated by acquiring the domain adapted image, the enrolled image and matched image are input in the same DenseNet-161. The spatial similarity of each image was evaluated in the classifier to confirm whether it is an authentic matching case or an imposter matching case. However, while evaluating the spatial similarity, misalignment or rotation, which were not removed during preprocessing, could be observed. These factors significantly affect the process of matching. To solve these problems, the enrolled image or matched image was matched through eight-way translation in this study. Then, the misalignment issue such as pixel translation was solved by designating the minimal matching value as the final matching score.

5. Experimental Results

In this section, we would explain experimental environments in Section 5.1, training of the domain adaptation model in Section 5.2, and training of finger-vein recognition model in Section 5.3. In addition, we would explain evaluation metrics in Section 5.4, and testing results and analyses with HKPolyU-DB after training with SDUMLA-HMT-DB (including ablation study) in Section 5.5. Finally, testing results and analyses with SDUMLA-HMT-DB after training with HKPolyU-DB (including ablation study) are presented in Section 5.6.

5.1. Experimental Environments

In this study, SDUMLA-HMT-DB [30] and HKPolyU-DB version 1 [31] were used. The HKPolyU database is divided into session 1 and session 2; only session 1 data were used in this study. HKPolyU-DB session 1 consists of 1872 images; two fingers of 156 persons were used for image acquisition, and six images were captured for each finger. SDUMLA-HMT-DB consists of 3816 images in which three fingers of each hand of 106 persons were used, and six images were captured for each finger. Each dataset was classified according to the finger used to acquire the image. HKPolyU-DB and SDUMLA-HMT-DB have a total of 312 classes and 636 classes, respectively. The number of classes is calculated by "the number of fingers" $\times$ "the number of hands" $\times$ "the number of persons". For example, because "the number of fingers", "the number of hands", and "the number of persons" in SDUMLA-HMT-DB are 3, 2, and 106, respectively, the number of classes becomes 636 ($3 \times 2 \times 106$) in SDUMLA-HMT-DB. To perform two-fold cross validation for training and testing, 156 classes were used for the training set and another 156 classes were used for the testing set for HKPolyU-DB, whereas 318 classes were used for the training set and another 318 classes were used for the testing set for SDUMLA-HMT-DB in 1st-fold validation. Specifically, the training and testing datasets did not include data from the same class. The training set and testing set were switched once for the experiment in the second-fold validation, and the average of the two accuracy values was used as the final value. In detail, as shown in Table 4, in the first-fold validation, the images of 318 classes (classes 1~318) were used for training whereas those of the remaining 318 classes (classes 319~636) were used for testing. In the second-fold validation, the images of 318 classes (classes 319~636) were used for training whereas those of the remaining 318 classes (classes 1~318) were used for testing. The sets used in each database are summarized in Table 4.

Table 4. Details of the experimental databases.

Database	Subset	Classes	Number of Original Images	Number of Augmented Images
HKPolyU-DB	Training	156	936	4680
	Test	156	936	-
SDUMLA-HMT-DB	Training	318	1908	9540
	Test	318	1908	-

We increased the number of training images by five times (including original training images) based on the data augmentation of image translation and cropping in the four directions (left, right, up, and down directions) by referring to [4]. Therefore, the total number of training images in HKPolyU-DB is 4680 (936 × 5) for each fold, and that in SDUMLA-HMT-DB is 9540 (1908 × 5) for each fold as shown in Table 4. With these augmented data, our models for domain adaptation and finger-vein recognition were successfully trained as shown in Figures 6 and 7.

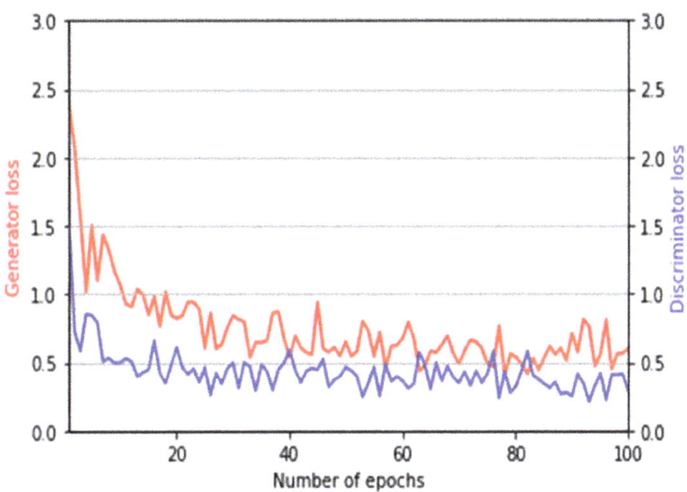

Figure 6. Graphs of training loss and accuracy by CycleGAN.

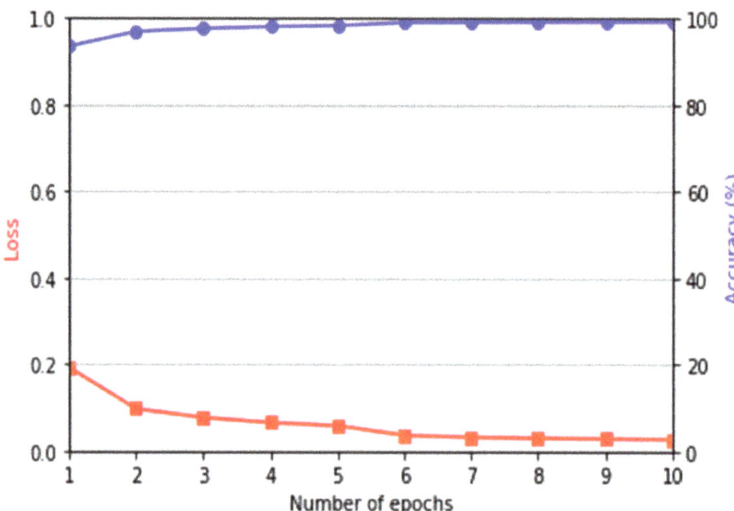

Figure 7. Graphs of the training loss and accuracy by DenseNet-161 using domain adapted data.

When we generated the images from HKPolyU-DB by CycleGAN, the test images of HKPolyU-DB were used for generation. Therefore, the number of generated images is 936 as shown in Table 4. When we generated the images from SDUMLA-HMT-DB by CycleGAN, the test images of SDUMLA-HMT-DB were used for generation. Therefore, the number of generated images is 1908 as shown in Table 4.

Training and testing were performed using a desktop computer equipped with an Intel® Core™ i7-3770K CPU @ 3.50GHz with 12GB RAM, and the graphics processing unit (GPU) card of NVIDIA Geforce GTX 1070 [32]. Moreover, compute unified device architecture (CUDA) version 9.0 [33] and CUDA deep neural network library (CUDNN) version 7.4.2 [34] were used. To execute the model and algorithm proposed in this study, Tensorflow framework version 1.15.1 [35] based on Python version 3.7.1 [36] was used.

5.2. Training of the Domain Adaptation Model

For the optimizer of the CycleGAN used for domain adaptation, the adaptive moment estimation (Adam) optimizer [37] was used. The initial learning rate was 0.0001; the exponential decay rate of the Adam optimizer was 0.9 for the first moment estimate and 0.999 for the second moment estimate. The learning rate strategies such as linear decay were not used. The model was trained for a total of 100 epochs. The discriminator was trained once for one mini-batch, whereas the generator was trained five times to solve the problem of the difficulty in training the generator of CycleGAN. Owing to this training strategy, the CycleGAN model used in this study was appropriately optimized for both the discriminator and generator. Figure 6 shows the loss graph of the generator and discriminator of the CycleGAN used in this study.

5.3. Training of Finger-Vein Recognition Model

A transfer learning strategy was used for training the finger-vein recognition model. The fully connected layer of the original network fine-tuned with the ImageNet was replaced with a fully connected layer with two-dimensional output, thus freezing the previous convolutional layer part and using the fully connected layer in the domain adapted image for training. Figure 7 shows the loss and accuracy graphs of the DenseNet-161 used in this study. These graphs imply that the DenseNet-161 model has been appropriately optimized.

5.4. Evaluation Metrics

An EER was used as the evaluation metric in this experiment. Each input determines genuine matching cases and imposter matching cases based on the matching score obtained during finger-vein recognition. Here, the rate of cases in which imposter matching cases have been categorized as genuine matching cases is the false acceptance rate (FAR), whereas the rate of cases in which the genuine matching cases are categorized as the imposter matching cases is the false rejection rate (FRR). The final EER is obtained at the threshold point where FAR and FRR are the same.

5.5. Testing with HKPolyU-DB after Training with SDUMLA-HMT-DB (Including Ablation Study)

In this section, the results of the experiment which proved the effect of the database that has been domain adapted from HKPolyU-DB to SDUMLA-HMT-DB are presented. As shown in Table 5, our CycleGAN was trained with the training data of HKPolyU-DB (input domain) and SDUMLA-HMT-DB (target domain), and the trained CycleGAN generated the domain adapted image (similar to the images of SDUMLA-HMT-DB) by using the testing data of HKPolyU-DB. Then, for testing, the generated images (similar to the images of SDUMLA-HMT-DB) were used as input to our finger-vein recognition model trained with the training data of SDUMLA-HMT-DB.

Table 5. Experimental scenario of our domain adaptation method (unit: %).

Training of CycleGAN	Image Generation by CycleGAN	Training of Finger-Vein Recognition Model	Testing of Finger-Vein Recognition Model
Using the training data of HKPolyU-DB (input domain) and SDUMLA-HMT-DB (target domain)	Using the testing data of HKPolyU-DB	Using the training data of SDUMLA-HMT-DB	Using the generated images by CycleGAN (similar to SDUMLA-HMT-DB)

For two-fold cross validation, the model for domain adaptation was trained using the training set. When both types of databases (HKPolyU-DB, SDUMLA-HMT-DB) were used during domain adaptation, the training set and the testing set were strictly separated for both databases in two-folds. Accordingly, the experiment was performed in an open-world setting in which the class of training data was different from the class of testing data.

Table 6 shows the comparison of the drop of finger-vein recognition performance for the same domain and cross-domain environment while the DenseNet-161 network is applied in the same manner without the CycleGAN-based domain adaptation proposed in this study.

Table 6. Comparisons of EER with same domain and cross-domain environment without our domain adaptation method (unit: %).

Training of Finger-Vein Recognition Model	Testing of Finger-Vein Recognition Model	EER
HKPolyU-DB	HKPolyU-DB	0.58
SDUMLA-HMT-DB	HKPolyU-DB	1.80

As shown in Table 6, when training and testing were conducted using HKPolyU-DB, the recognition rate was high with the EER of 0.58%. In contrast, when the model was trained using SDUMLA-HMT-DB and tested using HKPolyU-DB without the CycleGAN-based domain adaptation, the accuracy was significantly lower. As shown in Table 4, the amount of data used in SDUMLA-HMT-DB were considerably greater than that used on in HKPolyU-DB, and performance drop occurred even though they are both databases of the same finger-vein scope. The difference in data between the two domains is not visually noticeable; however, the heterogeneity between the two domains is definitely present.

Moreover, the qualities of images in HKPolyU-DB are relatively better than those of images in SDUMLA-HMT-DB, and the intra-class variance is lower. In other words, the training set is a much more complex case than the testing set; thus, the performance drop is not significant. However, compared with the same domain environment, the cross-validation environment experienced a considerable performance drop, and the domain adaptation method was used to solve this problem. Table 7 and Figure 8 show the accuracy of finger-vein recognition of the various domain adaptation methods. Here, genuine acceptance rate (GAR) is defined as 100–FRR (%). Therefore, we can find that the ratio of FRR to FAR is smaller in case that the ROC curve is positioned higher (closed to the left-top position of Figures 8 and 9), which means the ratio of GAR to FAR is higher. The experimental results showed that the accuracy is significantly higher when the proposed CycleGAN-based method is used compared to the cases when a domain adaptation method is not applied or other domain adaptation methods were used. This result implies that domain adaptation based on the proposed method sufficiently transferred the feature information of each domain.

Table 7. Comparisons of EERs of the proposed method and other domain adaptation methods in case of training with SDUMLA-HMT-DB and testing with HKPolyU-DB (unit: %).

Method	EER
No domain adaptation	1.80
StarGAN-v2 [38]	1.34
ComboGAN [39]	2.77
CycleGAN (proposed method)	0.85

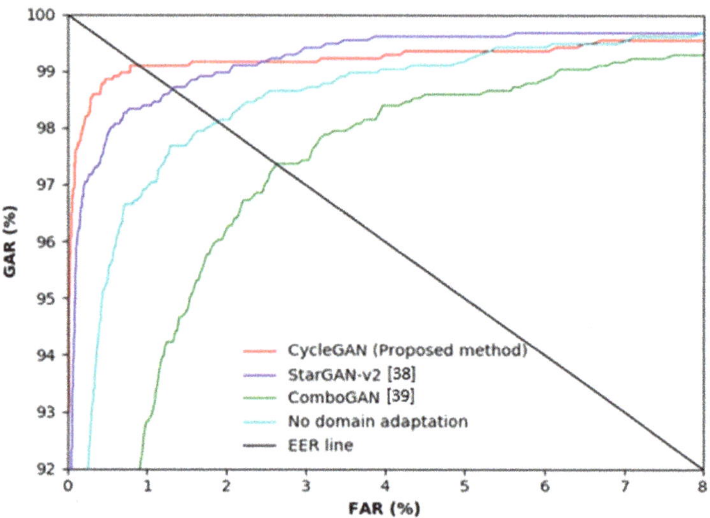

Figure 8. ROC curves of finger-vein recognition by the proposed method and other methods in case of training with the SDUMLA-HMT database and testing with the HKPolyU database.

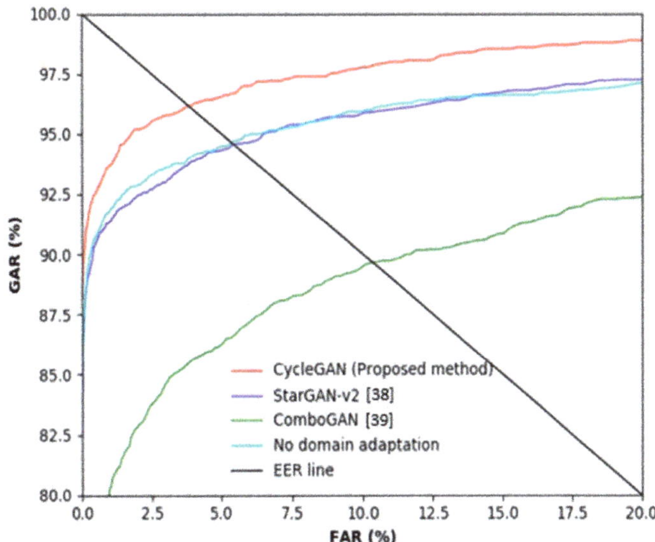

Figure 9. ROC curves of finger-vein recognition by the proposed method and other methods in case of training with HKPolyU-DB and testing with SDUMLA-HMT-DB.

Table 8 shows a comparison of the accuracy of the proposed method and the state-of-the-art methods. The experimental results highlighted that the proposed method had a higher recognition accuracy than the state-of-the-art methods.

Table 8. Comparisons of EER by the state-of-the-art methods and the proposed method in case of training with SDUMLA-HMT-DB and testing with HKPolyU-DB (unit: %).

Method	EER
Huang et al. [40]	9.46
Miura et al. [41]	6.49
Liu et al. [42]	5.01
Gupta et al. [43]	4.47
Miura et al. [44]	4.45
Dong et al. [45]	3.53
Liu et al. [46]	1.47
Xi et al. [47]	1.44
Joseph et al. [48]	1.27
Proposed method	0.85

5.6. Testing with SDUMLA-HMT-DB after Training with HKPolyU-DB (Including Ablation Study)

In this section, we performed the experiments again by exchanging HKPolyU-DB and SDUMLA-HMT-DB compared to the experiments of Section 5.5. Table 9 shows the result of performing training and testing with SDUMLA-HMT-DB and of performing training with HKPolyU-DB and testing with SDUMLA-HMT-DB. The performance drop is greater compared to the result shown in Table 6, which can be because the degree of noise, misalignment, and blur in the images in SDUMLA-HMT-DB are considerably greater than those of the images in HKPolyU-DB. Therefore, the recognition performance in the cross-domain environment is significantly low because of the unique trait of the domain transformed by noise or an image capturing device.

Table 9. Comparisons of EER with same domain and cross-domain environment without our domain adaptation method (unit: %).

Training of Finger-Vein Recognition Model	Testing of Finger-Vein Recognition Model	EER
SDUMLA-HMT-DB	SDUMLA-HMT-DB	2.17
HKPolyU-DB	SDUMLA-HMT-DB	4.42

Table 10 and Figure 9 show the accuracy of finger-vein recognition obtained by various domain adaptation methods. The experimental results showed that the accuracy is significantly higher when the proposed CycleGAN-based method is used compared to when a domain adaptation method is not applied or other domain adaptation methods were used. Thus, the feature information that can be obtained from SDUMLA-HMT-DB has been well adapted while partially maintaining the unique shape information of HKPolyU-DB. The results of StarGAN-v2 and ComboGAN are poorer than those of the proposed CycleGAN. Table 7 and Figure 8 present similar results. Fundamentally, a CycleGAN is a network designed for style transfer between two domains, whereas ComboGAN and StarGAN-v2 are designed for multi-domain transfer. Particularly, a StarGAN-v2 can not only simply discriminate between real or fake data using a style code but can also discriminate the type of domain generated. In a multi-domain focused architecture, performance is poorer as the discrepancy between domains is greater. Only a specific region cannot have high activation due to the trait of finger-vein data, and the heterogeneity in the shape information is noticeably significant even if the databases appear similar. Furthermore, ComboGAN not only mitigates the number of generators which increases with multi-domain transfer cases but also attempts to solve the problem of deteriorating performance caused by a greater difference in the domains of the existing StarGAN. However, the encoder and decoder separated by the number of domains recognize a specific database as one style as proposed by the ComboGAN, i.e., it failed to completely learn the domain distribution.

Table 10. Comparisons of EERs of the proposed method and other domain adaptation methods in case of training with HKPolyU-DB and testing with SDUMLA-HMT-DB (unit: %).

Method	EER
No domain adaptation	4.42
StarGAN-v2 [38]	4.43
ComboGAN [39]	8.96
CycleGAN (proposed method)	3.40

Table 11 shows the comparison of the accuracy between the proposed method and the state-of-the-art methods. The experimental result showed that the proposed method had a higher recognition accuracy than the state-of-the-art methods.

Figures 10 and 11 show examples of the image domains adapted using various methods. Figures 10a and 11a show the examples of the original image; the images on the left in (b)–(g) are the source images and those on the right are images generated through domain adaptation using the source images. That is, the left and right images of Figure 10b–g, respectively, show original images and domain adapted images from SDUMLA-HMT-DB and HKPolyU-DB using various methods ((b), (c) our method, (d), (e) ComboGAN, (f), (g) StarGAN-v2). By comparing the right images of (b) and (c) with those of (d)–(g), the right images of (b) and (c) by our method have more similar image characteristics (including the distinctiveness of vein patterns) to the original images of HKPolyU-DB (Figure 10a) compared to the right images of (d)–(g). In addition, as shown in Figure 11, by comparing the right images of (b) and (c) by our method with those of (d)–(g) by other methods, the right images of (b) and (c) have more similar image characteristics (including to the distinctiveness of vein patterns) to the original images of SDUMLA-HMT-DB (Figure 11a) compared to the right images of (d)–(g).

Table 11. Comparisons of EER of the state-of-the-art methods and proposed method in case of training with HKPolyU-DB and testing with SDUMLA-HMT-DB (unit: %).

Method	EER
Jalilian et al. [18]	3.57
Pham et al. [49]	8.09
Miura et. al. [44]	5.46
Miura et al. [41]	4.54
Yang et al. [50]	3.96
CycleGAN (proposed method)	3.40

As shown in all examples, the image generated by the proposed method using a CycleGAN has the best quality; the images generated by the StarGAN-v2 are somewhat blurry and exhibit dark noises while transferring the target domain style to a certain extent. Lastly, the image generated by the ComboGAN shows that the difference in data quantity between SDUMLA-HMT-DB and HKPolyU-DB as well as the separated encoder and decoder structure did not produce good performance. Unlike facial emotion data in which features are concentrated in specific regions, the information is not concentrated in specific regions in the finger-vein data; thus, it is difficult to assign a style. Therefore, the results in Figures 10 and 11 are produced if the generator structure is not concrete because the dataset is widely distributed.

Finally, the effect of the proposed method was analyzed by comparing the cases in which recognition errors were produced in all schemes in which the proposed method and domain adaptation were not applied and cases in which the model correctly recognized the images only using the proposed method. Figure 12 summarizes the error cases generated in the no adaptation method where SDUMLA-HMT-DB was used as the training set and in the proposed method where SDUMLA-HMT-DB was domain adapted to HKPolyU-DB. Figure 12a,b show the cases in which errors occurred even when domain adaptation was performed using the proposed method. Specifically, Figure 12a is an example of a false rejection case, and Figure 12b is the example of a false acceptance case. As shown in Figure 12a, a major pixel translation observed even when the enrolled image and the matched image were an authentic matching case. In Figure 12b, both images were not properly acquired because of the imbalance in lighting intensity of the NIR sensor used for acquiring the finger-vein images. Because of these problems, the finger-vein pattern appeared only in a limited region of the image, which resulted in an imposter matching case which appeared as an authentic matching case. In addition, correctly recognizing if the shape pattern, which is important information, is distributed in a similar manner, is a challenging task.

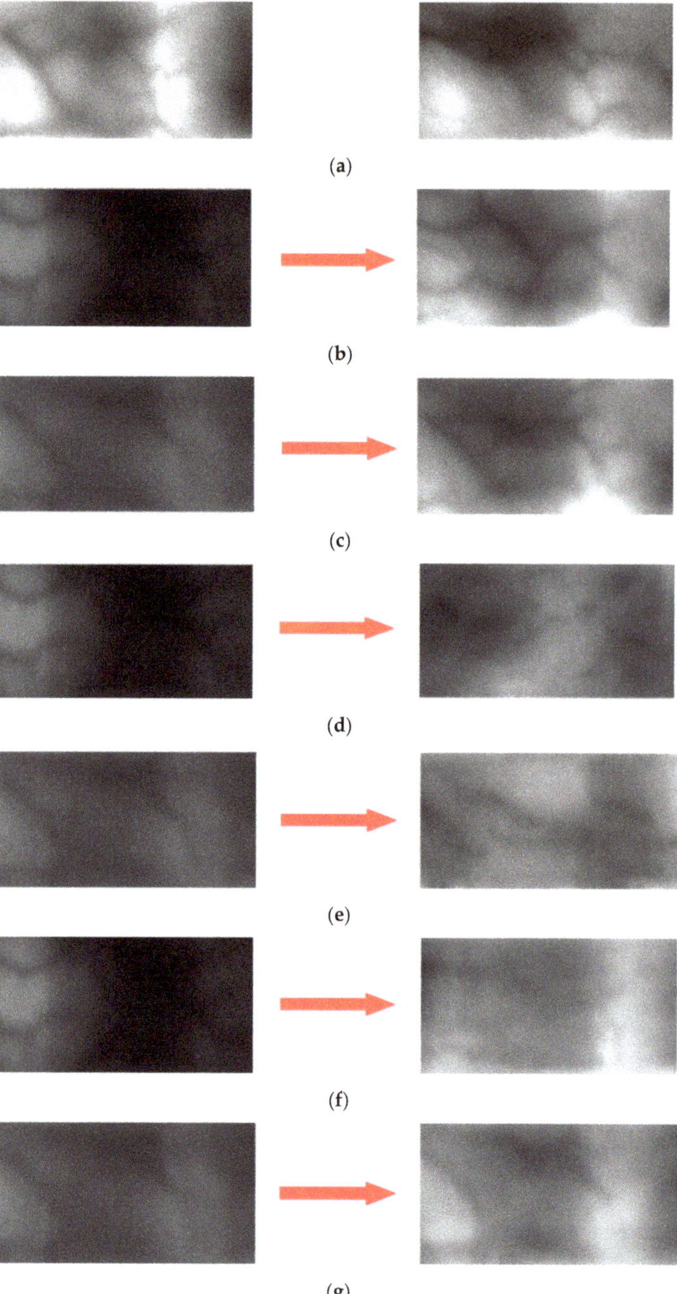

Figure 10. Examples of original images and domain adapted images: (**a**) Original image of HKPolyU-DB. Left and right images of (**b–g**) respectively, show original images and domain adapted images from SDUMLA-HMT-DB and HKPolyU-DB using various methods. (**b**,**c**) proposed method, (**d**,**e**) ComboGAN, (**f**,**g**) StarGAN-v2.

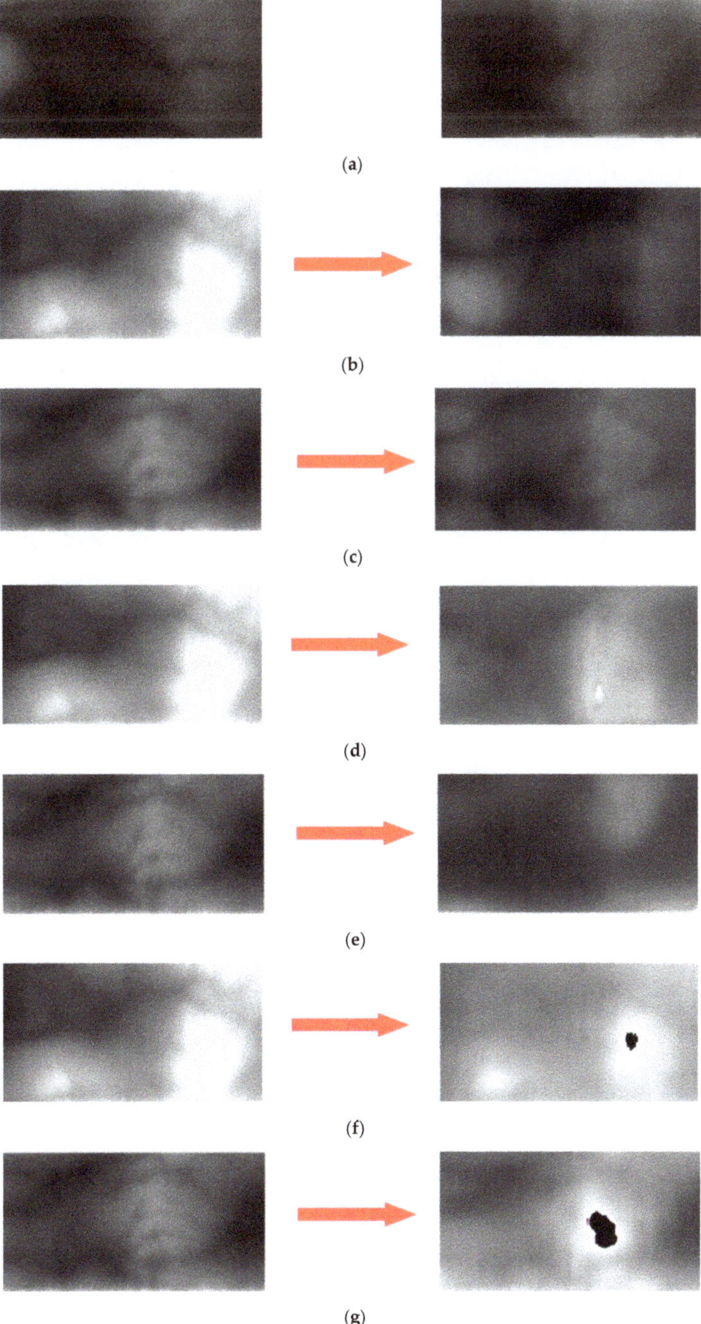

Figure 11. Examples of original images and domain adapted images: (**a**) Original image of SDUMLA-HMT-DB. Left and right images of (**b**–**g**), respectively, show original images and domain adapted images from HKPolyU-DB to SDUMLA-HMT-DB using various methods. (**b**,**c**) proposed method, (**d**,**e**) ComboGAN, (**f**,**g**) StarGAN-v2.

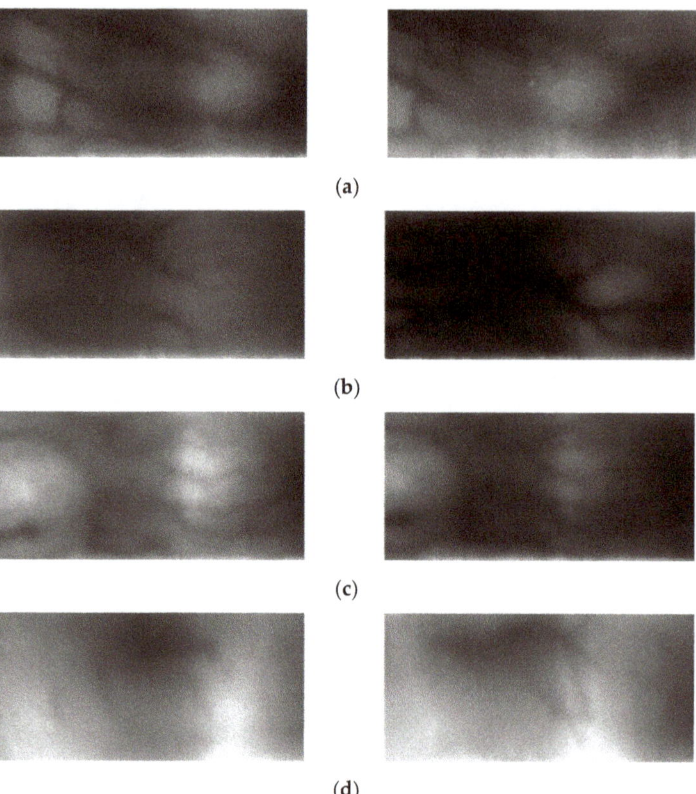

Figure 12. Examples of errors in case of testing with HKPolyU-DB: (**a**) False rejection case by both proposed method and no domain adaptation, (**b**) false acceptance case by both proposed method and no domain adaptation, (**c**) false rejection case by the no adaptation method, but correct recognition case by the proposed method, (**d**) false acceptance case by the no adaptation method, but correct rejection case by the proposed method. Left and right images of (**a–d**) show enrolled and matched images, respectively.

Figure 12c,d are the results of correct recognition when the proposed method was used in which 12c shows the falsely rejected case and 12d shows the falsely accepted case in a scheme where the domain adaptation method was not applied. Figure 12c is an authentic matching case; however, a problem was observed when the intensity of lighting varied during the image capturing trial. However, the data for which domain adaptation was performed are effective against the variance in lighting intensity as such information of the source domain, SDUMLA-HMT-DB, was also transferred. Figure 12d also shows that it is difficult to identify the overall finger-vein pattern because finger-vein information is acquired from a limited region; however, a good recognition performance was still observed when the proposed method was used appropriately using the scarcely available finger-vein pattern. Therefore, a robust performance was achieved for extracting the finger-vein valley through domain adaptation.

Unlike Figure 12, Figure 13 summarizes the error cases generated in the no adaptation method where HKPolyU-DB was used as the training set, and in the proposed method where HKPolyU-DB was domain adapted to SDUMLA-HMT-DB. The information was mostly not contained in the images properly for the data of SDUMLA-HMT-DB, which is similar to the data of HKPolyU-DB. In particular, the cases in Figure 13a,b only contained a small amount of finger-vein patterns, and the recognition was performed using the background information during testing. This problem cannot be easily solved by domain adaptation, and therefore, it was not successfully recognized in the case where the proposed method was used. Even though the case in Figure 13c is an authentic matching case, the pixel translation between the enrolled image and the matched image was significantly large, while the forms of the shades slightly varied. However, for the data generated using the proposed method, the finger-vein pattern of each domain was effectively transferred, thus producing robust performance for the finger-vein pattern of SDUMLA-HMT-DB along with the focused form of the finger-vein pattern. This shows that the network was optimized to generate variations in the vein pattern information by focusing on the vein pattern when training the CycleGAN. Figure 13d also shows that it is difficult to identify the overall finger-vein pattern because the finger-vein information is acquired from a limited region; however, a good recognition performance was still observed when the proposed method was used appropriately using the scarcely available finger-vein pattern.

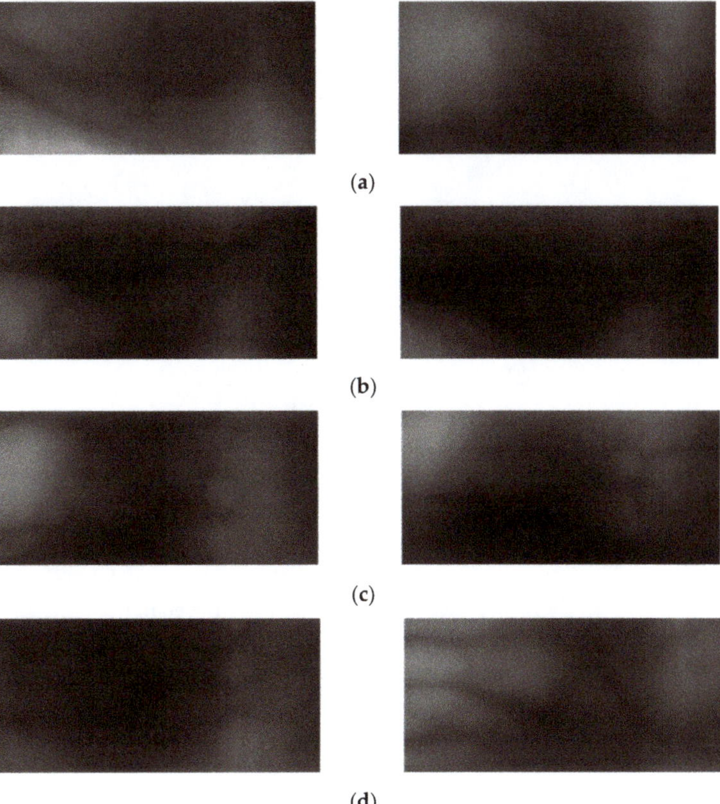

Figure 13. Examples of errors in the case of testing with SDUMLA-HMT-DB: (**a**) False rejection case by both the proposed method and no domain adaptation, (**b**) false acceptance case by both the proposed method and no domain adaptation, (**c**) false rejection case by the no adaptation method, but correct recognition case by the proposed method, (**d**) false acceptance case by the no adaptation method, but correct rejection case by the proposed method. Left and right images of (**a–d**) show enrolled and matched images, respectively.

6. Discussion

In this section, we briefly compared the previous and proposed methods with advantages and disadvantages, as shown in Table 12.

Table 12. Comparisons of the previous and proposed methods for hand-based biometrics.

Categories	Considering the Cross-Domain Problem	Method	Modality	Advantage	Disadvantage
Non-training-based	No	Wide line detector and pattern normalization [40]	Finger-vein	Simple and computationally efficient than training-based method	Performance is not good compared to training-based method
		Maximum curvature points [41]			
		Minutiae matching [42]			
		Multi-scale matched filter [43]			
		Repeated line tracking [44]			
		Personalized best patches map [45]			
		Superpixel-based [46]			
		Discriminative binary codes [47]			
		Fuzzy rule-based [48]			
		Local binary pattern [49]			
		Tri-branch vein structure [50]			
	Yes	Dimension reduction and orientation coding algorithm [7]	Palmprint		
		SIFT [8]	Dorsal hand-vein		
		Improved SIFT [9]			
		BGP and Gabor-HoG [10]	Fingerprint		
		Least square-based domain transformation function [11]			

Table 12. *Cont.*

Categories	Considering the Cross-Domain Problem	Method	Modality	Advantage	Disadvantage
Training-based	No	VGG-16 and CNN [20]		Preprocessing is not required	No consideration about the heterogeneous data problem
		Patch-based MobileNet [21]			
		CGAN [19]			Does not show good performance in cross-sensor environments
		FCN [18]	Finger-vein	Using compact information on recognition stage increases generality	Unreliable label data were used
	Yes	FLDA [12]	Face and fingerprint	Simple method for domain adaptation	Needs multiple modality data from same people
		Universal material translator wrapper [13]	Fingerprint	Uses a simple style transfer network	Generated images cannot deal with level 3 features
		DeepDomainPore network [14]		Can exploit level 3 features using low-resolution input	Long preprocessing time and ground truth required for source data
		PalmGAN [15]	Palmprint	Automatically generates label data for target domain	- Long preprocessing time and ground truth required for source data Segmentation method is unstable
		Auto-encoder [16]		Automatically generates label data for target domain Simple method for domain matching with good matching performance	
		DeepScatNet and RDF [17]	Finger-selfie		
		CycleGAN-based (Proposed method)	Finger-vein	High performance for domain adaptation Does not need ground truth for source data	Intensive training for CycleGAN is necessary

In case of five-fold or 10-fold cross validation, the number of training data becomes much larger, and the consequent accuracy of testing becomes higher than that by two-fold cross validation in most cases due to the sufficient training of model. However, it is very difficult to acquire the sufficient number of training data in real world cases. Considering these cases, we aim at measuring the testing accuracies even with insufficient training data based on two-fold cross validation in our experiments.

The proposed method failed the correct recognition in the following cases; (i) a major pixel translation observed even when the enrolled image and the matched image were an authentic matching case, (ii) both the enrolled and matched images were not properly acquired because of the imbalance in lighting intensity of the NIR sensor used for acquiring the finger-vein images, and (iii) the captured image only contained small amount of finger-vein patterns, and the recognition was performed using the background information during testing.

7. Conclusions

In this study, we propose a finger-vein recognition system in which domain adaptation is applied to solve the problem of the performance drop in a finger-vein recognition system when unobserved data are used. Domain adaptation was performed using CycleGAN, and the proposed domain adapted model proved to be effective using two databases—HKPolyU-DB and SDUMLA-HMT-DB. All cases found in the real environment include unobserved data; thus, a performance drop in similar circumstances is critical. As a finger-vein recognition system is used for security purposes, unstable performance depending on specific situations would decrease the reliability, thus making its application to real-world applications difficult. Using the proposed method, a stable finger-vein recognition system with improved generality can be applied to various real-world applications.

In this research, we focused on checking the possibility of domain adaptation of heterogeneous finger-vein databases. Therefore, we used the well-known CycleGAN and DenseNet-161 whose performances for style transfer of unpaired data and classification were already confirmed, respectively, in many previous researches of different applications. We performed only the fine-tuning of CycleGAN and DenseNet-161 with our experimental data. We would research the method of further customization of CycleGAN and DenseNet-161 to enhance the accuracies as future works.

In addition, we would research the advanced domain adaptation method, which can solve the cases of major pixel translation between the enrolled and matched images, the imbalance of lighting intensity in the captured image, and the small amount of finger-vein patterns contained in the captured image explained in Section 6. We would also evaluate the performance by five-fold or 10-fold cross validation in future work. In addition, a finger-vein recognition system with a more robust performance for unobserved data will be further studied in the future by improving the generality of the domain through multiple-domain adaptation, rather than simple domain adaptation between two databases. Furthermore, the efficacy of domain adaptation proposed in this study will also be researched for diverse biometric data such as palm and hand dorsal vein images, visible and NIR iris images, and visible and NIR face images.

Author Contributions: K.J.N. and K.R.P. designed finger-vein recognition system with heterogeneous databases by domain adaption based on CycleGAN, conducted and analyzed experiments, and wrote the original paper. J.C. and J.S.H. implement the preprocessing method and experiments. All authors have read and agreed to the published version of the manuscript.

Institutional Review Board Statement: Ethical review and approval were waived for this study because we used two open databases of SDUMLA-HMT-DB [30] and HKPolyU-DB [31].

Informed Consent Statement: Patient consent was waived because we used two open databases of SDUMLA-HMT-DB [30] and HKPolyU-DB [31].

Data Availability Statement: Trained models with algorithm can be available upon reasonable request according to the instructions in [25].

Acknowledgments: This work was supported in part by the Ministry of Science and ICT (MSIT), Korea, under the ITRC (Information Technology Research Center) support program (IITP-2020-2020-0-01789) supervised by the IITP (Institute for Information & communications Technology Promotion), in part by the National Research Foundation of Korea (NRF) funded by the MSIT through the Basic Science Research Program (NRF-2020R1A2C1006179), and in part by the NRF funded by the MSIT through the Basic Science Research Program (NRF-2019R1A2C1083813).

Conflicts of Interest: The authors declare no conflict of interest.

References

1. Lee, E.C.; Jung, H.; Kim, D. New Finger Biometric Method Using Near Infrared Imaging. *Sensors* **2011**, *11*, 2319–2333. [CrossRef] [PubMed]
2. Meng, X.; Yang, G.; Yin, Y.; Xiao, R. Finger Vein Recognition Based on Local Directional Code. *Sensors* **2012**, *12*, 14937–14952. [CrossRef] [PubMed]
3. Peng, J.; Wang, N.; El-Latif, A.A.A.; Li, Q.; Niu, X. Finger-vein Verification Using Gabor Filter and SIFT Feature Matching. In Proceedings of the 8th International Conference on Intelligent Information Hiding and Multimedia Signal Processing, Piraeus, Greece, 18–20 July 2012; pp. 45–48.
4. Song, J.M.; Kim, W.; Park, K.R. Finger-vein Recognition Based on Deep Densenet Using Composite Image. *IEEE Access* **2019**, *7*, 66845–66863. [CrossRef]
5. Kim, W.; Song, J.M.; Park, K.R. Multimodal Biometric Recognition Based on Convolutional Neural Network by the Fusion of Finger-vein And Finger Shape Using Near-Infrared (NIR) Camera Sensor. *Sensors* **2018**, *18*, 2296. [CrossRef] [PubMed]
6. Lu, Z.; Li, M.; Zhang, J. Automatic Illumination Control Algorithm for Capturing the Finger Vein Image. In Proceedings of the 13th World Congress on Intelligent Control and Automation, Changsha, China, 4–8 July 2018; pp. 881–886.
7. Jia, W.; Hu, R.-X.; Gui, J.; Zhao, Y.; Ren, X.-M. Palmprint recognition across different devices. *Sensors* **2012**, *12*, 7938–7964. [CrossRef] [PubMed]
8. Wang, Y.; Zheng, X.; Wang, C. Dorsal Hand Vein Recognition across Different Devices. In Proceedings of the In Chinese Conference on Biometric Recognition, Chengdu, China, 14–16 October 2016; pp. 307–316.
9. Wang, Y.; Zheng, X. Cross-device Hand Vein Recognition Based on Improved SIFT. *Int. J. Wavelets Multiresolution Inf. Process.* **2018**, *16*, 1840010. [CrossRef]
10. Alshehri, H.; Hussain, M.; Aboalsamh, H.A.; Zuair, M.A.A. Cross-Sensor Fingerprint Matching Method Based on Orientation, Gradient, And Gabor-Hog Descriptors with Score Level Fusion. *IEEE Access* **2018**, *6*, 28951–28968. [CrossRef]
11. Ghiani, L.; Mura, V.; Tuveri, P.; Marcialis, G.L. On the Interoperability of Capture Devices in Fingerprint Presentation Attacks Detection. In Proceedings of the First Italian Conference on Cybersecurity, Venice, Italy, 17–20 January 2017; pp. 66–75.
12. Kute, R.S.; Vyas, V.; Anuse, A. Cross Domain Association Using Transfer Subspace Learning. *Evol. Intell.* **2019**, *12*, 201–209. [CrossRef]
13. Gajawada, R.; Popli, A.; Chugh, T.; Namboodiri, A.; Jain, A.K. Universal Material Translator: Towards Spoof Fingerprint Generalization. In Proceedings of the 2019 International Conference on Biometrics, Crete, Greece, 4–9 June 2019; pp. 1–8.
14. Anand, V.; Kanhangad, V. Unsupervised Domain Adaptation for Cross-sensor Pore Detection in High-resolution Fingerprint Images. *arXiv* **2020**, arXiv:1908.10701v2. Available online: https://arxiv.org/abs/1908.10701 (accessed on 12 December 2020).
15. Shao, H.; Zhong, D.; Li, Y. PalmGAN for Cross-domain Palmprint Recognition. In Proceedings of the 2019 IEEE International Conference on Multimedia and Expo, Shanghai, China, 8–12 July 2019; pp. 1390–1395.
16. Shao, H.; Zhong, D.; Du, X. Cross-domain Palmprint Recognition Based on Transfer Convolutional Autoencoder. In Proceedings of the 2019 IEEE International Conference on Image Processing, Taipei, Taiwan, 22–25 September 2019; pp. 1153–1157.
17. Malhotra, A.; Sankaran, A.; Vatsa, M.; Singh, R. On Matching Finger-selfies Using Deep Scattering Networks. *IEEE Trans. Biometrics Behav. Identit. Sci.* **2020**, *2*, 350–362. [CrossRef]
18. Jalilian, E.; Uhl, A. Finger-vein Recognition Using Deep Fully Convolutional Neural Semantic Segmentation Networks: The Impact of Training Data. In Proceedings of the 2018 IEEE International Workshop on Information Forensics and Security, Hong Kong, China, 11–13 December 2018; pp. 1–8.
19. Dabouei, A.; Kazemi, H.; Iranmanesh, S.M.; Dawson, J.; Nasrabadi, N.M. ID Preserving Generative Adversarial Network for Partial Latent Fingerprint Reconstruction. In Proceedings of the 2018 IEEE 9th International Conference on Biometrics Theory, Applications and Systems, Redondo Beach, CA, USA, 22–25 October 2018; pp. 1–10.
20. Nogueira, R.F.; de Alencar Lotufo, R.; Machado, R.C. Fingerprint Liveness Detection Using Convolutional Neural Networks. *IEEE Trans. Inf. Forensics Secur.* **2016**, *11*, 1206–1213. [CrossRef]
21. Chugh, T.; Cao, K.; Jain, A.K. Fingerprint Spoof Buster: Use of Minutiae-centered Patches. *IEEE Trans. Inf. Forensics Secur.* **2018**, *13*, 2190–2202. [CrossRef]
22. Chui, K.T.; Lytras, M.D.; Vasant, P. Combined Generative Adversarial Network and Fuzzy C-Means Clustering for Multi-Class Voice Disorder Detection with an Imbalanced Dataset. *Appl. Sci.* **2020**, *10*, 4571. [CrossRef]
23. Isola, P.; Zhu, J.-Y.; Zhou, T.; Efros, A.A. Image-to-image Translation with Conditional Adversarial Networks. In Proceedings of the IEEE Conference on Computer Vision and Pattern Recognition, Honolulu, HI, USA, 21–26 July 2017; pp. 1125–1134.
24. Zhu, J.-Y.; Park, T.; Isola, P.; Efros, A.A. Unpaired Image-to-image Translation Using Cycle-consistent Adversarial Networks. In Proceedings of the IEEE International Conference on Computer Vision, Venice, Italy, 22–29 October 2017; p. 2223.
25. Dongguk CycleGAN-Based Domain Adaptation and DenseNet-Based Finger-Vein Recognition Models (DCDA&DFRM) with Algorithms. Available online: http://dm.dgu.edu/link.html (accessed on 9 August 2020).
26. Noh, K.J.; Choi, J.; Hong, J.S.; Park, K.R. Finger-vein Recognition Based on Densely Connected Convolutional Network Using Score-level Fusion with Shape and Texture Images. *IEEE Access* **2020**, *8*, 96748–96766. [CrossRef]
27. Gonzalez, R.C.; Woods, R.E. *Digital Image Processing*, 3th ed.; Prentice Hall: Upper Saddle River, NJ, USA, 2010.

28. Kumar, A.; Zhang, D. Personal Recognition Using Hand Shape and Texture. *IEEE Trans. Image Process.* **2006**, *15*, 2454–2461. [CrossRef] [PubMed]
29. Huang, G.; Liu, Z.; Van Der Maaten, L.; Weinberger, K.Q. Densely Connected Convolutional Networks. In Proceedings of the IEEE Conference on Computer Vision and Pattern Recognition, Honolulu, HI, USA, 21–26 July 2017; pp. 2261–2269.
30. Yin, Y.; Liu, L.; Sun, X. SDUMLA-HMT: A Multimodal Biometric Database. In Proceedings of the 6th Chinese Conference on Biometric Recognition, Beijing, China, 3–4 December 2011; pp. 260–268.
31. Kumar, A.; Zhou, Y. Human Identification Using Finger Images. *IEEE Trans. Image Process.* **2012**, *21*, 2228–2244. [CrossRef]
32. NVIDIA GeForce GTX 1070. Available online: https://www.nvidia.com/en-in/geforce/products/10series/geforce-gtx-1070/ (accessed on 10 July 2020).
33. CUDA. Available online: https://developer.nvidia.com/cuda-90-download-archive (accessed on 10 July 2020).
34. CUDNN. Available online: https://developer.nvidia.com/cudnn (accessed on 10 July 2020).
35. Tensorflow: The Python Deep Learning Library. Available online: https://www.tensorflow.org/versions/r1.15/api_docs/python/tf (accessed on 10 July 2020).
36. Python. Available online: https://www.python.org/downloads/release/python-371 (accessed on 10 July 2020).
37. Kingma, D.P.; Ba, J. Adam: A Method for Stochastic Optimization. In Proceedings of the 3rd International Conference on Learning Representations, San Diego, CA, USA, 7–9 May 2015; pp. 1–15.
38. Choi, Y.; Uh, Y.; Yoo, J.; Ha, J.-W. Stargan v2: Diverse Image Synthesis for Multiple Domains. In Proceedings of the IEEE/CVF Conference on Computer Vision and Pattern Recognition, Seattle, WA, USA, 14–19 June 2020; pp. 8188–8197.
39. Anoosheh, A.; Agustsson, E.; Timofte, R.; Van Gool, L. Combogan: Unrestrained Scalability for Image Domain Translation. In Proceedings of the IEEE Conference on Computer Vision and Pattern Recognition Workshops, Salt Lake City, UT, USA, 18–22 June 2018; pp. 783–790.
40. Huang, B.; Dai, Y.; Li, R.; Tang, D.; Li, W. Finger-vein Authentication Based on Wide Line Detector and Pattern Normalization. In Proceedings of the 20th International Conference on Pattern Recognition, Istanbul, Turkey, 23–26 August 2010; pp. 1269–1272.
41. Miura, N.; Nagasaka, A.; Miyatake, T. Extraction of Finger-vein Patterns Using Maximum Curvature Points in Image Profiles. *IEICE Trans. Inf. Syst.* **2007**, *E90-D*, 1185–1194. [CrossRef]
42. Liu, F.; Yang, G.; Yin, Y.; Wang, S. Singular Value Decomposition Based Minutiae Matching Method for Finger Vein Recognition. *Neurocomputing* **2014**, *145*, 75–89. [CrossRef]
43. Gupta, P.; Gupta, P. An Accurate Finger Vein Based Verification System. *Digit. Signal Process.* **2015**, *38*, 43–52. [CrossRef]
44. Miura, N.; Nagasaka, A.; Miyatake, T. Feature Extraction of Fingervein Patterns Based on Repeated Line Tracking and Its Application to Personal Identification. *Mach. Vis. Appl.* **2004**, *15*, 194–203. [CrossRef]
45. Dong, L.; Yang, G.; Yin, Y.; Liu, F.; Xi, X. Finger Vein Verification Based on a Personalized Best Patches Map. In Proceedings of the IEEE International Joint Conference on Biometrics, Clearwater, FL, USA, 29 September–2 October 2014; pp. 1–8.
46. Liu, F.; Yin, Y.; Yang, G.; Dong, L.; Xi, X. Finger Vein Recognition with Superpixel-based Features. In Proceedings of the IEEE International Joint Conference on Biometrics, Clearwater, FL, USA, 29 September–2 October 2014; pp. 1–8.
47. Xi, X.; Yang, L.; Yin, Y. Learning Discriminative Binary Codes for Finger Vein Recognition. *Pattern Recognit.* **2017**, *66*, 26–33. [CrossRef]
48. Joseph, R.B.; Ezhilmaran, D. An Efficient Approach to Finger Vein Pattern Extraction Using Fuzzy Rule-based System. In Proceedings of the 5th Innovations in Computer Science and Engineering, Hyderabad, India, 16–17 August 2017; pp. 435–443.
49. Pham, T.D.; Park, Y.H.; Nguyen, D.T.; Kwon, S.Y.; Park, K.R. Nonintrusive Finger-vein Recognition System Using NIR Image Sensor and Accuracy Analyses According to Various Factors. *Sensors* **2015**, *15*, 16866–16894. [CrossRef] [PubMed]
50. Yang, L.; Yang, G.; Xi, X.; Meng, X.; Zhang, C.; Yin, Y. Tri-branch Vein Structure Assisted Finger vein Recognition. *IEEE Access* **2017**, *5*, 21020–21028. [CrossRef]

sensors

MDPI

Article

Monocular Depth Estimation with Joint Attention Feature Distillation and Wavelet-Based Loss Function

Peng Liu [1,2,3], **Zonghua Zhang** [1,2,*], **Zhaozong Meng** [1,2] **and Nan Gao** [1,2]

[1] State Key Laboratory of Reliability and Intelligence of Electrical Equipment, Hebei University of Technology, Tianjin 300130, China; niatdlut@163.com (P.L.); zhaozong.meng@hebut.edu.cn (Z.M.); ngao@hebut.edu.cn (N.G.)

[2] School of Mechanical Engineering, Hebei University of Technology, Tianjin 300130, China

[3] Key Laboratory of Intelligent Data Information Processing and Control of Hebei Province, Tangshan University, Tangshan 063000, China

* Correspondence: zhzhang@hebut.edu.cn; Tel.: +86-1862-288-0015

Abstract: Depth estimation is a crucial component in many 3D vision applications. Monocular depth estimation is gaining increasing interest due to flexible use and extremely low system requirements, but inherently ill-posed and ambiguous characteristics still cause unsatisfactory estimation results. This paper proposes a new deep convolutional neural network for monocular depth estimation. The network applies joint attention feature distillation and wavelet-based loss function to recover the depth information of a scene. Two improvements were achieved, compared with previous methods. First, we combined feature distillation and joint attention mechanisms to boost feature modulation discrimination. The network extracts hierarchical features using a progressive feature distillation and refinement strategy and aggregates features using a joint attention operation. Second, we adopted a wavelet-based loss function for network training, which improves loss function effectiveness by obtaining more structural details. The experimental results on challenging indoor and outdoor benchmark datasets verified the proposed method's superiority compared with current state-of-the-art methods.

Keywords: monocular depth estimation; feature distillation; joint attention; loss function

check for **updates**

Citation: Liu, P.; Zhang, Z.; Meng, Z.; Gao, N. Monocular Depth Estimation with Joint Attention Feature Distillation and Wavelet-Based Loss Function. *Sensors* **2021**, *21*, 54. https://dx.doi.org/10.3390/s21010054

Received: 28 November 2020
Accepted: 21 December 2020
Published: 24 December 2020

Publisher's Note: MDPI stays neutral with regard to jurisdictional claims in published maps and institutional affiliations.

1. Introduction

Depth estimation is a fundamental computer vision task and is in high demand for manifold 3D vision applications, such as scene understanding [1], robot navigation [2,3], action recognition [4], 3D object detection [5], etc. Monocular depth estimation (MDE) is a more affordable solution for depth acquisition due to extremely low sensor requirements, compared with common depth sensors, e.g., Microsoft's Kinect or stereo images. However, MDE is ill-posed and inherently ambiguous due to one-too-many mapping from 2D to 3D and remains a very challenging topic.

Classical approaches often design hand-crafted features to deduce depth information, but hand-crafted features have no generality across different real-world scenes. Hence, classical approaches have considerable difficulty in acquiring reasonable accuracy. Deep convolutional neural network (DCNN) architectures could be considered as the effective reconstruction methods for many applications with ill-posed problem properties [6–8]. Powerful feature generalization and representation has become available recently through DCNN, which have been successfully introduced to MDE and demonstrated superior performances to the classical approaches [9].

Most DCNN-based MDE methods are based on encoder–decoder architecture. Standard DCNN originally designed for the image classification task are selected as encoders, e.g., ResNet [10], DenseNet [11], SENet [12], etc. These encoders gradually decrease the feature map spatial resolution by pooling while learning the rich feature representation.

Since feature map resolution increases during decoding, various deep-learning methods have been adopted to provide high-quality estimations, including skip connection [13–17], multiscale feature extraction [18–22], attention mechanism [23–26], etc. Although great improvements have been achieved for MDE methods, reconstructing the depth for fine-grain details still requires further improvements, as shown in Figure 1.

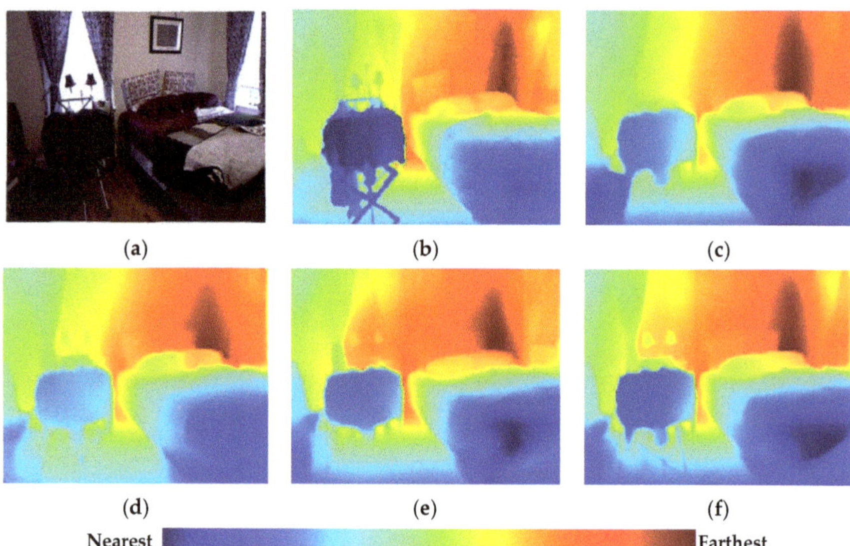

Figure 1. A depth estimation example: (**a**) RGB image; (**b**) ground truth depth map; and (**c–f**) depth maps by Chen et al. [20], Alhashim et al. [17], Hu et al. [22], and the proposed method. We set colors of all indoor depth maps in our work according to the distance as the color bar above.

The current methods struggle to precisely recover large-scale geometry regions (walls) and local detail regions with rich structural information (boundaries and small parts) simultaneously, because the methods still lack the sufficient flexibility and discriminative modulation ability to handle regions with different feature information during up-sampling. This insufficiency limits the feature representation and significantly reduces the estimation accuracy in many cases.

Another area for improvement is the loss function design. Several loss function terms are commonly combined to construct loss functions for predicting a better-quality depth. Various weight-setting methods for the loss function terms have been proposed to balance the training process [27–29], but how to enhance loss function effectiveness for fixed loss term combinations remains an open question.

Therefore, we proposed a new DCNN to settle this issue. We designed an attention-based feature distillation block (AFDB) to address the insufficiency above and integrate it into each up-sampling process in the decoder. To our best knowledge, this is the first time feature distillation has been introduced to MDE. The AFDB enriches feature representation through a series of distillation and residual asymmetric convolution (RAC) layers. We also propose a joint attention module (JAM) to adaptively and simultaneously rescale features depending on the channel and spatial contexts. The designed AFDB incorporates the proposed JAM, providing flexible and discriminative modulation to handle the features.

We also designed a wavelet-based loss function to enhance the loss function effectiveness by combining the multiple loss function with discrete wavelet transform (DWT). The estimated depth map is first divided into many patches using DWT at various frequencies, highlighting high-frequency information from depth map edge areas. The loss for each patch is then reasonably combined to generate the final loss. The experimental results verified that this loss function modification could significantly improve various metrics on benchmark datasets.

Our main contributions are summarized as follows:

- A novel AFDB was designed for the proposed DCNN-based MDE method by combining feature distillation and joint attention mechanisms to boost discriminative modulation for feature processing.
- A wavelet-based loss function was adopted to optimize the training by highlighting the structural detail losses and, hence, improve the estimation accuracy.
- The proposed network was superior to most state-of-the-art MDE methods on two public benchmark datasets: NYU-Depth-V2 and KITTI.

2. Related Works

We discuss and summarize supervised DCNN-based MDE methods in Section 2.1 and briefly review the related techniques, i.e., attention mechanism, feature distillation, and loss function design, in Sections 2.2–2.4, respectively.

2.1. Supervised DCNN-Based MDE Methods

The Supervised DCNN-based MDE methods utilize the DCNN to realize the nonlinear mapping from the RGB image to the depth map. The Supervised DCNN-based methods have become significantly efficient for MDE, with many publicly available RGB and depth map (RGBD) datasets, due to their powerful feature generalization and representation. Eigen et al. [30] proposed a multiscale deep network for MDE that included coarse and fine-scaled network pathways with skip connections between the corresponding layers. Laina et al. [31] used ResNet architecture and several up-projection operators to attain the final depth maps. Cao et al. [32] designed a fully convolutional deep residual network that explicitly considered the long tail distribution of the ground truth depth and regarded the MDE problem as a pixel-wise classification task.

Repeated pooling while learning the rich-feature representations for supervised DCNN-based models inevitably reduces the feature map spatial resolution, which poorly influences the fine-grain depth estimation. Li et al. [33] and Zheng et al. [34] integrated hierarchical depth features to settle this problem. They combined different resolution depth features with up-convolution to realize a coarse-to-fine process. Godard et al. [14] and Liu et al. [13] used skip connection to aggregate feature maps in lower layers, with same resolution feature maps in deeper layers. Other studies [18–22] have aggregated multiscale contexts to improve prediction performances. For example, Fu et al. [18] applied dilated convolution with multiple dilation rates to extract multiscale features and, subsequently, developed a full-image encoder to capture image level features, Zhao et al. [19] employed image super-resolution techniques to generate multiscale features, and Chen et al. [20] proposed an adaptive dense feature aggregation module to aggregate effective multiscale features to infer scene structures.

Several recent multitask learning methods [35–40] have been successfully introduced for MDE by estimating depth maps with other information, such as semantic segmentation labels, surface normals, super pixels, etc. For example, Eigen and Fergus [35] combined semantic segmentation, surface normal, and depth estimation cues to build a single DCNN. This single architecture simplifies implementing a system that requires multiple prediction tasks. Ito et al. [36] proposed a 3D representation for semantic segmentation and depth estimation from a single image. Lin et al. [37] proposed a hybrid DCNN to integrate semantic segmentation and depth estimation into a unified framework. Although multitask learning methods can boost estimation performances, the required multibranch design in the decoder increases the model parameters and reduces the running speed.

2.2. Attention Mechanism

The attention mechanism can enhance the network representation by increasing the model sensitivity to informative and important features. This has been widely adopted for

MDE. For example, Chen et al. [23] enhanced the feature discrimination by designing an attention-based context fusion network to extract image and pixel-level context information, Li et al. [24] applied a channel-wise attention mechanism to extract discriminative features for each resolution, Wang et al. [25] used joint attention mechanisms in their framework to improve the presentation for highest level of feature maps, Chen et al. [15] proposed spatial attention and global context blocks to extract features by blending cross-channel information, and Huynh et al. [41] proposed a guiding depth estimation to favor planar structures by incorporating a nonlocal coplanarity constraint with a nonlocal attention mechanism.

2.3. Feature Distillation

Feature distillation is a recently developed method that has been efficiently applied to super-resolution tasks. The method usually adopts channel splitting to distill feature maps and gain more efficient information. Hui et al. [42] first proposed a feature distillation network to aggregate long and short path features. Hui et al. [43] further advanced the concept and constructed a lightweight cascaded feature multi-distillation block by combining distillation with selective fusion operation. The selective fusion was implemented by their proposed contrast-aware attention layer. Liu et al. [44] recently proposed a lightweight residual feature distillation network using a shallow residual block and multiple feature distillation connections to learn more discriminative representations. The proposed model was the winning solution for the advances in image manipulation 2020 (AIM2020) constrained image super-resolution challenge [45].

2.4. Loss Function Design

Learning in DCNNs is essentially an optimization process, i.e., a neural network adjusts the network weights depending on the loss function value. Therefore, the loss function is important for generating the final estimation model. Many previous studies combined multiple loss terms to build the loss function. However, some loss terms can be ignored during training when many are included, and an adaptive weight adjustment strategy is also required to balance the contribution from each loss term, since they reduce at different rates. Jiang et al. [27] proposed an adaptive weight allocation method based on a Gaussian model for their proposed hybrid loss function. Liu et al. [28] proposed an effective adaptive weight adjustment strategy to adjust each loss term's weight during training. Lee et al. [29] proposed a loss rebalancing algorithm to initialize and rebalance weights for loss terms adaptively during training. Yang et al. [46] adopted DWT to reform the structural similarity (SSIM) loss [47] and achieved improved reconstructions. These methods were proposed to enhance the loss function effectiveness under fixed loss term combinations.

Although great improvements have been achieved for MDE methods, reconstructing the depth for fine-grain details still requires further improvements. Our proposed method employed a single-task encoder–decoder architecture that has fewer model parameters and faster running speed compared with the multitask learning architecture. We efficiently integrated feature distillation and joint attention mechanisms in the decoder to further boost the discriminative modulation for feature processing. We also combined multiple loss functions with DWT to enhance the loss function effectiveness.

3. Proposed Method

This section describes the proposed MDE method. Sections 3.1 and 3.2 discuss the network architecture and provide details for the proposed AFDB, respectively. Section 3.3 details the proposed wavelet-based loss function.

3.1. Network Architecture

Figure 2 shows the proposed network architecture. We use a standard encoder–decoder architecture with skip connections between same resolution layers. The encoder is modified from the standard DCNN that was originally designed for image classification by removing the final average pooling and fully connected layers. In the decoding stage, we first attached a 1×1 convolutional layer to the top of the encoder for feature reduction. We concatenated up-sampled feature maps in the decoder with feature maps from the encoder that have the same resolution to enrich the feature representation and provide flexible and discriminative modulation for the feature maps. The concatenated feature maps were refined using the proposed AFDB. After gradually recovering the feature maps back to the expected depth map resolution, the AFDB output was fed into a 3×3 convolutional layer to derive the final estimation.

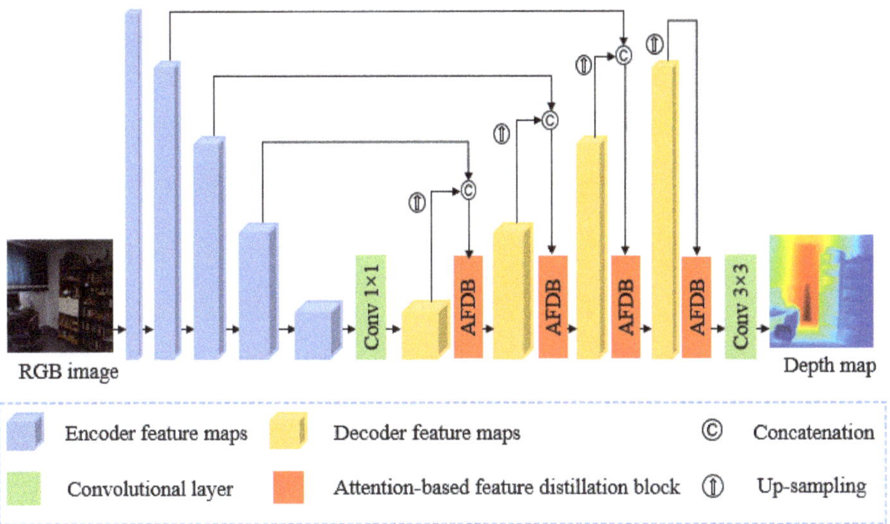

Figure 2. Proposed network architecture.

3.2. Attention-Based Feature Distillation

Figure 3 shows the proposed AFDB to enrich the feature representation and improve the flexible and discriminative modulation during up-sampling in the decoder. The first 1×1 convolutional layer reduces the concatenated feature map channels from the encoder and decoder with the same resolution. The subsequent block with a residual connection includes the progressive refinement, local fusion, and joint attention modules. The progressive refinement module enriches the feature representation through several distillation and feature refinement steps. The local fusion module is a commonly employed structure that includes concatenation and a 1×1 convolutional layer, providing local feature reduction and fusion for all branch outputs from the progressive refinement module. The JAM further enhances the feature discriminative modulation by fully considering the feature channel and spatial contexts.

The proposed AFDB was modified from the feature distillation block structure proposed by [44], incorporating two improvements. We replaced the shallow residual block of [44] with the RAC in the progressive refinement module, which efficiently enhanced the model robustness to rotational distortions in image classification [48]. We effectively integrated a channel attention branch in parallel to the original contrast aware attention layer, enhancing the discriminative modulation for the block.

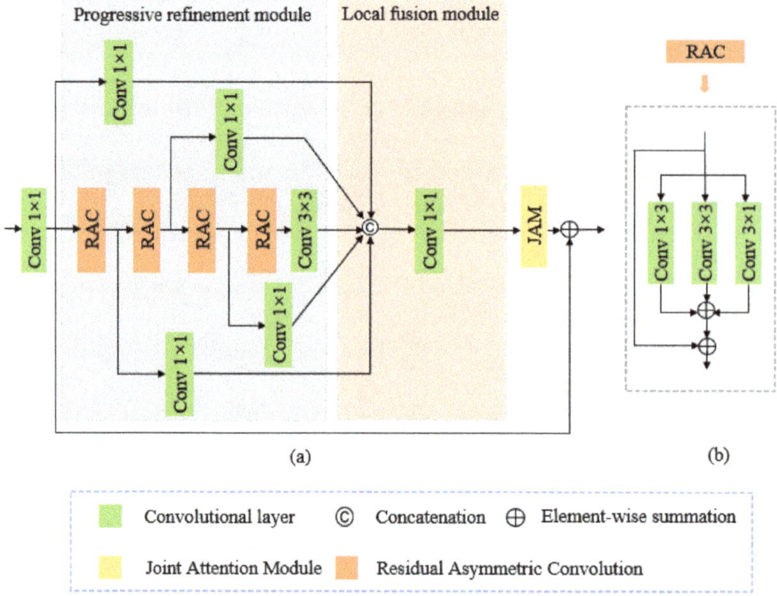

Figure 3. Proposed AFDB design with a four-step distillation example: (**a**) AFBD and (**b**) RAC structures.

3.2.1. Progressive Refinement Module

Figure 3a shows the proposed progressive refinement module structure. Each step uses a 1×1 convolutional layer to distill some features and an RAC layer to further refine the remaining features simultaneously. The RAC comprises an asymmetric convolution with skip connections, where the asymmetric convolution comprises three parallel layers with 3×3, 3×1, and 1×3 kernels. The outputs are summed to enrich the feature representation.

Given the input features F_{in} for the progressive refinement block and four-step distillation, the procedure can be described as

$$F_{\text{ref}_1}, F_{\text{dis}_1} = \text{Split}_1(F_{\text{in}}), \tag{1}$$

$$F_{\text{ref}_2}, F_{\text{dis}_2} = \text{Split}_2(F_{\text{ref}_1}), \tag{2}$$

$$F_{\text{ref}_3}, F_{\text{dis}_3} = \text{Split}_3(F_{\text{ref}_2}), \tag{3}$$

and

$$F_{\text{ref}_4}, F_{\text{dis}_4} = \text{Split}_4(F_{\text{ref}_3}), \tag{4}$$

where Split_i denotes the i-th channel splitting operation, which includes a 1×1 convolutional layer to generate the distilled features F_{dis_i} and a 3×3 convolutional layer to generate the refined features F_{ref_i}, which will be further processed by succeeding layers. Distilled feature channels are half the dimensionality of the original.

After the four-step operation, we use a 3×3 convolutional layer to further filter the last RCAB:

$$F_{\text{fil}} = W_{\text{fil}}^{3 \times 3}(F_{\text{ref}_4}), \tag{5}$$

where W denotes convolution.

The local fusion procedure can be expressed as

$$F_{\text{LF}} = W_{LF}^{1 \times 1}(\text{Concat}(F_{\text{fil}}, F_{\text{dis}_1}, F_{\text{dis}_2}, F_{\text{dis}_3}, F_{\text{dis}_4})), \tag{6}$$

where Concat denotes concatenation.

3.2.2. Joint Attention Module

Figure 4 shows the proposed JAM structure, inspired by lightweight joint attention modules [49] that infer attention maps along the channel and spatial dimensions simultaneously, to further enhance the feature discriminative modulation. We adopted a residual connection and joint attention mechanism to facilitate the gradient flow. The JAM produces a 3D attention map for the input feature maps by combining parallel channel and spatial attention branches. Thus, JAM can refine feature maps and enhance the feature representation while fully considering the channel and spatial contexts.

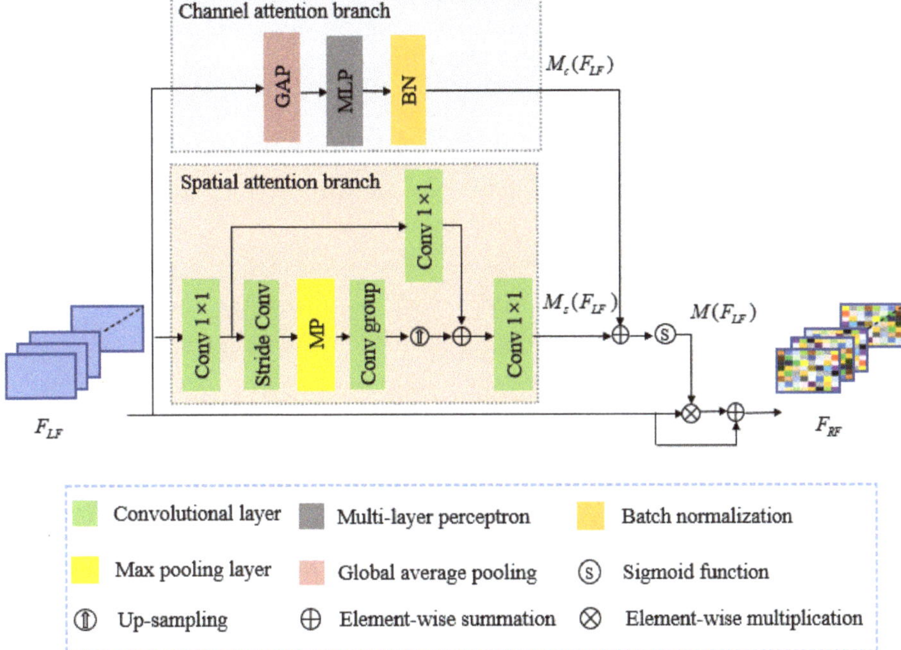

Figure 4. Proposed joint attention module (JAM) structure.

Figure 4 shows that, for a given input feature map F_{LF}, i.e., the local fusion module output, we simultaneously compute the channel attention $M_c(F_{LF})$ and spatial attention $M_s(F_{LF})$ in the channel and spatial attention branches, respectively. The joint 3D attention map $M(F_{LF})$ is then computed as

$$M(F_{LF}) = \sigma(M_c(F_{LF}) + M_s(F_{LF})), \tag{7}$$

where σ denotes the sigmoid function. The refined feature maps are

$$F_{RF} = F_{LF} + F_{LF} \otimes M(F_{LF}), \tag{8}$$

where $\otimes$ denotes element-wise multiplication.

The channel attention $M_c(F_{LF})$ exploits the inter-channel relationships for the feature maps, which mainly includes three steps (Figure 4):

1. Global average pooling on the input feature maps to fetch global information for each channel.

2. Multilayer perceptron with one hidden layer to predict the attention across the computed channels.
3. Batch normalization layer to adjust the scale with another spatial branch output.

The procedure can be described mathematically as

$$M_c(F_{LF}) = BN(MLP(GAP(F_{LF}))),\qquad(9)$$

where BN denotes the batch normalization, MLP denotes the multilayer perceptron, and GAP denotes the global average pooling.

Spatial attention $M_s(F_{LF})$ emphasizes or restrains the feature maps in different spatial locations, which mainly includes five steps (Figure 4):

1. 1×1 convolutional layer to compress the channel dimensions.
2. Stride convolution and max-pooling layers combined to enlarge the receptive field to receive more useful information.
3. Convolutional group with two 3×3 convolutional layers to catch the spatial context information and up-sampling layer to recover the spatial dimensions.
4. 1×1 convolutional shortcut and adding its output to the step 3 output to further enrich the spatial context information.
5. 1×1 convolutional layer to recover the channel dimensions.

Thus, the spatial attention is computed as

$$M_s(F_{LF}) = W_{s_3}^{1 \times 1}\left(Up\left(W_{s_2}^{3 \times 3}\left(W_{s_1}^{3 \times 3}\left(Mp\left(W_s^{stride}\left(W_{s_1}^{1 \times 1}(F_{LF})\right)\right)\right)\right)\right) + W_{s_2}^{1 \times 1}\left(W_{s_1}^{1 \times 1}(F_{LF})\right)\right),\qquad(10)$$

where Up denotes up-sampling, and Mp denotes max-pooling.

3.3. Wavelet-Based Loss Function

In order to balance the reconstructing depth maps by minimizing the difference between the ground truth while also penalizing the loss of high-frequency details that typically correspond to the object boundaries in the scene, four loss terms were combined in our loss function as follows:

1. Depth loss. Balance loss contributions for different distances. We calculate the BerHu loss [31] in logarithm space:

$$L_{dep} = \frac{1}{n}\sum_{i=1}^{n}\ln(|g_i - d_i|_b + \alpha_1),\qquad(11)$$

where

$$|x|_b = \begin{cases} |x|, & |x| \le c \\ \frac{x^2+c^2}{2c}, & |x| > c \end{cases},\qquad(12)$$

d_i and g_i are the predicted depth map value and corresponding ground truth for pixel index i, respectively, n is the total number of pixels in the current batch, $\alpha_1 = 5$ is a constant parameter; and we set $c = 0.2\max_{n}(|g_i - d_i|)$.

2. Gradient loss. Penalizes acute object boundary changes in both the x and y directions that show abundant fine-feature granularity:

$$L_{gra} = \frac{1}{n}\sum_{i=1}^{n}\ln\left(\left|\nabla_x^{sobel}(e_i)\right| + \left|\nabla_y^{sobel}(e_i)\right| + \alpha_2\right),\qquad(13)$$

where e is the L_1 Euclidean distance between the predicted depth map and the corresponding ground truth, ∇_x^{sobel} and ∇_y^{sobel} represent the horizontal and vertical Sobel operators that calculate the gradient information, and $\alpha_2 = 0.5$ is a constant parameter.

3. Normal loss. Minimize the angle between the predicted surface normal and corresponding ground truth to help emphasize the small details in the predicted depth map:

$$
L_{\text{nor}} = \frac{1}{n} \sum_{i=1}^{n} \left| 1 - \frac{\left\langle n_i^d, n_i^g \right\rangle}{\sqrt{\left\langle n_i^d, n_i^d \right\rangle} \sqrt{\left\langle n_i^g, n_i^g \right\rangle}} \right|,
\tag{14}
$$

where $n_i^d = \left[-\nabla_x(d_i), -\nabla_y(d_i), 1 \right]$ and $n_i^g = \left[-\nabla_x(g_i), -\nabla_y(g_i), 1 \right]$ are the surface normal for the predicted depth map and corresponding ground truth, respectively.

4. SSIM loss. Global consistency metric commonly employed for computer vision tasks:

$$
L_{\text{SSIM}} = 1 - \frac{\left(2\mu_d\mu_g + c_1\right)\left(2\delta_{dg} + c_2\right)}{\left(\mu_d^2 + \mu_g^2 + c_1\right)\left(\delta_d^2 + \delta_g^2 + c_2\right)},
\tag{15}
$$

where μ_d and μ_g are the predicted depth map and ground truth means, respectively, δ_d and δ_g are predicted depth map and ground truth standard deviations, respectively, δ_{dg} is the covariance between the predicted depth map and ground truth, and constants $c_1 = 2$ and $c_2 = 6$ follow [46].

Given the DWT invertibility, all depth maps features are preserved by the decomposition scheme. Importantly, DWT captures the depth map location and frequency information, which is helpful for penalizing the high-frequency detail loss that typically corresponds with the object texture. Thus, we propose combining the DWT and multiple loss terms. Figure 5 shows applying iterative DWT decomposes the depth map into different sub-band images, which can be expressed as

$$
I_{i+1}^{\text{LL}}, I_{i+1}^{\text{LH}}, I_{i+1}^{\text{HL}}, I_{i+1}^{\text{HH}} = \text{DWT}\left(I_i^{\text{LL}} \right),
\tag{16}
$$

where subscript i refers to output from the i-th DWT iteration, and I_0^{LL} is the original depth map.

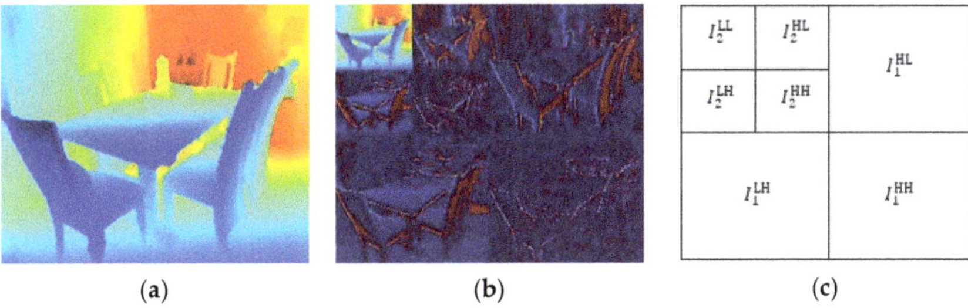

Figure 5. Discrete wavelet transform (DWT) process for depth maps, with two iterations for example: (**a**) original depth map, (**b**) depth map after 2 DWT iterations, and (**c**) labels for different image patches.

The four loss terms described above are calculated from the original depth map, I_0^{LL}, and sub-band images $I_i^{\text{LL}}, i = 1, \cdots, n$, where n is the number of DWT iterations. We supplemented some depth losses on the basis of the sub-band images $I_i^{\text{LH}}, I_i^{\text{HL}}$, and $I_i^{\text{HH}}, i = 1, \cdots, n$, i.e., loss information for high-frequency details that typically correspond to the object's horizontal edge, vertical edge, and corner in the depth map, which are very useful for fine-grain estimation. These loss terms can be expressed as

$$
L_{\text{W-dep}} = \sum_{i=0}^{n} L_{\text{dep}}\left(I_i^{\text{LL}} \right) + \sum_{i=1}^{n} \left(L_{\text{dep}}\left(I_i^{\text{LH}} \right) + L_{\text{dep}}\left(I_i^{\text{HL}} \right) + L_{\text{dep}}\left(I_i^{\text{HH}} \right) \right),
\tag{17}
$$

$$L_{W-gra} = \sum_{i=0}^{n} L_{gra}\left(I_i^{LL} \right), \tag{18}$$

$$L_{W-nor} = \sum_{i=0}^{n} L_{nor}\left(I_i^{LL} \right), \tag{19}$$

and

$$L_{W-SSIM} = \sum_{i=0}^{n} L_{SSIM}\left(I_i^{LL} \right), \tag{20}$$

and hence, the final loss function is

$$L_{total} = L_{W-dep} + L_{W-gra} + L_{W-nor} + L_{W-SSIM}. \tag{21}$$

Similar conclusions were found by [15] and [46]. Reference [46] extended the SSIM loss by combining it with DWT and showed that this simple modification could improve reconstruction for single-image dehazing. Reference [15] showed that simply allocating larger weights to edge areas in the loss function could boost performances in the border areas.

4. Experiments

Section 4.1 describes the experimental setup, including the datasets, evaluation metrics, and implementation details. Section 4.2 compares the experimental results with the current state-of-the-art methods on two public datasets: NYU-Depth-V2 [50] (indoor scenes) and KITTI [51] (outdoor scenes). Section 4.3 uses the NYU-Depth-V2 dataset to analyze the effectiveness and rationality of the AFDB and wavelet-based loss function. Finally, Section 4.4 uses cross-dataset validation on the iBims-1 [52] dataset to assess the proposed method's generality.

4.1. Experimental Setup

4.1.1. Datasets

The NYU-Depth-V2 dataset contains 464 indoor scenes captured by Microsoft Kinect devices. Following the official split, we used 249 scenes (approximately 50-K pair-wise images) for training and 215 scenes (654 pair-wise images) for testing.

The KITTI dataset was captured using a stereo camera and rotating LIDAR sensor mounted on a moving car. Following the commonly used Eigen split [30], we used 22-K images from 28 scenes for training and 697 images from different scenes for testing.

iBims-1 is a high-quality RGBD dataset comprising 100 high-quality images and corresponding depth maps particularly designed to test MDE methods. A digital single-lens reflex camera and high-precision laser scanner were used to acquire the high-resolution images and highly accurate depth maps for diverse indoor scenarios. We use iBims-1 for cross-dataset validation to assess the proposed method's generality.

4.1.2. Evaluation Metrics

The performance was quantitatively evaluated using standard metrics for these datasets, as shown below for the ground truth depth y_i^*, estimated depth y_i, and total pixels n in all evaluated depth maps.

- Absolute relative difference (Abs Rel):

$$\text{Abs Rel} = \frac{1}{n} \sum_i \frac{|y_i - y_i^*|}{y_i^*}. \tag{22}$$

- Squared relative difference (Sq Rel):

$$\text{Sq Rel} = \frac{1}{n} \sum_i \frac{\|y_i - y_i^{*2}\|}{y_i^*}. \tag{23}$$

- Mean Log10 error (log10):

$$\log 10 = \frac{1}{n} \sum_i \left| \log_{10} y_i - \log_{10} y_i^* \right|. \tag{24}$$

- Root mean squared error (RMS):

$$\text{RMS} = \sqrt{\frac{1}{n} \sum_i \left(y_i - y_i^* \right)^2}. \tag{25}$$

- Log10 root mean squared error (logRMS):

$$\text{logRMS} = \sqrt{\frac{1}{n} \sum_i \left(\log_{10} y_i - \log_{10} y_i^* \right)^2}. \tag{26}$$

- Threshold accuracy (TA):

$$\text{TA} = \frac{1}{n} \sum_i g(y_i, y_i^*), \tag{27}$$

where

$$g(y_i, y_i^*) = \begin{cases} 1, & \delta = \max\left(\frac{y_i^*}{y_i}, \frac{y_i}{y_i^*} \right) < \text{thr} \\ 0, & \text{otherwise} \end{cases}. \tag{28}$$

The threshold accuracy is the ratio of the maximum relative error δ below the threshold thr. Conditions $\delta < 1.25$, $\delta < 1.25^2$, and $\delta < 1.25^3$ were used in the experiment, denoted as δ_1, δ_2, and δ_3, respectively.

4.1.3. Implementation Details

The proposed model was implemented with the PyTorch [53] framework and trained using two Nvidia RTX 2080ti graphics processing units (GPUs). The encoders were both pretrained on the ImageNet dataset [54], and the other layers were randomly initialized. The Adam [55] optimizer was selected with $\beta_1 = 0.9$ and $\beta_2 = 0.999$, and the weight decay = 0.0001. We set the batch size = 16 and trained the model for 20 epochs.

For the NYU-Depth-V2 dataset, we first cropped each image to 228 × 304 pixels, and the offline data augmentation methods were as the same as those of the mainstream approaches [18,20,22], i.e., each training image was augmented with random scaling (0.8, 1.2), rotation (−5°, 5°), horizontal flip, rectangular window dropping, and color shift (multiplied by random value (0.8, 1.2)).

For the KITTI dataset, we masked out the sparse depth maps projected by the LIDAR point cloud and evaluated the predicted results only for valid points with ground depths. We capped the maximum estimation at the KITTI dataset maximum depth (80 m). The data augmentation methods were the same as those in [23].

4.2. Results

Table 1 shows the evaluation metrics comparing the proposed model with several state-of-the-art methods on NYU-Depth-V2. The DenseNet-161, ResNet-101, and SENet-154 encoders were selected to verify the proposed method's flexibility. Figure 6 visualizes the trade-off between the performance and model parameters. The results for the comparison methods were taken from their relevant literature.

Table 1. Model performance on NYU-Depth-V2. Best scores are highlighted in bold font. The attention-based feature distillation block (AFDB) distillation step = 5 and discrete wavelet transform (DWT) iteration = 3. Abs Rel: absolute relative difference and RMS: root mean squared error.

Method	Error (Lower is Better)			Accuracy (Higher is Better)		
	Abs Rel	RMS	Log10	δ_1	δ_2	δ_3
Eigen et al. [30]	0.212	0.873	-	0.611	0.887	0.969
Laina et al. [31]	0.127	0.573	0.055	0.811	0.953	0.988
Chen et al. [23]	0.138	0.496	-	0.826	0.964	0.990
Lee et al. [21]	0.131	0.538	-	0.837	0.971	0.994
Qi et al. [40]	0.128	0.569	-	0.834	0.960	0.990
Zhao et al. [19]	0.128	0.523	0.059	0.813	0.964	0.992
Li et al. [33]	0.134	0.540	0.056	0.832	0.965	0.989
Hao et al. [26]	0.127	0.555	-	0.841	0.966	0.991
Alhashim et al. [17]	0.123	0.465	0.053	0.846	0.974	0.994
Huang et al. [39]	0.122	0.459	0.051	0.859	0.972	0.993
Hu et al. [22]	0.115	0.530	0.050	0.866	0.975	0.993
Fu et al. [18]	0.115	0.509	0.051	0.828	0.965	0.992
Wang et al. [25]	0.115	0.519	0.049	0.871	0.975	0.993
Chen et al. [20]	**0.111**	0.514	**0.048**	**0.878**	0.977	0.994
Ours (DenseNet-161)	0.122	0.534	0.050	0.857	0.972	0.993
Ours (ResNet-101)	0.123	0.532	0.052	0.854	0.972	0.992
Ours (SENet-154)	0.113	**0.504**	**0.048**	**0.878**	**0.978**	**0.995**

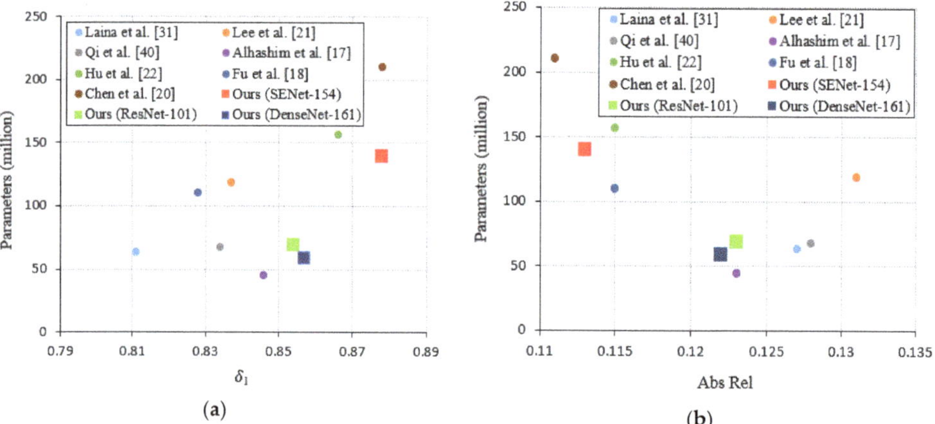

Figure 6. Model parameters and performance (**a**) with respect to δ_1 and (**b**) with respect to the absolute relative difference (Abs Rel).

Table 1 confirms that the proposed method achieved good performances for all the encoder architectures, with the SENet-154 encoder architecture providing the best performance. The proposed method also achieved a comparable or better performance compared with the current state-of-the-art methods.

Figure 6 shows that the proposed model achieved better a trade-off between the performance and model parameters, with only the Abs Rel metric being less than [20], but [20] has more parameters. The proposed method with the DenseNet-161 and ResNet-101 encoders achieved better performances compared with other methods with less than 100 M parameters.

Figure 7 compares the estimated depth maps, and more qualitative results are presented in Appendix A. The display pixels for all the estimated depth maps were the same as those for ground truth to provide easier comparisons. The proposed method achieved better geometric details and object boundaries than the other methods. Thus, the proposed method provides better fine-grain estimations.

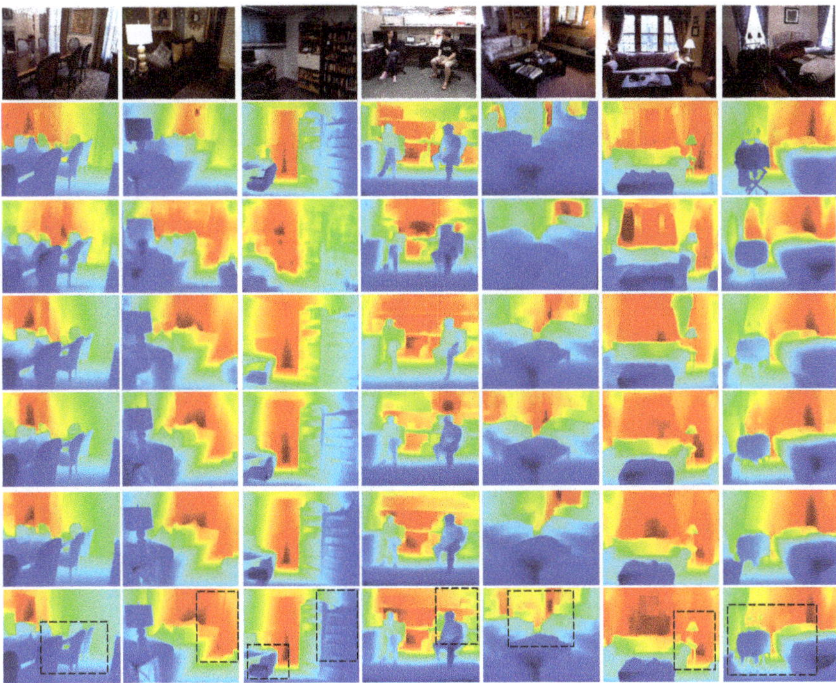

Figure 7. Qualitative evaluations on NYU-Depth-V2. Rows from top to bottom: original RGB images, ground truth depth maps, Laina et al. [31], Alhashim et al. [17], Hu et al. [22], Chen et al. [20], and the proposed method. Regions in black boxes highlight the better-predicted results. Color indicates depth, where red is far and blue is close.

Table 2 compares the proposed method on the KITTI test dataset using the SENet-154 encoder, with some quantitative comparisons in Figure 8 and more qualitative results in Appendix A. The proposed method outperforms most state-of-the-art methods and provides better object boundaries.

Table 2. Performance evaluation on the KITTI. The best scores are highlighted in bold font. Sq Rel: squared relative difference.

Method	Error (Lower is Better)				Accuracy (Higher is Better)		
	Abs Rel	RMS	Sq Rel	logRMS	δ_1	δ_2	δ_3
Eigen et al. [30]	0.190	7.156	1.515	0.270	0.692	0.899	0.967
Godard et al. [14]	0.148	5.927	1.515	0.247	0.802	0.922	0.964
Jiang et al. [27]	0.128	5.299	1.037	0.224	0.837	0.939	0.971
Li et al. [33]	0.104	4.513	0.697	0.164	0.868	0.967	0.990
Liu et al. [13]	0.106	4.274	0.686	0.176	0.878	0.968	0.986
Wang et al. [25]	0.096	4.327	0.655	0.171	0.893	0.963	0.983
Alhashim et al. [17]	0.093	4.170	0.589	0.171	0.886	0.965	0.986
Chen et al. [23]	0.083	3.599	0.437	0.127	0.919	0.982	**0.995**
Fu et al. [18]	0.072	**2.727**	0.307	**0.120**	0.932	**0.984**	0.994
Ours (SENet-154)	**0.071**	2.848	**0.306**	0.121	**0.933**	0.983	**0.995**

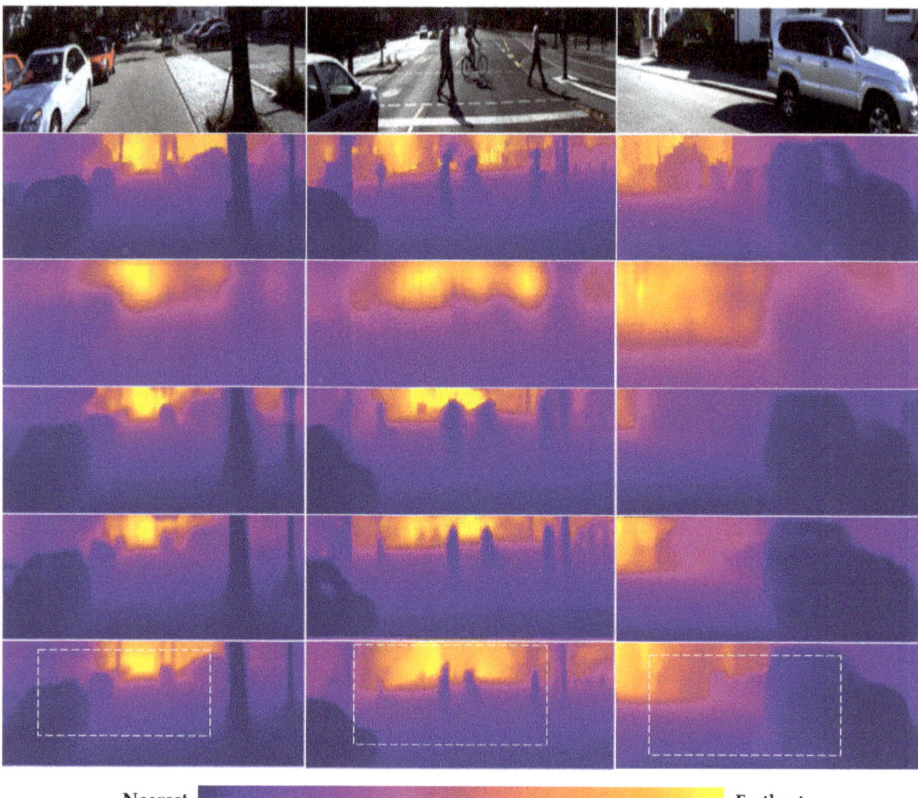

Nearest   Farthest

Figure 8. Qualitative evaluations on the KITTI dataset. Rows from top to bottom: original RGB images, ground truth depth maps, Eigen et al. [30], Godard et al. [14], Chen et al. [23], and the proposed method. Regions in the white boxes highlight the better-predicted results. The ground truth maps were interpolated from the sparse measurements for better visualization. Color indicates depth; yellow is far, and purple is close. We set the colors of all outdoor depth maps in our work according to the distance, as in the color bar above.

4.3. Algorithm Analysis

We conducted several experiments on NYU-Depth-V2 to investigate the effectiveness and rationality for the proposed AFDB and wavelet-based loss functions with the SENet-154 encoder.

4.3.1. AFDB

Figure 9 and Table 3 compare other feature distillation methods with the proposed AFDB. Distillation steps = 4, and DWT iterations = 2 for all evaluations. All metrics are improved for the proposed AFDB at the cost of a few more model parameters. The proposed feature distillation could better predict detailed depth map characteristics.

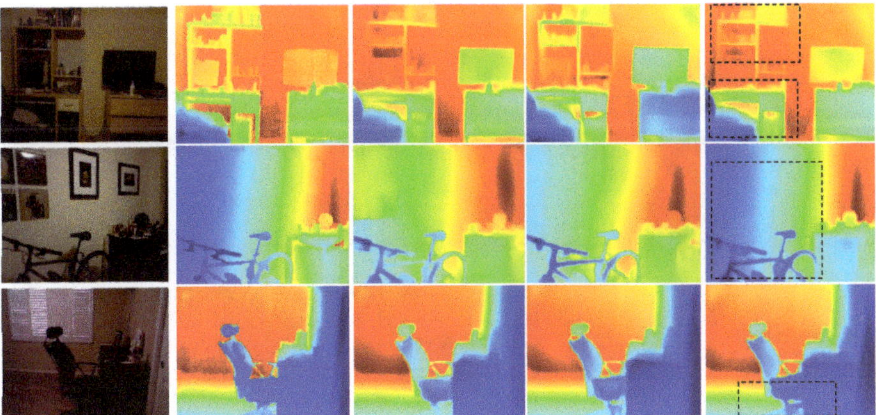

Figure 9. Feature distillation methods on NYU-Depth-V2. Columns from left to right: original RGB images, ground truth depth maps, Hui et al. [43], Liu et al. [44], and proposed approach. Regions in black boxes highlight the better-predicted results. Color indicates depth; red is far, and blue is close.

Table 3. Feature distillation performance on NYU-Depth-V2.

Method	Parameters	Error (Lower is Better)			Accuracy (Higher is Better)		
		Abs Rel	RMS	Log10	δ_1	δ_2	δ_3
Hui et al. [43]	127.6 M	0.121	0.515	0.050	0.863	0.973	0.992
Liu et al. [44]	133.1 M	0.114	0.517	0.049	0.871	0.976	0.993
AFDB	135.7 M	0.113	0.509	0.049	0.877	0.978	0.994

Table 4 shows the ablation effects, i.e., distillation step and JAM influences, for the prediction results and model performance. We used two DWT iterations to decompose the depth map. More distillation steps can improve the evaluation metrics but increases the model parameters. Almost all evaluation metrics worsened for six or more distillation steps, mainly because five-step distillation generates sufficient features for subsequent treatments, and more steps just increase the local feature fusion burdens. All metrics are improved for the proposed JAM at the cost of a few more model parameters.

Table 4. The AFDB performance under different settings. Method subscripts show the distillation steps (w/o means without). JAM: joint attention module.

Method	Parameters	Error (Lower is Better)			Accuracy (Higher is Better)		
		Abs Rel	RMS	Log10	δ_1	δ_2	δ_3
AFDB $_{3,\,JAM}$	134.4 M	0.117	0.511	0.050	0.870	0.974	0.994
AFDB $_{4,\,JAM}$	135.7 M	0.113	0.509	0.049	0.877	0.978	0.994
AFDB $_{5,\,JAM}$	139.2 M	0.113	0.504	0.048	0.878	0.978	0.995
AFDB $_{6,\,JAM}$	142.7 M	0.121	0.503	0.050	0.867	0.976	0.994
AFDB $_{4,\,w/o\,JAM}$	133.9 M	0.117	0.511	0.050	0.867	0.974	0.992

4.3.2. Loss Function

Table 5 shows the performance metrics for the proposed model with different loss functions for network training. We gradually added the loss terms described in Section 3.3 to assess the loss terms selection rationality using four-step distillation as the baseline. All evaluation metrics improved with increased loss terms. Thus, the proposed loss function selection method is effective and rational.

Table 5. Proposed method performance for different loss functions. SSIM: structural similarity. Each loss function is defined in Section 3.3.

Loss Function	Error (Lower is Better)			Accuracy (Higher is Better)		
	Abs Rel	RMS	Log10	δ_1	δ_2	δ_3
L_{dep}	0.121	0.534	0.051	0.857	0.970	0.992
$L_{dep}+L_{gra}$	0.117	0.525	0.050	0.865	0.975	0.993
$L_{dep}+L_{gra}+L_{nor}$	0.116	0.521	0.050	0.868	0.976	0.993
$L_{dep}+L_{gra}+L_{nor}+L_{SSIM}$	0.114	0.515	0.049	0.872	0.976	0.994

Table 6 shows the effects from DWT iterations using the wavelet-based loss function (Equation (21)) to train the network. Three DWT iterations are sufficient to obtain the optimal results. The increased iterations reduce the performance, because the depth map size gradually reduces with the increased iterations, and the detailed depth map features from the smallest scale become indistinct, which may adversely influence the estimation quality.

Table 6. DWT iteration effects on the model performance using the wavelet-based loss function.

DWT Iterations	Error (Lower is Better)			Accuracy (Higher is Better)		
	Abs Rel	RMS	Log10	δ_1	δ_2	δ_3
One	0.114	0.509	0.049	0.874	0.975	0.994
Two	0.113	0.509	0.049	0.877	0.978	0.994
Three	0.113	0.504	0.048	0.877	0.978	0.994
Four	0.114	0.509	0.049	0.873	0.976	0.994

4.4. Cross-Dataset Validation

We performed cross-dataset validation to assess the proposed method's generality. We used the iBims-1 dataset, because it contains different indoor scenarios and has higher-quality depth maps closer to real depth values compared with NYU-Depth-V2. Therefore, cross-dataset validation on the iBims-1 dataset could verify the model efficiency for different data distributions between training and testing sets. The corresponding evaluation metrics are also more objective and accurate due to the higher precision depth maps.

The proposed network was first trained on NYU-Depth-V2 to generate a pretrained model. Then, the pretrained model was used without fine-tuning to estimate the iBims-1 depth maps. Table 7 shows the corresponding evaluation metrics for iBims-1, and Figure 10 shows some qualitative comparisons. The settings for the compared methods were the same as for the proposed method. The pretrained models for the compared methods were generated by running their open-source codes.

Table 7. Cross-dataset validation trained on NYU-Depth-V2 and tested on the iBims-1 dataset.

Method	Error (Lower is Better)			Accuracy (Higher is Better)		
	Abs Rel	RMS	Log10	δ_1	δ_2	δ_3
Alhashim et al. [17]	0.346	2.772	0.199	0.179	0.547	0.827
Hu et al. [22]	0.360	2.815	0.208	0.162	0.497	0.816
Chen et al. [20]	0.349	2.750	0.200	0.162	0.531	0.849
Ours	0.329	2.665	0.184	0.192	0.601	0.876

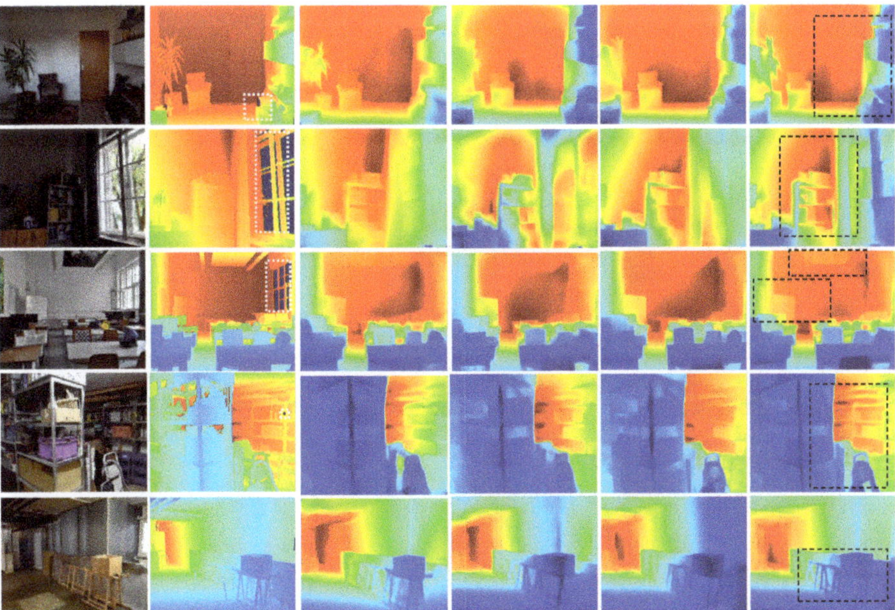

Figure 10. Cross-validation trained on NYU-Depth-V2 and tested on the iBims-1 datasets. Columns from left to right: original RGB images, ground truth depth maps, Alhashim et al. [17], Hu et al. [22], Chen et al. [20], and the proposed method. Regions in white boxes show missing or incorrect depth values from the ground truth data. Regions in black boxes highlight the better-predicted results. Colors indicate depth; red is far, and blue is close.

The test results of the pretrained models on iBims-1 were quite different from those on NYU-Depth-V2. In contrast to the earlier comparisons in Table 1, [17] has better performances than [20] and [22]. The proposed model achieved significantly better performances than the three comparative methods. Thus, the proposed method could better estimate the geometric details and object boundaries for these different scenes than the three current state-of-the-art methods.

5. Conclusions

This paper proposed a new DCNN for monocular depth estimation. Two improvements were realized compared with previous methods. We made a combination of joint attention and feature distillation mechanisms in the decoder to boost the feature discriminative modulation and proposed a wavelet-based loss function to emphasize the detailed depth map features. The experimental results on the two public datasets verified the proposed method's effectiveness. The experiments were also conducted to verify the proposed approach effectiveness and rationality. The generality for the proposed model was demonstrated using cross-dataset validation.

Future works will focus on applying the proposed MDE methods to 3D vision applications, such as augmented reality, simultaneous localization and mapping (SLAM), and indoor scene reconstruction.

Author Contributions: Funding acquisition, Z.Z.; methodology, P.L. and Z.M.; project administration, Z.Z. and N.G.; resources, N.G.; software, P.L.; validation, P.L.; writing—original draft, P.L. and Z.M.; and writing—review and editing, Z.Z. All authors have read and agreed to the published version of the manuscript.

Funding: This research was funded by National Key R&D Program of China (under Grant No. 2017YFF0106404) and the National Natural Science Foundation of China (under Grants No. 52075147 and 51675160).

Institutional Review Board Statement: Not applicable.

Informed Consent Statement: Not applicable.

Data Availability Statement: The data presented in this study are available on request from the corresponding author.

Conflicts of Interest: The authors declare no conflict of interest.

Appendix A

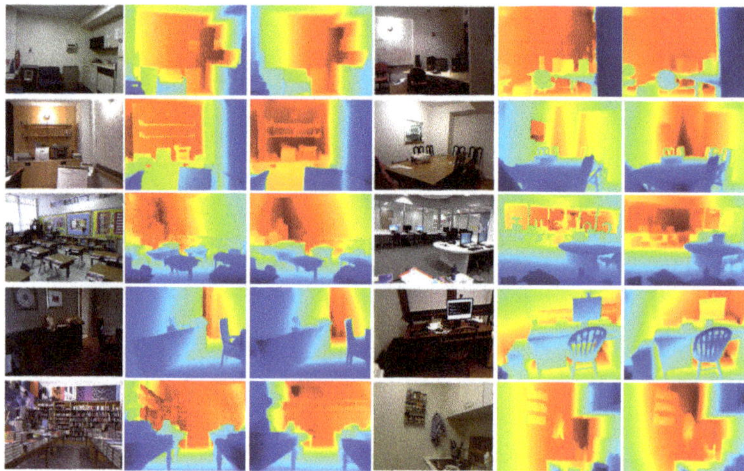

Figure A1. Example outcomes for the proposed method on NYU-Depth-V2. Columns from left to right: original RGB images, ground truth depth maps, and proposed model predicted depth maps. Colors indicate depth; red is far, and blue is close.

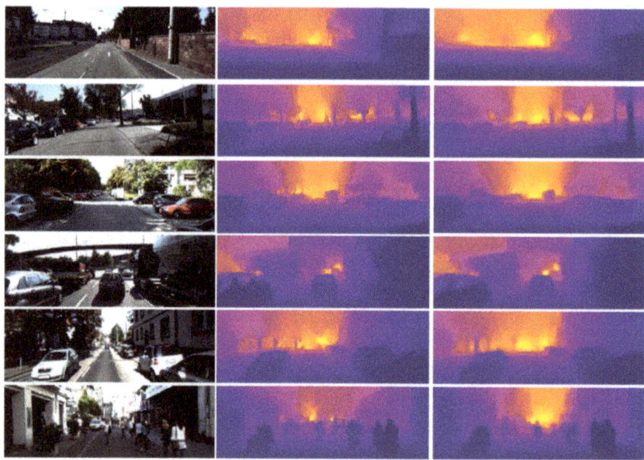

Figure A2. Example outcomes for the proposed method on KITTI. Columns from left to right: original RGB images, ground truth depth maps, and proposed model predicted depth maps. Colors indicate depth; yellow is far, and purple is close. Ground truth maps were interpolated from sparse measurements for better visualization.

References

1. Naseer, M.; Khan, S.; Porikli, F. Indoor scene understanding in 2.5/3D for autonomous agents: A survey. *IEEE Access* **2018**, *7*, 1859–1887. [CrossRef]
2. Othman, K.M.; Rad, A.B. A doorway detection and direction (3Ds) system for social robots via a monocular camera. *Sensors* **2020**, *20*, 2477. [CrossRef]
3. Ball, D.; Ross, P.; English, A.; Milani, P.; Richards, D.; Bate, A. Farm workers of the future: Vision-based robotics for broad-acre agriculture. *IEEE Robot. Autom. Mag.* **2017**, *24*, 97–107. [CrossRef]
4. Li, Z.; Dekle, T.; Cole, F.; Tucker, R. Learning the depths of moving people by watching frozen people. In Proceedings of the 2019 IEEE Conference on Computer Vision and Pattern Recognition (CVPR), Los Angeles, CA, USA, 15–21 June 2019; pp. 4521–4530.
5. Ku, J.; Mozifian, M.; Lee, J.; Harakeh, A.; Waslander, S.L. Joint 3d proposal generation and object detection from view aggregation. In Proceedings of the 2018 IEEE/RSJ International Conference on Intelligent Robots and Systems (IROS), Madrid, Spain, 1–5 October 2018; pp. 1–8.
6. Mateev, V.; Marinova, I. Machine learning in magnetic field calculations. In Proceedings of the 19th International Symposium on Electromagnetic Fields in Mechatronics, Electrical and Electronic Engineering (ISEF), Nancy, France, 29–31 August 2019; pp. 1–2.
7. Tsai, Y.S.; Hsu, L.H.; Hsieh, Y.Z.; Lin, S.S. The real-time depth estimation for an occluded person based on a single image and OpenPose method. *Mathematics* **2020**, *8*, 1333. [CrossRef]
8. Yang, C.H.; Chang, P.Y. Forecasting the demand for container throughput using a mixed-precision neural architecture based on CNN–LSTM. *Mathematics* **2020**, *8*, 1784. [CrossRef]
9. Khan, F.; Salahuddin, S.; Javidnia, H. Deep learning-based monocular depth estimation methods—A state-of-the-art review. *Sensors* **2020**, *20*, 2272. [CrossRef] [PubMed]
10. He, K.; Zhang, X.; Ren, S.; Sun, J. Deep residual learning for image recognition. In Proceedings of the 2016 IEEE Conference on Computer Vision and Pattern Recognition (CVPR), Las Vegas, NV, USA, 26–30 June 2016; pp. 770–778.
11. Huang, G.; Liu, Z.; Laurens, V.; Weinberger, K.Q. Densely connected convolutional networks. In Proceedings of the 2017 IEEE Conference on Computer Vision and Pattern Recognition (CVPR), Honolulu, HI, USA, 21–26 July 2017; pp. 2261–2269.
12. Hu, J.; Shen, L.; Albanie, S.; Sun, G.; Wu, E. Squeeze-and-excitation networks. *IEEE Trans. Pattern Anal. Mach. Intell.* **2020**, *42*, 2011–2023. [CrossRef] [PubMed]
13. Liu, J.; Li, Q.; Cao, R.; Tang, W.; Qiu, G. A contextual conditional random field network for monocular depth estimation. *Image Vis. Comput.* **2020**, *98*, 103922. [CrossRef]
14. Godard, C.; Aodha, O.M.; Brostow, G.J. Unsupervised monocular depth estimation with left-right consistency. In Proceedings of the 2017 IEEE Conference on Computer Vision and Pattern Recognition (CVPR), Honolulu, HI, USA, 21–26 July 2017; pp. 270–279.
15. Chen, T.; An, S.; Zhang, Y.; Ma, C.; Wang, H.; Guo, X.; Zheng, W. Improving monocular depth estimation by leveraging structural awareness and complementary datasets. In Proceedings of the 2020 European Conference on Computer Vision (ECCV), Glasgow, UK, 23–28 August 2020; pp. 90–108.
16. Lin, L.; Huang, G.; Chen, Y.; Zhang, L.; He, B. Efficient and high-quality monocular depth estimation via gated multi-scale network. *IEEE Access* **2020**, *8*, 7709–7718. [CrossRef]
17. Alhashim, I.; Wonka, P. High quality monocular depth estimation via transfer learning. *arXiv* **2018**, arXiv:1812.11941.
18. Fu, H.; Gong, M.; Wang, C.; Batmanghelich, K.; Tao, D. Deep ordinal regression network for monocular depth estimation. In Proceedings of the 2018 IEEE Conference on Computer Vision and Pattern Recognition (CVPR), Salt Lake City, UT, USA, 18–22 June 2018; pp. 2002–2011.
19. Zhao, S.; Zhang, L.; Shen, Y.; Zhao, S.; Zhang, H. Super-resolution for monocular depth estimation with multi-scale sub-pixel convolutions and a smoothness constraint. *IEEE Access* **2018**, *7*, 16323–16335. [CrossRef]
20. Chen, X.; Chen, X.; Zha, Z. Structure aware residual pyramid network for monocular depth estimation. In Proceedings of the 28th International Joint Conference on Artificial Intelligence (IJCAI), Macao, China, 10–16 August 2019; pp. 694–700.
21. Lee, J.H.; Kim, C.S. Monocular depth estimation using relative depth maps. In Proceedings of the 2019 IEEE Conference on Computer Vision and Pattern Recognition (CVPR), Long Beach, CA, USA, 16–20 June 2019; pp. 9729–9738.
22. Hu, J.; Ozay, M.; Zhang, Y.; Okatani, T. Revisiting single image depth estimation: Toward higher resolution maps with accurate object boundaries. In Proceedings of the Workshop on Applications of Computer Vision (WACV), Hilton Waikoloa Village, HI, USA, 8–10 January 2019; pp. 1043–1051.
23. Chen, Y.; Zhao, H.; Hu, Z. Attention-based context aggregation network for monocular depth estimation. *arXiv* **2019**, arXiv:1901.10137v1.
24. Li, R.; Xian, K.; Shen, C. Deep attention-based classification network for robust depth prediction. In Proceedings of the 14th Asian Conference on Computer Vision (ACCV), Perth, Australia, 4–6 December 2018; pp. 663–678.
25. Wang, J.; Zhang, G.; Yu, M.; Xu, T. Attention-based dense decoding network for monocular depth estimation. *IEEE Access* **2020**, *8*, 85802–85812. [CrossRef]
26. Hao, Z.; Li, Y.; You, S.; Lu, F. Detail preserving depth estimation from a single image using attention guided networks. In Proceedings of the Sixth International Conference on 3D Vision (3DV), Verona, Italy, 5–8 September 2018; pp. 304–313.
27. Jiang, J.; Ehab, H.; Zhang, X. Gaussian weighted deep modeling for improved depth estimation in monocular images. *IEEE Access* **2019**, *7*, 134718–134729. [CrossRef]

28. Liu, P.; Zhang, Z.; Meng, Z.; Gao, N. Joint attention mechanisms for monocular depth estimation with multi-scale convolutions and adaptive weight adjustment. *IEEE Access* **2020**, *8*, 184437–184450. [CrossRef]

29. Lee, J.H.; Kim, C.S. Multi-loss rebalancing algorithm for monocular depth estimation. In Proceedings of the 2020 European Conference on Computer Vision (ECCV), Glasgow, UK, 23–28 August 2020; pp. 785–801.

30. Eigen, D.; Puhrsch, C.; Fergus, R. Depth map prediction from a single image using a multi-scale deep network. In Proceedings of the 2014 IEEE Conference and Workshop on Neural Information Processing Systems (NIPS), Montreal, QC, Canada, 13–20 December 2014; pp. 2366–2374.

31. Laina, I.; Rupprecht, C.; Belagiannis, V.; Tombari, F.; Navab, N. Deeper depth prediction with fully convolutional residual networks. In Proceedings of the Fourth International Conference on 3D Vision (3DV), Stanford, CA, USA, 25–28 October 2016; pp. 239–248.

32. Cao, Y.; Wu, Z.; Shen, C. Estimating depth from monocular images as classification using deep fully convolutional residual networks. *IEEE Trans. Circuits Syst. Video Technol.* **2018**, *28*, 431–444. [CrossRef]

33. Li, B.; Dai, Y.; He, M. Monocular depth estimation with hierarchical fusion of dilated CNNs and soft-weighted-sum inference. *Pattern Recognit.* **2018**, *83*, 328–339. [CrossRef]

34. Zheng, Z.; Xu, C.; Yang, J.; Tai, Y.; Chen, L. Deep hierarchical guidance and regularization learning for end-to-end depth estimation. *Pattern Recognit.* **2018**, *83*, 430–442. [CrossRef]

35. Eigen, D.; Fergus, R. Predicting depth, surface normals and semantic labels with a common multi-scale convolutional architecture. In Proceedings of the 2015 IEEE International Conference on Computer Vision (ICCV), Santiago, Chile, 7–13 December 2015; pp. 2650–2658.

36. Ito, S.; Kaneko, N.; Sumi, K. Latent 3D volume for joint depth estimation and semantic segmentation from a single image. *Sensors* **2020**, *20*, 5765. [CrossRef]

37. Lin, X.; Sánchez-Escobedo, D.; Casas, J.R.; Pardàs, M. Depth estimation and semantic segmentation from a single RGB image using a hybrid convolutional neural network. *Sensors* **2019**, *19*, 1795. [CrossRef]

38. Yan, H.; Zhang, S.; Zhang, Y.; Zhang, L. Monocular depth estimation with guidance of surface normal map. *Neurocomputing* **2017**, *280*, 86–100. [CrossRef]

39. Huang, K.; Qu, X.; Chen, S.; Chen, Z.; Zhang, W.; Qi, H.; Zhao, F. Superb monocular depth estimation based on transfer learning and surface normal guidance. *Sensors* **2020**, *20*, 4856. [CrossRef]

40. Qi, X.; Liao, R.; Liu, Z.; Urtasun, R.; Jia, J. Geonet: Geometric neural network for joint depth and surface normal estimation. In Proceedings of the 2018 IEEE Conference on Computer Vision and Pattern Recognition (CVPR), Salt Lake City, UT, USA, 18–22 June 2018; pp. 283–291.

41. Huynh, L.; Nguyen-Ha, P.; Matas, J.; Rahtu, E.; Heikkila, J. Guiding monocular depth estimation using depth-attention volume. In Proceedings of the 2020 European Conference on Computer Vision (ECCV), Glasgow, UK, 23–28 August 2020; pp. 581–597.

42. Hui, Z.; Wang, X.; Gao, X. Fast and accurate single image super-resolution via information distillation network. In Proceedings of the 2018 IEEE Conference on Computer Vision and Pattern Recognition (CVPR), Salt Lake City, UT, USA, 16–22 June 2018; pp. 723–731.

43. Hui, Z.; Gao, X.; Yang, Y.; Wang, X. Lightweight image super-resolution with information multi-distillation network. In Proceedings of the 27th ACM International Conference on Multimedia (ACM Multimedia), Nice, France, 21–25 October 2019; pp. 2024–2032.

44. Liu, J.; Tang, J.; Wu, G. Residual feature distillation network for lightweight image super-resolution. *arXiv* **2020**, arXiv:2009.11551.

45. Zhang, K.; Danelljan, M.; Li, Y.; Timofte, R.; Liu, J.; Tang, J.; Wu, G.; Zhu, Y.; He, X.; Xu, W.; et al. AIM 2020 challenge on efficient super-resolution: Methods and results. *arXiv* **2020**, arXiv:2009.06943.

46. Yang, H.; Yang, C.H.; Tsai, Y.J. Y-Net: Multi-scale feature aggregation network with wavelet structure similarity loss function for single image dehazing. In Proceedings of the 2020 IEEE International Conference on Acoustics, Speech and Signal Processing (ICASSP), Barcelona, Spain, 4–8 May 2020; pp. 2628–2632.

47. Wang, Z.; Bovik, A.C.; Sheikh, H.R.; Simoncelli, E.P. Image quality assessment: From error visibility to structural similarity. *IEEE Trans. Image Process.* **2004**, *13*, 600–612. [CrossRef]

48. Ding, X.; Guo, Y.; Ding, G.; Han, J. ACNet: Strengthening the kernel skeletons for powerful CNN via asymmetric convolution blocks. In Proceedings of the 2019 IEEE International Conference on Computer Vision (ICCV), Seoul, Korea, 27–31 October 2019; pp. 1911–1920.

49. Park, J.; Woo, S.; Lee, J.Y.; Kweon, I.S. BAM: Bottleneck attention module. In Proceedings of the 2018 British Machine Vision Conference (BMVC), Newcastle, UK, 3–6 September 2018; pp. 147–163.

50. Silberman, N.; Hoiem, D.; Kohli, D.; Fergus, R. Indoor segmentation and support inference from RGBD images. In Proceedings of the European Conference on Computer Vision, Florence, Italy, 7–13 October 2012; pp. 746–760.

51. Geiger, A.; Lenz, P.; Stiller, C.; Urtasun, R. Vision meets robotics: The KITTI dataset. *Int. J. Robot. Res.* **2013**, *32*, 1231–1237. [CrossRef]

52. Koch, T.; Liebel, L.; Fraundorfer, F.; Korner, M. Evaluation of CNN-based single-image depth estimation methods. In Proceedings of the 2018 European Conference on Computer Vision (ECCV), Munich, Germany, 8–14 September 2018; pp. 331–348.

53. Paszke, A.; Gross, S.; Chintala, S.; Chanan, G.; Yang, E.; DeVito, Z. Automatic differentiation in PyTorch. In Proceedings of the Advances in Neural Information Processing Systems Workshops (NIPS), Long Beach, CA, USA, 4–9 December 2017; pp. 1–4.

54. Deng, J.; Dong, W.; Socher, R.; Li, L.; Li, K.; Li, F. Image-Net: A large-scale hierarchical image database. In Proceedings of the 2009 IEEE Conference on Computer Vision and Pattern Recognition (CVPR), Miami, FL, USA, 20–25 June 2009; pp. 248–255.
55. Kingma, D.P.; Ba, J. Adam: A method for stochastic optimization. In Proceedings of the 3rd International Conference on Learning Representations (ICLR), San Diego, CA, USA, 7–9 May 2015; pp. 1–15.

Article

Classification and Prediction of Typhoon Levels by Satellite Cloud Pictures through GC–LSTM Deep Learning Model

Jianyin Zhou [1], Jie Xiang [1,*] and Sixun Huang [1,2]

[1] College of Meteorology and Oceanography, National University of Defense Technology, Nanjing 211101, China; 15165381462@163.com (J.Z.); huangsixun2021@163.com (S.H.)
[2] State Key Laboratory of Satellite Ocean Environment Dynamics, Second Institute of Oceanography, State Oceanic Administration, Hangzhou 361005, China
* Correspondence: xiangjie2021@163.com

Received: 9 August 2020; Accepted: 6 September 2020; Published: 9 September 2020

Abstract: Typhoons are some of the most serious natural disasters, and the key to disaster prevention and mitigation is typhoon level classification. How to better use data of satellite cloud pictures to achieve accurate classification of typhoon levels has become one of classification the hot issues in current studies. A new framework of deep learning neural network, Graph Convolutional–Long Short-Term Memory Network (GC–LSTM), is proposed, which is based on the data of satellite cloud pictures of the Himawari-8 satellite in 2010–2019. The Graph Convolutional Network (GCN) is used to process the irregular spatial structure of satellite cloud pictures effectively, and the Long Short-Term Memory (LSTM) network is utilized to learn the characteristics of satellite cloud pictures over time. Moreover, to verify the effectiveness and accuracy of the model, the prediction effect and model stability are compared with other models. The results show that: the algorithm performance of this model is better than other prediction models; the prediction accuracy rate of typhoon level classification reaches 92.35%, and the prediction accuracy of typhoons and super typhoons reaches 95.12%. The model can accurately identify typhoon eye and spiral cloud belt, and the prediction results are always kept in the minimum range compared with the actual results, which proves that the GC–LSTM model has stronger stability. The model can accurately identify the levels of different typhoons according to the satellite cloud pictures. In summary, the results can provide a theoretical basis for the related research of typhoon level classification.

Keywords: deep learning; GC–LSTM model; typhoon; satellite image; prediction system

---

1. Introduction

Typhoons have the widest damage range among all the natural disasters, which is the most invading disaster of coastal areas. In particular, storms caused by typhoons have caused huge losses to coastal ships and the marine industry [1]. According to statistics, economic losses caused by typhoons account for 1%–3% of Gross Domestic Product (GDP) [2]. The prediction of typhoons has always been a scientific issue in this field [3]. During the typhoon outbreak, it is difficult to obtain typhoon data directly from conventional climate and ocean monitoring data, which makes it difficult to predict typhoons [4]. With the improvement of the satellite remote sensing technology, meteorological satellite cloud pictures can more accurately and stably monitor the weather changes in real-time in all weathers, becoming the main means of observing and predicting typhoons [5]. Research on using satellite cloud pictures has achieved a series of results in the process of typhoon generation and development [6,7]. The influence of typhoons on people's lives and property is closely related to its intensity. According to the satellite cloud picture analysis, different typhoon levels will cause different cloud clusters. Therefore, the levels of typhoons can be predicted according to the cloud cluster data transmitted by satellite cloud pictures [8]. The current research methods for typhoon level prediction are mainly divided into the subjective empirical method and the simulation analysis method [9]. The subjective empirical method requires professional knowledge; for example, the Dvorak analysis method is estimated by understanding the cloud system structure characteristics and specific parameters through empirical rules and constraints [10]. The simulation analysis method is the most used; atmospheric physical quantities such as different initial fields and boundary conditions are comprehensively considered to predict typhoon levels [11] The simulation analysis method depends on the accurate recognition of satellite cloud pictures, and the typhoon level prediction model has poor accuracy and large errors [12]. Therefore, the establishment of a high-precision typhoon level classification model is crucial for the study of typhoons.

As a classification and recognition method with strong generalization ability, deep learning overcomes the shortcomings of traditional methods that require prior knowledge to explicitly extract features. Of the latest years, many scholars worldwide have applied deep learning techniques to marine meteorological research. Krapivin et al. employed sequential analysis and seepage theory tools to analyze the process of the ocean–atmosphere coupling system; additionally, they adopted the SVM to detect the state features of this system; such a method helped to monitor the changes and directions of the ocean transition process and could predict significant changes in the state of the ocean–atmosphere system [13]. Varotsos et al. proposed an information modeling tracker for tropical cyclones based on the clustering algorithm to assess the instability of the atmosphere–ocean system; the synthesized functional prediction structure could be a reliable global ocean monitoring system, which could effectively reduce the risk of tropical cyclones [14]. Zhu et al. (2019) established a short-term heavy rain recognition model based on physical parameters and deep learning; this model could automatically predict the probability of heavy rain occurrence based on data from various monitoring stations [15]. Scher employed a deep neural network based on the principles of cyclic model dynamics; after training the model, the network could predict ocean weather several hours in advance [16]. Kaba et al. input astronomical factors, extraterrestrial radiation and climate variables, sunshine duration, cloud cover, minimum temperature, and maximum temperature as attributes to obtain a climate prediction model; the prediction accuracy of the model for the marine climate was 98% [17]. Xiao et al. used convolutional Long Short-Term Memory (LSTM) networks as building blocks and trained the blocks in an end-to-end manner to achieve accurate and comprehensive predictions of sea surface temperature in the short and medium term [18]. However, there is little research on the characteristics of using satellite cloud pictures to identify typhoon levels through deep learning technology [19]. The main reason is that traditional machine learning algorithms need to explicitly extract features for classification, but it is difficult to extract features related to typhoon level classification in satellite cloud pictures.

Therefore, the poor accuracy, complex satellite cloud picture feature extraction, and low recognition rate in typhoon prediction are the problems to be focused on here. By introducing a Graph Convolutional Network (GCN) and a Long Short-Term Memory (LSTM) neural network, a typhoon Graph Convolutional–LSTM (GC–LSTM) neural network is constructed. The GC–LSTM model uses satellite cloud picture data of 20 years as samples for deep learning, and is compared with traditional typhoon prediction models. The results are expected to provide accurate and fast weather information for relevant departments to help them make decisions and reduce human life and property losses caused by typhoons.

2. Materials and Methods

2.1. Satellite Cloud Images and Data Sources

The meteorological satellite cloud picture is to use the meteorological satellite instruments in space to photograph the earth's atmosphere, to find the weather through some rules, and to verify the weather in combination with the ground weather [20]. It can display various types of clouds on a single picture, characterize weather phenomena at different scales, and provide very useful telemetry data for weather analysis and forecasting. Generally, satellite cloud pictures can be divided into infrared satellite cloud pictures, visible light satellite cloud pictures, and cloud pictures processed and synthesized according to requirements. Figure 1 below is a processed RGB color cloud picture.

Figure 1. Meteorological satellite cloud picture.

Typhoons are the most common severe weather system. Violent winds above level 12, huge waves above 6–9 m, storm surges above 2–3 m, and heavy rains above 200–300 mm always accompany typhoons, which are harmful to marine ships, the marine engineering industry even people's lives and properties in coastal and inland areas. The embryo of a typhoon, that is, the initial cyclonic low-pressure circulation, has the following sources: (1) low-pressure disturbance in the tropical convergence zone, accounting for about 80–85%; (2) easterly wind belt disturbance—easterly wind wave, accounting for about 5–10%; (3) westerly wind belt disturbance degeneration, accounting for about 5%; (4) low-level vortices induced by high-level cold vortices, accounting for less than 5%. There is a very favorable weather situation in the Northwest Pacific: a strong tropical easterly wind belt, a strong and active equatorial westerly wind belt, an active tropical convergence belt, and a southwest-southeast monsoon convergence belt. Hence, typhoons are particularly prone to generate. Before a typhoon is formed, it undergoes an enhanced development process, which usually develops from an existing tropical cloud cluster of 3–4 d. Tropical cloud clusters and isolated cloud clusters in the zonal cloud belt if they can maintain existence for more than 3–4 d and can present a cyclonic low-pressure circulation. Once the surrounding long cloud belts can form one or several convections and be involved in the low pressure, after 1–2 d, the low pressure can develop into a tropical cyclone typhoon. Therefore, when identifying the satellite cloud picture, whether it meets the rules based on the cloud cluster features should be determined first. Then, whether it is a meteorological feature of a typhoon should be judged based on the overall cloud cluster. In this way, the satellite cloud pictures can effectively identify typhoons. However, traditional methods also depend on recognizing these features; without enhancement, features in the satellite cloud pictures cannot be compared accurately, thereby reducing the prediction accuracy.

The data come from the National Institute of Informatics (NII) of Japan. (1) The website is: http://agora.ex.nii.ac.jp/digital-typhoon/. The Japanese "Himawari" series of satellite cloud pictures are utilized. From 2010 to 2019, Japan successively launched the "Himawari" satellites. In particular, there are 16 "Himawari-8" geosynchronous weather satellites. For the band channel, the spatial resolution can reach up to 500 m; (2) Analysis area: The coverage range of satellite cloud picture data is the upper part of the Northwest Pacific (120° E–160° W); (3) Data time: The high-resolution satellite cloud picture data of this area have been downloaded, as well as the information of typhoon intensity; all the data were transmitted from the Himawari-8 satellite from 2010 to 2019 in Japan.

The model established here is based on Ubunt16.10. The processor is the computing node of the Beijing Supercomputing Center server. The hardware configuration is 2 channels and 32 cores, EPYC 7452 @2.35 GHz, and 256 GB memory. The deep learning framework used is open-source Keras. The dataset contains more than 1000 typhoon processes. The experiment uses infrared cloud pictures as data samples. The objective is to obtain all-weather meteorological data. According to the typhoon level index, different typhoon level labels are formulated, as shown in Table 1:

Table 1. Typhoon level standard label.

Typhoon Level	Maximum Wind Speed/kt	Maximum Wind Speed/(m/s)
Tropical depression	<34	<17
Typhoon	>34–<64	>17–<33
Strong typhoon	>64–<85	>33–<44
Super Typhoon	>85–<105	>44–<54

Cloud pictures processing: First, the median filtering is performed on the original infrared image to remove the noise in the cloud image, which effectively retains the edge information of the picture. Second, the nearest neighbor scaling is used to convert the cloud picture into a $24 \times 24 \times 1$ format as input information. Then, according to the criteria in Table 1, the classification is performed, where A represents tropical depression, B represents typhoon, C represents strong typhoon, and D represents super typhoon. Finally, a dataset of 3500 training samples and 600 test samples is constructed. There are, respectively, 1000 training sets and 200 test sets for tropical depression, typhoons, strong typhoons, and super typhoons. Some satellite cloud picture samples are shown in Figure 2.

（A）Tropical depression　　　（B）Typhoon

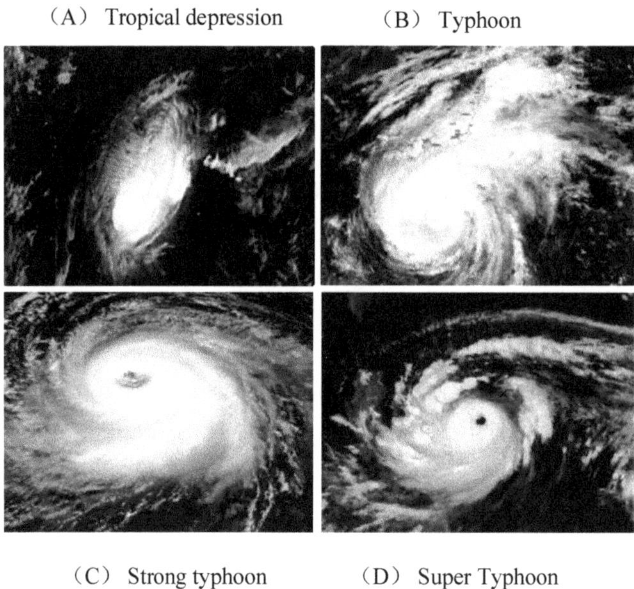

（C）Strong typhoon　　　（D）Super Typhoon

Figure 2. Some meteorological satellite cloud picture samples. Please add explanation of subfigures in caption.

2.2. Traditional Convolutional Neural Network

Deep learning Convolutional Neural Network (CNN) is a feed-forward neural network, which can quickly respond to nearby covered networks through artificial neurons, thus achieving a deep learning algorithm for rapid response to data [21]. It consists of a convolutional layer and a sampling layer alternately forming a network topology [22]. CNN uses the method of backpropagating neurons to realize the update network of each neuron information. The feature extraction process can be expressed as a score function $S(x_i, w, b)$. The cross-entropy loss function for the classification error of the i sample (x_i, y_i) is defined as:

$$L_i = -\ln e^{S_{y_i}} + \ln \sum e^{S_i} \tag{1}$$

In (1), S_{yi} represents the number of scores for the true classification of the i-th sample of the training set, and S_i indicates the ratio of the index of the current element to the sum of all the element indices. The output of sample (x_i, y_i) after passing through the network is $f(x)$, and the corresponding sample loss value is:

$$L_i(f(x), y) = -\ln f(x)_y \tag{2}$$

The error-sensitive items of the output layer of the CNN of the deep learning layer l are:

$$\delta_l = \frac{\partial L}{\partial a^l} = \nabla_a^l(x) - \ln f(x)_y = f(x) - y \tag{3}$$

where a^l represents the input of layer l. Finally, the backpropagation rule of CNN is used to update the weight of each neuron, which makes the overall error function of the model continuously decrease.

Figure 3 shows the convolution layer of CNN, which uses its convolution kernel to convolve with the input image and then outputs the feature image of this layer through the neuron activation function, which realizes the feature extraction of the image. The convolution process is defined as follows:

$$x_j^l = f(\sum_{i \in M_j} x_i^{l-1} \times k_{ij}^l + b_j^l) \tag{4}$$

where l is the number of convolutional layers in the model, k_{ij}^l is the number of convolution kernels, b_j^l is the additive bias, f is the activation function, and M_j is the input image. The specific structure process is shown in Figure 3:

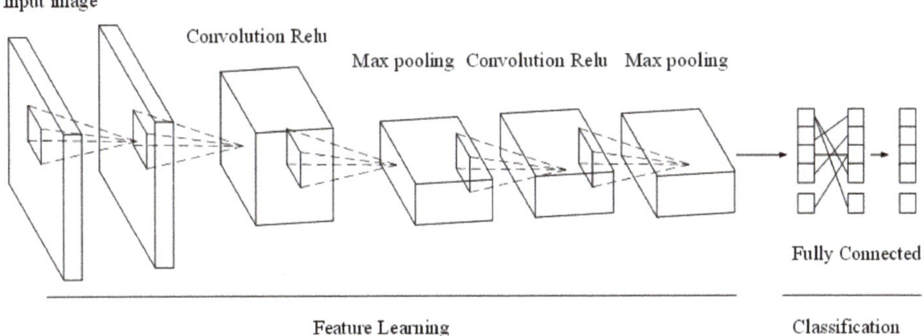

Figure 3. Structure of the deep learning convolution layer.

Figure 4 shows the data collection layer of the convolution nerve. The data collection layer is the process of reducing the resolution of the current input feature image and reducing the amount of calculation, thereby improving the network convergence speed. The data collection can be defined as:

$$x_j^l = f(\beta_j^l down(x_i^{l-1}) + b_j^l) \tag{5}$$

where $down(\bullet)$ represents the data collection function, β_j^l and b_j^l represent the product bias and additive bias, respectively, f is the activation function. Among them, the characteristic of the sampling layer C_x

is 2×2 sampling, and every 4 pixels are combined into 1 pixel. Weighted by the multiplicative bias w_{x+1}, the addictive bias b_{x+1} is output through the activation function S_{x+1}.

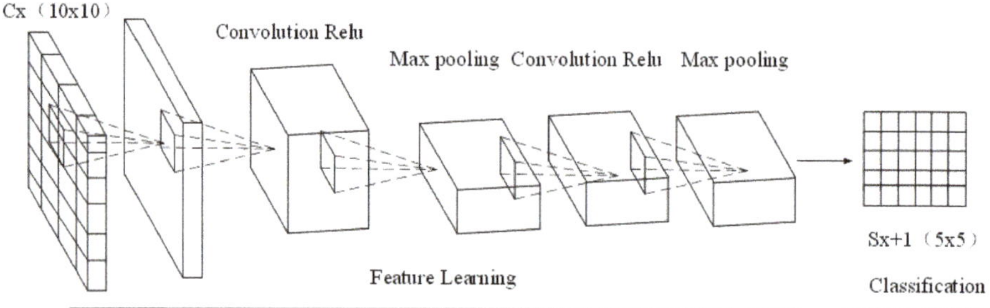

Figure 4. Structure of the deep learning collection layer.

2.3. Deep Learning GCN Algorithm

The network topology of traditional CNN is formed by alternately arranged convolution layers and sampling layers. If the input features are not prominent, the pooling layer will lose some image information while reducing the dimensionality, reducing the network learning capability. The traditional CNN directly utilizes and compares the pictures with the original typhoon pictures. If the resolution of the transmitted data CNN image is low, the CNN prediction accuracy of the typhoon levels will decrease more. In the constructed dataset of satellite cloud pictures, due to the sophisticated atmospheric factors during typhoon formation, the spiral radius of the cloud pictures is not apparent. Therefore, traditional CNN is not suitable for feature extraction of typhoon cloud images. In contrast, the GCN network is a deep learning algorithm specially established based on images. It can extract the original image of the satellite cloud pictures according to some rules, effectively extract the features of the image, and improve the local image resolution; then, the model compares the processed image with the trained image database so that the processed image prediction accuracy is higher.

GCN is an algorithm that adds a lot of image processing based on CNN [23]. Most real-world network data is represented in the form of graphs, such as social networks, protein interaction networks, and knowledge graphs; however, they do not have a regular spatial structure and can process image data that cannot be processed by CNN. The algorithm transfers the convolution method on the image to the graph, and proposes two methods based on space and spectrum decomposition. The space-based method is to establish a perceptual domain for each node (selecting the neighbor node of the node); in other words, the nodes in the graph are connected in the spatial domain to achieve a hierarchical structure, so as to perform convolution learning. Based on the spectral decomposition method, the Laplacian matrix is used to transform the feature vector into the spectral domain; then, the point in the spectral domain is multiplied and the inverse Fourier transform is performed to achieve the convolution on the graph. The specific structure is shown in Figure 5. GCN is a CNN that directly acts on graphs. GCN allows end-to-end learning of structured data, and extracts the features of network nodes by learning the structural features of the network. Here, the GCN is utilized to extract the network structure features at each moment.

In the recognition of the typhoon cloud pictures, the general model directly utilizes the pictures and compares them with the original typhoon pictures. If the resolution of the transmitted data image is low, the prediction accuracy of the typhoon levels will decrease more. The GCN network is a deep learning algorithm specially established based on images. It can extract the original image of the satellite cloud pictures according to some rules, effectively extract the features of the image, and improve the local image resolution; then, the model compares the processed image with the trained image database so that the processed image prediction accuracy is higher.

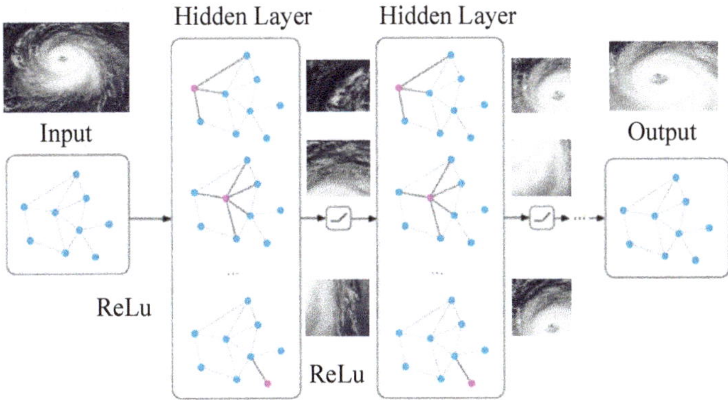

Figure 5. Schematic diagram of Graph Convolutional Network (GCN) structure.

The identification of satellite cloud pictures via GCN is as follows: (1) connecting a single sample of satellite cloud pictures to form a row vector; and (2) superimposing n vectors into the GCN system. Information can be transferred between nodes based on a correlation coefficient matrix. Therefore, the data-driven method constructs a correlation coefficient matrix, which contains the original satellite cloud picture's image and edge features. This model uses the data-driven method to establish a directed graph between markers, and GCN maps the category markers to the corresponding category classifier. Finally, the category relationship is modeled, and at the same time, the model learning ability is improved. A correlation coefficient matrix is constructed for GCN by balancing the node and its neighboring nodes for node feature updates, thereby effectively solving the overfitting and excessive smoothing problems that hinder GCN's performance.

2.4. LSTM Neural Network Algorithm

LSTM is a type of Recurrent Neural Network (RNN) network. It is often used to process and predict important events with very long intervals and delays in time series [24]. In typhoon prediction, accurate time prediction is very important. Therefore, the LSTM algorithm is chosen, which can accurately establish a time relationship graph. Based on the data transmitted from the traditional satellite and after feature extraction, the data will be arranged in chronological order. Second, according to the time interval of the previous typhoon, the algorithm can predict the time of the typhoon well. An LSTM unit contains an input gate, output gate, and forget gate. Among them, the input gate controls model input, the output gate controls model output, and the forget gate calculates the degree of forgetting of the memory module at the previous moment. The structure of the LSTM model is shown in Figure 6, the specific calculation is as follows:

$$f_t = \sigma(W_f \cdot [h_{t-1}, x_t] + b_f) \tag{6}$$

where f_t and i_t denote the forget gate and the input gate of the t step in the sentence sequence, respectively. In each sentence sequence, the forget gate controls the degree of forgetting the information of each word, and the input gate controls the degree to which each word information is newly written into long-term information.

$$i_t = \sigma(W_i \cdot [h_{t-1}, x_t] + b_i) \tag{7}$$

$$C = \tanh(W_c \cdot [h_{t-1}, x_t] + b_c) \tag{8}$$

$$C_t = f_t \times C_{t-1} + i_t \times C \tag{9}$$

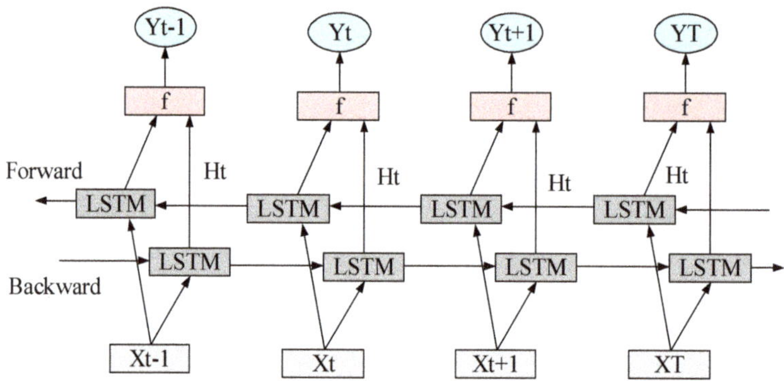

Figure 6. Schematic diagram of the Long Short-Term Memory (LSTM) model structure.

The two gates f_t and i_t use the Sigmoid function, the value range is [0, 1], and the value of tanh function is [−1, 1]. C_{t-1} is the state of the neuron at time $t - 1$, and C_t is the state of the neuron at time t.

$$h_t = o_t \times \tanh(C_t) \tag{10}$$

$$o_t = \sigma(w_o \cdot [h_{t-1}, x_t] + b_o) \tag{11}$$

where o_t is the output degree of the output-gate-controlled word long-term information h_t is the output of step t in the sentence sequence. The above equations show that the word information of the current step of LSTM is determined by the word information retained in the previous step and the word information saved after being filtered by the input gate at the current time. Here, the LSTM network is introduced to effectively mine the long-term data and utilize the raw data information, to effectively process the cloud picture data.

2.5. Construction of the GCN–LSTM Fusion Model

The traditional CNN uses a topology structure composed of a convolutional layer and a data acquisition layer. When the input image features are not obvious, the resolution of the input picture will be actively reduced in the pooling layer of the network, which loses some important information of the pictures, resulting in a decline in network learning capabilities; in the meantime, the corresponding model accuracy will continue to decline.

The complex atmospheric factors in the formation of a typhoon make the features within the spiral radius of the cloud picture not obvious. In the data of satellite cloud pictures, every detail is very concerned. However, the traditional CNN is not very suitable for the feature extraction of typhoon cloud pictures. Based on the traditional CNN, the cyclic convolution is utilized to enhance the feature extraction capability of the model. By taking the advantages of the LSTM network, a novel deep learning model GC–LSTM is proposed to extract features in atmospheric cloud pictures to achieve accurate prediction of typhoons.

The entire GC–LSTM model combines the advantages of the two models: LSTM network and GCN. LSTM is used to learn the timing information of the connected state of each node, and GCN is used to learn snapshots at every moment. The structural characteristics of the network make it capable of effectively processing high-dimensional, time-dependent, and sparse structural sequence data. This model is easy to build and train, and can adapt to different network applications. Also, the accurate prediction of typhoon levels can be achieved. The specific structural framework is shown in Figure 7.

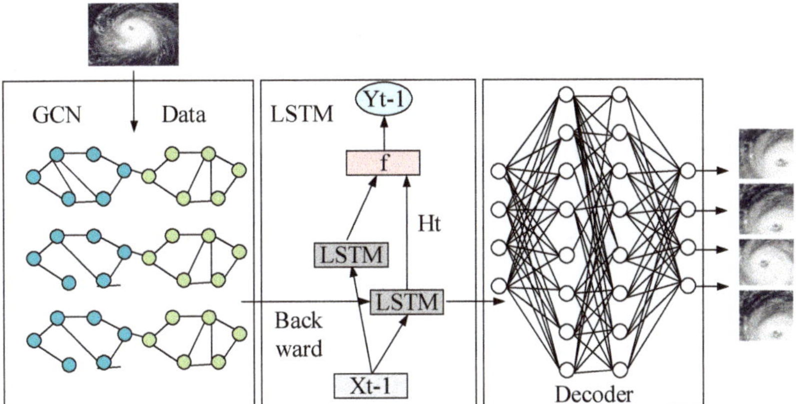

Figure 7. Overall framework of the GCN–LSTM fusion model.

2.6. Model Verification and Optimization

(1) Model accuracy (*ACC*): It is the part that passes the true correct rate. If the number of real typhoons in the *i* sample of all n satellite cloud picture samples is y, and the data predicted by the model is O_i, then the correct rate of the classification of the typhoon satellite cloud picture model is calculated as follows; if the number predicted by satellite cloud pictures is more consistent with the real number, the correct rate of model classification is greater.

$$ACC(y_i, O_i) = \frac{1}{n}\sum_{i=0}^{n} 1(y_i = O_i) \tag{12}$$

(2) Precision (*Pre*): It indicates the proportion of processed samples that are correctly divided into positive samples [25].

$$Pre = \frac{N_{TP}}{N_{TP} + N_{FP}} \tag{13}$$

where N_{TP} represents the number of satellite cloud pictures that should be correctly classified; N_{FP} represents the number of true correct classification after passing through the typhoon satellite cloud picture prediction model.

(3) Recall (*Rec*): It represents the proportion of positive samples in the original positive samples [26]. It indicates the proportion of the total number of correctly predicted numbers after the typhoon satellite cloud picture prediction model.

(4) Recognition Rate (*RR*): It is the ratio of the wrongly recognized image/the recognized image [27].

(5) Matching Speed (*MS*): It refers to the time from the completion of image acquisition to the completion of model prediction.

$$Rec = \frac{TP}{TP + FN} \times 100\% \tag{14}$$

where *TP* is the number of typhoon searches that are correctly identified by satellite cloud pictures, and *FN* is the number of typhoon searches that are not correctly identified by satellite cloud pictures.

The model performance is compared. M1 stands for the ANN model, M2 stands for the RNN model, M3 stands for the GCN model, M4 stands for the LSTM model, M5 stands for the GCN–LSTM, and M6 stands for the RNN–LSTM model. RNN–LSTM has been employed in research [28]. Lian et al. (2020) utilized the RNN–LSTM model for predicting the path of satellite cloud pictures; the results revealed the high prediction accuracy of this model [29]. Zhao et al. (2020) proposed a typhoon identification and typhoon center location method based on deep learning; the average accuracy of this method was 96.83%, and the average detection time of each sample was 6 ms, which met the real-time typhoon eye detection. At present, typhoon paths are often predicted by the single algorithms or the CNN + SVM/LSTM algorithm. Although some of these algorithms have high accuracy, the output results are unstable, or the operation requires a higher configuration, which reduces the recognition speed [30]. No one has systematically summarized and compared the problems of the fusion algorithms, nor utilized the combination of GCN–LSTM for typhoon type prediction. The optimal model parameters mainly determine the convolution number of the CNN and the proportion of neurons. Among them, the convolution number of the input data is set to Q1 (1×1), Q2 (3×3), Q3 (5×5), Q4 (7×7), Q5 (8×8), and Q6 (9×9); the proportion of different neurons is set between 0% and 90%, and increases by 10% each time; the optimal parameter setting of the model is judged by its accuracy.

3. Results

3.1. Performance Analysis of Different Models

The performance of different models is shown in Figure 8. In terms of accuracy, the accuracy of the training set is higher than that of the test set, which is 5% higher on average. Comparing different models, it is found that as the number of training sets increases, the accuracy of the model continues to increase, which is consistent with the actual situation. When the training data are more, the accuracy of the model is higher. Compared with the single algorithm, the accuracy of the model is lower than the fusion model. The single model with the highest accuracy rate is the LSTM model. Because of its data memory function, the overall accuracy is significantly higher. Compared with other models, the average accuracy rate is 88.21%, and the fusion model with the highest accuracy rate is the GCN–LSTM model, with an average accuracy rate of 91.51%. In terms of accuracy and recall rate, the results are consistent with the conclusion of the accuracy rate. The highest is the GCN–LSTM model, with an average recall rate of 91.04% and an average accuracy of 92.35%. In terms of the recognition rate, the model shows a higher advantage, and the average recognition rate is as high as 93.61%. In terms of matching speed, the model has a stronger ability to recognize satellite cloud pictures than other models. Due to the advantages of the GCN and the rapid processing of data by LSTM, its average processing speed is maintained at 25.5 ms. According to the above results, the constructed GCN–LSTM model has a higher recognition rate for satellite cloud pictures, and the accuracy is 13.3% higher than that of the traditional ANN network. The classification effect of satellite cloud pictures is also significantly enhanced. The model shows a strong advantage in typhoon recognition.

3.2. Determination of Optimal Model Parameters

It is illustrated in Figure 9 that by using different convolutional numbers based on the original model, the convergence speed of the model differs greatly. In the training data set, when the convolution kernel number is 1×1, the loss value of the model is 0.55. When the convolution kernel number is 3×3, the model convergence speed is the slowest, and the stability can only be achieved when the number of iterations is 350 times. When the convolution kernel number is 7×7, the model converges and reaches stability at 310 times, and the loss value is 0.018. When the convolution kernel number is 5×5, the optimal loss value for the model convergence at 320 times is 0.0094. When the convolution kernel number is 8×8, the optimal loss value of the model at 300 times of convergence is 0.035, the overall convergence iteration number of the test set is larger, and the optimal loss value is also larger than the training set. In summary, 5×5 is chosen as the optimal convolution kernel number.

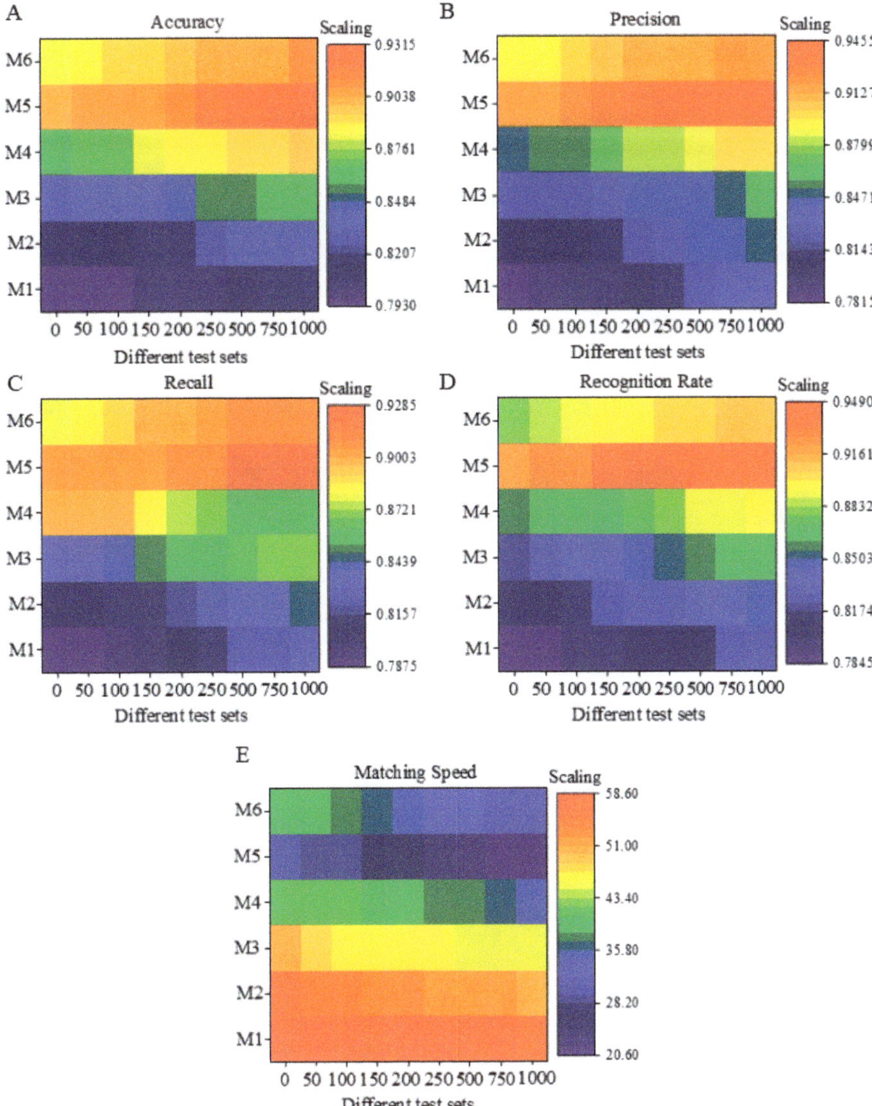

Figure 8. Performance comparison results of different models (Note: M1 is the ANN model, M2 is the Recurrent Neural Network (RNN) model, M3 is the GCN model, M4 is the LSTM model, M5 is the GCN–LSTM, and M6 is the RNN–LSTM model, where 0–200 is the test set and 200–1000 is the training set). Figures (**A–E**) show the performance results of different algorithms in accuracy, precision, recall, recognition rate, and processing speed under different test sets.

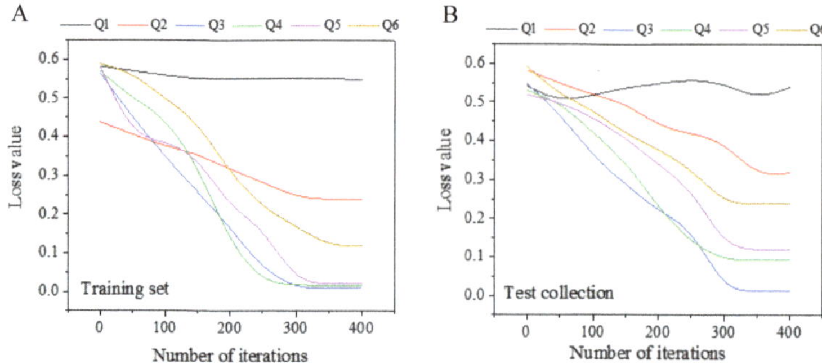

Figure 9. Feature extraction effect of different number of convolution kernels (Note: Q1 (1 × 1 convolution kernel), Q2 (3 × 3 convolution kernel), Q3 (5 × 5 convolution kernel), Q4 (7 × 7 convolution kernel), Q5 (8 × 8 convolution kernel), and Q6 (9 × 9 convolution kernel)). Figure (**A**) shows the feature extraction effect under different numbers of convolution kernels on the training set; Figure (**B**) shows the feature extraction effect under different numbers of convolution kernels on the test set.

It is illustrated in Figure 10 that compared to the convolution kernel numbers of 1 × 1 and 3 × 3, when the convolution kernel number is 5 × 5, the neural network has a high processing efficiency for the satellite cloud picture. The 5 × 5 convolution kernel has obvious feature extraction effects for typhoon eye, cloud wall, and spiral cloud belt (yellow area), but the 8 × 8 convolution kernel will increase typhoon similar redundancy and thus lose local features. Compared with the original image of the vortex cloud area, it is found that the model is more sensitive to the yellow area, but not sensitive to the fibrous cloud at the edge of the typhoon. Therefore, adding the model to the yellow area feature extraction is beneficial to the classification of the model. This is consistent with the truth that many scholars predict typhoons by locating typhoon eyes, segmenting dense cloud areas, and extracting spiral cloud belt features. This also proves the feasibility of classifying typhoon levels through deep learning and satellite cloud pictures.

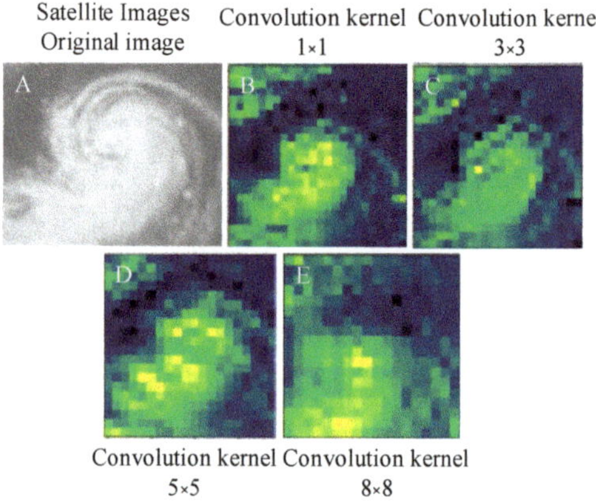

Figure 10. Feature extraction result of some convolution kernels.

By using different convolution kernel numbers based on the original model, the accuracy of the model is not much different, which is shown in Figure 11. In the training data set, when the convolution kernel number is 1 × 1, the model accuracy rate is 83.8%; when the convolution kernel number is 3 × 3, the model accuracy rate is 87.2%; when the convolution kernel number is 7 × 7, the model accuracy rate is 93.3%; when the convolution kernel number is 5 × 5, the model accuracy rate is 97.1%; when the convolution kernel number is 8×8, the model accuracy rate is 91.9%; the overall accuracy of the test set is lower than the training set. In summary, 5×5 is chosen as the optimal number of convolutions for the model.

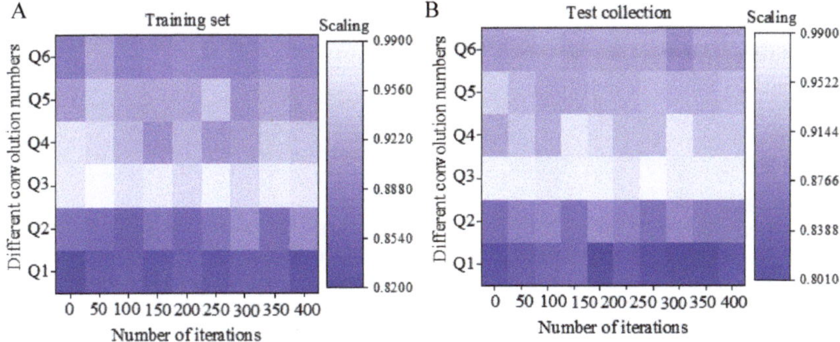

Figure 11. Model accuracy under different convolution kernels. Figure (**A**) shows the accuracy of the model under different convolution kernels on the training set; Figure (**B**) shows the accuracy of the model under different convolution kernels on the test set.

The accuracy of the model under different neuron ratios is shown in Figure 12. As the proportion of neurons continues to increase, the accuracy of the model shows a slight increase and then decreases. When the proportion of neurons is 30%, 50%, 80%, and 90%, the accuracy rate of the model drops; the lowest accuracy rate is 91% when the proportion of neurons is 30%, and the highest recognition rate is 92.5% when the proportion of neurons is 70%. The overall model accuracy of the test set is lower than that of the training set.

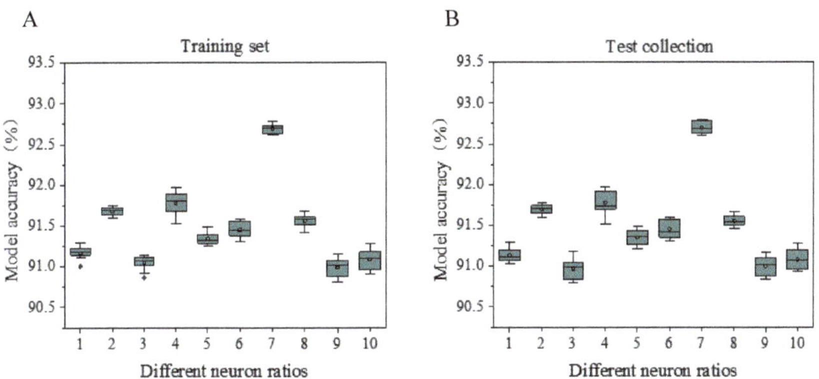

Figure 12. Model accuracy rate at different neuron ratios. Figure (**A**) shows the accuracy of the model under different neuron ratios on the training set; Figure (**B**) shows the accuracy of the model under different neuron ratios on the test set.

3.3. Application Analysis of Model Examples

It is shown in Table 2 that using the GC–LSTM model under the optimal parameters, the prediction accuracy rate of typhoons and super typhoons can reach 95.12%, but the prediction accuracy rate of tropical depressions is only 83.36% lower, which may be that the specific cloud characteristics and complete cloud structure are not formed during the typhoon formation process. Approximate results can also be seen in the results of some sample satellite cloud pictures. Therefore, the accuracy of the model is low. For strong typhoons, the accuracy rate is 93.24%, which is 2% higher than the traditional typhoon prediction model [31]. In summary, the GC–LSTM typhoon prediction model has high prediction performance, and the prediction accuracy of typhoons and super typhoons can reach 95.12%.

Table 2. Comprehensive evaluation of typhoon level prediction.

Categorical Data	Tropical Depression (0-)	Typhoon (1-)	Strong Typhoon (2-)	Super Typhoon (3-)
Tropical depression	83.36	12.67	9.59	3.28
Typhoon	1	95.12	0	0
Strong typhoon	1	1	93.24	7.24
Super Typhoon	0	0	1	95.12

Figure 13A,B shows the data prediction of the GCN–LSTM model from 2010 to 2019. After the results are compared with actual values, the annual average absolute error is obtained. The result shows that after 6 h of model training, the average absolute error remains between 1 and 15; after 12 h of training, the average absolute error of the model drops by 33.33% and remains between 1 and 10, which further shows that the accuracy of the GCN–LSTM model is improved greatly.

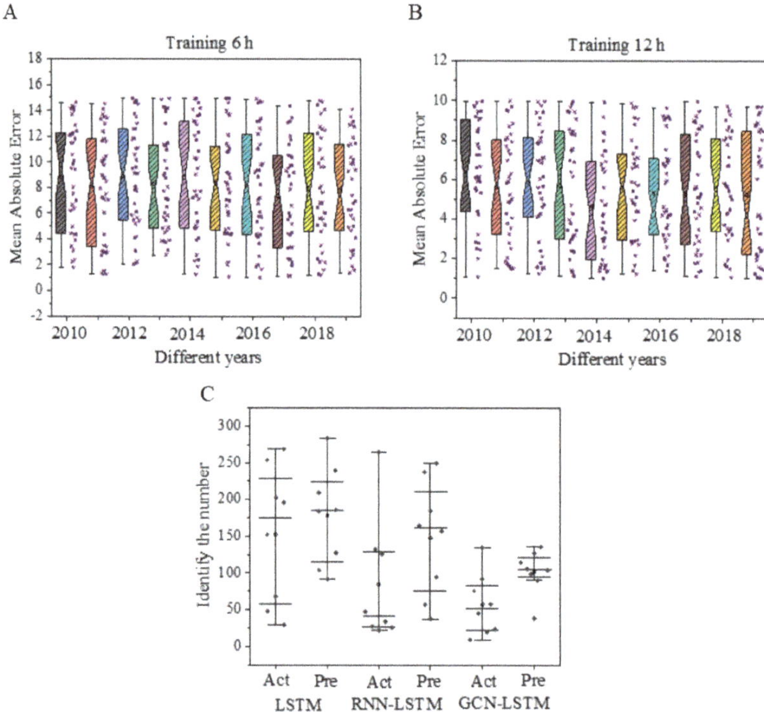

Figure 13. Model stability test results. Figure (**A,B**) illustrate the mean absolute errors of the GCN-LSTM model after 6 h and 12 h training on the data of 2010–2019, respectively. Figure (**C**) shows the identified and predicted number of typhoons of LSTM, RNN-LSTM, and GCN-LSTM models.

Figure 13C illustrates that LSTM, the best single model, and RNN–LSTM and GCN–LSTM, two best fusion models, are employed for prediction. Act represents the actual output result on the test dataset, and Pre represents the actual output result on the training dataset. All satellite cloud pictures, from 2010 to 2019, are output and analyzed according to different years. The results show that when a fixed amount of data is input, the LSTM test set's result is maintained between 30 and

253. Data fluctuates sharply in different years. Therefore, the model is unstable. Compared to the RNN–LSTM model, this model's stability also has a large deviation due to the need for a large number of neural networks to perform operations. The prediction result interval of GCN–LSTM is 38–104, with the least data fluctuation. The above results show that the proposed GCN–LSTM model is more stable.

4. Conclusions

Through the analysis of the original typhoon prediction model, problems such as poor prediction accuracy, low recognition rate, and complex feature extraction are found in the traditional typhoon prediction models. Therefore, based on the deep learning neural network framework, the advantages of GCN and LSTM are utilized to build the typhoon GC–LSTM model to effectively process the problems in the construction of satellite cloud picture models. The proposed model is significantly better than other prediction models in terms of algorithm performance. Compared with the traditional ANN model, it has improved by 13.3%. The prediction accuracy of typhoons and super typhoons through optimization of parameters has reached 95.12%. It can accurately identify typhoon eyes and spiral cloud belts, and its stability is better. This model can provide a theoretical basis for typhoon prediction related research. Although the construction process and the actual application effect of the model have been elaborated on as much as possible, due to the objective limitations, the following deficiencies are found: (1) only one GCN image processing neural network is used, and no other image legend recognition algorithm is used for processing; (2) for the actual application of the model, only the prediction effect of the model is comprehensively evaluated, but actual application analysis with a large amount of data is not involved. In the future, in-depth research will be continued in these two areas, with a view to truly applying this satellite cloud picture-based typhoon prediction model to actual analysis, thereby reducing the impact of typhoons on people's lives and property.

Author Contributions: Conceptualization, J.Z. and J.X.; methodology, S.H.; software, J.Z.; validation, J.Z., J.X. and S.H.; formal analysis, J.Z.; investigation, J.Z.; resources, J.Z.; data curation, J.Z.; writing—original draft preparation, J.Z.; writing—review and editing, J.Z.; visualization, J.Z.; supervision, S.H.; project administration, J.Z.; funding acquisition, J.X. All authors have read and agreed to the published version of the manuscript.

Funding: Foundation for Innovative Research Groups of the National Natural Science Foundation of China: 41475021, 91730304, 41575026.

References

1. Uson, M.A.M. Natural disasters and land grabs:The politics of their intersection in the Philippines following super typhoon Haiyan. *Can. J. Dev. Stud. Rev. Can. D'études Dév.* **2017**, *38*, 414–430. [CrossRef]
2. Chen, L.L.; Tseng, C.H.; Shih, Y.H. Climate-related economic losses in Taiwan. *Int. J. Glob. Warm.* **2017**, *11*, 449–463. [CrossRef]
3. Ding, X.; Chen, Y.; Pan, Y.; Reeve, D. Fast ensemble forecast of storm surge along the coast of China. *J. Coast. Res.* **2016**, *75*, 1077–1081. [CrossRef]
4. Corcione, V.; Nunziata, F.; Migliaccio, M. Megi typhoon monitoring by X-band synthetic aperture radar measurements. *IEEE J. Ocean. Eng.* **2017**, *43*, 184–194. [CrossRef]
5. Rüttgers, M.; Lee, S.; Jeon, S.; You, D. Prediction of a typhoon track using a generative adversarial network and satellite images. *Sci. Rep.* **2019**, *9*, 1–15. [CrossRef]
6. Su, X. Using Deep Learning Model for Meteorological Satellite Cloud Image Prediction. Available online: https://ui.adsabs.harvard.edu/abs/2017AGUFMIN13B0064S/abstract (accessed on 9 September 2020).
7. Zhao, L.; Chen, Y.; Sheng, V.S. A real-time typhoon eye detection method based on deep learning for meteorological information forensics. *J. Real Time Image Process.* **2020**, *17*, 95–102. [CrossRef]
8. Liou, Y.A.; Liu, J.C.; Liu, C.P.; Liu, C.C. Season-dependent distributions and profiles of seven super-typhoons (2014) in the Northwestern Pacific Ocean from satellite cloud images. *IEEE Trans. Geosci. Remote Sens.* **2018**, *56*, 2949–2957. [CrossRef]

9.   Chen, S.T. Probabilistic forecasting of coastal wave height during typhoon warning period using machine learning methods. *J. Hydroinform.* **2019**, *21*, 343–358. [CrossRef]

10.   Olander, T.L.; Velden, C.S. The Advanced Dvorak Technique (ADT) for estimating tropical cyclone intensity: Update and new capabilities. *Weather Forecast.* **2019**, *34*, 905–922. [CrossRef]

11.   Lu, J.; Feng, T.; Li, J.; Cai, Z.; Xu, X.; Li, L.; Li, J. Impact of assimilating Himawari-8-derived layered precipitable water with varying cumulus and microphysics parameterization schemes on the simulation of Typhoon Hato. *J. Geophys. Res. Atmos.* **2019**, *124*, 3050–3071. [CrossRef]

12.   Gao, S.; Zhao, P.; Pan, B.; Li, Y.; Zhou, M.; Xu, J.; Zhong, S.; Shi, Z. A nowcasting model for the prediction of typhoon tracks based on a long short term memory neural network. *Acta Oceanol. Sin.* **2018**, *37*, 8–12. [CrossRef]

13.   Krapivin, V.F.; Soldatov, V.Y.; Varotsos, C.A.; Cracknell, A.P. An adaptive information technology for the operative diagnostics of the tropical cyclones; solar–terrestrial coupling mechanisms. *J. Atmos. Sol. Terr. Phys.* **2012**, *89*, 83–89. [CrossRef]

14.   Varotsos, C.A.; Krapivin, V.F.; Soldatov, V.Y. Monitoring and forecasting of tropical cyclones: A new information-modeling tool to reduce the risk. *Int. J. Disaster Risk Reduct.* **2019**, *36*, 101088. [CrossRef]

15.   Zhu, Z.; Peng, G.; Chen, Y.; Gao, H. A convolutional neural network based on a capsule network with strong generalization for bearing fault diagnosis. *Neurocomputing* **2019**, *323*, 62–75. [CrossRef]

16.   Scher, S. Toward data-driven weather and climate forecasting: Approximating a simple general circulation model with deep learning. *Geophys. Res. Lett.* **2018**, *45*, 616–622. [CrossRef]

17.   Kaba, K.; Sarıgül, M.; Avcı, M.; Kandırmaz, H.M. Estimation of daily global solar radiation using deep learning model. *Energy* **2018**, *162*, 126–135. [CrossRef]

18.   Xiao, C.; Chen, N.; Hu, C.; Wang, K.; Xu, Z.; Cai, Y.; Xu, L.; Chen, Z.; Gong, J. A spatiotemporal deep learning model for sea surface temperature field prediction using time-series satellite data. *Environ. Model. Softw.* **2019**, *120*, 104502–104521. [CrossRef]

19.   Wang, H.; Shao, N.; Ran, Y. Identification of Precipitation-Clouds Based on the Dual-Polarization Doppler Weather Radar Echoes Using Deep–Learning Method. *IEEE Access* **2018**, *7*, 12822–12831. [CrossRef]

20.   Devika, G.; Ilayaraja, M.; Shankar, K. Optimal Radial Basis Neural Network (ORB-NN) For Effective Classification of Clouds in Satellite Images with Features. *Int. J. Pure Appl. Math.* **2017**, *116*, 309–329.

21.   Chang, P.; Grinband, J.; Weinberg, B.; Bardis, M.; Khy, M.; Cadena, G.; Su, M.Y.; Cha, S.; Filippi, C.; Bota, D. Deep-learning convolutional neural networks accurately classify genetic mutations in gliomas. *Am. J. Neuroradiol.* **2018**, *39*, 1201–1207. [CrossRef] [PubMed]

22.   Winkler, J.K.; Fink, C.; Toberer, F.; Enk, A.; Deinlein, T.; Hofmann-Wellenhof, R.; Thomas, L.; Lallas, A.; Blum, A.; Stolz, W. Association between surgical skin markings in dermoscopic images and diagnostic performance of a deep learning convolutional neural network for melanoma recognition. *JAMA Dermatol.* **2019**, *155*, 1135–1141. [CrossRef] [PubMed]

23.   Zhou, J.; Cui, G.; Zhang, Z.; Yang, C.; Liu, Z.; Wang, L.; Li, C.; Sun, M. Graph neural networks: A review of methods and applications. *arXiv* **2018**, arXiv:1812.08434.

24.   Fischer, T.; Krauss, C. Deep learning with long short-term memory networks for financial market predictions. *Eur. J. Oper. Res.* **2018**, *270*, 654–669. [CrossRef]

25.   Zhou, L.; Zhang, Z.; Chen, Y.C.; Zhao, Z.Y.; Yin, X.D.; Jiang, H.B. A deep learning-based radiomics model for differentiating benign and malignant renal tumors. *Transl. Oncol.* **2019**, *12*, 292–300. [CrossRef]

26.   Grabler, P.; Sighoko, D.; Wang, L.; Allgood, K.; Ansell, D. Recall and cancer detection rates for screening mammography: Finding the sweet spot. *Am. J. Roentgenol.* **2017**, *208*, 208–213. [CrossRef]

27.   Lee, H.J.; Ullah, I.; Wan, W.; Gao, Y.; Fang, Z. Real-time vehicle make and model recognition with the residual SqueezeNet architecture. *Sensors* **2019**, *19*, 982. [CrossRef]

28.   Chattopadhyay, A.; Hassanzadeh, P.; Subramanian, D.; Palem, K. Data-Driven prediction of a multi-scale Lorenz 96 chaotic system using a hierarchy of deep learning methods: Reservoir computing, ANN, and RNN-LSTM. *EarthArXiv* **2019**, *21*, 654–660.

29.   Lian, J.; Dong, P.; Zhang, Y.; Pan, J. A Novel Deep Learning Approach for Tropical Cyclone Track Prediction Based on Auto-Encoder and Gated Recurrent Unit Networks. *Appl. Sci.* **2020**, *10*, 3965. [CrossRef]

30. Heming, J.T.; Prates, F.; Bender, M.A.; Bowyer, R.; Cangialosi, J.; Caroff, P.; Coleman, T.; Doyle, J.D.; Dube, A.; Faure, G. Review of recent progress in tropical cyclone track forecasting and expression of uncertainties. *Trop. Cyclone Res. Rev.* **2019**, *8*, 181–218. [CrossRef]
31. Ouyang, H.-T. Input optimization of ANFIS typhoon inundation forecast models using a Multi-Objective Genetic Algorithm. *J. Hydro Environ. Res.* **2018**, *19*, 16–27. [CrossRef]

Article

NCC Based Correspondence Problem for First- and Second-Order Graph Matching[†]

Beibei Cui [1,2,*] and Jean-Charles Créput [2]

[1] College of Electrical Engineering, Henan University of Technology, Zhengzhou 450001, Henan, China

[2] CIAD, University Bourgogne Franche-Comté, UTBM, 90010 Belfort, France; jean-charles.creput@utbm.fr

[*] Correspondence: beibei_cui@sina.com

[†] This paper is an extended version of our paper published in Cui, B.; Créput, J.-C. Using Entropy and Marr Wavelets to Automatic Feature Detection for Image Matching. In Proceedings of the 2019 15th International Conference on Signal-Image Technology & Internet-Based Systems, Sorrento, Italy, 26–29 November 2019.

Received: 20 July 2020; Accepted: 1 September 2020; Published: 8 September 2020

Abstract: Automatically finding correspondences between object features in images is of main interest for several applications, as object detection and tracking, identification, registration, and many derived tasks. In this paper, we address feature correspondence within the general framework of graph matching optimization and with the principal aim to contribute. We proposed two optimized algorithms: first-order and second-order for graph matching. On the one hand, a first-order normalized cross-correlation (NCC) based graph matching algorithm using entropy and response through Marr wavelets within the scale-interaction method is proposed. First, we proposed a new automatic feature detection processing by using Marr wavelets within the scale-interaction method. Second, feature extraction is executed under the mesh division strategy and entropy algorithm, accompanied by the assessment of the distribution criterion. Image matching is achieved by the nearest neighbor search with normalized cross-correlation similarity measurement to perform coarse matching on feature points set. As to the matching points filtering part, the Random Sample Consensus Algorithm (RANSAC) removes outliers correspondences. One the other hand, a second-order NCC based graph matching algorithm is presented. This algorithm is an integer quadratic programming (IQP) graph matching problem, which is implemented in Matlab. It allows developing and comparing many algorithms based on a common evaluation platform, sharing input data, and a customizable affinity matrix and matching list of candidate solution pairs as input data. Experimental results demonstrate the improvements of these algorithms concerning matching recall and accuracy compared with other algorithms.

Keywords: normalized cross-correlation; Marr wavelets; entropy and response; graph matching; RANSAC

---

1. Introduction

Computer vision is an important research direction in current computer science since it occupies a pivotal position in human perception simulation. Automatically finding correspondences between object features in images is of interest for several applications, as target tracking [1], 3D object retrieval [2], pattern recognition [3], image stitching [4] and in many other fields.

Image matching is used to determine the geometric alignment of two or more images of the same scene taken by the same or different sensors from different viewpoints at the same or different times. We can distinguish dense correspondence, that determines correspondences at the pixel level, and sparse correspondence, that determines correspondences between a sparse set of higher lever features being first extracting from images. Most of the time such features represent invariant information at some location in the image, like corners, edges, gradients. Since we are interested in sparse correspondence, we study the standard methods for automatic extraction of the feature point sets from images. This becomes a critical step since it should avoid the presence of outliers and should allow discriminating objects easily. We propose a method for feature extraction step suitable to first-order matching.

Local feature descriptors, that is, providing detail feature detection and feature description information, play a fundamental and vital role in the process of feature correspondence, directly affecting the accuracy and objective score of graph matching (GM). High-quality local feature descriptors describe key points with uniqueness, repeatability, accuracy, compactness, and effective representation. These key points can keep robust and constant in terms of scaling, rotation, affine transformation, illumination, and occlusion [5]. Here we focus on the theoretical and mathematical descriptions of various local feature descriptors. In the feature points detection algorithm, local descriptors are typically used to describe image regions near feature points. Currently, methods for extracting feature points include Harris descriptor [6], Gilles descriptor [7], LoG descriptor [8], corner detector (CD) [9], Harrislaplace descriptor [10], SIFT descriptor [11], and so on. Among these feature extraction algorithms proposed in the literature, Marr wavelets which was originally used in [12] is favored for several properties: robustness (against distortion), rotationally invariant, noise insensitivity [13]. We choose to focus on the latter, whereas other methods will serve as a basis for comparative evaluation.

Given a pair of images, how to detect and extract feature points is the first step in image matching. Automatic feature extraction is the key point, and further related image matching can be performed based on the feature-to-feature correspondence. Given some standard nearest neighbor matching strategies, how to improve the reliability of the feature set is the problem to be solved here. To this end, we try to combine or enhance standard and easy-to-implement feature detection methods to make the final overall method (including feature detection and matching) competitive in terms of computation time and matching quality.

In order to obtain better matching results, the method proposed inserts a Laplace filter-based image preprocessing method before detecting feature points to increase the size of the candidate feature set. Since the Marr wavelet within scale-interaction method is more inclined to extract the edge information of the object, the Laplacian method can be used to enhance the edge details of the image. Then, a sparse feature point method based on entropy selection is proposed as a new filtering step. Filtering (also known as convolution) is also a very popular operation in the field of image processing, which can be applied to image encryption by changing pixel values [14]. This step combines the local entropy evaluation with the brightness deviation response as a new process for feature selection.

Entropy is a key concept in thermodynamics and statistical mechanics. It not only plays a special role in physical quantities, but also relates to the macro and micro aspects of nature, and determines the behavior of the macro system. Entropy is a well-defined quantity regardless of the type or size of the system under consideration. Entropy has many general properties, for instance, invariance, continuity, additivity, concavity, etc. [15].The probability distribution of entropy can be interpreted not only as a measure of uncertainty, but also as a measure of information [16]. Local entropy represents structured information, which is used to count the probability of occurrence of gray level in the sub-image, but independently around a single pixel. We claim that it is not sensitive to the influence of noise and can improve the accuracy of the image description. Based on the mesh division, the feature points in the sub-regions can be sorted according to their local entropy and selected trough deviation values. Then, entropy selection can not only effectively reduce the useless feature points for saving time, but also ensure the uniformity of feature point distribution. Mainly, the advantage of the proposed entropy and response algorithm is to realize a good compromise (trade-off) between accuracy and computation time together when compared to other standard approaches from literature. It is worth noting that both accuracy and computation time are essential criteria to compare and evaluate heuristic methods [17,18].

A necessary graph matching procedure based on the normalized cross-correlation similarity measure is applied to measure the effectiveness of the method in image matching. It entrusts quality assessment to the Random Sample Consensus Algorithm (RANSAC) program, which eliminates mismatched pairs and calculates true match recalls. The experiments were performed on standard image processing benchmarks. They showed how to increase the size of the feature set and matching accuracy with saving computation time.

The contributions of our paper are proposing two optimized algorithms: first-order and second-order for graph matching. Both of them are realized based on normalized cross-correlation (NCC) algorithm, which allows us to address graph matching or derived sub-problems with a closer relationship with experiments on integer quadratic programming (IQP) models in the Matlab platform. Especially for the first-order graph matching, we have proposed a new combination of feature points detection algorithms among Laplace filter, Marr wavelets, and the entropy-response based selection method.

This paper is organized as follows. Section 2 introduces the motivations and taxonomy of graph matching. The formulation of standard graph matching is explained in Section 3. In Section 4, the different steps of the proposed feature extraction and first-order graph matching procedures are respectively presented. A preliminary version of this section forward appears as part of our previous conference paper [19]. We also extended the NCC based second-order graph matching in Section 5. The evaluation of the two proposed algorithms and their comparison with some algorithms are exposed in Section 6. Section 7 presents the conclusion.

2. Background of Graph Matching

2.1. Motivations of Graph Matching

Graph matching plays an important role in processing many practical applications in computer vision, such as feature correspondence [20], action recognition [21], image classification [22,23], shape matching [24], image retrieval [25], and pattern recognition [26,27]. The goal of graph matching is to find the optimal mapping constraint between two sets of nodes in two corresponding images, which will preserve the relationship between the graphs as much as possible so that when vertices are labeled based on the correspondence, they look 'the most similar'.

In a more general case, this problem is expressed mathematically as a quadratic distribution problem, including finding a distribution that maximizes the objective function. Over the past decade, considerable effort has been invested in developing approximate methods to address more general QAP. Gold and Rangarajan [28] proposed a graduated assignment algorithm that combines graduated nonconvexity, two-way (assignment) constraints, and sparsity to solve a series of linear approximations to the cost function using Taylor expansion iterations. Leordeanu [29] presented an efficient approximation by using the spectral matching method (SM), which approximates the IQP problem to spectral relaxation. Cour et al. [17] introduced a new spectral matching with affine constraints (SMAC), which can not only provide a higher relaxation than SM but also keep the speed and scalability benefits of SM. Torresani et al. [20] designed a complex objective function which refers to dual decomposition (DD) that can be effectively optimized by double decomposition. Cho and Lee [18] presented reweighted random walk (RRWM) algorithms for graph matching. Later, Cho et al. [30] provided a max-pooling graph matching method (MPM), which not only resists deformation but also significantly tolerates outliers. This central idea of this algorithm is that the pairs with maximum scores are the correct matches.

Recently, some authors have proposed the use of high-dimensional relationships between super edges for high-order graph matching. The most commonly used is based on third-order research. The calculation of the high-order affinity matrix is generally based on the tuples of feature points, and it is achieved by comparing the corresponding edges and angle information of two sets of corresponding triangles. Another vital characteristic of high-order matching is that it is invariant to changes in scale and affine. Zass and Shashua [31] reformulated a high-order graph matching (HGM) in the view of probabilistic view of the probability setting of convex optimization representation. Chertok and Keller [32] proposed a general framework for solving higher-order assignment problems based on the core assumption that high order affinities are encoded in an affinity tensor. In this algorithm, they derive a marginalization scheme that can map triples to matrices or vectors. Olivier Duchenne et al. [33] derived a tensor-based high-order graph matching (TM) that invariant to affine, rigid, or transformations. This algorithm defined a tensor to represent the affinity assignment between the tuples of features. It is a multidimensional power iteration operation during which the solution will be projected onto the closest assignment matrix. Lee et al. [34] demonstrated a hypergraph matching via reweighted random walks (RRWHM) in a probabilistic manner. The algorithm uses a personalized jump and re-weighting scheme, which effectively reflects the one-to-one matching constraints in the random walk process. It can realize strong anti-deformation and anti-noise performance compared with other state-of-the-art methods. Ngoc [35] presented a general framework with a flexible tensor block coordinate ascent scheme for hypergraph matching. It is a crucial idea under a multilinear reconstruction using the original objective function, which can guarantee the third-order matching scores their algorithms increase monotonically. Another hypergraph matching algorithm based on tensor refining was proposed in [36], accompanied by an alternative adjustment approach to accelerate the convergence processing.

Despite decades of extensive research, graph matching is still a challenging problem for two main reasons: (1) generally, the objective function is non-convex and prone to local minima. (2) The constraints of space and time complexities. We set out to conduct research to solve these challenges.

2.2. Taxonomy of Graph Matching

The goal of the graph matching process is mainly focused on finding the correspondences between two characteristics, edges, and points, under certain constraints. A graph-based matching method treats a set of points as graphics. In the most common cases, the algorithm used to solve inexact matching problems can be classified into three categories: first-order, second-order, and high-order graph matching methods.

- First-order graph matching At the view of first-order graph matching, it mainly implicates vertex-to-vertex properties based on local feature descriptors, focuses primarily on unary information, regardless of edge-related associated information. This matching method was proposed initially to find the correspondence between two sets of points and transformation parameters at the same time. It uses a coordinate positioning and grayscale information to calculate the transformation parameters and uses a soft assignment algorithm to estimate the correspondence between two sets of points. Finally, it converts the alignment of the two-point sets into the optimal match between the two graphs. Although the result of the first-order matching is stable, it fails when there is ambiguity, such as local appearance and repeating texture. In the case of high noise and outlier registration, it performs poorly, which dramatically limits its range of applications. Therefore, first-order matching is more suitable for focusing on non-rigid motion types with a small local affine transformation.

- Second-order graph matching Second-order graph matching mainly enriches the vertex, and edge approaches; it is objective function established by a matrix representing the affinity properties between candidate pairs, that is to say, each node represents the correspondence between points, and the weights represent pairwise protocols between the corresponding potential pairs. A secure connection of the adjacency matrix can judge the correct assignment or not. Therefore, it better stands for some point of view between candidate pairs and overcomes the disadvantages of the first-order algorithm. Since the graph matching algorithm is usually based on the integer quadratic programming (IQP) formula, with an approximate solution, so the second-order graph matching problem is also the NP-hard problem. That means it can be formulated as an optimization problem; the purpose is to get the best match and receive a higher score based on the objective scoring function.

- High-order graph matching Hypergraph is a natural generalization of traditional graphs. Since pairwise assignments are sensitive to scale-invariant between two corresponding graphs, pairwise relationships are not enough to capture the entire geometrical structure. Unlike pairwise matching in which each link can have two vertices at its ends, each link in the hypergraph can have three or more vertices, which can have a more powerful tool to model complex structures for more high-level information. Therefore, the most crucial idea to solve the hypergraph problem is to search for higher-order constraints, rather than unary or pairwise constraints. Essentially, hypergraph matching is a combinatorial optimization problem. In the process of obtaining the final solution, it is not straightforward for us to find its optimal global solution based on a reasonable time. In recent years, more accessible methods use probability frames to explain hypergraph matching, of which tensor-based models are often used. Considering the complex structure of the data, we believe that the learning and construction of hypergraphs will hopefully become an increasingly promising research direction in the future.

3. General Formulation of Graph Matching Problem

This section introduces the general representation of traditional graph matching and the definition of the high-order graph matching problem. The main purpose is on finding the correspondence one-to-one mapping between two feature sets from two image sources. The goal is to maximize a function score among the set of correspondence pairs. In first-order matching, only local attribute descriptors are considered and evaluated, whereas, in the general case of graph matching, two order potentials between pairs (edges) of features must also be maximized to established the similarity between edges of features. On the other hand, the high-order GM method considers the invariant geometric information by considering the relationship between tuples of feature points. The input feature graph becomes a hyper-graph, where hyper-edges replace edges, that is subsets of k points, with the order $k \geq 2$, rather than only considering couple of points.

Suppose we are given a pair of graphs $G_P = (P, E^P)$ with N_P feature points for the reference graph G_P, and $G_Q = (Q, E^Q)$ with N_Q feature points for the query graph G_Q. P and Q are the two sets of feature points, and E^P and E^Q denote edge sets. We note $i, j \in P$ and $a, b \in Q$ as representing feature points. Therefore, the main problem is to find a suitable one-to-one mapping from one feature set to the other feature set as illustrated in the Figure 1. The pictures in Figure 1 are from PF-WILLOW dataset (https://www.di.ens.fr/willow/research/proposalflow/).

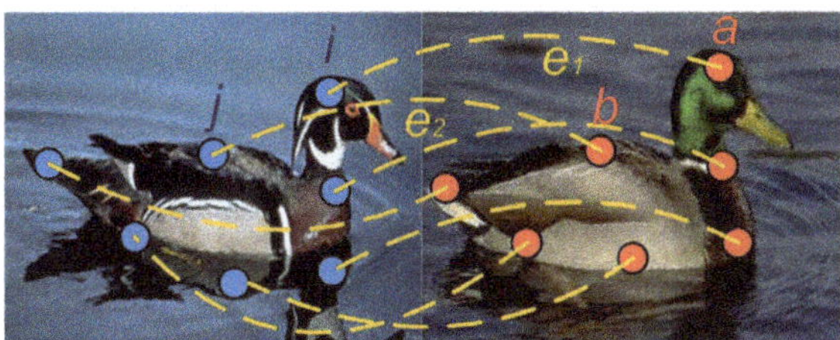

Figure 1. Feature point correspondence mapping.

Finding a mapping form P to Q can be equivalent to find an $N_P \times N_Q$ assignment matrix X, such that $X_{ia} = 1$ when point i is assigned to point a, and $X_{ia} = 0$ otherwise. Therefore, a one-to-one admissible solution must verify the following constraints in (1), that requires a binary solution, and (2) and (3), that express the two-ways constraints of a one-to-one mapping.

$$X \in \{0, 1\}^{N_P \times N_Q} \tag{1}$$

$$\forall i \sum_{a=1}^{N_P} X_{ia} \leq 1 \tag{2}$$

$$\forall a \sum_{i=1}^{N_Q} X_{ia} \leq 1 \tag{3}$$

Then, the problem of graph matching can be formulated as the maximization of the following general objective score function (4):

$$score(X) = \sum_{ia,jb} H_{ia,jb} X_{ia} X_{jb}, \tag{4}$$

where $H_{ia,jb}$ means the similarity or affinity measurement corresponding to the tuple of feature points i, j and a, b. The higher is the score $H_{ia,jb}$, the higher are the similarities between the two corresponding edges (i, j) and (a, b). The product $X_{ia} X_{jb}$ is equal to 1 if and only if points i, j are respectively mapped to points a, b.

Then, we need to know how to compute such a positive and symmetric similarity matrix H. Many cost functions may be used to compute affinity matrices for first-order and second-order GM. Note that $H_{ia,ia}$ represents first-order similarity term, between local attributes of points $i \in P$ and $a \in Q$. For example, the authors in [37,38] use the normalized cross-correlation (NCC) cost function, as we have used to validate our feature point extraction method in this thesis. Nevertheless, any point-to-point distance function can be used, as the Euclidean distance between SIFT descriptors, or sum of squared error data terms.

Duchenne et al. in [33] propose a general formula to compute the second-order affinity term $H_{ia,jb}$ as shown in (5), where f is a feature vector associated to each edge.

$$\forall ia, jb \; H_{ia,jb} = exp(-\gamma \|f_{i,j} - f_{a,b}\|^2) \tag{5}$$

Leordeanu et al. in [29], and as most often encountered, computes the Euclidean distance between the corresponding candidate point pairs i, j and a, b, to build the affinity term $H_{ia,jb}$, as shown in Equation (6). Here, σ_d is the sensitive controller of the deformation.

$$H(ia, jb) = \begin{cases} 4.5 - \dfrac{(d_{ij} - d_{ab})^2}{2\sigma_d^2} & if |d_{ij} - d_{ab}| < 3\sigma_d \\ 0 & otherwise \end{cases} \tag{6}$$

4. First-Order NCC Based GM

In this section, we introduce the implementation of the system image matching process, including feature point detection and extraction, feature point matching, and removing mismatching pairs. The schematic diagram of the whole process is shown in Figure 2. Regarding the interest point matching strategy, the proposed pipeline illustrated in Figure 2 can be summarized as (1) Laplacian filter is used for edge detection; (2) Marr wavelets are used to identify salient points in the image; (3) entropy based metric for selecting the most distinctive points of the image denoted as feature points; (4) feature points matching; (5) outlier removal through RANSAC.

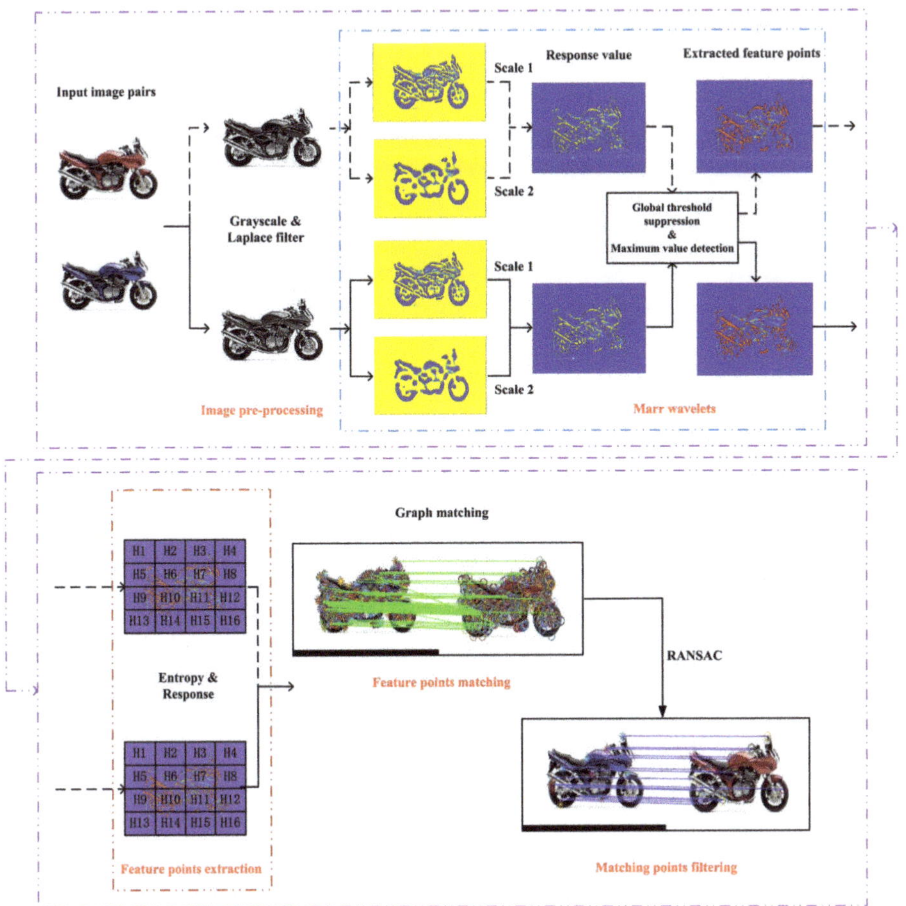

Figure 2. The basic flowchart of graph matching.

4.1. Image Pre-Processing

Laplacian is a second-order derivative operator that detects the zero-crossing in image intensity and usually produces more accurate edge detection results [39]. Laplace filter represents a discrete approximation to the mathematical Laplace operator. Its second-order partial derivative in the orthogonal direction of continuous space and the approximation of its mathematical equivalent are defined below [40]:

$$\nabla^2 f(x,y) = \frac{\partial^2 f(x,y)}{\partial x^2} + \frac{\partial^2 f(x,y)}{\partial y^2} \tag{7}$$

$$\nabla^2 f(x,y) \cong \{f(x+1,y) + f(x-1,y) + f(x,y+1) + f(x,y-1)\} - 4f(x,y) \tag{8}$$

$$I(X) = I(x,y) = f(x,y) - c \cdot \nabla^2 f(x,y) \tag{9}$$

where $f(x,y)$ is the original image, $I(x,y)$ is the processed image, c is a constant.

From formula (8), the digital mask filter w can be viewed as the following 3×3 set of filter coefficients as shown in Figure 3a. The process of the Laplacian filter sharpening is essentially a convolution process. Suppose the origin pixel of f is located in the upper left corner of the image f, and set the middle value of mask w as the center of kernel. Let w move at all possible positions so that the center kernel of w can coincide with each of pixels of f. The convolution operation is essentially the sum of the products of the corresponding positions of the two functions. The convolution between f and its corresponding mask filter w is shown in Figure 3b.

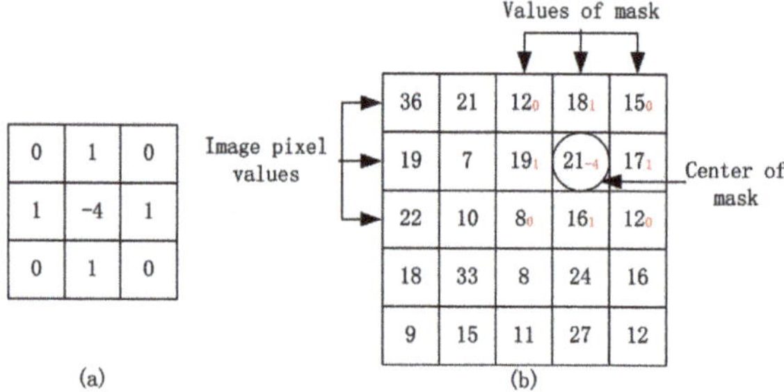

Figure 3. The process of the Laplacian filter sharpening: (**a**) the digital mask filter w, (**b**) the convolution operation.

The definition of two-dimensional convolution is as:

$$I(x,y) = f * w = \sum_{k,l} f(x+k, y+l) w(k,l) \tag{10}$$

4.2. Marr Wavelets within Scale-Interaction

Receptive field [41] is used to describe the stimulation pattern of the retina. The receiving field of high-level neuron cells in the visual pathway is synthesized from the receiving field of low-level neuron cells. Therefore, as the level increases, the range of the receptive field becomes more impressive. Ultra-complex neuron cell models [42] can respond to complex object features through powerful non-linear processing functions, and most ultra-complex neuron cells have sensitive termination characteristics at the ends of line segments, corner points, and line segments with high curvature. In other words, the response of ultra-complex neuron cells to light can be simulated by the difference in response of spatial filters with different bandwidths to light. The response of the received field to light can be represented by a spatial filter function, such as a Gaussian difference function or a Gabor wavelet function.

The scale-interaction model for feature detection is based on filtering using a class of self-similar Gabor functions or Gabor wavelets [43,44], which can achieve the minimum joint resolution in the spatial and frequency domains. This recommendation is made because it is unique in reaching the smallest possible value of the joint uncertainty [45]. The function of feature detection is defined as the following formula:

$$Q_{ij}(x,y,\theta) = f(W_i(x,y,\theta) - \gamma W_j(x,y,\theta)) \tag{11}$$

where γ is a normalizing factor, $W_i(x,y,\theta)$ and $W_j(x,y,\theta)$ are spatial filters. They go through the transformation of a nonlinear function f at location (x,y) with preferred orientation θ in two scales i and j respectively. If feature detection function Q_{ij} obtains a local maximum at the location (x,y), this location is considered to be a potential feature point position.

For further optimization, the Marr wavelets [46] were used instead of Gabor wavelets within the scale-interaction model, because of its isotropic [47,48]. Two-dimensional Marr wavelets and their corresponding feature detection function are defined as:

$$M_i(X) = \lambda_i(2 - \lambda_i^2 X^2) exp(-\frac{\lambda_i^2 X^2}{2}) \tag{12}$$

$$Q_{ij}(X) = |M_i(X) - \gamma M_j(X)| \tag{13}$$

$$R_{ij}(X) = I(X) * Q_{ij}(X) \tag{14}$$

where $X = (x, y)$, and $X^2 = (x^2 + y^2)$. $\lambda_i = 2^{-i}$ and i represents the scaling value of Marr wavelets. Convolve $Q_{ij}(X)$ with an grayscale image $I(X)$. If its response value $R_{ij}(X)$ obtains a maximum local value, then X is considered to be a potential feature point. Algorithm 1 is pseudo-code of Marr wavelets within scale-interaction. Figure 4 illustrates the process of extracting response value using Marr wavelets within scale-interaction. (b) and (c) are convolution results between the original picture and the mask filter. Then local maximum points are extracted on this basis.

Algorithm 1 Marr wavelets within scale-interaction algorithm

Input: $I(X), i, j$
Output: points
1: $I_i = MarrFilter(I(X), i)$
2: $I_j = MarrFilter(I(X), j)$
3: $I_{sub} \leftarrow |I_i - I_j|$
4: $local_{thr} \leftarrow max(max(I_{sub})) * r$
5: **if** $I_{sub}(i, j) < local_{thr}$ **then**
6: $\quad I_{localthr}(i, j) \leftarrow I_{sub}(i, j) \leftarrow 0$
7: **end if**
8: $points \leftarrow corner_{peaks} \leftarrow I_{localthr}$
9: **function** $MarrFilter(I(X), scale)$
10: $\quad \delta = 2^{scale}$
11: $\quad x = -(2 * fix(\delta)) : 1 : (2 * fix(\delta))$
12: $\quad y = -(2 * fix(2 * \delta)) : 1 : (2 * fix(2 * \delta))$
13: $\quad M_i(X) \leftarrow X \leftarrow meshgrid(x, y)$
14: $\quad I_{fil} \leftarrow I(X) * M_i(X)$
15: **end function**

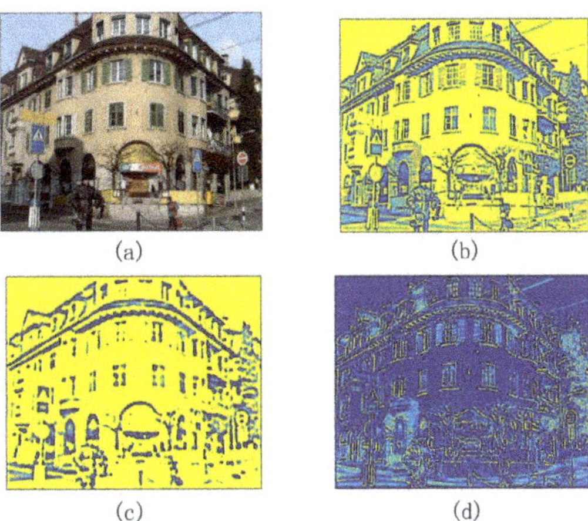

Figure 4. From the left to the right: (**a**) original image, (**b**) filtered result when $i = 1$, (**c**) filtered result when $i = 2$ and (**d**) response image.

4.3. Entropy and Response

Aiming at the problems of uneven feature distribution, too many feature points, and long matching time, a feature point extraction method based on local entropy and feature point response was proposed, called the entropy and response algorithm (ER). In this section, three main parts are explained.

4.3.1. Entropy Algorithm

In actual images, feature points often appear as sharp changes of gray values or inhomogeneity in grayscale distribution; that is, the local region of a feature point has a large amount of information. Entropy is a measure of information in an image, and local entropy is a measure of local area information of an image. Local entropy value under feature-rich region is much higher than the local entropy value under feature-poor region. Therefore, it is possible to determine which regions have more features by calculating the local information entropy of image, and then extract the feature points in these regions.

Information entropy [49,50] is the amount that represents the overall characteristics of the source in a common sense. It is considered from the statistical properties of the entire source to measure the expected value of a random variable. An image is essentially a source of information that can be described by information entropy. Let the gray image G have m gray levels, mesh division is performed to obtain $n \times n$ sub-regions. The whole information entropy H_i and the average entropy $\bar{H}$ of the image are calculated as follows:

$$H_i = \sum_{i=1}^{m} p_i log_2 p_i \tag{15}$$

$$\bar{H} = \frac{1}{n^2} \sum_{j=1}^{n^2} H_j \tag{16}$$

where p_i is the probability that the i_{th} gray level appears, that is, the ratio of the number of pixels whose gray value is i to the total number of pixels of the image. So the local entropy is counted for the probability of occurrence of gray level in the sub-image. Since the value of the information entropy is only related to the distribution of the local gray-scale pixels, but independent with a single pixel, so it is not sensitive to the influence of noise and can improve the accuracy of the image authenticity description. Here, we use the local information entropy of the image to extract feature points. Under the meshing strategy, the image is divided into $n \times n$ sub-regions. Therefore, the sub-average entropy value of each sub-grid can be calculated. Figure 5 is a schematic diagram of mesh division, n is set to 40.

(a) (b)

Figure 5. Schematic diagram of meshing: (**a**) image divided by 4×4 sub-regions, (**b**) image divided by 8×8 sub-regions.

4.3.2. Response Algorithm

After mesh division and the computation of each local entropy, we will get $n \times n$ sub-regions for the whole image, then detected feature points are mapped into the respective sub-areas. In this case, if we compute and sort the entropy values of all of these feature points extracted, then the first N feature points with larger entropy values can be selected to describe the whole image. However, this method only utilizes the entropy values of feature points, without considering its distribution in the image. Finally, feature points with high entropy values may mostly appear in the same local area, which will cause aggregation. So a block division and response algorithm are proposed to deal with this problem.

As mentioned before, if a pixel presents a sharp change in its neighborhood, this pixel will have a stronger deviation value from the mean value. Based on the Bresenham discrete circle [51] with the pixel point p_i as the center and three pixels as the radius, 16-pixel points on the discrete circumference are considered in correspondence with the central pixel point p_i. This is shown in Figure 6. These 16 pixels are assigned to dark and bright areas. The dividing criteria and deviation [52] are respectively defined as the following:

$$S_{bright} = \{x | I_{p_i,x} > I_{p_i} + t\} \tag{17}$$

$$S_{dark} = \{x | I_{p_i,x} \leq I_{p_i} - t\} \tag{18}$$

$$Dev = max(\sum_{x \in S_{bright}} |I_{p_i,x} - I_{p_i}| - t, \sum_{x \in S_{dark}} |I_{p_i} - I_{p_i,x}| - t) \tag{19}$$

where S_{bright} indicates bright area, S_{dark} indicates dark area. I_{p_i} is the gray value of center point p_i, $I_{p_i,x}$ represents the gray value of a pixel labeled x on a discrete circumference centered at pixel p_i, t is the set threshold. Dev is the sum of the deviation value among the gray values of the pixel p_i and its corresponding neighboring pixels located in the bright or dark area.

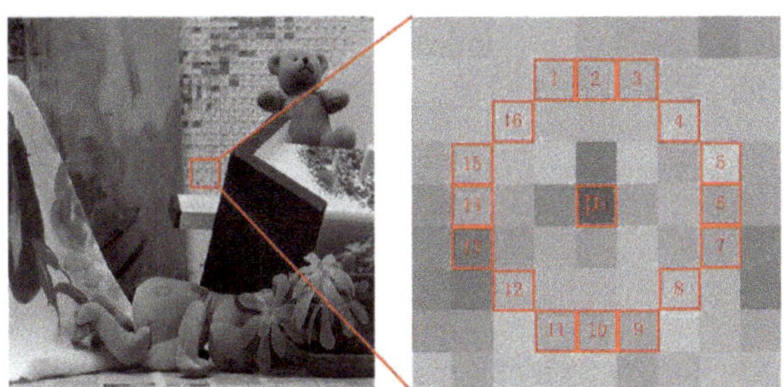

Figure 6. Bresenham discrete circle centered on pixel p_i.

4.3.3. Distribution Criterion

After the calculation of the entropy and response, a distribution criterion is proposed to extract the corresponding points that meet the requirements. In the region where the local entropy is bigger than the average of entropy $\bar{H}$, the feature points are extracted with the ratio r. The only one strongest response point is extracted in each remaining region where the local entropy is smaller than $\bar{H}$. Here, we choose the unified ratio method for the selection of r. Assuming that there are m regions whose local entropy is greater than $\bar{H}$, then $r_i(1, 2, ..., m)$ is the ratio of extracting feature points in the i_{th} region. For example, when $r = 10\%$, that is, in the m_{th} region with large local entropy, the feature points with the top 10% responses given by Dev are extracted. The appropriate value of r is set empirically.

The detail of the entire ER method is outlined in Algorithm 2. Based on the mesh division strategy, we first compute each of the local average entropy value of all these sub-regions. Then the feature points are sorted according to the computation of their deviation values in each of their sub-region. Finally, the distribution criterion can not only effectively reduce some of the useless feature points, but also ensure the uniformity of feature point distribution. As we can see, the step entropy and response can both develop a custom algorithm to identify feature points. Entropy strategy is only to calculate the reflection degree of an individual pixel, but the response is based on the deviation value of a set pixels between the center point, and it is adjacent points based on the Bresenham discrete circle principle, so that the mutation of response is more reflected for finding points with reliable contrast. Therefore, deviation is more capable of extracting more qualified feature points than entropy.

Algorithm 2 Entropy and response algorithms

Input: $I(X), points, m, n, t, r$
Output: $points_{select}$
1: **for** $j = 1 \to n^2$ **do**
2:      $points_j \leftarrow points$
3:      $H_j, \bar{H} \leftarrow Entropy(I(X), m, n)$
4:      $p_{sub(j)} \leftarrow Response(points_j, t)$
5:      **if** $H_j \geqslant \bar{H}$ **then**
6:          $points_{select} \leftarrow P_{(p_{sub(j)} * r)}$
7:      **else**
8:          $points_{select} \leftarrow p_{max}$
9:      **end if**
10:     $j = j + 1$
11: **end for**
12: **return** $points_{select}$
13: **function** $Entropy(I(X), m, n)$
14:     $p_{i(m*1)} \leftarrow Isub(i)_{(n*n)} \leftarrow I(X)$
15:     $H = \sum_{i=1}^{m} p_i log_2 p_i$
16:     $\bar{H} = \frac{1}{n^2} \sum_{j=1}^{n^2} H_j$
17: **end function**
18: **function** $Response(points, t)$
19:     $I_{p,x} \leftarrow circle_p \leftarrow points$
20:     $Dev = max(\sum_{x \in S_{bright}} |I_{p,x} - I_p| - t, \sum_{x \in S_{dark}} |I_p - I_{p,x}| - t)$
21:     $p_{sub} \leftarrow P_{(Dev_p)}$
22: **end function**

4.4. Definition of First-Order GM Problem Based on NCC

The purpose of graph matching [53] is to determine the correct attribute correspondences $P = (V^P, E^P)$ and $Q = (V^Q, E^Q)$ between two graphs P and Q, where V means vertex, and E represents edge. We customize corresponding mapping edges $e_1 = ij \in E^P, e_2 = ab \in E^Q$.

The objective of graph matching is to find the correct corresponding point pairs between two graphs P and Q among the feature points extracted. A unidirectional 'one-to-one' constraint is assumed, which requires one node in P to match at most one node in Q.

Cross-correlation is a standard method for estimating the similarity between two sets of data [54]. Normalized cross-correlation (NCC) is an essential application, which has been used widely for many signal processing applications due to its effective and direct representation in the frequency domain, and it is less sensitive to linear variations in the amplitude of two comparison signals [55]. We use the NCC algorithm to measure the similarity between two feature point p in graph P and q in graph Q. These ratios $Ra(p, q)$ of the calculated correlation values represent the degree of matching between the two sets of corresponding images. The NCC algorithm used to find similarity match between a window near feature point p and a window around feature point q is defined as:

$$Ra(p, q) = \frac{\sum_i [(Wp_i - \overline{Wp})(Wq_i - \overline{Wq})]}{\sqrt{\sum_i (Wp_i - \overline{Wp})^2} \sqrt{\sum_i (Wq_i - \overline{Wq})^2}} \tag{20}$$

where the summations are over all window coordinates, Wp_i and Wq_i are pixel intensity in windows for p and q respectively, each of the windows is sized as 5×5. Also, $\overline{Wp}$ and $\overline{Wq}$ are the corresponding mean of the window pixels. The coordinate of maximum values in this normalization cross-correlation is the position of the best matches for reference images.

Based on the NCC similarity measure, we use the nearest neighbor ratio (NNR) method to perform a rough match on the feature point set. The selection and matching process of the NNR algorithm with one-way 'one-to-one' constraint is described as follows: Based on the sample feature point pairs in the two images, first use the NCC algorithm to extract all the most significant corresponding pairs. These ratios are then compared with a fixed threshold. If the NCC ratio is higher than this fixed threshold, the corresponding point pair is considered a match. Otherwise, the pair of points are discarded. The fixed threshold is usually a constant not greater than 0.9. Since correct matching has stronger similarity than incorrect matching, this is a functional judging characterization for graph matching according to the NNR concept. Figure 7 shows a flowchart of data processing. The detected feature points of the two images are respectively put into two corresponding buffers. Each point p in graph P is used to calculate the ratio of NCC to all points q in graph Q, and then all ratios are sorted in descending order. Under the principle of NNR, some will be extracted as the best matching points, otherwise, they will be discarded.

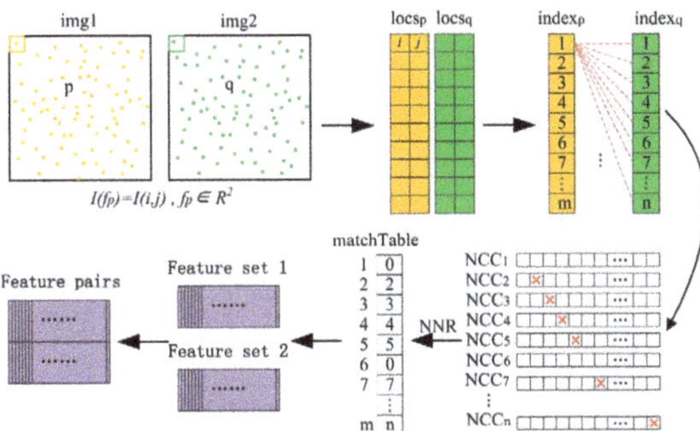

Figure 7. The basic flowchart of data processing.

4.5. Outlier Elimination

In this application, before using the Marr filter for convolution, we will use Laplacian for convolution as a preprocessing step. Due to the linear relationship of these two operations, this corresponds to a single convolution, which is a smoothed fourth-order derivative filter. Therefore, as each derivative amplifies the noise, the number of feature points increases. On the other hand, although the NNR method is easy to implement and sometimes well-matched, some points in these extracted feature point sets do not match, so a mismatched cleanup operation is required. Therefore, it is particularly important to find ways to reduce the mismatch caused by interference.

The RANSAC algorithm [56,57] is called the Random Sample Consensus Algorithm, which can well eliminate the existence of mismatches. The algorithm has strong robustness and the ability to correct data sets. The basic idea of the RANSAC algorithm is to use an iterative method to extract the sample set from the model. Find an optimized parametric model that can include more internal points in the data set and then test the extracted samples using the residual set. Points in the algorithm that fit the data set model are called interior points. Otherwise, they are called outliers. Therefore, the RANSAC algorithm can be used to find the best parameter model in the data set containing outliers through an iterative algorithm. The detailed implementation process of RANSAC is as follows:

(a) Randomly extracts non-collinear a pairs of feature points from the data set ($a = 4$ in experiment), then calculate their transformation homography matrix H, and record it as model M.

(b) Calculate projection error of each point in the dataset with model M_k. If the error is less than a predefined threshold τ, add it into the inner point set I_k.

(c) If the current number of elements in inner point set I_k is greater than the number in optimal inner point set I_{best}, then update I_{best} and re-estimate the model M_{best}.

(d) If the number of iteration is more than k, the operation will be exited; otherwise, the number of iterations is increased by 1, and the above steps are repeated.

The threshold τ is selected in accordance with n-dimensional chi-square distribution. χ is the cumulative chi-square distribution. Assume that the out-of-class points are white Gaussian noise with a mean of 0 and a variance of η. The number of iterations k will be updated instead of fixed until it is greater than the maximum number of iterations.

$$\tau^2 = \chi_n^{-1}(\mu)\eta^2 \tag{21}$$

$$k = \frac{log(1 - p_c)}{log(1 - \mu^a)} \tag{22}$$

where p_c is the confidence level, generally taking 0.95 to 0.99; μ is the ratio of inlier point; a is the minimum number of samples required to calculate for model. The pseudo-code of RANSAC is outlined in Algorithm 3.

Algorithm 3 Random Sample Consensus (RANSAC) algorithm

Input: p_c, k_{max}, τ, a, m
Output: M_{best}, I_{best}
 1: $k = 0, I_{max} = 0$
 2: **for** $k < k_{max}$ **do**
 3:     $\tau^2 = X_n^{-1}(\mu)\eta^2$
 4:     Use randomly sampled subset a to estimate M_k and I_k.
 5:     **if** $|I_k| > I_{max}$ **then**
 6:         $M_{best} = M_k, I_{best} = I_k$
 7:         $\mu = |I_{best}|/m, k_{max} = log(1 - p_c)/log(1 - \mu^a)$
 8:     **end if**
 9:     $k = k + 1$
10: **end for**

4.6. Parameters of the Proposed Algorithm

This section details the specific parameters used in the experiments. In the feature point detection part, the Marr wavelet algorithm is used to define the feature points as the maximum local value inside the scale-interaction image (with $\gamma = 1$). The two scales we chose are $i = 1$ and $j = 2$. Then the mesh division and feature points extraction are processed, the image is meshed by $n \times n$ to obtain n^2 sub-regions. Here, $n = 40$ is selected. The detected feature points are mapped into various sub-regions and sorted based on the deviation value Dev in each sub-region to which they belong. At the same time, each local information entropy H_i and average information entropy $\bar{H}$ are calculated. Assume that there are k sub-regions with local information entropy greater than the average information entropy, then feature points with maximum responsiveness of 30% are extracted from these k regions. In the feature point matching part, the NCC similarity measurement algorithm and NNR method are used to match the feature point set roughly. For the matching point filtering part, RANSAC can better remove the unmatched points, and finally get more accurate matching results. Its computational complexity is $O\left(max(N_P, N_Q)\right)$, where N_P is the number of features in the reference image, and N_Q is number of features in query image. The algorithm is suitable for performing real-time global methods. Besides, it can also achieve efficient parallel implementation in GPU systems. Algorithm 4 outlines the details of the entire process of the proposed algorithm.

Algorithm 4 The proposed algorithm

Input: Input images I_1 and I_2
 1: I_1 and I_2 processed under Laplace filter
 2: $(I_{1'}, I_{2'}) = MarrWaveletsFunction(I_1, I_2)$
 3: $(points1, points2) \leftarrow (corner_{peaks1}, corner_{peaks2}) \leftarrow (I_{1'}, I_{2'})$
 4: $(points_{select1}, points_{select2}) = EntropyResponseFunction(points1, points2)$
 5: $(points_{select1'}, points_{select2'}) = RANSACFunction(points_{select1}, points_{select2})$
 6: $MatchingPairs = NCCandNNR(points_{select1'}, points_{select2'})$

---

5. Second-Order NCC Based GM

The first-order GM provides convolution-based algorithms, whereas the second-order GM emphasizes geometric inter-feature relationships, transforming the correspondence problem to a purely geometric problem stated in a high dimensional space, generally modeled as an integer quadratic programming. This section presents our second application. We introduce in this section a new contribution with an application to second-order graph matching in the Matlab framework. The framework is based on the original Matlab application provided by Cho et al. [18]. This application constitutes a useful framework for graph matching as an IQP problem. It offers useful mathematical abstractions, and it allows us to develop and compare many algorithms based on a common evaluation platform, sharing input data, but also customizing affinity matrices and a matching list of candidate solution pairs as input data. This allows us to reuse these common data and context to start elaborate NCC algorithms for second-order graph matching application (As we discussed in detail in Section 4.4). This approach uses NCC algorithm to search for the indicator vector, then the matching score will be computed under IQP based formulation. By considering the second-order term, the algorithm determines the mapping between two graphs that should reflect the geometric similarity relationship between the pairwise matching features. All the algorithms are executed and compared based on the same experimental framework with common data from standard benchmarks in the domain.

We set P and Q the two sets of features of query graph $G_P = (P, E^P)$ and reference graph $G_Q = (Q, E^Q)$ respectively. We note $i, j \in P$ and $a, b \in Q$ as feature points, $ij \in E^P$ and $ab \in E^Q$ as edges. Also, $e_1 = (i, a)$ and $e_2 = (j, b)$ represent, when needed, candidate assignments. The main task it to find a suitable one-to-one mapping between P and Q. The feature points correspondence mapping is shown in the Figure 1. The yellow lines are correct matches.

Affinity matrix M, also known as affinity tensor, is used to organize the mutual similarities between sets of feature points. The measurement of affinity can be interpreted as a product of a solution vector x, that represents the set of candidate correspondences, by the matrix. The solution variable $x \in \{0, 1\}^{N_P N_Q}$ is an indicator vector such that $x_{ia} = 1$ means feature $i \in P$ matches with feature $a \in Q$, $x_{ia} = 0$ means no correspondence, and where N_P and N_Q are the respective set sizes of P and Q.

A graph matching score S between edges can be defined by the following equation:

$$S = \sum_{ij \sim ab} f(ij, ab) = \sum_{ia, jb} M(ia, jb) x_{ia} x_{jb} = x^T M x, \tag{23}$$

where $ij \sim ab$ means (i, a) and (j, b) are correspondence pairs, and x is the indicator vector. Then, the purpose of the graph matching IQP problem is computing solution x^* that maximizes the matching score as follows:

$$x^* = argmax(x^T M x), \tag{24}$$

$$s.t.\ x \in \{0, 1\}^{N_P N_Q}, \tag{25}$$

$$\forall i \sum_{a=1}^{N_Q} x_{ia} \leq 1, \tag{26}$$

$$\forall a \sum_{i=1}^{N_P} x_{ia} \leq 1. \tag{27}$$

The binary constraint is expressed by Equation (25), while (26) and (27) express the two-way constraints, that specify the solution to be a one-to-one mapping from P to Q. Note that by removing constraint (27), we obtain a many-to-one mapping, that is, a (partial) function from P to Q. In this chapter both constraints must be verified.

Affinity matrix M which consists of the relational similarity values between edges and nodes must is considered as an input of the problem. It can be noted that its size is defined by the total number of candidate assignment pairs considered. Then, the affinity matrix size may vary from $O\left((N_P N_Q)^2\right)$, in the case of full possible pairs, to $O\left((K \times N_P)^2\right)$ where K is some constant, in case of a restricted list of candidate pairs. Note that this list of candidate pairs must be added as part of the input to relate the entries of the affinity matrix to the feature points. The indicator variable x size varies also accordingly to the symmetric affinity matrix size. Its length corresponds to the column, of line, size of the matrix, and may vary from $N_P \times N_Q$ to $K \times N_P$ depending on the application.

Here, the matching score is completely retained as pairwise geometric only. The individual affinity $M(e_1, e_1)$ that represents first order affinity, is set to zero since there is no information about individual affinity. That is to say, all the diagonal values of the affinity matrix are zeros. The pairwise affinity $M(e_1, e_2) = M(ia, jb)$ between edges is given by:

$$M(ia, jb) = max(50 - d_{ia;jb}, 0), \tag{28}$$

where $d_{ia;jb}$ is the mutual projection error function used in [58] between two candidate assignments (i, a) and (j, b), that includes euclidean distance evaluation d_{ij} between locations of features i, j. The Table 1 summarizes notations and definitions used in this paper.

Table 1. Summarization of notations.

Notation	Purpose
G_P	Reference graph
G_Q	Query graph
P	Set of features in G_P
Q	Set of features in G_Q
N_P	Total number of data featrues of G_P
N_Q	Total number of data featrues of G_Q
C	Mapping constrains
L	A set of candidate assignments
i, j	Feature points in G_P
a, b	Feature points in G_Q
$e_1 = (i, a), e_2 = (j, b)$	Candidate assignments
M	Affinity matrix
$M(e_1, e_2)$	Pairwise affinity
$M(e_1, e_1)$	Individual affinity
S	Graph matching score
x	Indicator vector
x^*	Optimal solution
d_{ij}	Euclidean distances between the point i and j

6. Experimental Evaluation

6.1. First-Order GM Experiment

The experiments were performed on a CPU Intel(R) Core(TM) i5-4590 3.3 GHz. In this section, we evaluate the proposed feature point matching method by using the Visual Geometry Group dataset (https://www.robots.ox.ac.uk/~vgg/data/). The following will be divided into two parts for experimental description: extract feature points and perform feature matching. For quantitative evaluation, a set of experiments was performed. The proposed algorithm based on Laplace filter and Marr wavelets under entropy method (L_Marr_E) was compared with other classic conventional methods, corner detector [9], Gilles [7], Harris [6], LoG [8] and SIFT [11] algorithms.

6.1.1. Feature Points Extraction

Based on the above experimental theory, the following experimental verification was performed. First, we needed to verify the importance of the Laplacian filtering algorithm at the feature point extraction stage. Figure 8b,c show comparison graphs before and after adding a Laplacian filter. We can clearly see that using Laplace filters could greatly increase the number of feature points detected. Figure 9 shows the feature point detection results between different algorithms: CD, Gilles, Harris, LoG, SIFT, and L_Marr_E. Correspondingly, Table 2 details the specific number of feature points extracted of Figure 9. Compared with the conventional method, this method could extract more feature points.

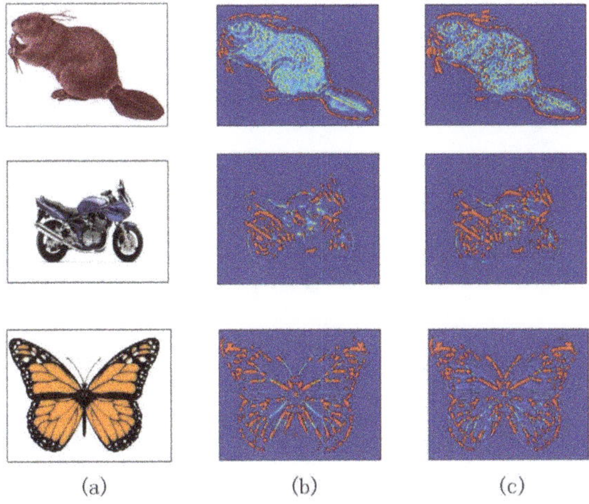

(a)   (b)   (c)

Figure 8. Quantitative experimental analysis of feature point extraction: (**a**) original image, (**b**) feature point extraction without Laplace filter, (**c**) feature point extraction with Laplace filter.

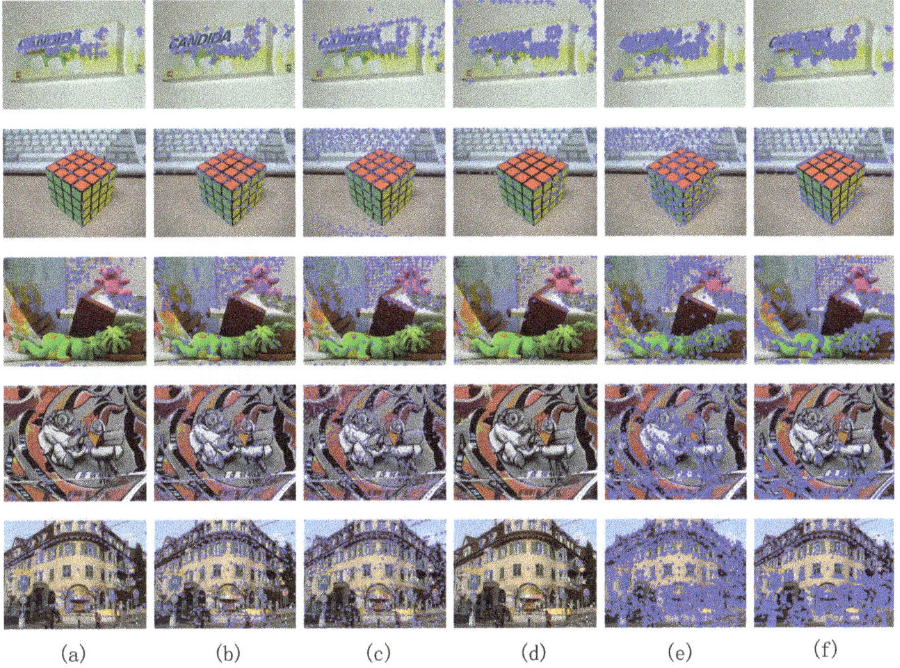

(a) (b) (c) (d) (e) (f)

Figure 9. Feature points extraction under different algorithms: (**a**) corner detector (CD) (**b**) Gilles (**c**) Harris (**d**) LoG (**e**) SIFT (**f**) L_Marr_E.

Table 2. Feature point extraction results under different kinds of algorithms.

Method	CD	Gilles	Harris	LOG	SIFT	L_Marr_E
img.1	72	73	104	174	226	264
img.2	70	246	457	165	980	654
img.3	144	274	341	110	763	1576
img.4	203	844	1133	153	3533	8402
img.5	320	627	696	140	3366	7806

6.1.2. Feature Points Matching

The following experiments verified the application of entropy-based refinement in feature matching. Combined with the fragile feature selection stage, ablation analysis provided some valuable intuitions for the pipeline in the feature matching stage. Then, the Figure 10 shows the comparison results, and we found that the RANSAC algorithm could thoroughly eliminate the mismatch points. Besides, the NCC algorithm based on Laplacian filter and Marr wavelet (L_Marr) not only had higher matching accuracy but also could increase the number of correct feature matches than the method without Laplace filter (Marr).

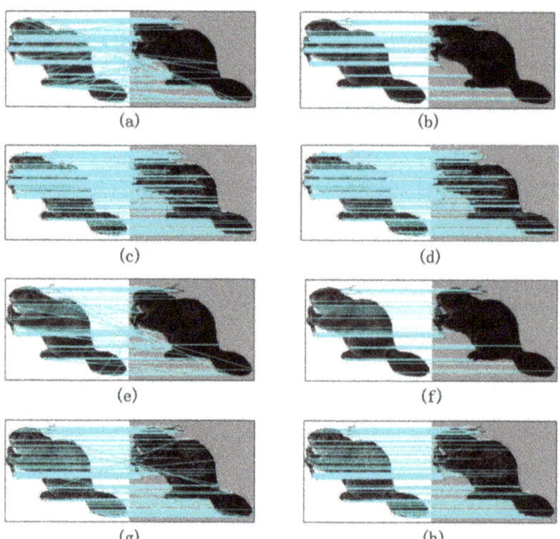

Figure 10. (**a,b**) show graph matching without Laplace filter before and after RANSAC optimization; (**c,d**) show graph matching under Laplace filter before and after RANSAC optimization; (**e,f**) show graph matching using entropy algorithm without Laplace filter before and after RANSAC optimization; (**g,h**) show graph matching using entropy algorithm under Laplace filter before and after RANSAC optimization.

Then we conducted another ablation study. We calculated the five groups of databases as used in Figure 9 and took their final average as shown in Table 3. It summarizes the number of image matching pairs, the number of recall, and time. The computation of graph matching Recall [59] can be defined as follows:

$$Recall = N_{rm} / N_{tm} \qquad (29)$$

where N_{rm} is defined as the number of detected true matches after RANSAC removes the mismatched points, and N_{tm} means the total number of correspondences. From the Table 3, we can find that the use of Laplace increases the recall rate, but at the same time it increases the computing time; however, the use of entropy can greatly improve operating efficiency under the premise of basically ensuring the recall value. Therefore, the two complement each other, the combination of Laplace and entropy can better realize the superiority of the proposed algorithm. As shown in the Table 3, the accuracy of the global L_Marr_E method remains very high, even if slightly lower than the accuracy of L_Marr method with no entropy, whereas a substantial computation time acceleration by a factor 5 is a benefit of the method with entropy element.

Table 3. Experimental comparison results for quantitative analysis of feature point matching.

Methods	N_{rm}	N_{tm}	Recall (%)	Time (s)
Marr	3323	4711	70.58	18.76
L_Marr	10407	10954	95.01	98.02
Marr_E	998	1455	68.59	10.33
L_Marr_E	3523	3740	94.20	21.45

We now selected the L_Marr_E method, which provided the best compromise in both accuracy and computation time in previous tests, as a competitor against other feature point selection methods from literature. We conducted another experiment based on test data from the University of Oxford's Object Category dataset, which contained 13 different image sets. Among them, seven data sets were selected for comparative analysis in this section. Table 4 shows the average of all image pairs selected from the dataset. The best recall results were obtained through L_Marr_E, which produced the most significant feature points in the two query images. It should be noted that when compared to standard feature point extraction methods, our method provided the best accuracy despite a slight increase of computation time. While entropy accelerated feature point selection in our framework with Laplacian and Marr wavelet, as shown in Table 3, it allowed on the contrary to improve accuracy against other independent standard feature point selection methods, as shown in Table 4. Even when compared to the famous SIFT method, our method remained competitive according to the trade-off between time and accuracy, accuracy being improved at a slight expense of computation time. Note that L_Marr_E also provided the highest number of feature points. Figure 11 shows the corresponding links between images under different algorithms: CD, Gilles, Harris, LoG, SIFT, and L_Marr_E. Therefore, the proposed algorithm ould obtain the better matching effect.

Table 4. Feature point matching results under different kinds of algorithms.

	Detector	1st Image	2nd Image	Time (s)	Recall (%)
1	CD	51	53	1.76	0.36
2	Gilles	124	134	2.24	0.16
3	Harris	208	226	2.8	0.35
4	LoG	300	300	4.27	0.27
5	SIFT	1275	1258	5.49	0.48
6	L_Marr_E	1998	1929	7.22	0.52

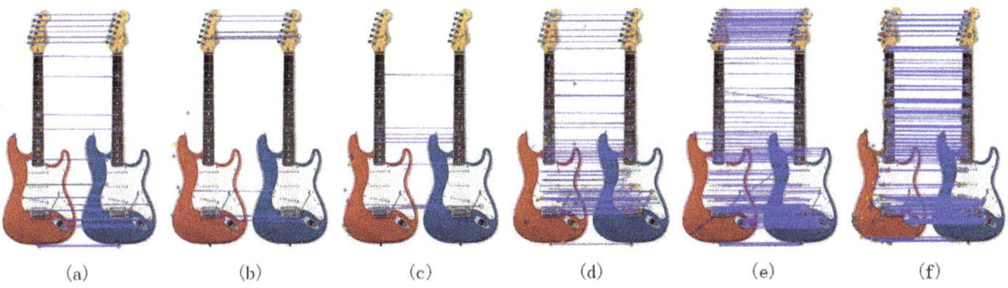

Figure 11. Feature points matching under different algorithms: (**a**) CD (**b**) Gilles (**c**) Harris (**d**) LoG (**e**) SIFT and (**f**) L_Marr_E.

6.2. Second-Order GM Experiment

In this subsection, experiments were conducted on a CPU Intel(R) Core(TM) i5-4590 3.3 GHz. We performed experiments on the CMU house image database and real image database. Accuracy, objective score and time were the most important main parameters in the field of graph matching. We needed to use them as criterion when comparing with other different algorithms. The datasets used in this paper, such as CALTECH and CMU house datasets, are the most popular databases for some state-of-the-art algorithms. Although there are still many different databases, in order to facilitate the comparative study of researchers who study graph matching, this article used these most popular databases for easier comparing research. The proposed NCC algorithm was compared with two state-of-the-art methods such as RRWM [18] and SM [29]. All of these different algorithms shared the same images, feature points, and affinity matrices as input data. Each test set had a ground-truth solution for accuracy evaluation.

6.2.1. Presentation

Performance evaluation could be done by computing the affinity score, but more importantly, by evaluating accuracy according to a ground-truth set of true assignment pairs, that reflected the application requirement of graph matching. We also evaluated the computation time. Based on the above IQP, the objective score could be obtained by formula (4). Accuracy could be obtained by dividing the actual correct number of matches detected by the maximum number of ground truth pairs that could be returned, and the formula was as follows:

$$Accuracy = x^* * V_{GTbool}/maxGT, \tag{30}$$

where x^* is the binary vector solution returned by the algorithm, $V_{GTbool} \in \{0,1\}^{N_P N_Q}$ is a binary vector representing the ground-truth pairs, and $maxGT$ is the maximum number of true assignments that could be returned when considering a many-to-one mapping, relaxing constraint formula (2). However, solutions returned by the algorithms had to necessarily verify a one-to-one mapping in this section. They verified both constraints formulas (2) and (3).

6.2.2. CMU House Image Matching

Experiments using the CMU house internal sequence dataset (http://vasc.ri.cmu.edu/idb/html/motion/) could evaluate sequence matching of the same object. A total of 110 pictures in this dataset were divided into different sequence gaps (from 10 to 100 with an interval of 10). Therefore, we ended up with ten sets of data pairs. Each pair of image sets consisted of an initial fixed-position picture (sequence 1) and its changing transformations. To evaluate the matching accuracy, 30 landmark feature extraction points were manually tracked and labeled as ground truth on all frames. In this typical classic CMU test, the proposed NCC method was competitive to the RRWM algorithm. The experimental results are shown in Figure 12, and the detailed information of the quantitative evaluation is recorded in Table 5. From Table 5 we can find that in this typical single object with only the angle conversion test, the proposed NCC algorithm was comparable to RRWM in terms of accuracy and score. NCC also cost less computing time than RRWM.

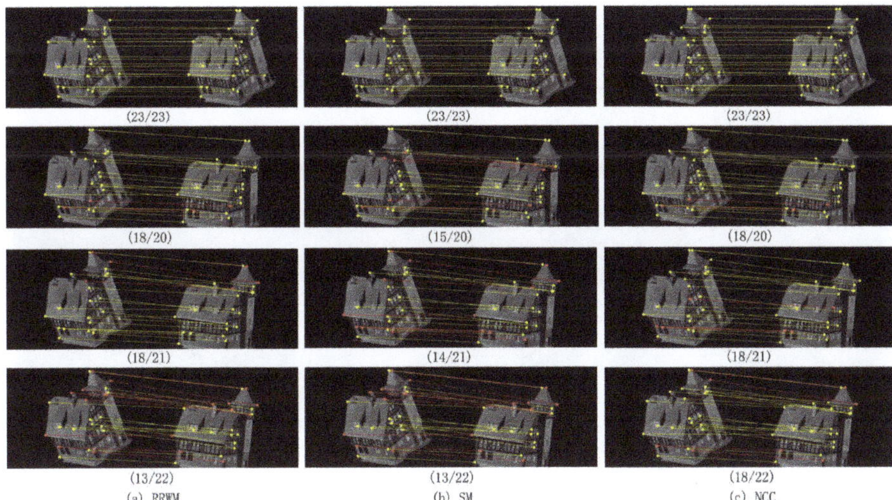

Figure 12. CMU house dataset matching results among (**a**) RRWM, (**b**) SM, and (**c**) NCC algorithms.

Table 5. Comparative evaluation on CMU database for reweighted random walk method (RRWM), spectral matching method (SM), and normalized cross-correlation (NCC).

	Methods	Accuracy	Score	Time (s)
1	RRWM	92.61	99.93	0.52
2	SM	89.21	94.44	0.07
3	NCC	94.88	95.31	0.23

6.2.3. Real Image Matching

In the experiments of real image matching, we used the CALTECH database (https://cv.snu.ac.kr/research/~RRWM/) customized by Cho et al. [18]. In this CALTECH database, it contained 30 different image pairs. All ground truths of these corresponding candidates were manually pre-labeled. Accuracy and objective scores were the main criteria for matching. Table 6 shows the average results of 30 pairs of images obtained by RRWM, SM, and NCC algorithms. As can be seen from the table, the performance of the NCC algorithm was not better than the RRWM and SM algorithms. Figure 13 shows the visualization of feature point connections. Correct and incorrect matches were marked with yellow and black lines, respectively.

Since the second-order graph matching based on IQP was an NP-hard problem, various approximate solutions were used to attempt to solve the pairwise similarity corresponding mapping problem. Leordeanu and Hebert provided a spectral matching (SM) algorithm [29] based on the main strength cluster of the adjacency matrix by finding its principal eigenvector. Reweighted random walks for graph matching (RRWM) algorithm was introduced by Cho et al. [18], and it combined mapping constraints with re-weighted jumping schemes. Both of these two methods implement an iteration loop to compute a principal eigenvector. This basic iteration loop technique belongs to the power iteration method. Although this method cannot guarantee to achieve the final global optimal solution during the operation, it can converge to the fixed point of the tensor. However, NCC algorithm is a single-threaded algorithm but not an iterative method. When the normalized cross-correlation algorithm calculation is over, the entire matching process is completed. It cannot run the optimization iterative until the solution x converges. Therefore, the NCC algorithm cannot achieve the expected optimization effect in the second-order matching.

Table 6. Comparative evaluation on CALTECH database for RRWM, SM, and NCC.

	Methods	Accuracy	Score	Time (s)
1	RRWM	61.13	99.92	0.16
2	SM	50.22	79.93	0.02
3	NCC	52.17	74.39	0.06

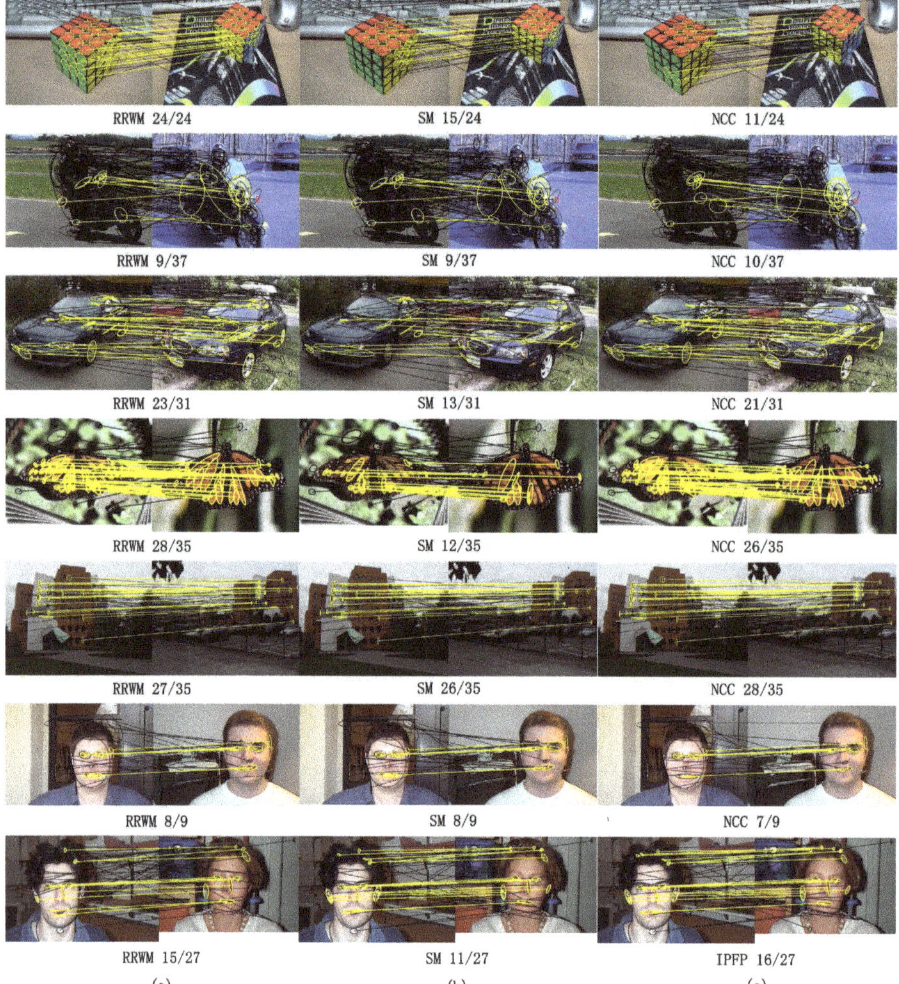

Figure 13. From the left to the right: (**a**) RRWM algorithm, (**b**) SM algorithm, and (**c**) NCC algorithm for graph matching. (The yellow lines represent the correct matching pairs, and the black lines represent the wrong matches.)

7. Conclusions

We study different declinations of feature correspondence problems by the use of the Matlab platform, in order to reuse and provide state-of-the-art solution methods, as well as experimental protocols and input data necessary with evaluation and comparison tools against existing sequential algorithms, most of the time developed in Matlab framework. While feature extraction methods are numerous, it is not straightforward that each could represent the objects to match from a query to a reference image adequately. In order to vary the feature set size while preserving a reasonable recall rate in graph matching, we have proposed a new combination of filters with an entropy-response based selection method. Laplace filter enhances the edges and details of an image. Secondly, Marr wavelets embedded in scale-interaction are used to detect feature points. Then, we use the entropy and brightness response to extract typical feature points. Most importantly, the entropy-based selection method greatly reduces the calculation time. Image matching is achieved by nearest neighbor search using a normalized cross-correlation similarity measure. Finally, the RANSAC process deletes the outlier correspondence to achieve matching optimization. The first-order comparison results show that our algorithm has a higher matching rate for reasonable computation time, despite the augmentation of the number of feature points under the Laplace algorithm. The second-order graph matching is also realized based on the NCC algorithm, which allows us to address graph matching or derived sub-problems with a closed relationship with experiments on IQP models in the Matlab platform. We found that the second-order NCC approach performs competitively to IQP models on CMU images on accuracy and score. The performance looks less satisfactory for real case images of the CALTECH database. In future work, we will improve implementation by exploiting natural parallelism of the method on the GPU platform.

Author Contributions: Conceptualization, B.C. and J.-C.C.; methodology, B.C.; software, B.C.; validation, B.C. and J.-C.C.; formal analysis, J.-C.C.; investigation, B.C. and J.-C.C.; resources, B.C.; data curation, B.C.; writing—original draft preparation, B.C.; writing—review and editing, J.-C.C.; visualization, B.C.; supervision, J.-C.C.; project administration, J.-C.C.; funding acquisition, B.C. All authors have read and agreed to the published version of the manuscript.

Funding: This work was supported by the China Scholarship Council (201604490109).

Conflicts of Interest: The authors declare no conflict of interest.

References

1. Tang, S.; Andres, B.; Andriluka, M.; Schiele, B. Multi-person tracking by multicut and deep matching. In Proceedings of the European Conference on Computer Vision, Amsterdam, The Netherlands, 8–16 October 2016; pp. 100–111.

2. Mohamed, W.; Hamza, A.B. Reeb graph path dissimilarity for 3D object matching and retrieval. *Vis. Comput.* **2012**, *28*, 305–318. [CrossRef]

3. Vu, N.S.; Caplier, A. Enhanced patterns of oriented edge magnitudes for face recognition and image matching. *IEEE Trans. Image Process.* **2011**, *21*, 1352–1365.

4. Juan, L.; Oubong, G. SURF applied in panorama image stitching. In Proceedings of the 2010 2nd International Conference on Image Processing Theory, Tools and Applications, Paris, France, 7–10 July 2010; pp. 495–499.

5. Leng, C.; Zhang, H.; Li, B.; Cai, G.; Pei, Z.; He, L. Local feature descriptor for image matching: A Survey. *IEEE Access* **2018**, *7*, 6424–6434. [CrossRef]

6. Harris, C.; Stephens, M. A combined corner and edge detector. *Citeseer* **1988**, *15*, 10–5244.

7. Gilles, S. Robust Description and Matching of Images. Ph.D. Thesis, Department of Engineering Science, University of Oxford, Oxford, UK, 1998.

8. Lindeberg, T. Feature detection with automatic scale selection. *Int. J. Comput. Vis.* **1998**, *30*, 79–116. [CrossRef]

9. Derpanis, K.G. *The Harris Corner Detector*; York University: Toronto, ON, Canada, 2004.

10. Mikolajczyk, K.; Schmid, C. Scale & affine invariant interest point detectors. *Int. J. Comput. Vis.* **2004**, *60*, 63–86.

11. Lowe, D.G. Object recognition from local scale-invariant features. In Proceedings of the Seventh IEEE International Conference on Computer Vision, Kerkyra, Greece, 20–27 September 1999; Volume 2, pp. 1150–1157.

12. Kutter, M.; Bhattacharjee, S.K.; Ebrahimi, T. Towards second generation watermarking schemes. In Proceedings of the 1999 International Conference on Image Processing, Kobe, Japan, 24–28 October 1999; Volume 1, pp. 320–323.
13. Tang, C.W.; Hang, H.M. A feature-based robust digital image watermarking scheme. *IEEE Trans. Signal Process.* **2003**, *51*, 950–959. [CrossRef]
14. Li, T.; Shi, J.; Li, X.; Wu, J.; Pan, F. Image Encryption Based on Pixel-Level Diffusion with Dynamic Filtering and DNA-Level Permutation with 3D Latin Cubes. *Entropy* **2019**, *21*, 319. [CrossRef]
15. Wehrl, A. General properties of entropy. *Rev. Mod. Phys.* **1978**, *50*, 221–260, doi:10.1103/revmodphys.50.221. [CrossRef]
16. Rényi, A. On measures of entropy and information. In *Proceedings of the Fourth Berkeley Symposium on Mathematical Statistics and Probability, Volume 1: Contributions to the Theory of Statistics*; The Regents of the University of California: Oakland, CA, USA, 1961.
17. Cour, T.; Srinivasan, P.; Shi, J. Balanced graph matching. In *Advances in Neural Information Processing Systems*; MIT Press: Philadelphia, PA, USA, 2007; pp. 313–320.
18. Cho, M.; Lee, J.; Lee, K.M. Reweighted random walks for graph matching. In Proceedings of the European Conference on Computer Vision, Heraklion, Greece, 5–11 September 2010; pp. 492–505.
19. Cui, B.; Creput, J.C. Using Entropy and Marr Wavelets to Automatic Feature Detection for Image Matching. In Proceedings of the 2019 15th International Conference on Signal-Image Technology & Internet-Based Systems (SITIS), Sorrento, Italy, 26–29 November 2019; doi:10.1109/sitis.2019.00084. [CrossRef]
20. Torresani, L.; Kolmogorov, V.; Rother, C. Feature correspondence via graph matching: Models and global optimization. In Proceedings of the European Conference on Computer Vision, Marseille, France, 12–18 October 2008; pp. 596–609.
21. Yao, B.; Fei-Fei, L. Action recognition with exemplar based 2.5 d graph matching. In Proceedings of the European Conference on Computer Vision, Florence, Italy, 7–13 October 2012; pp. 173–186.
22. Neuhaus, M.; Bunke, H. A graph matching based approach to fingerprint classification using directional variance. In Proceedings of the International Conference on Audio-and Video-Based Biometric Person Authentication, Hilton Rye Town, NY, USA, 20–22 July 2005; pp. 191–200.
23. Banerjee, B.; Bovolo, F.; Bhattacharya, A.; Bruzzone, L.; Chaudhuri, S.; Buddhiraju, K.M. A novel graph-matching-based approach for domain adaptation in classification of remote sensing image pair. *IEEE Trans. Geosci. Remote. Sens.* **2015**, *53*, 4045–4062. [CrossRef]
24. Berg, A.C.; Berg, T.L.; Malik, J. Shape matching and object recognition using low distortion correspondences. In Proceedings of the 2005 IEEE Computer Society Conference on Computer Vision and Pattern Recognition (CVPR'05), San Diego, CA, USA, 20–25 June 2005; pp. 26–33.
25. Zhu, G.; Doermann, D. Logo matching for document image retrieval. In Proceedings of the 2009 10th International Conference on Document Analysis and Recognition, Barcelona, Spain, 26–29 July 2009; pp. 606–610.
26. Conte, D.; Foggia, P.; Sansone, C.; Vento, M. Graph matching applications in pattern recognition and image processing. In Proceedings of the 2003 International Conference on Image Processing (Cat. No. 03CH37429), Barcelona, Spain, 14–17 September 2003.
27. Foggia, P.; Percannella, G.; Vento, M. Graph matching and learning in pattern recognition in the last 10 years. *Int. J. Pattern Recognit. Artif. Intell.* **2014**, *28*, 1450001. [CrossRef]
28. Gold, S.; Rangarajan, A. A graduated assignment algorithm for graph matching. *IEEE Trans. Pattern Anal. Mach. Intell.* **1996**, *18*, 377–388. [CrossRef]
29. Leordeanu, M.; Hebert, M. A spectral technique for correspondence problems using pairwise constraints. In Proceedings of the Tenth IEEE International Conference on Computer Vision (ICCV'05), Beijing, China, 17–21 October 2005; pp. 1482–1489.
30. Cho, M.; Sun, J.; Duchenne, O.; Ponce, J. Finding matches in a haystack: A max-pooling strategy for graph matching in the presence of outliers. In Proceedings of the IEEE Conference on Computer Vision and Pattern Recognition, Columbus, OH, USA, 23–28 June 2014; pp. 2083–2090.
31. Zass, R.; Shashua, A. Probabilistic graph and hypergraph matching. In Proceedings of the 2008 IEEE Conference on Computer Vision and Pattern Recognition, Anchorage, AK, USA, 23–28 June 2008; pp. 1–8.

32. Chertok, M.; Keller, Y. Efficient high order matching. *IEEE Trans. Pattern Anal. Mach. Intell.* **2010**, *32*, 2205–2215. [CrossRef]
33. Duchenne, O.; Bach, F.; Kweon, I.S.; Ponce, J. A tensor-based algorithm for high-order graph matching. *IEEE Trans. Pattern Anal. Mach. Intell.* **2011**, *33*, 2383–2395. [CrossRef]
34. Lee, J.; Cho, M.; Lee, K.M. Hyper-graph matching via reweighted random walks. In Proceedings of the CVPR 2011, Providence, RI, USA, 20–25 June 2011; pp. 1633–1640.
35. Nguyen, Q.; Gautier, A.; Hein, M. A flexible tensor block coordinate ascent scheme for hypergraph matching. In Proceedings of the IEEE Conference on Computer Vision and Pattern Recognition, Boston, MA, USA, 7–12 June 2015; pp. 5270–5278.
36. Zhou, J.; Wang, T.; Lang, C.; Feng, S.; Jin, Y. A novel hypergraph matching algorithm based on tensor refining. *J. Vis. Commun. Image Represent.* **2018**, *57*, 69–75. [CrossRef]
37. Yang, Z. Fast template matching based on normalized cross correlation with centroid bounding. In Proceedings of the 2010 International Conference on Measuring Technology and Mechatronics Automation, Changsha, China, 13–14 March 2010; Volume 2, pp. 224–227.
38. Saravanan, C.; Surender, M. Algorithm for face matching using normalized cross-correlation. *Int. J. Eng. Adv. Technol. (IJEAT) ISSN* **2013**, *2*, 2249–8958.
39. Tang, C.; Gao, T.; Yan, S.; Wang, L.; Wu, J. The oriented spatial filter masks for electronic speckle pattern interferometry phase patterns. *Opt. Express* **2010**, *18*, 8942–8947. [CrossRef]
40. van Vliet, L.J.; Young, I.T.; Beckers, G.L. A nonlinear Laplace operator as edge detector in noisy images. *Comput. Vis. Graph. Image Process.* **1989**, *45*, 167–195. [CrossRef]
41. Stewart, C.V. Robust parameter estimation in computer vision. *SIAM Rev.* **1999**, *41*, 513–537. [CrossRef]
42. Dobbins, A.; Zucker, S.W.; Cynader, M.S. Endstopping and curvature. *Vis. Res.* **1989**, *29*, 1371–1387. [CrossRef]
43. Manjunath, B.; Chellappa, R.; von der Malsburg, C. A feature based approach to face recognition. In Proceedings of the 1992 IEEE Computer Society Conference on Computer Vision and Pattern Recognition, Champaign, IL, USA, 15–18 June 1992; pp. 373–378.
44. Manjunath, B.; Shekhar, C.; Chellappa, R. A new approach to image feature detection with applications. *Pattern Recognit.* **1996**, *29*, 627–640. [CrossRef]
45. Daugman, J.G. Uncertainty relation for resolution in space, spatial frequency, and orientation optimized by two-dimensional visual cortical filters. *JOSA A* **1985**, *2*, 1160–1169. [CrossRef]
46. Bhattacharjee, S.K.; Kutter, M. Compression Tolerant Image Authentication. In Proceedings of the 1998 International Conference on Image Processing, Chicago, IL, USA, 7 October 1998; pp. 435–439.
47. Antoine, J.P.; Murenzi, R. Two-dimensional directional wavelets and the scale-angle representation. *Signal Process.* **1996**, *52*, 259–281. [CrossRef]
48. Ding, N.; Liu, Y.; Jin, Y.; Zhu, M. Image registration based on log-polar transform and SIFT features. In Proceedings of the 2010 International Conference on Computational and Information Sciences, Chengdu, China, 17–19 December 2010; pp. 749–752.
49. Kapur, J.N.; Sahoo, P.K.; Wong, A.K. A new method for gray-level picture thresholding using the entropy of the histogram. *Comput. Vis. Graph. Image Process.* **1985**, *29*, 273–285. [CrossRef]
50. Liu, M.Y.; Tuzel, O.; Ramalingam, S.; Chellappa, R. Entropy rate superpixel segmentation. In Proceedings of the Computer Vision and Pattern Recognition (CVPR), Providence, RI, USA, 20–25 June 2011; pp. 2097–2104.
51. Rosten, E.; Porter, R.; Drummond, T. Faster and better: A machine learning approach to corner detection. *IEEE Trans. Pattern Anal. Mach. Intell.* **2010**, *32*, 105–119. [CrossRef]
52. Kraft, M.; Schmidt, A.; Kasinski, A.J. High-Speed Image Feature Detection Using FPGA Implementation of Fast Algorithm. *VISAPP* **2008**, *8*, 174–179.
53. Lee, J.; Cho, M.; Lee, K.M. A graph matching algorithm using data-driven markov chain monte carlo sampling. In Proceedings of the 2010 20th International Conference on Pattern Recognition, Istanbul, Turkey, 23–26 August 2010; pp. 2816–2819.
54. Bourke, P. Cross correlation. *Cross Correl. Auto Correl. 2D Pattern Ident.* **2019**, *3*, 1996.
55. Yoo, J.C.; Han, T.H. Fast normalized cross-correlation. *Circuits Syst. Signal Process.* **2009**, *28*, 819. [CrossRef]
56. Raguram, R.; Chum, O.; Pollefeys, M.; Matas, J.; Frahm, J.M. USAC: A universal framework for random sample consensus. *IEEE Trans. Pattern Anal. Mach. Intell.* **2013**, *35*, 2022–2038. [CrossRef]
57. Derpanis, K.G. Overview of the RANSAC Algorithm. *Image Rochester* **2010**, *4*, 2–3.

58. Cho, M.; Lee, J.; Lee, K.M. Feature correspondence and deformable object matching via agglomerative correspondence clustering. In Proceedings of the 2009 IEEE 12th International Conference on Computer Vision, Kyoto, Japan, 29 September–2 October 2009; pp. 1280–1287.
59. Miksik, O.; Mikolajczyk, K. Evaluation of local detectors and descriptors for fast feature matching. In Proceedings of the 21st International Conference on Pattern Recognition (ICPR2012), Tsukuba, Japan, 11–15 November 2012; pp. 2681–2684.

Article

Crowd Counting with Semantic Scene Segmentation in Helicopter Footage

Gergely Csönde [1,*], Yoshihide Sekimoto [2] and Takehiro Kashiyama [2]

[1] Department of Civil Engineering, The University of Tokyo, 4-6-1 Komaba, Meguro, Tokyo 1538505, Japan
[2] Institute of Industrial Science, The University of Tokyo, 4-6-1 Komaba, Meguro, Tokyo 1538505, Japan; sekimoto@iis.u-tokyo.ac.jp (Y.S.); ksym@iis.u-tokyo.ac.jp (T.K.)
* Correspondence: csonde@iis.u-tokyo.ac.jp

Received: 29 July 2020; Accepted: 24 August 2020; Published: 27 August 2020

Abstract: Continually improving crowd counting neural networks have been developed in recent years. The accuracy of these networks has reached such high levels that further improvement is becoming very difficult. However, this high accuracy lacks deeper semantic information, such as social roles (e.g., student, company worker, or police officer) or location-based roles (e.g., pedestrian, tenant, or construction worker). Some of these can be learned from the same set of features as the human nature of an entity, whereas others require wider contextual information from the human surroundings. The primary end-goal of developing recognition software is to involve them in autonomous decision-making systems. Therefore, it must be foolproof, which is, it must have good semantic understanding of the input. In this study, we focus on counting pedestrians in helicopter footage and introduce a dataset created from helicopter videos for this purpose. We use semantic segmentation to extract the required additional contextual information from the surroundings of an entity. We demonstrate that it is possible to increase the pedestrian counting accuracy in this manner. Furthermore, we show that crowd counting and semantic segmentation can be simultaneously achieved, with comparable or even improved accuracy, by using the same crowd counting neural network for both tasks through hard parameter sharing. The presented method is generic and it can be applied to arbitrary crowd density estimation methods. A link to the dataset is available at the end of the paper.

Keywords: remote sensing; helicopter footage; deep learning; computer vision; image processing; crowd counting; semantic segmentation; multitask learning

1. Introduction

With the recent rapid developments in convolutional neural networks (CNNs), many image processing tasks that were very difficult a decade ago have become easier. Perhaps the most obvious example is the recognition of humanoid shapes in images. The detection and counting of humans in digital images, especially in crowded scenes, offer several applications, including traffic monitoring, safety surveillance, and finding stranded people following disasters.

At present, the task of image-based crowd counting can be divided into two main categories: direct and indirect methods. In the former case, all individuals are separately identified in the image, following which the total number of humans is obtained by counting those individuals directly. However, indirect methods produce additional abstract information, from which the total count can be achieved through a more elaborate procedure. In this study, we focus on indirect methods; specifically, density map estimator (DME) CNNs, although the work presented can also be applied to direct methods.

DME CNNs use an arbitrary three-channel RGB image as input and produce a single-channel crowd density map. Thereafter, the total count of humans can be obtained by integrating over the entire image. The advantage of this method is that, in addition to yielding the total count of humans, it provides a detailed map of their spatial distribution. The networks are usually fully convolutional, which means that the input can have an arbitrary size.

The typical footage type used for crowd counting is created with street-level cameras and drones flying at low altitudes. Although these image types are used extensively, they suffer from strong perspective distortions and obstructions, owing to the low camera angle. As a result, people appear in many different sizes and shapes in the imagery, and many methods can only detect a subset of these varieties, thereby yielding suboptimal density maps. Most works in the field are focusing on solving these issues at present.

Another issue exists that is unrelated to the previous problems and, to the best of the authors' knowledge, is not currently being investigated by researchers. Although generic human detection has its uses, it also exhibits limitations from certain practical viewpoints. If the task is to identify a subset of humans, identifying that all humans will produce faulty results. The simplest example is the differentiation between actual living humans and fake ones, such as dummies, statues, or posters, which inevitably appear in urban footage whether or not they are exhibited in conventional datasets.

The set of features used for identifying humans can also be used for identifying human subcategories, depending on the task; for example, people with a certain hair color, wearing a certain type of clothing, or carrying a bag. However, more abstract subcategories exist that are difficult or impossible to identify based on humanoid features alone. One such subcategory is pedestrians.

Counting pedestrians offers several applications. Such data can be used to reconstruct the traffic flow of people, which can be subsequently used for further research purposes or monitoring the use of infrastructural objects. Safety monitoring is another useful application, and the monitoring of curfews is also an interesting option. Owing to the recent Coronavirus outbreak, many countries have imposed a certain level of restrictions relating to time spent on the streets or social distancing.

For the abovementioned purposes, people inside buildings, on balconies, or even on rooftops are irrelevant, as are people tending to their gardens or working at construction sites. Moreover, as mentioned previously, inanimate human-like objects may also exist, such as billboards and other forms of street advertisements (Figure 1). From the perspective of human detection, there is no real difference between an "image of a human" or an "image of an image of a human". We must recognize the billboard itself to differentiate between the two. The latter is not specific to pedestrians, but it is a generic issue. Furthermore, seemingly random false positive detections may occur in the regions of an image where there are not supposed to be people. Finally, the density map may contain background noise, which distorts the accuracy.

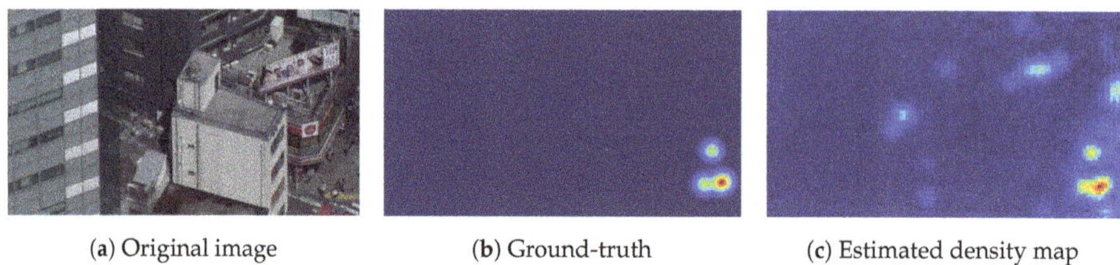

(**a**) Original image (**b**) Ground-truth (**c**) Estimated density map

Figure 1. Sample image exhibiting the problem of "fake human" detection. It is clear that the billboard with humans on top of the building produces false positive detections.

Changing the footage type is an unorthodox means of solving the problems of perspective distortions and occlusions. As opposed to conventional datasets, helicopter footage is free of perspective distortions and it exhibits fewer obstructions. This is because of the high altitude and steep camera angle. With such a setup, there is a very small size difference between humans in different parts of the image, and even when they are standing close to one another, large parts of their bodies remain visible. Moreover, one image can cover a much larger area from a higher altitude than from the street level.

In summary, helicopter footage contains several advantageous attributes for human detection. Most importantly, it exhibits features that are not covered by other conventional datasets, as demonstrated in [1], in which a new helicopter dataset was introduced as part of the investigation. In this study, we introduce a new large-scale urban helicopter dataset, known as Tokyo Hawkeye 2020 (TH2020) [2], which shares the same attributes, but has a much larger image and pedestrian count, and it also exhibits inanimate human-like objects in the urban environment.

We propose a solution that also happens to help with false detections in regions where ordinary humans cannot possibly exist in order to address the issue of differentiating between all human-like entities and actual pedestrians. We explicitly apply semantic segmentation to enhance the DME results. The goal in the semantic segmentation task is to sort the regions of the input image into several semantic categories. If C different categories exist, the network generates C confidence score maps: one for every category. Semantic segmentation was also used as binary classification in several other works, which differentiated between human and non-human regions in order to attempt to enhance the accuracy. Our approach differs from this: in the footage that we introduce, the ratio of human regions is very low when compared to the background. Therefore, we ignore humans in the segmentation and focus on classifying the background, as illustrated in Figure 2.

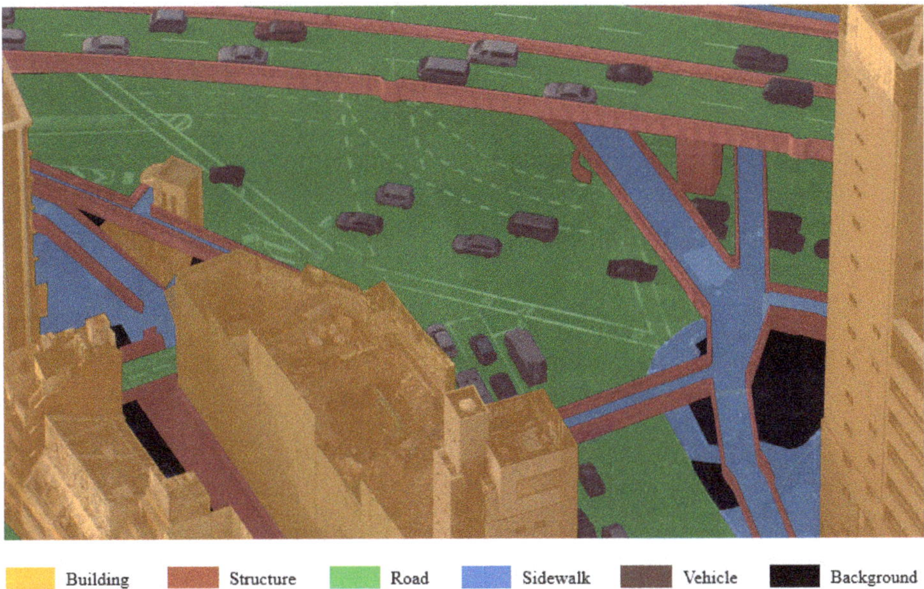

Figure 2. Semantic segmentation of background for helicopter footage, where humans are ignored.

Semantic segmentation for aerial footage has been well developed and it can achieve high accuracy. The segmentation can be used to identify the regions of images in which pedestrians can or cannot be found. We demonstrate that this approach can be used to improve the accuracy of pedestrian detection by masking regions where pedestrians cannot exist. Moreover, we show that the segmentation can be achieved by the same network as the density map estimation by performing optimization for a combinatorial loss (crowd density and segmentation) with minimal additional computational cost, while the density map accuracy does not decrease significantly or even improves. The former approach can be used with steep-angle, high-altitude footage, but is unfeasible with conventional crowd counting benchmark datasets. Although the latter approach would be applicable to arbitrary datasets, to the best of the authors' knowledge, no other dataset has the required annotation information. Because the introduced methods are independent of the underlying DME network, they can be applied to any arbitrary architecture. To support this statement, we present experimental results using three different state-of-the-art DME networks.

2. Related Work

2.1. Crowd Counting

The majority of recent works have focused on solving two problems that arise in state-of-the-art crowd counting methods: varying scales and perspective distortions. Several researchers have diretly addressed the varying target sizes [3–7]. Others have experimented with increasing or changing the receptive field of the output pixels [8–11]. Certain scholars have attempted to tackle perspective distortions by using focused attention mechanisms [12–16]. Learning residual errors and correcting density estimations with these errors has also proven to be a viable approach [17–19].

All of these methods serve the purpose of improving the accuracy of generic human detection. However, we need to include the surroundings of the people in the calculations if we wish to differentiate between people on roofs, balconies, or posters and those on the street. Although certain methods may do so unintentionally and implicitly to an extent, our aim is to address this explicitly using semantic segmentation.

The use of semantic segmentation in crowd counting is not new [20–24]. All of these works aimed to create segmentation to separate human regions from non-human regions. In many situations, this is a simple binary classification, but even when multiclass segmentation is used, the problem is eventually reduced to human versus non-human separation.

In our work, we ignore humans as a segmentation class and focus on the segmentation of the background. We classify the background from the functional perspective of how likely it is that pedestrians exist in that region. This approach can be used with steep-angle, high-altitude footage, but it is unfeasible with conventional crowd counting benchmark datasets. We elaborate further on this in Section 5.1.

Besides density map methods, there is another closely related paradigm called localized regression. The key difference lies with the target map. In this approach, every map pixel aims to represent the total count in a small, localized region of the input. One typical way to create such a target map is to apply a uniform square-shaped kernel to the head annotations [25]. The result will be a map of redundant localized counts. Reference [26] also changes the counting task to a classification task by predicting count intervals instead of specific counts. These methods achieve competitive accuracy.

While most research in crowd counting is based on image processing techniques, it is worth mentioning that there are efforts underway to also use the rest of the electromagnetic spectrum for this purpose. Some methods rely on on-person devices, such as smart phones or radio frequency identification cards. The problem with these is that they are working with the assumption that people would carry the device at all times, not to mention the serious privacy issues that are raised by use of such devices. For indoor counting in buildings, where there are plenty of Wi-Fi enabled devices installed, personal device-free Wi-Fi based methods yield high count accuracy [27,28] while preserving privacy. However, in open outdoor spaces, these methods are less viable due to the significantly lower number of installed devices and the lack of refractive surfaces.

2.2. Semantic Segmentation

The first significant breakthrough in CNN-based semantic segmentation was the realization that the task is essentially no different from classification, and any arbitrary classification network can easily be converted into a segmentation network [29]. Subsequently, various network architectures were rapidly developed in an attempt to improve the segmentation accuracy. For example, FuseNet [30] incorporates depth information into the training in order to improve the segmentation accuracy.

An important issue in the segmentation task is that the network is usually required to provide a larger receptive field when compared to object detection. Dilated convolutions can solve this problem.

DeepLabv3 [31] is a model based on ResNet [32], which uses dilated convolutions in its atrous spatial pyramid pooling layer.

Several architectures have been specifically developed for aerial semantic segmentation [33–37]. A common feature of the above methods is that they are multimodal or multispectral. The ISPRS aerial datasets [38,39] are typical benchmark datasets for aerial segmentation.

2.3. Existing Datasets

Numerous CNN-based human detection or counting methods have been developed and tested on standard datasets, such as the UCSD [40], INRIA [41], Caltech Benchmark [42], UCF_CC_50 [43], and ShanghaiTech [3] datasets. Although these image sets can be distinguished from one another by the average person count, they were all obtained from near-street-level cameras and from low angles. Therefore, these images contain many obstructions and perspective distortions. These issues pose substantially less difficulty in aerial footage.

Obstructions are caused by people walking alongside one another. In images that were taken from a steep angle, the obstruction is not very large because people do not walk over one another. Perspective distortion is caused by the relative difference between the distances of objects from the camera. The size of an object in an image is inversely proportional to its distance from the camera. Therefore, if the ratio of the distance of two objects is much larger (or smaller) than 1.0, the difference in their sizes in the image will also be very large. This ratio is closer to 1.0 in photographs taken from a high altitude. Figure 3 depicts these issues.

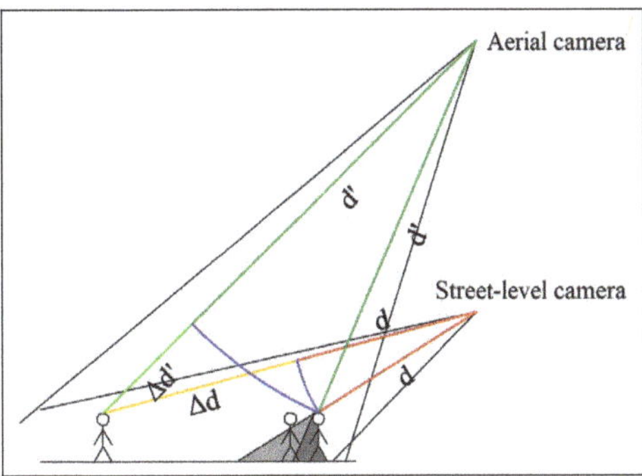

Figure 3. Illustration of differences between street-level and aerial footage. $(d + \Delta d)/d$ is much further from 1.0 than $(d' + \Delta d')/d'$. Moreover, the dark gray area is obstructed from both cameras, whereas the light gray area is only obstructed from the street-level camera. Note that the image is not to scale.

Of course, other aerial datasets are also available, which have been created either with airplanes or, more recently, using drones. Airplane footage is typically obtained from a high altitude, which is not suitable for the detection of small targets. The authors of [44] introduced a high-altitude, vertical-angle airplane dataset with acceptable resolution. The elevations are lower in the case of drone datasets, so human detection can be achieved. The angle is often vertical, which is advantageous for considering occlusion, but the human features are less recognizable. If the angle is low, perspective distortions occur, but a better viewpoint is provided for human features. Examples of drone datasets include the SDD [45], VisDrone2019 [46], and Okutama-Action [47] datasets.

Some of the above-mentioned datasets are image based, whereas others are obtained from videos. The image-based datasets were annotated manually. The common annotation strategy for video-based datasets is to annotate certain key frames, such as every tenth one, and then interpolate the annotation between frames. This vastly increases the dataset size, but not the unique object instance count.

Certain video-based datasets, such as SDD and Okutama–Action, use unique ID-based annotation, so that it can also be considered for tracking.

3. Materials and Methods

3.1. Segmentation-Based Region Masking

During our preliminary experiments, we identified two main issues with pedestrian counting. The first is that even the best DMEs often produce false positive predictions and contain background noise in locations where humans cannot be present, such as flat wall surfaces or treetops. The other is that, even though the annotation only contains pedestrians, the trained models learn to identify humans regardless of the context and, as a result, non-pedestrians and poster images of humans are also identified and counted.

Both of the above issues can be solved by identifying the regions in which pedestrians cannot be present. We refer to these as invalid regions. Consequently, we refer to regions in which pedestrians may be present as valid regions. The invalid region information can be used to modify a base crowd density map. We do so by integrating the density map estimation and semantic segmentation. We use the mean absolute error (MAE) primarily and root mean squared error (RMSE) secondarily for the density map quality measurement metrics.

In the following sections, we explain two methods for region-based masking, but, first, we introduce our dataset in detail because it is essential for our methods.

3.2. Dataset

We introduce a new helicopter-based pedestrian dataset. There are two main reasons for this. Firstly, a dataset from such a viewpoint exhibits features that are not covered by other conventional datasets, and models trained on those datasets show very weak performance on helicopter footage, as demonstrated in [1]. Secondly, the method that we introduce depends on the attributes of the dataset.

Because our dataset is an extended version of the TH2019 dataset [1], we named it TH2020. It has the same attributes as TH2019, except for the dataset size; there are roughly 20 times more images and 15 times more pedestrian annotations in our dataset. Specifically, TH2020 comprises 6237 static images with a resolution of 960×540, containing 120,772 pedestrian annotations from 10 different locations and 22 different sessions. The images were obtained from helicopter footage from a high altitude at a steep, but not vertical angle. Owing to the high altitude, the perspective distortions and scale differences are negligible, although the average target size is very small. It is important to note that, because the camera angle is not vertical, pedestrians are partially visible from the sides, thereby increasing the number of recognizable features as compared to those of a vertical camera angle, which is the most typical in airplane footage. Figure 4 presents some sample images.

The dataset includes two types of ground-truth annotations: a pedestrian annotation and a semantic segmentation annotation. The former only contains pedestrian head coordinates, including bikers, as there is no reason to differentiate between these from a practical viewpoint. Moreover, distinguishing bikers from pedestrians is very challenging; for example, if there are 10 people surrounding a bicycle, any one of these may be the rider. People on rooftops or on construction sites are excluded from the ground-truth. We generated a ground-truth density map from the head coordinate annotation using the conventional method. We created a matrix with the same dimensions as the input image. The matrix contained 1s at the head coordinates, and 0s otherwise. Thereafter, we convolved this matrix with a Gaussian kernel. We used a fixed kernel with a standard deviation of 15 instead of the adaptive method.

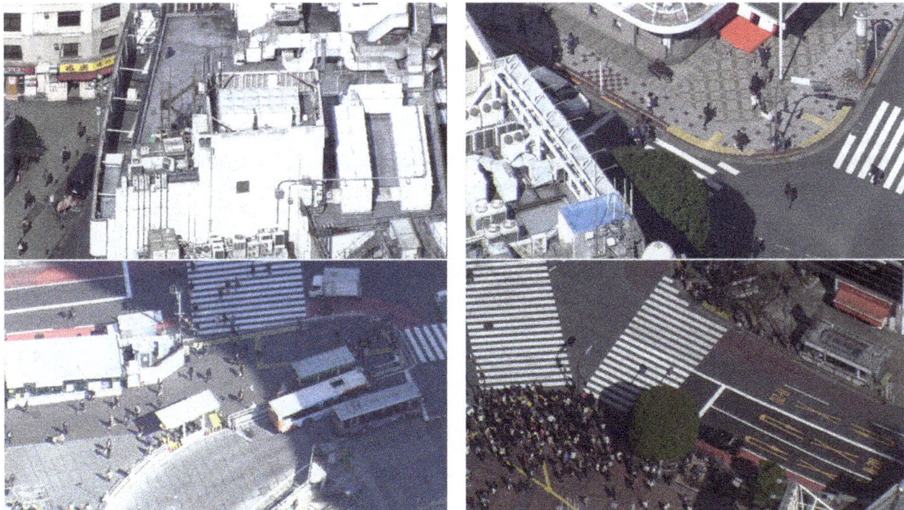

Figure 4. Sample images from TH2020 dataset.

For the segmentation, only 5040 images were annotated, owing to financial and time limitations. The annotation assigned one category to every pixel of the image. When we created the segmentation ground-truth, we were not aiming for an extensive annotation. Rather, we attempted to achieve simplicity and focused on a functional perspective for pedestrian detection. We wanted to ignore objects that were too small to occlude humans completely, so we considered those objects as part of their surrounding entities. Moreover, we required high-level classes, as we did not want the annotation to be too detailed. Therefore, we decided to annotate the following categories:

- Road: areas designated primarily for vehicle traffic; humans are often present.
- Sidewalk: areas designated for pedestrian traffic; humans are often present.
- Building: areas covered by construction that pedestrians can enter (houses, offices, etc.); humans may appear in windows, on balconies, and on rooftops.
- Structure: areas covered by construction, often with girder-like features, where entry is not possible (walls, columns, railings, etc.); humans are unlikely.
- Plant: areas covered by vegetation that can block visibility, such as trees with leaves and large bushes; humans are unlikely.
- Vehicle: areas covered by cars, buses, trucks, etc.; depending on the vehicle and lighting conditions, human heads may be present, but they are unlikely.
- Background: anything else; typically grass, rubble, railway tracks, etc.; humans may be present, but are unlikely.

With regard to privacy matters, it should be mentioned that the resolution of the footage is not enough to identify any individual; only coarse features, such as hair, clothing, or luggage, can be identified. Therefore, there are no privacy issues with the dataset, as opposed to conventional benchmark datasets, where there are clearly recognizable people.

We divided the dataset into training and evaluation parts scene-wise instead of applying a random split in order to avoid overfitting. The training set was the same for the segmentation and density map estimation, whereas the evaluation set for the segmentation was a subset of the density map estimation evaluation set, as not all images had segmentation ground-truth. Specifically, there were 4114 training images, 926 segmentation evaluation images, and 2123 density map estimation evaluation images.

3.3. Simple Masking with Separate CNNs

In this approach, we used an arbitrary DME CNN and an arbitrary semantic segmentation CNN, and trained both on the TH2020 dataset. For inference, we used the input image and generated an ordinary crowd density map with the unaltered DME CNN. Thereafter, we used the semantic segmentation CNN to generate a segmentation map for the same image. This segmentation map could have arbitrary classes. In our experiments, we used those that were explained in Section 3.2. Subsequently, we took the segmentation map and simplified it as a valid–invalid region map based on a predetermined table. Valid pixels had a value of 1, whereas invalid ones were 0. As the two CNNs were independent from one another, their output resolutions could differ. In such a case, we resampled the valid–invalid map to the DME output resolution using nearest-neighbor interpolation. As the final step, we masked the original crowd density map by multiplying it element-wise with the valid–invalid map. Alternatively, it would be possible to assign real numbers between 0 and 1 to every segmentation class and to use that as a mask, but, because this was our first attempt with masking, we aimed for simplicity. Therefore, we used the zero–one map. Figure 5 illustrates the flow.

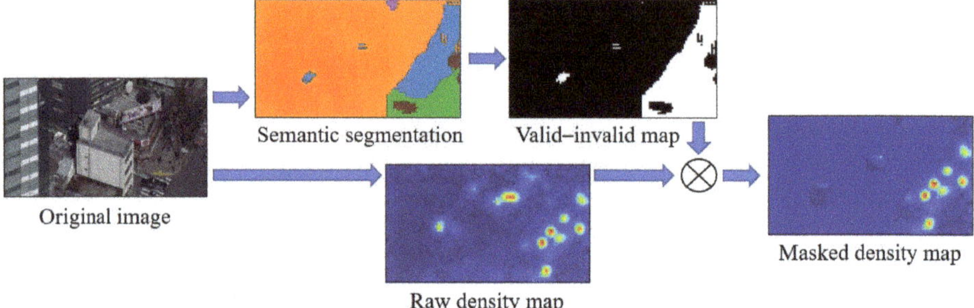

Figure 5. Flow of region-based masking procedure.

There is more than one means of creating a valid–invalid map from segmentation. The most obvious method is that, whereby for every pixel, the value for that segmentation class from the predetermined table is inserted. Although this is certainly a viable approach, it has two drawbacks. Firstly, the segmentation is not perfect, so it is possible that some estimated invalid regions will overlap with some pedestrians. Secondly, it is very common for pedestrians to appear at the edges of invalid regions because the camera angle is not vertical. For example, when someone walks in front of a building and very close to it, most of the body will overlap with the building.

We proposed max-pool masking to solve the above problems. We applied a max-pool layer to the valid–invalid map with a carefully selected kernel size (we used a kernel size of seven) and a single pixel stride. That is, an invalid pixel would only remain invalid if it did not have any valid pixels in a certain vicinity. In this manner, the sizes of the invalid regions were shrunk to counter the effects of segmentation errors and natural overlaps.

3.4. Dme Segmentation Multitask Training

Any arbitrary classification network can easily be converted into a segmentation network, as demonstrated in [29]. However, we decided not to limit our approach to classification, as many image processing CNNs are based on classification networks in any case. Furthermore, technically, any arbitrary fully CNN can be changed into a semantic segmentation network by replacing the final output layers and loss function, which is the approach we used.

We used crowd counting CNNs, increased the number of channels in the final output layer to C, and changed the loss function to a conventional segmentation loss, specifically, the softmax cross-entropy loss. However, it occurred to us that the output did not contain any parameters, which meant that the model was exactly the same as that used for crowd counting, so we could attempt to perform the two tasks simultaneously. Moreover, we realized that, because we were generating a crowd density map and segmentation map at the same time, we could use the segmentation map directly for masking the density map. Thus, our final architecture contained C + 2 output channels: one for crowd density map estimation, C for the segmentation categories, and one for the masked crowd density map. Figure 6 presents a graphical layout of our proposed method.

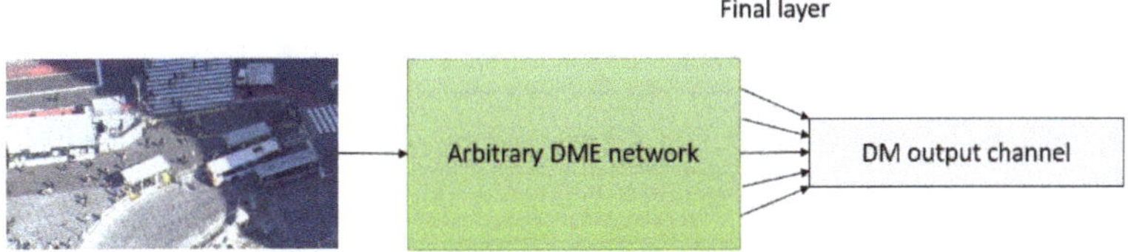

(**a**) Regular density map estimator (DME) with one output channel

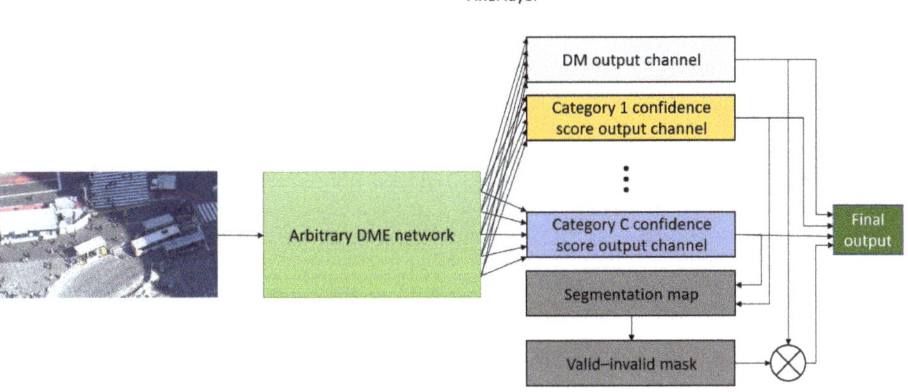

(**b**) Proposed multitask architecture with C + 2 output channels, where C is the number of segmentation categories

Figure 6. Visual representation of difference between regular DME network and our proposed multitask segmentation method.

For the loss function, we calculated the Euclidean norm over the density map per-pixel errors and softmax cross-entropy for the segmentation channels, and combined these two losses with a balancing factor α.

$$L_c = \frac{1}{N} \sum_{i=0}^{N} (d_i - \hat{d}_i)^2, \tag{1}$$

$$L_s = \frac{1}{N} \sum_{i=0}^{N} -log \left(\frac{e^{-p_i^g}}{\sum_{c=0}^{C} e^{-p_i^c}} \right), \tag{2}$$

$$L = L_c + \alpha L_s, \tag{3}$$

where N is the number of pixels in the mini-batch, d_i and $\hat{d}_i$ are the ground-truth and estimated per-pixel crowd density values, respectively, C is the number of categories, p_i^g is the predicted confidence score for the ground-truth category at pixel i, and p_i^c is the predicted confidence score for category c at pixel i.

The Euclidean norm could also be calculated over the masked crowd density map instead of the raw one, but, in this case, finding the correct alpha was very difficult, because the valid–invalid map could easily become stuck in the complete 0 or complete 1 state.

In this approach, we also evaluated the semantic segmentation quality, for which we used the mean intersection-over-union (mIoU).

We added the MTSM prefix (for MultiTask Segmentation Method) to distinguish between the original backbone crowd counting network and the multitask version. For example, if the backbone was CSRNet architecture, we referred to our modified version as MTSM-CSRNet.

Finally, it should be noted that the conversion method also works backwards; that is, a semantic segmentation network can be converted into a DME network, but we have not conducted such experiments yet and, thus, the accuracy of the backwards direction remains to be determined.

4. Results

4.1. Simple DME and Segmentation Methods

In our experiments, we used CSRNet [9], CAN [48], and SPNet [10] as the DME networks and DeepLabv3 as the semantic segmentation network, although, as we will demonstrate, there is no real difference. We selected a general-purpose segmentation network, because methods for aerial footage also require height information, which we did not have for our footage.

As we were using a custom dataset instead of a conventional benchmark dataset, we first had to establish a baseline accuracy value for all of the networks. The DME results can be viewed in the first line of Table 1, whereas Tables 2 and 3 display the DeepLabv3 results. In the segmentation task, we assigned valid values to the road, sidewalk, and background classes. The average pedestrian count in the evaluation set was approximately 30, which means that the relative MAE was slightly over 0.2. We were confident that the models were well trained, as these models produced similar relative errors on conventional datasets. In the case of DeepLabv3, the evaluation was not so simple. We could not compare the results to those of aerial segmentation datasets, as those also contain height information, which aids in achieving very high mIoU values. Our best option was a comparison with the Cityscapes dataset [49]. Deeplabv3 achieved an mIoU of 81.3 on Cityscapes. The accuracy achieved on our dataset was slightly worse than this, but the annotation was very different, so this difference did not mean that the trained model was not as effective as it could possibly be.

Moreover, we made several attempts with SANet [50], because the authors claimed to achieve high accuracy on conventional benchmark datasets, but we could not train the model. Even the lowest error we could achieve was almost three times higher than that of the other networks. Therefore, we excluded it from our investigation.

Table 1. Summary of crowd density estimation accuracies for models trained on TH2020. The mean intersection-over-union (mIoU) column is only included to indicate the accuracy of the segmentation mask used.

Method		CSRNet			CAN			SPNet		
		MAE	RMSE	mIoU	MAE	RMSE	mIoU	MAE	RMSE	mIoU
Simple DME		7.35	19.63		6.5	13.54		6.29	16.97	
Masked	Simple DME	7.26	19.9	0.7562	6.03	13.95	0.7562	6.14	17.15	0.7562
	MTSM	6.84	20.94	0.7332	6.9	22.3	0.7562	6.58	19.56	0.5821
Raw		6.84	20.85		6.82	22.2		6.56	19.56	

Table 2. Per-category intersection-over-unions (IoUs) and means for DeepLabv3 and our MTSM models trained on TH2020.

Network	Building	Structure	Plant	Sidewalk	Vehicle	Road	Background	Mean	Upsampled Mean
DeepLabV3	0.8494	0.3431	0.6552	0.7905	0.732	0.8632	0.6765	0.7014	0.7014
MTSM-CSRNet	0.9207	0.6263	0.6115	0.7571	0.7004	0.8584	0.6583	0.7332	0.7131
MTSM-CAN	0.9428	0.6693	0.642	0.7632	0.7072	0.8834	0.6858	0.7562	0.7358
MTSM-SPNet	0.9043	0.4162	0.4541	0.673	0.4674	0.8308	0.3288	0.5821	0.5778

Table 3. Valid–invalid IoUs and their means for DeepLabv3 and our MTSM models trained on TH2020.

Network	Invalid (Building-Structure-Plant-Vehicle)	Valid (Road-Sidewalk-Background)	Mean
DeepLabV3	0.8916	0.8846	0.8881
MTSM-CSRNet	0.8999	0.8979	0.8989
MTSM-CAN	0.9112	0.9115	0.9113
MTSM-SPNet	0.8533	0.8508	0.852

4.2. Simple Masking with Separate CNNs

We took our most accurate segmentation model, which happened to be MTSM-CAN, and used it to mask the density maps for all three DME networks. The MAE improved in all scenarios. The results are displayed in the second line of Table 1. Figure 7 presents a comparison of the different models.

4.3. DME Segmentation Multitask Training

For this method, we modified CSRNet, CAN, and SPNet according to Section 3.4. The DME results are displayed in the third and fourth lines of Table 1.

We reported the results for the models with the lowest MAE and reasonable segmentation quality. The latter means that the softmax cross-entropy loss was lower than 0.2. Tables 2 and 3 present the per-class and valid–invalid IoUs as compared to the same metrics of DeepLabv3. Figure 7 displays a comparison among the models.

The output of DeepLabv3 had the same resolution as the original image, whereas our backbone networks applied downsampling. To enable a fair comparison, we reported the mIoU for both the downsampled segmentation map and the downsampled segmentation map upsampled back to the original resolution.

CSRNet and CAN were demonstrated to be strongly adaptable to our MTSM model, with comparable and even improved accuracy. In the case of SPNet, both the counting and segmentation metrics decreased. The most important difference between SPNet and the other architectures is that SPNet is shallow when compared to the others.

It can be observed from Table 1 that the masking actually decreased the counting accuracy. In the following section, we investigate the reason for the varying masking efficiency.

4.4. Masking Efficiency

The level of improvement depends on the quality of both the density map estimation and semantic segmentation. The dependence on the segmentation quality is straightforward. If the invalid areas are larger in the estimation than in the ground-truth map, true positive detections can be masked out. However, nothing may change if they are smaller than the ground-truth.

The dependence on the crowd density estimation quality is slightly more complicated. For argument's sake, let us assume that the segmentation map is correct. This is a safe assumption, because we achieved very high invalid region IoUs (Table 3), and the possibility of errors is further reduced by the max-pool masking. Moreover, let us assume that the density map does not contain negative elements. Although many architectures do not enforce this, in our experiments the robust networks eliminated negative values (or reduced them to very close to 0).

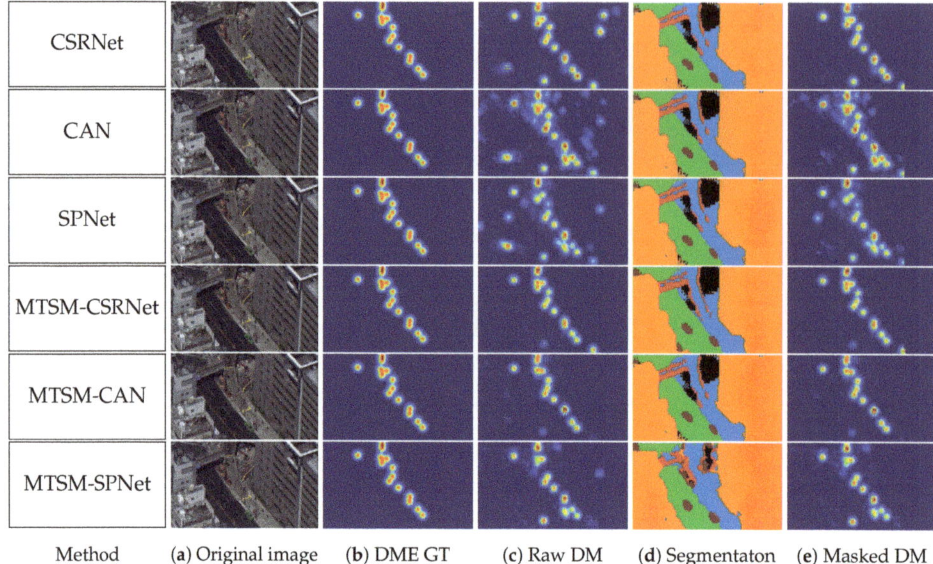

<table>
<tr><td>CSRNet</td></tr>
<tr><td>CAN</td></tr>
<tr><td>SPNet</td></tr>
<tr><td>MTSM-CSRNet</td></tr>
<tr><td>MTSM-CAN</td></tr>
<tr><td>MTSM-SPNet</td></tr>
</table>

Method	(**a**) Original image	(**b**) DME GT	(**c**) Raw DM	(**d**) Segmentaton	(**e**) Masked DM

Figure 7. Sample results for comparison across methods. The segmentation for the regular density map (DM) methods was generated by MTSM-CAN.

If the total count is overestimated, then there must be several false positive detections. As we assumed perfect segmentation, the masking can only remove false positives. If there is a removed false positive count that is more than twice the original error, the MAE will increase. However, this also means that the estimation has numerous false negatives and the network attempts to compensate for the error with false positives.

If the total count is underestimated and something is still masked out, thereby increasing the MAE, the model produces false negatives and false positives simultaneously, but the false negatives outweigh the false positives. In summary, the masking efficiency depends on whether or not the estimated density map is littered with both false positives and false negatives.

If there are many false positives or false negatives, then they will cause spike-like errors in the density map. These spikes significantly increase the L2 norm of the pixel errors (that is, the Euclidean loss). In contrast, if there is mostly only background noise, the L2 norm will be much smaller than the L1 norm. Therefore, the ratio of the two norms provides information regarding the amount of spike-like errors; that is, the number of false positives and false negatives in the estimation. We collected this information and plotted it against the MAE improvement, which can be observed from Table 4 and Figure 8.

Table 4. L2/L1 norm ratio as compared to change in accuracy after masking for investigated models.

Method	Average L2/L1 Norm Ratio	Change in MAE	Invalid Mask mIoU
CSRNet	0.0457	−0.09	0.9112
CAN	0.0315	−0.47	0.9112
SPNet	0.0457	−0.15	0.9112
MTSM-CSRNet	0.0509	0	0.8999
MTSM-CAN	0.0529	0.08	0.9112
MTSM-SPNet	0.0444	0.02	0.8533

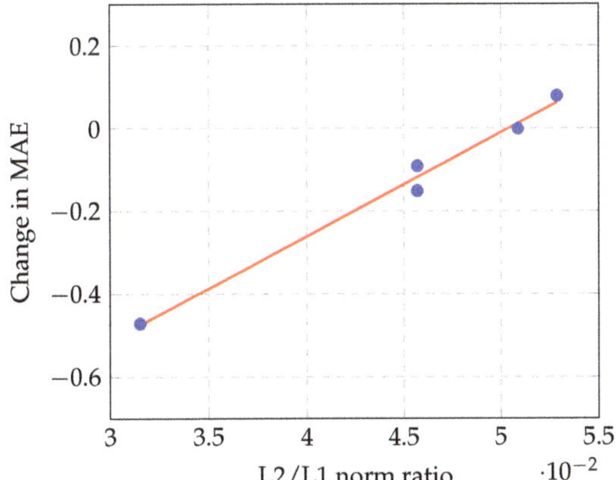

Figure 8. Correlation between DME quality and masking efficiency in case of high-quality segmentation mask.

4.5. Training Details

We downloaded open-source implementations for all models except for SPNet, which we implemented ourselves (excluding the final ReLU layer).

As DME networks are downscaling and sometimes upscaling, the output resolution differs from the original resolution. Thus, because the ground-truth has the same resolution as the original image, it must be resampled. We used bicubic interpolation for the density map and nearest-neighbor interpolation for the segmentation map. For the density map, we also had to multiply the pixel magnitudes by the resampling factor in order to preserve the total count. We applied a random horizontal flip as augmentation. As the footage had a well-defined upwards vertical direction, it was pointless to apply rotation or a vertical flip, and we did not want to complicate the training further. As all three DME networks have a VGG-16 [51] backbone, we initialized them using pre-trained VGG-16 parameter weights. There were no suitable pre-trained weights available for DeepLabv3; therefore, we had to train it from scratch. The parameters without pre-trained values were randomly initialized with a normal distribution ($\mu = 1, \sigma = 0.02$).

We trained the models on full images, because DeepLabv3 struggled with patch-wise training, and it was important for all of the models to be trained on the same dataset for comparison. We used one or four GPUs for the DME methods. In the case of four GPUs, the batch size was increased, but we did not observe any change in the accuracy, only in the execution speed. We used four GPUs for DeepLabv3 owing to the large memory requirement. Each GPU had 16 GB memory in which a batch of two images could fit, yielding a total batch size of 8. For the DME, we set the batch size to range from 12 to 16 per GPU (the maximum number that could fit). We used the Adam optimizer with an initial learning rate of 10^{-4} and a momentum of 0.95. The learning rate was directly halved every 20 epochs, starting from the 40th epoch.

The details of the MTSM training were mostly the same as those in the case of the baseline models. The baseline model parameters were used as initial values for MTSM-CSRNet and MTSM-CAN. As MTSM-SPNet did not perform effectively when initialized from the baseline model, we launched several training sessions from scratch in order to achieve the highest accuracy.

However, the most important aspect was to determine the appropriate value for α. The Euclidean loss and softmax cross-entropy are very different functions. Therefore, using a simple constant scaling factor would result in the gradient components having different weightings in different sections of the training. During our training attempts on MTSM-CSRNet and MTSM-CAN, we often observed that with a constant α, the training shifted towards either the crowd density or segmentation very rapidly. We used an initial α of 10^{-6} and, whenever the training became too one-sided, we shifted it towards the weaker loss, which usually meant increasing α. In the case of MTSM-SPNet, the training could be performed by calculating the loss function over the masked density map using a constant α of 10^{-5}. In practice, we used two balancing factors: one for each loss component, so the final loss would be closer to 1, because small fractions are difficult to read.

5. Discussion

5.1. Correlation between Density Map Quality and Masking Efficiency

Our data exhibited a linear correlation between the average L2/L1 norm ratio of the images and the change in the MAE caused by masking. This means that a model with a smaller average L2/L1 norm ratio can benefit more from the masking method. A factor that could not be observed from our experimental results is that models with very different MAEs may have the same average L2/L1 norm ratio. For example, if there are two models and one produces proportionally larger per-pixel errors than the other, then it will also have proportionally larger MAEs, while the average L2/L1 norm ratio will be the same. However, if the per-pixel errors are all proportionally larger, the per-pixel errors that are masked will also be proportionally larger, which is exactly the change in the MAE. In summary, based on the MAE value, the same average L2/L1 norm ratio may be correlated to different changes in the MAE. This suggests that the average L2/L1 norm ratio is probably linearly correlated to the change in the MAE relative to the MAE, but, as our experimental results had MAEs that were very close to one another, there was also a correlation with the absolute change.

Referring back to the results in Table 1, it can be observed that, although the MAE values improved, the RMSE increased slightly. This may occur if there are moderate improvements in images with small errors along with small degradations in images with large errors. If the dataset has a small variance in the per-image ground-truth count, a high RMSE indicates that images exist in the dataset that are greatly over- or underestimated. However, if the dataset has a large variance in the per-image ground-truth count, a high RMSE is inevitable, because every method exhibits a certain proportionality between the ground-truth count and absolute error; that is, if the ground-truth has high variance, the absolute error will also have high variance. This means that the RMSE is not a very useful metric for datasets with high ground-truth variation, such as ours, unless a method is developed that produces an error that is independent from the ground-truth count.

Moreover, it should be noted that the proposed masking method is not feasible with conventional crowd counting benchmark datasets for two reasons. Firstly, in conventional datasets, the ratio of the area that is covered by humans is very large, and often almost the entire image is covered with humans. Therefore, they cannot be ignored in the segmentation. Secondly, even if segmentation could be achieved by ignoring the humans, owing to the low camera angle and close distance, the heads would overlap excessively with invalid regions and, therefore, they would be masked out.

5.2. MTSM Model Parameters

Our aim was to delve deeper into the inner workings of the MTSM models. We were interested in understanding how these are realized, and how they differ from training two independent networks for crowd counting and segmentation apart from the obvious memory and/or computational requirements. We were dealing with two different tasks that could either be intertwined or not. The latter case means that inside the network, two parallel subnetworks are realized and the channels from the one do not affect the output of the other. However, if the two tasks are intertwined, there cannot be such separation and the parameters affect both outputs.

The existence of two separable subnetworks would mean that the current architectures for crowd counting are excessive and the same accuracy can be achieved with a substantially smaller model. However, our experiments demonstrated that this is not the case. If there were two subnetworks, the parameters in the final layer would have to be separable into two groups by affecting either one output or the other, which we investigated. For every output channel in the final layer, we calculated the absolute mean of its parameters over the input channels. Thereafter, we verified whether there were any input channels with parameters that were thousands of times smaller than the mean. If so, the given output channel would barely be affected by those input channels.

We found that only a handful of input channels exhibit such behavior, and most of these do not affect any output channels, regardless of the type. Therefore, we can confirm that the two tasks are strongly intertwined. In fact, there is no difference between them in the sense that both tasks are performed using the same feature set, and they are only divided by the form of the output. Our point is that well-designed architecture performs effectively in more than one image processing task, and not just in a very particular narrow field. Therefore, combining the task of crowd counting with other computer vision tasks to save on computation time offers development potential and future research possibilities.

Furthermore, whereas region-based masking is unfeasible for different types of datasets, the multitask segmentation method can be applied to any human dataset. Unfortunately, to the best of our knowledge, the TH2020 dataset is the only one that includes both human head annotation ground-truth and semantic segmentation ground-truth; therefore, we cannot conduct experiments on any other datasets.

5.3. Considerations about the Dataset

We designate the dataset as helicopter footage; however, the attributes of the dataset are what more significant. The most important of these is the high altitude. High enough so that perspective distortions and size differences can be markedly reduced with a steep camera angle while still covering large areas, but not so high that the targets become too small for detection. While technologically it is feasible to create such footage with drones, there are a few reasons why the use of helicopters is the only option currently.

In most countries, the use of airspace is very strictly regulated. Moreover, in many large metropolitan regions, commercial or public use of drones is prohibited. In regions with more flexible regulations, the use of drones is tied to special permissions but only allowed up to a very limited altitude. Certainly, if authorities are considering putting aerial surveillance systems in place, they may change these regulations for cost efficiency. However, it is still important to keep in mind technological requirements. Due to the high altitude, the camera equipment must have high zooming capabilities to achieve sufficient resolution. Moreover, precise aiming equipment is required to find and keep the target region in focus. These devices greatly increase the payload weight, thereby requiring larger drones with larger energy consumption and shorter flight times.

We also want to mention that there are practical applications where the use of drones is not possible at the current technological level. One such example is a search and rescue operation, where simultaneous crowd counting and semantic segmentation is a helpful tool to identify people in immediate danger. For the rescue, helicopter is the only option and splitting the operation into search and rescue parts to save money may cost time and, consequently, human lives.

6. Conclusions

We have introduced a new large-scale human dataset obtained from a helicopter. Based on the experimental results, the dataset itself is not particularly challenging owing to the lack of scale differences and perspective distortions. However, it offers a viewpoint that is not covered by other conventional crowd counting benchmark datasets. Moreover, the area covered by one image is much larger than that of conventional crowd counting datasets, meaning that the same region requires less computational resources to process.

The focus of our work was DME crowd counting. We have introduced a new method of extracting deeper semantic information regarding humans. Specifically, we have changed the task from crowd counting to pedestrian counting, which cannot be performed simply by detecting human features. The generated density map must use some information from the surroundings of the humans. We achieved this by applying semantic segmentation to the background, ignoring humans, and using these results to mask regions of the crowd density map where the segmentation indicates a non-pedestrian area.

Our experiments demonstrated that if the segmentation quality is high, the efficiency of this masking method is dependent on the quality of the density map estimation. In fact, the change in the accuracy exhibits a strong correlation with the ratio of the L2 norm and L1 norm of the per-pixel estimation error. In particular, estimations with lower L2/L1 ratios (that is, better estimation quality) benefit more from the masking method, whereas estimations with a high L2/L1 ratio may exhibit a decrease in accuracy.

We have also theorized that the density map and segmentation map can both be generated by the same network, resulting in a minimal increase in the memory and computational requirements compared to either task alone. We tested this theory with three different backbone networks, among which the two deeper architectures exhibited comparable or even improved accuracy. This means that there is no real difference between the image processing tasks in the sense that both tasks are performed using the same feature set, and well-designed models can perform effectively in a wide variety of problems. Therefore, combining crowd counting with other computer vision tasks is a valid research direction.

We attempted to replicate the work of other researchers appropriately where necessary, and to confirm the validity of open-source software resources. Furthermore, we executed our experiments with the utmost care to avoid any hazards that could falsify our results.

Author Contributions: Conceptualization, T.K. and Y.S.; methodology, G.C.; software, G.C.; validation, G.C.; formal analysis, G.C.; investigation, G.C.; resources, T.K. and Y.S.; data curation, G.C., T.K. and Y.S.; writing—original draft preparation, G.C.; writing—review and editing, T.K. and Y.S.; visualization, G.C.; supervision, Y.S.; project administration, Y.S.; funding acquisition, Y.S. All authors have read and agreed to the published version of the manuscript.

Funding: This research received no external funding.

Conflicts of Interest: The authors declare no conflict of interest.

References

1. Csönde, G.; Sekimoto, Y.; Kashiyama, T. Congestion Analysis of Convolutional Neural Network-Based Pedestrian Counting Methods on Helicopter Footage. *arXiv* **2019**, arXiv:1911.01672.
2. Tokyo Hawkeye 2020 Pedestrian Dataset. Available online: https://github.com/sekilab/TokyoHawkeye2020 (accessed on 27 August 2020).
3. Zhang, Y.; Zhou, D.; Chen, S.; Gao, S.; Ma, Y. Single-Image Crowd Counting via Multi-Column Convolutional Neural Network. In Proceedings of the IEEE Conference on Computer Vision and Pattern Recognition (CVPR), Las Vegas, NV, USA, 27–30 June 2016.
4. Sindagi, V.A.; Patel, V.M. CNN-Based cascaded multi-task learning of high-level prior and density estimation for crowd counting. In Proceedings of the 2017 14th IEEE International Conference on Advanced Video and Signal Based Surveillance (AVSS), Lecce, Italy, 29 August–1 September 2017; pp. 1–6. [CrossRef]
5. Sam, D.B.; Surya, S.; Babu, R.V. Switching Convolutional Neural Network for Crowd Counting. In Proceedings of the 2017 IEEE Conference on Computer Vision and Pattern Recognition (CVPR), Honolulu, HI, USA, 21–26 July 2017; pp. 4031–4039.
6. Marsden, M.; McGuinness, K.; Little, S.; O'Connor, N.E. ResnetCrowd: A residual deep learning architecture for crowd counting, violent behaviour detection and crowd density level classification. In Proceedings of the 2017 14th IEEE International Conference on Advanced Video and Signal Based Surveillance (AVSS), Lecce, Italy, 29 August–1 September 2017; pp. 1–7.
7. Liu, L.; Qiu, Z.; Li, G.; Liu, S.; Ouyang, W.; Lin, L. Crowd Counting With Deep Structured Scale Integration Network. In Proceedings of the IEEE International Conference on Computer Vision (ICCV), Seoul, Korea, 27 October–2 November 2019.

8. Liu, M.; Jiang, J.; Guo, Z.; Wang, Z.; Liu, Y. Crowd Counting with Fully Convolutional Neural Network. In Proceedings of the 2018 25th IEEE International Conference on Image Processing (ICIP), Athens, Greece, 7–10 October 2018; pp. 953–957.

9. Li, Y.; Zhang, X.; Chen, D. CSRNet: Dilated Convolutional Neural Networks for Understanding the Highly Congested Scenes. In Proceedings of the IEEE Conference on Computer Vision and Pattern Recognition (CVPR), Salt Lake City, UT, USA, 18–22 June 2018; pp. 1091–1100.

10. Chen, X.; Bin, Y.; Sang, N.; Gao, C. Scale Pyramid Network for Crowd Counting. In Proceedings of the 2019 IEEE Winter Conference on Applications of Computer Vision (WACV), Waikoloa Village, HI, USA, 7–11 January 2019; pp. 1941–1950.

11. Zou, Z.; Su, X.; Qu, X.; Zhou, P. DA-Net: Learning the Fine-Grained Density Distribution with Deformation Aggregation Network. *IEEE Access* **2018**, *6*, 60745–60756. [CrossRef]

12. Sindagi, V.A.; Patel, V.M. Generating High-Quality Crowd Density Maps Using Contextual Pyramid CNNs. In Proceedings of the IEEE International Conference on Computer Vision (ICCV), Venice, Italy, 22–29 October 2017.

13. Zhang, A.; Shen, J.; Xiao, Z.; Zhu, F.; Zhen, X.; Cao, X.; Shao, L. Relational Attention Network for Crowd Counting. In Proceedings of the IEEE/CVF International Conference on Computer Vision (ICCV), Seoul, Korea, 27 October–2 November 2019.

14. Yan, Z.; Yuan, Y.; Zuo, W.; Tan, X.; Wang, Y.; Wen, S.; Ding, E. Perspective-Guided Convolution Networks for Crowd Counting. In Proceedings of the IEEE/CVF International Conference on Computer Vision (ICCV), Seoul, Korea, 27 October–2 November 2019.

15. Guo, D.; Li, K.; Zha, Z.J.; Wang, M. DADNet: Dilated-Attention-Deformable ConvNet for Crowd Counting. In Proceedings of the 27th ACM International Conference on Multimedia, Nice, France, 21–25 October 2019; pp. 1823–1832. [CrossRef]

16. Shi, M.; Yang, Z.; Xu, C.; Chen, Q. Revisiting Perspective Information for Efficient Crowd Counting. In Proceedings of the IEEE Conference on Computer Vision and Pattern Recognition (CVPR), Long Beach, CA, USA, 15–21 June 2019.

17. Sindagi, V.A.; Yasarla, R.; Patel, V.M. Pushing the Frontiers of Unconstrained Crowd Counting: New Dataset and Benchmark Method. In Proceedings of the IEEE International Conference on Computer Vision (ICCV), Seoul, Korea, 27 October–2 November 2019; pp. 1221–1231.

18. Sam, D.B.; Babu, R.V. Top-Down Feedback for Crowd Counting Convolutional Neural Network. *arXiv* **2018**, arXiv:1807.08881.

19. Liu, L.; Wang, H.; Li, G.; Ouyang, W.; Lin, L. Crowd Counting using Deep Recurrent Spatial-Aware Network. *arXiv* **2018**, arXiv:1807.00601.

20. Shi, Z.; Mettes, P.; Snoek, C.G.M. Counting With Focus for Free. In Proceedings of the IEEE International Conference on Computer Vision (ICCV), Seoul, Korea, 27–28 October 2019; pp. 4200–4209

21. Jiang, S.; Lu, X.; Lei, Y.; Liu, L. Mask-aware networks for crowd counting. *IEEE Trans. Circuits Syst. Video Technol.* **2019**. [CrossRef]

22. Wan, J.; Luo, W.; Wu, B.; Chan, A.B.; Liu, W. Residual Regression with Semantic Prior for Crowd Counting. In Proceedings of the IEEE Conference on Computer Vision and Pattern Recognition (CVPR), Seoul, Korea, 27–28 October 2019; pp. 4036–4045.

23. Sindagi, V.A.; Patel, V.M. Inverse Attention Guided Deep Crowd Counting Network. In Proceedings of the16th IEEE International Conference on Advanced Video and Signal Based Surveillance (AVSS), Taipei, Taiwan, 18–21 September 2019.

24. Gao, J.; Wang, Q.; Li, X. PCC Net: Perspective Crowd Counting via Spatial Convolutional Network. *IEEE Trans. Circuits Syst. Video Technol.* **2019**. [CrossRef]

25. Paul Cohen, J.; Boucher, G.; Glastonbury, C.A.; Lo, H.Z.; Bengio, Y. Count-ception: Counting by Fully Convolutional Redundant Counting. In Proceedings of the IEEE International Conference on Computer Vision (ICCV) Workshops, Venice, Italy, 22–29 October 2017; pp. 18–26.

26. Xiong, H.; Lu, H.; Liu, C.; Liu, L.; Cao, Z.; Shen, C. From Open Set to Closed Set: Counting Objects by Spatial Divide-and-Conquer. In Proceedings of the IEEE/CVF International Conference on Computer Vision (ICCV), Seoul, Korea, 27 October–3 November 2019; pp. 8362–8371.

27. Zou, H.; Zhou, Y.; Yang, J.; Spanos, C.J. Device-free occupancy detection and crowd counting in smart buildings with WiFi-enabled IoT. *Energy Build.* **2018**, *174*, 309–322. [CrossRef]

28. Liu, S.; Zhao, Y.; Xue, F.; Chen, B.; Chen, X. DeepCount: Crowd Counting with WiFi via Deep Learning. *arXiv* **2019**, arXiv:1903.05316.

29. Long, J.; Shelhamer, E.; Darrell, T. Fully Convolutional Networks for Semantic Segmentation. In Proceedings of the IEEE Conference on Computer Vision and Pattern Recognition (CVPR), Boston, MA, USA, 7–12 June 2015; pp. 3431–3440.

30. Hazirbas, C.; Ma, L.; Domokos, C.; Cremers, D. FuseNet: Incorporating Depth into Semantic Segmentation via Fusion-Based CNN Architecture. In *Computer Vision—ACCV 2016*; Lai, S.H., Lepetit, V., Nishino, K., Sato, Y., Eds.; Springer International Publishing: Cham, Switzerland, 2017; pp. 213–228.

31. Chen, L.; Papandreou, G.; Schroff, F.; Adam, H. Rethinking Atrous Convolution for Semantic Image Segmentation. *arXiv* **2017**, arXiv:1706.05587.

32. He, K.; Zhang, X.; Ren, S.; Sun, J. Deep Residual Learning for Image Recognition. In Proceedings of the 2016 IEEE Conference on Computer Vision and Pattern Recognition (CVPR), Las Vegas, NV, USA, 27–30 June 2016; pp. 770–778.

33. Paisitkriangkrai, S.; Sherrah, J.; Janney, P.; Van-Den Hengel, A. Effective Semantic Pixel Labelling With Convolutional Networks and Conditional Random Fields. In Proceedings of the IEEE Conference on Computer Vision and Pattern Recognition (CVPR) Workshops, Boston, MA, USA, 7–12 June 2015; pp. 36–43.

34. Marmanis, D.; Wegner, J.D.; Galliani, S.; Schindler, K.; Datcu, M.; Stilla, U. Semantic Segmentation of Aerial Images with an Ensemble of CNSS. In *ISPRS Congress*; Halounova, L., Schindler, K., Limpouch, A., Šafář, V., Pajdla, T., Mayer, H., Elberink, S.O., Mallet, C., Rottensteiner, F., Skaloud, J., et al., Eds.; Copernicus Publications: Göttingen, Germany, 2016; pp. 473–480.

35. Fu, G.; Liu, C.; Zhou, R.; Sun, T.; Zhang, Q. Classification for High Resolution Remote Sensing Imagery Using a Fully Convolutional Network. *Remote Sens.* **2017**, *9*, 498. [CrossRef]

36. Maggiori, E.; Tarabalka, Y.; Charpiat, G.; Alliez, P. High-Resolution Aerial Image Labeling With Convolutional Neural Networks. *IEEE Trans. Geosci. Remote Sens.* **2017**, *55*, 7092–7103. [CrossRef]

37. Audebert, N.; Saux, B.L.; Lefèvre, S. Beyond RGB: Very high resolution urban remote sensing with multimodal deep networks. *ISPRS J. Photogramm. Remote Sens.* **2018**, *140*, 20–32. [CrossRef]

38. Lyu, Y.; Vosselman, G.; Xia, G.S.; Yilmaz, A.; Yang, M.Y. UAVid: A semantic segmentation dataset for UAV imagery. *ISPRS J. Photogramm. Remote Sens.* **2020**, *165*, 108–119. [CrossRef]

39. ISPRS Aerial Dataset with 2D Semantic Labeling. Available online: http://www2.isprs.org/commissions/comm3/wg4/semantic-labeling.html (accessed on 26 June 2020).

40. Chan, A.; Liang, Z.S.; Vasconcelos, N. Privacy preserving crowd monitoring: Counting people without people models or tracking. In Proceedings of the IEEE Conference on Computer Vision and Pattern Recognition, Anchorage, AK, USA, 23–28 June 2008; pp. 1–7.

41. Dalal, N.; Triggs, B. Histograms of Oriented Gradients for Human Detection. In Proceedings of the 2005 IEEE Computer Society Conference on Computer Vision and Pattern Recognition (CVPR'05), San Diego, CA, USA, 20–25 June 2005; pp. 886–893. [CrossRef]

42. Dollár, P.; Wojek, C.; Schiele, B.; Perona, P.; Darmstadt, T. Pedestrian detection: A benchmark. In Proceedings of the IEEE Conference on Computer Vision and Pattern Recognition, Miami, FL, USA, 20–25 June 2009.

43. Idrees, H.; Saleemi, I.; Seibert, C.; Shah, M. Multi-source Multi-scale Counting in Extremely Dense Crowd Images. In Proceedings of the 2013 IEEE Conference on Computer Vision and Pattern Recognition, Portland, OR, USA, 23–28 June 2013; pp. 2547–2554. [CrossRef]

44. Bahmanyar, R.; Vig, E.; Reinartz, P. MRCNet: Crowd Counting and Density Map Estimation in Aerial and Ground Imagery. *arXiv* **2019**, arXiv:1909.12743.

45. Robicquet, A.; Sadeghian, A.; Alahi, A.; Savarese, S. Learning Social Etiquette: Human Trajectory Understanding in Crowded Scenes. In *Computer Vision—ECCV 2016*; Leibe, B., Matas, J., Sebe, N., Welling, M., Eds.; Springer International Publishing: Cham, Switzerland, 2016; pp. 549–565.

46. Zhu, P.; Wen, L.; Bian, X.; Haibin, L.; Hu, Q. Vision Meets Drones: A Challenge. *arXiv* **2018**, arXiv:1804.07437.

47. Barekatain, M.; Marti, M.; Shih, H.F.; Murray, S.; Nakayama, K.; Matsuo, Y.; Prendinger, H. Okutama-Action: An Aerial View Video Dataset for Concurrent Human Action Detection. In Proceedings of the IEEE Conference on Computer Vision and Pattern Recognition (CVPR) Workshops, Honolulu, HI, USA, 21–26 July 2017; pp. 28–35.

48. Liu, W.; Salzmann, M.; Fua, P. Context-Aware Crowd Counting. In Proceedings of the IEEE/CVF Conference on Computer Vision and Pattern Recognition (CVPR), Long Beach, CA, USA, 15–21 June 2019; pp. 5099–5108.

49. Cordts, M.; Omran, M.; Ramos, S.; Rehfeld, T.; Enzweiler, M.; Benenson, R.; Franke, U.; Roth, S.; Schiele, B. The Cityscapes Dataset for Semantic Urban Scene Understanding. In Proceedings of the IEEE Conference on Computer Vision and Pattern Recognition (CVPR), Las Vegas, NV, USA, 27–30 June 2016; pp. 3213–3223.

50. Cao, X.; Wang, Z.; Zhao, Y.; Su, F. Scale Aggregation Network for Accurate and Efficient Crowd Counting. In Proceedings of the European Conference on Computer Vision (ECCV), Munich, Germany, 8–14 September 2018; pp. 734–750.

51. Simonyan, K.; Zisserman, A. Very Deep Convolutional Networks for Large-Scale Image Recognition. In Proceedings of the International Conference on Learning Representations, San Diego, CA, USA, 7–9 May 2015.

Article

An Input-Perceptual Reconstruction Adversarial Network for Paired Image-to-Image Conversion

Aamir Khan [1], **Weidong Jin** [1,2,*], **Muqeet Ahmad** [3], **Rizwan Ali Naqvi** [4] and **Desheng Wang** [1]

[1] School of Electrical Engineering, Southwest Jiaotong University, Chengdu 611756, China;
    aamir@my.swjtu.edu.cn (A.K.); wds@my.swjtu.edu.cn (D.W.)
[2] China-ASEAN International Joint Laboratory of Integrated Transport, Nanning University,
    Nanning 530000, China
[3] School of Information Science and Technology, Southwest Jiaotong University, Chengdu 611756, China;
    muqeetahmad@my.swjtu.edu.cn
[4] Department of Unmanned Vehicle Engineering, Sejong University, Seoul 05006, Korea;
    rizwanali@sejong.ac.kr
* Correspondence: wdjin@home.swjtu.edu.cn

Received: 17 June 2020; Accepted: 23 July 2020; Published: 27 July 2020

Abstract: Image-to-image conversion based on deep learning techniques is a topic of interest in the fields of robotics and computer vision. A series of typical tasks, such as applying semantic labels to building photos, edges to photos, and raining to de-raining, can be seen as paired image-to-image conversion problems. In such problems, the image generation network learns from the information in the form of input images. The input images and the corresponding targeted images must share the same basic structure to perfectly generate target-oriented output images. However, the shared basic structure between paired images is not as ideal as assumed, which can significantly affect the output of the generating model. Therefore, we propose a novel Input-Perceptual and Reconstruction Adversarial Network (IP-RAN) as an all-purpose framework for imperfect paired image-to-image conversion problems. We demonstrate, through the experimental results, that our IP-RAN method significantly outperforms the current state-of-the-art techniques.

Keywords: image-to-image conversion; image de-raining; label to photos; edges to photos; generative adversarial network (GAN)

---

1. Introduction

The main objectives of image-to-image conversion tasks are the discovery of suitable latent space and understanding of features maps from source to target images. These tasks have multiple applications in computer graphics, image processing, and computer vision. Image processing applications include: (i) image in-painting, where damaged parts of an image are restored [1,2], (ii) image de-raining where rain-streaks are removed from an input image to get rain-free image [3,4], (iii) image super-resolution where high-quality images are generated from similar degraded images [5–10]. Additional applications exist, however they are not constrained to image denoising [11–13], style transfer [14], image segmentation [15] and image colorization [16,17].

Recently, researchers have developed convolutional neural networks (CNNs) for multiple image-to-image conversion problems. These models mostly come in the form of an encoder-decoder structure where the encoder encodes an input image to some latent space, the decoder decodes from the latent space to the required output image and then they punish the network with a loss function to pick up the mapping between two image domains. Many different loss functions and distinct motivations [5,18] established these models. CNNs utilize reconstruction or pixel-wise losses [5,17,19,20] to generate output images, which are the most upfront techniques. For example, in pixel space, the least absolute or the least-squares losses used to estimate the difference between the ground-truth and generated images. Pixel-wise computation can construct sensible photos. However, in many cases, these losses just capture low-frequency instead of high-frequency components of images, leading to some critical flaws concerning the outputs, e.g., image blurring and image artifacts [7].

Recent years have witnessed that the procedures using the concept of generative adversarial networks (GANs) [21] have accomplished remarkable results in image-to-image conversion tasks. GANs, introduced by Goodfellow et al., is made up of a generator network G and a discriminator network D, targeting to model the real images distribution by synthesizing generated samples, which are very similar to real images. GAN-based models need more memory and computational time in the training process than simple CNN based models as they need to train two networks, i.e., the Discriminator network and the Generator network [22]. Whereas in the testing process, there is only one network, i.e., the generator network. Therefore, the memory and the computational time of GAN based models in the testing process are nearly similar to CNN based models. The significant advantage of using GAN based model is that it generates sharper and more realistic images than CNN based models [23–25]. Hence, the algorithms using the concepts of GANs and conditional GANs (cGANs) [26] have turned out to be a common approach for numerous image-to-image conversion tasks [8,23]. Based on cGAN, pix2pix-cGAN [23] became a representative method aimed at solving the paired image-to-image conversion problems, the objective of which is to map the conditional distribution of the real images conditioned on the given input images [25,27–29].

The critical part of image-to-image conversion tasks is that they have to map high-resolution input grids into high-resolution output grids. Additionally, the issue we consider is that the input and the output have dissimilar surface nature, but both must render the same basic structure to ensure perfect outcomes. There are two popular methods to find out the basic structure of an image, i.e., perceptual features based method [6] and moments based method [30]. The key challenges with methods of moments (MoM) [31] for training deep generative networks are in describing millions of sufficient distinct moments and identiflying an objective function for learning the desirable moments [31,32]. On the other hand, the use of features from deep neural networks (VGG-16) pre-trained on ImageNet dataset [33,34] has led to important advancements in computer vision. Perceptual features have been widely used in piece of works such as super-resolution [6], style transfer [14], and transfer learning [35]. The image generation model is considered to learn from the information in the form of input images, which plays a significant part in the image-to-image conversion task to achieve desire targeted outputs. In paired image datasets, the input structure is roughly matched with the output structure and can significantly affect the production of the image generation models. For example, Figure 1 shows that the window frames are not accurately labeled in the corresponding input images. Hence, the image generation model requires further information to capture targeted high-resolution output grids against each given missing high-resolution input grids. Despite considerable progress, we note that the previous approaches have not examined optimized additional input information for imperfect paired datasets.

To overcome the problem of imperfect paired datasets and to attain desired results, we opted to feed this extra information in the form of input-perceptual loss (i.e., calculated between imperfect paired images) into the objective function of the proposed model. It is an essential issue, as the perfect paired dataset is expensive and hard to collect. This work introduces a trade-off between collecting large-amount of the perfect paired dataset and an optimized training for the image-to-image conversion network.

The remainder of the study is as follows: We discuss the previous research of the image-to-image conversion with details in Section 2. The IP-RAN methodology, objective, and network architecture are explained in Section 3. In Section 4, we present the experiments, results, and analysis of different loss functions and generator configurations. Section 5 presents the conclusions and future work.

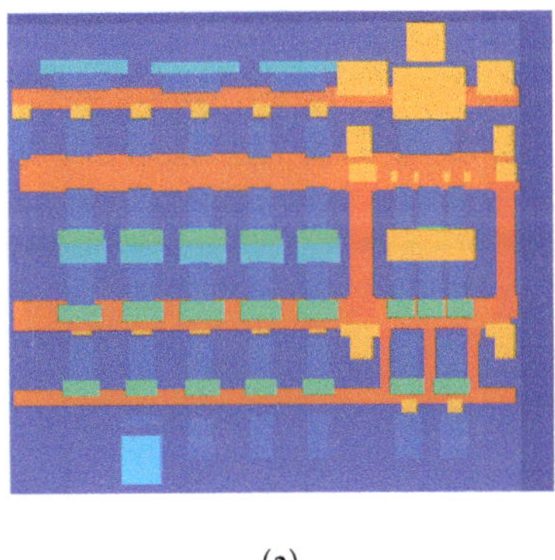

(a)

(b)

Figure 1. Example of the label to architectural photos. (**a**) shows an input labeled image. (**b**) shows marked objects in a ground-truth image.

2. Related Work

In previous years, the training of deep convolutional neural networks using back-propagation algorithms with per-pixel loss functions has solved a broad range of feed-forward image-to-image conversion tasks [18,36]. Various techniques of image-to-image conversion employ only pixel-level losses or pixel-level losses preceded by several additional losses [6,23]. Image segmentation techniques generate dense scene labels by operating networks in a fully convolutional way over a single input image [20,37–40]. Image de-raining techniques try to eliminate rain strikes in uncontrolled weather images [41,42]. Image super-resolution techniques generate a high-resolution image for a given its low-resolution matching part [5,6]. Image in-painting is designed to retrieve the missing portion of the given damaged image [1,43,44]. Other examples of image-to-image conversion techniques modeled on feed-forward CNNs exist, however, they are not constrained to depth estimations [37,45,46] and image colorization [19], etc.

A series of GAN-family [16,26,47,48] networks was introduced in a short time for an enormous variety of problems since Goodfellow introduced the influential concept of Generative Adversarial Nets (GAN) [21] in 2014. GANs also showed promising results in numerous applications for computer vision, for example, image generation, representation learning [48], image editing [49], etc. Specifically, various extended GANs accomplished good results at several image generation applications such as style transfer [24], super-resolution [7], image inpainting [1], text2image [50], and like many other domains including videos [51] and 3D data [52]. These studies also consist of but are not constrained to the PGN introduced for video prediction [53], the iGAN introduced for interactive application [54], the SRGAN added for super-resolution [7], and the ID-CGAN presented for image de-raining [3].

Moreover, some of these works based on GANs are dedicated to developing an improved generative model, for example, WGAN(-GP) [55,56], Energy-based GAN [57], Progressive GAN [58], SN-GAN [59] and E-GAN [60]. A conditional image generation based on GANs has also been actively studied recently. Some advanced GAN models continuously improved the quality of particular tasks, e.g., InfoGAN [16], cGANs [26], and LAPGAN [61] have been introduced to image translation recently for their easy execution and outstanding results. The cGANs [26] hold category labels as conditional data for the generation of particular images. Some of the works have included GANs into their designs to enhance the efficiency of conventional tasks, e.g., for small entity (or object) detection, the PGAN [62] was adopted. Specifically, Li et al. [62] developed an innovative perceptual-discriminator network, which includes a perception block and an adversarial block. Wang et al. [25] used different layers of discriminator network to measures perceptual losses. Sung et al. [63] introduced new paired input conditions for the replacement of conditional adversarial networks to improve the image-to-image translation tasks.

Additionally, some modifications of the GANs [29,64–66] examined cross-domain image conversions over discovering the linear mapping relationship among various image domains. In particular, primal GAN intentions to investigate the mapping relationships between input images and target images, although a double (or opposite) GAN does the opposite task. Such GANs shape a closed-loop and enable the translation and reconstruction of images from either domain. These designs can also be used to execute image conversion operations in the lack of paired examples by merging cycle consistency loss and GAN loss. However, paired data is available for training in specific applications, Ge et al. [29], Zhu et al. [64], Yi et al. [65], and Kim et al. [66] ignore that paired data often achieves less than paired methods [23]. It is therefore still essential at this point to study paired data training, particularly for performance motivated circumstances and implementations like the photo-realistic picture synthesis [7], high-resolution image synthesis [8], real-world image painting [67], etc.

In GANs based works, generator networks are the same as the aforementioned encoder-decoder structure in CNNs. As the training of deep CNNs suffer from vanishing gradient problem. Therefore, many previous works [3,4,25] used skip-connections in the generator to pass the gradient easily to prior layers of the encoder. Unfortunately, these skip-connections directly carry unwanted information from the inputs to the resultant images, hence affecting the visual quality of the constructing images. In the demand to develop a visually appealing image-to-image conversion model, we have to consider the following facts into the optimization method:

- The principle, to perfectly map targeted output images must not be affected by the texture of the given input images, which should be the essential pillar in the formation of a generator structure.
- The visual quality of constructed images should also be considered in the optimization method rather than just relying on qualitative performance metric values. This principle can guarantee that the generated images look visually appealing and realistic.

Under the above criteria, we present the Input-Perceptual and Reconstruction Adversarial Networks (IP-RAN) for image-to-image conversion tasks. The IP-RAN consists of an encoder-decoder network G; for converting an input image to the desired output image, a discriminator network D; to flag the real or fake photos and an input-perceptual loss network P; to calculate fundamental structure difference between an input image and the ground-truth image. We employ the input-perceptual, the traditional reconstruction L1, and the generative adversarial losses in the objective function. Initially, this work utilized the input-perceptual loss to calculate the missing information of the basic structure in the input images according to the target images. Then, this study used similar to many traditional losses the L1 loss for penalizing generated images to be near to the targeted images. Meanwhile, we used the generative adversarial losses to estimate the distribution of converted images, i.e., to punish the generated distribution for converging into the target distribution of output, which generally results in the production of more visually pleasing images. The contributions of this study are as follows:

- This study introduces a novel approach to deal with imperfect paired datasets and the method of feeding extra information into the objective function in the form of input-perceptual losses calculated between the input images and the target images for imperfect paired datasets.
- We introduce an optimized method based on pix2pix-cGAN and conditional GANs (cGANs) frameworks for existing imperfect pair datasets.
- We also analyzed the primary two different configurations of the generator structure, and the results show the proposed approach is better than previous methods.
- We achieve both qualitative and quantitative results by using IP-RAN, which indicates that the adopted technique produces better results than the baseline models.

Table 1 shows a comparison between the proposed and existing methods.

Table 1. Comparison between state-of-the-art and proposed method.

Methods	Advantages	Disadvantages
CNNs (Reconstruction L1 and L2 losses) based methods [1,17]	Need less computation as one network is to trained.	Need big datasets to train
	Fast and easy to train	Produce blurry results
Simple GAN (Adversarial Loss) based methods [21,26]	Can be trained with small datasets	More computation than CNNs as two different networks to be trained
	Produces sharp and realistic images	GAN networks are difficult to train
		There is an image artifacts problem
Adversarial, reconstruction and perceptual losses with skip-connections in generator network based methods [3,4,23,25,29,63]	Achieve good quality results than CNNs and simple GAN by combining two loss functions	Skip-connections affect the quality of generated images by directly passing unwanted input information to the output of the network.
	Skip-connections in generator configuration reduce vanishing gradient problem	
Proposed method	This method adds extra information to the objective function to optimize the results.	Need to calculate input-perceptual losses which increase training time
	Use the Resnet bottleneck structure in the generator configuration to reduce the vanishing gradient problem.	
	Achieves excellent results visually and quantitatively	

3. Methodology

In this work, we have two sets of paired training images, i.e., a set of input images $\{x_i\}_{i=1}^{N} \in X$ and a set of target output images $\{y_i\}_{i=1}^{N} \in Y$. We train the generative network G that the fake generated images $G(x)$ to be same as the real targeted images, and alongside we train a discriminative network D to distinguish the fake generated images $G(x)$ from the real targeted images. The generator network learns the mapping from an input domain to a real-world domain by minimizing adversarial losses, aiming to deceive the discriminator network. The generator has sub-networks: an encoder *Enc*, residual blocks *Res*, and a decoder network *Dec*. The encoder network contains a sequence of convolutional layers, which convert an input image into encoded feature space $Enc(x)$. Later, the output of encoder network, $Enc(x)$, becomes the input of residual blocks [68]. The output of the residual layers, $Res(Enc(x))$, is the activation maps which feed to the decoder network *Dec*. At that moment, a sequence of fractionally-stride convolutionary layers decode the converted features into the fake generated image $G(x)$. Equation (1) expresses the output of the generator network:

$$G(x) = Dec(Res(Enc(x)))\tag{1}$$

The whole network architecture is shown in Figure 2 and is called the Input-Perceptual Pixel-Reconstruction Generative Adversarial Networks (IP-RAN).

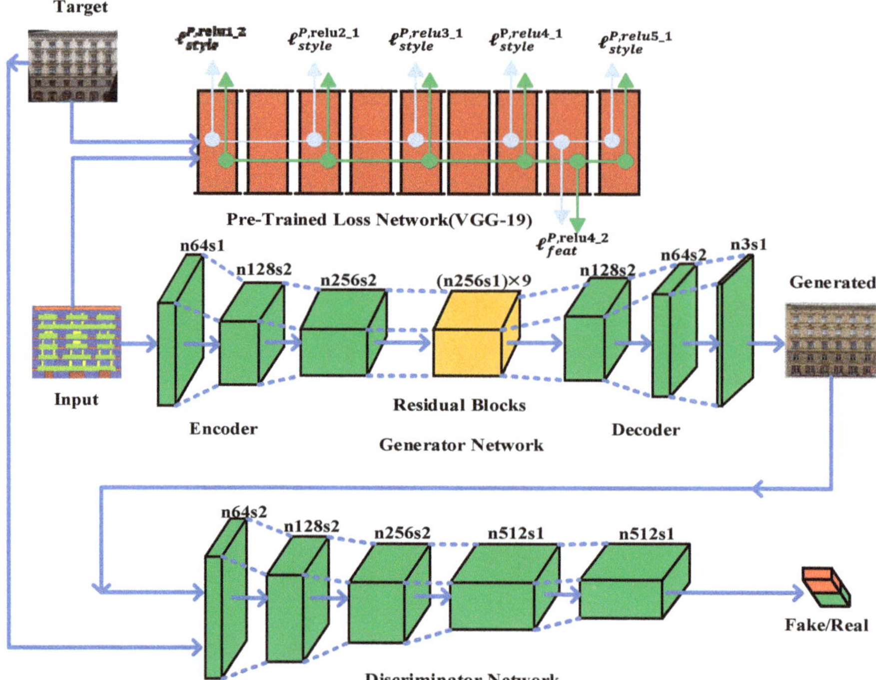

Figure 2. IP-RAN framework. IP-RAN consists of generator network, *G*, input-perceptual loss network, *P*, and discriminator network, *D*. The generator network, *G*, is intended to generate translated images from given input images. It is composed of an encoder-decoder structure that includes two down-sampling layers of stride-2 convolution, several residual blocks, and two up-sampling layers stride-2 of transposed convolution. Input-perceptual loss network, *P*, is the pre-trained VGG-19 and used to extract features from hidden layers to calculate the perceptual loss. The discriminator network, *D*, consists of convolutional-BatchNorm-LeakyRelu layers, and its output is used to distinguish generated images from real images.

3.1. Objective

The input-perceptual loss calculated between high-resolution input grids and targeted high-resolution output grids, which decrease the effect of less information in the input images and useful against imperfect paired datasets. Equation (2) expresses input-perceptual loss:

$$\mathcal{L}_P(P) = \varphi_c l_f + \varphi_s l_s \tag{2}$$

where l_f is the feature reconstruction as given in Equation (3), and l_s is the style reconstruction losses as given in Equation (4), are the two parts of the perceptual loss function, as Johnson et al. described in [6]. Input-perceptual losses are utilized to measure fundamental structural differences such as common patterns, texture, colors, etc., between the high-resolution input grids and the high-resolution target grids.

Let $P_i(x)$ be the activation maps for the ith layer of the network P when processing the image x. If i is a convolutional layer then $P_i(x)$ will be an activation map having a shape of $C_i \times H_i \times W_i$. The feature reconstruction loss can be calculated as Euclidean distance between activation maps as follows:

$$l_f = \ell_{feat}^{P,i}(x,y) = \frac{1}{C_i H_i W_i} \|P_i(x) - P_i(y)\|_2^2 \tag{3}$$

where P_i denotes the non-linear CNN transformation at the ith layers of the loss network, P. The ℓ^P_{feat} loss aims to measure the discrepancy between high-level features of the given images.

The style reconstruction loss can be computed as squared Frobenius norm for the discrepancy between the Gram matrices of the input and the targeted images as follows:

$$l_s = \ell^{P,i}_{style}(x, y) = \left\|\mathbb{G}^P_i(x) - \mathbb{G}^P_i(y)\right\|^2_F \tag{4}$$

where $\mathbb{G}^P_i(x)$ is the Gram matrix of ith layer activation maps of a given image x extracted from network P. $\mathbb{G}^P_i(x)$ is defined as the components of the $C_i \times C_i$ matrix is given by:

$$\mathbb{G}^P_i(x)_{c,c^*} = \frac{1}{C_i H_i W_i} \sum_{h=1}^{H_i} \sum_{w=1}^{W_i} P_i(x)_{h,w,c} P_i(x)_{h,w,c^*} \tag{5}$$

where $P_i(x)$ interpret as giving C_i-dimensional activation maps for each point on $H_i \times W_i$ grid, and the Gram matrix, $\mathbb{G}^P_i(x)$, relates to non-centric covariance of the C_i-dimensional activation maps, processing each grid site as an autonomous sample. Therefore, it gathers details about the features that appear to be working together. The Gram matrix can also be determined accurately by transforming $P_i(x)$ into a matrix ϕ of shape $C_i \times H_i W_i$; then $\mathbb{G}^P_i(x) = \frac{\phi\phi^T}{C_i H_i W_i}$.

Generative adversarial loss [21], which trains G and D together as the two-player mini-max game with loss function $\mathcal{L}_{GAN}(G, D)$. The generator network G attempts to produce an image $G(x)$ that appears similar to the image in the target domain Y, while the discriminator network D attempts to differentiate between them. In particular, we train the discriminator network, D, to maximize the likelihood of classifying the correct label to the targeted image and the generated image $G(x)$, while training G is to minimize the likelihood of classifying the correct label to the generated image $G(x)$. The mini-max game can be formulated as:

$$\min_G \max_D \mathbb{E}_{y \in Y}[\log(D(y))] + \mathbb{E}_{x \in X}[\log(1 - D(G(x)))] \tag{6}$$

GANs-based models have revealed the significant ability to learn generative models, particularly for image generation tasks [16,53,55]. Therefore, we also implement the GANs learning process to resolve image conversion tasks. As illustrated in Figure 2, the image generation network G is used to produce output image, $G(x)$, against the input image, $x \in X$. In the meantime, each input image x_i has a correspondent target image y_i. We assume that all target images, y, follow the distribution $y \in Y$, and the generated images, $G(x)$, are motivated to have matching distribution as targeted images y, i.e., $G(x) \sim Y$. Besides, to accomplish the generative adversarial learning approach, a discriminative network, D, is added, and the adversarial loss function can be expressed as follows:

$$\min_G \max_D V(G, D) = \mathbb{E}_{y \in Y}[\log(D(y))] + \mathbb{E}_{x \in X}[\log(1 - D(G(x)))] \tag{7}$$

We use least squares loss (LSGAN) as discussed in [69], which offers a non-saturated and smooth gradient for discriminator network D. Adversarial loss, $\mathcal{L}_{GAN}(G, D)$, is expressed as:

$$\mathcal{L}_{GAN}(G, D) = \mathbb{E}_{y \in Y}\left[(D(y) - 1)^2\right] + \mathbb{E}_{x \in X}\left[D(G(x))^2\right] \tag{8}$$

The generative adversarial loss turns as per the numerical measurement to punish the variance between the distributions of generated images and ground-truth images.

The basic GAN framework is unstable as it trains two competing neural networks. In [64], the author noted that one cause for instability is that there are un-unique solutions during the training of the generator. As shown in Figure 3, several artifacts introduced by the standard GAN structure can be observed which significantly impacts the visual quality of the output image. Previous methods have found that it is useful to combine GAN objectives with more traditional losses such as L2 loss [1] in such way that the work of the discriminator remains unchanged as in Equation (8), but the task of the generator is not only to deceive the discriminator but also to make generated image closer to the targeted ground truth image according to L2. In our method, we used L1 distance instead of L2, because L1 encourages blur reduction:

$$\mathcal{L}_{L1}(G) = \mathbb{E}_{x,y}\big[\|y - G(x)\|_1\big] \tag{9}$$

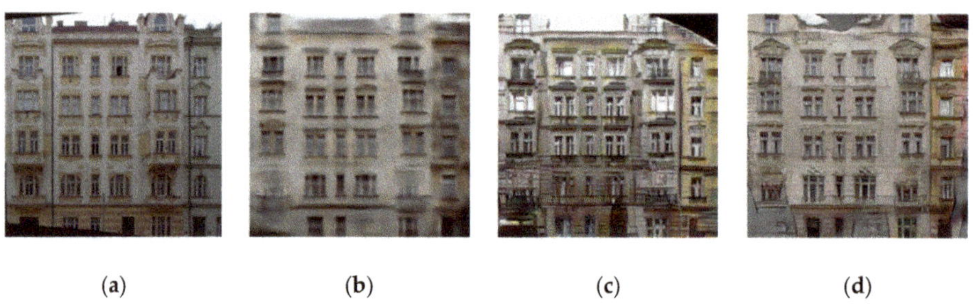

(a)	(b)	(c)	(d)

Figure 3. (a) ground-truth image, (b) image generated by conventional CNN using L1 loss function, (c) image generated by standard GAN using adversarial loss and (d) image constructed by the proposed method with Input-Perceptual and Reconstruction Adversarial losses

The adversarial loss helps the generator and protect from the blurry effect of L1 loss as well as remain close to the targeted output images. The final objective for the generator network is expressed as:

$$\mathcal{L}_{G_T} = \varphi_g \mathcal{L}_{GAN}(G) + \varphi_{L1} \mathcal{L}_{L1}(G) + \varphi_P \mathcal{L}_P(P) \tag{10}$$

where $\mathcal{L}_{G_T}$ represents the total generator network loss which is the sum of the generator's adversarial loss, $\mathcal{L}_{GAN}(G)$, L1 reconstruction loss, $\mathcal{L}_{L1}(G)$, and the input-perceptual, $\mathcal{L}_P(P)$.

3.2. Network Architecture

Figure 2 demonstrates the proposed structure consisting of three CNNs networks, i.e., the generator network, G, the input-perceptual loss network, P, and the discriminative network, D.

Recently, many solutions [3,23,25] to these problems used skip-connections in the generator network to shuttle the information directly from input to output throughout the network and to solve the vanishing gradient problem. On the one hand, skip-connections are useful in resolving the vanishing gradient problem. Still, for image-to-image conversion problems, these skip-connections are carrying unwanted information from the input throughout the network and influencing the performance of the results critically, see Figure 2. We utilize the ResNet [68] framework same as Johnson et al. [6], with an encoder-decoder structure instead of skip-connections between encoder-decoder layers to avoid unwanted information coming from the input and to produce visually pleasing results. Our generator network includes two downsampling layers of stride-2 convolution, nine residual blocks, and two upsampling layers with stride-2 of transposed convolution and utilizes instance normalization [70], for specifications, see Table 2. The input-perceptual loss network, P, uses VGG-19 pre-trained on the ImageNet dataset [33,34]. We extract features from six layers (Relu-1 of block1, Relu-1 of block2, Relu-1 of block3, Relu-1 of block4, Relu-1 of block5) for style loss l_s and Relu-2 of block4 for feature loss l_f of pre-trained VGG-19 to calculate input-perceptual losses.

In this work, we use 70 × 70 Markovian PatchGANs [7,23,71] for the discriminator network D to classify whether 70 × 70 overlapping patches of images are real or fake. Patch-level discriminator has fewer parameters than a full-image discriminator and can operate in a fully convolutionary fashion on images of arbitrary size [23].

Table 2. Generator Network of IP-RAN.

	Operation	Pre-Reflection Padding	Kernel Size	Stride	Non-Linearity	Feature Maps
Encoder entry 2	Convolution	3	7	1	ReLU	64
	Convolution		3	2	ReLU	128
	Convolution		3	2	ReLU	256
Residual Blocks	Residual block	1	3	1	ReLU	256
	Residual block	1	3	1	ReLU	256
	Residual block	1	3	1	ReLU	256
	Residual block	1	3	1	ReLU	256
	Residual block	1	3	1	ReLU	256
	Residual block	1	3	1	ReLU	256
	Residual block	1	3	1	ReLU	256
	Residual block	1	3	1	ReLU	256
	Residual block	1	3	1	ReLU	256
Decoder	Deconvolutional		3	2	ReLU	128
	Deconvolutional		3	2	ReLU	256
	Convolutional	3	7	1	Tanh	256

4. Experiments and Results

In this section, we first discuss the specifications of the datasets, proposed model, and training parameters. We compared the IP-RAN with the standard approaches and current state-of-the-art methods. We also discuss the information on the experiments and performance measures used to test the proposed method.

4.1. Datasets

Experiments are carried out on several datasets to evaluate the performance of IP-RAN and other state-of-the-art methods. We use three public paired datasets which are as follows:

- CMP facades dataset [72] is used to train for architectural "Labels to Photos" task.
- Dataset provided by ID-CGAN [3] is used to train for the "Image De-raining" task.
- Dataset formed by pix2pix [23] is used to train for the "Edges to Photos" task. The original dataset has come from [54] and [73], and the use of the HED edge detector [74] to extract edges. All images are scaled to 256×256.

4.2. Model and Parameter Details

In this subsection, we discuss the model and the parameter details. In the case of GAN loss ($\mathcal{L}_{GAN}$), we replace the criterion of negative log-likelihood with a least-square loss [69] for the network's training stabilization. This least-square loss is found more stable throughout the training procedure and produces higher quality results. In general, for $\mathcal{L}_{GAN}(G, D)$, we set that G, train to minimize $E_{x \sim p_{data}(x)}\left[(D(G(x)) - 1)^2\right]$ and D, train to minimize $E_{y \sim p_{data}(y)}\left[(D(y) - 1)^2\right] + E_{x \sim p_{data}(x)}\left[(D(G(x))^2\right]$. Furthermore, we divide the discriminator's criterion by 2 when optimizing D, which slows the learning rate of D proportional to G. We apply the Adam optimizer [75] and use minibatch Stochastic Gradient Decent (SGD), setting a learning rate of $\alpha = 0.0002$, $\beta 1 = 0.5$. Relu activation function, with slope value of 0.2, is used in the generator network, G, except the last layer used *tanh*. The Batch size is set to one for all of the experiments. The training parameters are set as $\varphi_g = 1$, $\varphi_{L1} = 10$, $\varphi_s = 1$ and $\varphi_c = 0$ for labels to photos task, $\varphi_g = 1$, $\varphi_{L1} = 10$, $\varphi_s = 1$ and $\varphi_c = 1 \times 10^{-6}$ for edges to photos task, and $\varphi_g = 1 \times 10^{-9}$, $\varphi_{L1} = 10$, $\varphi_s = 1$ and $\varphi_c = 1 \times 10^{-6}$ for image de-raining task.

4.3. Evaluation Criteria

For a performance demonstration of image-to-image conversion tasks, we performed qualitative and quantitative tests to determine the quality of the generated images. We directly present input and generated images for qualitative assessments. We apply quantitative measures on test sets to assess the performance of different model and configurations such as, Peak Signal to Noise Ratio (PSNR), Structural Similarity Index (SSIM) [76], Visual Information Fidelity (VIF) [77], and Universal Quality Index (UQI) [78]. These quantitative measures valuation are based on the luminance channel of the image. FID score [79] determines the distance between the real data distribution and the generated data distribution.

4.4. Analysis of Different Loss Functions

We train models to separate the effect of different variations of loss functions on the architectural CMP facades "label to photos" dataset. We perform tests to compare the impact of each part of Equation (10). Figure 4 shows the qualitative results of the variations mentioned below on labels to photos problem.

- L1, by setting $\varphi_g = 0$ and $\varphi_P = 0$ in Equation (10), causes to generate blurry outputs.
- The cGAN, by setting $\varphi_{L1} = 0$ and $\varphi_P = 0$ in Equation (10), leads to much sharper outputs but brings visual artifacts.
- L1 and cGAN together, by setting $\varphi_P = 0$ in Equation (10) causes sensible results but still far from the targeted outputs.
- The results of the proposed loss function in Equation (10), show a significant improvement in quality and similarity to the targeted results.

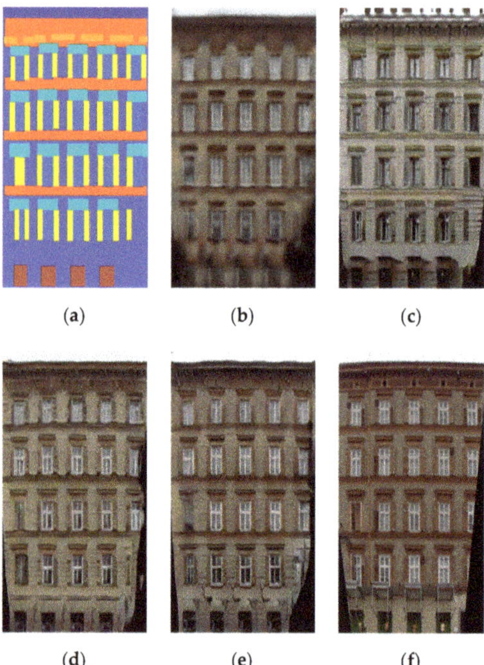

Figure 4. Shows label input against different loss functions that produce different architectural photo results. (**a**) input label image, (**b**) result of L1 ($\mathcal{L}_{L1}(G)$) alone, (**c**) result of cGAN ($\mathcal{L}_{GAN}(G)$) alone, (**d**) result of L1+cGAN ($\mathcal{L}_{GAN}(G) + \mathcal{L}_{L1}(G)$), (**e**) result of the IP-RAN, and (**f**) target output photo.

In Table 3, we compared the abovementioned cases quantitatively using the PSNR, SSIM, UQI, VIF, and FID scores on the labels to photos dataset. L1 achieves higher scores in PSNR, SSIM, UQI, and VIF, but the output results are blurred images and are very poor in FID-score. Hence, pointing out that the results are visually unpleasant. We observed from Figure 4 and Table 3 that for blurry images PSNR, SSIM, UQI, and VIF evaluation scores perform inferiorly. Table 3 shows that cGAN alone achieves poor scores in PSNR, SSIM, UQI, and VIF, which indicating that results are less similar to the targeted output. However, it has got a good FID-score as compare to L1 that shows results have a recognizable structure. Table 3 shows that the IP-RAN achieves the best possible scores in PSNR, SSIM, UQI, VIF, and FID. Hence, the results are similar to the targeted output as well as have a recognizable structure, and they are visually pleasing.

Table 3. Quantitative results compared with different loss functions.

	PSNR(dB)	SSIM	UQI	VIF	FID
L1	**13.43**	**0.2837**	**0.8186**	**0.0627**	176.74
cGAN	11.86	0.1996	0.7722	0.0399	111.00
L1+cGAN=CGAN	12.80	0.2399	0.8035	0.0480	113.53
IP-RAN	12.84	0.2426	0.8052	0.0488	**110.29**

4.5. Analysis of Different Generator Configuration

The encoder-decoder structure does not have skip-connections among the layers. The U-Net structure has skip-connections between encoder layers and decoder layers, as shown in Figure 5. We have trained both structures on image de-raining dataset and labels to photos dataset with similar loss function using pix2pix-cGANs [23] architecture. We conducted tests to compare both structures.

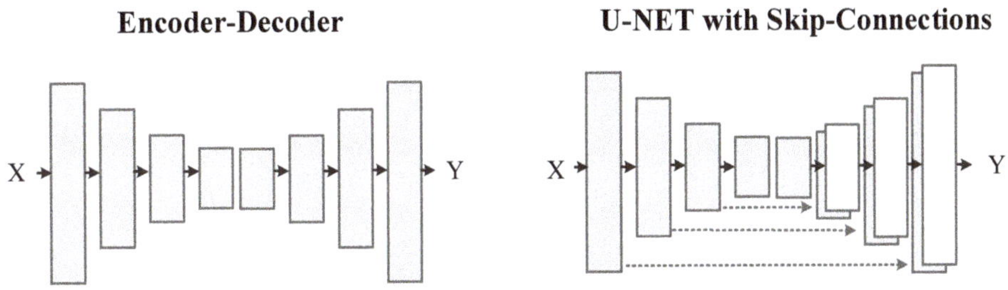

Figure 5. Different structures of image-to-image generation networks.

Figure 6 shows the encoder-decoder structure achieves excellent results without losing any information than the U-Net structure. Skip-connections passing unwanted information of the input images, which have a severe influence on generated images, leads to corrupted results and poorly achieved their targets. In the image de-raining task, the generator structure with skip-connections poorly converts between the rain to de-raining images. In Figure 6c the first four rows, where rain-streaks still can be found in resultant images. The resultant images inherit this unwanted information via skip-connections from the corresponding input images. Figure 6c the last four rows, where resultant images contain bluish and greenish color effects, which are directly coming from the input labeled images via skip-connections.

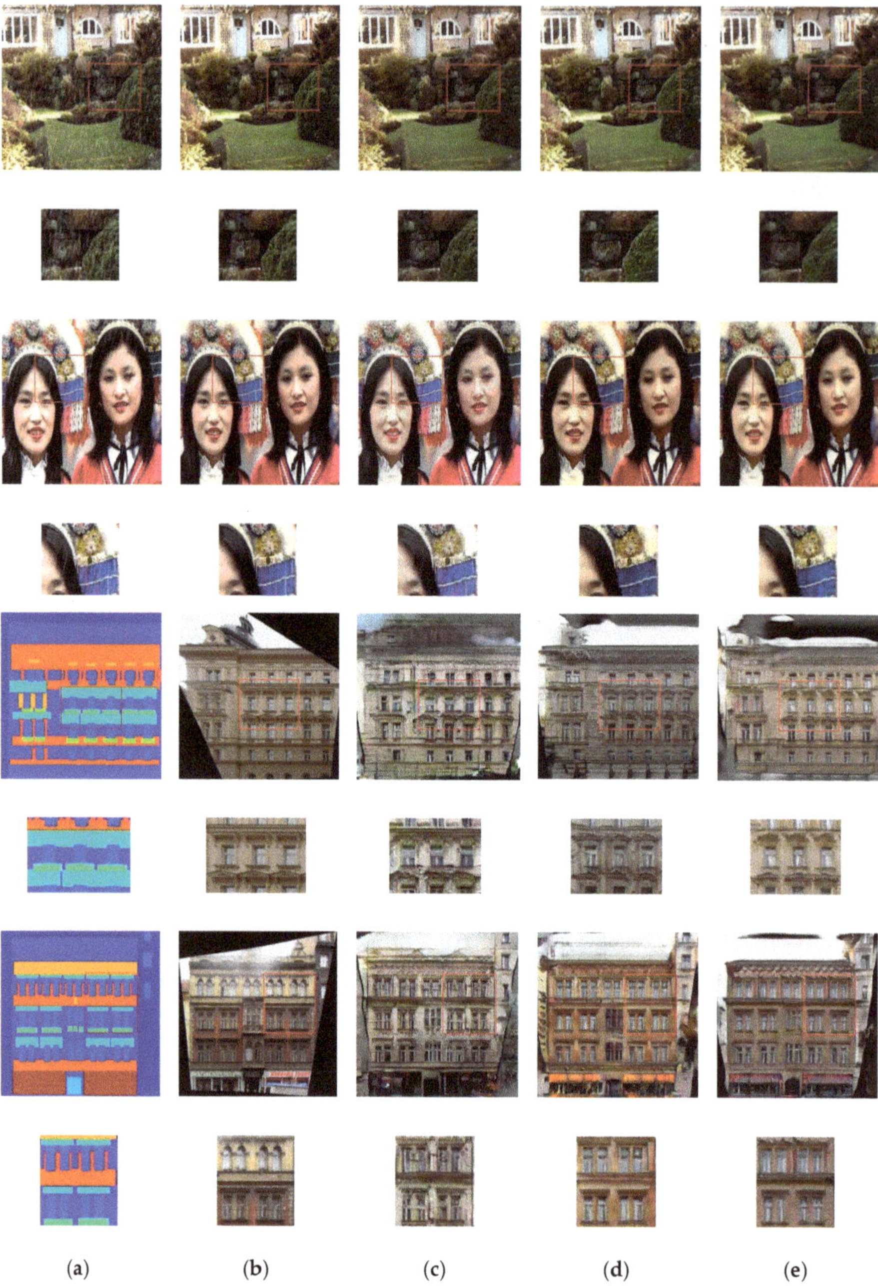

(a)   (b)   (c)   (d)   (e)

Figure 6. Sample results in the first four rows of rainy to de-raining images and last four rows of labels to architectural photos. For good visual comparison, the smaller images below the test images represent specific regions-of-interest. (**a**) input images, (**b**) targeted photos, (**c**) U-Net with skip-connections, (**d**) Encoder-Decoder and (**e**) IP-RAN.

4.6. Comparison with Baseline

For comparison purposes, we selected the following latest state-of-the-art approaches for image-to-image conversion problems:

- Pix2Pix-cGAN [23]: Pix2pix is designed for paired image datasets based on the cGAN architecture. Pix2Pix utilizes L1 reconstruction loss and adversarial loss to train its model for the conversion of input images to output images.
- UTN-GAN [29]: UTN-GAN introduced a GAN-based unsupervised transformation network with hierarchical representations learning and weight-sharing technique. The reconstruction network learns the hierarchical representations of the input image, and the mutual high-level representations are shared with the translation network to realize the target-domain oriented image translation.
- PAN [25]: PAN can learn a mapping function to transform input images to targeted output images. PAN consists of a image transformer network and a discriminator network. In PAN, the discriminator measures perceptual losses on different layers and identifies between real and fake images. PAN uses perceptual adversarial losses to train the generator model.
- iPANs [63]: iPANs used U-NET as image transformation network and perceptual similarity network as a discriminator network. iPANs introduced new paired input conditions for the replacement of conditional adversarial networks to improve the image-to-image translation tasks. In this method the ground-truth images which are identical images are the real pair, whereas the generated images and ground-truth images are the fake pair.
- ID-CGAN [3]: ID-CGAN introduced to handle the image de-raining task by combining the pixel-wise least-squares reconstruction loss, conditional generative adversarial losses, and perceptual losses. ID-CGAN used cGAN structure to map from rainy images to de-rainy images. ID-CGAN consists of a dense generator to transform from an input image to its counter-part output images. ID-CGAN used the pre-trained VGG-16 network to calculate the perceptual losses between generated and ground-truth images.

4.6.1. Comparison with Pix2Pix-cGAN, PAN, UTN-GAN and iPANs

We attempt to transform semantic labels to architectural photos. This inverse conversion is a complicated process and distinct from the tasks of image segmentation. Pix2Pix-cGAN and UTN-GAN used adversarial and reconstruction losses, and PAN and iPANs used adversarial and perceptual losses to produce labels to architectural photos as shown in Figure 7. After the comparison, we observe the adopted approach captures further information and generates realistic and more similar images to the targeted photos with less deformation. Furthermore, the quantitative assessment in Table 4 also demonstrates that the IP-RAN can attain substantially improved results.

Creating a real-world object from the corresponding input edges is one of the image-to-image conversion tasks as well. We train the IP-RAN on the dataset given by [23] to convert edges-to-shoes and compare its results by the outcomes of pix2pix-cGAN, PAN, UTN-GAN and iPANs. Figure 8 shows shoe photos generated from given input edges by the proposed method, pix2pix-cGAN, PAN, UTN-GAN and iPANs, while Table 5 presents the quantitative measures on the test set results. By observing and comparing the constructed shoe photos, we find that the IP-RAN, pix2pix-cGAN and PAN accomplished promising results, so far, it's difficult to express which of these is better. On the measurement score of UQI and FID, the IP-RAN performed slightly weak compared to pix2pix-cGAN and PAN, yet superior in the other quantitative measures.

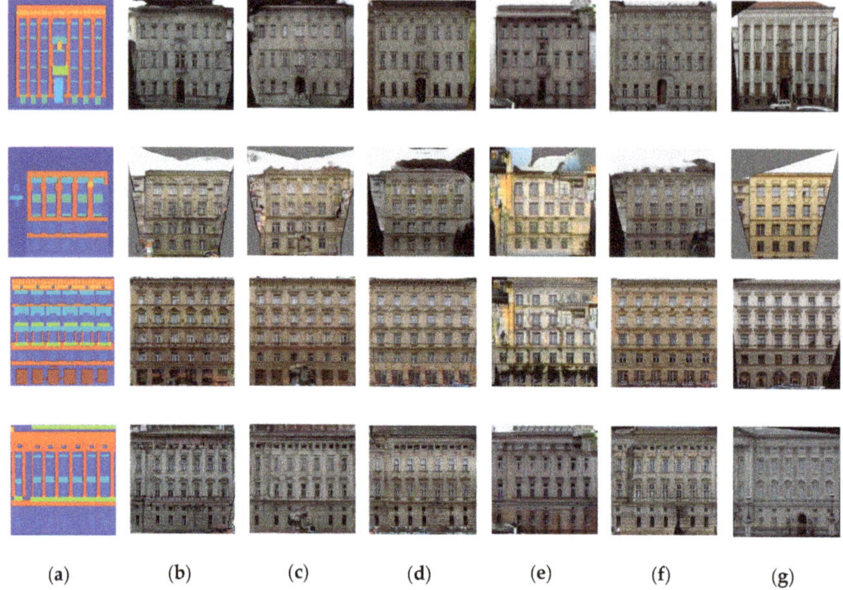

(a) (b) (c) (d) (e) (f) (g)

Figure 7. Samples results from paired labels to architectural photos. (**a**) input images, (**b**) results of pix2pix-cGAN, (**c**) results of UTN-GAN, (**d**) results of PAN, (**e**) results of iPANs, (**f**) results of the IP-RAN, and (**g**) targeted photos.

(a) (b) (c) (d) (e) (f)

Figure 8. Samples result from edges to shoes. (**a**) input images, (**b**) results of the IP-RAN, (**c**) results of the pix2pix-cGAN, (**d**) results of PAN, (**e**) results of UTN-GAN, and (**f**) results of iPANs

Table 4. Quantitative results of labels to architectural photos, bold results show good scores.

	PSNR(dB)	SSIM	UQI	VIF	FID
Pix2Pix-cGAN	**13.37**	**0.2559**	**0.8195**	**0.0541**	113.53
UTN-GAN	12.78	0.2362	0.8016	0.0481	111.86
PAN	12.82	0.2370	0.8030	0.0477	112.47
iPANs	11.46	0.1765	0.7603	0.0382	140.70
IP-RAN	12.84	0.2426	0.8052	0.0488	**110.29**

Table 5. Quantitative results of Edges to Shoes, bold results show good scores.

	PSNR(dB)	SSIM	UQI	VIF	FID
Pix2Pix-cGAN	19.33	0.7569	**0.9220**	0.2092	**59.93**
UTN-GAN	15.41	0.6588	0.8255	0.1786	104.9
PAN	19.11	0.7389	0.9187	0.2034	62.13
iPANs	15.71	0.6671	0.8444	0.1778	117.1
IP-RAN	**19.42**	**0.7608**	0.9179	**0.2153**	62.15

4.6.2. Comparison with UTN-GAN, ID-CGAN and iPANs

ID-CGAN and iPANs try to resolve the image de-raining problem. They aim to eliminate rain streaks from a given input rainy photos. Assuming un-predictable weather situations, the image de-raining or de-snowing alone is a challenging image-to-image conversion problem.

We try to resolve a single image de-raining task by the IP-RAN using a similar configuration to ID-CGAN. We train our adopted scheme on the image de-raining dataset provided by ID-CGAN [3]. This dataset contains 700 synthesizing images for training, whereas 100 artificial and 50 real-world rainy images are presented for testing purposes. Figure 9 shows the sample results of synthetic test images. As per the collection of ground-truth images are available against the set of synthetic test photos, we measure and report the quantitative outcomes in Table 6. Furthermore, we assess UTN-GAN, ID-CGAN, iPANs and IP-RAN on natural rainy images, and the results are shown in Figure 10.

Figure 9. *Cont.*

(a)   (b)   (c)   (d)   (e)

Figure 9. Sample results of synthetic test images. For good visual comparison, the smaller images below the test images represent specific regions-of-interest. (**a**) input images, (**b**) results of UTN-GAN, (**c**) results of ID-CGAN, (**d**) results of iPANs and (**e**) results of the IP-RAN.

Table 6. Quantitative results of image de-raining, bold results show good scores.

	PSNR(dB)	SSIM	UQI	VIF	FID
UTN-GAN	21.81	0.7325	0.9056	0.2939	127.4
ID-CGAN	**24.42**	0.8490	**0.9433**	0.3708	76.71
iPANs	22.44	0.7687	0.9252	0.3101	112.72
IP-RAN	23.69	**0.8518**	0.9412	**0.3740**	**75.90**

From Figures 9 and 10, we can observe that ID-CGAN, iPANs and the IP-RAN have accomplished great results in image de-raining tasks. The findings of the iPANs look slightly better, but contain some artifacts and blurriness. However, by examining the results carefully, the adopted scheme eliminates more rain-streaks with a lesser amount of color distortion. Moreover, as specified in Table 6, for a synthetic set of test images, the introduced method's evaluation scores and the resultant images are far more comparable with the corresponding ground-truth photos than with the results of the other methods. In the single image de-raining problem, the adopted method can accomplish more improved results than UTN-GAN, ID-CGAN, iPANs; one of the possible reasons is that these methods used skip-connection in their generator network. These skip-connection passes useful as well as unwanted information directly from the input image to the output images throughout the network and influence the results. Even though ID-CGAN achieved highest score in the PSNR and UQI metrics, still rain-streak can be seen in the resultant images of ID-CGAN. On the other hand, the adopted method tries to resolve the problem through the proposed loss function using an encoder-decoder generator structure. The novel training scheme of IP-RAN can benefit the generator to learn better-quality mapping from the input images to the output images, leading to improved performance.

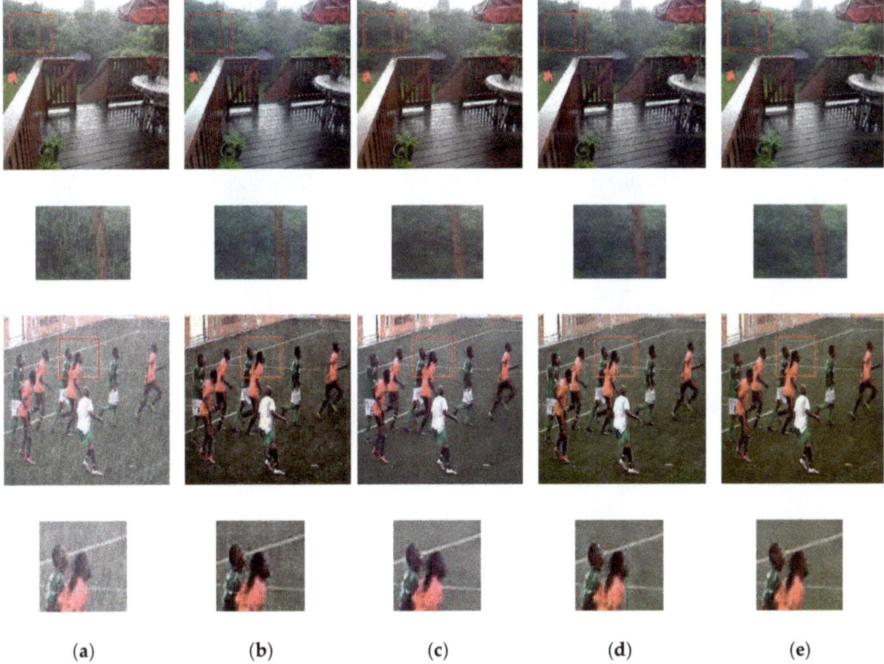

(a)   (b)   (c)   (d)   (e)

Figure 10. Sample results of real-world rainy images. For good visual comparison, the smaller images below the test images represent specific regions-of-interest. (**a**) input images, (**b**) results of UTN-GAN, (**c**) results of ID-CGAN, (**d**) results of iPANs and (**e**) results of the IP-RAN.

5. Conclusions

We have introduced a novel cGAN-based scheme to overcome the lack of information in input labels for imperfect paired datasets. In this work, we propose a novel Input-Perceptual and Reconstruction Adversarial Network (IP-RAN) for paired image-to-image conversion tasks as a general-purpose framework. We merge the input-perceptual loss with the adversarial and the per-pixel reconstruction Euclidean losses as an innovative loss function for imperfect paired datasets. Also, we analyze two popular generator configurations and evaluated their results quantitatively and qualitatively. A generator without skip-connections produced much better and visually pleasing results than a generator with skip-connections. We conducted extensive experiments on multiple datasets to assess the efficiency of the IP-RAN. The adopted scheme outperforms the state-of-the-art works for image-to-image conversion problems. The experimental results of several image-to-image conversion tasks illustrated that the proposed framework is efficient and capable of practical imperfect paired image-to-image conversion applications. In this study, we explored input-perceptual losses to feed the extra information of imperfect paired datasets for only paired image-to-image conversion tasks. Future work is required to examine the impact of input-perceptual losses for unpaired image-to-image conversion applications.

Author Contributions: Methodology, A.K. and W.J.; software and coding, A.K, and D.W.; Experimentation and formal analysis, A.K.; writing—original draft preparation, A.K.; writing—review and editing, A.K., M.A., R.A.N. and D.W.; supervision, W.J.; funding acquisition, W.J. All authors have read and agreed to the published version of the manuscript.

Funding: This research was funded by National Natural Science Foundation of China, grant number 61134002.

Conflicts of Interest: The authors declare no conflict of interest.

References

1. Pathak, D.; Krahenbuhl, P.; Donahue, J.; Darrell, T.; Efros, A.A. Context encoders: Feature learning by inpainting. In Proceedings of the IEEE Conference on Computer Vision and Pattern Recognition, Las Vegas, NV, USA, 27–30 June 2016.
2. Bertalmio, M.; Sapiro, G.; Caselles, V.; Ballester, C. Image Inpainting. Available online: http://rbrad.ulbsibiu.ro/teaching/courses/docs/acs/bertalmi.pdf (accessed on 2 May 2020).
3. Zhang, H.; Sindagi, V.; Patel, V.M. Image de-raining using a conditional generative adversarial network. *IEEE Trans. Circuits Syst. VideoTechnol.* **2019**. [CrossRef]
4. Ramwala, O.A.; Paunwala, C.N.; Paunwala, M.C. Image De-Raining for Driver Assistance Systems using U-Net based GAN. In Proceedings of the 2019 IEEE International Conference on Signal Processing, Information, Communication & Systems (SPICSCON), Dhaka, Bangladesh, 28–30 November 2019.
5. Dong, C.; Loy, C.C.; He, K.; Tang, X. Image super-resolution using deep convolutional networks. *IEEE Trans. Pattern Anal. Mach. Intell.* **2015**, *38*, 295–307. [CrossRef] [PubMed]
6. Johnson, J.; Alahi, A.; Fei-Fei, L. Perceptual losses for real-time style transfer and super-resolution. *arXiv* **2016**, arXiv:1603.08155.
7. Ledig, C.; Theis, L.; Huszár, F.; Caballero, J.; Cunningham, A.; Acosta, A.; Aitken, A.; Tejani, A.; Totz, J.; Wang, Z. Photo-Realistic Single Image Super-Resolution Using a Generative Adversarial Network. Available online: http://iie.fing.edu.uy/~{}mdelbra/DL2018/papers/11_2017_Ledig_CVPR.pdf (accessed on 3 May 2020).
8. Wang, T.-C.; Liu, M.-Y.; Zhu, J.-Y.; Tao, A.; Kautz, J.; Catanzaro, B. High-Resolution Image Synthesis and Semantic Manipulation with Conditional Gans. Available online: https://davidwatkinsvalls.com/files/papers/1711.11585.pdf (accessed on 3 May 2020).
9. Nasrollahi, K.; Moeslund, T.B. Super-resolution: A comprehensive survey. *Mach. Vis. Appl.* **2014**, *25*, 1423–1468. [CrossRef]
10. Dai, Q.; Cheng, X.; Qiao, Y.; Zhang, Y. Crop Leaf Disease Image Super-Resolution and Identification with Dual Attention and Topology Fusion Generative Adversarial Network. *IEEE Access* **2020**, *8*, 55724–55735. [CrossRef]
11. Elad, M.; Aharon, M. Image Denoising via Sparse and Redundant Representations over Learned Dictionaries. *IEEE Trans. Image Process.* **2006**, *15*, 3736–3745. [CrossRef]
12. Kumar, R.; Maji, S.K. A Novel Framework for Denoised High Resolution Generative Adversarial Network–DHRGAN. In Proceedings of the 2020 7th International Conference on Signal Processing and Integrated Networks (SPIN), Noida, India, 27–28 February 2020.
13. Matsui, T.; Ikehara, M. GAN-Based Rain Noise Removal from Single-Image Considering Rain Composite Models. *IEEE Access* **2020**, *8*, 40892–40900. [CrossRef]
14. Gatys, L.A.; Ecker, A.S.; Bethge, M. Image Style Transfer Using Convolutional Neural Networks. Available online: http://liaoqing.me/course/AI%20Project/[2016%20CVPR]Image%20Style%20Transfer%20Using%20Convolutional%20Neural%20Networks.pdf (accessed on 4 May 2020).
15. Khan, M.W. A survey: Image segmentation techniques. *Int. J. Future Comput. Commun.* **2014**, *3*, 89. [CrossRef]
16. Chen, X.; Duan, Y.; Houthooft, R.; Schulman, J.; Sutskever, I.; Abbeel, P. Infogan: Interpretable Representation Learning by Information Maximizing Generative Adversarial Nets. Available online: http://resources.dbgns.com/study/GAN/InfoGAN.pdf (accessed on 4 May 2020).
17. Zhang, R.; Isola, P.; Efros, A.A. Colorful Image Colorization. Available online: https://richzhang.github.io/colorization/resources/colorful_eccv2016.pdf (accessed on 4 May 2020).
18. Chen, Y.; Lai, Y.-K.; Liu, Y.-J. Transforming Photos to Comics Using Convolutional Neural Networks. Available online: https://core.ac.uk/download/pdf/82967487.pdf (accessed on 4 May 2020).
19. Cheng, Z.; Yang, Q.; Sheng, B. Deep colorization. In Proceedings of the IEEE International Conference on Computer Vision, Santiago, Chile, 7–13 December 2015.
20. Long, J.; Shelhamer, E.; Darrell, T. Fully Convolutional Networks for Semantic Segmentation. Available online: https://computing.ece.vt.edu/~{}f15ece6504/slides/L13_FCN.pdf (accessed on 4 May 2020).
21. Goodfellow, I.; Pouget-Abadie, J.; Mirza, M.; Xu, B.; Warde-Farley, D.; Ozair, S.; Courville, A.; Bengio, Y. Generative Adversarial Nets. Available online: https://chyson.net/papers/Generative%20Adversarial%20Nets.pdf (accessed on 4 May 2020).

22. Cheng, K.; Tahir, R.; Eric, L.K.; Li, M. An analysis of generative adversarial networks and variants for image synthesis on MNIST dataset. *Multimed. Tools Appl.* **2020**, 1–28. [CrossRef]

23. Isola, P.; Zhu, J.-Y.; Zhou, T.; Efros, A.A. Image-to-Image Translation with Conditional Adversarial Networks. Available online: https://gangw.web.illinois.edu/class/cs598/papers/CVPR17-img2img.pdf (accessed on 4 May 2020).

24. Chen, X.; Xu, C.; Yang, X.; Song, L.; Tao, D. Gated-gan: Adversarial gated networks for multi-collection style transfer. *IEEE Trans. Image Process.* **2018**, *28*, 546–560. [CrossRef]

25. Wang, C.; Xu, C.; Wang, C.; Tao, D. Perceptual Adversarial Networks for Image-to-Image Transformation. *IEEE Trans. Image Process.* **2018**, *27*, 4066–4079. [CrossRef] [PubMed]

26. Mirza, M.; Osindero, S. Conditional Generative Adversarial Nets. Available online: http://resources.dbgns.com/study/GAN/conditional_gan.pdf (accessed on 4 May 2020).

27. Kupyn, O.; Budzan, V.; Mykhailych, M.; Mishkin, D.; Matas, J. Deblurgan: Blind Motion Deblurring Using Conditional Adversarial Networks. Available online: http://www.gwylab.com/pdf/deblur-gan.pdf (accessed on 4 May 2020).

28. Regmi, K.; Borji, A. Cross-View Image Synthesis Using Conditional Gans. Available online: https://openaccess.thecvf.com/content_cvpr_2018/papers_backup/Regmi_Cross-View_Image_Synthesis_CVPR_2018_paper.pdf (accessed on 4 May 2020).

29. Ge, H.; Yao, Y.; Chen, Z.; Sun, L. Unsupervised transformation network based on GANs for target-domain oriented image translation. *IEEE Access* **2018**, *6*, 61342–61350. [CrossRef]

30. Wu, C.-H.; Horng, S.-J.; Lee, P.-Z. A new computation of shape moments via quadtree decomposition. *Pattern Recognit.* **2001**, *34*, 1319–1330. [CrossRef]

31. Ravuri, S.; Mohamed, S.; Rosca, M.; Vinyals, O. Learning implicit generative models with the method of learned moments. *arXiv* **2018**, arXiv:1806.11006.

32. Santos, C.N.D.; Mroueh, Y.; Padhi, I.; Dognin, P. Learning Implicit Generative Models by Matching Perceptual Features. Available online: https://www.researchgate.net/profile/Cicero_Dos_Santos2/publication/332264118_Learning_Implicit_Generative_Models_by_Matching_Perceptual_Features/links/5d1e0ff6a6fdcc2462c0cccb/Learning-Implicit-Generative-Models-by-Matching-Perceptual-Features.pdf (accessed on 7 May 2020).

33. Simonyan, K.; Zisserman, A. Very deep convolutional networks for large-scale image recognition. *arXiv* **2014**, arXiv:1409.1556.

34. Russakovsky, O.; Deng, J.; Su, H.; Krause, J.; Satheesh, S.; Ma, S.; Huang, Z.; Karpathy, A.; Khosla, A.; Bernstein, M. Imagenet large scale visual recognition challenge. *Int. J Comput. Vis.* **2015**, *115*, 211–252. [CrossRef]

35. Huh, M.; Agrawal, P.; Efros, A.A. What makes ImageNet good for transfer learning? *arXiv* **2016**, arXiv:1608.08614.

36. Voulodimos, A.; Doulamis, N.; Doulamis, A.; Protopapadakis, E. Deep learning for computer vision: A brief review. *Comput. Intell. Neurosci.* **2018**, *2018*, 13. [CrossRef]

37. Eigen, D.; Fergus, R. Predicting Depth, Surface Normals and Semantic Labels with a Common Multi-Scale Convolutional Architecture. Available online: http://iie.fing.edu.uy/~{}mdelbra/DL2017/papers/09_2015_Eigen_ICCV.pdf (accessed on 8 May 2020).

38. Farabet, C.; Couprie, C.; Najman, L.; LeCun, Y. Learning hierarchical features for scene labeling. *IEEE Trans. Pattern Anal. Mach. Intell.* **2012**, *35*, 1915–1929. [CrossRef]

39. Noh, H.; Hong, S.; Han, B. Learning Deconvolution Network for Semantic Segmentation. Available online: http://www-prima.imag.fr/Prima/jlc/Courses/2018/PRML/Noh_Learning_Deconvolution_Network_ICCV_2015_paper.pdf (accessed on 8 May 2020).

40. Garcia-Garcia, A.; Orts-Escolano, S.; Oprea, S.; Villena-Martinez, V.; Garcia-Rodriguez, J. A review on deep learning techniques applied to semantic segmentation. *arXiv* **2017**, arXiv:1704.06857.

41. Fu, X.; Huang, J.; Ding, X.; Liao, Y.; Paisley, J. Clearing the skies: A deep network architecture for single-image rain removal. *IEEE Trans. Image Process.* **2017**, *26*, 2944–2956. [CrossRef] [PubMed]

42. Fu, X.; Huang, J.; Zeng, D.; Huang, Y.; Ding, X.; Paisley, J. Removing Rain from Single Images via a Deep Detail Network. Available online: https://xueyangfu.github.io/paper/2017/cvpr/cvpr2017.pdf (accessed on 8 May 2020).

43. Ružić, T.; Pižurica, A. Context-aware patch-based image inpainting using Markov random field modeling. *IEEE Trans. Image Process.* **2014**, *24*, 444–456. [CrossRef] [PubMed]

44. Qin, C.; Chang, C.-C.; Chiu, Y.-P. A novel joint data-hiding and compression scheme based on SMVQ and image inpainting. *IEEE Trans. Image Process.* **2013**, *23*, 969–978. [CrossRef]

45. Eigen, D.; Puhrsch, C.; Fergus, R. Depth Map Prediction from a Single Image Using a Multi-Scale Deep Network. Available online: http://datascienceassn.org/sites/default/files/Depth%20Map%20Prediction%20from%20a%20Single%20Image%20using%20a%20Multi-Scale%20Deep%20Network.pdf (accessed on 9 May 2020).

46. Liu, F.; Shen, C.; Lin, G. Deep Convolutional Neural Fields for Depth Estimation from a Single Image. Available online: https://www.cv-foundation.org/openaccess/content_cvpr_2015/app/3B_078.pdf (accessed on 9 May 2020).

47. Berthelot, D.; Schumm, T.; Metz, L. Began: Boundary equilibrium generative adversarial networks. *arXiv* **2017**, arXiv:1703.10717.

48. Radford, A.; Metz, L.; Chintala, S. Unsupervised representation learning with deep convolutional generative adversarial networks. *arXiv* **2015**, arXiv:1511.06434.

49. Brock, A.; Lim, T.; Ritchie, J.M.; Weston, N. Neural photo editing with introspective adversarial networks. *arXiv* **2016**, arXiv:1609.07093.

50. Reed, S.; Akata, Z.; Yan, X.; Logeswaran, L.; Schiele, B.; Lee, H. Generative adversarial text to image synthesis. *arXiv* **2016**, arXiv:1605.05396.

51. Vondrick, C.; Pirsiavash, H.; Torralba, A. Generating Videos with Scene Dynamics. Available online: https://pdfs.semanticscholar.org/7188/6726f0a1b4075a7213499f8f25d7c9fb4143.pdf (accessed on 9 May 2020).

52. Wu, J.; Zhang, C.; Xue, T.; Freeman, B.; Tenenbaum, J. Learning a Probabilistic Latent Space of Object Shapes via 3d Generative-Adversarial Modeling. Available online: https://core.ac.uk/download/pdf/141473151.pdf (accessed on 9 May 2020).

53. Lotter, W.; Kreiman, G.; Cox, D. Unsupervised learning of visual structure using predictive generative networks. *arXiv* **2015**, arXiv:1511.06380.

54. Zhu, J.-Y.; Krähenbühl, P.; Shechtman, E.; Efros, A.A. Generative Visual Manipulation on the Natural Image Manifold. Available online: https://www.philkr.net/media/zhu2016generative.pdf (accessed on 9 May 2020).

55. Arjovsky, M.; Chintala, S.; Bottou, L. Wasserstein generative adversarial networks. In Proceedings of the International Conference on Machine Learning, Sydney, Australia, 6–11 August 2017.

56. Gulrajani, I.; Ahmed, F.; Arjovsky, M.; Dumoulin, V.; Courville, A.C. Improved Training of Wasserstein Gans. Available online: http://www.cs.utoronto.ca/~{}bonner/courses/2020s/csc2547/papers/adversarial/improved-training-of-WGANs,-gulrajani,-nips2017.pdf (accessed on 9 May 2020).

57. Zhao, J.; Mathieu, M.; LeCun, Y. Energy-based generative adversarial network. *arXiv* **2016**, arXiv:1609.03126.

58. Karras, T.; Aila, T.; Laine, S.; Lehtinen, J. Progressive growing of gans for improved quality, stability, and variation. *arXiv* **2017**, arXiv:1710.10196.

59. Miyato, T.; Kataoka, T.; Koyama, M.; Yoshida, Y. Spectral normalization for generative adversarial networks. *arXiv* **2018**, arXiv:1802.05957.

60. Wang, C.; Xu, C.; Yao, X.; Tao, D. Evolutionary generative adversarial networks. *IEEE Trans. Evol. Comput.* **2019**, *23*, 921–934. [CrossRef]

61. Denton, E.L.; Chintala, S.; Fergus, R. Deep Generative Image Models Using a Laplacian Pyramid of Adversarial Networks. Available online: https://research.fb.com/wp-content/uploads/2016/11/deep-generative-image-models-using-a-laplacian-pyramid-of-adversarial-networks.pdf (accessed on 13 May 2020).

62. Li, J.; Liang, X.; Wei, Y.; Xu, T.; Feng, J.; Yan, S. Perceptual generative adversarial networks for small object detection. In Proceedings of the IEEE Conference on Computer Vision and Pattern Recognition, Honolulu, HI, USA, 21–26 July 2017.

63. Sung, T.L.; Lee, H.J. Image-to-Image Translation Using Identical-Pair Adversarial Networks. *Appl. Sci.* **2019**, *9*, 2668. [CrossRef]

64. Zhu, J.-Y.; Park, T.; Isola, P.; Efros, A.A. Unpaired Image-to-Image Translation Using Cycle-Consistent Adversarial Networks. Available online: https://pdfs.semanticscholar.org/c43d/954cf8133e6254499f3d68e45218067e4941.pdf (accessed on 13 May 2020).

65. Yi, Z.; Zhang, H.; Tan, P.; Gong, M. Dualgan: Unsupervised Dual Learning for Image-to-Image Translation. Available online: https://www.cs.sfu.ca/~{}haoz/pubs/yi_iccv17_dualGAN.pdf (accessed on 14 May 2020).

66. Kim, T.; Cha, M.; Kim, H.; Lee, J.K.; Kim, J. Learning to Discover Cross-Domain Relations with Generative Adversarial Networks. Available online: http://axon.cs.byu.edu/Dan/673/papers/kim.pdf (accessed on 14 May 2020).

67. Chen, Q.; Koltun, V. Photographic Image Synthesis with Cascaded Refinement Networks. Available online: https://www.cqf.io/papers/Photographic_Image_Synthesis_ICCV2017.pdf (accessed on 15 May 2020).

68. He, K.; Zhang, X.; Ren, S.; Sun, J. Deep Residual Learning for Image Recognition. Available online: http://www.cs.sjtu.edu.cn/~{}shengbin/course/cg/Papers%20for%20Selection/Deep%20Residual%20Learning%20for%20Image%20Recognition.pdf (accessed on 15 May 2020).

69. Mao, X.; Li, Q.; Xie, H.; Lau, R.Y.; Wang, Z.; Paul Smolley, S. Least Squares Generative Adversarial Networks. Available online: https://www.researchgate.net/profile/Haoran_Xie/publication/322060458_Least_Squares_Generative_Adversarial_Networks/links/5bfad008a6fdcc538819cf3e/Least-Squares-Generative-Adversarial-Networks.pdf (accessed on 15 May 2020).

70. Ulyanov, D.; Vedaldi, A.; Lempitsky, V. Instance normalization: The missing ingredient for fast stylization. *arXiv* **2016**, arXiv:1607.08022.

71. Li, C.; Wand, M. Precomputed Real-Time Texture Synthesis with Markovian Generative Adversarial Networks. Available online: https://arxiv.org/pdf/1604.04382v1.pdf (accessed on 15 May 2020).

72. Tyleček, R.; Šára, R. Spatial Pattern Templates for Recognition of Objects with Regular Structure. Available online: https://pdfs.semanticscholar.org/3edc/81db7c70d9123ea04829a98fc9fd62b29b1d.pdf (accessed on 16 May 2020).

73. Yu, A.; Grauman, K. Fine-Grained Visual Comparisons with Local Learning. Available online: https://aronyu.io/vision/papers/cvpr14/aron-cvpr14.pdf (accessed on 16 May 2020).

74. Xie, S.; Tu, Z. Holistically-Nested Edge Detection. Available online: https://pages.ucsd.edu/~{}ztu/publication/iccv15_hed.pdf (accessed on 16 May 2020).

75. Kingma, D.P.; Ba, J. Adam: A method for stochastic optimization. *arXiv* **2014**, arXiv:1412.6980.

76. Wang, Z.; Bovik, A.C.; Sheikh, H.R.; Simoncelli, E.P. Image quality assessment: From error visibility to structural similarity. *IEEE Trans. Image Process.* **2004**, *13*, 600–612. [CrossRef]

77. Sheikh, H.R.; Bovik, A.C. Image information and visual quality. *IEEE Trans. Image Process.* **2006**, *15*, 430–444. [CrossRef]

78. Wang, Z.; Bovik, A.C. A universal image quality index. *IEEE Signal Process. Lett.* **2002**, *9*, 81–84. [CrossRef]

79. Heusel, M.; Ramsauer, H.; Unterthiner, T.; Nessler, B.; Hochreiter, S. Gans Trained by a Two Time-Scale Update Rule Converge to a Local Nash Equilibrium. Available online: http://papers.neurips.cc/paper/7240-gans-trained-by-a-two-time-scale-update-rule-converge-to-a-local-nash-equilibrium.pdf (accessed on 16 May 2020).

Article

Global-and-Local Context Network for Semantic Segmentation of Street View Images

Chih-Yang Lin [1], Yi-Cheng Chiu [2], Hui-Fuang Ng [3,*], Timothy K. Shih [2,*] and Kuan-Hung Lin [2]

[1] Department of Electrical Engineering, Yuan Ze University, Taoyuan 32003, Taiwan;
    andrewlin@saturn.yzu.edu.tw
[2] Department of Computer Science & Information Engineering, National Central University,
    Taoyuan City 32001, Taiwan; chialinwu001@gmail.com (Y.-C.C.); paul92035@gmail.com (K.-H.L.)
[3] Department of Computer Science, University Tunku Abdul Rahman, Kampar 31900, Malaysia
* Correspondence: nghf@utar.edu.my (H.-F.N.); timothykshih@gmail.com (T.K.S.)

Received: 12 February 2020; Accepted: 18 May 2020; Published: 21 May 2020

Abstract: Semantic segmentation of street view images is an important step in scene understanding for autonomous vehicle systems. Recent works have made significant progress in pixel-level labeling using Fully Convolutional Network (FCN) framework and local multi-scale context information. Rich global context information is also essential in the segmentation process. However, a systematic way to utilize both global and local contextual information in a single network has not been fully investigated. In this paper, we propose a global-and-local network architecture (GLNet) which incorporates global spatial information and dense local multi-scale context information to model the relationship between objects in a scene, thus reducing segmentation errors. A channel attention module is designed to further refine the segmentation results using low-level features from the feature map. Experimental results demonstrate that our proposed GLNet achieves 80.8% test accuracy on the Cityscapes test dataset, comparing favorably with existing state-of-the-art methods.

Keywords: semantic segmentation; global context; local context; fully convolutional networks

---

1. Introduction

Image semantic segmentation is a computer vision task in which each pixel of an image is assigned to a corresponding object label. Semantic segmentation provides richer information, including object's boundary and shape, than other object segmentation techniques such as object detection which produces only object bounding boxes. Therefore, semantic segmentation of street view images is important to the development of vision sensors for autonomous vehicle systems that require rich information for scene understanding [1,2]. Many state-of-the-art semantic segmentation methods are based on fully convolutional networks (FCN) [3]. FCN is an end-to-end, pixel-wise fully convolutional neural network for image semantic segmentation which replaces the fully connected layer in a conventional convolutional neural network (CNN) with convolutional layers and then restores the output to the original image size by up-sampling the final convolutional layer using a deconvolution operation [4]. However, consecutive pooling operations at the encoding stage usually lead to the reduction of feature resolution, which in turn will degrade final segmentation performance. As shown in the first row of Figure 1, small objects such as poles and traffic light are missing due to loss of detailed information resulted from pooling operations. To address this issue, *dilated* convolution (also called *atrous* convolution) has been proposed to replace the down-sampling and up-sampling operations in order to gain a wider field of view while preserving full spatial information [5–7].

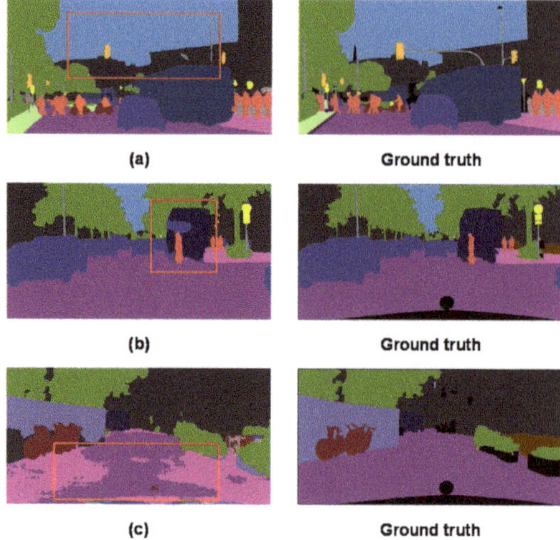

Figure 1. Some semantic segmentation issues observed in the Cityscapes dataset. (**a**) Missing small and thin objects. (**b**) Incorrect segmentation of similar parts. (**c**) Incomplete segmentation of large objects.

Local context information is an important component in semantic segmentation for segmenting complex street view scenes that contain different objects with similar features. The second row in Figure 1 shows a portion of the 'truck' object that has been mistakenly classified as a 'car' object because it has a similar appearance to cars. Consequently, local context information should be explicitly encoded in the semantic segmentation models [6]. Another important factor affecting segmentation is the existence of objects in multiple scales. With regard to this, different scales of an image are generated to extract multi-scale features, and the features are then merged to predict the results [8–10]. In [11,12], dilated convolution layers with different rates were used to capture object features at multiple scales, and the experimental results indicated that using multiple field-of-views can improve overall segmentation accuracy.

Convolutional operations process one local neighborhood at a time as in a conventional CNN [13,14], which may cause incomplete segmentation of large objects. As shown in the third row of Figure 1, with local convolutional operations, parts of the road near roadside area were wrongly predicted as sidewalk due to the lack of global context information. To resolve this, global contextual information has been utilized in scene segmentation and been shown to be effective in reducing false segmentation [15]. However, a systematic way to integrate both global and local contextual information in a single network has not yet been fully investigated.

In this paper, we propose a global-and-local context network (GLNet) for semantic segmentation of street view images for autonomous vehicle systems, which incorporates both global and local contextual information in a single network. An overview of the proposed GLNet is shown in Figure 2. The global context module captures semantic of spatial interdependencies while the local context module uses dilated convolution of different rates to extract dense multi-scale features that are important for a network to adapt to different object sizes. GLNet fuses the outputs of the two modules to capture both global and local contextual information to help the network recognize the relationships between objects in a scene, helping eliminate false alarms. Furthermore, CNN low-level features are used through channel weighting to restore the edges and fine details of the segmentation results.

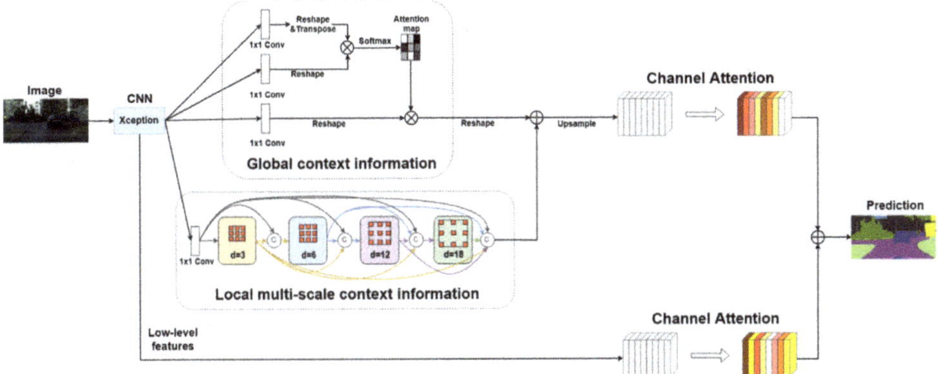

Figure 2. An overview of the proposed GLNet. First, a CNN is applied on the input image and the feature map from the last convolution layer of CNN is kept. Next, global module and local module are deployed on the feature map to capture both global and multi-scale context information. Finally, the channel attention module restores edges and fine details of the segmented objects using low-level features.

2. Related Work

Many recent image semantic segmentation methods have been based on fully convolutional networks (FCN) [3], which train on an end-to-end network for pixel-wise prediction, achieving state-of-the-art result. For instances, encoder-decoder networks such as SegNet [16] and U-Net [17] have been proposed to enhance semantic segmentation results. The encoder gradually reduces the spatial dimension of feature maps to extract salient features. On the other hand, the decoder increasingly up-samples the feature map back to the original spatial resolution of the input image and produces pixel-wise segmentation result. However, the encoding process usually causes the reduction of feature resolution, which in turn degrades final segmentation performance.

The use of contextual information is important to pixel-level prediction tasks. Objects in complex scenes exhibit large-scale changes, which introduces great challenges for advanced feature representations. In [11,12], dilated convolutions were used to expand the receptive field and to encode multi-scale context information. Zhao et al. [18] proposed a pyramid scene parsing network to capture multi-scale context information by aggregating feature maps of different resolutions through a pyramid pooling module. DeepLabv3+ [19] used parallel Atrous Spatial Pyramid Pooling (ASPP) which connects parallel dilated convolutions of different rates on the feature map to effectively encode multi-scale information. Zhou et al. [20] designed a multi-scale deep context convolutional network for semantic segmentation which combines the feature maps from different levels of network. Although multi-scale context information help to capture different scales objects, it cannot leverage the relationship between objects in a global view, which is also essential to scene segmentation.

DANet [21] uses self-attention to capture long-range global context by exploring orthogonal relationships in both spatial and channel dimensions using non-local operator [15]. Convolution is essentially a local operation. Non-local operations extract long-range dependencies directly from any two positions in an image. The position attention module of DANet selectively aggregates the features of each location by weighted summation of all locations. CFNet [22] adds an extra global average pooling path to determine whole scene statistics. However, both models do not incorporate local multi-scale features in their networks. Self-attention is also used in OCNet [23] to learn pixel-level object context information to enhance context aggregation. Attempt to incorporate additional depth information for improving semantic segmentation has also been explored in [24] and [25]. Our proposed method incorporates rich global spatial information and dense local multi-scale context information to model the relationship between objects in a scene to reduce segmentation errors.

3. Method

In this section, we describe the proposed GLNet network in detail. The overall network structure of GLNet is shown in Figure 2. First, an Xception [26] network with 65 layers is used as the backbone to extract features from the input image. Next, the feature map from the last convolution layer of the Xception network is input into both the global module and the local module and the outputs from the two modules are fused to generate the segmentation result. Lastly, the channel attention module refines the segmentation result by utilizing the low-level features from the Xception network.

3.1. Global Module

The global module is intended to capture the spatial dependencies of the feature map. To this end, we have adopted the non-local operations proposed in [15] and the details of the global module are shown in Figure 3. In the global module, the response at a feature location is weighted by features at all locations in the input feature map. The weights are determined based on the feature similarity between two corresponding locations. Thus, any two positions with similar features can contribute to each other regardless of the distance between them.

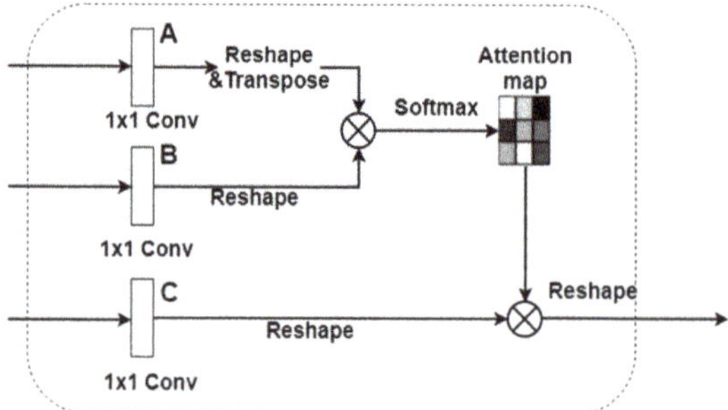

Figure 3. The global module. The number of channels of the CNN feature map is first reduced to 1/8 of the input channels before generating the attention map, and the weighted result is reverted back to the shape of the input feature map.

The last layer of the Xception network is first convolved with a 1×1 convolution to reduce the number of channels from 2048 to 256, or 1/8 of the original number of channels. This produces a $H \times W \times nc$ feature map, where H, W are the height and width of the feature map, and nc denotes the number of channels. Three copies of the feature map (**A**, **B**, **C**) are generated and reshaped to $N \times nc$, where $N = H \times W$. As shown in Figure 3, feature map **A** is then transposed and multiplied by feature map **B**, and the resulting $N \times N$ matrix is passed through softmax function to produce the spatial attention map. Finally, feature map **C** is multiplied by the attention map and the result is reshaped back to $H \times W \times nc$. As a result, the output of the global module is a feature map weighted by the global spatial interdependencies among pixel locations.

3.2. Local Module

Objects in a scene usually exhibit large-scale changes, posing a difficult challenge for feature representation in semantic segmentation since multi-scale information must be properly encoded. DeepLab [19] handled this problem using an Atrous Spatial Pyramid Pooling (ASPP) module in which dilated convolutions with different rates are applied to the input feature map and the results are fused to account for different object scales. However, the ASPP module is not dense enough to deal with large object scale changes in complex scenes.

To resolve this, we propose a local module based on DenseASPP [12] to extract dense multi-scale features that are important for a network to adapt to large-scale variations. The local module combines the advantages of parallel and cascaded dilated convolutions to obtain larger and denser receptive field than ASPP, thus achieving superior multi-scale representation of objects in a scene. The details of the local module are depicted in Figure 4. Similar to the global module, the feature map from the Xception network is first convolved with 1×1 convolution to reduce the number of channels to nc (1/8 of the input channels) before it is fed to the local module. Next, the input feature map is sequentially convolved with dilated convolution with increasing dilation rates from 3, 6, 12, to 18. To lessen network complexity, dilation rate 24 is not included here as in DenseASPP. The input to each dilated convolution layer is formed by concatenating the input feature map with the outputs from previous convolutions and then convolved with 1×1 convolution to maintain the number of channels at nc. Lastly, the outputs from the four dilated convolutions are concatenated together with the input feature map and result is again reduced to nc channels before outputting to the next processing step. Compared with DenseASPP, the total number of parameters of the local module is reduced from 6.48 M to 2.01 M. It can be seen from Figure 4 that the output of the local module is a feature map that contains dense multi-scale feature information for the input image.

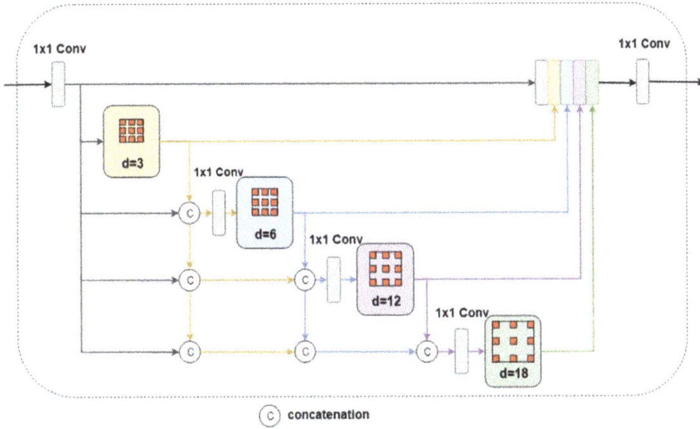

Figure 4. The local module. The output of each dilated convolutional layer is concatenated with the input feature map and convolved with 1×1 convolution to maintain fixed channel number before going into the next dilated convolution layer.

3.3. Channel Attention Module

The Global-Local module extracts high level multi-scale semantic information. However, details of objects may be lost during the down sampling process in the initial stage of the CNN network. The low-level feature maps produced by CNN before the max pooling layer should preserve detailed information including edges and other fine details of the objects in the image. To recover the finer details of objects, we propose a channel attention module which fuses the output from the Global-Local module and the low-level feature map of CNN, as shown in Figure 5.

The output of the Global-Local module is up-sampled to match the dimension of the low-level feature map of CNN before the fusion process. In addition, Squeeze-and-Excitation (SE) block [27] is applied to both feature maps to obtain cross channel information and learn the channel-wise attention. The SE block generates a weight between 0 and 1 per channel through sigmoid function, and the weights are then used to perform dynamic feature reweighting via channel-wise multiplication between the weights and the feature maps. As a result, the channel attention module also explores channel dependency in the feature map, which has been proven useful for image classification and segmentation tasks [28].

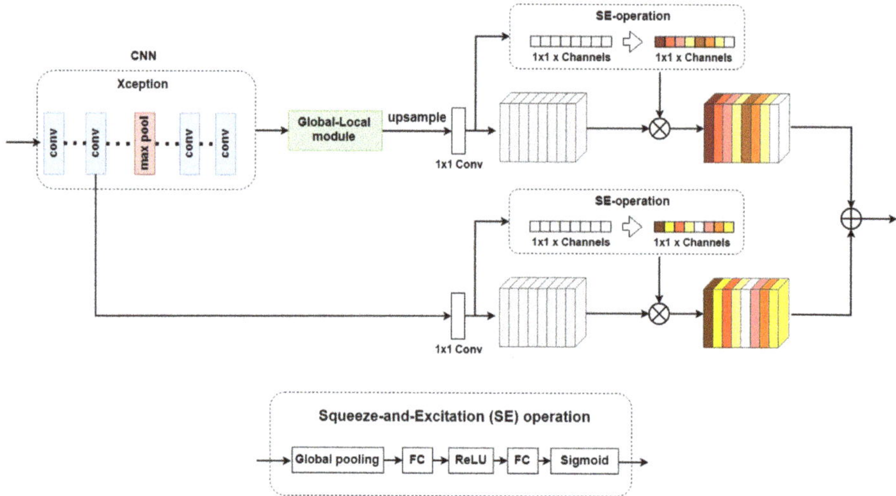

Figure 5. The Channel attention module. The output of the Global-Local module is up-sampled to match the size of the CNN feature map before the fusion process, and the SE operation explores channel dependency in the feature map.

4. Experiments

In this section, we describe the implementation details and discuss the experimental results of the proposed GLNet network. Comprehensive experiments and ablation studies on the Cityscapes dataset [29] were carried out to evaluate the proposed network, and the performance improvements of the modules are highlighted.

4.1. Implementation Details

The implementation of GLNet is built upon the TensorFlow framework. The pre-trained weights of Xception-65 [26] were drawn from the Imagenet [30]. The original image size was 1024 × 2048, while the input image size was randomly cropped to 768 × 768. We adopted the learning rate policy from [6,18] where the current learning rate is calculated as follows:

$$\left(1 - \frac{iter}{max\_iter}\right)^{power},$$ (1)

where *iter* and *max_iter* represent the current training step and the overall training iterations, respectively. An initial learning rate of 0.01 was used throughout the experiments, whereas *max_iter* was set at 90,000 steps and *power* is set to 0.9. Furthermore, the batch size was fixed at 8 in our experiments.

The loss function is defined as:

$$loss = \sum_{n=1}^{N} l_{mce}((X_n), Y_n),$$ (2)

where X_n are the training images and Y_n are the corresponding ground truth, and l_{mce} denotes the multi-class cross-entropy loss for predictions.

The training data came from the Cityscapes dataset, which is comprised of 5000 annotated urban street scenes (2975 for training, 500 for validation and 1525 for testing) for pixel-level semantic labeling. Intersection over union (IoU) and pixel accuracy averaged across the 19 classes in the dataset were used as performance metrics for evaluating the methods. The GLNet takes about 6 h of training time using a single NVIDIA Tesla® V100 GPU with batch size fixed at 8.

4.2. Ablation Study

An extensive ablation study on the Cityscapes dataset is conducted to determine the effectiveness of the various modules and the design choice of the proposed GLNet.

4.2.1. Effect of Global Context Information

Before examining the overall performance of the proposed GLNet network, we first investigate the effect of incorporating global context information on semantic segmentation, which is one of the main objectives of this study. Figure 6 shows the sample segmentation results when incorporating global context information versus results without the global context information. The first column is the input images. The second column shows visualizations of the attention maps generated by the Global-Local module, which are the results of fusing the outputs from both the global and the local modules (see Figure 2). The third column shows the same attention map, excluding the output from the global module.

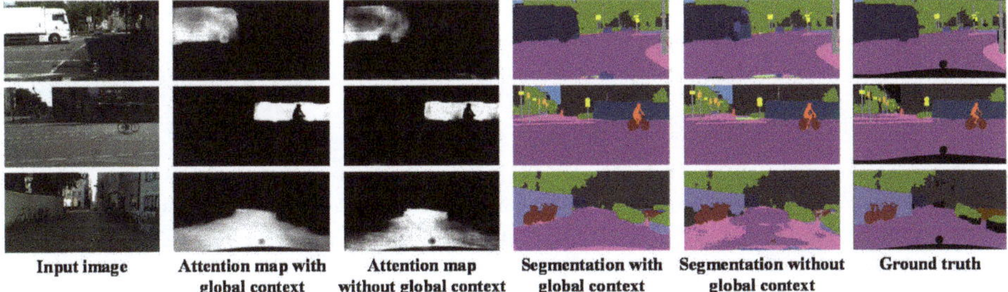

Figure 6. Visualization results of incorporating global context information versus without global context information.

It can be seen in Figure 6 that the attention maps that incorporated global context information are less noisy than those without global context information. The segmentation results in column 4 and 5 confirm that without the global context information, it is more problematic to segment larger objects, such as the truck in the first row, the bus in the second row, and the road in the last row. This indicates that using information from local context alone and ignoring the global scene context will result in misclassified pixels.

4.2.2. GLNet Modules

In this experiment, we tested the effects of the various modules in GLNet on the overall performance of the network. The results are summarized in Table 1 and the visualization results are shown in Figure 7. The first test was to compare the proposed local module with the ASPP module used in DeepLabv3+ [19]. The local module increased the mean IoU by 0.5% over the ASPP module on the Cityscapes dataset. The first and second columns in Figure 7 show a slight decrease in false classification of pixels using the proposed local module. The above results suggest that dense multi-scale features extracted by the local module have positive influence on the overall segmentation results.

Next, we tested the effect of applying the channel attention module on the segmentation performance, which fuses low-level feature using SE block. With the inclusion of SE block, the mean IoU increased about 1%, from 77.6% to78.4%. Finally, when all three modules in GLNet were activated, the mean IoU improved to 80.1%, 3% higher than using only the ASPP module of DeepLabv3+ [19]. By visually examining the segmentation results in Figure 7, we can see that the proposed GLNet architecture has significantly fewer segmentation errors, especially for large objects and for the background area. This is due to the inclusion of global information in the feature extraction process.

Table 1. Performance evaluation on Cityscapes val. set.

BaseNet	ASPP	Local	Global	SE	Mean IoU%
Xception-65	✓				77.1%
Xception-65		✓			77.6%
Xception-65		✓		✓	78.4%
Xception-65		✓	✓	✓	80.1%

Note: ASPP denotes ASPP module [19], Local represents local module, Global represents global module, and SE implies SE block [24].

Figure 7. Visualization results on Cityscapes val. set. Refer to Table 1 for descriptions of corresponding modules.

4.2.3. Global Module

As mentioned in Section 3, non-local block [15] was chosen to gather global attention in the proposed global module. It is worth to experiment with other attention mechanisms for capturing global attention in the feature map. The Convolutional Block Attention Module (CBAM) proposed recently by Woo et al. [31] and the Pyramid Pooling Module in PSPNet [18] are selected here to compare with the non-local block method. CBAM applies average pooling and max pooling along the channel axis of the feature map and then the result is passed through a 7×7 convolution to capture both cross-channel and spatial attentions [31]. Pyramid pooling module harvests different sub-region representations and concatenates them into a feature representation that carries both local and global context information [18].

The results of applying different methods for extracting global context information in the global module are shown in Table 2. The results indicate that using non-local block offers a slight edge over CBAM on the Cityscapes dataset. Pyramid pooling does not perform as well and has the lowest IoU score. One possible explanation of the low performance for pyramid pooling is that in the original PSPNet implementation, the number of output channels after concatenation is 3072, which has been reduced to 256 channels in order to fit our network architecture. The reduction in channel number might result in loss of feature information. Visualization results in Figure 8 confirm that non-local block can handle large objects better than CBAM and pyramid pooling. For example, as shown in the upper left column in Figure 8, CBAM and pyramid pooling misclassified portion of the truck object as car pixels, whereas non-local block was able to segment the truck object correctly. Therefore, non-local block is recommended to be used in the proposed GLNet network.

Table 2. Performance on Cityscapes val. set of applying different methods in the global module.

BaseNet	Non-Local	CBAM	Pyramid Pooling	Mean IoU%
Xception-65			✓	70.3%
Xception-65		✓		78.9%
Xception-65	✓			80.1%

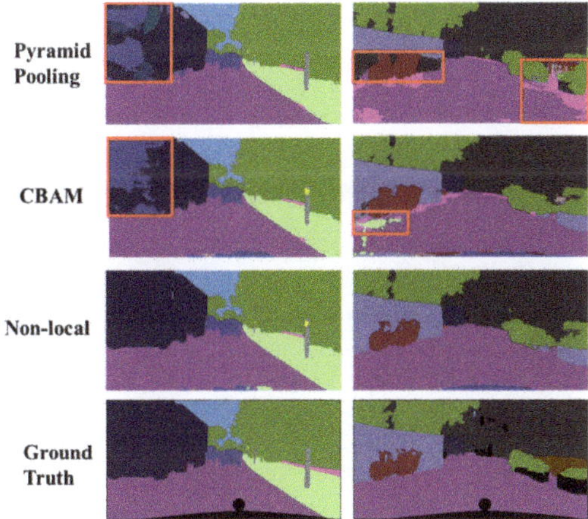

Figure 8. Visualization results on Cityscapes val. set of applying different methods in the global module.

4.2.4. Fusion vs. Concatenation

Another experiment was to explore the use of concatenation operation versus fusion operation for combining high-level feature and low-level feature and the results are shown in Table 3. The results demonstrate that the performance of using fusion operation is slightly better than using a concatenation operation. In-depth analysis of the segmentation results of individual object classes reveal that concatenation operation actually produced slightly better segmentation of small objects and fine details, thus providing nicer visualization results. Conversely, fusion operation generated better segmentation results on larger objects such as buses and trucks. Since large objects occupy more pixels and contribute more to the IoU calculation, as a result, the mean IoU under fusion operation is higher.

Table 3. Performance on Cityscapes val. set using concatenation operation versus fusion operation for combining features.

BaseNet	Concatenation	Fusion	Mean IoU%
Xception-65	✓		79.9%
Xception-65		✓	80.1%

4.3. Performance Comparison

4.3.1. Performance on Cityscapes Dataset

We have benchmarked our proposed GLNet with state-of-the-art methods on the Cityscapes test set and validation set and the results are shown in Tables 4 and 5. All methods are trained with the fine dataset of Cityscapes. For the test set, GLNet achieves a mean IoU of 80.8, the highest among all the methods tested. Out of the 19 classes in the dataset, GLNet obtains or shares the best scores for 13 classes. Similar results are reported on the validation set, where GLNet has the best overall segmentation performance. Other methods, for example PSPNet [18], extract context information through global average pooling, and concatenates the information in the final output feature map. Although such an approach produces satisfactory segmentation results, but some fine details are missing and it generates false edges in the results. The works in [11,12,19] use Atrous Spatial Pyramid Pooling (ASPP) module to cope with different object scales. However, they do not carry enough global context information to handle large objects. In contrast to the previous methods, the proposed GLNet incorporates global and local context information, together with low-level feature to tackle complex scenes that contain a mixture of fine and large objects.

Table 4. Performance comparison on Cityscapes test set.

Method	Mean IoU	Road	Sidewalk	Building	Wall	Fence	Pole	Traffic Light	Traffic Sign	Vegetation	Terrain	Sky	Person	Rider	Car	Truck	Bus	Train	Motorcycle	Bicycle
FCN-8s [3]	65.3	97.4	78.4	89.2	34.9	44.2	47.4	60.1	65.0	91.4	69.3	93.7	77.1	51.4	92.6	35.3	48.6	46.5	51.6	66.8
DeepLab-v2 [6]	70.4	97.9	81.3	90.3	48.8	47.4	49.6	57.9	67.3	91.9	69.4	94.2	79.8	59.8	93.7	56.5	67.5	57.5	57.7	68.8
RefineNet [32]	73.6	98.2	83.3	91.3	47.8	50.4	56.1	66.9	71.3	92.3	70.3	94.8	80.9	63.3	94.5	64.6	76.1	64.3	62.2	70
DUC [33]	77.6	98.5	85.5	92.8	58.6	55.5	65	73.5	77.9	93.3	72	95.2	84.8	68.5	95.4	70.9	78.8	68.7	65.9	73.8
ResNet-38 [34]	78.4	98.5	85.7	93.1	55.5	59.1	67.1	74.8	78.7	**93.7**	**72.6**	95.5	86.6	69.2	95.7	64.5	78.8	74.1	69	76.7
PSPNet [18]	78.4	98.6	86.2	92.9	50.8	58.8	64.0	**75.6**	79.0	93.4	72.3	95.4	86.5	71.3	95.9	68.2	79.5	73.8	69.5	77.2
DenseASPP [12]	80.6	**98.7**	**87.1**	**93.4**	**60.7**	**62.7**	65.6	74.6	78.5	93.6	72.5	95.4	86.2	**71.9**	**96.0**	**78.0**	90.3	80.7	69.7	76.8
GLNet	**80.8**	**98.7**	86.7	**93.4**	56.9	60.5	**68.3**	75.5	**79.8**	**93.7**	**72.6**	**95.9**	**87.0**	71.6	**96.0**	73.5	**90.5**	**85.7**	**71.1**	**77.3**

Table 5. Performance comparison on Cityscapes val set.

Method	Mean IoU	Road	Sidewalk	Building	Wall	Fence	Pole	Traffic Light	Traffic Sign	Vegetation	Terrain	Sky	Person	Rider	Car	Truck	Bus	Train	Motorcycle	Bicycle
FCN-8s [3]	57.3	93.5	75.7	87.2	33.7	41.7	36.4	40.5	57.1	89.0	52.7	91.8	64.3	29.9	89.2	34.2	56.4	34.0	19.7	62.2
ICNet [10]	67.2	97.3	79.3	89.5	49.1	52.3	46.3	48.2	61.0	90.3	58.4	93.5	69.9	43.5	91.3	64.3	75.3	58.6	43.7	65.2
DeepLab-v2 [6]	69.0	96.7	76.7	89.4	46.2	49.3	43.6	55.0	64.8	89.5	56.0	91.6	73.3	53.2	90.8	62.3	79.6	65.8	58.0	70.2
PSPNet [18]	76.5	98.0	84.4	91.7	57.8	62.0	54.6	67.4	75.2	91.4	63.2	93.4	79.1	60.6	94.4	77.2	84.6	79.4	63.3	75.1
DeepLab-v3+ [19]	77.1	98.2	85.1	92.6	56.4	61.9	65.5	68.6	78.8	92.5	61.9	95.1	81.8	61.9	94.9	72.6	84.6	71.3	64.0	77.1
DenseASPP [12]	79.5	**98.6**	**87.0**	**93.2**	**59.9**	**63.3**	64.2	71.4	80.4	**93.1**	**64.6**	94.9	81.8	63.8	**95.6**	**84.0**	**90.8**	79.9	66.4	78.1
GLNet	**80.1**	98.4	86.7	93.1	59.5	62.7	**68.4**	**73.0**	**81.7**	92.9	64.4	**95.3**	**84.0**	**65.4**	95.3	82.6	90.6	**81.0**	**67.9**	**79.7**

Recently, DANet [21] reached a new state-of-the-art performance level of 81.5 on the Cityscapes dataset. However, DANet relies on extra strategies such as data augmentation, multiple feature maps of different grid sizes, and segmentation map fusion to further improve its performance [21]. Such improvement strategies can also be used in other methods. Therefore, DANet has not been included here for comparison. Without the extra steps, DANet achieved a mean IoU of 77.57, which is lower than our score.

The main difference between the proposed GLNet and existing segmentation methods is that GLNet explicitly models global contextual information, local contextual information, and low-level features in a single network and systematically combines the information for semantic segmentation of complex street view scenes. The benefit of incorporating all three pieces of information in one network enables GLNet to simultaneously dealing with large objects and background, and paying attention to fine details. Figure 9 shows visualization results of DenseASPP and GLNet, which are the top two performers in Table 4. Although the performance improvement of GLNet over DenseASPP is incremental in terms of mean IoU, it can be seen from Figure 9 that GLNet generates much better visualization results compared to DenseASPP. GLNet was able to segment large objects such as sidewalk and vegetation as well as small objects such as pedestrians and light poles equally well, whereas DenseASPP produced unsatisfactory results.

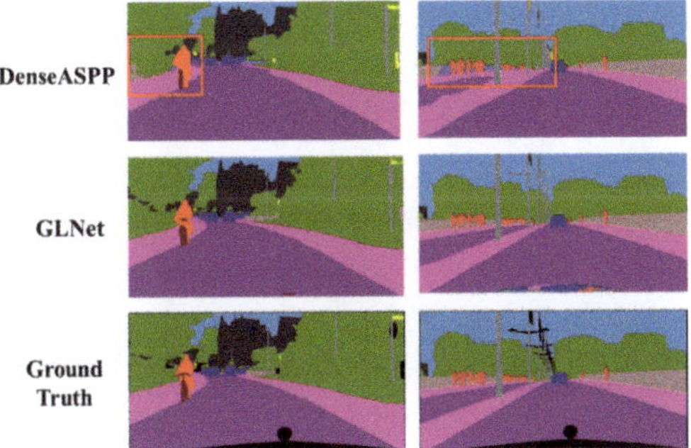

Figure 9. Visualization results of DenseASPP and GLNet on Cityscapes val. set.

4.3.2. Network Size and Inference Speed

Network size in terms of number of parameters and inference speed in frame per second (FPS) of the proposed GLNet and several other methods are listed in Table 6. The tests were running on NVIDIA GeForce GTX 1080 with Tensorflow Cuda 9.0. The test images were 1024 × 2048 raw images from the Cityscapes validation set. It can be seen from Table 6 that the number of parameters of the GLNet is slightly fewer than the DeepLab methods [6,19], and it is much lower than FCN [3] and PSPNet [18]. The inference speed of GLNet is comparable to DeepLab and PSPNet with 1.24 FPS. The difference in inference speed is partially owing to the fact that different backbone networks are used in different models. For instance, Xception-65 is used as the backbone in GLNet while DeepLab uses ResNet-101 as the backbone network.

Table 6. Network size and inference speed.

Method	Params	FPS
FCN-8s [3]	134.4 M	2.70
DeepLab-v2 [6]	43.9 M	1.38
PSPNet [18]	67.6 M	1.13
ICNet [10]	6.7 M	5.58
DeepLab-v3+ [19]	41.2 M	2.08
GLNet	39.8 M	1.24

5. Conclusions

We have proposed the GLNet which integrates global spatial information and dense local multi-scale context information in a single model for semantic segmentation of complex street view scenes. The global context module captures semantic of spatial interdependencies whereas the local context module extracts dense multi-scale features, and the output of the global-local module is fused with low-level features to recover fine scene details. Experimental results on the Cityscapes dataset have demonstrated superior performance of the proposed GLNet over existing state-of-the-art methods. This study highlights the importance of incorporating both global and local contextual information in image semantic segmentation. We hope this insight can contribute to future semantic segmentation works.

Author Contributions: C.-Y.L. provided conceptualization and algorithm design. Y.-C.C. conducted the implementation and testing. H.-F.N. helped in data analysis and writing. T.K.S. guided and supervised the project. K.-H.L. conducted experiments and testing for revisions. All authors have read and agreed to the published version of the manuscript.

Funding: This work was supported in part by Ministry of Science and Technology (MOST), Taiwan, under the Grant MOST 107-2221-E-155-048-MY3, MOST 108-2634-F-008-001, and 109-2634-F-008-007.

Conflicts of Interest: The authors declare no conflict of interest.

References

1. Sharma, S.; Ball, J.; Tang, B.; Carruth, D.; Doude, M.; Islam, M.A. Semantic Segmentation with Transfer Learning for Off-Road Autonomous Driving. *Sensors* **2019**, *19*, 2577. [CrossRef] [PubMed]
2. Sáez, Á.; Bergasa, L.M.; López-Guillén, E.; Romera, E.; Tradacete, M.; Gómez-Huélamo, C.; Del Egido, J. Real-Time Semantic Segmentation for Fisheye Urban Driving Images Based on ERFNet †. *Sensors* **2019**, *19*, 503. [CrossRef] [PubMed]
3. Long, J.; Shelhamer, E.; Darrell, T. Fully convolutional networks for semantic segmentation. In Proceedings of the IEEE Conference on Computer Vision and Pattern Recognition, Boston, MA, USA, 7–12 June 2015; pp. 3431–3440.
4. Noh, H.; Hong, S.; Han, B. Learning deconvolution network for semantic segmentation. In Proceedings of the IEEE International Conference on Computer Vision, Santiago, Chile, 11–18 December 2015; pp. 1520–1528.
5. Chen, L.-C.; Papandreou, G.; Kokkinos, I.; Murphy, K.; Yuille, A.L. Semantic image segmentation with deep convolutional nets and fully connected crfs. *arXiv* **2014**, arXiv:1412.7062.
6. Chen, L.-C.; Papandreou, G.; Kokkinos, I.; Murphy, K.; Yuille, A.L. DeepLab: Semantic Image Segmentation with Deep Convolutional Nets, Atrous Convolution, and Fully Connected CRFs. *IEEE Trans. Pattern Anal. Mach. Intell.* **2018**, *40*, 834–848. [CrossRef] [PubMed]
7. Yu, F.; Koltun, V. Multi-scale context aggregation by dilated convolutions. *arXiv* **2015**, arXiv:1511.07122.
8. Chen, L.-C.; Yang, Y.; Wang, J.; Xu, W.; Yuille, A.L. Attention to scale: Scale-aware semantic image segmentation. In Proceedings of the IEEE Conference on Computer Vision and Pattern Recognition, Las Vegas, NV, USA, 26 June–1 July 2016; pp. 3640–3649.
9. Eigen, D.; Fergus, R. Predicting depth, surface normals and semantic labels with a common multi-scale convolutional architecture. In Proceedings of the IEEE International Conference on Computer Vision, Santiago, Chile, 11–18 December 2015; pp. 2650–2658.
10. Zhao, H.; Qi, X.; Shen, X.; Shi, J.; Jia, J. ICNeT for real-time semantic segmentation on high-resolution images. In Proceedings of the European Conference on Computer Vision, Munich, Germany, 8–14 September 2018; pp. 405–420.
11. Chen, L.-C.; Papandreou, G.; Schroff, F.; Adam, H. Rethinking atrous convolution for semantic image segmentation. *arXiv* **2017**, arXiv:1706.0558.
12. Yang, M.; Yu, K.; Zhang, C.; Li, Z.; Yang, K. DenseASPP for semantic segmentation in street scenes. In Proceedings of the IEEE Conference on Computer Vision and Pattern Recognition, Salt Lake City, UT, USA, 18–23 June 2018; pp. 3684–3692.
13. Krizhevsky, A.; Sutskever, I.; Hinton, G.E. Imagenet classification with deep convolutional neural networks. In Proceedings of the International Conference on Neural Information Processing Systems, Lake Tahoe, NV, USA, 3–8 December 2012; pp. 1097–1105.
14. LeCun, Y.; Bottou, L.; Bengio, Y.; Haffner, P. Gradient-based learning applied to document recognition. *Proc. IEEE* **1998**, *86*, 2278–2324. [CrossRef]
15. Wang, X.; Girshick, R.; Gupta, A.; He, K. Non-local neural networks. In Proceedings of the IEEE Conference on Computer Vision and Pattern Recognition, Salt Lake City, UT, USA, 18–23 June 2018; pp. 7794–7803.
16. Badrinarayanan, V.; Kendall, A.; Cipolla, R. Segnet: A deep convolutional encoder-decoder architecture for image segmentation. *IEEE Trans. Pattern Anal. Mach. Intell.* **2017**, *39*, 2481–2495. [CrossRef] [PubMed]
17. Ronneberger, O.; Fischer, P.; Brox, T. U-net: Convolutional networks for biomedical image segmentation. In Proceedings of the International Conference on Medical Image Computing and Computer-assisted Intervention (MICCAI), Munich, Germany, 5–9 October 2015; pp. 234–241.
18. Zhao, H.; Shi, J.; Qi, X.; Wang, X.; Jia, J. Pyramid scene parsing network. In Proceedings of the IEEE Conference on Computer Vision and Pattern Recognition, Honolulu, HI, USA, 21–26 July 2017; pp. 2881–2890.
19. Chen, L.-C.; Zhu, Y.; Papandreou, G.; Schroff, F.; Adam, H. Encoder-decoder with atrous separable convolution for semantic image segmentation. In Proceedings of the European Conference on Computer Vision, Munich, Germany, 8–14 September 2018; pp. 801–818.
20. Zhou, Q.; Yang, W.; Gao, G.; Ou, W.; Lu, H.; Chen, J.; Latecki, L. Multi-scale deep context convolutional neural networks for semantic segmentation. *World Wide Web* **2019**, *22*, 555–570. [CrossRef]

21. Fu, J.; Liu, J.; Tian, H.; Li, Y.; Bao, Y.; Fang, Z.; Lu, H. Dual attention network for scene segmentation. In Proceedings of the IEEE Conference on Computer Vision and Pattern Recognition, Long Beach, CA, USA, 16–20 June 2019; pp. 3146–3154.

22. Zhang, H.; Zhang, H.; Wang, C.; Xie, J. Co-occurrent features in semantic segmentation. In Proceedings of the IEEE Conference on Computer Vision and Pattern Recognition, Long Beach, CA, USA, 16–20 June 2019; pp. 548–557.

23. Yuan, Y.; Wang, J. OCNet: Object context network for scene parsing. *arXiv* **2018**, arXiv:1809.00916.

24. Hu, X.; Yang, K.; Fei, L.; Wang, K. ACNET: Attention based network to exploit complementary features for rgbd semantic segmentation. In Proceedings of the IEEE Conference on Image Processing (ICIP), Taipei, Taiwan, 22–25 September 2019; pp. 1440–1444.

25. Yang, K.; Wang, K.; Bergasa, L.M.; Romera, E.; Hu, W.; Sun, D.; Sun, J.; Cheng, R.; Chen, T.; López, E. Unifying terrain awareness for the visually impaired through real-time semantic segmentation. *Sensors* **2018**, *18*, 1506. [CrossRef] [PubMed]

26. Chollet, F. Xception: Deep learning with depthwise separable convolutions. In Proceedings of the IEEE Conference on Computer Vision and Pattern Recognition, Honolulu, HI, USA, 21–26 July 2017; pp. 1251–1258.

27. Hu, J.; Shen, L.; Sun, G. Squeeze-and-excitation networks. In Proceedings of the IEEE Conference on Computer Vision and Pattern Recognition, Salt Lake City, UT, USA, 18–23 June 2018; pp. 7132–7141.

28. Zhang, H.; Dana, K.; Shi, J.; Zhang, Z.; Wang, X.; Tyagi, A.; Agrawal, A. Context encoding for semantic segmentation. In Proceedings of the IEEE Conference on Computer Vision and Pattern Recognition, Salt Lake City, UT, USA, 18–23 June 2018; pp. 7151–7160.

29. Cordts, M.; Omran, M.; Ramos, S.; Rehfeld, T.; Enzweiler, M.; Benenson, R.; Franke, U.; Roth, S.; Schiele, B. The cityscapes dataset for semantic urban scene understanding. In Proceedings of the IEEE Conference on Computer Vision and Pattern Recognition, Las Vegas, NV, USA, 26 June–1 July 2016; pp. 3213–3223.

30. Russakovsky, O.; Deng, J.; Su, H.; Krause, J.; Satheesh, S.; Ma, S.; Huang, Z.; Karpathy, A.; Khosla, A.; Bernstein, M. Imagenet large scale visual recognition challenge. *Int. J. Comput. Vis.* **2015**, *115*, 211–252. [CrossRef]

31. Woo, S.; Park, J.; Lee, J.-Y.; Kweon, I.S. CBAM: Convolutional block attention module. In Proceedings of the European Conference on Computer Vision, Munich, Germany, 8–14 September 2018; pp. 3–19.

32. Lin, G.; Milan, A.; Shen, C.; Reid, I. Refinenet: Multi-path refinement networks for high-resolution semantic segmentation. In Proceedings of the IEEE Conference on Computer Vision and Pattern Recognition, Honolulu, HI, USA, 21–26 July 2017; pp. 1925–1934.

33. Wang, P.; Chen, P.; Yuan, Y.; Liu, D.; Huang, Z.; Hou, X.; Cottrell, G. Understanding convolution for semantic segmentation. In Proceedings of the IEEE Winter Conference on Applications of Computer Vision, Lake Tahoe, NV, USA, 12–15 March 2018; pp. 1451–1460.

34. Wu, Z.; Shen, C.; Van Den Hengel, A. Wider or deeper: Revisiting the resnet model for visual recognition. *Pattern Recog.* **2019**, *90*, 119–133. [CrossRef]

Article

Deep Binary Classification via Multi-Resolution Network and Stochastic Orthogonality for Subcompact Vehicle Recognition

Joongchol Shin [†], Bonseok Koo [†], Yeongbin Kim and Joonki Paik *

Department of Image, Chung-Ang University, Seoul 06974, Korea; mbstel275@gmail.com (J.S.); izg2sd@gmail.com (B.K.); sawors2010@gmail.com (Y.K.)
* Correspondence: paikj@cau.ac.kr; Tel.: +82-10-7123-6846
† These authors contributed equally to this work.

Received: 17 March 2020; Accepted: 7 May 2020; Published: 9 May 2020

Abstract: To encourage people to save energy, subcompact cars have several benefits of discount on parking or toll road charge. However, manual classification of the subcompact car is highly labor intensive. To solve this problem, automatic vehicle classification systems are good candidates. Since a general pattern-based classification technique can not successfully recognize the ambiguous features of a vehicle, we present a new multi-resolution convolutional neural network (CNN) and stochastic orthogonal learning method to train the network. We first extract the region of a bonnet in the vehicle image. Next, both extracted and input image are engaged to low and high resolution layers in the CNN model. The proposed network is then optimized based on stochastic orthogonality. We also built a novel subcompact vehicle dataset that will be open for a public use. Experimental results show that the proposed model outperforms state-of-the-art approaches in term of accuracy, which means that the proposed method can efficiently classify the ambiguous features between subcompact and non-subcompact vehicles.

Keywords: vehicle recognition; multi resolution network; optimization

---

1. Introduction

Typically, subcompact cars are defined by the engine displacement, width, and height under 1000 cc, 1.6 m, and 2.0 m, respectively. To satisfy these specifications, the subcompact car has a unique shape such as shorter-bonnet and hatchback. In addition, there are various environmental benefits because the subcompact cars have a small displacement engine and a light weight. To encourage people to drive subcompact cars, many countries provide several benefits—discounts on tall road charge and parking fee. Since classification of subcompact cars from other requires labor-intensive human investigation, an automatic vehicle classification system is needed. In general, vehicle classification methods can be classified into two approaches: one uses infrared sensors to measure physical dimensions of a vehicle such as length, height, and width. The other uses a single camera and image processing algorithms to recognize the visual characteristics of vehicles [1,2]. Despite of accuracy and robustness, the infrared sensor-based system is too expensive to be installed in many places. Thus, we propose an image recognition system to reduce the installation and maintenance cost. To classify the visual feature in images, Dalal et al. extracted the histogram of oriented gradients (HOG) and classify the HOG using support vector machine (SVM) [3]. To the best of authors' knowledge, the HOG-based SVM is the most popular approach to recognize objects before deep learning has become popular. To enhance HOG features that are affected by rotation, or distance, and occlusion, various approaches were proposed. Llorca et al. proposed vehicle manufacturer recognition by detecting the vehicle-logo, but a subcompact car can not be completely classified using only manufacturer information [4].

Clady et al. recognized the vehicle type by separating objects and the background in interactively selected regions [5]. This method is robust to the variance in the distance. However, the region should be passively selected. Mohottala et al. created vehicle images using computer graphics (CG), and then classify the type of vehicle using eigenvalues [6].

Although this approach can easily obtain the vehicle data, it cannot avoid error in real vehicle data. Michael and Daniel classified the eigenvalues of vehicle classes using neural networks [7]. Since they used an artificial neural network, classification accuracy was acceptable only without occlusion. Huttunen et al. adaptively recognized the vehicle classes using a deep neural network [8]. This method can recognize the multi-class vehicles such as sedan, truck, and bus. However, in subcompact car classification, it has overffiting while learning the subcompact vehicle class because it is difficult to discriminate the subcompact vehicle from others. Simonyan et al. proposed the deep convolutional neural networks called VGG16 and VGG19 [9]. Since the VGG networks can be pre-trained via a large-scale image dataset [10], it can have a very deep hidden-layer to recognize the vehicle. However, it cannot robustly classify the subcompact vehicles because of both obscure features and environmental variables as shown in Figure 1. He et al. proposed the more deep residual networks [11]. This network can be designed more deeply such as 50, 101, 151 layers because of the residual learning. However, in the binary classification, the VGG networks are also deep enough. Xie et al. applied the split-transform-merge strategy to deep residual networks [12]. This strategy can effectively recognize various features, but it can not adaptively crop the image region. Karpathy et al. proposed the multiple convolutional neural networks with center clipping and image fusion for video classification [13]. It can recognize the obscure objects and actions in video, but it cannot localize objects that are not in the center of the image.

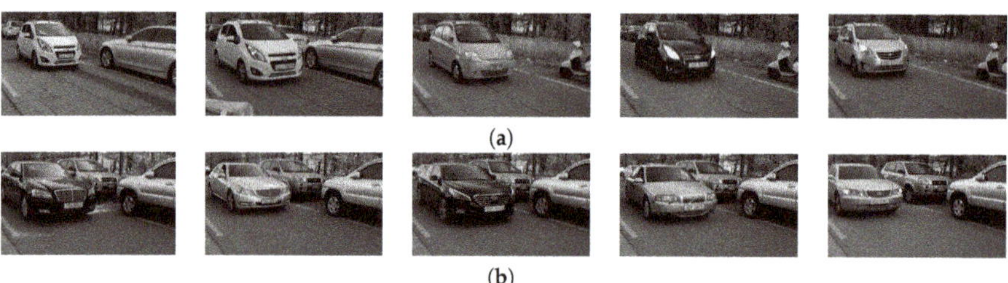

(a)

(b)

Figure 1. (a) Subcompact vehicles and (b) sedans. It is not easy to differentiate two classes using small features such as head lamp or rear-view mirror. On the other hand, there are differences in bigger features such as bonnet and overall shape of vehicles.

To solve this isolated problem, we proposed a novel multi-resolution network and stochastic orthogonal learning method. More specifically, the proposed method include three functional steps: (i) we emphasize the features using retinex model-based image-enhancement [14], (ii) we track the bonnet region using an optimized correlation filters [15], and (iii) we engage this region and image using the proposed multi-resolution network. In addition, we learned our muti-resolution network by considering stochastic orthogonality of probabilities between subcompact and general vehicles. We also build a subcompact vehicle dataset including 1500 training data and 2000 test-images. Experimental results show that the proposed method outperforms state-of-the-arts approaches in terms of accuracy by over 12.25%. This paper is organized as follows. In Section 2, we describe the related works. The proposed multi-resolution network is presented in Section 3 followed by experimental results in Section 4, and Section 5 concludes this paper.

2. Related Works

2.1. Support Vector Machine

To classify the features in images, the SVM can be applied by minimizing as following Equation [16]

$$\therefore \arg\min_{\vec{w},b} \frac{1}{2}\|\vec{w}\|^2 - \sum_{i=1} \alpha_i \left(y_i \left(\vec{w} \bullet \vec{x}_i + b\right) - 1\right), \tag{1}$$

where $\vec{w}$ and b represent weight and bias of hyper-plane to classify the features, α is an operator to find the support vectors, and y denotes the label such as positive or negative. The optimized hyper-plane of the support vector machine works well, but it should be estimated low-dimensional features such as histograms of gradients and scale-invariant features [3,17] to apply imaging systems.

2.2. Neural Network

The neural networks can classify the non-linear features because the each node in hidden-layer discriminates the complicated patterns as shown Figure 2. Each node includes the weight, bias, and activate function such as sigmoid, and relu. These parameters can be easily estimated by simple cost function and chain rule as

$$E_{total} = \sum \frac{1}{2}(y - f(x))^2, \tag{2}$$

and

$$\begin{aligned}
\frac{\partial E_{total}}{\partial w_*} &= \frac{\partial E_{total}}{\partial out_{o1}} * \frac{\partial out_{o1}}{\partial net_{o1}} * \frac{\partial net_{o1}}{\partial w_*}, \\
\frac{\partial E_{total}}{\partial b_*} &= \frac{\partial E_{total}}{\partial out_{o1}} * \frac{\partial out_{o1}}{\partial net_{o1}} * \frac{\partial net_{o1}}{\partial b_*},
\end{aligned} \tag{3}$$

where f returns the results of neural networks, w_* and b_* are weight and bias in $*$-th node. Therefore, each parameter can be estimated as

$$\begin{aligned}
w_*(t+1) &= w_*(t) - \frac{\partial E_{total}}{\partial w_*}, \\
b_*(t+1) &= b_*(t) - \frac{\partial E_{total}}{\partial b_*}.
\end{aligned} \tag{4}$$

However, the neural net also has limitation to apply large-scale image classification.

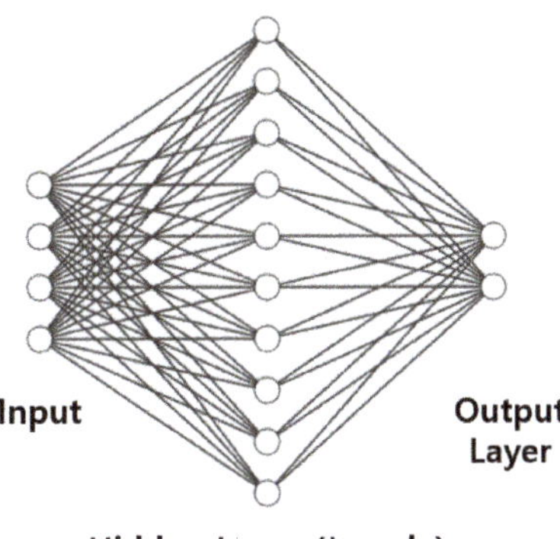

Figure 2. The architecture of neural network.

2.3. Convolutional Neural Network (CNN)

Since an image has various features such as gradients, color, and intensity information, the convolution operators in the hidden layer are effective to extract the image features [9]. Furthermore, these convolution operator can also be optimized by chain-rule. For example, we visualize the extracted image features in convolution layer using CNN feature simulator [18]. Note that the convolution operator can extract the large scale features and textures as shown Figure 3. In other words, the CNN can not only classify the multi-class image but also recognize the detail textures. Therefore, the CNN can be applied to various field using the transfer learning method such as medical imaging [19], intelligent transportation system [20,21], and remote sensing [22]

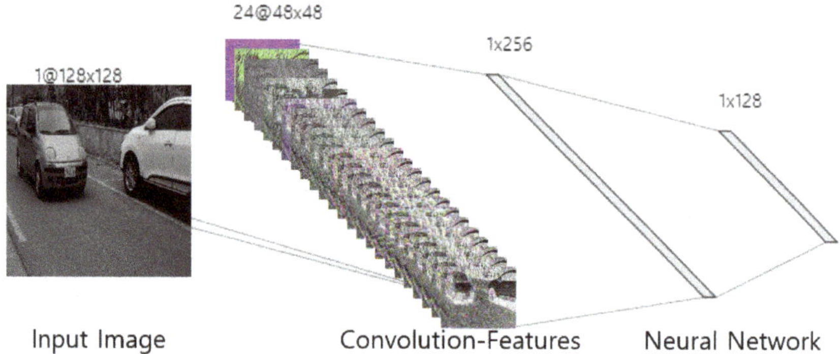

Figure 3. The convolutional neural network and convolution features.

3. Proposed Method

3.1. Subcompact Vehicle Dataset

To train and test the proposed network, we collected vehicle images using a digital camera (gray-scale) at a parking gate in Seoul, South Korea. Since the camera was installed under the charge machine, images were captured from an angle viewed from below as shown in Figure 4. Furthermore, we collected vehicle images for 1 year and 6 months to reflect various environmental variables such as day light, back light, dust, and night. The collected images were classified into five types including subcompact sedan, subcompact van, subcompact truck, sedan and sport utility vehicle (SUV), and truck and van according to the design and shape. The dataset was split into training and test sets including 1500 and 2000 images, respectively. The goal of this work is the binary classification (subcompact vehicle or not). To this end, each set of the proposed dataset is divided into subcompact and non-subcompact vehicles as shown in Figure 5.

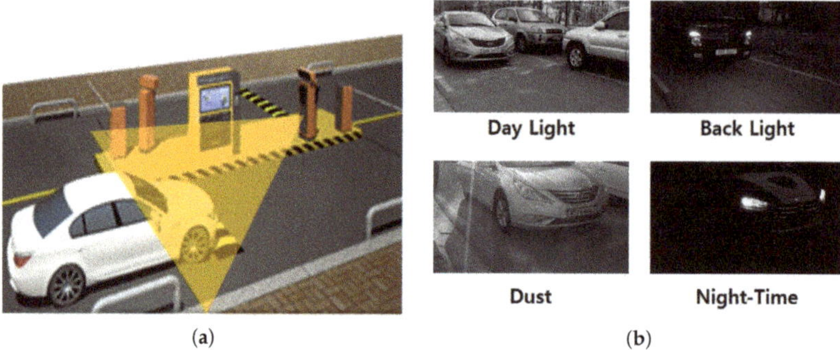

Figure 4. (**a**) Camera installation and (**b**) four different illumination conditions.

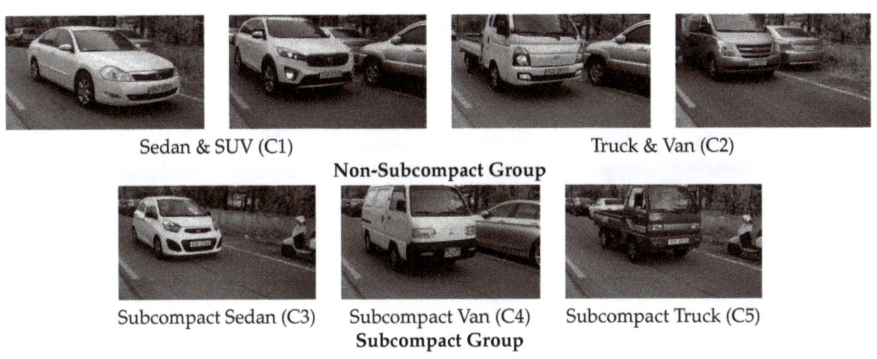

Sedan & SUV (C1) Truck & Van (C2)

Non-Subcompact Group

Subcompact Sedan (C3) Subcompact Van (C4) Subcompact Truck (C5)

Subcompact Group

Figure 5. An example of five vehicle classes in the proposed dataset.

3.2. Pre-Convolution Layer

The proposed network consists of pre-convolution and multi-resolution network layers. In the pre-convolution layer, we resize the original 1920 × 1080 px images to 400 × 300 px, and we amplify the intensity and increase the local-contrast using a simple retinex-based image enhancement algorithm as [14]

$$H(x) = \frac{I(x)}{\max\left(l(x)\,\varepsilon\right)},\tag{5}$$

where H represents high-resolution image, I the input gray-image in the dataset, and ε is a very small positive number to avoid division by zero. The illuminance map l can be estimated as a smoothing term [23]

$$l = med(I) - med\left(|med(I) - I|\right),\tag{6}$$

where *Med* represents the local-median filter [24]. Figure 6a,c show that the environmental variables can be normalized, and at the same time local-contrast in the shadow region is also enhanced. To process a low-resolution image, the vehicle region should be localized. In this paper, we detect the region using a correlation filter which has low-computational complexity and efficient localization performance [25]. To reflect a feature of texture, we applied multi-channel correlation filter (MCCF) with the histogram of oriented gradients, which can be defined as ridge regression

$$E(w) = \frac{1}{2}\sum_i^N\sum_j^D \left(y_i(j) - \sum_{k=1}^K w^{(k)T}x_i^{(k)}\left[\Delta\tau_j\right]\right)^2 + \frac{\lambda}{2}\sum_{k=1}^K \left(w^{(k)}\right)^2,\tag{7}$$

where $y_i(j)$ is the desired and shifted response in i-th sample $y_i = [y_i(1),....,y_i(D)]^T$, $x_i[\Delta\tau_j]$ is a set of cyclically shifted vehicle images in the training dataset. N represents the number of training images, K is the channels of feature map including HOG 34-channels, and w represents the correlation filter. The response map y for a vehicle coordinate in the frequency domain has a Gaussian-shaped distribution centering on a pre-annotated region. Since both input patch and response map are circular matrices for cyclic convolution, the correlation filter w can be simply expressed in the frequency domain as

$$\hat{w}^* = \left(\lambda I + \sum_i^N \hat{X}_i^T \hat{X}_i\right)^{-1} \sum_{i=1}^N \hat{X}^T \hat{y}_i,\tag{8}$$

where, $\hat{w}$ represents a variable w in the frequency domain, $*$ and T respectively represent the complex conjugate and transpose of a matrix. The optimized correlation filter can estimate the coordinates of the vehicle region via maximum response-region and distribution as shown in Figure 6d–f. We cropped high-resolution images, based on this picking coordinates to obtain low-resolution images L (280 × 200).

Note that the proposed pre-convolution layer should be processed in Central processing unit (CPU) for efficient memory allocation.

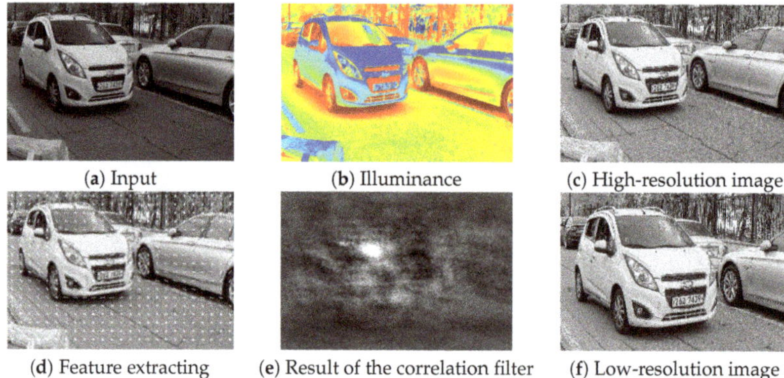

Figure 6. Step-by-step results in the pre-convolution layer (a–f).

3.3. Multi-Resolution Network

We changed the size of both high- and low-resolution images to 224×224 to recognize vehicle type. Note that the scale of the proposed subcompact dataset is gray. Therefore, we generate the 3 zero-min channels with the average color of ImageNet [10], and concatenate 3 zero-min channels to create a pseudo color as

$$H^R = H - 0.4850,\ H^G = H - 0.4580,\ H^B = H - 0.4076, \tag{9}$$

and

$$L^R = L - 0.4850,\ L^G = L - 0.4580,\ L^B = L - 0.4076, \tag{10}$$

where H and L are single gray-scale ($224 \times 224 \times 1$), and H^* and L^* have pseudo RGB channel ($224 \times 224 \times 3$) as shown as high- and low-resolution in Figure 7. We correspondingly defined both high- and low-resolution network with 13 convolution layers, 5 max-pooling layers, and 3 fully connected layers as shown in Figure 7. A 3×3 filter is used in each convolution layer, ReLU is used for an activation function, and 2×2 max-pooling filters are used to maximize the receptive field [9]. Each fully connected layer has 4096 perceptrons, except for the last five layers. Finally, the soft-max operator returns the probability using returned five values. In this paper, the proposed multi-resolution network is combined to our pre-convolution layer described in Section 3.2.

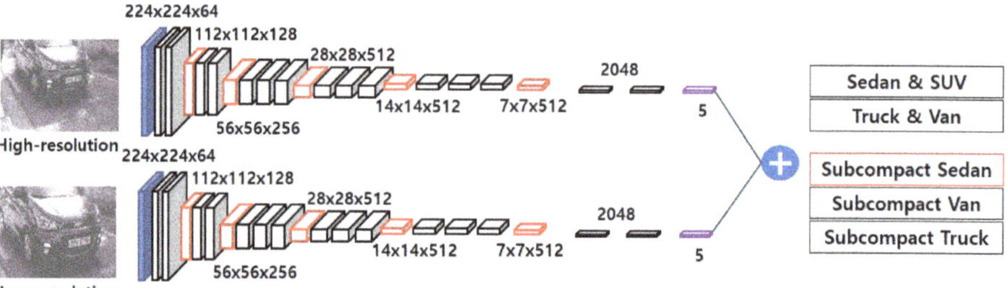

Figure 7. The architecture of the proposed multi-resolution network. Blue, black, and red cubes represent the resized input, convolution, and max-pool layers, respectively. Black and purple boxes are fully-connected and softmax layers, respectively.

3.4. Orthogonal Learning

To combine the results of low- and high-resolution networks, we define the average least square loss as

$$L_{least} = \frac{1}{2B} \sum_{n=0}^{B} \sum_{i=1}^{N} (g_i{}^n - P_i(H_n^*))^2 + (g_i - P_i(L_n^*))^2,$$ (11)

where H_n and L_n respectively represent the $n-$th high and low resolution image, B denotes the size of batch, and N is the vehicle type between C1 to C5. g represents the one-hot vector of size $B \times 5$, which is pre-labeled in our dataset described in Section 3.1. To reduce the correlation between two groups, we also define the orthogonal loss as [26]

$$L_o = \frac{1}{B} \sum_{n=0}^{B} \sum_{i=1}^{2} (P_i(H_n^*, L_n^*) P_3(H_n^*, L_n^*) + P_i(H_n^*, L_n^*) P_4(H_n^*, L_n^*) + + P_i(H_n^*, L_n^*) P_5(H_n^*, L_n^*)).$$ (12)

In binary classification, the error can be reduced when sum of multiplication between subcompact and other group is closed to zero as shown in Figure 8. Therefore, the proposed total loss can be defined as

$$L_{total} = L_{least} + L_o.$$ (13)

To reduce the total loss L_{total}, the set of parameters including convolution kernel, bias, and perceptron weight are updated via stochastic gradient decent optimization [9]. The learning and dropout rates are set to 0.0001 and 0.5, respectively. For supervised learning, we train the model using 1500 labeled training data given in Section 3.1. For transfer learning, all of convolution layers are pretrained by ImageNet data [10]. We trained the proposed model using 4500 iterations, and 15 batches are engaged to the proposed multi-resolution network for each learning. Figure 9 shows the proposed learning procedure. Finally, to distinguish subcompact vehicle, the probability values of the optimized model are estimated using the thresholding operator as

$$D_n = \begin{cases} False & \arg\max \frac{1}{2}(P(H_n^*) + P(L_n^*)) \in \{C1, C2\} \\ True & \arg\max \frac{1}{2}(P(H_n^*) + P(L_n^*)) \in \{C3, C4, C5\} \end{cases}$$ (14)

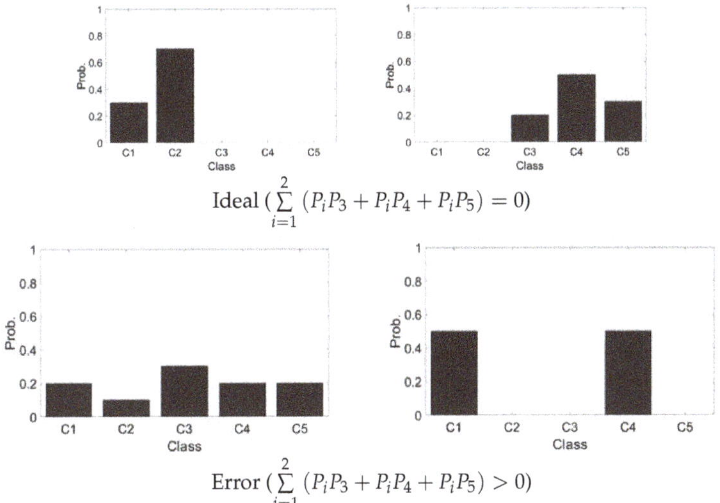

Figure 8. An example of orthogonal learning.

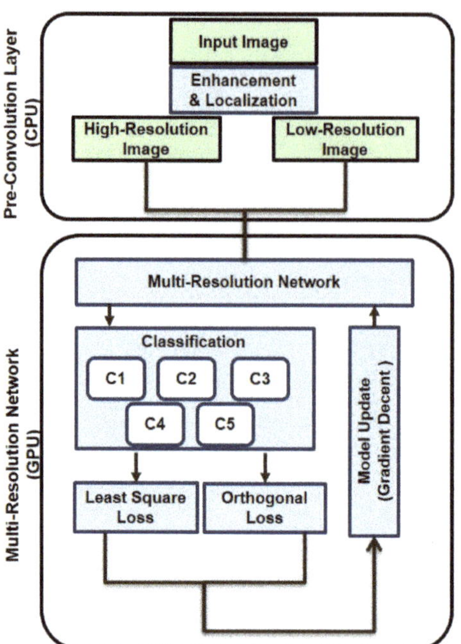

Figure 9. Dual procedure of the proposed method using both CPU and GPU.

4. Experimental Results

4.1. Quantitative Evaluation

To evaluate the proposed method, we compared experimental results with 2000 test data and state-of-the-arts classification models including HOG based recognition model ($HOG + SVM$) proposed by Dalal et al. [3], MCCF combined HOG recognition ($MCCF + HOG + SVM$) [15], multi-resolution image based HOG recognition ($Retinex + MCCF + HOG + SVM$), deep neural network based Huttunen's method (DNN) [8], convolutional neural network (CNN) with 16 layers proposed by Simonyan et al. [9], retinex CNN $Retinex + CNN$, based on proposed pre-convolution layer ($Retinex + MCCF + CNN$), and the proposed multi-resolution network without orthogonal learning ($Retinex + MCCF + MRN$). All the algorithms were implemented in visual studio 2015 and Python 3.5 using on a desktop PC with i7 CPU, 64 GB RAM, and NVIDIA RTX 2080ti graphics processing unit (GPU). We also quantitatively measured accuracy (Acc.), precision, recall, and false-positive rate (FPR) as

$$precision = \frac{TP}{TP + FP}, \tag{15}$$

$$Recall = \frac{TP}{TP + FN}, \tag{16}$$

$$FPR = \frac{FP}{TP + FP}, \tag{17}$$

and

$$Acc. = 100 \times \frac{TP + TN}{2000}, \tag{18}$$

where TP, TN, FP, and FN respectively represent the true-positive, true-negative, false-positive, and false negative. Table 1 shows several evaluation results using state-of-the arts and the proposed methods. Since *HOG* uses handcraft-based features, its recognition performance is limited for ambiguous features. *HOG + SVM* method results in many mis-classification cases represented by *FN* and *FP* as shown in Table 1.

Table 1. Quantitative comparison with state-of-the art approaches.

Method	TP	FN	TN	FP	Precision	Recall	FPR	Acc.
HOG + SVM	165	335	912	588	0.2191	0.3300	0.7809	53.85%
MCCF + HOG + SVM	43	457	1477	23	0.6515	0.0860	0.3485	76.00%
Retinex + MCCF + HOG + SVM	37	463	1455	45	0.4512	0.0740	0.5488	74.60%
DNN	182	318	1377	123	0.5967	0.3650	0.4033	77.95%
CNN	198	302	1457	43	0.8216	0.3960	0.1784	82.75%
Retinex + CNN	410	90	1474	26	0.9404	0.8200	0.0596	94.20%
Retinex + MCCF + CNN	373	127	1444	56	0.8695	0.7460	0.1305	90.85%
Retinex + MCCF + MRN	417	83	1478	22	0.9499	0.8340	0.0501	94.75%
Proposed MRN	423	77	1477	23	0.9484	0.8460	0.0516	95.00%

MCCF + HOG + SVM method can improve the false-positive case because MCCF based localization effectively removes unnecessary information such as background, but FN case can be increased. Since retinex-based image enhancement enhance too much textures, *MCCF + HOG + SVM* outperforms *Retinex + MCCF + HOG + SVM* in every sense. The deep neural network (*DNN*) can effectively increase the TP compared with SVM-based methods, but false-positive rate was slightly higher than *MCCF + HOG + SVM*. Simonyan's convolutional neural network model can improve both TP and TN, so accuracy was highly increased over 4% than both *DNN* and *SVM* based methods, Especially, false-positive rate rapidly decreases compared with DNN and SVM based method, but accuracy is not enough because of the imbalance between true positive and false negative. Since the enhanced textures in a shadow region can compensate for the imbalance, the retinex-based CNN model (*Retinex + CNN*) outperforms the vanilla *CNN* in terms of the recall, accuracy. As a result of the localization error of *MCCF*, *Retinex + MCCF + CNN* can generate errors such as FN and FP, but the combined version *Retinex + MCCF + MRN* outperforms *CNN* models in all of evaluation terms. This is mean that the proposed *MRN* can adaptively reflect between localized information and enhanced textures to recognize the sub-compact vehicle. Furthermore, the proposed orthogonal learning method has TP values higher than *Retinex + MCCF + MRN* because it can generate the uncorrelated group-probability vectors. Note that the precision, recall, fpr, and accuracy are better by 0.1268, 0.45, 1.268, and 12.25% than convolutional neural network (*CNN*). In conclusion, the proposed approach can effectively classify the ambiguous objects because it designed and optimized with consideration of the group-error and ambiguous features. In addition, the multi-class recognition performance is compared with the *CNN* model as shown in Figure 10. The *CNN* method misclassifies between sedan and subcompact vehicles many times. Furthermore, subcompact truck and van are sometimes mis-recognized by the *CNN* method. However, the proposed method can not only reduce the mis-recognized case but also improve the accuracy by 10.85%. In addition, we conducted ablation study using validation check as shown in Figure 11. When we train the MRN without pseudo color (9), the performance can be degraded as shown black line in Figure 11 because the MRN was pre-trained by true color image dataset. The center crop-based MRN(Center crop) means that the MCCF operator in proposed pre-convolution layer was replaced to center cropping method [13]. Since the center cropping method can not adaptively localize the object, it can not outperform than proposed MRN. Furthermore, if MRN was optimized by only least square loss (11), the extracted features are not suitable for binary classification. Thus, when MRN was learned without proposed orthogonal loss (12), the performance of the MRN can not reach the orthogonal learning-based MRN as shown in Figure 11.

Label	C1	C2	C3	C4	C5	Total
C1	1263	37	0	42	0	1342
C2	5	152	1	0	0	158
C3	163	16	57	4	2	242
C4	5	58	1	58	2	124
C5	2	58	2	0	72	134
Accuracy	80.10%					2000

Convolutional Neural Network [9]

Label	C1	C2	C3	C4	C5	Total
C1	1263	58	0	21	0	1342
C2	5	151	0	0	2	158
C3	37	9	182	10	4	242
C4	2	18	0	101	3	124
C5	1	10	0	1	122	134
Accuracy	90.95%					2000

Multi-Resolution Network

Figure 10. Multi-classification comparison.

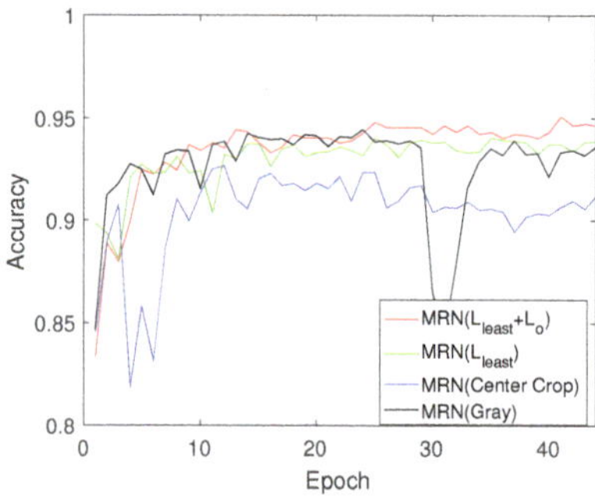

Figure 11. Ablation study: MRN ($L_{least} + L_o$) is the proposed orthogonal learning based multi-resolution network (MRN), MRN (L_{least}) is the least square loss based MRN, MRN (Center Crop) is the center cropping method to generate the low-resolution image, and MRN (Gray) is a single channel based MRN.

4.2. Baseline Comparison

To compare the efficient baseline networks, we evaluate the accuracy with several efficient networks such as VGG16 [9], residual network50(resnet50) [11], and resinext50 [12]. Table 2 shows the maximum binary, and multi class accuracy values. Since general networks do not consider ambiguous in binary classification problem, the MRN outperforms than several based line networks in terms of both binary and multi-class accuracy. The residual network based MRN slightly lower than the VGG16 based MRN because the VGG16 have already deep layers in binary classification. However, resnext based MRN outperforms the VGG16 based MRN in terms of binary accuracy because the split-transform-merge strategy can effectively apply to recognize the ambiguous binary objects. Figure 12 shows the accuracy for each training epoch. Note that the proposed networks outperform than state-of-the-arts baseline-networks in most epoch. In addition, we recorded computational complexity and allocated GPU-memory on average to verify the computational efficiency.

Table 2. Effects on the accuracy for different baseline networks.

Method	Baseline	Tool	Accuracy (Binary)	Accuracy (Multi)	Proc. Time (ms)	GPU-Memory (GB)
CNN	VGG16	Tensorflow	0.8275	0.8010	65 ms	1.3 GB
CNN	Resnet50	Pytorch	0.8740	0.8435	80 ms	1.0 GB
CNN	Resnext50	Pytorch	0.8890	0.8575	86 ms	1.0 GB
MRN	VGG16	Tensorflow	0.9500	0.9095	70 ms	1.6 GB
MRN	Resnet50	Pytorch	0.9290	0.8810	100 ms	1.2 GB
MRN	Resnext50	Pytorch	0.9530	0.8955	106 ms	1.3 GB

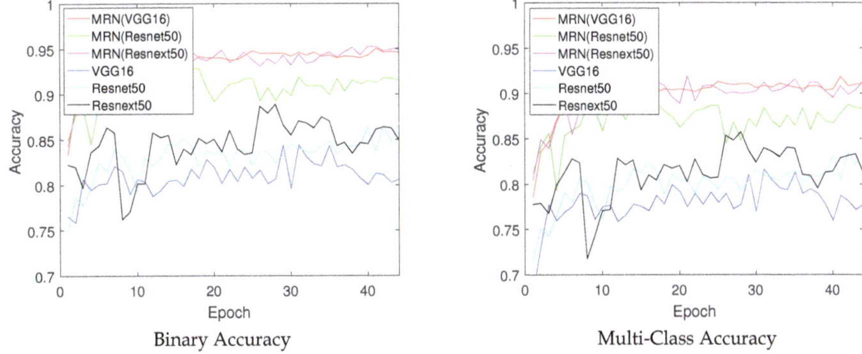

Figure 12. Accuracy evaluation according to each epoch.

4.3. Visualization

To verify the performance of the proposed orthogonal learning, we visualized the output values of last layer in the both CNN [9] and the proposed MRN by projecting to two-dimensional space using t-stochastic neighbor embedding (t-SNE) [27]. In Figure 13a, the visualization is not easy because of the correlation between subcompact vehicle and other groups. However, Figure 13b shows that the points are clustered to easily classify, which means that the proposed orthogonal learning can remove the group-correlation. As a result, proposed orthogonal learning can improve the performance of deep binary classification.

In Figure 14, we visualized the classified label and localized regions, where the localized bonnet region is the input to the low-resolution network, and the entire image is engaged to the high-resolution network. The resulting label (subcompact and non-subcompact vehicle) based on the average value of the two probabilities is reflected at the end of red-box. The proposed MRN can not only classify in various illuminations but also distinguish the ambiguous vehicle types.

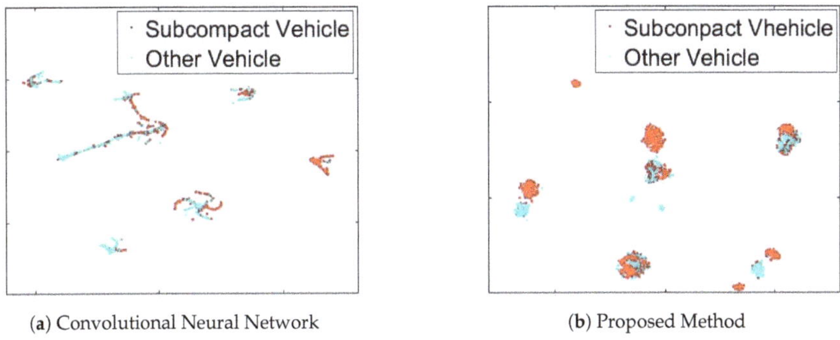

(**a**) Convolutional Neural Network　　　　　(**b**) Proposed Method

Figure 13. Probability Visualization using t-Stochastic Neighbor Embedding (t-SNE) [27].

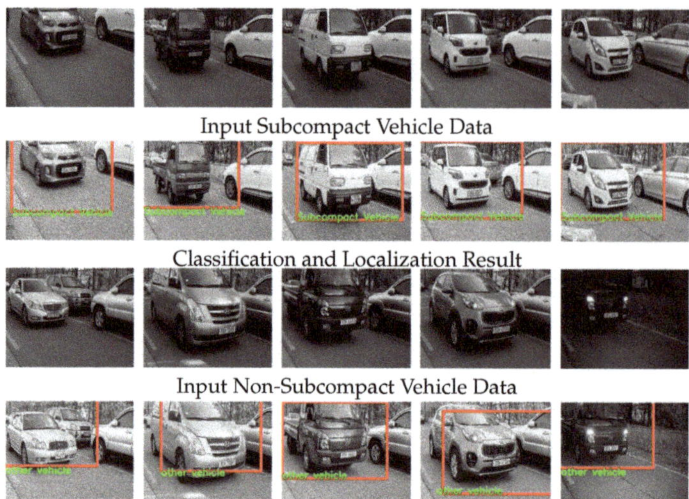

Input Subcompact Vehicle Data

Classification and Localization Result

Input Non-Subcompact Vehicle Data

Classification and Localization Result

Figure 14. Classification and localization result.

5. Conclusions

To recognize ambiguous features between the subcompact and other vehicles, we collected a novel set of subcompact vehicle images, and proposed a pre-convolution layer that is combined with the multi-resolution network with an orthogonal learning method. The proposed method can not only enhance the textures using retinex-based enhancement but also adaptively cropped the bonnet region using correlation computation. As a result, our MRN can avoid over-fitting by ambiguous features between vehicle types, and outperforms the existing CNN(VGG16) method by 12.25%. Therefore, the proposed method can be applied to various traffic management systems such as toll and parking gates for automatic charging system. In the future work, we will expand the proposed MRN by combining the license plate detection system. The source code is available at https://github.com/JoongcholShin.

Author Contributions: Conceptualization, J.S. and Y.K.; methodology, J.S., B.K.; software, J.S., B.K., and Y.K.; validation, J.S. B.K., and Y.K.; formal analysis, J.S., B.K., and Y.K.; investigation, J.S., B.K., and Y.K.; resources, J.P.; data curation, J.S., B.K., and Y.K.; writing—original draft preparation, J.S., B.K., and Y.K.; writing—review and editing, J.P.; visualization, J.S and B.K; supervision, B.K., Y.K., and J.P.; project administration, J.P.; funding acquisition, J.P. All authors have read and agreed to the published version of the manuscript.

Funding: This research was partly supported by Institute for Information & communications Technology Promotion(IITP) grant funded by the Korea government(MSIT) (2017-0-00250, and by Intelligent Defense Boundary Surveillance Technology Using Collaborative Reinforced Learning of Embedded Edge Camera and Image Analysis) and supported by the Chung-Ang University Research Scholarship Grants in 2020.

Acknowledgments: We acknowledge support given by Nexpa System Co., Ltd.

Conflicts of Interest: The authors declare no conflict of interest.

References

1. Ding, J.; Cheung, S.; Tan, C.; Varaiya, P. Signal processing of sensor node data for vehicle detection. In Proceedings of the 7th International IEEE Conference on Intelligent Transportation Systems (IEEE Cat. No.04TH8749), Washington, WA, USA, 3–6 October 2004; pp. 70–75.
2. Sifuentes, E.; Casas, O.; Pallas-Areny, R. Wireless Magnetic Sensor Node for Vehicle Detection With Optical Wake-Up. *IEEE Sens. J.* **2011**, *11*, 1669–1676. [CrossRef]

3. Dalal, N.; Triggs, B. Histograms of oriented gradients for human detection. In Proceedings of the 2005 IEEE Computer Society Conference on Computer Vision and Pattern Recognition (CVPR'05), San Diego, CA, USA, 20–25 June 2005; Volume 1, pp. 886–893.

4. Llorca, D.F.; Arroyo, R.; Sotelo, M.A. Vehicle logo recognition in traffic images using HOG features and SVM. In Proceedings of the 16th International IEEE Conference on Intelligent Transportation Systems (ITSC 2013), The Hague, The Netherlands, 6–9 October 2013; pp. 2229–2234.

5. Clady, X.; Negri, P.; Milgram, M.; Poulenard, R. Multi-class Vehicle Type Recognition System. In *Artificial Neural Networks in Pattern Recognition*; Prevost, L., Marinai, S., Schwenker, F., Eds.; Springer: Berlin/Heidelberg, Germany, 2008; pp. 228–239.

6. Han, D.; Hwang, J.; Hahn, H.s.; Cooper, D.B. Vehicle Class Recognition Using Multiple Video Cameras. In *Computer Vision—ACCV 2010 Workshops*; Koch, R., Huang, F., Eds.; Springer: Berlin/Heidelberg, Germany, 2011; pp. 246–255.

7. Madden, M.G.; Munroe, D.T. Multi-Class and Single-Class Classification Approaches to Vehicle Model Recognition from Images. In Proceedings of the 16th Irish Conference on Artificial Intelligence and Cognitive Science (AICS '05), Portstewart, UK, 7–9 September 2005.

8. Huttunen, H.; Yancheshmeh, F.S.; Ke Chen. Car type recognition with Deep Neural Networks. In Proceedings of the 2016 IEEE Intelligent Vehicles Symposium (IV), Gothenburg, Sweden, 19–22 June 2016; pp. 1115–1120.

9. Simonyan, K.; Zisserman, A. Very Deep Convolutional Networks for Large-Scale Image Recognition. In Proceedings of the International Conference on Learning Representations, San Diego, CA, USA, 7–9 May 2015.

10. Russakovsky, O.; Deng, J.; Su, H.; Krause, J.; Satheesh, S.; Ma, S.; Huang, Z.; Karpathy, A.; Khosla, A.; Bernstein, M.; Berg, A.C.; Fei-Fei, L. ImageNet Large Scale Visual Recognition Challenge. *Int. J. Comput. Vis.* **2015**, *115*, 211–252. [CrossRef]

11. He, K.; Zhang, X.; Ren, S.; Sun, J. Deep Residual Learning for Image Recognition. In Proceedings of the IEEE Conference on Computer Vision and Pattern Recognition (CVPR), Las Vegas, NV, USA, 27–30 June 2016.

12. Xie, S.; Girshick, R.; Dollar, P.; Tu, Z.; He, K. Aggregated Residual Transformations for Deep Neural Networks. In Proceedings of the IEEE Conference on Computer Vision and Pattern Recognition (CVPR), Honolulu, HI, USA, 21–26 July 2017.

13. Karpathy, A.; Toderici, G.; Shetty, S.; Leung, T.; Sukthankar, R.; Fei-Fei, L. Large-Scale Video Classification with Convolutional Neural Networks. In Proceedings of the 2014 IEEE Conference on Computer Vision and Pattern Recognition, Columbus, OH, USA, 24–28 June 2014; pp. 1725–1732.

14. Jobson, D.J.; Rahman, Z.; Woodell, G.A. Properties and performance of a center/surround retinex. *IEEE Trans. Image Process.* **1997**, *6*, 451–462. [CrossRef] [PubMed]

15. Galoogahi, H.K.; Sim, T.; Lucey, S. Multi-channel Correlation Filters. In Proceedings of the 2013 IEEE International Conference on Computer Vision, Sydney, NSW, Australia, 1–8 December 2013; pp. 3072–3079.

16. Burges, C.J.C. A tutorial on support vector machines for pattern recognition. *Data Min. Knowl. Discov.* **1998**, *2*, 121–167. [CrossRef]

17. Lowe, D.G. Object Recognition from Local Scale-Invariant Features. In Proceedings of the International Conference on Computer Vision ICCV, Corfu, Kerkyra, Greece, 20–27 September 1999.

18. Ozbulak, U. PyTorch CNN Visualizations. 2019. Available online: https://github.com/utkuozbulak/pytorch-cnn-visualizations (accessed on 2 March 2020).

19. Xiong, B.; Zeng, N.; Li, Y.; Du, M.; Shi, W.; , G.M.; Yang, Y. Determining the Online Measurable Input Variables in Human Joint Moment Intelligent Prediction Based on the Hill Muscle Model. *Sensors* **2020**, *20*, 1185. [CrossRef] [PubMed]

20. Liu, T.; Xu, H.; Ragulskis, M.; Cao, M.; Ostachowicz, W. A Data-Driven Damage Identification Framework Based on Transmissibility Function Datasets and One-Dimensional Convolutional Neural Networks: Verification on a Structural Health Monitoring Benchmark Structure. *Sensors* **2020**, *20*, 1059. [CrossRef] [PubMed]

21. Li, Y.; Song, B.; kang, X.; Guizani, M. Vehicle-Type Detection Based on Compressed Sensing and Deep Learning in Vehicular Networks. *Sensors* **2020**, *18*, 4500. [CrossRef] [PubMed]

22. Zhang, J.; Lu, C.; Wang, J.; Yue, X.G.; Lim, S.J.; Al-Makhadmeh, Z.; Tolba, A. Training Convolutional Neural Networks with Multi-Size Images and Triplet Loss for Remote Sensing Scene Classification. *Sensors* **2020**, *20*, 1188. [CrossRef] [PubMed]

23. Jiang, X.; Yao, H.; Zhang, S.; Lu, X.; Zeng, W. Night video enhancement using improved dark channel prior. In Proceedings of the 2013 IEEE International Conference on Image Processing, Melbourne, VIC, Australia, 15–18 September 2013; pp. 553–557.

24. Gonzalez, R.C.; Wood, R.E. *Digital Image Processing*, 3rd ed.; Prentice Hall: Englewood Cliffs, NJ, USA, 2008.

25. Bolme, D.S.; Beveridge, J.R.; Draper, B.A.; Lui, Y.M. Visual object tracking using adaptive correlation filters. In Proceedings of the 2010 IEEE Computer Society Conference on Computer Vision and Pattern Recognition, San Francisco, CA, USA, 13–18 June 2010; pp. 2544–2550.

26. Kim, J.; Park, Y.; Kim, G.; Hwang, S.J. SplitNet: Learning to Semantically Split Deep Networks for Parameter Reduction and Model Parallelization. In Proceedings of the 34th International Conference on Machine Learning Research (PMLR), Sydney, NSW, Australia, 6–11 August 2017; Volume 70, pp. 1866–1874.

27. van der Maaten, L.; Hinton, G. Visualizing Data using t-SNE. *J. Mach. Learn. Res.* **2008**, *9*, 2579–2605.

Article

Shedding Light on People Action Recognition in Social Robotics by Means of Common Spatial Patterns

Itsaso Rodríguez-Moreno *, José María Martínez-Otzeta, Izaro Goienetxea, Igor Rodriguez-Rodriguez and Basilio Sierra

Department of Computer Science and Artificial Intelligence, University of the Basque Country, Manuel Lardizabal 1, 20018 Donostia-San Sebastián, Spain; josemaria.martinezo@ehu.eus (J.M.M.-O.); izaro.goienetxea@ehu.eus (I.G.); igor.rodriguez@ehu.eus (I.R.-R.); b.sierra@ehu.eus (B.S.)
* Correspondence: itsaso.rodriguez@ehu.eus

Received: 26 March 2020; Accepted: 23 April 2020; Published: 24 April 2020

Abstract: Action recognition in robotics is a research field that has gained momentum in recent years. In this work, a video activity recognition method is presented, which has the ultimate goal of endowing a robot with action recognition capabilities for a more natural social interaction. The application of Common Spatial Patterns (CSP), a signal processing approach widely used in electroencephalography (EEG), is presented in a novel manner to be used in activity recognition in videos taken by a humanoid robot. A sequence of skeleton data is considered as a multidimensional signal and filtered according to the CSP algorithm. Then, characteristics extracted from these filtered data are used as features for a classifier. A database with 46 individuals performing six different actions has been created to test the proposed method. The CSP-based method along with a Linear Discriminant Analysis (LDA) classifier has been compared to a Long Short-Term Memory (LSTM) neural network, showing that the former obtains similar or better results than the latter, while being simpler.

Keywords: action recognition; social robotics; common spatial patterns

---

1. Introduction

Social robotics aims at providing robots with artificial social intelligence to improve human–machine interaction and to introduce them in complex human contexts [1]. An effective social interaction between humans and robots requires these robots to understand and adapt to the human behaviour. Using visual perception for human activity recognition will aid the robot to provide better responses and thus enhance its social capabilities. The robot will be able to understand when the user wants to engage with it by recognising the action she/he performs.

Human activity recognition in videos is a task which consists in recognising certain actions from a series of observations. This field of research has received great attention since 1980 due to the amount of applications for which it is useful, such as health sciences, human-computer interaction, surveillance or sociology [2]. For example, in the field of surveillance [3], the automatic detection of suspicious actions allows an alert to be sent and some measures to be taken to deal with the danger. Another example is the use of action recognition for rehabilitation, which involves recognising the action the patients perform and being able to determine if they are performing it correctly or incorrectly. The principal field where this task is studied is in computer vision, based on videos. The visual features of a video provide basic information of the events or actions that occur.

Understanding what is happening in a video is really challenging, and different features can be taken into account when analysing a video sequence. For example, Video Motion Detection is a constrained approach which consists in detecting the movement in a static background. On the other hand, Video Tracking focuses on associating objects in consecutive frames, which can be difficult if the objects are moving fast in relation to the frames per second rate. Moreover, if the object in the scene must be recognised (already a challenging task), an additional complexity is added to the problem.

In the last few years many attempts to solve these problems have been made using different techniques such as Optical Flow, Hidden Markov Models (HMM) or, more recently, deep learning [4,5]. For example, the authors of [6,7] use Histograms of Optical Flow to perform recognition. However, in [8,9] the authors use the depth information obtained by depth cameras (Microsoft Kinect or Inter Realsense), due to the fact that depth images provide additional useful information about movement. The work of [10] must also be mentioned, as it is a reference for methods that use deep learning for this task. The authors propose a two-stream architecture incorporating spatial and temporal networks, which has been used in many subsequent methods.

Considering the computational cost and the complexity that come from the need of combining temporal and spatial information, the video classification problem progresses slowly when compared with image classification.

In this paper, a new approach for video action recognition is presented. The Common Spatial Pattern algorithm is used, a method normally applied in Brain Computer Interface (BCI) for EEG systems [11]. Videos are recorded and processed with OpenPose [12] software in order to obtain a sequence of skeleton data. This skeleton data corresponds to the position of the joints of the person performing the action of the video. A sequence of skeleton data is extracted from the video, and this data can be treated as a multidimensional signal. It is then filtered according to the Common Spatial Patterns (CSP) algorithm and characteristics extracted from these filtered data are used as features for a classifier. Linear Discriminant Analysis and Random Forest (RF) classifiers have been tested to build the models from the features extracted in the previous step. Variance, maximum, minimum and interquartile range (IQR) of the filtered signals have been taken as features to feed the aforementioned classifiers. The spatial filter generated by CSP is employed as a dimensionality reduction approach and can also be interpreted in EGG data analysis as a technique that sheds light on the relationships between the filtered signals, in a similar manner to Principal Component Analysis [13] (PCA), from which it is derived. While no direct visual interpretation is possible when applied to skeleton data, this dimensionality reduction technique allows for extracting the signal components which maximally discriminate between classes.

In Figure 1 an interaction example of a person with the robot is displayed. On the left, the skeleton superposed over the actual person that is interacting with the robot is shown. The skeleton contains the (X,Y) position of 25-keypoints, which include body, head and feet information. On the right, another point of view can be seen, with the expected response of the robot. A more detailed explanation about the employed human pose estimation system and the skeleton definition is provided in Section 4.1.

To apply CSP, as a first step, the skeleton of the person appearing in each frame is extracted using OpenPose, and the (X,Y) position of each of the 25 joints that OpenPose detects are used as input data to the CSP. Therefore, in the presented method, input videos are represented as frame sequences and the temporal sequence of each skeleton joint is treated as an input signal (channel) to the CSP. In Figure 2, the following data acquisition process is shown.

In order to validate the proposed CPS-based approach, an experiment is performed where it is compared with Long Short-Term Memory [14] neural networks, yielding better results.

The rest of the paper is organised as follows. First, in Section 2 some related works are mentioned in order to introduce the topic. In Section 3 a theoretical framework is presented to explain the proposed algorithm in detail.

In Section 4 the used dataset and related skeleton capture system, as well as the experimentation carried out, are explained thoroughly, and the obtained results are shown, including a comparison between the presented approach and a Keras [15] implementation of a LSTM network. A brief introduction to LSTMs is also presented in this section. The paper concludes with the Section 5, where the conclusions from the presented work are presented and some future work is pointed out.

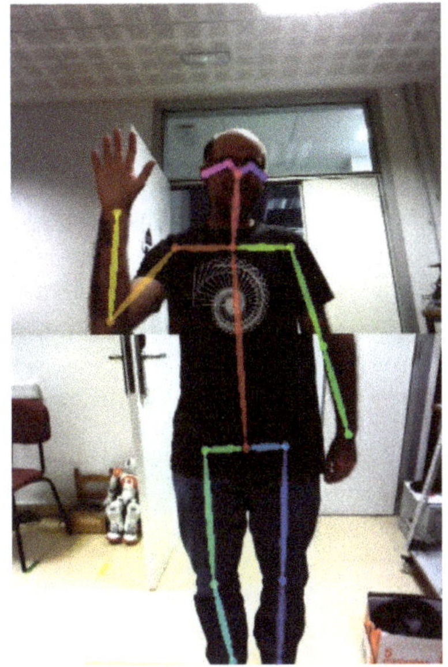

(**a**) Image captured by the robot.

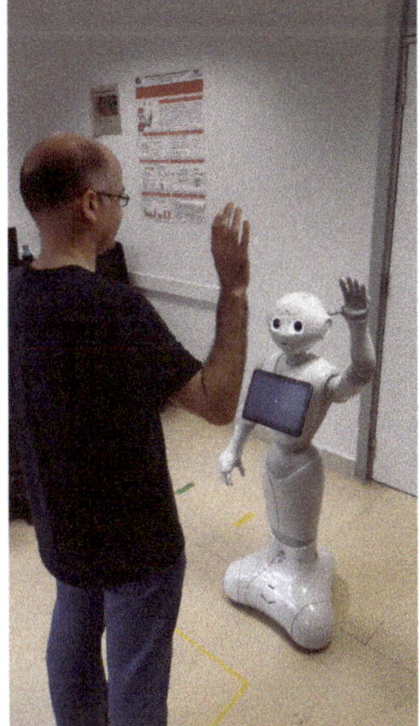

(**b**) Expected reaction of the robot.

Figure 1. Interaction example.

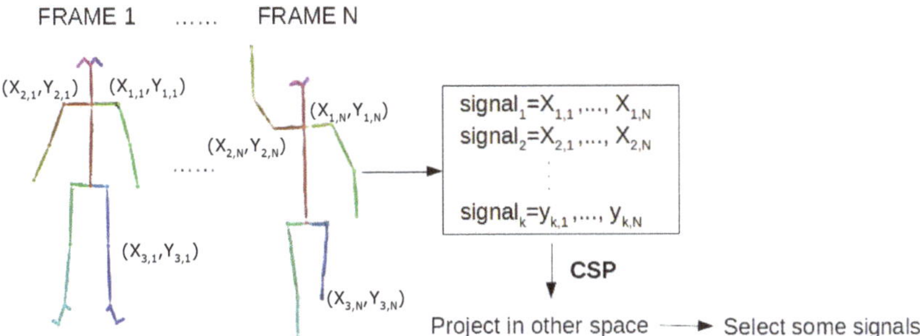

Figure 2. Proposed approach.

2. Related Work

As activity recognition has been an active research area lately, many different strategies have been developed to deal with this problem. There are several ways to extract visual features, both static image features and temporal visual features, and subsequently use them to perform the recognition. Temporal visual features are a combination of static image features and time information, so through these features temporal video information is achieved.

In [16] the authors use a temporal template as the basis of their representation, continuing with their work presented in [17]. This temporal template consists of a static vector-image where the value of the vector at each point represents a function of the motion properties at the corresponding spatial location in an image sequence. The authors of [18] demonstrate that local measurements in terms of spatio-temporal interest points (local features) can be used to recognise complex motion patterns. In [19] the authors present a hybrid hierarchical model, where video sequences are represented as collections of spatial and spatio-temporal features. These features are obtained by extracting both static and dynamic interest points, and the model is able to combine static and motion image features, as well as to perform categorisation of human actions in a frame-by-frame basis. Laptev et al. [20] contribute to the recognition of realistic videos and use movie scripts for automatic annotation of human actions in videos. Due to the promising performance that they achieve in image classification [21–24], they employ spatio-temporal features and spatio-temporal pyramids, extending the spatial pyramids presented in [22].

Many other methods make use of the optical flow to solve this issue. Optical flow is the motion of objects between consecutive frames, caused by the relative movement between an observer and a scene. Therefore, optical flow methods try to calculate the motion between two image frames which are taken at times t and $t + \Delta t$ at every position, assuming that the intensity of objects does not change during the movement. The authors of [6] use Histograms of Oriented Gradients (HOG) for human pose representations and time series of Histogram of Oriented Optical Flow (HOOF) to characterise human motion. In [7], the authors also use HOOF features for frame representation, which are independent to the scale of the moving person and to the direction of motion. There are many approaches which are based on histograms [25–27]. The authors of [28] introduce a motion descriptor based on the direction of optical flow, using the Lucas–Kanade algorithm [29] to compute it. In [30], the authors defend that to deal with the video-based action recognition problem temporally represented video information is needed. In their work, optical flow vectors are grouped according to their angular features and then summed and integrated with a new velocity concept.

It should also be mentioned that the interest of using depth data captured by depth cameras for the action recognition problem has grown, due to the advances in imaging technology to capture depth information in real time, and there are many approaches which use this extra information to make the recognition [8,9,31,32].

Some works focus on using skeleton data to perform activity recognition. In [33], the authors present a representation for action recognition, for which they use a human pose estimator and extract heatmaps for the human joints in each frame. Ren et al. [34] proposed a method for encoding geometric relational features into colour texture images, where temporal variations of different features are converted into the colour variations of their corresponding images. They use a multistream CNN model to classify the images.

As a result of the great performance that deep learning methods have achieved in image classification, these techniques have also been applied to video-based activity recognition. Taking these two publications [10,35] as a starting point, deep learning has continued to be used for activity recognition, mainly with Convolutional Neural Networks (CNN) and LSTMs. Wang et al. in their work [36] presented a very deep two-stream CNN in order to improve the results of recent architectures, getting closer to image domain deep models. In [37], trajectory-pooled deep-convolutional descriptor (TDD) is introduced, where the authors first train two-stream CNNs and then use them as feature extractors to achieve convolutional spatial and temporal feature maps from the learned networks. In the work of Feichtenhofer et al. [38], authors show that it is important to associate spatial feature maps of a particular area to temporal feature maps for that corresponding region. Authors of [39] proposed an action recognition method by processing the video data using Convolutional Neural Networks and deep bidirectional LSTM (DB-LSTM) networks. The use of deep learning for video recognition is still a work in progress, and even though the obtained results are not as good as those obtained in image recognition, better results are being achieved.

3. CSP-Based Approach

The core motivation of the presented method is to treat temporal sequences of skeleton joints as signals to be later processed with the CSP algorithm. In this section the CSP algorithm and the proposed approach, which makes use of that algorithm, are introduced.

3.1. CSP

In the last few years, the Common Spatial Pattern algorithm (first mentioned in [40] as Fukunaga-Koontz Transform) has been widely used in Brain Computer Interface (BCI) applications for electroencephalography (EEG) systems [41–43]. It is a mathematical technique used in signal processing and it consists in finding an optimum spatial filter which reduces the dimensionality of the original signals. CSP was presented as an extension of Principal Component Analysis. Considering just two different classes, a CSP filter maximises the variance of filtered signals of EEG of one of the targets while it minimises the variance for the other, in this way maximising the difference of the variances between the classes.

The feature extraction is organised in the following way:

Let X_1 and X_2 denote two sets of n signals where a signal is a sequence of values read from a sensor. First the covariance matrices are computed as in (1).

$$R_1 = \frac{X_1 X_1^T}{trace(X_1 X_1^T)}; \quad R_2 = \frac{X_2 X_2^T}{trace(X_2 X_2^T)} \tag{1}$$

Then, the eigen decomposition of the composite spatial covariance matrix is computed as in (2), where λ is the diagonal matrix of eigenvalues and U is the normalised eigenvectors matrix. To scale the principal components, the whitening transformation is used (3), obtaining an identity matrix as covariance and variance 1 for each variable.

$$R_1 + R_2 = U\lambda U^T \tag{2}$$

$$P = \sqrt{\lambda^{-1}} U^T \tag{3}$$

R_1 and R_2 covariance matrices are transformed using P (4). After that, taking into account that the sum of two corresponding eigen values is 1 ($\psi_1 + \psi_2 = I$), the eigen decomposition is computed in order to find their common eigenvectors (5).

$$S_1 = PR_1 P^T; \quad S_2 = PR_2 P^T \tag{4}$$

$$S_1 = V\psi_1 V^T; \quad S_2 = V\psi_2 V^T \tag{5}$$

The CSP filters are obtained as in (6), which maximises the separation between both classes. Using W as a projection matrix (just the first q and the last q vectors), each trial can be projected, obtaining a filtered signal matrix as in (7).

$$W = P^T V \tag{6}$$

$$Z = W^T X \tag{7}$$

The feature vector to be created for classification purposes is shown in (8), where $var_p(Z_i)$ is the variance of the row p of the i-th trial of Z. The feature vector value for the p-th component of the i-th trial is the logarithm of the normalised variance. The feature vector has $2q$ dimensionality, where q indicates how many vectors of the spatial filter are used in the projection. Exactly, q first and q last vectors are used, which yield the smallest variance for one class and simultaneously, the largest variance for the other class.

$$f_p^i = log\left(\frac{var_p(Z_i)}{\sum_{p=1}^{2q} var_p(Z_i)}\right) \tag{8}$$

The purpose of this algorithm is to filter the data so their variance could be used to discriminate two populations, that is, to separate the signals belonging to two different classes. This algorithm can be useful in action recognition, where actions belonging to different classes have to be separated. From each video a group of signals is extracted (in the proposed approach, the coordinates of the joints' positions), and then, the CSP algorithm filters the signals in a way that maximum variance difference is obtained for two different classes. Features from the filtered data obtained by CSP are therefore used as input to a classification algorithm to discriminate instances that belong to different classes.

3.2. Proposed Approach

Even though the CSP algorithm has been used mainly with EEG problems, in this paper a new application is presented; the use of CSP filters for feature extraction in the human action recognition task. In the presented method, each video represents a trial and each skeleton joint is treated as an EEG channel, so the videos are taken as time series where the joints of the extracted skeletons are the channels which change over time.

In Brain–Computer Interface, some electrodes are placed along the scalp and they are used to record the electrical activity of the brain. Therefore, the signals are obtained from the electrodes and then the CSP is applied using the electroencephalography signals.

However, in the proposed approach, the signals used to feed the CSP are obtained in another way. The full process can be seen in Figure 2, where the signals are composed with keypoints of the skeleton of the actor who is performing the action to recognise. Each trial is a video where the signals are the values of the skeleton position over time. Once the skeletons are processed and, hence, the signals are formed, the CSP is computed in order to separate the classes according to their variance.

The main focus of the experimentation is the use of the variance of the signals after applying the Common Spatial Pattern algorithm as input to the classification algorithms. However, in addition to the variance, much more information can be extracted from these transformed signals, which may be useful when performing the classification. Hence, some experiments are performed with just the information of the variances and other experiments also with information about the maximum, minimum and the interquartile range ($IQR = Q3 - Q1$) of the signal. Once the features are extracted from the transformed signals, Linear Discriminant Analysis and Random Forest classifiers are used to perform the classification. The Linear Discriminant Analysis [44] tries to separate the different classes by finding a linear combination of features which describe each of the targets. Random Forest [45] is a Bagging (Bootstrap Aggregating) multiclassifier composed of decision trees.

4. Experimental Results

4.1. Robotic Platform and Human Pose Estimation

The robotic platform employed in the performed experiments is a Pepper robot developed by Softbank Robotics (https://www.softbankrobotics.com/emea/en/pepper). Pepper is a human-like torso that is fitted onto a holonomous wheeled platform. It is equipped with full-colour RGB LEDs, three cameras and several sensors located in different parts of its body that allow for perceiving the surrounding environment with high precision. In this work, only the information provided by the two identical RGB cameras, with a resolution of 320×240 pixels, situated on the forehead of the robot has been used (see Figure 3). The images of both cameras have been combined to obtain a wider field of view and better capture the person in front of the robot, thus obtaining an image of 320×480 resolution. An example of the combined image is shown in Figure 1a.

In order to obtain the data to apply CSP, as a first step, the skeleton of the person appearing in the scene has to be obtained. For this purpose, it has been decided to extract the skeletons using OpenPose [12], one of the most popular bottom-up approaches for multiperson human pose estimation. As with many bottom-up techniques, OpenPose first detects parts (keypoints) belonging to every person in the image and then assigns those parts to distinct individuals. The assignment is made using a nonparametric representation of association scores via Part Affinity Fields (PAFs), a set of 2d vectors fields that encode the location and orientation of limbs over the image. OpenPose can detect human body, feet, hands, and facial keypoints (135 keypoints in total) on single images. Due to the high computational cost that estimating all the keypoints requires, in this work only the BODY_25

(COCO [46] + feet) model has been used for human pose estimation. It returns the (X,Y) positions in the image of the extracted 25-keypoints, including head, body, and feet (see Figure 4).

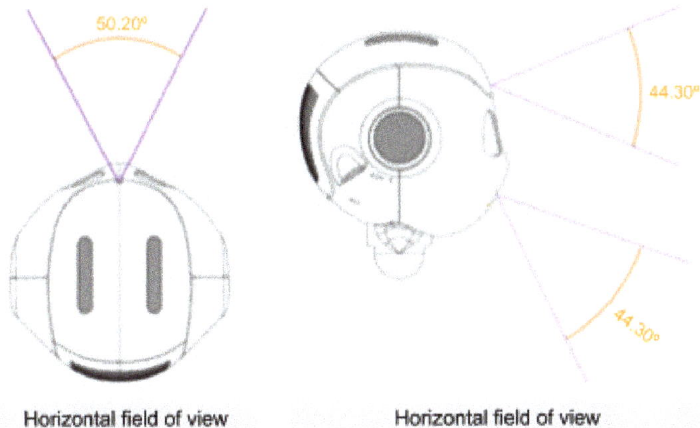

Figure 3. Pepper's RGB cameras position and orientation.

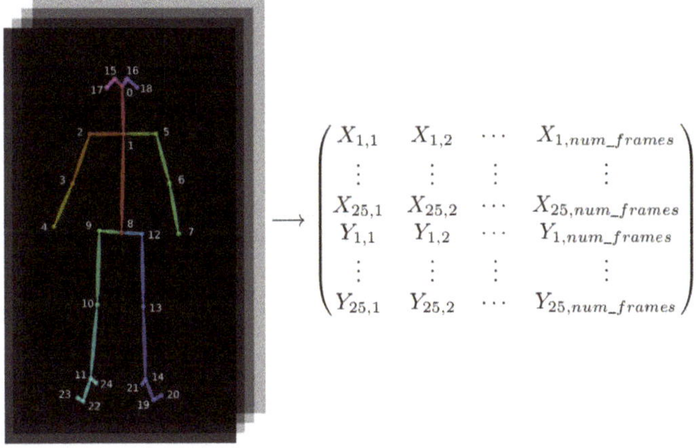

$$\longrightarrow \begin{pmatrix} X_{1,1} & X_{1,2} & \cdots & X_{1,num\_frames} \\ \vdots & \vdots & \vdots & \vdots \\ X_{25,1} & X_{25,2} & \cdots & X_{25,num\_frames} \\ Y_{1,1} & Y_{1,2} & \cdots & Y_{1,num\_frames} \\ \vdots & \vdots & \vdots & \vdots \\ Y_{25,1} & Y_{25,2} & \cdots & Y_{25,num\_frames} \end{pmatrix}$$

Figure 4. Skeleton's joint positions and matrix representation of the extracted signals.

4.2. Dataset

The videos in the database have been recorded using the combined image obtained from Pepper's forehead cameras. It consists of 272 videos with six action categories and around 45 clips belong to each category, performed by 46 different people. The robot adjusts the orientation of its head according to the location of the face of the person appearing in its field of view.

All the participants in this study gave their consent in being recorded for this research purpose. No raw video data has been stored, and only minimum information about joints' spatial coordinates has been maintained. All this data is anonymised, with no information about sex, age, race, or any other condition of the participants.

The action categories and video information can be seen in Table 1.

Table 1. Characteristics of each action category.

Category	#Video	Resolution	FPS
COME	46	320 × 480	10
FIVE	45	320× 480	10
HANDSHAKE	45	320 × 480	10
HELLO	44	320 × 480	10
IGNORE	46	320 × 480	10
LOOK AT	46	320 × 480	10

These are the six categories that the robot must differentiate:

1. COME: gesture for telling the robot to come to you.
2. FIVE: gesture of "high five".
3. HANDSHAKE: gesture of handshaking with the robot.
4. HELLO: gesture for indicating hello to the robot.
5. IGNORE: ignore the robot, pass by.
6. LOOK AT: stare at the robot in front of it.

Examples of skeletons extracted from videos of the six different classes are shown in Figure 5. It can be seen in the examples that all the videos follow the same pattern: the actor appears in the scene, approaches the robot and finally, the action is performed.

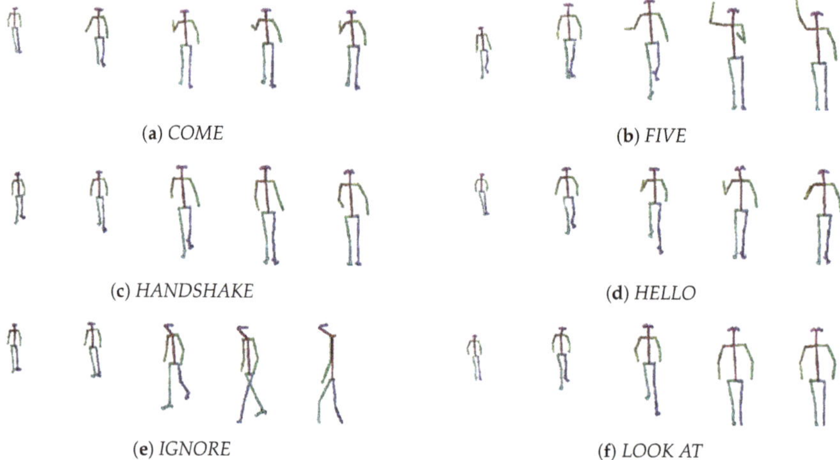

(a) *COME*

(b) *FIVE*

(c) *HANDSHAKE*

(d) *HELLO*

(e) *IGNORE*

(f) *LOOK AT*

Figure 5. Frame sequence examples for different categories.

In this case, the actions that have to be recognised are centred in the actor who performs them. Therefore, the skeleton of the actor has been extracted in every frame of each video. OpenPose returns the (X,Y) positions of 25-keypoints (joints). After obtaining the skeleton information for every frame of each video, fifty different signals are created to represent each video, where each signal will be the position of a skeleton keypoint over time. This way, there will be 50 signals (25 for the X position of the joints and another 25 for the Y position) with the same length as the original video (one skeleton per frame). The skeleton appearance and the matrix extracted from skeletons can be seen in Figure 4.

Some joints could be missing from the captured skeletons when the actor does not fit entirely in the camera range. In these cases, the missing joint values are estimated by a linear interpolation, using the previous and next values for that joint. The interpolation is done to avoid missing values and assuming that consecutive values of joints positions follow a smooth curve. The process of interpolation for the signal of one video can be seen graphically in Figure 6, where Figure 6a,c show the 25 X and 25 Y signals before interpolation and Figure 6b,d the 25 X and 25 Y signals after interpolating them.

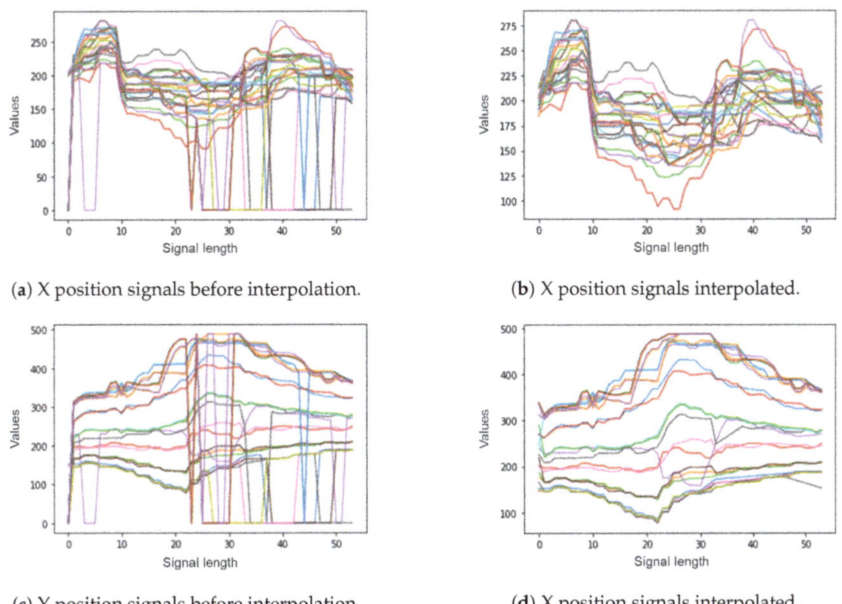

(**a**) X position signals before interpolation.

(**b**) X position signals interpolated.

(**c**) Y position signals before interpolation.

(**d**) X position signals interpolated.

Figure 6. Linear interpolation example.

Furthermore, the length of all the input data must be the same to apply the proposed method, therefore, it might be necessary to apply a preprocessing step to the videos. As the duration of the original videos differ, it has been decided to convert all the videos to the length of the longest clip.

As mentioned before, OpenPose provides the skeletons of the people of the scene for each frame of the video. It could happen that in some frames no person is detected and no skeleton is formed. Analysing this dataset, it can be noticed that full skeletons are only missed at the beginning of some of the videos and it has been decided to repeat the first skeleton encountered as many times as necessary.

After performing these changes, 50 signals with maximum video's length are obtained. These signals are then used to feed the CSP.

4.3. Long Short-Term Memory (LSTM) Neural Networks

LSTMs are a category of recurrent neural networks (RNNs) which belong to the growing field of deep learning paradigms. RNNs are artificial neural networks in which connections between units form a directed cycle. Due to this architecture, recurrent neural networks possess an internal state that stores information about past inputs. This endows the recurrent networks with the ability to process sequences of inputs and exhibits a dynamic temporal behaviour in response to those sequences.

Training RNNs to learn long-term dependencies by gradient-descent methods used to be difficult due to the vanishing or exploding gradient problem [47,48]. In recent years, sophisticated optimisation techniques, specialised network designs, and new weight initialisation methods have addressed this problem with great success [49]. LSTM design introduces gates that control how much of the past and the current state has to get through to the next time step.

In a RNN, the following terms are defined:

- x_t: input vector at time step t.
- $h_t = \phi(Wx_t + Uh_{t-1})$: hidden state at time step t. W and U are weight matrices applied to the current input and to the previous hidden state, respectively. ϕ is an activation function, typically sigmoid (σ), tanh, or ReLU.
- $o_t = \text{softmax}(Vs_t)$: output vector at time step t. V is a weight matrix.

In LSTMs accounting for the capability of forgetting selectively, the node's state is needed, so the terms are typically the following:

- x_t: input vector at time step t.
- $f_t = \sigma(W_f x_t + U_f h_{t-1})$: activation vector of the forget gate at time step t.
- $i_t = \sigma(W_i x_t + U_i h_{t-1})$: activation vector of the input gate at time step t.
- $o_t = \sigma(W_o x_t + U_o h_{t-1})$: activation vector of the output gate at time step t.
- $c_t = f_t \circ c_{t-1} + i_t \circ \tanh(W_c x_t + U_c h_{t-1})$: cell state vector at time step t.
- $h_t = o_t \circ \tanh(c_t)$: hidden state at time step t.

W_f, W_i, W_o, and W_c are weight matrices applied to the current input, while U_f, U_i, U_o, and U_c are applied to the previous hidden state. The $\circ$ operator represents the Hadamard product.

4.4. Results

Once the data have been processed, the previously explained CSP algorithm is performed. The used CSP method is implemented to work with just two classes, therefore all the tests have been carried out using pairs of classes, although multiclass classification is possible using pairwise classification approaches, such as One versus One (OVO) as a class binarization technique [50].

In Table 2 the obtained results by Linear Discriminant Analysis (LDA) classifier can be seen, and in Table 3 the results obtained by RF classifier are shown, where best results are highlighted in boldface. Both tables present the accuracy values obtained for every pair of classes of the database, using 10-fold cross validation for the evaluation. Parameter q indicates that only $2 \times q$ feature vectors are considered, where $2 \times q$ are the q first and q last vectors, when sorted by variance. Therefore, a feature vector of $2 \times q$ dimensionality is obtained after applying CSP, and that feature vector is the input to LDA or RF classifiers. In each table the accuracy values obtained with two different types of feature vectors are shown; variance when only the variances of the transformed signals are used to form the feature vectors and variance, max, min, IQR when apart from the variances, maximum, minimum, and IQR values are also represented in the feature vectors.

Table 2. Results obtained applying Common Spatial Patterns (CSP) with different q values and using LDA as classifier.

	Variance			Variance, Max, Min, IQR		
Pair of Categories	$q = 5$	$q = 10$	$q = 15$	$q = 5$	$q = 10$	$q = 15$
COME-FIVE	0.7579 ± 0.13	0.8124 ± 0.12	0.7667 ± 0.17	0.7578 ± 0.12	$\mathbf{0.8344 \pm 0.14}$	0.7667 ± 0.16
COME-HANDSHAKE	$\mathbf{0.8668 \pm 0.10}$	0.8019 ± 0.12	0.6910 ± 0.17	$\mathbf{0.8667 \pm 0.13}$	0.7900 ± 0.12	0.6567 ± 0.16
COME-HELLO	$\mathbf{0.5334 \pm 0.16}$	0.5000 ± 0.09	0.5000 ± 0.14	0.4778 ± 0.16	0.4444 ± 0.09	0.4778 ± 0.15
COME-IGNORE	$\mathbf{0.9779 \pm 0.05}$	0.9667 ± 0.05	0.9667 ± 0.05	0.9667 ± 0.05	0.9667 ± 0.05	0.9444 ± 0.06
COME-LOOK_AT	0.8678 ± 0.09	$\mathbf{0.8900 \pm 0.09}$	0.8789 ± 0.11	0.8678 ± 0.10	0.8356 ± 0.14	0.8033 ± 0.14
FIVE-HAND	$\mathbf{0.9557 \pm 0.06}$	0.9333 ± 0.06	0.9223 ± 0.05	0.9333 ± 0.11	0.9000 ± 0.11	0.9000 ± 0.08
FIVE-HELLO	$\mathbf{0.8208 \pm 0.14}$	0.7986 ± 0.15	0.7764 ± 0.17	0.7750 ± 0.18	0.7528 ± 0.18	0.7319 ± 0.21
FIVE-IGNORE	0.9668 ± 0.07	0.9668 ± 0.07	0.9556 ± 0.11	$\mathbf{0.9667 \pm 0.07}$	0.9556 ± 0.11	0.9556 ± 0.11
FIVE-LOOK_AT	$\mathbf{0.9667 \pm 0.05}$	0.9556 ± 0.06	0.9556 ± 0.06	0.9556 ± 0.08	0.9556 ± 0.08	0.9011 ± 0.17
HANDSHAKE-HELLO	0.7431 ± 0.19	0.7861 ± 0.14	0.8097 ± 0.10	0.7111 ± 0.24	0.7889 ± 0.21	0.8000 ± 0.10
HANDSHAKE-IGNORE	0.9889 ± 0.04	$\mathbf{1.0000 \pm 0.00}$	1.0000 ± 0.00	$\mathbf{1.0000 \pm 0.00}$	0.9889 ± 0.04	0.9889 ± 0.04
HANDSHAKE-LOOK_AT	$\mathbf{0.8235 \pm 0.18}$	0.7789 ± 0.16	0.7567 ± 0.12	0.8122 ± 0.17	0.7467 ± 0.17	0.7456 ± 0.12
HELLO-IGNORE	0.9333 ± 0.14	0.9221 ± 0.14	0.9333 ± 0.14	$\mathbf{0.9556 \pm 0.14}$	0.9444 ± 0.14	0.9444 ± 0.11
HELLO-LOOK_AT	0.8445 ± 0.11	0.8334 ± 0.12	0.8556 ± 0.14	0.8556 ± 0.09	0.8000 ± 0.10	$\mathbf{0.8667 \pm 0.10}$
IGNORE-LOOK_AT	$\mathbf{0.9889 \pm 0.04}$	0.9889 ± 0.04	0.9889 ± 0.04	0.9778 ± 0.05	0.9678 ± 0.05	0.9678 ± 0.05
MEAN	**0.8691**	0.8623	0.8506	0.8586	0.8448	0.8301

Table 3. Results obtained applying CSP with different q values and using RF as classifier.

	Variance			Variance, Max, Min, IQR		
Pair of Categories	$q = 5$	$q = 10$	$q = 15$	$q = 5$	$q = 10$	$q = 15$
COME-FIVE	0.6800 ± 0.29	0.6022 ± 0.24	0.5811 ± 0.19	$\mathbf{0.7133 \pm 0.21}$	0.6244 ± 0.23	0.5922 ± 0.21
COME-HANDSHAKE	0.7000 ± 0.20	0.6900 ± 0.29	0.6344 ± 0.29	$\mathbf{0.7556 \pm 0.16}$	0.6678 ± 0.32	0.6344 ± 0.32
COME-HELLO	$\mathbf{0.5111 \pm 0.22}$	0.3889 ± 0.21	0.4222 ± 0.17	0.4889 ± 0.22	0.4222 ± 0.20	0.3889 ± 0.20
COME-IGNORE	$\mathbf{0.9233 \pm 0.12}$	0.8900 ± 0.17	0.8800 ± 0.18	$\mathbf{0.9233 \pm 0.12}$	0.8911 ± 0.15	0.8578 ± 0.20
COME-LOOK_AT	$\mathbf{0.8133 \pm 0.23}$	0.7800 ± 0.20	0.7456 ± 0.25	0.8122 ± 0.23	0.8122 ± 0.24	0.7789 ± 0.24
FIVE-HANDSHAKE	$\mathbf{0.8889 \pm 0.17}$	0.7778 ± 0.15	0.6444 ± 0.17	0.8444 ± 0.17	0.7667 ± 0.12	0.6667 ± 0.17
FIVE-HELLO	$\mathbf{0.6264 \pm 0.22}$	0.5500 ± 0.22	0.5028 ± 0.23	$\mathbf{0.6264 \pm 0.22}$	0.5361 ± 0.23	0.5236 ± 0.24
FIVE-IGNORE	0.9444 ± 0.14	0.9344 ± 0.14	0.9344 ± 0.14	$\mathbf{0.9556 \pm 0.11}$	0.9456 ± 0.11	0.9233 ± 0.14
FIVE-LOOK_AT	0.9000 ± 0.19	0.8889 ± 0.21	0.8233 ± 0.23	$\mathbf{0.9111 \pm 0.21}$	0.9000 ± 0.21	0.8556 ± 0.25
HANDSHAKE-HELLO	0.6875 ± 0.18	0.5708 ± 0.14	0.6111 ± 0.20	$\mathbf{0.6889 \pm 0.19}$	0.5819 ± 0.16	0.6556 ± 0.15
HANDSHAKE-IGNORE	$\mathbf{0.9789 \pm 0.04}$	0.9578 ± 0.07	0.9133 ± 0.12	$\mathbf{0.9789 \pm 0.04}$	0.9578 ± 0.07	0.9244 ± 0.11
HANDSHAKE-LOOK_AT	0.7344 ± 0.26	$\mathbf{0.7556 \pm 0.29}$	0.6789 ± 0.29	0.7456 ± 0.26	0.7456 ± 0.28	0.6678 ± 0.25
HELLO-IGNORE	0.9000 ± 0.14	0.8889 ± 0.17	0.8667 ± 0.21	$\mathbf{0.9111 \pm 0.15}$	0.8889 ± 0.17	0.8667 ± 0.21
HELLO-LOOK_AT	0.7667 ± 0.22	0.6556 ± 0.32	0.6556 ± 0.35	$\mathbf{0.7889 \pm 0.23}$	0.7556 ± 0.29	0.7333 ± 0.28
IGNORE-LOOK_AT	0.9222 ± 0.12	$\mathbf{0.9333 \pm 0.14}$	0.9222 ± 0.14	$\mathbf{0.9333 \pm 0.09}$	0.9111 ± 0.15	$\mathbf{0.9333 \pm 0.14}$
MEAN	0.7985	0.7509	0.7211	**0.8052**	0.7605	0.7335

Looking at the results of Table 2, it can be observed that best outcomes are achieved when $q = 5$, that is, taking 10 values per video is enough to perform the classification. An accuracy higher than 80% is attained for most of the category pairs. Regarding the categories, some of them are better distinguished than others. For example, good results are obtained when classifying the class ignore with all other classes, so it can be supposed that the features obtained for the category ignore are quite different from the rest. However, videos that belong to the pair of classes come and hello are more difficult to differentiate, which can be easily deduced looking at the skeletons of both classes. Concerning the feature vector type, the results indicate that there is no need to use more information than the variances of the transformed signals to obtain better results; the accuracy values obtained with the variances are higher. Nevertheless, the obtained results indicate that the presented approach yields a good classification accuracy.

The results of Table 3 show that RF classifier performs worse than LDA, obtaining lower accuracy values in general. In this case, the feature vector type which uses the variance, max, min, and IQR values achieves better outcomes. Regarding both the q value and the categories, the conclusions presented for the results obtained by LDA classifier are maintained.

In order to assess the effectiveness of the presented method when compared with another technique, a Long Short-Term Memory network has been chosen, as this type of neural network has been widely used for video action recognition tasks. The LSTM network has been implemented in Python using the Keras library. The input shape is bidimensional (number of frames, number of joints), and the output space is of 64 units. Then another dense layer for classification is added, of size 2, as this is the number of classes for each individual problem. The Adam optimisation algorithm [51] has been used, as well as categorical cross-entropy as loss function. It has been trained during 100 epochs, with a batch size of 25. The comparison is made between the aforementioned LSTM and the proposed approach with the configuration which has achieved highest accuracy, in this instance, variance $q = 5$ with LDA classifier. The results are shown in Table 4, where best results are highlighted in boldface.

LSTM achieves accuracy values between 70% and 90% for most of the pairs. In this case, the accuracy obtained for come-hello pair has been improved notoriously. However, the results obtained for the rest of the classes are not that significant.

The results show that the presented method performs better than LSTM. More precisely, it outperforms LSTM results for 9 of 15 category pairs. Moreover, the mean value of all the tested pairs has been calculated for each technique, and it can be concluded that the proposed approach obtains higher accuracy values. Therefore, the CSP-based method not only achieves better results in most classifications but the average of the values obtained is higher.

Table 4. Comparison between the proposed approach and LSTM approach.

Pair of Categories	CSP (Variance and $q = 5$) + LDA	LSTM
COME-FIVE	0.7579 ± 0.13	$\mathbf{0.8628 \pm 0.11}$
COME-HANDSHAKE	$\mathbf{0.8668 \pm 0.10}$	0.7739 ± 0.16
COME-HELLO	0.5334 ± 0.16	$\mathbf{0.7336 \pm 0.17}$
COME-IGNORE	$\mathbf{0.9779 \pm 0.05}$	0.9575 ± 0.06
COME-LOOK_AT	$\mathbf{0.8678 \pm 0.09}$	0.7849 ± 0.10
FIVE-HANDSHAKE	$\mathbf{0.9557 \pm 0.06}$	0.8125 ± 0.14
FIVE-HELLO	0.8208 ± 0.14	$\mathbf{0.9125 \pm 0.07}$
FIVE-IGNORE	0.9668 ± 0.07	$\mathbf{0.9789 \pm 0.04}$
FIVE-LOOK_AT	$\mathbf{0.9667 \pm 0.05}$	0.8889 ± 0.11
HANDSHAKE-HELLO	$\mathbf{0.7431 \pm 0.19}$	0.7108 ± 0.21
HANDSHAKE-IGNORE	$\mathbf{0.9889 \pm 0.04}$	0.9764 ± 0.05
HANDSHAKE-LOOK_AT	0.8235 ± 0.18	$\mathbf{0.8350 \pm 0.12}$
HELLO-IGNORE	0.9333 ± 0.14	$\mathbf{0.9789 \pm 0.04}$
HELLO-LOOK_AT	$\mathbf{0.8445 \pm 0.11}$	0.5733 ± 0.18
IGNORE-LOOK_AT	$\mathbf{0.9889 \pm 0.04}$	0.9775 ± 0.05
MEAN	$\mathbf{0.8691}$	0.8505

Furthermore, the other three configurations tested above with LDA classifier (variance$-q = 10$, variance$-q = 15$ and variance, max, min, IQR$-q = 5$) also outperform the results obtained by the LSTM method.

variance q = 5		*variance* q = 10		*var, max, min, IQR* q = 5		*variance* q = 15		LSTM
0.8691	>	0.8622	>	0.8586	>	0.8506	>	0.8505

5. Conclusions

In this paper a new approach for activity recognition in video sequences is presented, in which Common Spatial Pattern signal processing has been applied to the skeleton joints data of people performing different activities. Features extracted from the transformed data have been used as input to Linear Discriminant Analysis and Random Forest classifiers, in order to perform action recognition. Two different sets of features have been selected: {Variance} and {Variance, Max, Min, IQR}. The results show that CSP processing followed by LDA classifier over variance features compares favourably to a Long Short-Term Memory model trained with the same data. From a database of six actions (fifteen possible pairs of actions), CSP and LDA obtains better results than LSTM in 9 of 15 category pairs.

Another advantage of the proposed method is the relative simplicity of LDA compared to LSTM networks and the lack of need for hyperparameter tuning. The set of features is also small, since only variance is used in the model that achieves best results.

As further work, it is planned to extend the range of human activities. Implementation of a real-time system could be of interest, for example, in social robotics.

Author Contributions: Research concept and supervision of technical writing: B.S.; software implementation, concept development, and technical writing: I.R.-M.; results validation and supervision of technical writing: J.M.M.-O. and I.G.; methodological analysis: I.R.-R. All authors have read and agreed to the published version of the manuscript.

Funding: This work has been partially funded by the Basque Government, Spain, grant number IT900-16, and the Spanish Ministry of Science (MCIU), the State Research Agency (AEI), and the European Regional Development Fund (FEDER), grant number RTI2018-093337-B-I00 (MCIU/AEI/FEDER, UE).

Acknowledgments: We gratefully acknowledge the support of NVIDIA Corporation with the donation of the Titan Xp GPU used for this research.

Conflicts of Interest: The authors declare no conflict of interest.

References

1. Breazeal, C. *Designing Sociable Robots*; Intelligent Robotics and Autonomous Agents, MIT Press: Cambridge, MA, USA, 2004.

2. Ke, S.R.; Thuc, H.; Lee, Y.J.; Hwang, J.N.; Yoo, J.H.; Choi, K.H. A review on video-based human activity recognition. *Computers* **2013**, *2*, 88–131. [CrossRef]

3.   Vishwakarma, S.; Agrawal, A. A survey on activity recognition and behavior understanding in video surveillance. *Vis. Comput.* **2013**, *29*, 983–1009. [CrossRef]

4.   Poppe, R. A survey on vision-based human action recognition. *Image Vis. Comput.* **2010**, *28*, 976–990. [CrossRef]

5.   Herath, S.; Harandi, M.; Porikli, F. Going deeper into action recognition: A survey. *Image Vis. Comput.* **2017**, *60*, 4–21. [CrossRef]

6.   Chen, C.C.; Aggarwal, J. Recognizing human action from a far field of view. In Proceedings of the 2009 Workshop on Motion and Video Computing (WMVC), Snowbird, UT, USA, 8–9 December 2009; pp. 1–7.

7.   Chaudhry, R.; Ravichandran, A.; Hager, G.; Vidal, R. Histograms of oriented optical flow and Binet-Cauchy kernels on nonlinear dynamical systems for the recognition of human actions. In Proceedings of the 2009 IEEE Conference on Computer Vision and Pattern Recognition, Miami, FL, USA, 20–25 June 2009; pp. 1932–1939.

8.   Oreifej, O.; Liu, Z. Hon4d: Histogram of oriented 4d normals for activity recognition from depth sequences. In Proceedings of the IEEE Conference on Computer Vision and Pattern Recognition, Portland, OR, USA, 23–28 June 2013; pp. 716–723.

9.   Liu, M.; Liu, H.; Chen, C. Robust 3D action recognition through sampling local appearances and global distributions. *IEEE Trans. Multimed.* **2018**, *20*, 1932–1947. [CrossRef]

10.   Simonyan, K.; Zisserman, A. Two-stream convolutional networks for action recognition in videos. In *Advances in Neural Information Processing Systems*; The MIT Press, Cambridge, MA, USA, 2014; pp. 568–576.

11.   Astigarraga, A.; Arruti, A.; Muguerza, J.; Santana, R.; Martin, J.I.; Sierra, B. User adapted motor-imaginary brain-computer interface by means of EEG channel selection based on estimation of distributed algorithms. *Math. Probl. Eng.* **2016**, *2016*. [CrossRef]

12.   Cao, Z.; Hidalgo, G.; Simon, T.; Wei, S.E.; Sheikh, Y. OpenPose: Realtime multi-person 2D pose estimation using Part Affinity Fields. *arXiv* **2018**, arXiv:1812.08008.

13.   Hotelling, H. Analysis of a complex of statistical variables into principal components. *J. Educ. Psychol.* **1933**, *24*, 417. [CrossRef]

14.   Hochreiter, S.; Schmidhuber, J. Long Short-Term Memory. *Neural Comput.* **1997**, *9*, 1735–1780. [CrossRef]

15.   Chollet, F.. Keras. 2015. Available online: https://keras.io (accessed on 24 April 2020).

16.   Bobick, A.F.; Davis, J.W. The recognition of human movement using temporal templates. *IEEE Trans. Pattern Anal. Mach. Intell.* **2001**, *23*, 257–267. [CrossRef]

17.   Bobick, A.; Davis, J. An appearance-based representation of action. In Proceedings of the 13th International Conference on Pattern Recognition, Vienna, Austria, 25–19 August 1996; Volume 1; pp. 307–312.

18.   Schuldt, C.; Laptev, I.; Caputo, B. Recognizing human actions: A local SVM approach. In Proceedings of the 17th International Conference on Pattern Recognition, Cambridge, UK, 26 August 2004, Volume 3; pp. 32–36.

19.   Niebles, J.C.; Fei-Fei, L. A hierarchical model of shape and appearance for human action classification. In Proceedings of the Computer Vision and Pattern Recognition, CVPR'07, Minneapolis, MN, USA, 17–22 June 2007; pp. 1–8.

20.   Laptev, I.; Marszalek, M.; Schmid, C.; Rozenfeld, B. Learning realistic human actions from movies. In Proceedings of the Computer Vision and Pattern Recognition, Anchorage, AK, USA, 23–28 June 2008; pp. 1–8.

21.   Bosch, A.; Zisserman, A.; Munoz, X. Representing shape with a spatial pyramid kernel. In Proceedings of the 6th ACM International Conference on Image And Video Retrieval, Amsterdam, The Netherlands, 9–11 July 2007; pp. 401–408.

22.   Lazebnik, S.; Schmid, C.; Ponce, J. Beyond bags of features: Spatial pyramid matching for recognizing natural scene categories. In Proceedings of the 2006 IEEE Computer Society Conference on Computer Vision and Pattern Recognition (CVPR'06), New York, NY, USA, 17–22 June 2006; pp. 2169–2178.

23.   Marszałek, M.; Schmid, C.; Harzallah, H.; Van De Weijer, J. Learning object representations for visual object class recognition. In Proceedings of the Visual Recognition Challange Workshop, in Conjunction with ICCV, Rio de Janeiro, Brazil, 14–20 October 2007.

24.   Zhang, J.; Marszałek, M.; Lazebnik, S.; Schmid, C. Local features and kernels for classification of texture and object categories: A comprehensive study. *Int. J. Comput. Vis.* **2007**, *73*, 213–238. [CrossRef]

25. Efros, A.A.; Berg, A.C.; Mori, G.; Malik, J. Recognizing action at a distance. In Proceedings of the Ninth International Conference on Computer Vision, Nice, France, 13–16 October 2003; pp. 726–733.

26. Tran, D.; Sorokin, A. Human activity recognition with metric learning. In Proceedings of the European Conference on Computer Vision, Marseille, France, 12–18 October 2008; Springer: Berlin/Heidelberg, Germany, 2008; pp. 548–561.

27. Ercis, F. Comparison of Histogram of Oriented Optical Flow Based Action Recognition Methods. Ph.D. Thesis, Middle East Technical University, Ankara, Turkey, 2012.

28. Lertniphonphan, K.; Aramvith, S.; Chalidabhongse, T.H. Human action recognition using direction histograms of optical flow. In Proceedings of the Communications and Information Technologies (ISCIT), 2011 11th International Symposium on Communications & Information Technologies (ISCIT 2011), Hangzhou, China, 12–14 October 2011; pp. 574–579.

29. Lucas, B.D.; Kanade, T. An Iterative Image Registration Technique With an Application To Stereo Vision; In Proceedings of the 7th International Joint Conference on Artificial Intelligence, Vancouver, BC, Canada, 24–28 August 1981; pp. 674–679.

30. Akpinar, S.; Alpaslan, F.N. Video action recognition using an optical flow based representation. In Proceedings of the International Conference on Image Processing, Computer Vision, and Pattern Recognition (IPCV), Las Vegas, NV, USA, 21–24 September 2014; p. 1.

31. Satyamurthi, S.; Tian, J.; Chua, M.C.H. Action recognition using multi-directional projected depth motion maps. *J. Ambient. Intell. Humaniz. Comput.* **2018**, *9*, 1–7. [CrossRef]

32. Yang, X.; Zhang, C.; Tian, Y. Recognizing actions using depth motion maps-based histograms of oriented gradients. In Proceedings of the 20th ACM International Conference on Multimedia, Nara, Japan, 29 October–2 November 2012; pp. 1057–1060.

33. Choutas, V.; Weinzaepfel, P.; Revaud, J.; Schmid, C. PoTion: Pose MoTion Representation for Action Recognition. In Proceedings of the IEEE Conference on Computer Vision and Pattern Recognition (CVPR), Salt Lake City, UT, USA , 18–22 June 2018.

34. Ren, J.; Reyes, N.H.; Barczak, A.; Scogings, C.; Liu, M. An Investigation of Skeleton-Based Optical Flow-Guided Features for 3D Action Recognition Using a Multi-Stream CNN Model. In Proceedings of the 2018 IEEE 3rd International Conference on Image, Vision and Computing (ICIVC), Chongqing, China, 27–29 June 2018; pp. 199–203.

35. Karpathy, A.; Toderici, G.; Shetty, S.; Leung, T.; Sukthankar, R.; Fei-Fei, L. Large-scale video classification with convolutional neural networks. In Proceedings of the IEEE Conference on Computer Vision and Pattern Recognition, Columbus, OH, USA, 23–28 June 2014; pp. 1725–1732.

36. Wang, L.; Xiong, Y.; Wang, Z.; Qiao, Y. Towards good practices for very deep two-stream ConvNets. *arXiv* **2015**, arXiv:1507.02159.

37. Wang, L.; Qiao, Y.; Tang, X. Action recognition with trajectory-pooled deep-convolutional descriptors. In Proceedings of the IEEE Conference on Computer Vision and Pattern Recognition, Boston, MA, USA, 7–12 June 2015; pp. 4305–4314.

38. Feichtenhofer, C.; Pinz, A.; Zisserman, A. Convolutional two-stream network fusion for video action recognition. In Proceedings of the IEEE Conference on Computer Vision and Pattern Recognition, Las Vegas, NV, USA, 26 June–1 July 2016; pp. 1933–1941.

39. Ullah, A.; Ahmad, J.; Muhammad, K.; Sajjad, M.; Baik, S.W. Action Recognition in Video Sequences using Deep Bi-Directional LSTM With CNN Features. *IEEE Access* **2018**, *6*, 1155–1166. [CrossRef]

40. Fukunaga, K.; Koontz, W.L. Application of the Karhunen-Loève Expansion to Feature Selection and Ordering. *IEEE Trans. Comput.* **1970**, *100*, 311–318. [CrossRef]

41. Ramoser, H.; Muller-Gerking, J.; Pfurtscheller, G. Optimal spatial filtering of single trial EEG during imagined hand movement. *IEEE Trans. Rehabil. Eng.* **2000**, *8*, 441–446. [CrossRef] [PubMed]

42. Wang, Y.; Gao, S.; Gao, X. Common spatial pattern method for channel selection in motor imagery based brain-computer interface. In Proceedings of the 2005 IEEE Engineering in Medicine and Biology 27th Annual Conference, Shanghai, China, 17–18 January 2006; pp. 5392–5395.

43. Novi, Q.; Guan, C.; Dat, T.H.; Xue, P. Sub-band common spatial pattern (SBCSP) for brain-computer interface. In Proceedings of the 2007 3rd International IEEE/EMBS Conference on Neural Engineering, Kohala Coast, HI, USA, 2–5 May 2007; pp. 204–207.

44. Fisher, R.A. The use of multiple measurements in taxonomic problems. *Ann. Eugen.* **1936**, *7*, 179–188. [CrossRef]

45. Ho, T.K. Random decision forests. In Proceedings of the 3rd international conference on document analysis and recognition, Montreal, QC, Canada, 14–16 August 1995; Volume 1; pp. 278–282.

46. Lin, T.Y.; Maire, M.; Belongie, S.; Hays, J.; Perona, P.; Ramanan, D.; Dollár, P.; Zitnick, C.L. Microsoft COCO: Common objects in context. In Proceedings of the European Conference on Computer Vision, Zurich, Switzerland, 6–12 September 2014; pp. 740–755.

47. Hochreiter, S.; Bengio, Y.; Frasconi, P.; Schmidhuber, J. Gradient flow in recurrent nets: The difficulty of learning long-term dependencies. In *A Field Guide to Dynamical Recurrent Neural Networks*; Kremer, S.C., Kolen, J.F., Eds.; IEEE Press: Piscataway, NJ, USA. 2001.

48. Bengio, Y.; Simard, P.; Frasconi, P. Learning long-term dependencies with gradient descent is difficult. *IEEE Trans. Neural Netw.* **1994**, *5*, 157–166. [CrossRef]

49. Talathi, S.S.; Vartak, A. Improving performance of recurrent neural network with relu nonlinearity. *arXiv* **2015**, arXiv:1511.03771.

50. Mendialdua, I.; Martínez-Otzeta, J.M.; Rodriguez-Rodriguez, I.; Ruiz-Vazquez, T.; Sierra, B. Dynamic selection of the best base classifier in one versus one. *Knowl.-Based Syst.* **2015**, *85*, 298–306. [CrossRef]

51. Kingma, D.P.; Ba, J. Adam: A method for stochastic optimization. *arXiv* **2014**, arXiv:1412.6980.

Article

A Double-Branch Surface Detection System for Armatures in Vibration Motors with Miniature Volume Based on ResNet-101 and FPN

Tao Feng [1], Jiange Liu [1], Xia Fang [1,*], Jie Wang [1] and Libin Zhou [2]

[1] School of Mechanical Engineering Sichuan University, Chengdu 610041, Sichuan, China;
    2017223025092@stu.scu.edu.cn (T.F.); liujiange666@163.com (J.L.); wangjie@scu.edu.cn (J.W.)
[2] College of Letters and Science, University of Wisconsin Madison, Madison, WI 53707, USA;
    Lzhou228@wisc.edu
* Correspondence: 18215575946@163.com

Received: 6 March 2020; Accepted: 14 April 2020; Published: 21 April 2020

Abstract: In this paper, a complete system based on computer vision and deep learning is proposed for surface inspection of the armatures in a vibration motor with miniature volume. A device for imaging and positioning was designed in order to obtain the images of the surface of the armatures. The images obtained by the device were divided into a training set and a test set. With continuous experimental exploration and improvement, the most efficient deep-network model was designed. The results show that the model leads to high accuracy on both the training set and the test set. In addition, we proposed a training method to make the network designed by us perform better. To guarantee the quality of the motor, a double-branch discrimination mechanism was also proposed. In order to verify the reliability of the system, experimental verification was conducted on the production line, and a satisfactory discrimination performance was reached. The results indicate that the proposed detection system for the armatures based on computer vision and deep learning is stable and reliable for armature production lines.

Keywords: armature; computer vision; deep learning; surface inspection

---

1. Introduction

A vibration motor is a source of excitation. Small vibration motors are used in digital products like cell phones to provide a vibration sense. Large vibration motors are used in metallurgy and mining to screen ingredients [1]. In terms of the vibration motors used in digital products, the quality of the motor is an important factor that has an impact on the user experience. In the process of motor production, the armature is assembled into a shell with magnets and bearings [2], so incipient faults in any part of the machinery could produce a chain reaction and lead to its defects [3–5]. However, due to its miniature volume, it is difficult to detect these defects.

Much effort has been made in fault diagnosis of rotating machinery. Asr et al. [6] designed a feature extraction method using empirical mode decomposition and fed the extracted features into a non-native Bayesian classifier for intelligent fault diagnosis of rotating machinery. Georgoulas et al. [7] applied symbolic dynamic entropy features to extract features of gearbox signals and applied a support vector machine to recognize the health conditions.

However, the inspection conducted after assembly is always dramatically influenced by the mechanical system. It is difficult to explore the form of the signal. Therefore, each part should be carefully examined before assembly. For a vibration motor, the armature is the core component. If the quality of the armature is ensured, many problems will be avoided during motor rotation. In the traditional way, surface inspection of the armature is done manually, which is costly and can be easily disturbed by many subjective factors [8]. Therefore, it is important for a factory to apply intelligent inspection to detect the surface of the armature.

With the rapid development of computer vision, more and more image technologies have been used in industrial environments to detect surface quality. Cabral et al. [9] realized the intelligent detection of various glass products by traditional computer vision methods such as edge detection operator and Hough circle. In reference [10], image processing algorithms including Canny edge extraction, histogram equalization, and image morphology closed operation are utilized to extract and locate a joint contour in a complicated background image. Firmin et al. [11] proposed a novel method for detection and quantification of corrosion on a pipe using digital image processing techniques to extract saturation value. All of these are the applications of traditional computer vision, which can only achieve some simple industrial applications. It is difficult to use these applications for the detection of complex industrial objects while obtaining good results.

Therefore, with the development of hardware performance, the deep learning algorithm represented by convolutional neural networks (CNNs) has gradually become the core of the machine learning algorithm [12]. To date, image feature extraction by a CNN has shown great success in the process of image classification [13,14], object detection [15], semantic segmentation [16,17], and high-resolution image reconstruction [18]. A CNN has also been successfully applied to inspect industrial surfaces recently. Masci et al. [19] adopted max-pooling convolutional neural networks to detect the defects in steel. Song et al. [20] proposed a novel residual squeeze-and-excitation network to discover anomalies and inspect the quality of adhesives on battery cell surfaces. Weimer et al. [21] discussed the structural design of the CNN to realize automatic feature extraction in industrial surfaces and verified the proposed elements through the dataset of industrial surface defects.

For surface inspection in this paper, we consider the armatures in micro vibration motors with miniature volume. Considering the rough surface and tiny features, we used more abundant semantic features extracted by the CNN network to detect the surface. The inspection standard comes from the armature manufacturer.

The rest of this paper is organized as follows. Firstly, we introduce the locating device we designed to take photos of the armature. Then, we discuss the method, in which the collected images of armatures were divided into a training set and a test set, and we introduce the classification standard. Finally, we propose a network structure designed for our dataset and a training method to make the network performs better.

2. Related Works

2.1. Image Acquisition

Figure 1 shows the inspected armature. We can see that the armature is made of an iron core wrapped with copper wire. The length of the whole armature is about 1 cm. The entire armature is symmetrical about 120° around the central axis. To detect as few images as possible to reduce the detection time, we can convert the detection of the armature to the detection of three major surfaces.

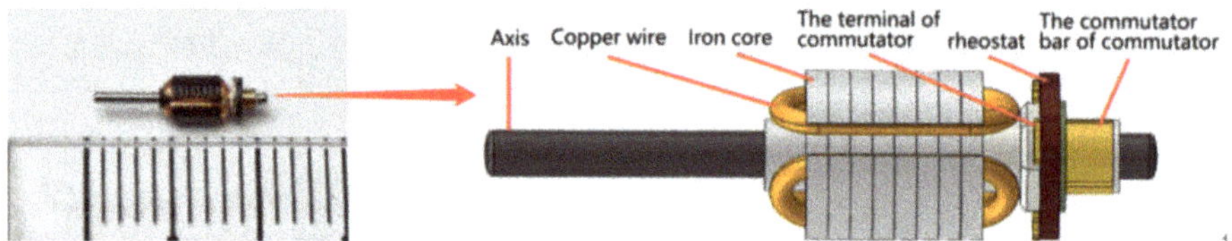

Figure 1. The inspected armature.

In industrial inspection, the workpiece is mostly irregular. In order to reduce the difficulty of positioning and detection, it is necessary to obtain a fixed imaging view through electromechanical structure. Figure 2 depicts the positioning and imaging device. The left side of Figure 2 is the entire device, and the right side is a zoomed-in view of the loading platform. The armature was placed on a loading platform and fixed by a small magnetic shaft hole. The loading platform was driven by a stepper motor to rotate and change the shooting angle of the armature. Because the reflectivity of the copper wire is greater than that of the iron core, when the beam of the digital fiber sensor shines on the iron core, as shown in Figure 1, the amount of reflected light is lower than that in the groove on the iron core shown in Figure 1, which is wrapped with copper wire. We used this phenomenon to find the position of the groove on the iron core. We used this position as the initial position for photographs. Then, taking advantage of the symmetry, a stepper motor with a closed loop system was used to drive the loading platform to precisely rotate 120° two times. In the end, we obtained the photos of three major surfaces, which can reflect the surface defects of a whole armature.

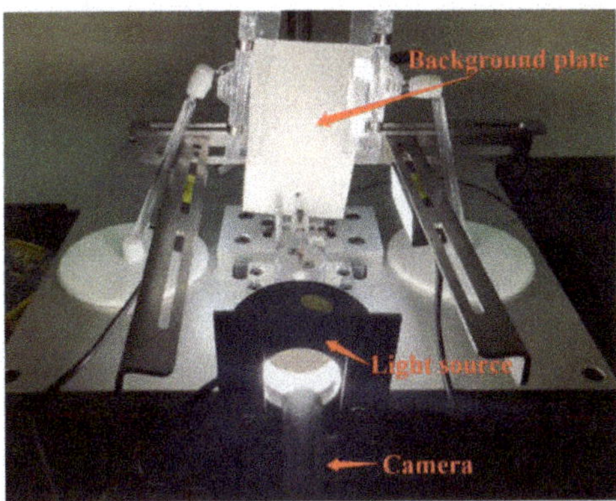

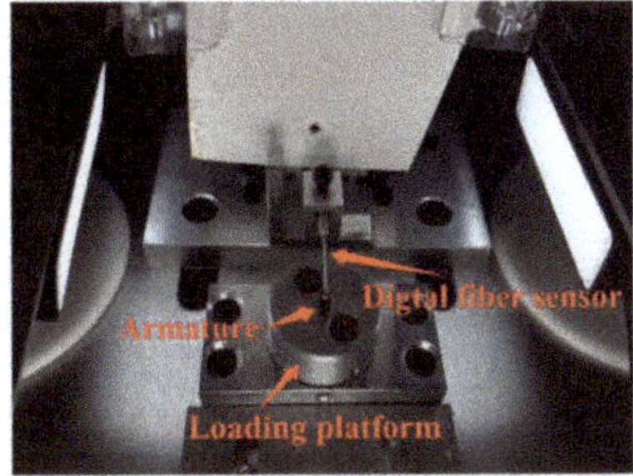

Figure 2. The positioning and imaging device.

Thus, for each armature we obtained three pictures. Whether the armature is defective is determined by detecting the three pictures. To avoid the influence of external illumination, a telephoto lens with fixed focus was selected. The shooting device was an industrial CCD camera with a resolution of 1920×960 pixels. The final processed and transmitted image format was JPG with a resolution of 540×480. Figure 3 gives the picture of three faces of an armature. From Figure 3, we can see that we obtained a good background and foreground. The edge between the armature and the background is well segmented. We can use simple morphological detection to identify the effective region of the armatures to eliminate unnecessary interference.

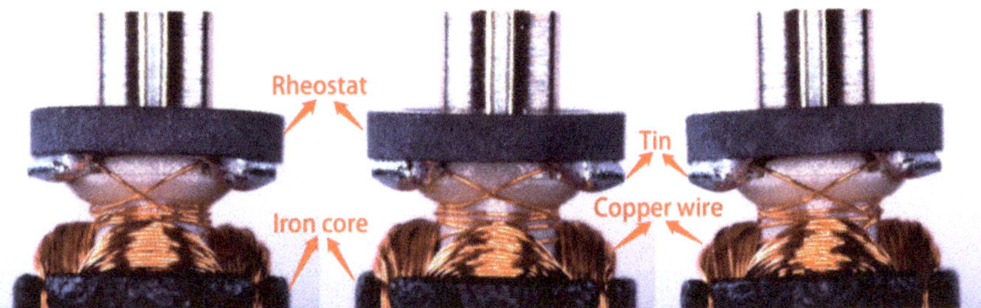

Figure 3. The picture of an armature we obtained from three faces.

2.2. Surface Defects on the Armatures

Figure 3 shows the full perspective of an armature. For our article, we only needed to examine a portion of the image because the pictures taken by the above device had a fixed scale and perspective.

Through the template matching algorithm in OpenCV, we could easily match the areas we needed from the template library we created. Figure 4 displays the final inspected area with a resolution of 350×120 pixels. We can determine from this figure that there are six kinds of surface defects that need to be inspected. The details of the surface defects are described in Table 1.

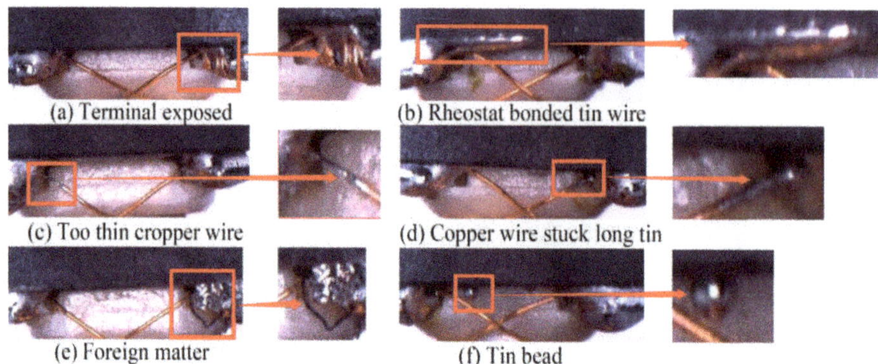

Figure 4. The final inspected area in our study and some of the surface defects.

Table 1. Types of surface defects.

Terminal exposed (Figure 4a)	The terminal connecting the copper wire and the resistor is not covered with tin. It is not strong enough to ensure the service life.
Rheostat bonded tin (Figure 4b)	A long piece of tin is attached to the rheostat, which will affect the use of the motor.
Too thin cropper wire (Figure 4c)	Because of the high temperature, the copper wire connecting the terminals is thinned, which will cause the copper wire to break easily.
Copper wire stuck, long tin (Figure 4d)	The attached tin on the copper wire is too long. The tin is not as hard as the copper wire, which will also cause the copper wire to break easily.
Foreign matter (Figure 4e)	There are foreign bodies sticking in this area, which will cause noise in the process of rotation.
Tin bead (Figure 4f)	In the process of rotation, the tin beads easily fall off, which will have a serious impact on the service life.

2.3. Data Calibration and Manual Observation

Each armature with a defective surface had one or more of the above-mentioned defects. Therefore, a defective armature cannot be classified in detail. In the process of labeling, only two classifications were tagged. Positive samples contained no surface defect, negative samples contained surface defects. Photos of the armatures were collected using our shooting and positioning device. We returned the collected photos to the company's quality control group. The quality control group divided these photos into two categories according to our classification task. Each armature picture was judged separately by two experts. The third expert re-judged the controversial samples. The determination of whether it is a positive sample was based on the judgment of the third expert. This dataset was determined by the quality control team. We regarded it as the correct data classification, which was our ground truth.

Although the inspected area is narrowed, we can see that the negative is still messy and varied. The armatures are tiny, so the defective areas are even smaller. It is a challenge for the eyes to distinguish between good and defective armatures. Due to the small size of the armatures, they must be studied under the microscope, which always contributes to eyestrain and high staff turnover. It is difficult for an enterprise to retain experienced workers. The demand for the enterprise is that the whole process of armature detection, including feeding, photographing, preprocessing, discrimination, and sorting, takes less than 3 s. Based on the above reasons, there are various mistakes in manual observation. For example, in the process of workers' detection, the detection standards are confused, and some minor defects are easy to ignore. As a result, based on the demand of detecting an armature in three seconds, the armatures in the dataset were observed and classified by five workers. After comparing the classification results with the correct classification results of the quality control group, we calculated the manual observation accuracy of each worker and found the average as the manual observation accuracy. In the end, the manual observation accuracy was 91.3%.

2.4. Traditional Computer Vision Method

In traditional computer vision, the information such as edges, textures, and colors are extracted and summarized. In our study, we used Sobel edge detection (Figure 5b), binarization (Figure 5d), and Canny edge detection (Figure 5e) to extract edges and textures information. Harris detection was used to detect the corners (Figure 5a). Because the main features of the inspected area were copper wire and tin, we took advantage of histogram backprojection to extract the areas that are similar to the tin in color (Figure 5c). In addition, the copper wire and tin are different in values between the R channel and B channel. Therefore, we subtracted the R and B channels to highlight the position of the copper wire (Figure 5f).

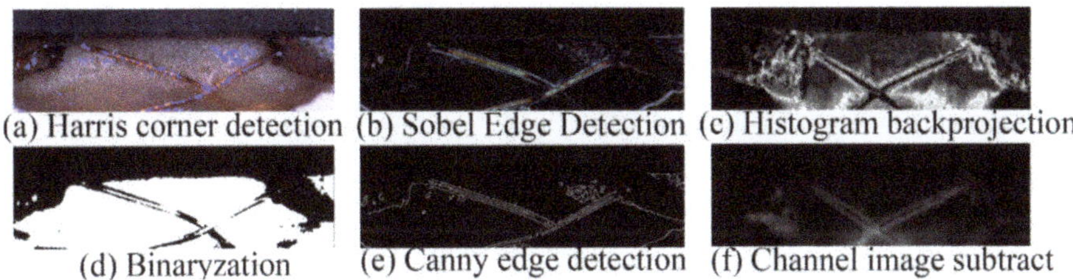

Figure 5. The feature maps extracted by traditional computer vision.

As can be seen from Figure 5, all methods suffer from a degree of information loss due to the effects of areas other than copper wires and tin, so the RGB image keeps the features integrated. We need a more effective feature extraction method for feature extraction of RGB images.

3. Methodology

For the armatures in our paper, the defective regions are small, but the features needed are complex and diverse. Furthermore, because of the concentricity error between the armatures and loading platform, the pictures of the armatures have a certain degree of angular difference. Therefore, we need to extract stronger semantic and more abstract features to inspect the surface of the armatures to resist interference.

3.1. ResNet-101 + FPN Network

Thus, we need some way to extract stronger semantic and more abstract image features. A convolutional neural network (CNN) is a kind of backpropagation neural network with deep structure including convolution calculation. It is one of the representative algorithms of deep learning. Generally, CNN has standard structure that the output of the features extractor which consists of stacked convolutional and pooling layers, can be directly input into the classifier. And this typical structure without any additional branch can be called a plain network like Alex, VGG [13,14]. For the plain network, a high-level feature map of each image can be obtained through a deep-network structure.

The small features become smaller or even disappear after downsampling. However, the information of high-resolution feature maps without downsampling is not abundant enough. It is not conducive to the extraction and characterization of complex small features, which is required for our dataset.

Inspired by the idea of feature pyramids in feature pyramid network (FPN) [22], our novel network as shown in Figure 6 has made some improvements on the basic design of the plain network, which comprises two specific branches: one for extracting feature maps, named the backbone branch, the other used as a network architecture called feature pyramid network (FPN) to fuse feature maps of multiple scales.

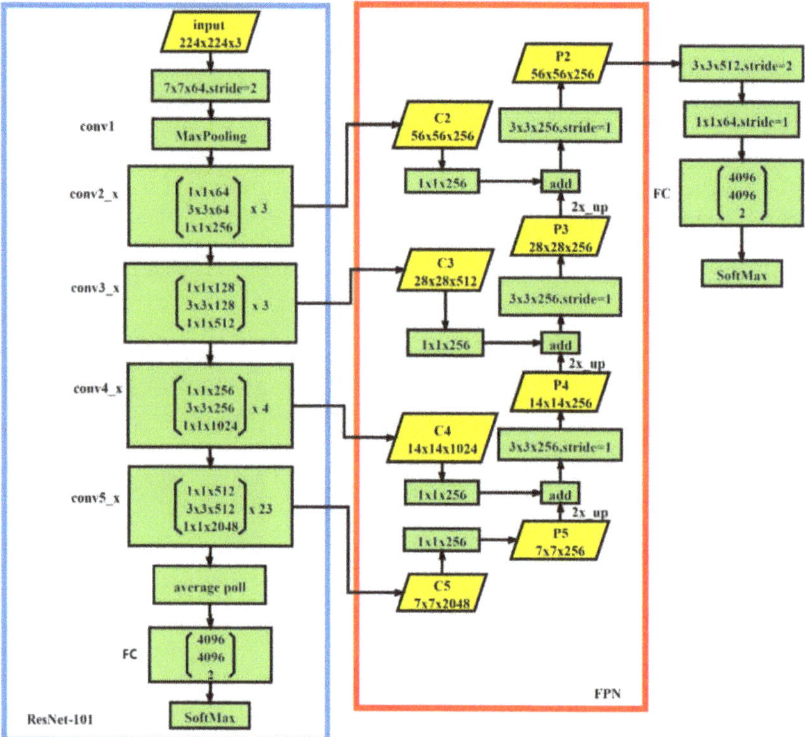

Figure 6. Final network structure of the ResNet101 + feature pyramid network (FPN).

In the process of forward propagation calculation, the backbone branch outputs feature maps of various resolutions. The FPN branch combines low-resolution, semantically strong features with high-resolution, semantically weak features via a top-down pathway, bottom-up pathway, and lateral connections.

Backbone branch: For the classification task, to extract features better, higher layers of representation amplify aspects of the input that are important for discrimination and suppress irrelevant variations [23]. As a result, the network must have enough depth. In order to ensure that the network has enough depth to extract rich semantic information to represent complex features, we used ResNet as the backbone network, since it is connected via a shortcut to learn residual mapping, which is better at transferring gradient information to prevent gradient vanishing and gradient degradation [24,25]. As a result, it can reach thousands of layers compared with a common plain network. In the end, allowing for the time and limited computing resources, we chose ResNet-101, a stack of 101-layer residual units, as our backbone network.

FPN branch: Our goal is to build a feature pyramid with high-level semantics throughout. The construction of our pyramid involves a bottom-up pathway, a top-down pathway, and lateral connections, as introduced in the following.

The top-down pathway provided the basic feature maps for our FPN. We collected the intermediate feature maps, which is the result of feedforward computation of the backbone network. There are often many layers producing intermediate feature maps of the same resolution, and we say these layers are in the same network stage. In the end, we chose the output of the last layer of each stage as our reference set of feature maps, which we enriched to create our pyramid. This choice is natural, since the deepest layer of each stage should have the strongest features.

For our backbone network, ResNet-101, we used the feature activation output by each stage's last residual block. We denote the output of these last residual blocks as {C2, C3, C4, C5}, and they have strides of {4, 8, 16, 32} pixels with respect to the input image. We did not include the output of the conv1 in the pyramid due to its large memory footprint.

The bottom-up pathway outputs higher resolution features by upsampling spatially coarser, but semantically stronger, feature maps from the upper level feature maps collected by the top-down pathway. These features are then enhanced with features from the bottom-up pathway via lateral connections. Each lateral connection merges feature maps of the same spatial size from the bottom-up pathway and the top-down pathway. To start the iteration, we simply attached a 1×1 convolutional layer on C5 to produce the coarsest resolution map. Finally, we appended a 3×3 convolution to each merged map to generate the final feature map, which was to reduce the aliasing effect of upsampling. This final set of feature maps is called {P2, P3, P4, P5}, corresponding to {C2, C3, C4, C5}, which are respectively of the same spatial sizes.

The result is a feature pyramid that has rich semantics at all levels. The feature pyramid is built quickly from a single input image [22], so P2~P5 all have rich semantics. Additionally, P2 has the highest resolution and fused more features with different scales. As a result, we only took the P2 layer feature map for prediction. We added two convolutional layers after P2 to reduce the size and dimension of the feature map for further information extraction and reduction of computation. After the two convolutional layers, we still added two fully connected layers to summarize the feature information. In the end, we connected SoftMax for binary classification.

3.2. Focal Loss

In the case of unbalanced input, the network will converge toward a large quantity of data. It will not care more about the small quantity of data with a large amount of information. For the data in this paper, negative samples have one or more of the above defects. The number of each kind of defective armature is also unbalanced. If the traditional cross-entropy loss function is used, it will lead to poor learning. To solve this problem, the focal loss was proposed by RetinaNet as follows [26]:

$$FL(p_t) = -\alpha_t(1 - p_t)^{\gamma} log(p_t) \tag{1}$$

α_t balances the importance of positive/negative examples. The focusing parameter γ smoothly adjusts the rate at which easy examples are down-weighted.

$p \in [0, 1]$ is the model's estimated probability for the class with true label, and p_t is defined as follows:

$$p_t = \begin{cases} p, & \text{if label is true} \\ 1 - p, & \text{otherwise} \end{cases} \tag{2}$$

We can see from Equation (2) that focal loss adds a modulating factor to the original cross-entropy loss function. When samples are misclassified and p_t is very small, the modulating factor is close to 1, and the loss function has no influence. When p_t approaches 1, the factor goes to 0, and the loss for well-classified examples is down-weighed. Therefore, focal loss can mine unbalanced data and weaken the contribution of the easy-to-classify armatures to loss. In this way, we can better mine the data distribution of negative samples and achieve a better learning effect.

3.3. Feature Library Matching

In the surface detection of the industrial product, the appearance of positive samples is relatively similar, and the appearance of the defective workpieces is relatively different. It is difficult to map all negative samples with different appearances into the same category. Therefore, we took advantage of the previously trained ResNet-101+FPN network, which has a strong representational ability for positive and negative samples, to extract the vector from its end and establish an a priori feature library according to the samples in the training set. We expected to get better results by matching the distance between the sample and a priori feature library to judge the category of the sample. The whole process is shown in Figure 7.

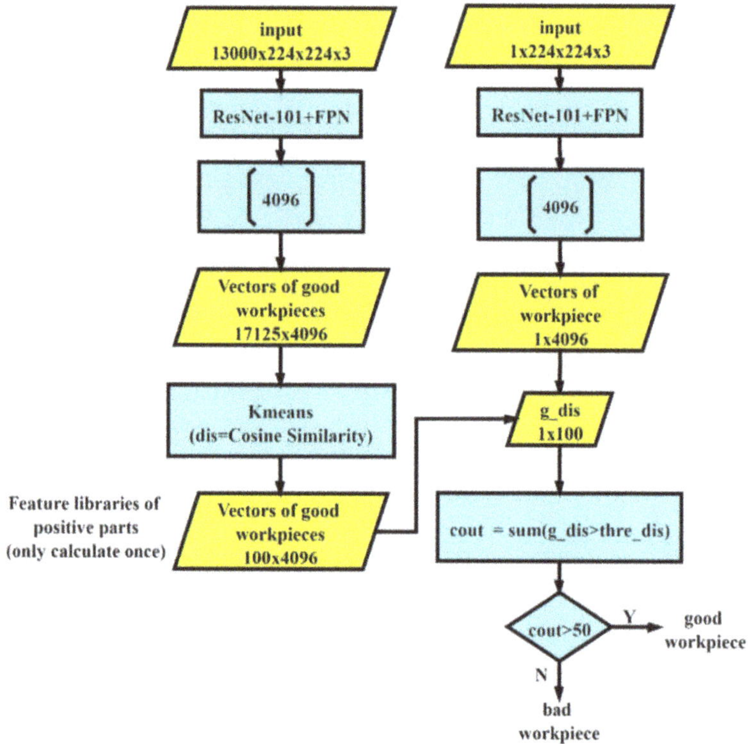

Figure 7. The method of feature library matching.

The flat quadrilaterals in the picture represent the data, and the rectangles represent calculations. We used the above-mentioned trained network to infer the training set. Vectors with a size of 4096 extracted from the first fully connected layer were collected and saved. They formed a feature library of the good samples. We inferred the test samples and obtained a feature vector with a size of 4096. We could match the feature vector with the feature library to calculate the similarity between the test samples and the good samples. As a vector similarity measurement, cosine similarity is widely used in text similarity calculation [27] and face recognition [28]. Cosine similarity is defined as follows:

$$similarity = cos(\theta) = \frac{\sum_{i=1}^{n}(x_i \times y_i)}{\sqrt{\sum_{i=1}^{n}(x_i)^2} \times \sqrt{\sum_{i=1}^{n}(y_i)^2}} \tag{3}$$

x_i and y_i represent the corresponding vectors. The result ranges from −1 to 1. A value of −1 means that the two vectors are pointing in opposite directions, a value of 1 means they are pointing in the same direction, and 0 usually means they are independent. A value in between them indicates similarity or dissimilarity. It can be seen that the range of the values is fixed and will not change like the Euclidean distance, which is helpful for us to find a better threshold value. Equation (3) indicates that it pays more attention to the similarity of the two vectors in the direction, rather than the absolute difference in value. For the comparison of feature vectors, it has a better effect. Therefore, we distinguished the samples by measuring the cosine similarity between feature vectors in the library and feature vectors of the samples.

It would be very time-consuming to compare the cosine similarity between the test workpiece and each vector in the feature library because of the inner product calculation. Therefore, we carried out K-means clustering to cluster the feature libraries of good parts into 100 groups by cosine similarity. The 100 cluster center vectors were selected as a new good feature library, and we only needed to compute the feature library once. We calculated the cosine similarity between the feature vectors of a test armature and each feature vector in the good feature library as *g_dis*, which has 100 dimensions. We set a threshold *thre_dis*. By comparing the size of *g_dis* and *thre_dis*, we can calculate the count, which is the number meeting the threshold condition. If the size of *count* is greater than 50 (the vote was more than half), it will output Y. By the voting mechanism, we could infer whether the sample was good or not.

3.4. The Double-Branch Discrimination Mechanism

To ensure the quality of the workpiece, it is necessary for mass production to reduce the probability of the defective samples being misjudged as good, so we introduced a double-branch discrimination mechanism to synthesize the results of the two methods. The entire double-branch discrimination mechanism is shown in Figure 8. A branch is determined by the result of the SoftMax layer. The other branch is determined by the method feature library matching. Only when both branches output Y at the same time will the system identify the sample as good. It is equivalent to improving the standard of discrimination. In this way, we expected to achieve better results in defective part discrimination.

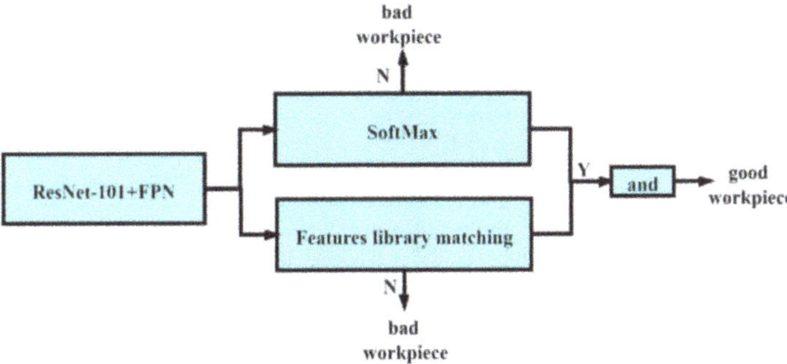

Figure 8. The double-branch discrimination mechanism.

4. Experiment

4.1. Dataset

The image data were captured by the device above. We returned the collected photos to the company's quality control group. The quality control group divided these photos into two categories according to our classification task. In the rest of this study, we tested and validated our approach based on this dataset and its classification. Because there are many types of defective armatures and the distribution is uneven, we tried and trained on different data volumes. We found that the larger the quantity of data, the more robust the final model is. To ensure a good learning effect for the data distribution of the defective samples and improve the robustness of the model, we studied a large quantity of data. Finally, the training set contained 26,000, and the test set contained 11,106. The ratio of the number between the training set and the test set is approximately 7:3, which followed the common ratio of 7:3 in machine learning.

4.2. Implementation Details

Data Augmentation: To fully connect the layers and enhance robustness, we resized the images to $240 \times 240 \times 3$ pixels and randomly cropped them to $224 \times 224 \times 3$ pixels as the input in each batch. Moreover, in the training process, we randomly flipped and rotated the loaded data and adjusted the brightness. To prevent the small changes in image quality making the judgment of the network change, we finally normalized the input data [29].

Experiment Environment: We built the network environment through TensorFlow, a deep learning framework compatible with Windows. We created a graph model and trained on a 1080 Ti graphics card with 12 GB video memory.

Training Strategy: Because the shortcut connection in the residual network can help the information to propagate easily [25,30], when the structure of FPN was added, the network became more complex, and information was harder to backpropagate. Thus, the method of training a backbone network, a side-branch structure, and an up-sampling structure together cannot ensure that the backbone network contains good semantic information for FPN to extract. To obtain a better backbone network and train our network better, we used a step-by-step method of training.

4.3. Training and Results

Firstly, we trained our network directly. The initial learning rate was set to 0.001 and was attenuated 10 times after the 80th epoch. In order to prevent the results from falling into a local optimal solution and gradient oscillation [31], we used the momentum optimizer to optimize loss and update the weight. We set the momentum to 0.01. After 300 iterations, we obtained the curve of loss and accuracy as the number of iterations increased, as shown in Figure 9.

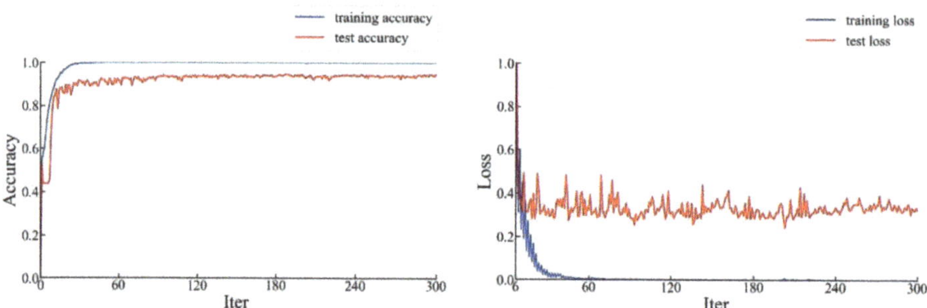

Figure 9. The accuracy and loss curve of the ResNet-101+FPN transfer learning from the ImageNet classification model.

As can be seen in Figure 9, the network tends to converge, and the highest accuracy in the whole process was 96.5%, while the loss was 0.317. Compared with ResNet-101, the accuracy was significantly improved.

The model with the highest accuracy on the test set was saved. Then, we tested it on test set and obtained the confusion matrix as shown in Table 2.

Table 2. The confusion matrix of the ResNet-101+FPN model (one stage).

		Predicted Value		Total
		Positive	Negative	
Observed Value	Positive	6788	153	6941
	Negative	221	3944	4165

Secondly, we extracted the partial parameters of ResNet-101 from the above trained ResNet101+FPN. We used the parameters to initialize and trained the ResNet-101 to obtain a more expressive backbone network. After 300 iterations, we obtained the curve of loss and accuracy as the number of iterations increased, as shown in Figure 10.

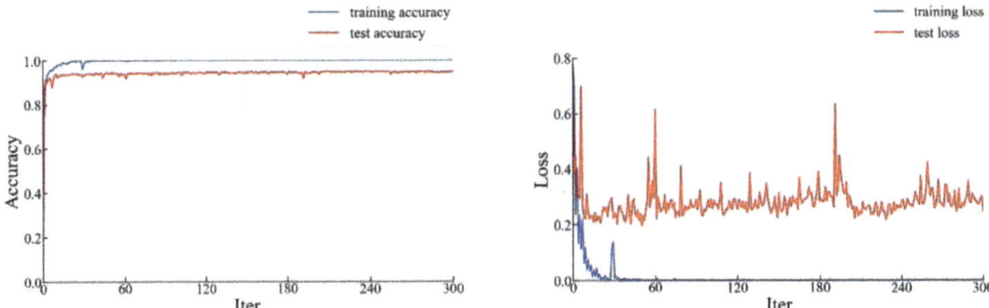

Figure 10. The accuracy and loss curve of the ResNet-101 initialized by trained the model of ResNet-101+FPN.

As can be seen from Figure 10, the highest prediction accuracy in the whole process was 96.8%, while the loss was 0.213. Compared with ResNet-101 transfer learning from the ImageNet classification model, our initialization method significantly improved the performance of ResNet-101. We realized the idea of training a good backbone network. We saved the model that performed best in accuracy to use in the follow-up work.

In the end, we used the trained backbone network to train ResNet101-FPN in two steps. In the first step, we used the more expressive ResNet-101 model above as the model of the backbone network. We adjusted the learning rate of ResNet-101 layers to 0 to fix the weight parameters of each layer. In this way, we only trained related layers of FPN. After 300 iterations, we obtained the curve of loss and accuracy as the number of iterations increased, as shown in Figure 11.

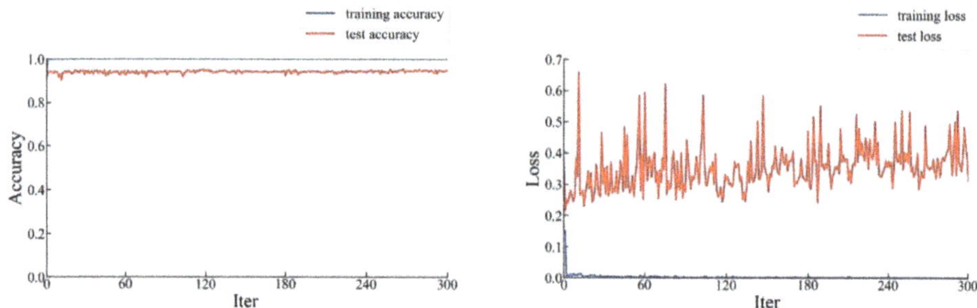

Figure 11. The accuracy and loss curve of the ResNet-101+FPN in the first step.

As can be seen in Figure 11, the highest prediction accuracy in the whole process is 97.2 %, while the loss was 0.293.

In the second step, we loaded the parameters obtained in the previous step into our network. We trained the entire network including ResNet-101 and FPN. After 300 iterations, we obtained the curve of loss and accuracy as the number of iterations increased, as shown in Figure 12.

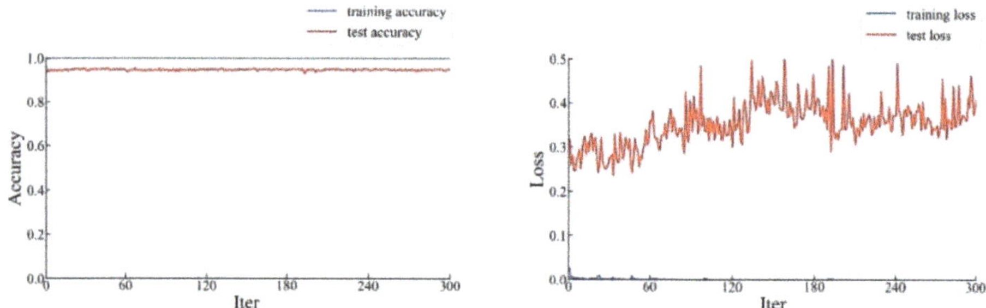

Figure 12. The accuracy and loss curve of the ResNet-101+FPN in the second step.

As can be seen from Figure 12, the highest prediction accuracy in the whole process was 97.1%, while the loss was 0.192. In the first step of training, the accuracy increased by 0.1% compared with the result of training our network directly. Although the accuracy did not increase again in the second step, the loss decreased significantly, indicating that the overall effect of the network was improved to some extent. Therefore, compared with directly training ResNet-101+FPN, our distributed training method achieved a significant improvement in terms of accuracy and loss.

The model with the highest accuracy on the test set was saved. Then, we tested to get the confusion matrix shown in Table 3.

Table 3. The confusion matrix of the ResNet-101+FPN model (two stage).

		Predicted Value		Total
		Positive	**Negative**	
Observed Value	Positive	6823	118	6941
	Negative	209	3956	4165

Through the above steps, we obtained a backbone network with a strong capability for feature extraction and a multi-scale fusion network. This was also better than directly training our network.

4.4. Establishing the Double-Branch Discrimination System and Results

Above all, we obtained a good network model. Since the range of cosine similarity is between -1 and 1, we searched the *thre_dis* with the step size of 0.05 and verified the branch of feature library matching on the test set to find the best threshold. Finally, when the *thre_dis* is 0.35, this branch has the highest accuracy in the test. The accuracy is 97.3%. After testing on the test set, we obtained the confusion matrix shown in Table 4.

Table 4. The confusion matrix of the branch implemented by feature library matching.

		Predicted Value		Total
		Positive	Negative	
Observed Value	Positive	6835	106	6941
	Negative	196	3969	4165

It can be seen that the method of creating a priori feature library can improve the accuracy.

To further reduce the number of the defective samples being misjudged as good, we combined the two branches in the way described before. Then, we tested the test set to get the confusion matrix shown in Table 5.

Table 5. The confusion matrix of the double-branch discrimination mechanism.

		Predicted Value		Total
		Positive	Negative	
Observed Value	Positive	6797	144	6941
	Negative	162	4003	4165

The accuracy of the test set dropped from 97.2% to 97.1%. As can be seen from Table 3, 4074 armatures were rejected by the first branch. Among them, 3956 were rejected correctly, and 118 were rejected incorrectly. As can be seen from Table 4, 4075 armatures were rejected by the second branch. Among them, 3969 were rejected correctly, and 106 were rejected incorrectly. As can be seen from Table 5, after combining the above two branches, 4147 were rejected. Among them, 4003 were rejected correctly, and 144 were rejected incorrectly. Through comparison, we find that more negative simples were correctly identified after combining the above two branches. Since both branches are required to be Y at the same time, the identification standards for good parts become stricter. This can guarantee better quality of the armatures.

4.5. Comparison and Discussions

The evaluation indexes used in this paper are accuracy, precision, recall, and f1-score, commonly used for classification problems. Our paper mainly considered the comparison of accuracy and precision. The accuracy can reflect the overall performance of the model. The accuracy rate can reflect the proportion of negative samples that are misjudged as positive samples, which is very important to ensure the quality of industrial products.

We trained and tested the same dataset on the Support Vector Machine (SVM), a machine-learning model, and the deep-learning model mentioned in our paper. Table 6 provides a comprehensive comparison of the index.

Table 6. A comprehensive comparison of the index.

	Accuracy	Recall	Precision	F1-Score	Time Per Image/s
SVM	83.2%	85.3%	87.5%	86.4%	0.574
VGG-16	92.4%	93.3%	94.6%	93.9%	0.035
ResNet-101	93.6%	95.1%	95.0%	95.0%	0.084
ResNet-101+FPN (trained directly)	96.6%	97.8%	96.8%	97.3%	0.095
ResNet-101+FPN (two stage trained)	97.1%	98.3%	97.0%	97.6%	0.095
Feature library matching	97.3%	98.5%	97.2%	97.8%	0.186
Double-branch discrimination mechanism	97.2%	97.9%	97.7%	97.7%	0.205

In summary, from Table 6, the armatures detection method we propose achieved good results. Compared with other structures, the structure of ResNet-101+FPN we designed is improved in all aspects. We also propose a training method to make our network performance better.

We can see that the proposed feature library matching method improved in all respects based on the parameters of our classification network. Although the accuracy and recall rate were reduced by 0.1%, the precision was improved by 0.5% after merging the two branches. It is a significant improvement in precision for the mass production under the condition that the accuracy is almost unchanged. With the precision improved, the probability of the defective samples being misjudged as good reduced, which ensured the quality of the workpiece and was in line with the requirements proposed by the enterprise. From the perspective of the time, there is little difference between the time of feature library matching and the time of the double-branch discrimination mechanism. Therefore, we chose double-branch discrimination mechanism to be applied to the production line.

4.6. Validation in Actual Production

In order to validate the reliability of the discrimination system in actual production, we inspected the armatures produced by the company in one day. The total number of the armatures inspected was about 360,000. Among them, the proportion of defective parts was about 2%. Due to the imbalance of positive and negative workpieces and most of the armatures with relatively good surfaces on the production line, the accuracy reached 98.9% in the actual production line. The result was also calculated by comparing the classification result of the discrimination system with the classification result of the quality control group. The whole process of the detection of the armatures, including feeding, photographing, preprocessing, discrimination, and sorting, is automated and takes less than 3 s, which is the enterprise's required time to identify an armature. Therefore, the armature inspection system proposed in this paper can meet the requirements of enterprise, both in terms of detection accuracy and detection time. The discrimination system can be stably used in the armature production line.

5. Conclusions

In this paper, a novel and complete system based on computer vision and deep learning is proposed for surface inspection of armatures in vibration motors with miniature volume. Concerning the characteristics of the data samples, we designed the structure of ResNet-101+FPN, which uses ResNet-101 as the backbone network and integrates the idea of FPN structure to utilize its representation capacity more effectively for small features. The network achieved satisfying performance on our dataset by accurately predicting the categories of armatures. In addition, we proposed a training method to make the network designed by us perform better. To guarantee the quality of the motor, we had to reduce the proportion of the defective samples being identified as good. Therefore, based on the trained network, we proposed a feature library matching method. After combining the classification network and the feature library matching method, the miscalculation of defective parts was significantly decreased. When testing on the production line, the discrimination could also ensure robustness and universality. The whole process of the detection of the armature takes less than 3 s, which is the enterprise's required time to identify an armature. Therefore, the system proposed in this paper can meet the requirements of enterprise, both in terms of effect and detection time. The discrimination system can be stably used in the armature production line.

Author Contributions: The authors T.F. and J.L. contributed equally to this work. Conceptualization, T.F. and J.L.; Data curation, T.F.and J.L.; Funding acquisition, J.W.; Investigation, T.F. and J.W.; Methodology, T.F. and J.L.; Project administration, X.F. and J.W.; Software, T.F., J.L. and L.Z.; Supervision, X.F. and J.W.; Validation, T.F., X.F. and L.Z.; Writing–original draft, T.F.; Writing–review & editing, X.F. and L.Z. All authors have read and agreed to the published version of the manuscript.

Funding: This presented work was supported by Sichuan Science and Technology Program (No. 2019YFG0359).

Conflicts of Interest: The authors declare no conflict of interest.

References

1. Meng, D. Research and Application of Vibration Motor Fault Diagnosis. Master's Thesis, Xi'an University of Architecture and Technology, Xi'an, China, 2017.
2. Chang, L.; Luo, F. Application of wavelet analysis in fault detection of cell phone vibration motor. In Proceedings of the 2009 International Asia Conference on Informatics in Control, Automation and Robotics, Bangkok, Thailand, 1–2 February 2009; pp. 473–476.
3. Worden, K.; Staszewski, W.J.; Hensman, J. Natural computing for mechanical systems research: A tutorial overview. *Mech. Syst. Signal Process.* **2011**, *25*, 4–111. [CrossRef]
4. Glowacz, A.; Glowacz, W.; Glowacz, Z.; Kozik, J. Early fault diagnosis of bearing and stator faults of the single-phase induction motor using acoustic signals. *Measurement* **2018**, *113*, 1–9. [CrossRef]
5. Guo, L.; Li, N.; Jia, F.; Lei, Y.; Lin, J. A recurrent neural network based health indicator for remaining useful life prediction of bearings. *Neurocomputing* **2017**, *240*, 98–109. [CrossRef]
6. Asr, M.Y.; Ettefagh, M.M.; Hassannejad, R.; Razavi, S.N. Diagnosis of combined faults in Rotary Machinery by Non-Naive Bayesian approach. *Mech. Syst. Signal Process.* **2017**, *85*, 56–70. [CrossRef]
7. Georgoulas, G.; Karvelis, P.; Loutas, T.; Stylios, C.D. Rolling element bearings diagnostics using the Symbolic Aggregate approXimation. *Mech. Syst. Signal Process.* **2015**, *60*, 229–242. [CrossRef]
8. Pesante-Santana, J.; Woldstad, J.C. Quality Inspection Task in Modern Manufacturing. In *International Encyclopedia of Ergonomics and Human Factors*; Karwowski, W., Ed.; Taylor and Francis: London, UK, 2001.
9. Cabral, J.D.D.; De Araújo, S.A. An intelligent vision system for detecting defects in glass products for packaging and domestic use. *Int. J. Adv. Manuf. Technol.* **2014**, *77*, 485–494. [CrossRef]
10. Long, Z.; Zhou, X.; Zhang, X.; Wang, R.; Wu, X. Recognition and Classification of Wire Bonding Joint via Image Feature and SVM Model. *IEEE Trans. Compon. Packag. Manuf. Technol.* **2019**, *9*, 998–1006. [CrossRef]
11. Bondada, V.; Pratihar, D.K.; Kumar, C.S. Detection and quantitative assessment of corrosion on pipelines through image analysis. *Procedia Comput. Sci.* **2018**, *133*, 804–811. [CrossRef]
12. Staar, B.; Lütjen, M.; Freitag, M. Anomaly detection with convolutional neural networks for industrial surface inspection. *Procedia CIRP* **2019**, *79*, 484–489. [CrossRef]
13. Krizhevsky, A.; Sutskever, I.; Hinton, G.E. Imagenet classification with deep convolutional neural networks. In Proceedings of the Advances in Neural Information Processing Systems, Lake Tahoe, NV, USA, 3–6 December 2012; pp. 1097–1105.
14. Simonyan, K.; Zisserman, A. Very deep convolutional networks for largescale image recognition. *arXiv* **2014**, arXiv:1409.1556.
15. Ji, Y.; Zhang, H.; Wu, Q.M.J. Salient object detection via multi-scale attention CNN. *Neurocomputing* **2018**, *322*, 130–140. [CrossRef]
16. Long, J.; Shelhamer, E.; Darrell, T. Fully convolutional networks for semantic segmentation. In Proceedings of the IEEE Conference on Computer Vision and Pattern Recognition, Boston, MA, USA, 7–12 June 2015; pp. 3431–3440.

17. Ren, S.; He, K.; Girshick, R.; Sun, J. Faster r-cnn: Towards real-time object detection with region proposal networks. In Proceedings of the Advances in Neural Information Processing Systems, Montreal, QC, Canada, 7–12 December 2015; pp. 91–99.

18. Huang, F.; Yu, Y.; Feng, T. Automatic extraction of impervious surfaces from high resolution remote sensing images based on deep learning. *J. Vis. Commun. Image Represent.* **2019**, *58*, 453–461. [CrossRef]

19. Masci, J.; Meier, U.; Ciresan, D.; Schmidhuber, J.; Fricout, G. Steel defect classification with max-pooling convolutional neural networks. In Proceedings of the 2012 International Joint Conference on Neural Networks (IJCNN), Brisbane, Australia, 10–15 June 2012; pp. 1–6.

20. Song, Z.; Yuan, Z.; Liu, T. In Residual Squeeze-and-Excitation Network for Battery Cell Surface Inspection. In Proceedings of the 2019 16th International Conference on Machine Vision Applications (MVA), Tokyo, Japan, 27–31 May 2019; pp. 1–5.

21. Weimer, D.; Scholz-Reiter, B.; Shpitalni, M. Design of deep convolutional neural network architectures for automated feature extraction in industrial inspection. *CIRP Ann.* **2016**, *65*, 417–420. [CrossRef]

22. Lin, T.-Y.; Dollar, P.; Girshick, R.; He, K.; Hariharan, B.; Belongie, S. Feature pyramid networks for object detection. In Proceedings of the IEEE Conference on Computer Vision and Pattern Recognition, Honolulu, HI, USA, 21–26 July 2017; pp. 2117–2125.

23. LeCun, Y.; Bengio, Y.; Hinton, G. Deep learning. *Nature* **2015**, *521*, 436–444. [CrossRef]

24. Pascanu, R.; Mikolov, T.; Bengio, Y. On the difficulty of training recurrent neural networks. In Proceedings of the International Conference on Machine Learning, Atlanta, GA, USA, 16–21 June 2013; pp. 1310–1318.

25. He, K.; Zhang, X.; Ren, S.; Sun, J. Deep residual learning for image recognition. In Proceedings of the IEEE Conference on Computer Vision and Pattern Recognition, Las Vegas, NV, USA, 27–30 June 2016; pp. 770–778.

26. Lin, T.-Y.; Goyal, P.; Girshick, R.; He, K.; Dollar, P. Focal loss for dense object detection. In Proceedings of the IEEE International Conference on Computer Vision, Venice, Italy, 22–29 October 2017; pp. 2980–2988.

27. Nguyen, H.V.; Bai, L. Cosine similarity metric learning for face verification. In Proceedings of the Asian Conference on Computer Vision, Queenstown, New Zealand, 8–12 November 2010; pp. 709–720.

28. Al-Anzi, F.; AbuZeina, D. Toward an enhanced Arabic text classification using cosine similarity and Latent Semantic Indexing. *J. King Saud Univ. Comput. Inf. Sci.* **2017**, *29*, 189–195. [CrossRef]

29. Kotsiantis, S.; Kanellopoulos, D.; Pintelas, P. Data preprocessing for supervised leaning. *Int. J. Comput. Sci.* **2006**, *1*, 111–117.

30. Venables, W.N.; Ripley, B.D. *Modern Applied Statistics with S-PLUS*; Springer Science & Business Media: Berlin/Heidelberg, Germany, 2013.

31. Ruder, S. An overview of gradient descent optimization algorithms. *arXiv* **2016**, arXiv:1609.04747.

Article

Multi-Modality Medical Image Fusion Using Convolutional Neural Network and Contrast Pyramid

Kunpeng Wang [1,2], Mingyao Zheng [3], Hongyan Wei [3], Guanqiu Qi [4] and Yuanyuan Li [3,*]

[1] School of Information Engineering, Southwest University of Science and Technology, Mianyang 621010, China; wkphnzk@163.com
[2] Robot Technology Used for Special Environment Key Laboratory of Sichuan Province, Mianyang 621010, China
[3] College of Automation, Chongqing University of Posts and Telecommunications, Chongqing 400065, China; ZMYzhengmingyao@126.com (M.Z.); weihy12@126.com (H.W.)
[4] Computer Information Systems Department, State University of New York at Buffalo State, Buffalo, NY 14222, USA; qig@buffalostate.edu
* Correspondence: liyy@cqupt.edu.cn

Received: 7 March 2020; Accepted: 8 April 2020; Published: 11 April 2020

Abstract: Medical image fusion techniques can fuse medical images from different morphologies to make the medical diagnosis more reliable and accurate, which play an increasingly important role in many clinical applications. To obtain a fused image with high visual quality and clear structure details, this paper proposes a convolutional neural network (CNN) based medical image fusion algorithm. The proposed algorithm uses the trained Siamese convolutional network to fuse the pixel activity information of source images to realize the generation of weight map. Meanwhile, a contrast pyramid is implemented to decompose the source image. According to different spatial frequency bands and a weighted fusion operator, source images are integrated. The results of comparative experiments show that the proposed fusion algorithm can effectively preserve the detailed structure information of source images and achieve good human visual effects.

Keywords: medical image fusion; convolutional neural network; image pyramid; multi-scale decomposition

---

1. Introduction

In the clinical diagnosis of modern medicine, various types of medical images play an indispensable role and provide great help for the diagnosis of diseases. To obtain sufficient information for accurate diagnosis, doctors generally need to combine multiple different types of medical images from the same position to diagnose the patient's condition, which often causes great inconvenience. If multiple types of medical images are only analyzed by doctor's space concepts and speculations, the analysis accuracy is subjectively affected, even parts of image information may be neglected. Image fusion techniques provide an effective way to solve these issues [1]. As the variety of medical imaging devices increases, the obtained medical images from different modalities contain complementary as well as redundant information. Medical image fusion techniques can fuse multi-modality medical images for more reliable and accurate medical diagnosis [2,3].

This paper proposes a CNN-based medical image fusion method. First, CNN-based model generates a weight map for any-size source image. Then, Gaussian pyramid decomposition is performed on the generated weight map, and the contrast image pyramid decomposition is applied to source images for obtaining the corresponding multi-scale sub-resolution images. Next, a weighted fusion operator based on the measurement of regional characteristics is used to set different thresholds for the top layer and the remaining layers of sub-decomposed images to obtain the fused sub-decomposed images. Finally, the fused image is obtained by the reconstruction of contrast pyramid. This paper has three main contributions as follows:

(1)   In training process of CNN, source images can be directly mapped to the weight map. Thus, it can also achieve the measurement of activity level and weight distribution in an optimal way to overcome the difficulties in design by learning network parameters in the training process.
(2)   Human visual system is sensitive to the changes of image contrast. Thus, this paper proposes a multi-scale contrast pyramid decomposition based image fusion solution, which can selectively highlight the contrast information of fused image to achieve better human visual effects.
(3)   The proposed solution uses a weighted fusion operator based on the measurement of regional characteristics. In the same decomposition layer, the fusion operators applied to different local regions may be different. Thus, the complementary and redundant information of fused image can be fully explored to achieve a better fusion effect and highlight important detailed features.

The remainder of this paper is organized as follows. Section 2 discusses the related works of medical image fusion. Section 3 demonstrates the proposed CNN-based medical image fusion solution in detail. Section 4 presents the comparative experiments and compares corresponding results. Section 5 concludes this paper.

2. Related Works

Researchers have proposed many medical image fusion methods in recent years [4,5]. Mainstream medical image fusion methods include decomposition-based and learning-based image fusion methods [6,7]. As a commonly used decomposition-based medical image fusion method, multi-scale transform (MST) generally has three steps in the fusion process: decomposition, fusion, and reconstruction. Pyramid-based method, wavelet, and multi-scale geometric analysis (MAG) based method are commonly used in MST[8]. In MAG-based methods, nonsubsampled contourlet transform (NSCT) [9,10] and nonsubsampled shearlet transform (NSST) [11] based methods have high efficiency in image representation. In addition to image transformation, the analysis of high- and low-frequency coefficients is also a key issue of MST-based fusion methods. Traditionally, the activity level of high-frequency coefficient is usually based on its absolute value. It is calculated in a pixel- or window-based way, and then uses a simple fusion rule, such as the selection of the maximum or weighted average, to obtain the fused coefficient. Averaging the coefficients of different source images was the most popular low-frequency fusion strategy in early research. In recent years, more advanced image transformations and more complex fusion strategies have been developed [12–17]. Liu proposed an integrated sparse representation (SR)- and MST-based medical image fusion framework [18]. Zhu proposed an NSCT based multi-modality decomposition method for medical images, which uses the phase consistency and local Laplacian energy to fuse high- and low-pass sub-bands, respectively [9]. Yin proposed a multi-modality medical image fusion method in NSST domain, which introduced pulse coupled neural network (PCNN) for image fusion [19]. To improve the fusion quality of multi-modality images, a novel multi-sensor image fusion framework based on NSST and PCNN was proposed by Li [20].

In the past decade, learning-based methods have been widely used in medical image fusion. Especially, SR- and deep learning-based fusion methods are most widely used [21,22]. In the early stage, SR-based fusion methods used a standard sparse coding model based on a single image component and local image blocks [23–25]. In the original spatial domain, source images were segmented into a set of overlapping image blocks for sparse coding. Most existing SR-based fusion methods attempt to improve their performances is the following ways: adding detailed constraints [5], designing more efficient dictionary learning strategy [26], using multiple sub-dictionaries in representation [27,28], etc. As an SR-based model, Kim proposed a dictionary learning method based on joint image block clustering for multi-modality image fusion. Zhu proposed a medical image fusion method based on cartoon-texture decomposition (CTD), and used an SR-based fusion strategy to fuse the decomposed coefficients [29]. Liu proposed an adaptive sparse representation (ASR) model for simultaneous image fusion and denoising [28]. All the above-mentioned methods propose complex fusion rules or different SR-based models. However, these specific rules cannot be applicable to every type of medical image fusion [27].

With the rapid development of artificial intelligence, deep learning-based image fusion methods have become a hot research topic [30–33]. As a main representative of artificial intelligence, deep learning is developed on the basis of traditional artificial neural networks. It can learn data characteristics autonomously, establish a human-like learning mechanism by simulating the neural network of human brain, and then analyze and learn the related data, such as images and texts [34,35]. CNN as a classical deep learning model can achieve the encoding of direct mapping from source images to weight map during the training process [29,36]. Thus, both activity-level measurements and weight distribution can be achieved together in an optimal way by learning network parameters. In addition, CNN's local connection and weight sharing feature can further improve the performance of image fusion algorithms, while reducing the complexity of entire network and the number of weights. At present, CNN plays an increasingly important role in medical image fusion. Xia integrated multi-scale transform and CNN into a multi-modality medical image fusion framework, which uses the deep stacked neural network to divide source images into high- and low-frequency components to do corresponding image fusion [37]. Liu proposed a CNN-based multi-modality medical image fusion algorithm, which applies image pyramids to the medical image fusion process in a multi-scale manner [38].

The calculation of weight map, which fuses the pixel activity information from different sources, is one of the most critical issues in existing deep learning based image fusion [38]. Most existing fusion methods use a two-step solution that contains activity-level measurement and weight assignment. In traditional transform-domain fusion methods, the absolute value of decomposition coefficient is used to measure its activity first. Then, the fusion rule, such as "choose-max" or "weighted-average", is used to select the maximum or weighted average [39]. According to the obtained measurements, the corresponding weights are finally assigned to different sources. To improve the fusion performance, many complicated decomposition methods and detailed weight assignment strategies have been proposed in recent years [28,40–45]. However, it is not easy to design an ideal activity level measurement or weight assignment strategy, which can consider all key issues [37].

3. The Proposed Medical Image Fusion Solution

As shown in Figure 1, the proposed medical image fusion framework has three main steps. First, it uses Siamese network model to generate the same-size weight map W for any-size source image A and B, respectively. Then, Gaussian pyramid decomposition is applied to the generated weight map W to obtain corresponding multi-scale sub-decomposed image G_W, which is used to determine the fusion operator in coefficient fusion process. $G_{W,l=N}^{l,k}$ and $G_{W,0 \leq l < N}^{l,k}$ are the top layer and the remaining layers of sub-decomposed image. It applies the contrast pyramid to the decomposition of source image A and B. The multi-scale sub-decomposed images C_A and C_B are obtained for the subsequent coefficient fusion process. $C_{A,l=N}^{l,k}$ and $C_{B,l=N}^{l,k}$ are the top layer of sub-decomposed image C_A and C_B, respectively. $C_{A,0 \leq l < N}^{l,k}$ and $C_{B,0 \leq l < N}^{l,k}$ are used to represent the remaining layers of sub-decomposed image C_A and C_B, respectively. Finally, different thresholds are set for the top layer and the remaining layers of sub-decomposed images, respectively. A weighted fusion operator based on the measurement of regional characteristics is used to fuse the different regions in the same decomposition layer to obtain the fused sub-decomposed image C_F. The final fused image F is obtained by the reconstruction of contrast pyramid.

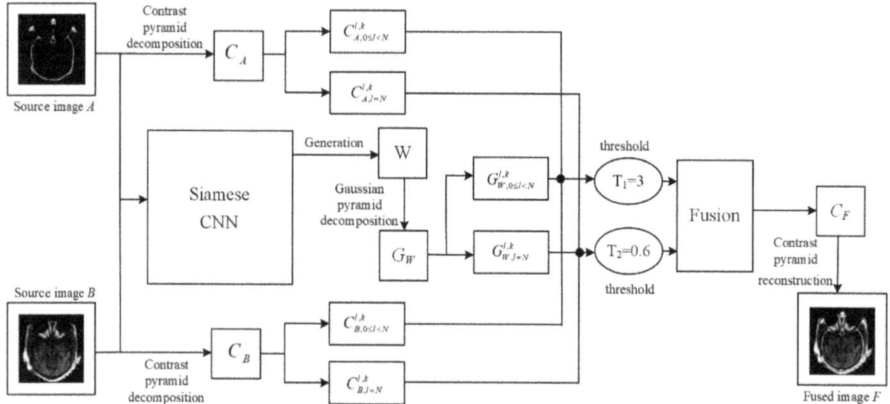

Figure 1. The proposed medical image fusion framework.

3.1. Generation of CNN-Based Weight Map

3.1.1. Network Construction

To obtain the weight map of pixel activity information from multiple source images, the proposed method uses CNN to achieve the measurement of optimal pixel activity level and weight distribution. This paper uses siamese network to improve the efficiency of CNN training. Siamese network has two branches. Each branch contains three convolutional layers and one max-pooling layer. The first two layers are convolutional layers. The first layer is used for the simple feature extraction of input image. In the second layer, the number of feature maps increases. The features of output map in the upper convolutional layer are extracted. The third layer is a max-pooling layer. It removes unimportant samples from feature map to further reduce the number of parameters. As a convolution layer, the fourth layer extracts more complex features from the output map of the pooling layer. To reduce memory consumption, it uses a lightweight network structure to reduce the training complexity. Specifically, the feature map of each branch's final output is concatenated first. Then, the concatenated ones are directly connected to a two-dimensional vector by a fully connected layer. To predict the probability distribution of different characteristics, the two-dimensional vector obtained by mapping is sent to a bi-directional softmax layer, and then classified by probability value. This paper uses the siamese network training architecture shown in Figure 2.

To achieve the classification in CNN network, this paper uses softmax classifier to obtain the classification probability by Equation (1).

$$f(p_i) = \frac{e^{p_i}}{\sum_{j=1}^{n} e^{p_j}} \tag{1}$$

If one p_i is larger than all the other p, then its mapping component is close to 1, and the others are close to 0, which normalizes all input vectors. The batch size is set to 128, thus the softmax loss function is obtained as Equation (2).

$$L = \sum_{i=0}^{batchsize} -\log f(p_i) \tag{2}$$

Taking the softmax loss function as the optimization goal, stochastic gradient descent is used to minimize the loss function. As the initial parameter settings, the momentum and weight decay are set to 0.9 and 0.0005, respectively. Thus, Equations (3) and (4) are used to update the weights.

$$v_{i+1} = 0.9 \cdot v_i - 0.0005 \cdot \alpha \cdot w_i - \alpha \cdot \frac{\partial L}{\partial w_i} \tag{3}$$

$$w_{i+1} = w_i + v_{i+1} \tag{4}$$

where v_i is the dynamic variable, w_i is weight after ith iteration, α is the learning rate, L represents the loss function, and $\frac{\partial L}{\partial w_i}$ is the loss derivative of weight w_i.

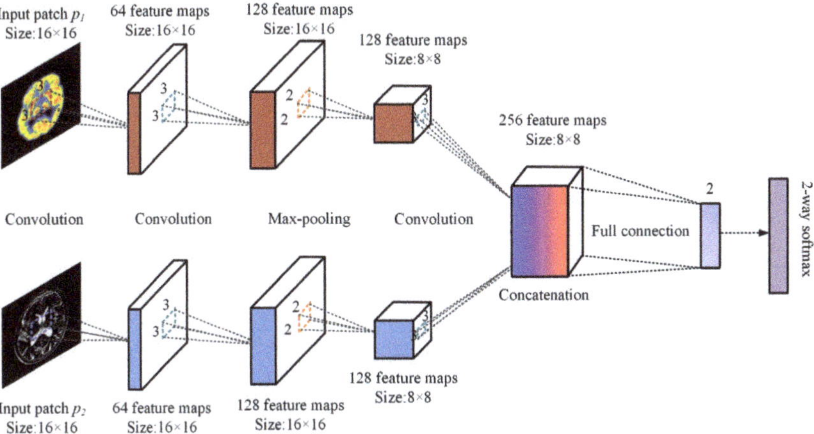

Figure 2. Siamese network training architecture.

3.1.2. Networking Training

It selects a high-quality multi-modality medical image set from http://www.med.harvard.edu/aanlib/home.html as training samples. It applies a Gaussian filter to each image to obtain corresponding five different-level fuzzy versions. Specifically, a Gaussian filter with a standard deviation of 2 and a cutoff value of 7×7 is used. Gaussian filter is used to blur the original image to obtain the first blurred image. In the following Gaussian filtering, the previous output image is used as the next input image. For instance, the output image of first Gaussian filtering is used as the input image of the second Gaussian filtering. Then, for each blurred and clear image, it randomly samples 20 pairs of 16×16 image blocks. p_c and p_b represent a pair of clear and blurred image blocks. When $p_1 = p_c$ and $p_2 = p_b$, it is defined as a positive example (marked as 1), where p_1 and p_2 are the inputs for the first and second branch, respectively. Oppositely, when $p_1 = p_b$ and $p_2 = p_c$, it is defined as a negative example (marked as 0). Therefore, the training set is ultimately composed of positive and negative examples. After the sample is generated, the weight of each convolutional layer is initialized by using Xavier algorithm, which adaptively determines the initialization scale based on the number of input and output neurons. The deviation of each layer is initialized to 0. The inclination rates of all layers are equal, and their initial values are set to 0.0001. When the loss reaches a steady state, the inclination rates are manually reduced to 10% of previous values. After about ten iterations, it can complete the network training.

3.1.3. The Generation of Weight Map W

In the image testing and fusion process, to process any-size source images, it converts the fully connected layer into two equivalent convolutional layers of equal kernel size. When the conversion is completed, any-size image A and B to be fused can be processed as a whole to generate a dense prediction map S. Every prediction S_i is a two-dimensional vector, and the value of each dimension is between 0 and 1. If one dimension is larger than another, this dimension can be normalized to 1, and the other one is set to 0. It simplifies the weight of corresponding image block with an output dimension value of 1. For two adjacent predictions in S, the steps of corresponding image blocks overlap. For overlapping areas, the weights are averaged. The output is the average weight of the overlapping image blocks. In the above way, it is possible to input any-size image A and B into the network, and generate the corresponding same-size weight map W.

3.2. Pyramid Decomposition

This paper uses both contrast pyramid and Gaussian pyramid to decompose source images. It builds the contrast pyramid first. Then, when the Gaussian pyramid is established, G^0 is the zeroth layer (bottom layer), and the l'th layer G^l can be constructed in the following manner. As shown in Equation (5), it convolves G^{l-1} by a window function $w(m, n)$ with low-pass characteristics first, and then downsamples the convolutional result by the interlaced every other row and column.

$$G^l = \sum_{m=-2}^{2} \sum_{n=-2}^{2} w(m, n) G^{l-1}(2i + m, 2j + n), \quad 0 < l \leq N, 0 \leq i < C, 0 \leq j < R_l \tag{5}$$

where $w(m, n)$ is the window function, C_l and R_l are the number of columns and the number of rows in the l'th-layer sub-image of the Gaussian pyramid, respectively, and N is the total number of the pyramid layers.

(1)   Separability: $w(m, n) = w(m)w(n), m \in [-2, 2], n \in [-2, 2]$;

(2)   Normalization: $\sum_{m=-2}^{2} w(n) = 1$;

(3)   Symmetry: $w(n) = w(-n)$; and

(4)   Equal contribution of odd and even terms: $w(-2) + w(2) + w(0) = w(-1) + w(1)$.

According to the above constraints, it can construct $w(0) = 3/8$, $w(1) = w(-1) = 1/4$, and $w(2) = w(-2) = 1/16$. Then, according to Constraint 1, it can get the window function $w(m, n)$ by calculation, as shown in Equation (6).

$$w = \frac{1}{256} \begin{bmatrix} 1 & 4 & 6 & 4 & 1 \\ 4 & 16 & 24 & 16 & 4 \\ 6 & 24 & 36 & 24 & 6 \\ 4 & 16 & 24 & 16 & 4 \\ 1 & 4 & 6 & 4 & 1 \end{bmatrix}. \tag{6}$$

At this point, the image Gaussian pyramid is constructed by $G^0, G^1, \cdots, G^N$.

After the construction of a Gaussian pyramid image by halving the size of each layer one by one, the interpolation method is used to interpolate and expand the Gaussian pyramid. Thus, the expanded lth-layer image G^l and the $l - 1$th-layer image G^{l-1} have the same size and the operation is shown as follows:

$$G^l_*(i, j) = 4 \sum_{m=-2}^{2} \sum_{n=-2}^{2} w(m, n) G^l \left[\frac{m+i}{2}, \frac{n+j}{2} \right], 0 < l \leq N, 0 < i < C_l, 0 < j < R_l \tag{7}$$

$$G^l \left[\frac{m+i}{2}, \frac{n+j}{2} \right] = \begin{cases} G^l \left(\frac{m+i}{2}, \frac{n+j}{2} \right), & \text{when } \frac{m+i}{2}, \frac{n+j}{2} \text{ are integer} \\ 0, & \text{otherwise} \end{cases} \tag{8}$$

where G^l_* is an expansion version of image Gaussian pyramid G^l. According to the above formulas, an expansion sequence is obtained by interpolating and expanding each layer of Gaussian pyramid, respectively.

According to the above formulas, an expansion sequence is obtained by interpolating and expanding each layer of Gaussian pyramid respectively. The decomposition of image contrast is shown as Equation (9).

$$\begin{aligned} C^l &= \frac{G^l}{G^l_*} - I, & N > l \geq 0 \\ C^N &= G^N, & l = N \end{aligned} \tag{9}$$

where C^l is the contrast pyramid, G^l is the Gaussian pyramid, I is the image decomposed by contrast pyramid, and l is the decomposition level, which composes the contrast pyramid $C^0, C^1, \cdots, C^N$ of source image.

Source image A and B are decomposed into corresponding sub-images by contrast pyramid, respectively. For the weight map generated in CNN network, it is decomposed into sub-images by Gaussian pyramid. Different thresholds are set for the top layer and the remaining layers of the obtained sub-images respectively in the fusion processing.

3.3. Fusion Rules

In the fusion process, to obtain better visual characteristics, richer details, and outstanding fusion effects, this paper adopts new fusion rules and the weighted average fusion operators based on regional characteristics. The fusion rules and operators are shown as follows:

(1) After the contrast pyramid decomposition, it calculates the energy E^l_A and E^l_B of corresponding local regions in each decomposition level l of source image A and B, respectively.

$$\begin{aligned} E^l_A(x,y) &= \sum_m \sum_n C^l_A(x+m, y+n)^2 \\ E^l_B(x,y) &= \sum_m \sum_n C^l_B(x+m, y+n)^2 \end{aligned} \tag{10}$$

where $E^l(x,y)$ represents the local area energy centered at (x,y) on the lth layer of contrast pyramid, C^l is the lth-layer image of contrast pyramid, and m and n represent the size of local area.

(2) Calculate the similarity of corresponding local regions in two source images.

$$M^l(x,y) = \frac{2 \sum_m \sum_m C^l_A(x+m, y+n) \, C^l_B(x+m, y+n)}{E^l_A(x,y) \, E^l_B(x,y)} \tag{11}$$

where E^l_A and E^l_B are calculated by Equation (10). The range of similarity is $[-1,1]$, and a value close to 1 indicates high similarity.

(3) Determine the fusion operators. Define a similarity threshold T (when $0 \leq l < N$, $T_1 = 3$; when $l = N$, $T_2 = 0.6$). When $M^l(x,y) < T$, it obtains:

$$\left. \begin{aligned} \text{when } E^l_A(x,y) \geq E^l_B(x,y), C^l_F(x,y) &= C^l_A(x,y); \\ \text{when } E^l_A(x,y) < E^l_B(x,y), C^l_F(x,y) &= C^l_B(x,y); \end{aligned} \right\} \tag{12}$$

when $M^l(x,y) \geq T$, weight map W based weighted mean model is:

$$\left. \begin{array}{l} \text{when } E_A^l(x,y) \geq E_B^l(x,y), \\ C_F^l(x,y) = W_{\max}^l(x,y)\,C_A^l(x,y) + W_{\min}^l(x,y)\,C_B^l(x,y); \\ \text{when } E_A^l(x,y) < E_B^l(x,y), \\ C_F^l(x,y) = W_{\min}^l(x,y)\,C_A^l(x,y) + W_{\max}^l(x,y)\,C_B^l(x,y); \end{array} \right\} \tag{13}$$

where C_F^l is the lth layer of sub-image after fusion.

$$\begin{array}{l} W_{\min}^l(x,y) = G_W^l(x,y) \\ W_{\max}^l(x,y) = 1 - W_{\min}^l(x,y) \end{array} \tag{14}$$

Finally, the integration strategy can be summarized as a whole by Equation (15).

$$C_F^l(x,y) = \left\{ \begin{array}{l} C_A^l(x,y), \\ if\ M^l(x,y) < T \& E_A^l(x,y) \geq E_B^l(x,y); \\ C_B^l(x,y), \\ if\ M^l(x,y) < T \& E_A^l(x,y) < E_B^l(x,y); \\ W_{\max}^l(x,y)\,C_A^l(x,y) + W_{\min}^l(x,y)\,C_B^l(x,y), \\ if\ M^l(x,y) \geq T \& E_A^l(x,y) \geq E_B^l(x,y); \\ W_{\max}^l(x,y)\,C_A^l(x,y) + W_{\min}^l(x,y)\,C_B^l(x,y), \\ if\ M^l(x,y) \geq T \& E_A^l(x,y) \geq E_B^l(x,y); \end{array} \right. \tag{15}$$

According to the above algorithm, when the similarity between the corresponding local regions of source image A and B is less than threshold T, it means that the "energy" difference of two local regions is large. At this time, the central pixel of the region with a larger "energy" is selected as the central pixel of corresponding region in the fused image. Conversely, when the similarity is greater than or equal to threshold T, it means that the "energy" of the region is similar in two source images. At this time, the weighted fusion operator is used to determine the contrast or gray value of the central pixel of the region in the fused image.

Since the central pixel with large local energy represents a distinct feature of source image, the local image features generally do not only depend on a certain pixel. Therefore, the weighted fusion operator based on region characteristics is used, which is more reasonable than other determination methods of fused pixel based on the simple selection or the weight of an independent pixel.

Finally, the decomposed sub-image C_F^l obtained after fusion is inversely transformed by contrast pyramid, which is also called image reconstruction. According to Equation (16), the accurate image reconstruction by contrast pyramid can be obtained.

$$\begin{array}{ll} G^l = \left(C^l + F\right) \odot G_*^l, & N > l \geq 0 \\ G^N = C^N, & l = N \end{array} \tag{16}$$

where $\odot$ denotes Hadamard product (also known as the element-wise multiplication).

The fused image F can be obtained by calculating the above-mentioned image reconstruction formula. Algorithm 1 shows the main steps of the proposed medical image fusion solution.

Algorithm 1 Proposed NSST-based multi-sensor image fusion framework.

Input:

    source image A and B;

    Parameters: pyramid decomposition level l, the number of pyramid levels N, similarity threshold T

Output:

    the fused image F

1: It inputs two any-size source images A and B to the trained siamese network.
2: It generates a dense prediction map S, where each prediction has two dimensions.
3: **for** any prediction S_i **do**
4:     It does normalization processing to obtain a corresponding image block weight with a dimension value of 1.
5: **end for**
6: **for** an overlapping region of two adjacent predictions S_j and S_{j+1} **do**
7:     It does the averaging process to obtain the mean value of the overlapping image block weights.
8:     It outputs the same size weight map W as source image.
9: **end for**
10: **for** each source image A, B, and weight map W **do**
11:     It does pyramid decomposition respectively to obtain a contrast sub-images C_A, C_B and a Gaussian sub-image G_W.
12:     **for** each decomposition level l obtained by the contrast pyramid decomposition of source image **do**
13:         It calculates the energy $E^l_{A,B}(x,y)$ of its corresponding local area.
14:         It determines the similarity of fusion mode $M^l(x,y)$.
15:         It defines a similarity threshold T (when $0 \leq l < N$, T_1=3; when $l = N$, T_2=0.6) to determine the strategy of coefficient fusion.
16:     **end for**
17: **end for**
18: The fused image F is obtained by the inverse pyramid transform of sub-image C^l_F after fusion.

---

4. Comparative Experiments and Analysis

4.1. Experiment Results and Analysis

The following comparative experiments were used to prove that the proposed CNN-based algorithm has good performance in medical image fusion. Eight different image fusion methods were used to fuse MR-CT, MR-T1-MR-T2, MR-PET, and MR-SPECT images, respectively. These eight methods are MST-SR [18], NSCT-PC [9], NSST-PCNN [19], ASR [28], CT [29], KIM [26], CNN-LIU [38], and the proposed solution.

Figure 3 shows the results of MR-CT image fusion experiments. In Figure 3c, the fused image obtained by MST-SR method has a general visualization performance, and the image contrast is high by analyzing the partially enlarged image. As shown in Figure 3d,e, the fused images by NSCT-PC and NSST-PCNN have a high brightness. According to the partial enlargements marked in green and red dashed frame, both methods have the poor performance in the preservation of image details. In Figure 3f,g, the fused images obtained by ASR and CT have low brightness. According to the analysis of details, the detailed information of image edge is not obvious, which is not good for human eye observation. As shown in Figure 3h, the fused image obtained by KIM has low sharpness and poor visual effect. Comparing Figure 3i,j, as well as the partially magnified images, it is difficult to visually distinguish the quality of the fused image obtained by CNN-LIU and the proposed method.

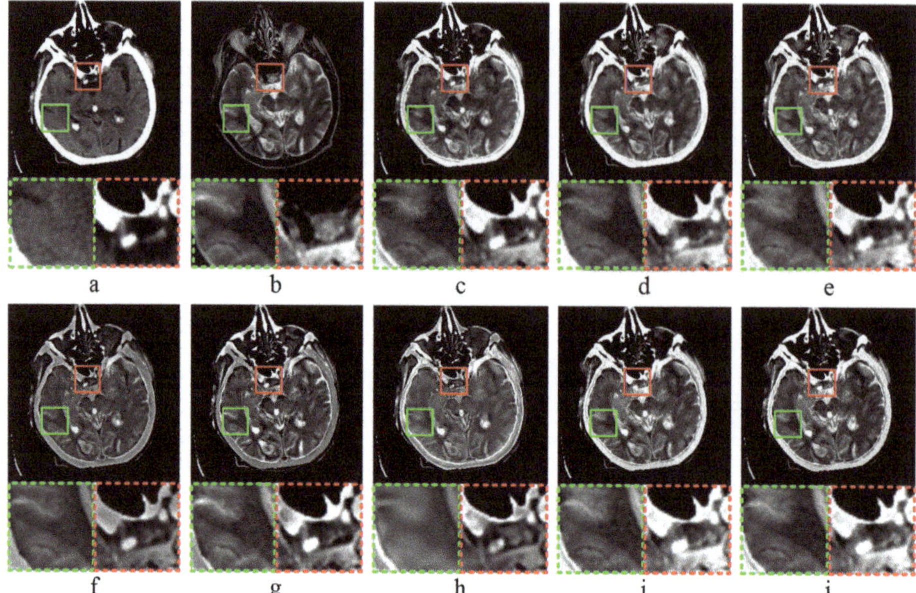

Figure 3. MR-CT image fusion experiments: (**a,b**) source images; and (**c–j**) the fused image obtained by MST-SR, NSCT-PC, NSST-PCNN, ASR, CT, KIM, CNN-LIU, and the proposed method, respectively. Two partially enlarged images marked in green and red dashed frames correspond to the regions surrounded by green and red frames in the fused image.

The results of MR-T1-MR-T2 image fusion experiments are shown in Figure 4. Comparing the fused result (Figure 4c) with source image (Figure 4a,b), the fused image obtained by MST-SR method has a low similarity to source image (Figure 4a), and does not well retain the detailed structure information of source image (Figure 4a). As shown in Figure 4d, the fused image obtained by NSCT-PC method is too smooth in some areas, and the detailed image texture is not sufficiently obvious. In Figure 4e, the fused image obtained by NSST-PCNN method has high brightness, and does not well preserve the detailed features of source images. ASR method obtains the fused image with low contrast and a lot of noises, as shown in Figure 4f. The fused image shown in Figure 4g was obtained by CT method, and has high edge brightness, which weakens the detailed texture information of image edges. According to Figure 4h, the fused image obtained by KIM method has low sharpness, and is blurred. As shown in Figure 4i,j, CNN-LIU and the proposed method reach the almost same visual performance of human eyes.

Figures 5 and 6 show the results of MR-PET image fusion experiments. In both Figures 5c and 6c, the fused images obtained by MST-SR method have high darkness, which is not conducive to human visual observation. According to Figure 5d,e, as well as the partially magnified areas, the fused images obtained by NSCT-PC and NSST-PCNN method have high brightness, and the detailed image information is not clear. Figures 5f,g and 6f,g show that the fused images obtained by ASR and CT methods have low brightness. It means that these two methods have poor performance in the preservation of image details. Comparing Figure 5i,j, the fused image of KIM method has low sharpness, which means the image is blurred. As shown in Figure 5i, the fused image obtained by CNN-LIU method has a low contrast, and the detailed edge information is not obvious. Both Figures 5j and 6j show that the proposed fusion method can preserve the detailed information of source images well, which is conducive to the observation of medical images and the diagnosis of diseases.

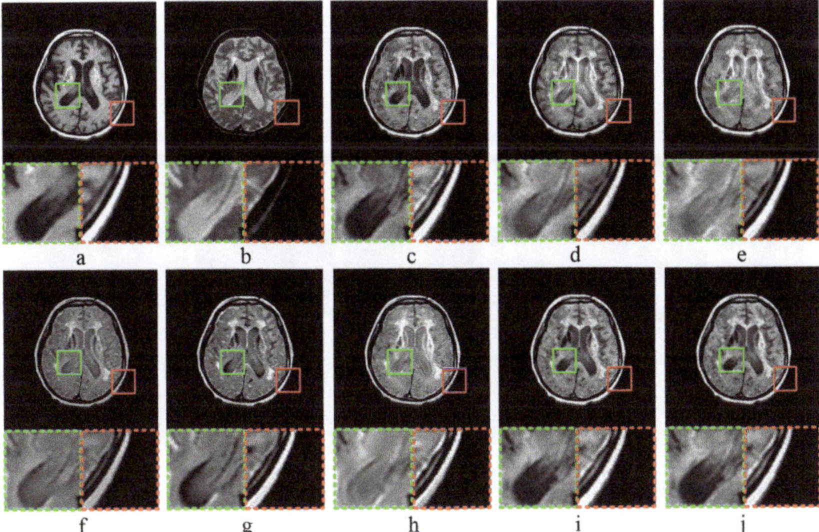

Figure 4. MR-T1 CMR-T2 image fusion experiments: (**a**,**b**) source images; and (**c**–**j**) the fused image obtained by MST-SR, NSCT-PC, NSST-PCNN, ASR, CT, KIM, CNN-LIU, and the proposed method, respectively. Two partially enlarged images marked in green and red dashed frames correspond to the regions surrounded by green and red frames in the fused image.

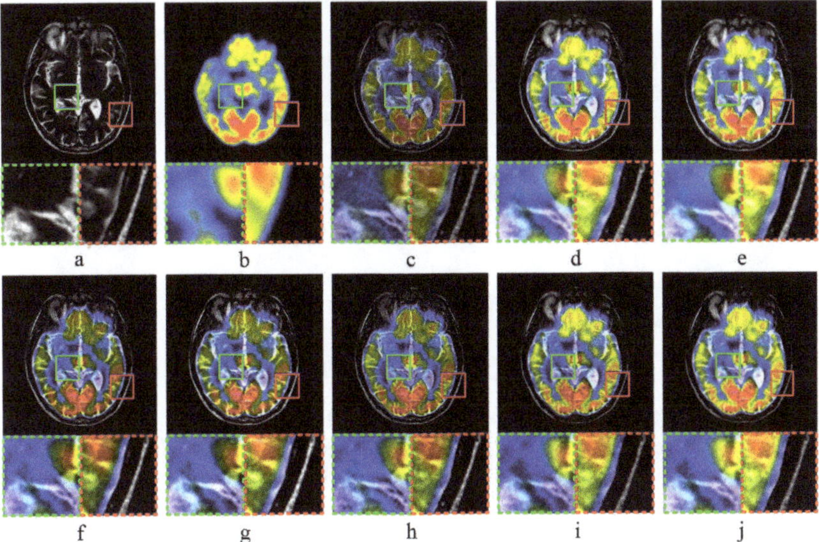

Figure 5. MR-PET image fusion Experiment 1: (**a**,**b**) source images; and (**c**–**j**) the fused image obtained by MST-SR, NSCT-PC, NSST-PCNN, ASR, CT, KIM, CNN-LIU, and the proposed method, respectively. Two partially enlarged images marked in green and red dashed frames correspond to the regions surrounded by green and red frames in the fused image.

Figures 7 and 8 show the results of MR-SPECT image fusion experiments. In Figure 7c, the fused image of MST-SR method has a low contrast and unclear edge details. As shown in Figure 8d,e, some edge regions are too smooth in the fused images obtained by NSCT-PC and NSST-PCNN methods, and the edge details are not clear. In Figures 7g and 8g, the images obtained by CT method have the high contrast, and CT method performs poorly on the detail retention of source images. The fused images shown in Figures 7f,h and 8f,h, which were obtained by ASR and KIM method, respectively, have the low brightness and poor visualization performance. As shown in Figures 7i,j and 8i,j, the fused images obtained by both CCN-LIU and the proposed method have the high brightness and good visualization performance. Comparing all the fused results in Figures 7 and 8, the fused images obtained by the proposed fusion method have the high similarity with source images, which can preserve the detailed structures of source images well and achieve good fusion performance.

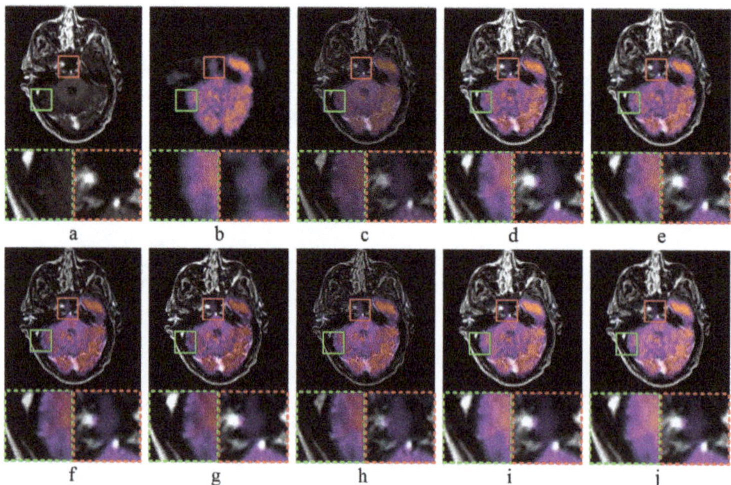

Figure 6. MR-PET image fusion Experiment 2: (**a**,**b**) source images; and (**c–j**) the fused image obtained by MST-SR, NSCT-PC, NSST-PCNN, ASR, CT, KIM, CNN-LIU, and the proposed method, respectively. Two partially enlarged images marked in green and red dashed frames correspond to the regions surrounded by green and red frames in the fused image.

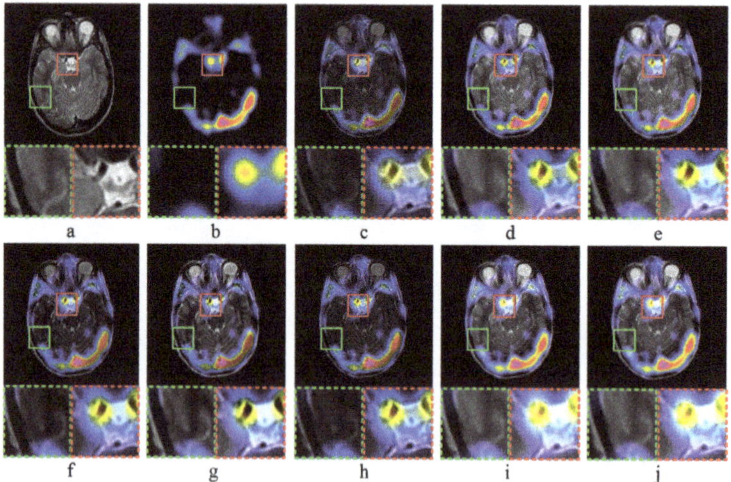

Figure 7. MR-SPECT image fusion Experiment 1: (**a**,**b**) source images; and (**c–j**) the fused image obtained by MST-SR, NSCT-PC, NSST-PCNN, ASR, CT, KIM, CNN-LIU, and the proposed method, respectively. Two partially enlarged images marked in green and red dashed frames correspond to the regions surrounded by green and red frames in the fused image.

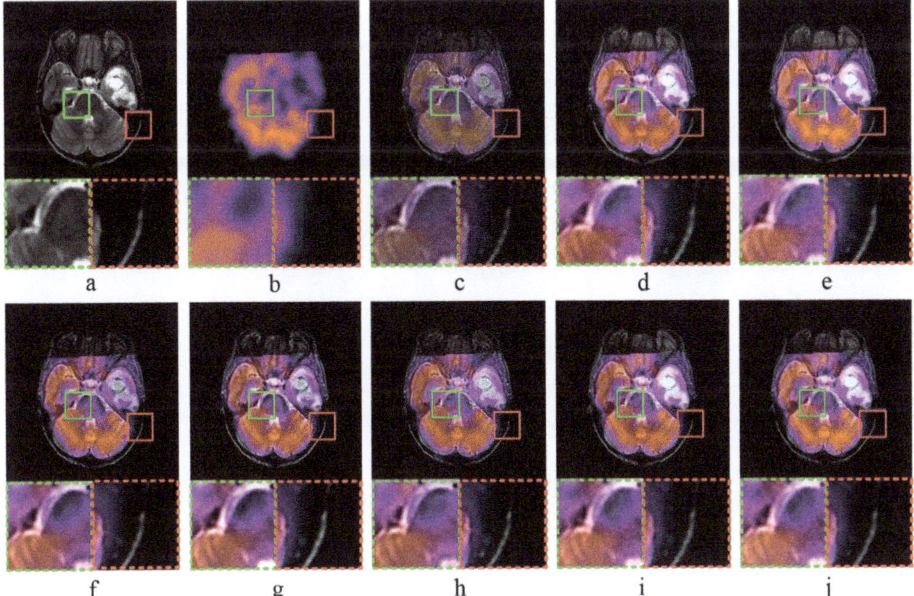

Figure 8. MR-SPECT image fusion Experiments 2: (**a**,**b**) source images; and (**c–j**) the fused image obtained by MST-SR, NSCT-PC, NSST-PCNN, ASR, CT, KIM, CNN-LIU, and the proposed method, respectively. Two partially enlarged images marked in green and red dashed frames correspond to the regions surrounded by green and red frames in the fused image.

4.2. Evaluation of Objective Metrics

For image fusion, a single evaluation metric lacks objectivity. Therefore, it is necessary to do a comprehensive analysis by using multiple evaluation metrics. In this study, four objective evaluation metrics, namely Q^{TE} [46,47], $Q^{AB/F}$[29,48], Q^{MI} [47], and Q^{VIF} [29,49], were used to evaluate the performances of different fusion methods. Q^{TE} is the Tsallis entropy of the fused image. The entropy value represents the amount of average information contained in the fused image. $Q^{AB/F}$ as a gradient-based quality indicator is mainly used to measure the edge information of fused images. Q^{MI} is the mutual information indicator, which is used to measure the amount of information contained in the fused image. Q^{VIF} is the information ratio between the fused image and source images to evaluate the human visualization performance of the fused image. The objective evaluation results of medical image fusion are shown in Figure 9. Among all the fusion results, the proposed method achieves good performance in all four objective evaluations. It confirms that the proposed method can preserve the detailed structure information of source images well and realize good human visual effects.

Table 1 shows the values of four objective metrics for eight fusion methods. The proposed method achieves the highest Q^{TE} value. Comparing with the seven other fusion methods, the fused image obtained by the proposed method has the highest Tsallis entropy, and contains more information than the others. According to the analysis of $Q^{AB/F}$, the fused images obtained by NSCT-PC, NSST-PCNN, CNN-LIU, and the proposed method have high $Q^{AB/F}$, which means these fused images perform well in the preservation of edge details. The fused image obtained by KIM has low $Q^{AB/F}$, which indicates that KIM does not have good performance in the preservation of edge information. For Q^{MI}, the proposed method is a little bit higher than the others. It means more information of source images is retained in the fused image, and the preservation ability of source image details is strong. The proposed method has the highest Q^{VIF}. Comparing with CNN-LIU, the proposed method has a higher information ratio between the fused image and source images, and achieves a better human visual effect as well.

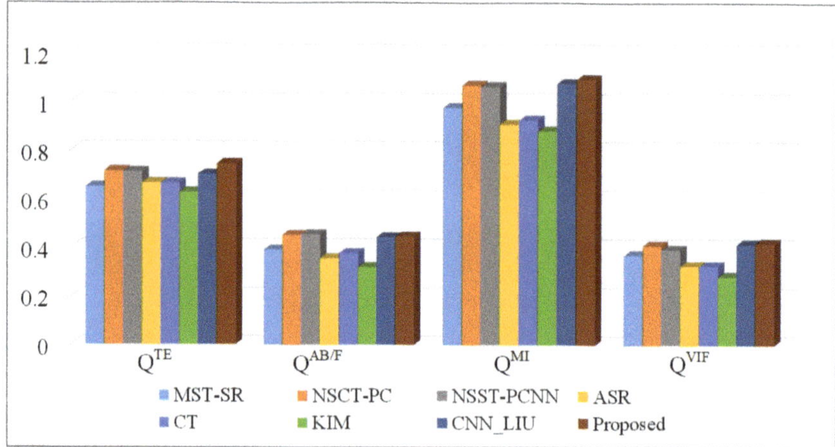

Figure 9. Objective evaluation results of eight fusion methods.

Table 1. Objective evaluations of medical image fusion comparative experiments.

	Q^{TE}	$Q^{AB/F}$	Q^{MI}	Q^{VIF}	Average Processing Time
MST-SR	0.6495	0.3911	0.9764	0.3693	15.0541
NSCT-PC	0.7150	0.4515	1.0681	0.4092	3.7743
NSST-PCNN	0.7113	**0.4537**	1.0610	0.3924	6.1595
ASR	0.6643	0.3550	0.9072	0.3258	35.2493
CT	0.6631	0.3769	0.9240	0.3258	14.5846
KIM	0.6257	0.3188	0.8792	0.2820	59.1929
CNN-LIU	0.7003	0.4421	1.0745	0.4145	14.5846
Proposed	**0.7445**	0.4449	**1.0925**	**0.4181**	12.8667

4.3. Threshold Discussion

In this study, a similarity threshold T was defined to fuse the multi-scale sub-decomposed images. For the top layer of sub-decomposed images, the threshold was set to 0.6. For the remaining layers of sub-decomposed images, the threshold was set to 3. Table 2 shows the values of five objective metrics for the fusion framework with different thresholds. According to Q^{MI}, the proposed method is a little bit lower than others. However, for Q^{TE}, $Q^{AB/F}$, and Q^{VIF}, the proposed method is higher than the others. It means more average information and edge information is contained in the fused image, and it has a higher information ratio between the fused image and source images. In addition, these three methods have close values in terms of time consumption. Overall, the proposed method performs better on five objective metrics.

Table 2. Objective evaluations of the fusion framework with different thresholds.

	Q^{TE}	$Q^{AB/F}$	Q^{MI}	Q^{VIF}	Average Processing Time
Threshold = 0.6	0.6973	0.4248	1.1165	0.4151	12.7256
Threshold = 3	0.7289	0.4255	**1.2540**	0.4068	12.6652
Proposed	**0.7445**	**0.4449**	1.0925	**0.4181**	12.8667

5. Conclusions

This paper proposes a CNN-based medical image fusion solution. The proposed method implements the measurement of activity level and weight distribution by CNN training to generate a weight map including the integrated pixel activity information. To obtain better visual effects, the multi-scale decomposition method based on contrast pyramid is used to fuse corresponding image components in different spatial frequency bands. Meanwhile, the complementary and redundant information of fused images is explored by the local similarity strategy in adaptive fusion mode. Comparative experiment results show that the fused images by proposed method have high visual quality and objective indicators. In the future, we will continue to explore the great potential of deep learning techniques and apply them to other types of multi-modality image fusion, such as infrared-visible and multi-focus image fusion.

Author Contributions: K.W. designed the proposed algorithm and wrote the paper; M.Z. and H.W. participated in the algorithm design, algorithm programming, and testing the proposed method; G.Q. participated in the algorithm design and paper writing processes; and Y.L. participated in the algorithm design, provided technical support, and revised the paper. All authors have read and approved the final manuscript.

Funding: This work was jointly supported by the National Natural Science Foundation of China under Grants 61803061 and 61906026; the National Nuclear Energy Development Project of China (Grant No. 18zg6103); Science and Technology Research Program of Chongqing Municipal Education Commission (Grant No. KJQN201800603); Chongqing Natural Science Foundation Grant cstc2018jcyjAX0167; the Common Key Technology Innovation Special of Key Industries of Chongqing science and Technology Commission under Grant Nos. cstc2017zdcy-zdyfX0067, cstc2017zdcy-zdyfX0055, and cstc2018jszx-cyzd0634; and the Artificial Intelligence Technology Innovation Significant Theme Special Project of Chongqing science and Technology Commission under Grant Nos. cstc2017rgzn-zdyfX0014 and cstc2017rgzn-zdyfX0035.

Conflicts of Interest: The authors declare no conflict of interest.

References

1. Ganasala, P.; Kumar, V. Feature-Motivated Simplified Adaptive PCNN-Based Medical Image Fusion Algorithm in NSST Domain. *J. Digit. Imaging* **2016**, *29*, 73–85. [CrossRef] [PubMed]

2. Li, S.; Kang, X.; Fang, L.; Hu, J.; Yin, H. Pixel-level image fusion: A survey of the state of the art. *Inf. Fusion* **2017**, *33*, 100–112. [CrossRef]

3. James, A.P.; Dasarathy, B.V. Medical image fusion: A survey of the state of the art. *Inf. Fusion* **2014**, *19*, 4–19. [CrossRef]

4. Zhu, Z.; Chai, Y.; Yin, H.; Li, Y.; Liu, Z. A novel dictionary learning approach for multi-modality medical image fusion. *Neurocomputing* **2016**, *214*, 471–482. [CrossRef]

5. Li, H.; He, X.; Tao, D.; Tang, Y.; Wang, R. Joint medical image fusion, denoising and enhancement via discriminative low-rank sparse dictionaries learning. *Pattern Recognit.* **2018**, *79*, 130–146. [CrossRef]

6. Li, Y.; Sun, Y.; Zheng, M.; Huang, X.; Qi, G.; Hu, H.; Zhu, Z. A novel multi-exposure image fusion method based on adaptive patch structure. *Entropy* **2018**, *20*, 935. [CrossRef]

7. Qi, G.; Wang, J.; Zhang, Q.; Zeng, F.; Zhu, Z. An Integrated Dictionary-Learning Entropy-Based Medical Image Fusion Framework. *Future Internet* **2017**, *9*, 61. [CrossRef]

8. Shen, J.; Zhao, Y.; Yan, S.; Li, X. Exposure Fusion Using Boosting Laplacian Pyramid. *IEEE Trans. Cybern.* **2014**, *44*, 1579–1590. [CrossRef]

9. Zhu, Z.; Zheng, M.; Qi, G.; Wang, D.; Xiang, Y. A Phase Congruency and Local Laplacian Energy Based Multi-Modality Medical Image Fusion Method in NSCT Domain. *IEEE Access* **2019**, *7*, 20811–20824. [CrossRef]

10. Li, Y.; Sun, Y.; Huang, X.; Qi, G.; Zheng, M.; Zhu, Z. An Image Fusion Method Based on Sparse Representation and Sum Modified-Laplacian in NSCT Domain. *Entropy* **2018**, *20*, 522. [CrossRef]

11. Xu, L.; Gao, G.; Feng, D. Multi-focus image fusion based on non-subsampled shearlet transform. *IET Image Process.* **2013**, *7*, 633–639. [CrossRef]

12. Qu, X.B.; Yan, J.W.; Xiao, H.Z.; Zhu, Z.Q. Image Fusion Algorithm Based on Spatial Frequency-Motivated Pulse Coupled Neural Networks in Nonsubsampled Contourlet Transform Domain. *Acta Autom. Sin.* **2008**, *34*, 1508–1514. [CrossRef]

13. Bhatnagar, G.; Wu, J.; Liu, Z. Directive Contrast Based Multimodal Medical Image Fusion in NSCT Domain. *IEEE Trans. Multimed.* **2013**, *15*, 1014–1024. [CrossRef]

14. Das, S.; Kundu, M.K. A Neuro-Fuzzy Approach for Medical Image Fusion. *IEEE. Trans. Biomed. Eng.* **2013**, *60*, 3347–3353. [CrossRef]

15. Liu, Z.; Yin, H.; Chai, Y.; Yang, S.X. A novel approach for multimodal medical image fusion. *Expert Syst. Appl.* **2014**, *41*, 7425–7435. [CrossRef]

16. Wang, L.; Li, B.; Tian, L. Multimodal Medical Volumetric Data Fusion Using 3-D Discrete Shearlet Transform and Global-to-Local Rule. *IEEE. Trans. Biomed. Eng.* **2014**, *61*, 197–206. [CrossRef]

17. Yang, Y.; Que, Y.; Huang, S.; Lin, P. Multimodal Sensor Medical Image Fusion Based on Type-2 Fuzzy Logic in NSCT Domain. *IEEE Sens. J.* **2016**, *16*, 3735–3745. [CrossRef]

18. Liu, Y.; Liu, S.; Wang, Z. A general framework for image fusion based on multi-scale transform and sparse representation. *Inf. Fusion* **2015**, *24*, 147–164. [CrossRef]

19. Yin, M.; Liu, X.; Liu, Y.; Chen, X. Medical Image Fusion with Parameter-Adaptive Pulse Coupled Neural Network in Nonsubsampled Shearlet Transform Domain. *IEEE Trans. Instrum. Meas.* **2019**, *68*, 49–64. [CrossRef]

20. Yin, L.; Zheng, M.; Qi, G.; Zhu, Z.; Jin, F.; Sim, J. A Novel Image Fusion Framework Based on Sparse Representation and Pulse Coupled Neural Network. *IEEE Access* **2019**, *7*, 98290–98305. [CrossRef]

21. Zhang, Q.; Liu, Y.; Blum, R.S.; Han, J.; Tao, D. Sparse representation based multi-sensor image fusion for multi-focus and multi-modality images: A review. *Inf. Fusion* **2018**, *40*, 57–75. [CrossRef]

22. Qi, G.; Zhang, Q.; Zeng, F.; Wang, J.; Zhu, Z. Multi-focus image fusion via morphological similarity-based dictionary construction and sparse representation. *CAAI TIT.* **2018**, *3*, 83–94. [CrossRef]

23. Wang, K.; Qi, G.; Zhu, Z.; Cai, Y. A Novel Geometric Dictionary Construction Approach for Sparse Representation Based Image Fusion. *Entropy* **2017**, *19*, 306. [CrossRef]

24. Yang, B.; Li, S. Multifocus Image Fusion and Restoration with Sparse Representation. *IEEE Trans. Instrum. Meas.* **2010**, *59*, 884–892. [CrossRef]

25. Yang, B.; Li, S. Pixel-level image fusion with simultaneous orthogonal matching pursuit. *Inf. Fusion* **2012**, *13*, 10–19. [CrossRef]

26. Kim, M.; Han, D.K.; Ko, H. Joint patch clustering-based dictionary learning for multimodal image fusion. *Inf. Fusion* **2016**, *27*, 198–214. [CrossRef]

27. Li, S.; Yin, H.; Fang, L. Group-Sparse Representation with Dictionary Learning for Medical Image Denoising and Fusion. *IEEE. Trans. Biomed. Eng.* **2012**, *59*, 3450–3459. [CrossRef]

28. Liu, Y.; Wang, Z. Simultaneous image fusion and denoising with adaptive sparse representation. *IET Image Process.* **2015**, *9*, 347–357. [CrossRef]

29. Zhu, Z.; Yin, H.; Chai, Y.; Li, Y.; Qi, G. A novel multi-modality image fusion method based on image decomposition and sparse representation. *Inf. Sci.* **2018**, *432*, 516–529. [CrossRef]

30. Shen, D.; Wu, G.; Suk, H.I. Deep Learning in Medical Image Analysis. *Annu. Rev. Biomed. Eng.* **2017**, *19*, 221–248. [CrossRef]

31. Litjens, G.; Kooi, T.; Bejnordi, B.E.; Setio, A.A.A.; Ciompi, F.; Ghafoorian, M.; van der Laak, J.A.W.M.; van Ginneken, B.; Sánchez, C.I. A Survey on Deep Learning in Medical Image Analysis. *Med. Image Anal.* **2017**, *42*, 60–88. [CrossRef] [PubMed]

32. Zhu, Z.; Qi, G.; Li, Y.; Wei, H.; Liu, Y. Image Dehazing by An Artificial Image Fusion Method based on Adaptive Structure Decomposition. *IEEE Sens. J.* **2020**, *42*, 1–11. [CrossRef]

33. Qi, G.; Chang, L.; Luo, Y.; Chen, Y.; Zhu, Z.; Wang, S. A Precise Multi-Exposure Image Fusion Method Based on Low-level Features. *Sensors* **2020**, *20*, 1597. [CrossRef]

34. Qi, G.; Zhu, Z.; Erqinhu, K.; Chen, Y.; Chai, Y.; Sun, J. Fault-diagnosis for reciprocating compressors using big data and machine learning. *Simul. Model. Pract. Theory* **2018**, *80*, 104–127. [CrossRef]

35. Li, D.; Dong, Y. Deep Learning: Methods and Applications. *Found. Trends Signal Process.* **2014**, *7*, 197–387. [CrossRef]

36. Qi, G.; Wang, H.; Haner, M.; Weng, C.; Chen, S.; Zhu, Z. Convolutional neural network based detection and judgement of environmental obstacle in vehicle operation. *CAAI TIT.* **2019**, *4*, 80–91. [CrossRef]

37. Xia, K.J.; Yin, H.S.; Wang, J.Q. A novel improved deep convolutional neural network model for medical image fusion. *Cluster Comput.* **2018**, *22*, 1515–1527. [CrossRef]

38. Liu, Y.; Chen, X.; Cheng, J.; Peng, H. A medical image fusion method based on convolutional neural networks. In Proceedings of the 20th International Conference on Information Fusion (Fusion), Xi'an, China, 10–13 July 2017. [CrossRef]
39. Li, H.; Li, X.; Yu, Z.; Mao, C. Multifocus image fusion by combining with mixed-order structure tensors and multiscale neighborhood. *Inf. Sci.* **2016**, *349-350*, 25–49. [CrossRef]
40. Shen, R.; Cheng, I.; Basu, A. Cross-Scale Coefficient Selection for Volumetric Medical Image Fusion. *IEEE. Trans. Biomed. Eng.* **2013**, *60*, 1069–1079. [CrossRef]
41. Singh, R.; Khare, A. Fusion of multimodal medical images using Daubechies complex wavelet transform-A multiresolution approach. *Inf. Fusion* **2014**, *19*, 49–60. [CrossRef]
42. Zhu, Z.; Qi, G.; Chai, Y.; Li, P. A Geometric Dictionary Learning Based Approach for Fluorescence Spectroscopy Image Fusion. *Appl. Sci.* **2017**, *7*, 161. [CrossRef]
43. Bhatnagar, G.; Wu, Q.M.J.; Liu, Z. A new contrast based multimodal medical image fusion framework. *Neurocomputing* **2015**, *157*, 143–152. [CrossRef]
44. Li, H.; Yu, Z.; Mao, C. Fractional differential and variational method for image fusion and super-resolution. *Neurocomputing* **2016**, *171*, 138–148. [CrossRef]
45. Li, H.; Liu, X.; Yu, Z.; Zhang, Y. Performance improvement scheme of multifocus image fusion derived by difference images. *Signal Process.* **2016**, *128*, 474–493. [CrossRef]
46. Cvejic, N.; Canagarajah, C.; Bull, D. Image fusion metric based on mutual information and Tsallis entropy. *Electron. Lett.* **2006**, *42*, 626–627. [CrossRef]
47. Liu, Z.; Blasch, E.; Xue, Z.; Zhao, J.; Laganiere, R.; Wu, W. Objective Assessment of Multiresolution Image Fusion Algorithms for Context Enhancement in Night Vision: A Comparative Study. *IEEE Trans. Pattern Anal. Mach. Intell.* **2012**, *34*, 94–109. [CrossRef]
48. Petrović, V. Subjective tests for image fusion evaluation and objective metric validation. *Inf. Fusion* **2007**, *8*, 208–216. [CrossRef]
49. Sheikh, H.; Bovik, A. Image information and visual quality. *IEEE Trans. Image Process.* **2006**, *15*, 430–444. [CrossRef]

Article

Target Recognition in Infrared Circumferential Scanning System via Deep Convolutional Neural Networks

Gao Chen * and Weihua Wang

National Key Laboratory of Science and Technology on ATR, National University of Defense Technology, Changsha 410073, China; atrwwh@126.com
* Correspondence: chengao18a@nudt.edu.cn; Tel.: +86-182-8029-7980

Received: 5 March 2020; Accepted: 27 March 2020; Published: 30 March 2020

Abstract: With an infrared circumferential scanning system (IRCSS), we can realize long-time surveillance over a large field of view. Recognizing targets in the field of view automatically is a crucial component of improving environmental awareness under the trend of informatization, especially in the defense system. Target recognition consists of two subtasks: detection and identification, corresponding to the position and category of the target, respectively. In this study, we propose a deep convolutional neural network (DCNN)-based method to realize the end-to-end target recognition in the IRCSS. Existing DCNN-based methods require a large annotated dataset for training, while public infrared datasets are mostly used for target tracking. Therefore, we build an infrared target recognition dataset to both overcome the shortage of data and enhance the adaptability of the algorithm in various scenes. We then use data augmentation and exploit the optimal cross-domain transfer learning strategy for network training. In this process, we design the smoother L1 as the loss function in bounding box regression for better localization performance. In the experiments, the proposed method achieved 82.7 mAP, accomplishing the end-to-end infrared target recognition with high effectiveness on accuracy.

Keywords: infrared circumferential scanning system; target recognition; deep convolutional neural networks; data augmentation; transfer learning; bounding box regression; loss function

---

1. Introduction

Objects with a temperature above absolute zero can continuously emit electromagnetic radiation into outer space. At room temperature, these radiations mainly concentrate in the infrared band. The infrared radiation emitted or reflected by the target is captured by thermal imagers to obtain high-contrast imaging results, which are usually grayscale images. Compared with visible cameras, the thermal infrared imager can work at night and has a specific ability to distinguish the true and false targets, because it relies on the difference in temperature and emissivity between the target and the ground. Compared with radar images, infrared images are more recognizable to human eyes because of the shorter wavelength of infrared radiation. Additionally, the thermal imager has passive characteristics, only receiving the target radiation, and does not need to transmit a signal, which means high concealment. With the advantages of all-day work, high concealment, and sensitivity, thermal infrared imagers are widely used to collect information in a variety of complex environments, and to achieve 24-h surveillance, especially in the defense system, which can provide supports for battlefield decision-making in modern warfare [1,2].

The infrared circumferential scanning system (IRCSS), equipped with a long linear infrared focal plane array (IRFPA), performs circum-sweep motion under precise servo control to realize circum-sweep imaging at multiple pitching angles [3]. Moreover, the horizontal coverage is close to 360 degrees, greatly expanding the detection range of infrared detectors. The typical imaging resolution of IRCSS can reach $768 \times 120,000$, which is much higher than that of the conventional forward-looking infrared system [4]. With the rapid development of image processing technology, the IRCSS has the capabilities of searching over a large field of view, long-range target automatic recognition, multi-target high-precision tracking, and integration with mechanical control systems.

When the field of view expands, there will be more interference in the imaging results of the IRCSS, such as clouds in the low-altitude background, and mountains and trees in the ground background, all of which make target recognition more challenging. However, most of the previous works only finished the target detection [5,6]. Our method is to upgrade the function of the IRCSS. Target recognition involves two levels of understanding of the target. Firstly, where is the target? We need to perform detection, deciding whether the target is in range, and localize it if it is. Secondly, what is the target? We need to perform identification, classifying the target. In the past, traditional hand-crafted methods took different algorithms for two subtasks—for example, threshold segmentation [7] for target localization, a histogram of oriented gradients (HOG) [8] for features extraction, and a support vector machine (SVM) [9] for classification. The representative work in this paradigm is the deformable part models (DPM) [10]. However, independence between algorithms becomes an obstacle to further enhance recognition performance. Since AlexNet [11] became the preferred option in ILSVRC 2012, there has been a strong interest in deep learning for computer vision, especially the deep convolutional neural network (DCNN). The DCNN has the structure of stacking various blocks, such as convolution layers, activation functions, pooling layers, and fully connected layers, which endows the network with a powerful feature extraction ability to automatically and adaptively obtain deep semantic information. In the target recognition field, DCNN-based methods utilize a network to accomplish two subtasks, realizing end-to-end processing. In learning-based models, DCNN-based methods need large-scale datasets, including images and annotations, to train the networks with a strong generalization ability. Owing to the access to large-scale datasets for the public, such as ImageNet [12], COCO [13], VOC [14], the performance of DCNN-based methods on visible images has been dramatically improved. However, in the infrared field, publicly available datasets are mostly used for target tracking, which are usually sequences of infrared images. Unfortunately, these datasets are unsuitable for infrared target recognition.

In this paper, as shown in Figure 1, we propose a DCNN-based method to address target recognition in the IRCSS. In the experiments, results demonstrated the effectiveness of the proposed method on accuracy with 82.7 mAP. The contributions in this paper can be summarized as follows:

1. We realize end-to-end target recognition on high-resolution imaging results of the IRCSS via the DCNN.
2. We build an infrared target recognition dataset to both overcome the shortage of data and enhance the adaptability of the algorithm in various scenes, including two types of targets in seven types of scenes with two types of aspect orientations, four types of sizes and twelve types of contrasts.
3. We design a loss function called the smoother L1 in the bounding box regression for better localization performance.

The rest of the paper is organized as follows. Section 2 reviews related works about target recognition and tracking algorithms. In Section 3, we describe the methodology of this paper. Section 4 is the experiment part, including experiment details, results, and analysis. Section 5 provides the conclusion and a plan for future work.

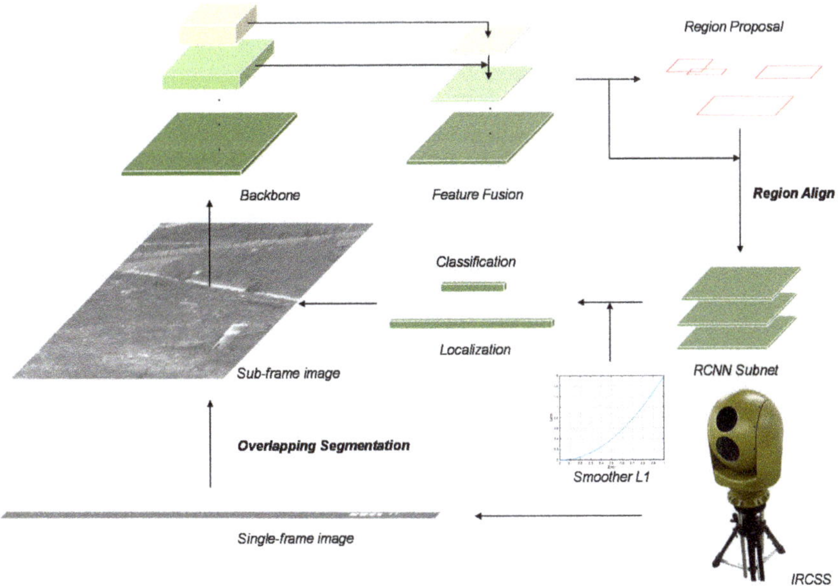

Figure 1. The overall architecture of the proposed method to address target recognition in the infrared circumferential scanning system (IRCSS). We perform the overlapping segmentation on the single-frame image of IRCSS. After getting the sub-frame image, it is sent to the recognition network. The backbone structure is detailed in Section 4. The feature fusion follows the design in the feature pyramid network (FPN) [15]. The region proposal network (RPN) and the region convolutional neural network (RCNN) subnet follows the design of the Faster RCNN [16]. The region proposals are processed by RoI align [17]. In the bounding box regression for target localization, the loss function is the smoother L1.

2. Related Work

2.1. Target Recognition and Tracking in Infrared Images

Thermal infrared sensors are not influenced by illumination variations and shadows, and objects can be distinguished from the background as the background is normally colder [18]. Considering these advantages and the demand for realistic, computer vision tasks in infrared images have emerged, such as target recognition and tracking.

Target recognition can be divided into two subproblems: target detection and identification—in other words, localizing the target in the image and figuring out its category. Target identification can be further divided into feature extraction and classifier design. The sliding window approach was the most straightforward way to localized the target [19]. It slid a window over the image to obtain image patches, and the target recognition model was then used to classify each patch covered by the window. To overcome the limitations of expensive computation of the sliding window approach, the selective search [20] approach was used to segment the image into original regions using the algorithm in [21] and then grouped similar regions based on color, texture, size, and shape compatibility. This process was repeated until the number of iterations was reached. For target identification, several methods have been proposed. In [22], a sparse representation-based classification (SRC) algorithm was proposed for infrared target recognition. In [23], the HOG and bag-of-words (BoW) was applied to further improve performance. With respect to the IRCSS, previous methods have finished the detection of targets. A time-domain multi-frame cumulative difference algorithm was proposed to detect the dim and small target in the large field of view and complicated background [6]. In [5], a rough-to-meticulous target detection algorithm was proposed for panorama infrared images. In the rough detection phase, the integrating processing of morphological filtering and interframe differences was utilized to pick up suspected targets most rapidly from high-resolution images and suspected target image slices were generated. In the meticulous detection phase, permanent false alarm adaptive threshold method and feature fusion were adopted to eliminate false alarm and generate a trajectory for the real targets.

Based on the annotation of a target only on the first frame of the video, target tracking aims to estimate a moving trajectory [24]. A discriminative correlation filter (DCF)-based tracker learns a correlation filter from annotations to discriminate the target from the background [25]. Even after several years, this branch is still flourishing in the tracking field. With respect to infrared tracking, in VOT-TIR2017 [26], which is a challenge on tracking in thermal infrared sequences, the winner DSLT [27] applied optical flow and extended Struck [28] with the ability to learn from dense samples and high dimensional features. The top accuracy tracker SRDCFir [29] introduced a spatial regularization function that penalized filter coefficients residing outside the target region to alleviate the periodic assumption when using circular correlation. Another branch of infrared tracking fuses infrared images and visible images. In the corresponding VOT-RGBT2019 challenge [30], the DCF-based tracker JMMAC [30] designed a robust RGBT (RGB and thermal) tracker that combined motion cues and appearance cues. The motion cue was inferred from key-point-based camera motion estimation and a Kalman filter applied to object motion. The appearance cues are generated by an extension of the efficient convolution operators (ECO) model [31]. In this paper, we are interested in target recognition. In the next part, we introduce an overview of DCNN-based target recognition methods.

2.2. DCNN-Based Target Recognition

Feature extraction plays an important role in both target recognition and tracking. The traditional hand-crafted features have been used in various modalities images [32]. Over the past few years, DCNN-based method has outperformed the traditional approaches in various computer vision domains, such as image classification, target recognition, and semantic segmentation, because of the strong ability of feature extraction.

DCNN-based methods utilize a single network to accomplish two subtasks in target recognition. According to the number of stages, they can be divided into two branches: two-stage methods and one-stage methods.

The two-stage methods, which are also known as classification-based methods, divide the recognition into two stages. In the first stage, the network selects target region proposals from the predefined boxes over the image and refines their position coordinates. This process can be regarded as binary classification. Each anchor is classified as the target-in or target-out. In the second stage, each proposal is classified and refined again. In general, benefiting from a region proposal network and two-time position refinement, two-stage methods have relatively higher precision. Faster RCNN [16] firstly finalized the above-mentioned workflow. After that, the feature pyramid network (FPN) [15] was proposed to fuse semantic information from deep layers and location information from shallow layers. Cascade RCNN [33] utilized three thresholds for better region proposals in the first stage. DetNAS [34] adopted the neural architecture search [35] to find the optimal architecture of the recognition network. CBNet [36] proposed a strategy of compositing connections between the adjacent backbones to build a more powerful backbone network than ResNet [37] and ResNeXt [38], which achieved the best 53.3 mAP on the COCO benchmark [13].

Meanwhile, the one-stage method can directly predict coordinates and categories of the targets by a multi-tasks loss function, which is also called the regression-based method. The basic architecture consists of the backbone network and detection subnet. Owing to less computation, one-stage methods are mostly proposed to achieve a faster speed of recognition. If the limit is 20FPS, YOLO [39] is the first to realize real-time target recognition. As an improved version, YOLO v3 [40] significantly improved the precision while maintaining the speed and has been widely used in realistic circumstances. SSD [41] was used to make predictions on feature maps of different scales. EfficientDet [42] utilized

EfficientNet [43] and BiFPN to develop a family of networks, among which EfficientDet-D7 achieved 51.0 mAP on the COCO benchmark [13].

Most DCNN-based target recognition algorithms are mostly proposed for RGB images. When used in the infrared system, diversities between characteristics of images may cause more problems, making recognition more challenging. In [44], a DCNN-based detector was designed in the infrared small unmanned aerial vehicle (SUAV) surveillance system by the laterally connected multi-scale feature fusion approach and densely paved predefined boxes. In [45], SVM and DCNN classification for infrared target recognition were compared. Some literature only takes CNN as feature extractors. In [46], CNN cooperated with the difference of Gaussian (DoG) to recognize the target. In [47], a compact and fully CNN was trained with synthetic data because of the shortage of infrared data. The trained network was used to address target recognition in an infrared defense system. In this paper, we propose a two-stage method to accomplish end-to-end target recognition in the IRCSS.

3. Methodology

We propose a DCNN-based two-stage method for target recognition in the IRCSS. Figure 1 shows the overall architecture. To be specific, owing to the large size of the single-frame image in the IRCSS, we perform overlapping segmentation on it to obtain sub-frame images. We then build an infrared target recognition dataset to both overcome the shortage of data and enhance the adaptability of the algorithm in various scenes. Furthermore, we adopt data augmentation to expend the dataset and exploit the optimal cross-domain transfer learning strategy for the train. In the network, we design a novel loss function in bounding box regression for target localization, called the smoother L1.

3.1. Sub-Frame Images of the IRCSS

An IRCSS consists of two parts: an infrared detector with a long linear IRFPA and a mechanical structure. Under precise control by a servomotor, the detector performs uniform rotation to obtain circumferential images. Compared with the simple staring thermal imager, the system can provide a large field of view and continuous circumferential images, which can be applied in environmental monitoring and night navigation.

As shown in Figure 2, the size of the circumferential image obtained by the IRCSS is much larger than that of the single-frame image obtained by the traditional staring infrared detector or the visible camera, up to $768 \times 40,000$. Directly handling the single-frame image, the efficiency of the algorithm can be slow because of the massive amount of data. In order to solve the contradiction between data quantity and algorithm efficiency, the method of sub-frame images is proposed.

Figure 2. The single-frame image obtained by the IRCSS. In this paper, its size is $768 \times 40,000$. To make it clearer to display, we zoom in on a helicopter target.

The single-frame circumferential image is divided into several blocks to reduce the amount of data processed by the algorithm, and the target recognition is carried out on each image block, called a sub-frame image. According to the sequence of obtaining the single-frame and sub-frame image, the methods can be divided into direct and indirect acquisition. The direct acquisition means that in the imaging process, the sub-frame image is directly obtained through the rotation of the IRCSS with an equal angle, so the complete single-frame image is no longer stored. The indirect acquisition means that a single-frame is firstly obtained, and the sub-frame is then obtained through segmentation.

The existing data are complete circumferential images, so sub-frame images are obtained by indirect acquisition. To be specific, we divide a circumferential image into several 768×768 image blocks. Figure 3 shows the acquisition of subframe images. In the case of direct segmentation, some targets can be segmented into different blocks, causing problems in recognition. In order to solve this, there is an overlapping area between contiguous blocks during segmentation, such that one target is complete in at least one block. The size of the overlapping area is selected according to the maximum size of the target.

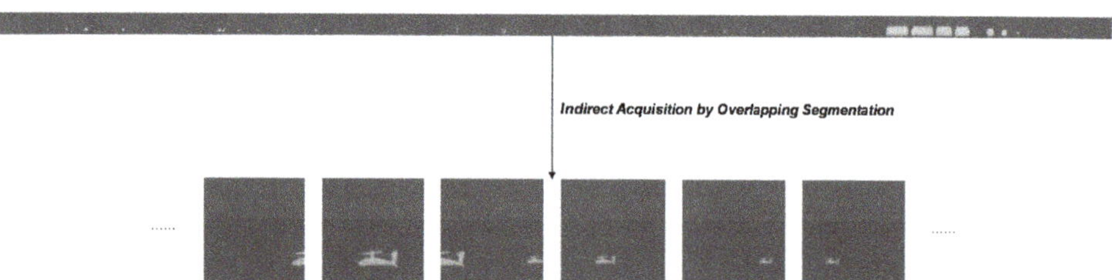

Figure 3. The indirect acquisition of sub-frame images by overlapping segmentation. One target is complete in at least one sub-frame image.

3.2. Infrared Target Recognition Dataset

As a type of data-driven algorithms, the DCNN needs a large amount of training data to ensure generalization performance, so that it can cope with the changes of the target itself and the scene. In the field of object recognition, training data refer to the images that contain targets and the annotations that describe the location and category of each target. It is usually expected to collect the data of different states of the targets in as many scenes as possible. However, the commonly used target recognition datasets are composed of visible (RGB) images, such as ImageNet, COCO, and VOC. In the infrared field, most of the datasets are used for target tracking, such as the VOT-TIR challenge [26,48], which consists of small sequences of infrared images containing targets. In these sequences, the target is of a single type, like pedestrians or vehicles, and the size and brightness of the target is almost unchanged, while the scene is also almost unchanged. If the training set and test set are determined by dividing a sequence randomly, the diversity between them can be too limited to guarantee the generalization of the network, although the performance on the test dataset can be noteworthy. Meanwhile, as shown in Figure 3, the background of the target in the existing data is too simple; if training is based only on this, the adaptability to the scene of the algorithm will be weak. In order to ensure that targets can be detected in infrared images of different scenes, we built an infrared target recognition dataset, including aspect orientation, size, contrast, and scene changes.

As shown in Figure 4, we separated targets from the existing infrared data, including two-aspect orientations and 12 contrasts of each types of targets, and selected frames as the background from image sequences of seven types of scenes [26,49], including a road, trees, a desert, grassland, a mountain, buildings, and cars, so that there was still diversity among backgrounds of the same scene.

When embedded in the background, each target was scaled to four sizes to simulate different distances from the detector. Thus, each type of scene contains 192 images, and the dataset has a total of 1344 images, some of which are shown in Figure 5. Compared with ImageNet and COCO, the dataset we built is too small, so we performed data augmentation, and details are shown in Section 4.

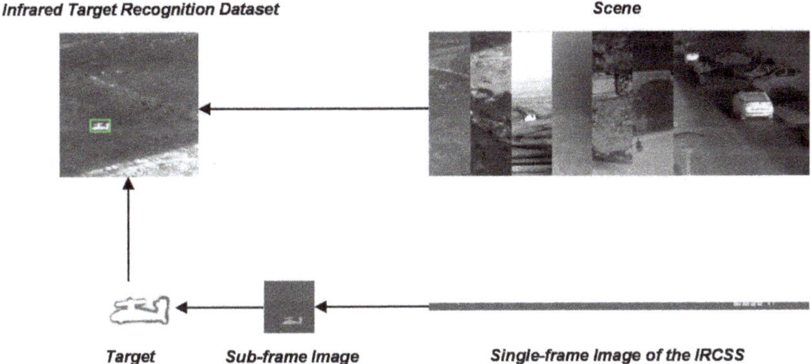

Figure 4. The process of building the infrared target recognition dataset. After getting the sub-frame image, we separated the target. The target was then embedded into the scene images.

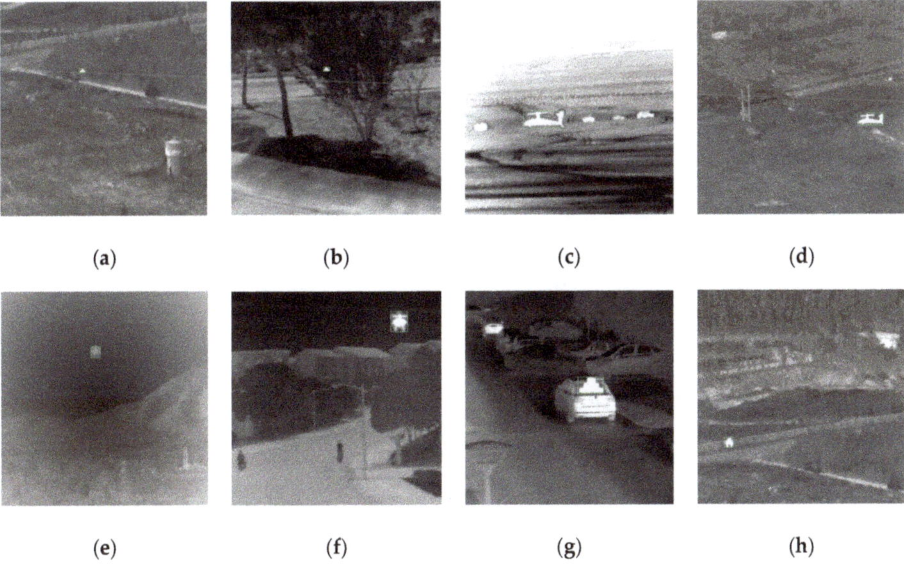

Figure 5. Some of the infrared target recognition dataset, including changes of target type, aspect orientation (front and side), size, contrast, and scene (road (**a**), (**h**), tree (**b**), desert (**c**), grassland (**d**), mountain (**e**), building (**f**), and car (**g**)).

If we train on our dataset from scratch, it will easily lead to overfitting. The results in Section 4 verify this assumption. Therefore, the cross-domain transfer learning is utilized. The network used for target recognition includes a backbone network for obtaining feature maps of the input image. After training, the features extracted by the shallow layer of the backbone are common for different targets, which generally are structural features, such as edges and angles [50]. Therefore, the weight obtained from training on a large dataset, called the source domain, can be transferred to the backbone network, and we continue training with our customized dataset, called the target domain, to finetune the weight. In this way, we can not only enhance the generalization ability of the DCNN-based algorithm in the target domain but also avoid overfitting.

Different source domain can produce different initial weights, and their finetuning effect on the target domain can also be different. In [51], the distribution of the relative size of the target in the

ImageNet and COCO dataset was statistically analyzed. In ImageNet, the median relative size of the target is 0.556, while it is 0.106 in COCO, which means there are more small targets in the COCO dataset. Additionally, in COCO, the relative sizes of the maximum 10% target and the minimum 10% target differ by 20 times, which is much more than that of ImageNet, which means targets in the COCO dataset have a more extreme scale variation. In Section 4, we exploited the optimal cross-domain transfer learning strategy by experiments.

3.3. Smoother L1

When training a DCNN for target recognition, we define a multi-task loss function to solve both classification and localization:

$$L = L_{cls} + L_{loc}. \tag{1}$$

The target location is realized by bounding box regression, and the objective function is a distance function between the prediction and the ground truth, which is also the target of regression, of the network:

$$L_{loc} = \sum_i \sum_{D \in \{x,y,w,h\}} distance\left(p_D^i, t_D^i\right). \tag{2}$$

Here, i represents the index of a region proposal participating in the regression, and we drop the superscript unless it is needed; D represents the four dimensions of a box coordinate, which is the abscissa and ordinate of the center of box and the width and height of the box; p represents the prediction; t represents the target. The specific definitions are as follows:

$$\begin{cases} p_x = \omega_x^T \phi(r) \\ p_y = \omega_y^T \phi(r) \\ p_w = \omega_w^T \phi(r) \\ p_h = \omega_h^T \phi(r) \end{cases} \tag{3}$$

$$\begin{cases} t_x = \frac{G_x - R_x}{R_w} \\ t_y = \frac{G_y - R_y}{R_h} \\ t_w = \log\left(\frac{G_w}{R_w}\right) \\ t_h = \log\left(\frac{G_h}{R_h}\right) \end{cases} \tag{4}$$

where ω_D (where D is one of x, y, w, h) represents the network parameters to be learned; $\phi(\cdot)$ represents the calculation of the DCNN, r represents the region proposal, and $\phi(r)$ represents the features calculated by the DCNN of a region proposal; R_D represents the coordinate of a region proposal; G_D represents the coordinate of the corresponding ground-truth box.

In training, the gradient descent is utilized to minimize the distance between prediction and target, which can also be called loss. We take 1 as the boundary of error. Hence, each sample can be classified as an inlier (< 1) or outlier (> 1). For the definition of loss functions, the L2 loss is adopted in the RCNN [52].

$$\text{L2 Loss} = x^2. \tag{5}$$

Because of the unlimited gradient of the L2 norm, the learning rate needs to be set very carefully in training to avoid the gradient explosion caused by outliers. In order to enhance the robustness of the loss function, smooth L1 is first utilized in Fast RCNN [53], which connects L2 with L1 by taking 1 as the boundary.

$$\text{L1 Loss} = |x| \tag{6}$$

$$\text{smooth L1} = \begin{cases} 0.5x^2, |x| < 1 \\ |x| - 0.5, others \end{cases}. \tag{7}$$

Thus, both the undifferentiability of L1 at 0 and the sensitivity of L2 to outliers are solved. However, during training, we realized that the value of gradient participates in the update of network parameters rather than the value of loss function. Consequently, we should pay more attention to the gradient when designing the loss function. In this paper, we propose a smoother L1.

As shown in Figure 6, compared with the smooth L1, the gradient of the smoother L1 changes from a linear function to a power function for inliers and remains as a constant for outliers. By this design, the nonlinearity, which can be regarded as the core of deep learning, of the network is enhanced. On the other hand, the transition of the gradient between outliers and inliers becomes smoother to avoid the large change of the gradient during the training, which is generally considered harmful to training.

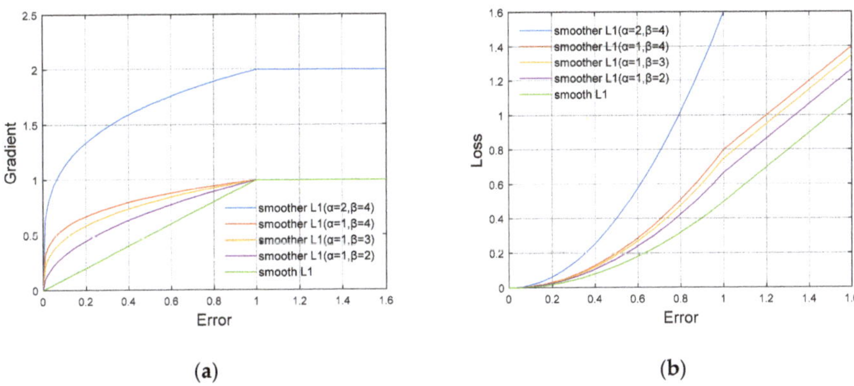

(a)   (b)

Figure 6. The curves of the smoother L1 with different α and β. We set α to control the gradient of outliers, and β to control the changing trend of the gradient of inliers. (**a**) The gradient curve of the smoother L1. (**b**) The loss curve of the smoother L1.

Meanwhile, according to the ablation studies, smoother L1 inherently alleviates the imbalance between classification loss and localization loss. The gradient formula is as follow:

$$\frac{\partial \text{smoother L1}}{\partial x} = \begin{cases} \alpha |x|^{1/\beta}, |x| < 1 \\ \alpha, others \end{cases}, \tag{8}$$

where α controls the gradient of outliers, and β controls the changing trend of the gradient of inliers. The larger or smaller the β, the closer it is to L1 loss or the smooth L1. L1 loss can be regarded as the smoother L1 as $\alpha = 1$, $\beta \to \infty$, and the smooth L1 can be regarded as the smoother L1 as $\alpha = 1$, $\beta = 1$. For the optimal setting of α and β, we did a coarse grid search in the experiment. We integrated the gradient formula to obtain the formula of the smoother L1:

$$\text{smoother L1} = \begin{cases} \frac{\alpha \cdot \beta}{\beta+1} |x|^{\frac{\beta+1}{\beta}}, |x| < 1 \\ \alpha \cdot |x| - \frac{\alpha}{\beta+1}, others \end{cases}. \tag{9}$$

4. Experiments

4.1. Implementation Details

We adopt data augmentation to expand the quantity of images. As shown in Figure 7, we perform horizontal flipping, Gaussian noise, rotation for each image. The dataset eventually contains 4032 images. It is divided into 2822 images for training, 403 images for validation, and 807 images for testing. All experiments are implemented on a Lenovo Linux PC with an Nvidia RTX2060 GPU and

Intel i7-9750 CPU. If not specifically noted, the batch size is set to 2 and every epoch contains 1411 iterations. We train all networks for 12 epochs, with the learning rate increasing linearly to 0.0025 in the first 500 iterations and decreasing by 0.1 after 8 and 11 epochs, respectively.

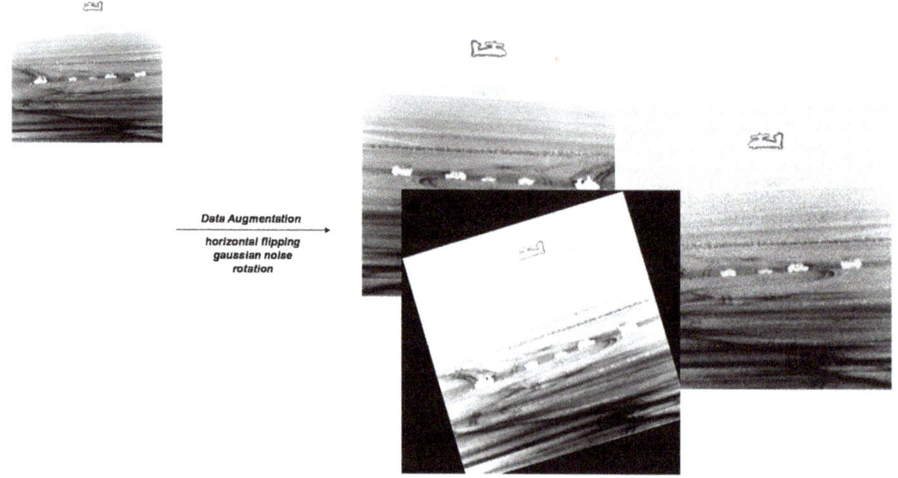

Figure 7. The data augmentation for each image, including horizontal flipping, Gaussian noise, rotation.

The evaluation metrics of recognition are standard COCO-style average precision (AP) [13], which is a mixture metric of widely used precision and recall. Specifically, as defined in Table 1, for the predicted bounding boxes of the same category, selecting an Intersection-over-Union (IoU) threshold T_α and then setting the confidence of each box to another threshold T_β, we classify each box as true positive (TP), false positive (FP), true negative (TN), and false negative (FN). Some ground truth may not have the corresponding prediction boxes; if only the prediction is judged, some FN may be missed. Therefore, following the formulas below, we calculate all the precision and recall metrics to draw the precision-recall curve. AP is the area under the curve. At last, we calculate the average value of all the AP values for all the classes in the dataset, denoted as AP_α. In the paper, we choose mAP (the average AP over 10 IoU thresholds 0.5:0.05:0.95), which is the primary metric, AP_{50} (the AP on IoU threshold 0.5), and AP_{75} (the AP on IoU threshold 0.75). The higher the AP, the better the performance.

$$precision = \frac{TP}{TP + FP} \tag{10}$$

$$recall = \frac{TP}{TP + FN} = \frac{TP}{GT}. \tag{11}$$

Table 1. The definition of true positive (TP), false positive (FP), true negative (TN), and false negative (FN) [54].

	Confidence > T_β	Confidence < T_β
IoU > T_α	TP	FN
IoU < T_α or Repetitive recognition *	FP	TN

* Corresponding to the same ground truth; if the IoU of multiply predicted boxes is larger than the threshold, only the bounding box with the largest IoU is considered the TP; others are considered FP.

With respect to the network, if not specifically noted, the backbone part is ResNet50 [37] that is introduced in Table 2, the feature fusion part is FPN [15], the region proposal network and the RCNN subnet follows the design of the Faster RCNN [6], and the region proposals are processed by RoI align [17]. All the hyper-parameters follow the settings of the Faster RCNN [16].

Table 2. The architecture of the backbone part in the network: ResNet50 [37].

Layer Name	Stage0	Stage1	Stage2	Stage3	Stage4
Operation *	7×7, 64, 2, 3 maxpool $(3 \times 3, 2, 1)$	$\begin{bmatrix} 1 \times 1,\ 64 \\ 3 \times 3,\ 64 \\ 1 \times 1,\ 256 \end{bmatrix} \times 3$	$\begin{bmatrix} 1 \times 1,\ 128 \\ 3 \times 3,\ 128 \\ 1 \times 1,\ 512 \end{bmatrix} \times 4$	$\begin{bmatrix} 1 \times 1,\ 256 \\ 3 \times 3,\ 256 \\ 1 \times 1,\ 1024 \end{bmatrix} \times 6$	$\begin{bmatrix} 1 \times 1,\ 512 \\ 3 \times 3,\ 512 \\ 1 \times 1,\ 2048 \end{bmatrix} \times 3$

* 7×7, 64, 2, 3 means convolution kernel = 7×7, num = 64, stride = 2, and padding = 3. For all the 1×1, stride = 1, padding = 0. For the first 3×3 in Stage 2, 3, 4, stride = 2, padding = 1; for other 3×3, stride = 1, padding = 1. Every convolution is followed by batch normalization and ReLU, except the last one in every stage is only followed by batch normalization. In the block, there is a shortcut that directly connects the input with the output, and batch normalization follows the addition.

4.2. Comparison of Methods

We performed a comparison with SSD [41], RetinaNet [55], Faster RCNN [16], and Faster RCNN+FPN [15] to evaluate the recognition performance of our proposed method in Table 3.

Table 3. Comparison of the different methods on the test dataset.

Method	Test		
	mAP	AP$_{50}$	AP$_{75}$
SSD(VGG)	72.5	90.5	85.8
RetinaNet	78.3	97.2	90.5
Faster RCNN	79.7	97.9	91.6
Faster RCNN+FPN	81.5	98.0	93.7
Ours	**82.7**	98.1	95.2

Our method achieves 82.7 mAP on the test dataset. Compared with the one-stage methods (SSD [41] and RetinaNet [55]), our two-stage recognition method obtained a significant improvement. Compared with the Faster RCNN [16], we achieved a 3.0-point-higher mAP. When adding the FPN [15], we still improved the mAP by 1.2 points, which was thanks to the new loss function—the smoother L1. From the improvement of AP$_{50}$ and AP$_{75}$, we knew that our method could achieve target localization with higher precision.

4.3. Exploiting the Optimal Cross-Domain Transfer Learning Strategy

4.3.1. Weight Initialization

In the experiments, we compared the recognition performance of different weight initialization, including Xavier [56], the pre-trained weight on COCO [13] and ImageNet [12] dataset. The result is shown in Table 4.

Table 4. Comparison of different methods of weight initialization on the validation dataset.

Weight Initialization	Validation		
	mAP	AP$_{50}$	AP$_{75}$
Xavier	\	\	\
ImageNet	80.1	95.9	94.0
COCO	83.7	97.0	97.0

When the network was initialized by Xavier, the trained detector failed to detect any targets on the validation dataset under the same training setting with others. As shown in Figure 8, regardless of the method of weight initialization, the classification loss could be reduced to quite a low level.

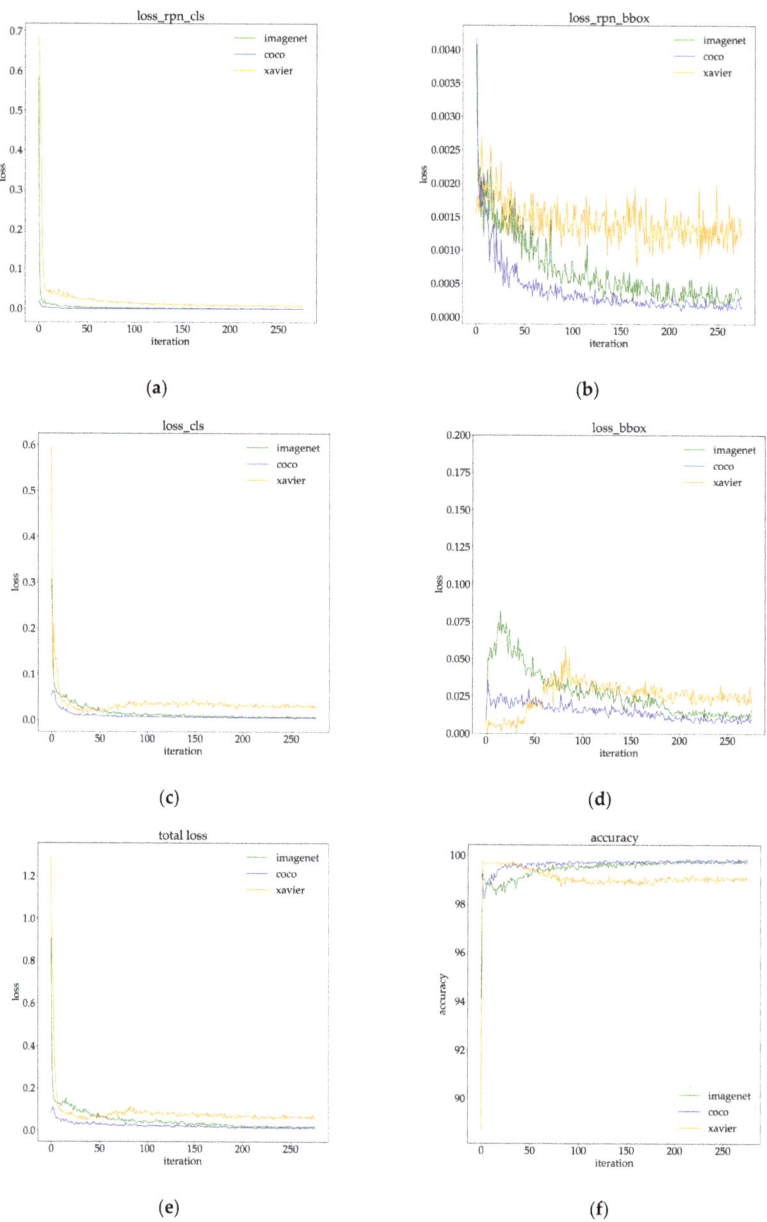

Figure 8. The visualization of training: (**a**) classification loss in RPN; (**b**) localization loss in RPN; (**c**) classification loss; (**d**) localization loss in bounding box regression; (**e**) total loss; (**f**) accuracy.

However, the Xavier initialization could not bring about a decrease in localization loss. Meanwhile, because classification loss was much larger than localization loss, the overall loss still showed a downward trend. We conjectured that the too-small size of the training dataset led to serious overfitting. On the other hand, even though there were significant differences between source domain datasets (COCO and ImageNet) and our customized dataset, transfer learning still worked well. We observed that the pre-trained weight of COCO brought down the loss to a lower level and had a better recognition performance in all metrics than that of ImageNet. We thought that the COCO pre-trained weight was more suitable for recognition network initialization because of more small targets and a broader range of target sizes in the COCO dataset, as mentioned in Section 3.2.

4.3.2. Frozen Stages

We compared the effect of different frozen stages in the network on time consumption and recognition performance. Table 5 shows the result.

Table 5. Comparison of different frozen stages in the network on the validation dataset.

Frozen Stages	Time Consumption	Validation		
		mAP	AP_{50}	AP_{75}
None	1 h 55 min	83.6	97.5	96.3
1	1 h 37 min	83.7	97.0	97.0
1, 2	1 h 30 min	82.2	97.0	95.9
1, 2, 3	1 h 18 min	80.7	96.5	95.2
1, 2, 3, 4	1 h 11 min	80.1	96.5	95.4

We observed that the training time decreased as the number of frozen stages increased, and the recognition results also decreased. Once a stage was frozen, it would no longer participate in the update of network parameters, thus saving the time. At the same time, the network could not have a better adaptation to the input image and thus could not extract more discriminative features. There was an 18-minute difference in time consumption if the first stage was frozen (as opposed to no freezing), but the recognition results were similar.

Therefore, in the other experiments, the network was initialized by the COCO pre-trained weight, of which the first stage was frozen during the training.

4.4. Ablation Studies on the Smoother L1

For the best setting in the smoother L1, the ablation studies are shown in Table 6.

Table 6. Ablation studies of the smoother L1 on the validation dataset.

Smooth L1		82.7			
β \ α	1	1.5	2	2.5	3
2	83.2	83.2	83.7	83.6	82.9
3	82.4	83.4	83.4	83.4	83.4
4	82.9	83.2	83.1	82.8	82.9

We know that α controls the gradient of outliers, and β controls the changing trend of the gradient of inliers according to Section 3.3. From another perspective, the change of α could be regarded as rebalancing the classification loss and localization loss. Furthermore, the change in β could be regarded as rebalancing the localization loss of inliers and outliers. As shown in Figure 9, we observed that the smoother L1 caused the localization loss to increase, which alleviated the imbalance between classification loss and localization loss. Benefiting from a more symmetrical multi-task loss function, the network equipped with the smoother L1 could bring a 1.0-point-higher AP than the smooth L1 baseline.

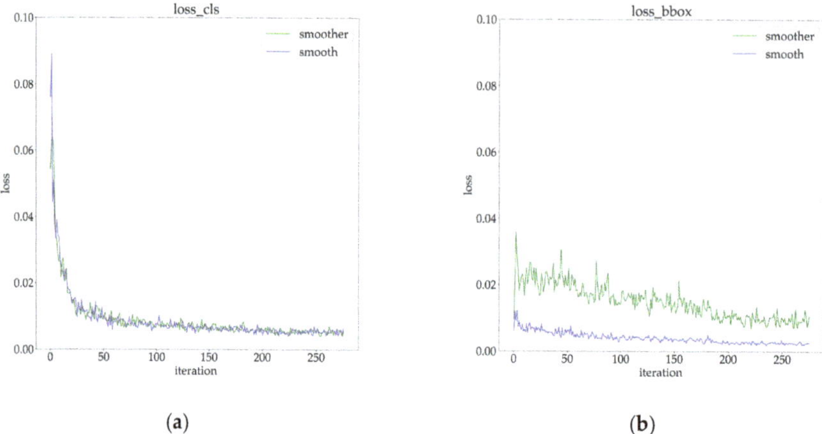

Figure 9. The visualization of training with the smoother L1 ($\alpha = 2$, $\beta = 2$) and the smooth L1: (**a**) classification loss; (**b**) localization loss in bounding box regression.

4.5. Scene Adaptability of the DCNN-Based Method

Target recognition tasks are faced with various scenes, but the dataset cannot contain all types. In order to observe the adaptability to different infrared scenes of our DCNN-based method, and to explore whether the trained network learns the scene information or target information that we prefer, the following experiment was conducted. Six types of scenes were taken for training, and the remaining type was taken as the test dataset. A total of seven experiments were conducted.

As shown in Table 7, when processing the scene that did not appear during the training, the network could still detect the target, but the performance of recognition decreased. Some of the recognition results are shown in Figure 10. We noticed that the network could recognize not only the specified target but also the original object in the scene image, such as the aircraft in the grassland and mountain scenes, and armored vehicles in the desert. It was apparent that the network could not correctly classify them because of the absence of their annotations, and they were considered as false alarms when calculating the metrics of recognition. We conjectured that the reason why the network could recognize these unknown targets in unknown scenes was that they had similar contours to the specified target; i.e., the network performed reorganization by information from the target itself instead of by speculating the location and category of the target based on the scene information. Because the object in the infrared image often lacks texture information, the characteristic of relying on contour information is noteworthy, and the designed network should pay more attention to learning contour information when recognizing the infrared target.

Table 7. Recognition results of the deep convolutional neural network (DCNN)-based method on different test scenes.

Test Scene	mAP	AP$_{50}$	AP$_{75}$
Grassland	67.2	97.3	77.8
Mountain	79.5	97.6	95.4
Road	80.9	96.8	92.6
Trees	74.5	93.6	89.8
Desert	76.0	93.5	92.4
Buildings	78.3	92.1	91.3
Cars	76.5	94.3	92.7

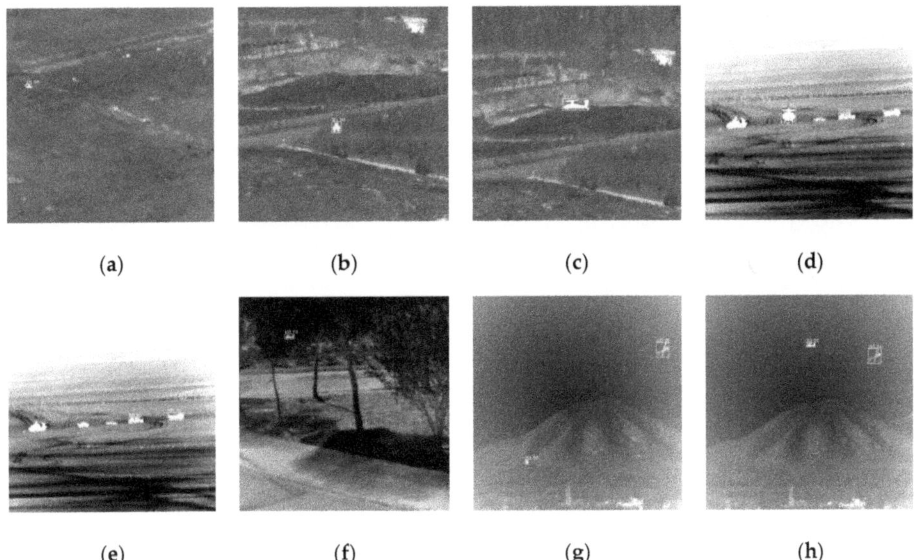

Figure 10. Some of the recognition results. The target is in the grassland (**a**), road (**b**), (**c**), desert (**d**), (**e**), trees (**f**), and mountains (**g**), (**h**). In (**d**), (**e**), (**g**), and (**h**), the original armored vehicles and aircraft in the scene were also recognized.

5. Conclusions and Prospect

In this paper, we propose a DCNN-based method to address end-to-end targets recognition in the IRCSS. The recognition accuracy reaches 82.7 mAP, proving the feasibility of the method. In order to solve the contradiction between the data quantity caused by the large size imaging results and the operation efficiency of the algorithm, direct and indirect acquisition of sub-frame images is proposed, and the indirect acquisition method with overlapping segmentation is selected according to the existing data in this paper. At the same time, we build an infrared target recognition dataset to both deal with the shortage of recognition data in the infrared field and enhance the adaptability of the algorithm in various scenes. During the training, on the one hand, the optimal cross-domain transfer learning strategy is exploited, including the analysis of the effect of ImageNet and COCO pre-trained weights on the recognition results and the optimal number of network frozen stages. On the other hand, through observation and analysis of the classification and localization loss, we design a smoother L1 loss function in bounding box regression and existing loss functions can be unified as specific values. Without significantly increasing the amount of calculation, it effectively improves recognition performance.

Some prospects for future work are given. Firstly, due to the low resolution of infrared images, we seek a special super-resolution algorithm as a pre-processing process in target recognition. Secondly, in the process of cross-domain transfer learning, we can adopt domain adaptation to further alleviate the performance degradation caused by the diversity between visible and infrared images. Thirdly, the method proposed in this paper has the background of practical demand, so we will continue to study how the method can be deployed on the embedded hardware platform to realize real-time automatic recognition of targets with high effectiveness both on accuracy and efficiency.

Author Contributions: Conceptualization: G.C.; data curation: W.W.; methodology: G.C. and W.W.; project administration: W.W.; software: G.C.; validation: G.C.; visualization: G.C.; writing—original draft: G.C.; writing—review and editing: W.W. All authors have read and agreed to the published version of the manuscript.

Funding: This research received no external funding.

Acknowledgments: The authors would like to thank the editor and the anonymous reviewers for their valuable comments and suggestions.

Conflicts of Interest: The authors declare that there is no conflict of interest.

References

1. Vollmer, M.; Möllmann, K.-P. *Infrared Thermal Imaging: Fundamentals, Research and Applications*; John Wiley & Sons: Hoboken, NJ, USA, 2017.
2. De Visser, M.; Schwering, P.B.; De Groot, J.F.; Hendriks, E.A. Passive ranging using an infrared search and track sensor. *Opt. Eng.* **2006**, *45*. [CrossRef]
3. Fan, H. A high performance IRST system based on 1152 × 6 LWIR detectors. *Infrared Technol.* **2010**, *32*, 20–24.
4. Fan, Q.; Fan, H.; Lin, Y.; Zhang, J.; Lin, D.; Yang, C.; Zhu, L.; Li, W. Multi-object extraction methods based on long-line column scanning for infrared panorama imaging. *Infrared Technol.* **2019**, *41*, 118–126.
5. Weihua, W.; Zhijun, L.; Jing, L.; Yan, H.; Zengping, C. A Real-time Target Detection Algorithm for Panorama Infrared Search and Track System. *Procedia Eng.* **2012**, *29*, 1201–1207. [CrossRef]
6. Hu, M. *Research on Detection Technology of Dim and Small Targets in Large Field of View and Complicated Background*; National University of Defense Technology: Changsha, China, 2008.
7. Otsu, N. A threshold selection method from gray-level histograms. *IEEE Trans. Syst. Man Cybern.* **1979**, *9*, 62–66. [CrossRef]
8. Dalal, N.; Triggs, B. Histograms of Oriented Gradients for Human Detection. In Proceedings of the 2005 IEEE Computer Society Conference on Computer Vision and Pattern Recognition (CVPR'05), San Diego, CA, USA, 20–25 June 2005; pp. 886–893.
9. Suykens, J.A.; Vandewalle, J. Least squares support vector machine classifiers. *Neural Process. Lett.* **1999**, *9*, 293–300. [CrossRef]
10. Felzenszwalb, P.F.; Girshick, R.B.; McAllester, D. Cascade Object Detection with Deformable Part Models. In Proceedings of the 2010 IEEE Computer Society Conference on Computer Vision and Pattern Recognition, San Francisco, CA, USA, 13–18 June 2010; pp. 2241–2248.
11. Krizhevsky, A.; Sutskever, I.; Hinton, G.E. Imagenet Classification with Deep Convolutional Neural Networks. Advances in Neural Information Processing Systems. *IEEE Trans. Pattern Anal. Mach. Intell.* **2012**, *1*, 1097–1105.
12. Deng, J.; Dong, W.; Socher, R.; Li, L.-J.; Li, K.; Li, F.F. Imagenet: A Large-Scale Hierarchical Image Database. In Proceedings of the 2009 IEEE Conference on Computer Vision and Pattern Recognition, Miami, FL, USA, 20–25 June 2009; pp. 248–255.
13. Lin, T.-Y.; Maire, M.; Belongie, S.; Hays, J.; Perona, P.; Ramanan, D.; Dollár, P.; Zitnick, C.L. Microsoft Coco: Common Objects in Context. In Proceedings of the European Conference on Computer Vision, Zurich, Switzerland, 6–12 September 2014; pp. 740–755.
14. Everingham, M.; Van Gool, L.; Williams, C.K.; Winn, J.; Zisserman, A. The pascal visual object classes (voc) challenge. *Int. J. Comput. Vis.* **2010**, *88*, 303–338. [CrossRef]
15. Lin, T.-Y.; Dollár, P.; Girshick, R.; He, K.; Hariharan, B.; Belongie, S. Feature Pyramid Networks for Object Detection. In Proceedings of the IEEE Conference on Computer Vision and Pattern Recognition, Honolulu, HI, USA, 21–26 July 2017; pp. 2117–2125.
16. Ren, S.; He, K.; Girshick, R.; Sun, J. Faster R-CNN: Towards Real-Time Object Detection with Region Proposal Networks. Advances in Neural Information Processing Systems. *IEEE Trans. Pattern Anal. Mach. Intell.* **2017**, *39*, 1137–1149. [CrossRef] [PubMed]
17. He, K.; Gkioxari, G.; Dollár, P.; Girshick, R. Mask R-CNN. In Proceedings of the IEEE International Conference on Computer Vision, Venice, Italy, 22–29 October 2017; pp. 2961–2969.
18. Zhang, L.; Gonzalez-Garcia, A.; van de Weijer, J.; Danelljan, M.; Khan, F.S. Synthetic data generation for end-to-end thermal infrared tracking. *IEEE Trans. Image Process.* **2018**, *28*, 1837–1850. [CrossRef]
19. Maji, S.; Malik, J. Object Detection Using a Max-Margin Hough Transform. In Proceedings of the 2009 IEEE Conference on Computer Vision and Pattern Recognition, Miami, FL, USA, 20–25 June 2009; pp. 1038–1045.
20. Uijlings, J.R.; Van De Sande, K.E.; Gevers, T.; Smeulders, A.W. Selective search for object recognition. *Int. J. Comput. Vis.* **2013**, *104*, 154–171. [CrossRef]

21. Felzenszwalb, P.F.; Huttenlocher, D.P. Efficient graph-based image segmentation. *Int. J. Comput. Vis.* **2004**, *59*, 167–181. [CrossRef]

22. Patel, V.M.; Nasrabadi, N.M.; Chellappa, R. Sparsity-motivated automatic target recognition. *Appl. Opt.* **2011**, *50*, 1425–1433. [CrossRef] [PubMed]

23. Khan, M.N.A.; Fan, G.; Heisterkamp, D.R.; Yu, L. Automatic Target Recognition in Infrared Imagery Using Dense Hog Features and Relevance Grouping of Vocabulary. In Proceedings of the IEEE Conference on Computer Vision and Pattern Recognition Workshops, Columbus, OH, USA, 23–28 June 2014; pp. 293–298.

24. Blackman, S.; Blackman, S.S.; Popoli, R. *Design and Analysis of Modern Tracking Systems*; Artech House Books: London, UK, 1999.

25. Henriques, J.F.; Caseiro, R.; Martins, P.; Batista, J. High-speed tracking with kernelized correlation filters. *IEEE Trans. Pattern Anal. Mach. Intell.* **2014**, *37*, 583–596. [CrossRef]

26. Kristan, M.; Leonardis, A.; Matas, J.; Felsberg, M.; Pflugfelder, R.; Cehovin Zajc, L.; Vojir, T.; Hager, G.; Lukezic, A.; Eldesokey, A. The Visual Object Tracking Vot2017 Challenge Results. In Proceedings of the IEEE International Conference on Computer Vision Workshops, Venice, Italy, 22–29 October 2017; pp. 1949–1972.

27. Yu, X.; Yu, Q.; Shang, Y.; Zhang, H. Dense structural learning for infrared object tracking at 200+ Frames per Second. *Pattern Recognit. Lett.* **2017**, *100*, 152–159. [CrossRef]

28. Hare, S.; Golodetz, S.; Saffari, A.; Vineet, V.; Cheng, M.-M.; Hicks, S.L.; Torr, P.H. Struck: Structured output tracking with kernels. *IEEE Trans. Pattern Anal. Mach. Intell.* **2015**, *38*, 2096–2109. [CrossRef]

29. Danelljan, M.; Hager, G.; Shahbaz Khan, F.; Felsberg, M. Learning Spatially Regularized Correlation Filters for Visual Tracking. In Proceedings of the IEEE International Conference on Computer Vision, Santiago, Chile, 7–13 December 2015; pp. 4310–4318.

30. Kristan, M.; Matas, J.; Leonardis, A.; Felsberg, M.; Pflugfelder, R.; Kamarainen, J.-K.; Cehovin Zajc, L.; Drbohlav, O.; Lukezic, A.; Berg, A. The Seventh Visual Object Tracking Vot2019 Challenge Results. In Proceedings of the IEEE International Conference on Computer Vision Workshops, Seoul, Korea, 27–28 October 2019.

31. Danelljan, M.; Bhat, G.; Shahbaz Khan, F.; Felsberg, M. ECO: Efficient Convolution Operators for Tracking. In Proceedings of the IEEE Conference on Computer Vision and Pattern Recognition, Honolulu, HI, USA, 21–26 July 2017; pp. 6638–6646.

32. Kang, M.; Ji, K.; Leng, X.; Xing, X.; Zou, H. Synthetic aperture radar target recognition with feature fusion based on a stacked autoencoder. *Sensors* **2017**, *17*, 192. [CrossRef]

33. Cai, Z.; Vasconcelos, N. Cascade R-CNN: Delving into High Quality Object Detection. In Proceedings of the IEEE Conference on Computer Vision and Pattern Recognition, Salt Lake City, UT, USA, 18–23 June 2018; pp. 6154–6162.

34. Chen, Y.; Yang, T.; Zhang, X.; Meng, G.; Xiao, X.; Sun, J. DetNAS: Backbone Search for Object Detection. In Proceedings of the Advances in Neural Information Processing Systems, Vancouver, BC, Canada, 8–14 December 2019; pp. 6638–6648.

35. Zoph, B.; Le, Q.V. Neural architecture search with reinforcement learning. *arXiv* **2016**, arXiv:1611.01578.

36. Liu, Y.; Wang, Y.; Wang, S.; Liang, T.; Zhao, Q.; Tang, Z.; Ling, H. Cbnet: A novel composite backbone network architecture for object detection. *arXiv* **2019**, arXiv:1909.03625.

37. He, K.; Zhang, X.; Ren, S.; Sun, J. Deep Residual Learning for Image Recognition. In Proceedings of the IEEE Conference on Computer Vision and Pattern Recognition, Las Vegas, NV, USA, 27–30 June 2016; pp. 777–778.

38. Xie, S.; Girshick, R.; Dollár, P.; Tu, Z.; He, K. Aggregated Residual Transformations for Deep Neural Networks. In Proceedings of the IEEE Conference on Computer Vision and Pattern Recognition, Honolulu, HI, USA, 21–26 July 2017; pp. 1492–1500.

39. Redmon, J.; Divvala, S.; Girshick, R.; Farhadi, A. You Only Look Once: Unified, Real-Time Object Detection. In Proceedings of the IEEE Conference on Computer Vision and Pattern Recognition, Las Vegas, NV, USA, 27–30 June 2016; pp. 779–788.

40. Redmon, J.; Farhadi, A. Yolov3: An incremental improvement. *arXiv* **2018**, arXiv:1804.02767.

41. Liu, W.; Anguelov, D.; Erhan, D.; Szegedy, C.; Reed, S.; Fu, C.-Y.; Berg, A.C. SSD: Single Shot Multibox Detector. In Proceedings of the European Conference on Computer Vision, Amsterdam, The Netherlands, 8–16 October 2016; pp. 21–37.

42. Tan, M.; Pang, R.; Le, Q.V. Efficientdet: Scalable and efficient object detection. *arXiv* **2019**, arXiv:1911.09070.

43. Tan, M.; Le, Q.V. Efficientnet: Rethinking model scaling for convolutional neural networks. *arXiv* **2019**, arXiv:1905.11946.

44. Zhang, Y.; Zhang, Y.; Shi, Z.; Zhang, J.; Wei, M. Design and Training of Deep CNN-Based Fast Detector in Infrared SUAV Surveillance System. *IEEE Access* **2019**, *7*, 137365–137377. [CrossRef]

45. Yardimci, O.; Ayyıldız, B.Ç. Comparison of SVM and CNN Classification Methods for Infrared Target Recognition. In Proceedings of the Automatic Target Recognition XXVIII, Orlando, FL, USA, 30 April 2018; p. 1064804.

46. Tanner, I.L.; Mahalanobis, A. Fundamentals of Target Classification Using Deep Learning. In Proceedings of the Automatic Target Recognition XXIX, Baltimore, MD, USA, 14 May 2019; p. 1098809.

47. d'Acremont, A.; Fablet, R.; Baussard, A.; Quin, G. CNN-based target recognition and identification for infrared imaging in defense systems. *Sensors* **2019**, *19*, 2040. [CrossRef]

48. Li, C.; Liang, X.; Lu, Y.; Zhao, N.; Tang, J. RGB-T object tracking: Benchmark and baseline. *Pattern Recognit.* **2019**, *96*, 106977. [CrossRef]

49. Science Data Bank: A Dataset for Dim-Small Target Detection and Tracking of Aircraft in Infrared Image Sequences. 2019. Available online: www.csdata.org/p/387/ (accessed on 27 March 2020).

50. Yosinski, J.; Clune, J.; Bengio, Y.; Lipson, H. How transferable are features in deep neural networks? Advances in neural information processing systems. *IEEE Trans. Pattern Anal. Mach. Intell.* **2014**, *2*, 3320–3328.

51. Singh, B.; Davis, L.S. An Analysis of Scale Invariance in Object Detection Snip. In Proceedings of the IEEE Conference on Computer Vision and Pattern Recognition, Salt Lake City, UT, USA, 18–23 June 2018; pp. 3578–3587.

52. Girshick, R.; Donahue, J.; Darrell, T.; Malik, J. Rich Feature Hierarchies for Accurate Object Detection and Semantic Segmentation. In Proceedings of the IEEE Conference on Computer Vision and Pattern Recognition, Columbus, OH, USA, 23–28 June 2014; pp. 580–587.

53. Girshick, R. Fast R-CNN. In Proceedings of the 2015 IEEE International Conference on Computer Vision (ICCV), Santiago, Chile, 7–13 December 2015; pp. 1440–1448.

54. Bishop, C.M. *Pattern Recognition and Machine Learning*; Springer: Berlin/Heidelberg, Germany, 2006.

55. Lin, T.-Y.; Goyal, P.; Girshick, R.; He, K.; Dollár, P. Focal Loss for Dense Object Detection. In Proceedings of the IEEE International Conference on Computer Vision, Venice, Italy, 22–29 October 2017; pp. 2980–2988.

56. Glorot, X.; Bengio, Y. Understanding the Difficulty of Training Deep Feedforward Neural Networks. In Proceedings of the Thirteenth International Conference on Artificial Intelligence and Statistics, Sardinia, Italy, 13–15 May 2010; pp. 249–256.

Article

Ultrasound Image-Based Diagnosis of Malignant Thyroid Nodule Using Artificial Intelligence

Dat Tien Nguyen, Jin Kyu Kang, Tuyen Danh Pham *, Ganbayar Batchuluun and Kang Ryoung Park

Division of Electronics and Electrical Engineering, Dongguk University, 30 Pildong-ro 1-gil, Jung-gu, Seoul 04620, Korea; nguyentiendat@dongguk.edu (D.T.N.); kangjinkyu@dgu.edu (J.K.K.); ganabata87@gmail.com (G.B.); parkgr@dgu.edu (K.R.P.)
* Correspondence: phamdanhtuyen@gmail.com; Tel.: +82-10-9264-4449; Fax: +82-2-2277-8735

Received: 18 February 2020; Accepted: 23 March 2020; Published: 25 March 2020

Abstract: Computer-aided diagnosis systems have been developed to assist doctors in diagnosing thyroid nodules to reduce errors made by traditional diagnosis methods, which are mainly based on the experiences of doctors. Therefore, the performance of such systems plays an important role in enhancing the quality of a diagnosing task. Although there have been the state-of-the art studies regarding this problem, which are based on handcrafted features, deep features, or the combination of the two, their performances are still limited. To overcome these problems, we propose an ultrasound image-based diagnosis of the malignant thyroid nodule method using artificial intelligence based on the analysis in both spatial and frequency domains. Additionally, we propose the use of weighted binary cross-entropy loss function for the training of deep convolutional neural networks to reduce the effects of unbalanced training samples of the target classes in the training data. Through our experiments with a popular open dataset, namely the thyroid digital image database (TDID), we confirm the superiority of our method compared to the state-of-the-art methods.

Keywords: ultrasound image; malignant thyroid nodule; artificial intelligence; deep learning; weighted binary cross-entropy loss

---

1. Introduction

Traditional disease diagnosis/treatment methods are mostly based on doctors' expert knowledge on any given condition. However, this diagnostic method has a big limitation, that is, its performance is much more dependent on the experiences and personal knowledge of doctors. As a result, the diagnostic performance varies and is limited. With the development of digital technology, image-based diagnosis techniques have been widely used to help doctors investigate problems with organs that are underneath the skin and/or deep inside the human body [1–11]. For example, doctors have used X-ray imaging to capture lung and/or bone images that can help to indicate whether a disease/injury exists in these organs [9,10]. To diagnose issues with the human brain, the Computer-Tomography (CT) and/or Magnetic Resonance Imaging (MRI) techniques have been widely used [2,3]. With the help of imaging techniques, the diagnosis performance can be much more enhanced. However, the use of captured images is still dependent on personal knowledge and experiences of doctors. To overcome this problem, Computer-Aided Diagnosis systems (CAD) have been developed to assist doctors in the diagnosis and treatment processes [1–10]. As indicated by its name, the CAD systems can serve as an additional expert in the double screening process that aims to enhance the human diagnostic performance based on a computer program [11]. This kind of system uses and processes one or more captured medical images of some organs such as X-ray, CT, and MRI scans, and yields its decision that can assist doctors in diagnosing diseases. Due to its purpose, CAD systems have been widely developed and used in real-life applications such as for diagnosing the brain [2,3,7,12], breasts [4,8,13–16], lungs [10], and thyroid diseases [17–35].

The thyroid is an important organ located in the human neck that produces and secretes two important hormones, namely triiodothyronine and thyroxine, which are responsible for the regulation of metabolism in the human body. Due to its important role in the human body, diagnosing and treatment of thyroid disease has become important [17–35]. As reported in the previous studies, one important problem commonly experienced in the thyroid region is the appearance of nodules that cause thyroid cancer. Thyroid nodules are abnormal lumps that appear on the thyroid region of the human body. They could be caused by many factors, including iodine deficiency, overgrowth of normal thyroid tissue, or thyroid cancer. Thyroid nodules are usually classified into two categories based on their characteristics namely, benign cases (which are noncancerous nodules), and malign cases (which can cause thyroid cancer) [36]. In both the benign and malign cases, the appearance of thyroid nodules can cause problems with patient health. With the appearance of nodules, the thyroid region can be malfunctioned. Although the benign case has little effects on patient health, it can cause aesthetic problems and/or make it difficult for the patients to breathe and/or swallow. The malign case can cause thyroid cancer. Fortunately, most detected thyroid nodules are benign cases as reported in the previous studies [19,36]. However, diagnosing and treating malign cases is still very important.

There have been several methods of diagnosing thyroid nodules such as physical examination, thyroid function tests, and Fine Needle Aspiration (FNA) biopsy. The physical exam is normally done at the first stage of the diagnosis process in which the patients are asked to perform several physical tests on the thyroid region such as swallowing to check the shape, size, and the movement of nodules. However, this method is just a primary test and normally does not give deep information about the nodules' condition. To gain a deep look inside the thyroid problem, thyroid function tests or FNA are normally invoked. In the thyroid function test method, the level (amount) of the two hormones (thyroxine and triiodothyronine, which are produced by the thyroid region) is measured to see whether there is any abnormality in thyroid functionality. FNA can also be applied in diagnosing thyroid nodules to produce good diagnosis results. However, these methods required are labor-intensive, invasive, and costly. As an alternative, image-based thyroid nodule diagnosis has been used in various applications. This method uses high-frequency sound waves (ultrasound wave) to produce images of the thyroid region. As a result, this method provides rich information of thyroid nodules such as the shape and structure of nodule as well as the condition of the nodules.

Using the ultrasound thyroid nodule images, there have been several previous studies on CAD for the thyroid nodule detection and classification problems. In contrast to the conventional thyroid diagnosis methods mentioned above, the CAD methods for thyroid nodules use ultrasound thyroid nodule images as inputs and produce thyroid nodule regions and/or the status of nodules (benign or malign) [17–36]. Similar to normal image processing systems, CAD systems for thyroid nodule use several image processing techniques to extract information from input images for detection/classification purposes. Based on the methods for extracting information from images, the previous studies can be categorized into three groups: the group using handcrafted feature extraction methods, the group using deep feature extraction methods, and the group that is a fusion of the two.

Handcrafted-based image feature extraction methods have been widely applied for a long time, especially with the simple image-based systems and/or before the appearance of deep learning-based techniques. As indicated by its name, this kind of method uses several handcrafted image feature extraction methods that are designed by experts based on their knowledge of specific problems to extract efficient features from input images for image-based processing systems. For the thyroid nodule CAD, the handcrafted feature-based method has also been used previously [22,24,31]. Chang et al. [24] used up to 78 texture features extracted from ultrasound thyroid nodule images for the thyroid nodule classification problem. Based on the extracted image features, they used Support Vector Machines (SVMs) to classify input images into several categories such as nodule versus non-nodule and follicles versus fibrosis. Sudarshan et al. [22] used wavelet transform to analyze the input ultrasound thyroid images for the thyroid nodule classification problem. A similar approach, Raghavendra et al. [31] used the segmentation-based fractal texture analysis technique to analyze ultrasound thyroid images under different threshold values for the classification problem. Ouyang et al. [26] found that linear and non-linear classifiers yield similar classification results for the thyroid nodule classification based on handcrafted image features. Since the handcrafted image feature extractors were designed and selected by expert knowledge of authors, they only reflect some limited aspects of the problem. As a result, the classification performance is limited.

With the development of technology, such as the back propagation algorithm, neural network, and Graphics Processing Units (GPUs), the deep learning-based technique has recently been applied to solve many problems in medical image processing systems [1,2,10,12,21,36]. For the thyroid nodule detection/classification problem, the deep learning-based method has gained a lot of success. As indicated by its name, the deep learning-based method, such as Convolutional Neural Network (CNN), automatically learns the useful texture features for the detection/classification problem instead of using handcrafted (fixed) feature extraction methods. As a result, the deep learning-based method can produce more superior results than handcrafted-based methods. In a study by Zhu et al. [21], they proposed a method for thyroid nodule classification using CNN systems. In their study, they fine-tuned the residual network (ResNet18-based network) and obtained good classification results using a public dataset. Similar to the work by Zhu et al., the work by Chi et al. [23] also used the CNN network to classify ultrasound thyroid nodule images into benign and malign nodules. However, different from the study by Zhu et al. [21], Chi et al. [23] used the GoogLeNet for a classification purpose. In addition, they trained their CNN model using two datasets to reduce the effect of the over-fitting problem and the variation of input images. In a study by Sundar et al. [28], the authors proposed a general framework for thyroid nodule classification using the CNN network, including the fine-tuning, training from scratch, and the use of pretrained networks for image feature extraction. With their proposed methods, they performed various experiments using two popular CNN architectures, including a relative shallow network based on VGG16-Net architecture, and a deep network based on Inception (GoogLeNet) architecture. In some other studies, the thyroid nodule classification can also be done by a detection-and-classification approach as shown in a study by Song et al. [27]. In that study, Song et al. used a detection network such as multiscale single-shot detection network (multiscale SSD) or Yolo network to roughly detect the position of thyroid nodules. Additionally, then, they performed the nodule classification using the detection results of the first step. This method has the advantage that noise and non-nodule regions can be removed before performing the classification step. However, it is difficult to find small nodules, and the network architecture is very complex.

As a fusion of the two mentioned approaches, there exist studies that combine the handcrafted and deep learning methods to enhance the classification performance. In a study by Nguyen et al. [36], they found that the information in the frequency domain can be useful for discriminating easy samples of benign and malign cases, and the deep learning-based method can be useful for discriminating harder samples (ambiguous samples). Based on this observation, they proposed a method that applies a cascade classifier scheme for the thyroid nodule classification problem. As a result of their study, they showed that the combination of handcrafted and deep features is efficient for enhancing classification accuracy compared to the use of individual feature extraction method. The following are more detailed differences between previous study [36] and our research. First, one CNN of ResNet was used in a previous study [36], but multiple CNNs of ResNet and InceptionNet are used in our research. Second, only the binary cross-entropy loss was used in a previous study [36] whereas only the weighted binary cross-entropy loss was newly adopted in our research. The weighted binary cross-entropy loss function is efficient for reducing the overfitting problem caused by the unbalanced training samples of the target classes in the training data. Third, the final classification of thyroid nodule was performed based on the one output score of ResNet in a previous study [36]. However, the outputs of multiple CNNs of ResNet and InceptionNet are combined by score level fusion in our research.

There is a common limitation in the aforementioned studies: they did not fully consider the problems associated with deep learning-based methods, such as the imbalance of training image samples, the depth of the network, and the variation of the size of objects. For example, the classification model can produce biased results if the training data have an imbalance of samples in target classes, or it is difficult to construct a very deep network that can capture features of both the small and large sizes of objects. In Table 1, we summarized the previous studies for the thyroid nodule classification problem in comparison with our proposed method. To overcome this limitation, we propose a novel approach for the thyroid nodule classification problem by modifying the loss function of a conventional CNN network and a combination of multiple CNN networks to enhance the learning ability of the deep learning method. In comparison with the methods in the previous studies, our proposed method is novel in the following four ways:

- We propose the use of multiple CNN-based models to analyze input ultrasound thyroid images deeply for the classification problem. Since each CNN model has its own architecture and characteristics of learning the characteristics of input images, the use of multiple CNN-based models can help to extract richer information compared to using an individual model.
- In order to solve the problem of unbalanced data samples between the benign and malign classes in the training data, we propose the use of a weighted binary cross-entropy loss function instead of the conventional binary cross-entropy loss function. As the name suggests, we assign a higher weight value to data samples of class (benign or malign), which have a smaller number of data than the other. This procedure helps increase the focus of the training process on this class rather than the other class. As a result, it helps to reduce the effects of the overfitting problem of the CNN networks when training with unbalanced data.
- We combine the outputs of multiple CNN-based models to enhance the classification performance using several bagging methods, including MIN, MAX, and AVERAGE combination rules which take the minimum, maximum, and average results of the multiple CNN-based models, respectively.
- We make our algorithm available to the public through [37], so that other researchers can make fair comparisons.

Table 1. Summary of the previous studies on the ultrasound thyroid nodule image classification problem.

Category	Method	Strength	Weakness
Handcrafted-based Methods	- Classification is implemented using extracted image features via human-designed methods [22,24,26,31]	- Easy to implement. - Does not require high-performance hardware devices	- Low classification accuracy
Deep learning-based methods	- Fine-tuning an existing CNN network for classification [21,23,28] - Extracts image features using a pretrained CNN network while classification is implemented using an SVM [28] - Combines detection and classification based on a CNN network [27]	- Utilizes the power of deep learning and transfer learning methods - Higher accuracy than handcrafted-based methods	- There is room for enhancing classification performance
Fusion of deep and handcrafted-based methods	- Extracts image features from both spatial and frequency domains for classification problem [36]	- Applies a cascade classifier scheme to enhance classification performance using handcrafted and deep features	- More complicated and takes longer processing time than using a single method (FFT-based or CNN-based methods)
	- Extracts image information from both spatial and frequency domains for classification problem - Combines classification results by multiple CNN models to enhance classification performance - Reduces the effect of unbalanced training samples of CNN network by using weighted cross-entropy loss function. (Proposed method)	- Analyzes the ultrasound thyroid images using different architectures of CNN network - Enhances the classification results compared to the use of single CNN architecture	- Requires strong hardware equipment to run multiple CNN networks - Takes longer processing time than the previous studies.

The remainder of our paper is organized as follows. In Section 2, we provide detailed descriptions of our proposed method for diagnosis of malignant thyroid nodules using the artificial intelligence technique. In Section 3, we validate the performance of the proposed method using a public ultrasound thyroid dataset, namely the Thyroid Digital Image Database (TDID) dataset [20]; compare the findings with the previous studies; and provide a discussion about our results. Finally, we present the conclusion of our study in Section 4.

2. Proposed Method

2.1. Overview of the Proposed Method

In Figure 1, we show some examples of ultrasound thyroid images in the TDID dataset [20]. As shown in these examples, the captured ultrasound thyroid images contained two main regions: the background (dark region) and the thyroid (brighter regions). Focusing on the thyroid region, the benign and malign cases exhibited several differences: the malign case images contained nodules with round-like shape and exhibited the calcification phenomenon. Based on this observation, we proposed a new thyroid nodule classification method as shown in Figure 2.

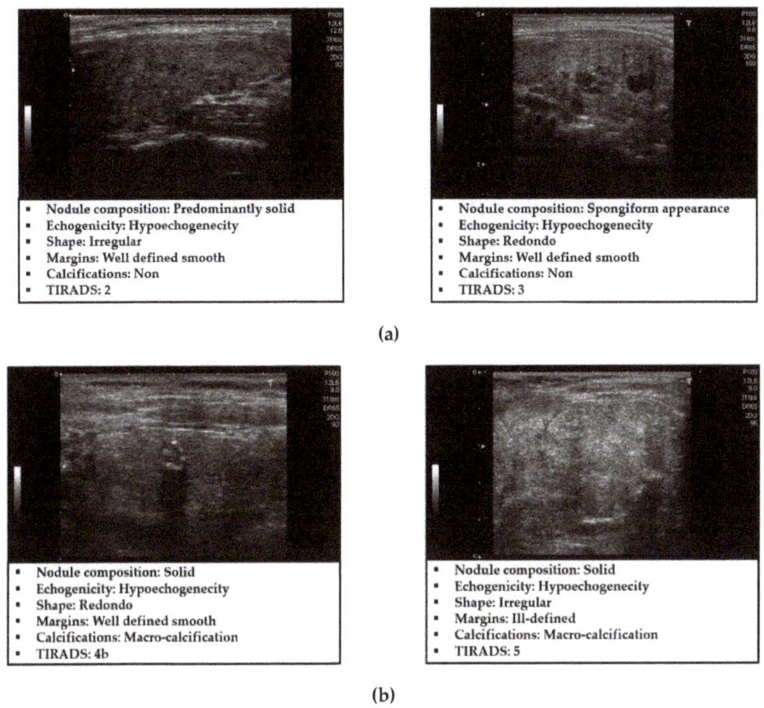

Figure 1. Example of captured ultrasound thyroid images in the thyroid digital image database (TDID) dataset [20]: (**a**) benign cases and (**b**) malign cases.

In Figure 2, we depicted the overall flow-chart of our proposed method for thyroid nodule classification using ultrasound images. As shown in this figure, our proposed method receives an ultrasound image of thyroid region and outputs a suggestion for doctors (radiologists) about whether the image contains a benign or malign case of a thyroid nodule. To perform its functionality, an input ultrasound thyroid image is first passed through a thyroid region detection method to filter-out the background and noise region before feeding it to our main algorithm. This step is necessary and important to enhance the classification performance because the background and noise provide redundant information, and consequently, they can have negative effects on the classification system. The detail description of this step is mentioned in Section 2.2. As a subsequent step, we performed a coarse classification step to classify the input ultrasound thyroid image into one of three categories, including 'benign', 'malign', and 'ambiguous benign–malign', using the image of the thyroid region (image obtained after filtering out the background and noise) based on a handcrafted-based method extracted in the frequency domain. This classification step was used to detect the easy benign or malign samples, reduce processing time, and shift the focus of the deep learning-based model on the more difficult samples. When the coarse classifier classified the input images as 'ambiguous benign–malign' cases, the input image was then further processed (classified) by a deep neural network based on the CNN method. The detail descriptions of these steps are included in the Section 2.3 for the coarse classifier, and Section 2.4 for the fine classifier.

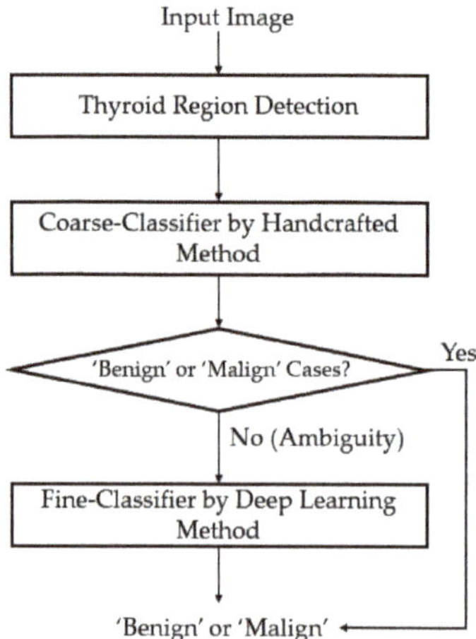

Figure 2. Flow chart of the proposed method.

2.2. Preprocessing of Captured Thyroid Images

As shown in Figures 1 and 2, the captured ultrasound thyroid images contained two main parts, that is, the background (boundary parts with low illumination and some additional artifacts) and the thyroid region (the inner brighter part that captures the details of the thyroid region). It is easy to see that the background regions contain no information about whether an image contains benign or malign cases of thyroid nodules. Besides, it also contains some artifact information that was added to an image as indicators for the radiologist, such as the patient information or capturing system configuration, during the image acquisition process. Due to this reason, the background region should be removed before passing images to the main classification system. This step is a preprocessing step and has been well-studied in a previous study [36]. In our study, we used a popular algorithm for removing background regions as shown in the studies by Zhu et al. [21] and Nguyen et al. [36]. Steps for localizing the thyroid region and removing the background regions are roughly described in Figure 3.

As shown in Figure 1, the thyroid region is normally displayed as the largest brighter region in the captured ultrasound thyroid image. Although several brighter regions exist in an ultrasound thyroid image, such as the illumination indicator and text for specifying capturing system configuration, the size of these regions is much smaller than that of the thyroid region. Based on this observation, we first performed an image binarization method to detect all brighter regions in the captured image using an optimal threshold value. In our study, we used a binarization method proposed by Otsu's et al. [38], which takes an input image and performs binarization adaptively by selecting the most suitable threshold value. A result of this binarization step is given in Figure 3b using the input image of Figure 3a. As shown in Figure 3b, although there were some brighter regions detected, the thyroid region had the largest size. Based on this truth, we detected the thyroid region by selecting the largest object in the binarized image and discarding the other regions as shown in Figure 3c. Finally, the detected thyroid region was determined by taking the bounding-box in the input image (in Figure 3a) based on the selected region of Figure 3c. An example of a resultant image of this step is given in Figure 3d using the input image of Figure 3a. As we can see from this example, the thyroid region was well localized using our localization method.

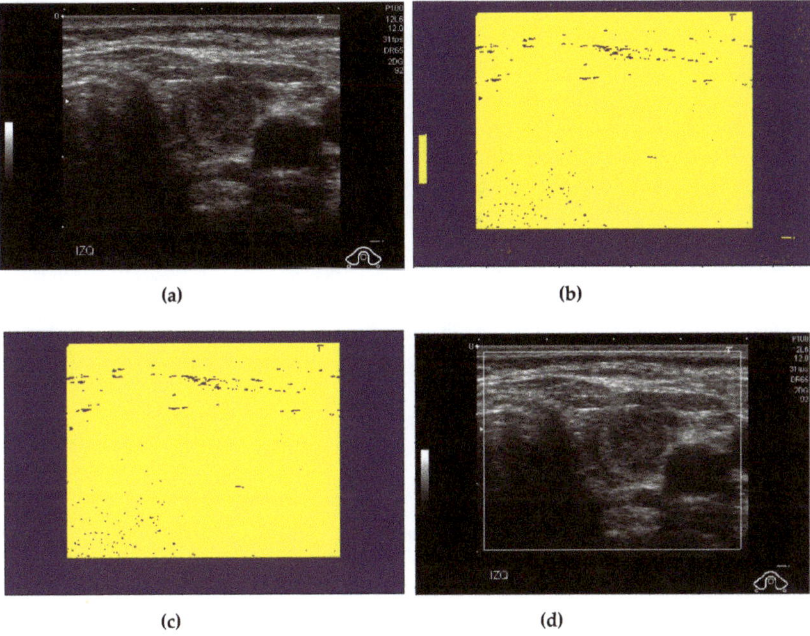

Figure 3. Example result of the thyroid region detection algorithm used in our study: (**a**) an input ultrasound thyroid image; (**b**) the binarized image; (**c**) the thyroid region detection by selecting the largest object; and (**d**) the final detection results.

2.3. Coarse Classifier Based on Information Extracted in the Frequency Domain

As shown in Figure 2, our proposed method is based on a cascade structure of classifiers using handcrafted-based and CNN-based methods. As the first stage of our proposed classification method, we performed a coarse classification based on the information extracted from images in the frequency domain, as suggested by Nguyen et al. [36]. The purpose of this classifier is to preclassify samples that are easily classified using information extracted in the frequency domain. As exploited by Nguyen et al. [36], there are differences between benign and malign case images in the frequency domain caused by the appearance of nodules and calcification phenomenon in the thyroid regions. That is, the appearance of nodules and calcification makes the captured ultrasound thyroid image of malign cases brighter, and the change in pixel values is faster around these nodule regions than other regions. Based on this observation, we used the Fast Fourier Transform (FFT) method to extract this difference and classify an ultrasound thyroid image into one of three categories: 'benign', 'malign', and 'ambiguous benign–malign,' as shown in Figure 4. As shown in Figure 4a, the thyroid region image was first transformed from the spatial domain to the frequency domain using the FFT method to extract the distribution of image energy in the frequency domain. With this extracted image in the frequency domain, Nguyen et al. proposed an image feature extraction method that uses the ratio between some selected frequency components and the total frequency components as shown in Equation (1) [36]. In this equation, P_s indicates the total power spectrum of image frequency components inside a selected frequency region, and the P indicates the total power spectrum of all frequency components of an image [36]. As indicated by Nguyen et al. [36], there could be several methods for selecting the frequency region in which we used to measure P_s (the selected frequency components) such as the use of frequency components inside a circle, horizontal, vertical, or a combination of them [36] around/through the DC component (zero-frequency component) of an image. However, as indicated by their work with the TDID dataset, they showed that the frequency components inside a circle around the DC component works better than other methods. Therefore, we selected to use the circle shape in our study as shown in Figure 4a (red circle).

$$\text{Score} = \frac{P_s}{P} \tag{1}$$

We compared the extracted image feature in the frequency domain with two threshold values, i.e., TH_LOW and TH_HIGH in Figure 4b for classifying the input ultrasound thyroid image into one of the three categories. These threshold values are experimentally obtained based on the training dataset. As a result, if the extracted image feature (Score in Figure 4b) is lower than TH_LOW, it is regarded as the 'benign' case image; if the extracted image feature is higher than the TH_HIGH image, it is regarded as the 'malign' case image. Otherwise, it is considered to belong to the 'ambiguous benign–malign' category in which we are not sure which class it should belong to. For this case, the final classification was done based on our second classifier that was based on the deep learning technique.

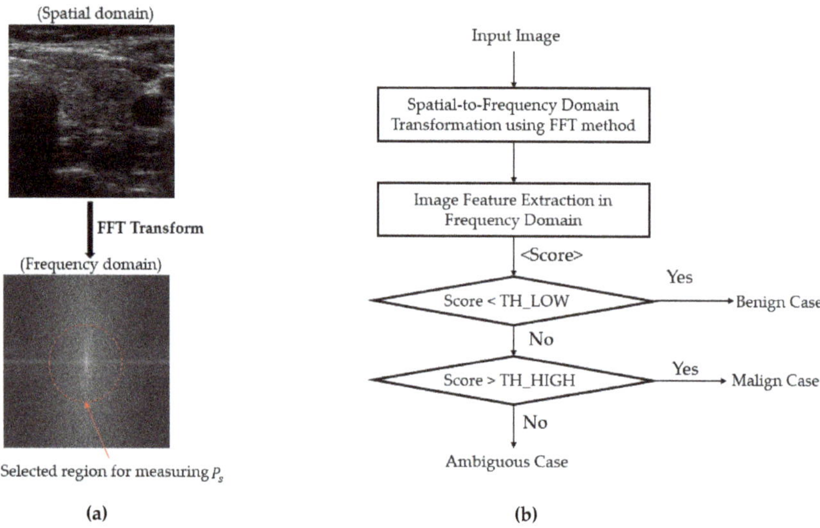

Figure 4. Coarse classifier based on information extracted in the frequency domain using the Fast Fourier Transform (FFT)-based method: (**a**) a thyroid image represented in spatial and frequency domain with a selected circle frequency regions and (**b**) the flowchart for classifying thyroid images into 'benign', 'malign', or 'ambiguous' region in our study.

2.4. Fine Classifier Based on a Combination of Multiple CNN Models

2.4.1. Introduction to the Deep Learning Framework

The deep learning-based method implies the use of a deep (many layers) neural network for a regression or classification problem. Although this is not a new technique, this method has recently attracted lots of attention from researchers because of the development of GPUs that are used to speed-up the processing of the network, and lots of superior (state-of-the-art) performances of digital signal processing systems have been reported [39–50]. This kind of signal processing technique has been successfully and widely used in many fields including image processing [39–48] and natural language processing [49,50]. In Figure 5, we show the general architecture of a CNN network, which is a special kind of deep learning-based technique and has been successfully used for the image classification problem. As shown in this figure, a CNN network is composed of two main components, including a feature extraction component based on convolution operation, and a classification component based on a multilayer perceptron (neural) network. This structure allows us to learn efficient representation (image texture features) of an input image using the filtering technique through the application of convolution operation. With the extracted image features, it is possible to learn a classifier to classify input image into predesigned classes. All of the network parameters (weights and biases of convolution filters and multilayer perceptron) can be trained and automatically obtained by a training process using a back propagation technique and training data. This is the key to make the learning-based method outperform the handcrafted-based method for the image-based classification problem. In addition, the use of convolution operation with a weight-sharing scheme allows us to construct a deeper network than the conventional neural network.

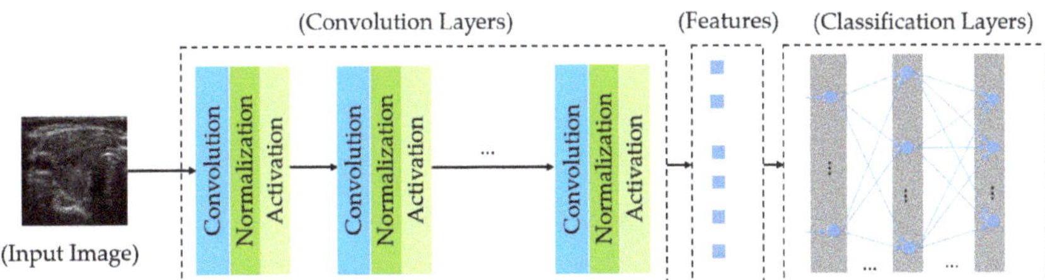

Figure 5. A general architecture of a Convolutional Neural Network (CNN) network for the image classification problem.

According to the type of applications, a suitable CNN network was used. Lots of CNN architectures have been proposed for various image-based systems such as image classification [39–43], object detection [44,45], 3D image reconstruction [46], and image feature extraction [47,48]. Although the CNN network has been successfully used in many image-based processing systems, it still has several limitations caused by its characteristics and internal structure. As mentioned in the previous studies [39–43], there are two main problems frequently associated with a CNN network. The first problem is caused by the depth of the network. To learn from data efficiently, we normally need to construct a deep network that contains many weight layers. However, the deep network is normally difficult to train due to the vanishing gradient problem [41]. The second problem is caused by the huge number of parameters that need to be learnt through the training process. For an image classification system, many CNN networks have been used, such as AlexNet, VGGNet, ResNet, DenseNet, and InceptionNet. According to their structures, the AlexNet contained about 62 million parameters, the VGGNet-16 contained about 138 million parameters, the VGGNet-19 contained about 143 million parameters, etc. To learn these huge amounts of parameters requires a strong hardware power (Central Processing Unit (CPU), GPU) as well as a large amount of training data. These problems can have strong negative effects on the performance of medical image-based systems because we normally require high performance systems using less training data. This is because it is difficult to collect a large number of medical images owing to special characteristics of this kind of images: they require expensive data acquisition devices and the cooperation of patients.

As explained above, the conventional CNN networks such as AlexNet or VGGNet were constructed by chaining weight layers (convolution and dense layers) to extract image features and learn classifiers for the classification problem. This is a basic CNN architecture and it works fine for a not-too-deep network. However, there is a problem, called vanishing gradients, which can occur when the depth of the network increases, and this problem makes the network difficult to train and consequently degrades the classification performance [41]. To solve this problem, He et al. [41] propose a new method for not only constructing a very deep CNN network, but also making it easier to train, namely the residual network (ResNet). In Figure 6, we described the methodology of the ResNet network building block. By using a new kind of connection, called skip connection, this new type of CNN architecture can make the network skip some training layers when the input and output of these layers are close to the identification function. As a result, the network is deeper and easier to train compared to the conventional CNN networks. In our study, we used this type of CNN architecture to construct a very deep network for learning texture feature of input ultrasound thyroid images. In detail, we used a ResNet50-based network that contained a total of 50 weight layers for our classification problem as explained in Section 2.1 and Figure 2.

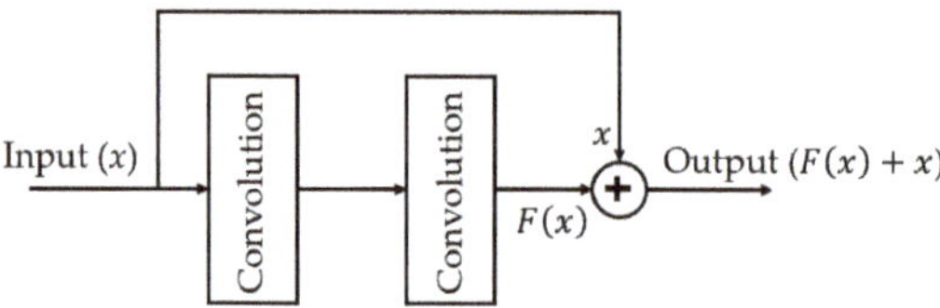

Figure 6. Methodology for constructing the residual convolution block.

Although the conventional CNN networks have successfully been used to capture image texture features, their performances are still affected by the large variation of the size of objects that appears in input images. As a result, choosing a right kernel size for the convolution mask is difficult to achieve, and normally a single optimal size of convolution mask does not exist. To solve this problem, Szegedy et al. [42] proposed a new network structure that applies multiple sizes of the convolution mask to extract image features from the input image. This new network structure is done by stacking its building blocks, namely inception blocks, as shown in Figure 7. As shown in this figure, instead of using a single convolution operation between a previous layer and the next layer as has been used in conventional CNN (Figure 7a), the inception block performed various convolution operations with various kernel sizes as shown in Figure 7b. Figure 7b shows the naïve inception block to demonstrate the methodology of the inception method in which the output feature maps are obtained by concatenating the outputs of several convolution and pooling layers [42]. Obviously, we could extract texture information at various object sizes (scales) by using multiple convolution operations at different sizes of the convolution kernel. As we could observe from this figure, the feature map at the output of the inception block was much richer in information than the conventional convolution block. This structure is not our contribution, but was proposed by the authors of the inception network [42].

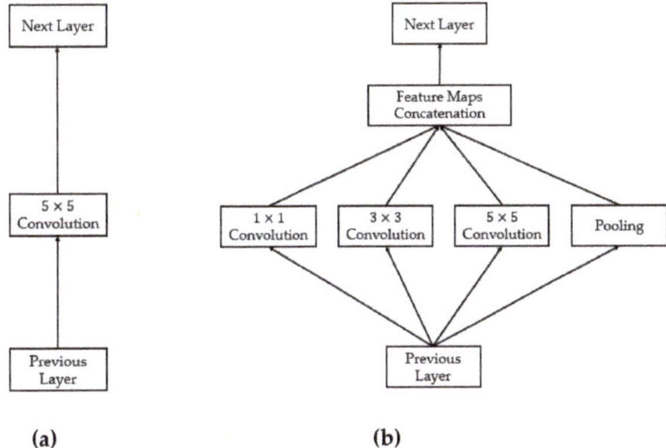

Figure 7. Comparison between: (**a**) the conventional convolution block versus (**b**) the naïve inception block.

2.4.2. Proposed Method for Thyroid Nodule Classification Using Multiple CNN Models

As the second classifier used in our proposed method, the deep learning-based method was applied in the case of the first classifier producing an 'ambiguous benign–malign' case as its answer. This result indicates that the input thyroid ultrasound images were difficult to classify based on the first classifier and needs to be processed further by the second classifier. In our study, we proposed the use of a combination of multiple CNN models for the classification purpose as shown in Figure 8.

As shown in this figure, we tended to enhance the classification performance of single CNN model by combining the classification results of multiple models that have different network architectures. For this purpose, we used two efficient CNN architectures, including the residual network and inception network, as explained in Section 2.4.1 and Figures 6 and 7. As shown in the previous studies, the residual network works well for thyroid nodule classification [21,36]. Therefore, we used this network in our study. Besides, in our study for the thyroid nodule classification problem, the nodule's size was varied according to the condition (status) of the thyroid nodules. To reduce the effect of this variation on our classification algorithm, we also used the inception network to learn the characteristics of thyroid nodules. In Tables 2 and 3, we show the detailed descriptions of the ResNet50-based and Inception-based network architectures used in our study. By combining the results of these two networks, we could enhance the classification result for CAD for the thyroid nodule classification system.

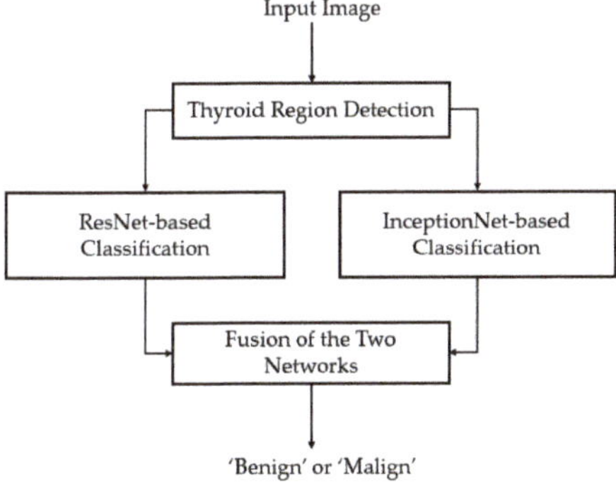

Figure 8. Flow-chart of the deep learning-based system constructed by combining classification results of multiple CNN networks.

Table 2. ResNet50-based CNN architecture used in our experiments.

Layer	Input Shape	Output Shape	Number of Parameters
Convolution Layers by ResNet-50 Network	(224, 224, 3)	(7, 7, 2048)	23,587,712
Global Average Pooling	(7, 7, 2048)	2048	0
Batch Normalization	2048	2048	8192
Dropout	2048	2048	0
Output Layer (Dense layer)	2048	2	4098

Table 3. Inception-based CNN architecture used in our experiments.

Layer	Input Shape	Output Shape	Number of Parameters
Convolution Layers by Inception Network	(224, 224, 3)	(5, 5, 2048)	21,802,784
Global Average Pooling	(5, 5, 2048)	2048	0
Batch Normalization	2048	2048	8192
Dropout	2048	2048	0
Output Layer (Dense layer)	2048	2	4098

To combine the results produced by ResNet50-based and Inception-based network, we used three combination methods, including the MIN, MAX, and SUM rules as shown in Equations (2)–(4). These combination methods have been widely used to combine classification scores of multiple biometric models or multiple classifiers for a single biometric [51–53]. In detail, at the output of each network we could obtain a classification score that presents the probability of the input image belonging to the benign or malign class. We referred to S_1 and S_2 as the decision scores produced by the ResNet50-based model and the Inception-based model, respectively. Then the MIN rule was performed by selecting the smallest score between these two scores; the MAX rule was performed by selecting the largest score between the two scores; and the SUM rule was performed by taking the average score of the two scores. If the final combination score was larger than the predetermined threshold, the input image was classified as the benign class. If not, it was regarded as the malign class. This optimal threshold was experimentally determined with training data.

$$MIN = \min(S_1, S_2) \tag{2}$$

$$MAX = \max(S_1, S_2) \tag{3}$$

$$SUM = \frac{S_1 + S_2}{2} \tag{4}$$

2.4.3. Weighted Binary Cross-Entropy Loss Function for Compensating the Imbalance of Training Samples

For the image classification problem, the previous studies mostly used the cross-entropy function to measure the loss (the difference between ground-truth and predicted labels) function [39–42]. In our specific case of thyroid nodule classification, the cross-entropy function is reduced to binary cross-entropy because we only have two classes of benign and malign. As a result, the formula for the loss function using Binary Cross-Entropy (BCE) is shown as Equation (5). In this equation, y and *(1-y)* indicate the ground-truth labels of two classes (benign and malign); and p and *(1-p)* indicate the predict labels (probability) of these classes. This is a very nice loss function that incorporates the probability theory into its calculation. However, this function only works well if the number of training data of the two classes is balanced because it considers the losses caused by each class equally. In the case of imbalance in the training data of the two classes such as the medical image processing system that is normally faced with the problem of data collection due to special characteristics of medical images, the binary cross-entropy function can produce bias in the trained classifier. To solve this problem, our proposed method uses a modified version of the binary cross-entropy, called the Weighted Binary Cross-Entropy (wBCE), as shown in Equation (6). As shown in this equation, we assigned different weight values to the losses caused by samples in each class in the binary cross-entropy function. As a result, the weighted cross-entropy function makes the training process to focus more on the class, which has a small number of samples than the other classes, and consequently reduces the bias in the trained classifier. The weight values can be determined experimentally based on the actual number of samples in each class in the training dataset. In our experiments with the TDID dataset, the optimal weights of w_0 (0.7) and w_1 (0.3) were determined with the training data, and this result corresponds to the fact that the number of malign samples is much larger than the number of benign samples. In addition, we show that the weighted binary cross-entropy function is sufficient to reduce bias in the classification results in Section 3.3.

$$BCE = -ylog(p) - (1 - y)\log(1 - p) \tag{5}$$

$$wBCE = -w_0 ylog(p) - w_1(1 - y)\log(1 - p) \tag{6}$$

The weighted cross-entropy loss function is not a new method to deal with the imbalanced data problem in a deep learning-based classification system. There exist several similar studies focusing on the problem such as the use of focused anchors loss [54], focal loss [55], and class-balanced loss [56]. In these studies, the main idea is that they can down-weight the well-classified examples to make the classification networks focus on the hard sample ones (focal loss, focused anchors loss); or assign weights to samples according to the volume of classes (class-balanced loss). As stated in our paper, the medical image processing systems normally face a common problem caused by the lack of training data. Due to this problem, the imbalanced data problem is normally occurred and consequently reduces the performance of the classification system. Therefore, the use of weighted cross-entropy loss function in our study can be seen as a simple application of this type of technique that is applicable to enhance the performance of the medical image processing systems.

3. Experimental Results

Based on the proposed method explained in Section 2, in this section, we present various experiments with a public ultrasound Thyroid Nodule Image Dataset (TDID dataset) to measure the classification performance of our proposed method. The experimental results are given in the subsections as follows.

3.1. Dataset and Experimental Setups

Although studies for the ultrasound image-based thyroid nodule classification problem exist [21–25,29], most of the datasets used in these studies are private. In addition, it is very difficult to collect a large amount of data owing to the lack of time and the special characteristics of the medical problems, in which expensive image collection systems and the patient's cooperation are required. Therefore, we decided to use a public thyroid nodule image dataset, namely the Thyroid Digital Image Database (TDID), which was collected and published by Pedraza et al. [20] at the Universidad Nacional de Colombia. This dataset has been widely used in the previous studies for the thyroid nodule classification problem [21,28,36]. Therefore, we can not only evaluate the classification performance of our proposed method, but also compare it with lots of the previous studies to investigate the efficiency of our study.

The TDID dataset was published in 2015 and contains ultrasound thyroid images of 298 patients. For each patient, one or more ultrasound images of the thyroid region were collected in the RGB format with the image size of 560 pixels × 360 pixels. As a result, we extracted a total of 450 thyroid nodule images for our experiments. To assess the condition of the thyroid region, a Thyroid Imaging Reporting And Data System (TI-RADS) score is given for each image that was evaluated by radiologists. The TI-RADS score is defined as a standard to evaluate the condition of thyroid nodules and can take one among seven possible values of {1, 2, 3, 4a, 4b, 4c, and 5}. Among these possible values, the TI-RADS score of 1, 2, and 3 indicate that the thyroid nodules are normal (TI-RADS score of 1), benign (TI-RADS score of 2), and no suspicious ultrasound features (TI-RADS score of 3), respectively. As indicated by their meaning, ultrasound thyroid images with TI-RADS scores of 1, 2, and 3 were grouped together to indicate that they belong to the benign case. The other four possible values of 4a, 4b, 4c, and 5 indicate that the thyroid nodule has one, two, three, and five suspicious features, respectively. Due to their meaning, the thyroid nodule images with these four TI-RADS scores were normally grouped together to indicate the malign case of thyroid nodule. In our study, we also used the TI-RADS score to preclassify thyroid images into either the benign or the malign category for the classification problem (ground-truth labels).

As explained in Section 2, our proposed method was based on a learning framework to determine the best classifier for the classification problem. Therefore, we divided the TDID dataset into the training and testing dataset for this purpose. In detail, we used a five-fold cross-validation scheme to train and measure the performance of our classification system. As a result, we randomly divided the benign and malign case data into five parts. Among these five parts, four parts were assigned as the training data, and the remaining part was assigned as the test data in the 1st fold validation. This process was repeated five times to train and measure the performance of our proposed method as a five-fold cross validation scheme. Then, the average testing accuracy of five folds was determined as a final testing accuracy. In Table 4, we show the detail information of our experimental data in TDID dataset. Although a validation set is usually used during the training process of a neural network, we did not use a validation set in our experiments. The reason is that the number of images in the TDID dataset was small consisting of 450 images. Even we could split this dataset into training/validation/testing sets, this division method consequently reduced the size of training and testing sets, which could result in the insufficiency of training a neural network, and we used only training and testing sets in our experiments like previous methods [21,28,36]. To train the CNN models mentioned in Section 2.4.2, we performed the fine-tuning technique to reduce the effects of under- or overfitting problem. The parameters for the training process are given in Table 5.

Table 4. Description of the TDID dataset used in our experiments (each number means the number of patients).

Benign Case		Malign Case		Total
Training Data	Testing Data	Training Data	Testing Data	
41	11	196	50	298

Table 5. Parameters for training CNN models in our study.

Optimizer	Number of Epochs	Batch Size	Initial Learning Rate	Stop Criteria
Adam	30	32	0.0001	End of Epochs

3.2. Criteria for Classification Performance of a Thyroid Nodule Classification Method

To measure the performance of a thyroid classification system, there are three popular metrics that have been in use, including the sensitivity, specificity, and the overall classification accuracy [21,23,28,36,57]. Similar to the previous studies, we also used these three performance measurements in our experiments to measure the performance of our proposed method as well as to compare our classification performance with the previous studies. Formulas for these measurements are given in Equations (7)–(9). Since CAD for a thyroid nodule classification system normally focuses on two different aspects of the classification problem, that is, the correct classification of benign case images and a correct classification of malign case images, the specificity and sensitivity measurements were used to measure the accuracy of these aspects. First, the sensitivity was measured as the ratio between the true positive (*TP*) samples (samples that are malign case images are correctly classified as the malign ones) over the total number of the malign cases image (TP + false negative (FN)) in a test dataset as shown in Equation (7). Second, the specificity is the measurement of true negative (TN) samples (samples that are benign cases are correctly classified as benign ones) over the total number of benign case images (TN + false positive (FP)) in a test dataset, as shown in Equation (8). As their definition and measurement methods, the sensitivity reflects the ability of a classification system in correctly detecting malign cases, while the specificity reflects the ability of a classification system to correctly detect (classify) benign cases. To access an overall (average) ability of the classification system, the third measurement (overall accuracy) was used and measured by the total number of correct classification/detection samples (true positive and true negative samples) over the total number of samples in a test dataset as shown in Equation (9). As indicated by the above explanations, high values of specificity, sensitivity, and accuracy were expected for a good classification system. In our experiments, we measured these criteria by using our proposed method with the TDID public dataset for performance measurement and comparison with other studies.

$$Sensitivity = \frac{TP}{TP + FN} \tag{7}$$

$$Specificity = \frac{TN}{TN + FP} \tag{8}$$

$$Accuracy = \frac{TP + TN}{TP + TN + FP + FN} \tag{9}$$

3.3. Classification Results Based on Multiple Artificial-Intelligence Models

As our first experiment, we measured the classification performance of the proposed deep learning-based network for the thyroid nodule classification problem. For this purpose, we first performed experiments using individual CNN network, i.e., ResNet50-based architecture as shown in Table 2, and Inception-based architecture as shown in Table 3. As shown in Section 2.4.3, we set the weight values of weighted binary cross-entropy w_0 and w_1 as 0.70 and 0.30, respectively, because the number of training samples in the benign cases is much smaller than the number of training samples in the malign case in the TDID dataset as shown in Table 4. In addition, to demonstrating the efficiency of the weighted binary cross-entropy loss function over the conventional binary cross-entropy loss function, we additionally performed an experiment in the case of equal weight values, i.e., w_0 of 0.50 and w_1 of 0.50. The detailed experimental results are given in Table 6 for both the ResNet50-based network and the Inception-based network. As shown in Table 6, using the ResNet50-based network, we obtained an overall accuracy of about 87.778% with the sensitivity of 91.356% and specificity of 64.018% in the case of using the conventional binary cross-entropy loss function. These experimental results are little different from those reported by Nguyen et al. [36]. The reason is caused by the unstableness of the training process in which the network parameters were randomly initialized at the beginning of the training process at some new layers as shown in Tables 2 and 3. Using the weighted binary cross-entropy loss function, we obtained an overall accuracy of 82.412% with a sensitivity of 83.950% and specificity of 72.524%. As we can see from these experimental results, the difference between the sensitivity and specificity in the case of using conventional binary cross-entropy loss function was about 27.338% (91.356%–64.018%). This result demonstrates that there was a bias in the classification result using the conventional binary cross-entropy loss function. Using the proposed weighted binary cross-entropy loss function, the difference between the sensitivity and specificity was much more reduced to about 11.426% (83.950%–72.524%). This result indicates that the bias was much more reduced by using the weighted cross-entropy loss function compared to the conventional binary cross-entropy loss function.

Table 6. Classification performance of the individual CNN network using the TDID dataset (unit: %).

Method	ResNet50-Based Network			Inception-Based Network		
	Accuracy	Sensitivity	Specificity	Accuracy	Sensitivity	Specificity
Using BCE [39–42]	87.778	91.356	64.018	81.506	83.406	68.760
Using wBCE (proposed method)	82.412	83.950	72.524	80.792	81.842	74.016

A similar phenomenon also occurred in our experiments with the Inception-based network. Using the conventional binary cross-entropy loss function, we obtained an overall classification accuracy of 81.506% with a sensitivity of 83.406% and a specificity of 68.760%. Using the proposed weighted binary cross-entropy loss function, we obtained an overall classification accuracy of about 80.792% with a sensitivity of 81.842% and a specificity of 74.016%. Similar to the experimental results by ResNet50-based network, the difference between the sensitivity and specificity was about 14.646% (83.460%–68.760%) in the case of using the conventional binary cross-entropy loss function that was much larger than the 7.826% (81.842%–74.016%) obtained in the case of using weighted binary cross-entropy loss function. Through these experimental results with the ResNet50-based network and Inception-based network, we could see that the proposed weighted binary cross-entropy loss function was more efficient for reducing the overfitting problem by reducing the difference between the sensitivity and specificity of the testing dataset.

Based on the trained models obtained by training the ResNet50-based network and Inception-based network, we further performed experiments by combining the classification results of these two models to investigate the enhancement ability of the combined network compared to the individual model. As explained in Section 2.4.2, we used three combination methods, including the MIN, MAX, and SUM rule, to combine the results of ResNet50-based network and Inception-based network as shown in Equations (2)–(4). The detailed experimental results are given in Table 7. Again, we performed experiments for the two cases of with and without the proposed weighted binary cross-entropy loss function. As shown in Table 7, we obtained the overall classification accuracy of 83.938%, 90.603%, and 82.677% for the case of using MIN, MAX, and SUM rule, respectively, with the use of the conventional binary cross-entropy loss function. The highest overall classification accuracy of 90.603% that was obtained using the MAX combination rule was much higher than the 87.778% obtained by using only ResNet50-based network or 81.506% using the Inception-based network. This result demonstrates that the combination of the results of the two networks helped to enhance the classification performance of our problem. In addition, the difference between the sensitivity and specificity using the MAX rule was about 36.728% (95.446%–58.718%). Similarly, we obtained an overall accuracy of 75.200%, 91.192%, and 78.709% for the case of using MIN, MAX, and SUM rule, respectively, with the use of the proposed weighted binary cross-entropy loss function. Again, the best classification accuracy was obtained using the MAX combination rule with the accuracy of about 91.192%. This classification accuracy was the highest accuracy among those obtained by only ResNet50-based model, Inception-based model even using conventional binary cross-entropy loss function or the proposed weighted binary cross-entropy loss function. In addition, the difference between the sensitivity and specificity was reduced to 29.396% (95.083%–65.687%), which was smaller than the 36.728% obtained using the conventional binary cross-entropy loss function. Through these experimental results, we could conclude that the combination of the multiple CNN networks could help to enhance the classification accuracy of the thyroid nodule classification, and the MAX rule outperformed the MIN and SUM rule for combining the results of individual models. In addition, the weighted binary cross-entropy loss function was efficient for reducing the overfitting problem caused by the unbalanced training samples of the target classes in the training data.

Table 7. Classification performance by combining the two CNN networks using MIN, MAX, and SUM rules (unit: %).

Method	MIN Rule			MAX Rule			SUM Rule		
	Accuracy	Sensitivity	Specificity	Accuracy	Sensitivity	Specificity	Accuracy	Sensitivity	Specificity
Using BCE [39–42]	83.938	85.868	71.142	90.603	95.446	58.718	82.677	83.894	74.219
Using wBCE (proposed method)	75.200	74.859	77.226	91.192	95.083	65.687	78.709	79.167	75.967

3.4. Classification Results by the Proposed Method

Based on our experimental results in Section 3.3, we finally performed the experiments to measure the performance of our proposed method as explained in Section 2.1 and Figure 2. The detailed experimental results are given in Table 8. Similar to the experiments in Section 3.3, we performed our experiments for two cases of with and without the proposed weighted binary cross-entropy loss function. For the case of the conventional binary cross-entropy loss function, we obtained overall classification accuracies of 86.928%, 90.603%, and 86.073% for the cases using MIN, MAX, and SUM rules, respectively. Compared with the classification results in Tables 6 and 7, we see that the proposed method enhanced the classification results for the cases of MIN and SUM rules. For the case of MAX rule, the proposed method produced the same classification accuracy as the combination of multiple CNN models, which was still much higher than the performance of individual CNN models.

For the case of using the proposed weighted binary cross-entropy loss function, our proposed method produced classification accuracies of 83.517%, 92.051%, and 85.286%, for the MIN, MAX, and SUM rules, respectively. These classification accuracies were higher than those produced by individual CNN models and the combination of them as shown in Tables 6 and 7. Especially, the highest classification accuracy of about 92.051% obtained by using the proposed method with a MAX combination rule was the highest classification result we obtained in all of our experiments in Tables 6–8. Compared to the case of using our proposed method but with the conventional binary cross-entropy loss function, the classification accuracy using our proposed method was also higher (92.051% versus 90.603%). This result again confirmed that our proposed method with the weighted binary cross-entropy loss function was efficient for reducing the overfitting problem, and consequently, enhancing the classification accuracy.

Table 8. Classification performance of our proposed method using the TDID dataset with MIN, MAX, and SUM rules (unit: %).

Method	MIN Rule			MAX Rule			SUM Rule		
	Accuracy	Sensitivity	Specificity	Accuracy	Sensitivity	Specificity	Accuracy	Sensitivity	Specificity
Using BCE [39–42]	86.928	89.331	71.142	90.603	95.446	58.718	86.073	87.831	74.219
Using wBCE (proposed method)	83.517	84.466	77.226	92.051	96.072	65.687	85.286	86.748	75.967

3.5. Performance Comparisons of Proposed Method with the State-of-the Art Methods

As explained in Section 1, there have been several previous studies that proposed their methods for solving the thyroid nodule classification problem. As one of the earliest studies, Zhu et al. [21] used the ResNet18-based network for the problem. To reduce the effect of overfitting, the transfer learning technique was applied, and they reported a classification accuracy of about 84.00% using the TDID dataset. To deal with the change in nodule sizes, Chi et al. [23] used the GoogLeNet, another name for the Inception network, for the problem. Using the method by Chi et al. [23], Nguyen et al. [36] evaluated the classification performance with the TDID dataset and reported an accuracy of about 79.36% in their experiments. In the study by Sundar et al. [28], they additionally performed experiments with the VGG16-based network for the thyroid nodule classification problem using their dataset. Using the VGG16-based network, a classification accuracy of 77.57% was obtained using the TDID dataset [36].

These mentioned studies have a similar characteristic in that they used a single CNN network with or without the transfer learning technique for the ultrasound image-based thyroid nodule classification problem. As a result, the performance of these studies depended extensively on the architecture of the selected CNN network as well as the training data. Most recently, Nguyen et al. [36] proposed a method based on a cascade classifier architecture that employs both handcrafted and deep learning-based methods. In that study, they first classified the input images using information extracted in the frequency domain. After that, the ambiguous samples were further processed by a deep learning-based network. The advantage of the study by Nguyen et al. [36] is that they combined information in both the frequency and spatial domains for the classification problem. However, they did not consider the difference in deep learning network architectures as well as the imbalance of image samples in the target classes as we did in this study. Nguyen et al. [36] reported a high classification accuracy of about 90.88% using their proposed method with the TDID dataset. Compared to the mentioned classification results by the previous studies, our proposed method produced much better classification accuracy. As shown in Section 3.4, our proposed method produced a classification accuracy of 92.051%. In Table 9, we summarized the previous classification performances in comparison with our proposed method. From the result in this table, we could conclude that our proposed method outperformed the previous studies for the ultrasound image-based thyroid nodule classification problem.

Table 9. Comparison of the overall accuracy of the previous studies and our proposed method with the TDID dataset (unit: %).

Methods		Accuracy
Zhu et al. [21]		84.00
Chi et al. [23]		79.36
Sundar et al. [28]	VGG16	77.57
	GoogLeNet	79.36
Nguyen et al. [36]		90.88
Proposed Method		92.05

3.6. Analysis and Discussion

As shown in Table 9, our proposed method outperformed all of the methods presented in the previous studies using the TDID dataset. To get a deep visualization about the performance of our proposed method compared to a previous study by Nguyen et al. [36], we show some example classification results performed by both studies in Figure 9. In Figure 9a, we show the cases in which the ground-truth benign case images were incorrectly classified in the study by Nguyen et al. [36]. However, using our proposed method, we correctly classified them as benign cases. As we can observe from these images, although it is hard to label them as benign or malign case based on human perception as well as the system by Nguyen et al. [36], our proposed method can still recognize them as benign case images. Similar to Figure 9a but with examples of the malign case, Figure 9b shows the example classification results of malign case images. As shown in this figure, our proposed method also correctly classified them as the malign cases, while the method by Nguyen et al. [36] produced incorrect classification labels. By human perception, we could find that these images contain nodules with the calcification phenomenon (white blob region inside a round region (nodule)) that indicates that they should be malign case images. However, the method by Nguyen et al. [36] made an incorrect decision. This example shows that our proposed method was more effective than the method used in the study by Nguyen et al. [36]. Through this example and our experimental results in Section 3.5, we concluded that our proposed method was more effective than the previous studies for the thyroid nodule classification problem using ultrasound images.

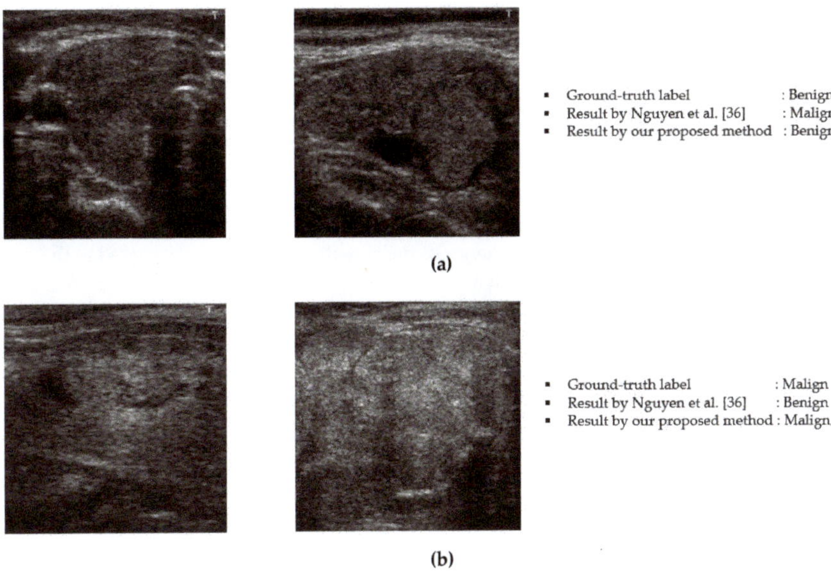

Figure 9. Example results obtained by our proposed method: (**a**) example results of the benign case and (**b**) example results of the malign case.

As the final experiment in our study, we measured the processing time of our proposed method to evaluate the real system applicability of our algorithm. For this purpose, we used a desktop computer with an Intel Core i7-6700 CPU, working clock of 3.4 GHz with 64 GB of RAM memory. To speed up the deep learning networks, we used a GPU, namely GeForce Titan X, to run the inference of the two deep learning models [58]. To implement our algorithm, we used Python programming language with the Tensorflow library for implementing the deep CNN networks [59]. The consequent experimental results are given in Table 10. As shown in Section 2, our proposed method mainly consists of three main steps, which include the preprocessing step, a coarse classification by FFT-based method, and a fine classification by the combination of ResNet and InceptionNet. As shown in Table 10, it took about 11.4646 ms for the preprocessing step (thyroid region extraction and normalization), 5.093 ms for classifying the input image using FFT-based method, 17.525 ms for running the ResNet50-based network, and 23.178 ms for running the Inception-based network. As shown in Figure 2 and Section 2.1, our proposed method could produce the final decision in two scenarios. First, with the easy input samples, our proposed method only used the preprocessing and FFT-based steps to produce its decision. For difficult (complex) samples, our proposed method must invoke the fine classification steps based on deep learning networks. As a result, it takes at least 16.739 ms (11.646 + 5.093) and at most 57.442 ms (11.646 + 5.093 + 17.525 + 23.178) to produce a final prediction by our proposed method. In other words, our proposed method could operate at a speed ranging from 17.4 (1000 ÷ 57.442) to 59.7 (1000 ÷ 16.739) fps. Averagely, we could conclude that our proposed method could operate at a speed of about 38 fps. Through experimental result, we see that our proposed method was suitable for real-system application using a desktop computer.

Table 10. Processing time of our proposed method (unit: ms).

Preprocessing Step.	FFT-Based Classification	ResNet50-Based Classification	Inception-Based Classification	Total
11.646	5.093	17.525	23.178	57.442

4. Conclusions

In this study, we enhanced the classification performance of the ultrasound imaging-based thyroid nodule classification system by analyzing captured images of the thyroid region in two domains, i.e., the spatial domain using the deep learning-based method, and frequency domain using the fast Fourier-based method. Compared to the previous studies, we used two different CNN architectures, which were different in depth and network structures in this study to analyze an ultrasound thyroid image. As a result, the input ultrasound thyroid image was better analyzed compared to the single network as used in the previous studies. Finally, by combining the classification results of two CNN networks, we enhanced the overall performance of the classification system compared to the previous studies. In addition, we applied the weighted binary cross-entropy loss function for learning the CNN models instead of the conventional cross-entropy loss function to reduce the effects of the unbalanced training samples in the training procedure, and consequently reduce the effect of the under/overfitting problem. Through experiments with the public TDID dataset, we proved that our proposed method could give more accurate predictions/suggestions for doctors (radiologists) when diagnosing thyroid nodule problems than the previous studies.

Author Contributions: D.T.N. and T.D.P. designed and implemented the overall system, made the code and performed experiments, and wrote this paper. J.K.K., G.B., and K.R.P. helped with parts of weighted binary cross-entropy loss, experiments, and gave comments during experiments. All authors have read and agreed to the published version of the manuscript.

Acknowledgments: This work was supported in part by the National Research Foundation of Korea (NRF) funded by the Korean Government, Ministry of Science and ICT (MSIT), under Grant NRF-2017R1C1B5074062, in part by the NRF funded by the MSIT through the Basic Science Research Program under Grant NRF-2020R1A2C1006179, and in part by the NRF funded by the MSIT, through the Bio and Medical Technology Development Program under Grant NRF-2016M3A9E1915855.

Conflicts of Interest: The authors declare no conflict of interest.

References

1. Vuong, Q.H.; Ho, M.T.; Vuong, T.T.; La, V.P.; Ho, M.T.; Nghiem, K.C.P.; Tran, B.X.; Giang, H.H.; Giang, T.V.; Latkin, C.; et al. Artificial intelligence vs. natural stupidity: Evaluating AI readiness for the Vietnamese medical information system. *J. Clin. Med.* **2019**, *8*, 168. [CrossRef] [PubMed]

2. Havaei, M.; Davy, A.; Warde-Farley, D.; Biard, A.; Courville, A.; Bengio, Y.; Pal, C.; Jodoin, P.-M.; Larochelle, H. Brain tumor segmentation with deep neural networks. *Med. Image Anal.* **2017**, *35*, 18–31. [CrossRef] [PubMed]

3. Cheng, C.-H.; Liu, W.-X. Identifying degenerative brain disease using rough set classifier based on wavelet packet method. *J. Clin. Med.* **2018**, *7*, 124. [CrossRef] [PubMed]

4. Xian, M.; Zhang, Y.; Cheng, H.D.; Xu, F.; Zhang, F.; Ding, J. Automatic breast ultrasound image segmentation: A survey. *Pattern Recognit.* **2018**, *79*, 340–355. [CrossRef]

5. Milletar, F.; Ahmadi, S.-A.; Kroll, C.; Plate, A.; Rozanski, V.; Maiostre, J.; Levin, J.; Dietrich, O.; Ertl-Wagner, B.; Botzel, K.; et al. Hough-CNN: Deep learning for segmentation of deep brain regions in MRI and ultrasound. *Comput. Vis. Image Underst.* **2017**, *16*, 92–102. [CrossRef]

6. Owais, M.; Arsalan, M.; Choi, J.; Park, K.R. Effective diagnosis and treatment through content-based medical image retrieval (CBMIR) by using artificial intelligence. *J. Clin. Med.* **2019**, *8*, 462. [CrossRef]

7. Zhang, W.; Li, R.; Deng, H.; Wang, L.; Lin, W.; Ji, S.; Shen, D. Deep convolutional neural networks for multi-modality isointense infant brain image segmentation. *Neuroimage* **2015**, *108*, 214–224. [CrossRef]

8. Moon, W.K.; Chang, S.-C.; Huang, C.-S.; Chang, R.-F. Breast tumor classification using fuzzy clustering for breast elastography. *Ultrasound Med. Biol.* **2011**, *37*, 700–708. [CrossRef]

9. Hrzic, F.; Stajduhar, I.; Tschauner, S.; Sorantin, E.; Lerga, J. Local-entropy based approach for x-ray image segmentation and fracture detection. *Entropy* **2019**, *21*, 338. [CrossRef]

10. Bhandary, A.; Prabhu, G.A.; Rajinikanth, V.; Thanaraj, K.P.; Satapathy, S.C.; Robbins, D.E.; Shasky, C.; Zhang, Y.D.; Tavares, J.M.R.; Raja, N.S.M. Deep-learning framework to detect lung abnormality—A study with chest x-ray and lung CT scan images. *Pattern Recogn. Lett.* **2020**, *129*, 271–278. [CrossRef]

11. Jung, N.Y.; Kang, B.J.; Kim, H.S.; Cha, E.S.; Lee, J.H.; Park, C.S.; Whang, I.Y.; Kim, S.H.; An, Y.Y.; Choi, J.J. Who could benefit the most from using a computer-aided detection system in full-field digital mammography? *World, J. Surg. Oncol.* **2014**, *12*, 168. [CrossRef] [PubMed]

12. Kamnitsas, K.; Ledig, C.; Newcombe, V.F.J.; Simpson, J.P.; Kane, A.D.; Menon, D.K.; Rueckert, D.; Glocker, B. Efficient multi-scale 3D CNN with fully connected CRF for accurate brain lesion segmentation. *Med. Image Anal.* **2017**, *36*, 61–78. [CrossRef] [PubMed]

13. Moon, W.K.; Huang, Y.-S.; Lee, Y.-W.; Chang, S.-C.; Lo, C.-M.; Yang, M.-C.; Bae, M.S.; Lee, S.H.; Chang, J.M.; Huang, C.-S.; et al. Computer-aided tumor diagnosis using shear wave breast elastography. *Ultrasonics* **2017**, *78*, 125–133. [CrossRef] [PubMed]

14. Acharya, U.R.; Ng, W.L.; Rahmat, K.; Sudarshan, V.K.; Koh, J.E.; Tan, J.H.; Hagiwara, Y.; Yeong, C.H.; Ng, K.H. Data mining framework for breast lesion classification in shear wave ultrasound: A hybrid feature paradigm. *Biomed. Signal Process. Control* **2017**, *33*, 400–410. [CrossRef]

15. Moon, W.K.; Cheng, I.-L.; Chang, J.M.; Shin, S.U.; Lo, C.-M.; Chang, R.-F. The adaptive computer-aided diagnosis system based on tumor sizes for the classification of breast tumors detected at screening ultrasound. *Ultrasonics* **2017**, *76*, 70–77. [CrossRef]

16. Xu, Y.; Wang, Y.; Yuan, J.; Cheng, Q.; Wang, X.; Carson, P.L. Medical breast ultrasound image segmentation by machine learning. *Ultrasonics* **2019**, *91*, 1–9. [CrossRef]

17. Koundal, D.; Gupta, S.; Signh, S. Computer aided thyroid nodule detection system using medical ultrasound images. *Biomed. Signal Process. Control* **2018**, *40*, 117–130. [CrossRef]

18. Tessler, F.N.; Middleton, W.D.; Grant, E.G.; Hoang, J.K.; Berland, L.L.; Teefey, S.A.; Cronan, J.J.; Beland, M.D.; Desser, T.S.; Frates, M.C.; et al. ACR thyroid imaging, reporting and data system (TI-RADS): White paper of the ACR TI-RADS committee. *J. Am. Coll. Radiol.* **2017**, *14*, 587–595. [CrossRef]

19. Ma, J.; Wu, F.; Zhu, J.; Xu, D.; Kong, D. A pretrained convolutional neural network based method for thyroid nodule diagnosis. *Ultrasonics* **2017**, *73*, 221–230. [CrossRef]

20. Pedraza, L.; Vargas, C.; Narvaez, F.; Duran, O.; Munoz, E.; Romero, E. An open access thyroid ultrasound-image database. In Proceedings of the 10th International Symposium on Medical Information Processing and Analysis, Cartagena de Indias, Colombia, 28 January 2015; pp. 1–6.

21. Zhu, Y.; Fu, Z.; Fei, J. An image augmentation method using convolutional network for thyroid nodule classification by transfer learning. In Proceedings of the 3rd IEEE International Conference on Computer and Communication, Chengdu, China, 13–16 December 2017; pp. 1819–1823.

22. Sudarshan, V.K.; Mookiah, M.R.K.; Acharya, U.R.; Chandran, V.; Molinari, F.; Fujita, H.; Ng, K.H. Application of wavelet techniques for cancer diagnosis using ultrasound images: A review. *Comput. Biol. Med.* **2016**, *69*, 97–111. [CrossRef]

23. Chi, J.; Walia, E.; Babyn, P.; Wang, J.; Groot, G.; Eramian, M. Thyroid nodule classification in ultrasound images by fine-tuning deep convolutional neural network. *J. Digit. Imaging* **2017**, *30*, 477–486. [CrossRef] [PubMed]

24. Chang, C.-Y.; Chen, S.-J.; Tsai, M.-F. Application of support-vector-machine-based method for feature selection and classification of thyroid nodules in ultrasound images. *Pattern Recognit.* **2010**, *43*, 3494–3506. [CrossRef]

25. Luo, S.; Kim, E.H.; Dighe, M.; Kim, Y. Thyroid nodule classification using ultrasound elastography via linear discriminant analysis. *Ultrasonics* **2011**, *51*, 425–431. [CrossRef] [PubMed]

26. Ouyang, F.-S.; Guo, B.-L.; Ouyang, L.-Z.; Liu, Z.-W.; Lin, Z.-W.; Meng, W.; Huang, X.-Y.; Chen, H.-X.; Hu, Q.-G.; Yang, S.-M. Comparison between linear and nonlinear machine-learning algorithms for the classification of thyroid nodule. *Eur. J. Radiol.* **2019**, *113*, 251–257. [CrossRef] [PubMed]

27. Song, W.; Li, S.; Liu, J.; Qin, H.; Zhang, B.; Zhang, S.; Hao, A. Multitask cascade convolution neural networks for automatic thyroid nodule detection and recognition. *IEEE J. Biomed. Health Inform.* **2019**, *23*, 1215–1224. [CrossRef]

28. Sundar, K.V.S.; Rajamani, K.T.; Sai, S.-S.S. Exploring image classification of thyroid ultrasound images using deep learning. In Proceedings of the International Conference on ISMAC in Computational Vision and Bio-Engineering, Palladam, India, 16–17 May 2018; pp. 1635–1641.

29. Song, J.; Chai, Y.J.; Masuoka, H.; Park, S.-W.; Kim, S.-J.; Choi, J.Y.; Kong, H.-J.; Lee, K.E.; Lee, J.; Kwak, N.; et al. Ultrasound image analysis using deep learning algorithm for the diagnosis of thyroid nodules. *Medicine* **2019**, *98*, e15133. [CrossRef]

30. Wang, L.; Yang, S.; Yang, S.; Zhao, C.; Tian, G.; Gao, Y.; Chen, Y.; Lu, Y. Automatic thyroid nodule recognition and diagnosis in ultrasound imaging with the Yolov2 neural network. *World, J. Surg. Oncol.* **2019**, *17*, 12. [CrossRef]

31. Raghavendra, U.; Acharya, U.R.; Gudigar, A.; Tan, J.H.; Fujita, H.; Hagiwara, Y.; Molinari, F.; Kongmebol, P.; Ng, K.H. Fusion of spatial gray level dependency and fractal texture features for the characterization of thyroid lessons. *Ultrasonics* **2017**, *77*, 110–120. [CrossRef]

32. Xia, J.; Chen, H.; Li, Q.; Zhou, M.; Chen, L.; Cai, Z.; Fang, Y.; Zhou, H. Ultrasound-based differentiation of malignant and benign thyroid nodules: An extreme learning machine approach. *Comput. Methods Programs Biomed.* **2017**, *147*, 37–49. [CrossRef]

33. Choi, W.J.; Park, J.S.; Kim, K.G.; Kim, S.-Y.; Koo, H.R.; Lee, Y.-J. Computerized analysis of calcification of thyroid nodules as visualized by ultrasonography. *Eur. J. Radiol.* **2015**, *84*, 1949–1953. [CrossRef]

34. Prochazka, A.; Gulati, S.; Holinka, S.; Smutek, D. Path-based classification of thyroid nodules in ultrasound images using direction independent features extracted by two-threshold binary decomposition. *Comput. Med. Imaging Graph.* **2019**, *71*, 9–18. [CrossRef] [PubMed]

35. Acharya, U.R.; Chowriappa, P.; Fujita, H.; Bhat, S.; Dua, S.; Koh, J.E.W.; Eugence, J.W.J.; Kongmebhol, P.; Ng, K.H. Thyroid lesion classification in 242 patient population using Gabor transform features from high resolution ultrasound images. *Knowl. Based Syst.* **2016**, *107*, 235–245. [CrossRef]

36. Nguyen, D.T.; Pham, D.T.; Batchuluun, G.; Yoon, H.S.; Park, K.R. Artificial intelligence-based thyroid nodule classification using information from spatial and frequency domains. *J. Clin. Med.* **2019**, *8*, 1976. [CrossRef] [PubMed]

37. Enhanced Ultrasound Thyroid Nodule Classification (US-TNC-V2) Algorithm. Available online: http://dm.dongguk.edu/link.html (accessed on 28 December 2019).

38. Otsu, N. A threshold selection method from gray-level histogram. *IEEE Trans. Syst. Man Cybern.* **1979**, *9*, 62–66. [CrossRef]

39. Krizhevsky, A.; Sutskever, I.; Hinton, G.E. ImageNet classification with deep convolutional neural networks. In Proceedings of the Neural Information Processing Systems, Lake Tahoe, NV, USA, 3–8 December 2012; pp. 1097–1105.

40. Simonyan, K.; Zisserman, A. Very deep convolutional neural networks for large-scale image recognition. *arXiv* **2014**, arXiv:1409.1556. Available online: https://arxiv.org/abs/1409.1556v6 (accessed on 20 September 2019).

41. He, K.; Zhang, X.; Ren, S.; Sun, J. Deep residual learning for image recognition. *arXiv* **2015**, arXiv:1512.03385. Available online: https://arxiv.org/abs/1512.03385v1 (accessed on 20 September 2019).

42. Szegedy, C.; Liu, W.; Jia, Y.; Sermanet, P.; Reed, S.; Anguelov, D.; Erhan, D.; Vanhoucke, V.; Rabinovich, A. Going deeper with convolutions. *arXiv* **2014**, arXiv:1409.4842v1. Available online: https://arxiv.org/abs/1409.4842v1 (accessed on 20 September 2019).

43. Srivastava, N.; Hinton, G.; Krizhevsky, A.; Sutskever, I.; Salakhutdinov, R. Dropout: A simple way to prevent neural networks from overfitting. *J. Mach. Learn. Res.* **2014**, *15*, 1929–1958.

44. Ren, S.; He, K.; Girshick, R.; Sun, J. Faster R-CNN: Towards real-time object detection with region proposal networks. *arXiv* **2015**, arXiv:1506.01497. Available online: https://arxiv.org/abs/1506.01497 (accessed on 20 September 2019). [CrossRef]

45. Redmon, J.; Farhadi, A. YOLOv3: An incremental improvement. *arXiv* **2018**, arXiv:1804.02767. Available online: https://arxiv.org/abs/1804.02767 (accessed on 20 September 2019).

46. Chu, M.P.; Sung, Y.; Cho, K. Generative adversarial network-based method for transforming single RGB image into 3D point cloud. *IEEE Access* **2018**, *7*, 1021–1029. [CrossRef]

47. Nguyen, D.T.; Yoon, H.S.; Pham, D.T.; Park, K.R. Spoof detection for finger-vein recognition system using NIR camera. *Sensors* **2017**, *17*, 2261. [CrossRef]

48. Nguyen, D.T.; Pham, D.T.; Lee, M.B.; Park, K.R. Visible-light camera sensor-based presentation attack detection for face recognition by combining spatial and temporal information. *Sensors* **2019**, *19*, 410. [CrossRef] [PubMed]

49. Young, T.; Hazarika, D.; Poria, S.; Cambria, E. Recent trends in deep learning based natural language processing. *arXiv* **2017**, arXiv:1708.02709. Available online: https://arxiv.org/abs/1708.02709 (accessed on 20 December 2019).

50. Otter, D.W.; Medina, J.R.; Kalita, J.K. A survey of the usages of deep learning in natural language processing. *arXiv* **2018**, arXiv:1807.10854. Available online: https://arxiv.org/abs/1807.10854 (accessed on 20 December 2019).

51. Islam, M.R. Feature and score fusion based multiple classifier selection for iris recognition. *Comput. Intell. Neurosci.* **2014**, 380585. [CrossRef]
52. Vishi, K.; Mavroeidis, V. An evaluation of score level fusion approaches for fingerprint and finger-vein biometrics. *arXiv* **2018**, arXiv:1805.10666. Available online: https://arxiv.org/abs/1805.10666 (accessed on 9 March 2020).
53. Nguyen, D.T.; Park, Y.H.; Lee, H.C.; Shin, K.Y.; Kang, B.J.; Park, K.R. Combining touched fingerprint and finger-vein of a finger, and its usability evaluation. *Adv. Sci. Lett.* **2012**, 85–95. [CrossRef]
54. Baloch, B.K.; Kumar, S.; Haresh, S.; Rehman, A.; Syed, T. Focused anchors loss: Cost-sensitive learning of discriminative features for imbalanced classification. In Proceedings of the machine learning research, Nagoya, Japan, 17–19 November 2019; pp. 822–835.
55. Lin, T.-Y.; Goyal, P.; Girshick, R.; He, K.; Dollar, P. Focal loss for dense object detection. *arXiv* **2017**, arXiv:1708.02002. Available online: https://arxiv.org/abs/1708.02002 (accessed on 9 March 2020).
56. Cui, Y.; Jia, M.; Lin, T.-Y.; Song, Y.; Belongie, S. Class-balanced loss based on effective number of samples. *arXiv* **2019**, arXiv:1901.05555. Available online: https://arxiv.org/abs/1901.05555 (accessed on 9 March 2020).
57. Carvajal, D.N.; Rowe, P.C. Research and statistics: Sensitivity, specificity, predictive values, and likelihood ratios. *Pediatr. Rev.* **2010**, *31*, 511–513. [CrossRef]
58. NVIDIA TitanX GPU. Available online: https://www.nvidia.com/en-us/geforce/products/10series/titan-x-pascal/ (accessed on 20 September 2019).
59. Tensorflow Deep-Learning Library. Available online: https://www.tensorflow.org/ (accessed on 20 September 2019).

Article

Presentation Attack Face Image Generation Based on a Deep Generative Adversarial Network

Dat Tien Nguyen, Tuyen Danh Pham, Ganbayar Batchuluun *, Kyoung Jun Noh and Kang Ryoung Park

Division of Electronics and Electrical Engineering, Dongguk University, 30 Pildong-ro 1-gil, Jung-gu, Seoul 04620, Korea; nguyentiendat@dongguk.edu (D.T.N.); phamdanhtuyen@gmail.com (T.D.P.); kjn0908@naver.com (K.J.N.); parkgr@dongguk.edu (K.R.P.)
* Correspondence: ganabata87@dongguk.edu; Tel.: +82-10-4098-7081

Received: 28 January 2020; Accepted: 19 March 2020; Published: 25 March 2020

Abstract: Although face-based biometric recognition systems have been widely used in many applications, this type of recognition method is still vulnerable to presentation attacks, which use fake samples to deceive the recognition system. To overcome this problem, presentation attack detection (PAD) methods for face recognition systems (face-PAD), which aim to classify real and presentation attack face images before performing a recognition task, have been developed. However, the performance of PAD systems is limited and biased due to the lack of presentation attack images for training PAD systems. In this paper, we propose a method for artificially generating presentation attack face images by learning the characteristics of real and presentation attack images using a few captured images. As a result, our proposed method helps save time in collecting presentation attack samples for training PAD systems and possibly enhance the performance of PAD systems. Our study is the first attempt to generate PA face images for PAD system based on CycleGAN network, a deep-learning-based framework for image generation. In addition, we propose a new measurement method to evaluate the quality of generated PA images based on a face-PAD system. Through experiments with two public datasets (CASIA and Replay-mobile), we show that the generated face images can capture the characteristics of presentation attack images, making them usable as captured presentation attack samples for PAD system training.

Keywords: generative adversarial network; presentation attack detection; artificial image generation; presentation attack face images

---

1. Introduction

1.1. Introduction to Face-Based Biometric System

Biometric models, such as fingerprints, faces, irises, and finger-vein models, have been widely used in high-performance systems for recognizing/identifying a person [1,2]. In addition, these recognition systems offer more convenience to users than conventional recognition methods, such as token- and knowledge-based methods [1]. However, with the development of digital technology, biometric recognition systems are facing an increasing threat from attackers using fake samples to successfully circumvent recognition systems.

Face-based recognition systems are popular biometric recognition systems and have been used for a long time to recognize people [3–5]. This type of biometric is based on the fact that facial appearance can be used to easily distinguish people. To prevent attackers, presentation attack detection for face recognition (face-PAD) systems have been proposed; these typically use a collection of real and presentation attack (PA) face images to train a detection model [6–15]. The performance of such face-PAD systems has been shown to be strongly dependent on the training data, in which PA images are captured by simulating several limited types of attacking methods, such as the use of a photo, video display, or mask. The real images that are captured using real human faces represented in front of capturing devices (camera) and PA images inherit differences because of different subjects such as the distribution of illumination, reflection, and noises. However, with the development of technology, the presentation attack face images are become closer to real face images and can possibly deceive the face recognition system, making it fail. In addition, the attack methods are very diverse according to PA instrument (PAI) and attack procedure, such as the use of a three-dimensional (3D) masks instead of two-dimensional (2D) masks, the use of a high quality photo/video instead of a low quality photo/video, or the use of different types of photo or different equipment for displaying videos. As a result, it is difficult to collect a large amount of PA face sample images that simulate all possible types of attacking methods to train the systems. Consequently, the performance of face-PAD systems can be reduced and biased if faced with a new type of attacking method that has not been simulated in the training data during detector training. This is still an open issue and must be studied in more detail to enhance the security of face recognition systems.

1.2. Problem Definition

As explained in Section 1.1, face-PAD systems are necessary for enhancing the security level of face-based recognition systems. However, a high-performance face-PAD system requires a huge amount of training data (real and PA images) in which the PA images can simulate all possible attack methods and scenario. Unfortunately, this kind of data is hard to collect in a real system because the attack methods and presentation attack instruments are diverse and can change and become more sophisticated as the technology develops. To solve this problem, our study aims to artificially generate PA images that are close to the captured PA images by learning the presentation attack characteristics of available captured PA images and the fusion of the real and these PA images. Our study makes the following four novel contributions:

- This is the first attempt to generate PA face images based on a deep-learning framework. By learning the characteristics of real and PA images in a training dataset, our method can efficiently generate PA images, which are difficult to collect using conventional image collection methods due to the diversity of attack methods.
- By training our CycleGAN-based generation network using both captured real and PA images, we learn the characteristics of PA images in addition to the fusion of real and PA images. This approach can consequently help to fill the gap of missing PA samples caused by the diversity of attack methods.
- We propose a new measurement method to evaluate the quality of generated images for biometric recognition systems based on the use of a conventional face-PAD system and the dprime measurement.
- The code and pre-trained models for PA image generation are available as a reference to other researchers [16].

The remainder of this paper is organized as follows: In Section 2, we summarize works related to our study. In Section 3, the proposed method is described in detail along with several necessary preprocessing steps. Using the proposed method in Section 3, we performed various experiments using two public datasets (including CASIA [7] and Replay-mobile [9]) to evaluate the generated PA images, and the results are given in Section 4. Finally, we conclude our work and discuss future work in Section 5.

2. Related Works

As explained in Section 1, researchers have paid much attention to developing face-PAD systems to detect PA samples from face recognition systems to enhance their security [6–15]. Initially, they used several handcrafted image feature extraction methods to extract image features and detect PA samples by applying some classification method based on the extracted image features [6,8,10,11]. For example, color information [10], texture information extracted by local binary pattern (LBP) or dynamic local ternary pattern (DLTP) [6,11], and the defocus phenomenon [8] have been used for face-PADs. In [17], Benlamoudi et al. proposed a method that combined multi-level local binary pattern (MLLBP) and multi-level binarized statistical image features (MLBSIF) for face-PAD. In addition, they compared the detection accuracy of their proposed method with other six handcrafted-based methods using CASIA dataset. However, their detection performances were not sufficient because they were designed by expert knowledge of researchers alone, which can only reflect some limited aspects of the face-PAD problem. Recently, with the development of learning-based methods, especially deep-learning, the detection performance of face-PAD systems has significantly enhanced by using image features extracted by convolutional neural networks (CNNs) instead of the handcrafted image features. Nguyen et al. [15] used a stacked CNN-RNN network to learn deep representation of input face sequences. By combining the deep and handcrafted image features, they showed that the detection performance of sequence-based face-PAD system is greatly enhanced compared to the use of only handcrafted features. Recently, Liu et al. [18] proposed a deep tree learning method for face-PAD systems. By using the zero-shot learning technique, they made the face-PAD system more generalized for unknown attack methods. However, a common limitation of deep-learning-based methods is that they require a large amount of data to efficiently train the detection models, which is usually difficult to obtain because it requires much labor and cost. An additional problem with common face-PAD systems is that attack methods are diverse. As shown in previous studies [6–15], attackers can use various methods to attack a face recognition system depending on the PAI and approach used. Therefore, it is practically impossible to collect sufficient data to simulate all possible attack cases to train a detection model. One possible solution to this problem is an automatic PA face image synthesis and generation method.

With the development of deep-learning frameworks, image generation has attracted many researchers in the computer vision research community [19–32]. A major method for generating images is the generative adversarial network (GAN), which was proposed by Goodfellow et al. [19]. This type of image generation method has successfully been applied to many computer vision tasks, such as image editing [20,21], super-resolution interpolation [22–24], image de-blurring [25], data generation [26–28], image-to-image translation [29–31], and attention prediction [32]. The key to the success of the GAN is that it trains two deep CNNs (the discriminator and generator) in an adversarial manner. Specifically, the discriminator network is responsible for discriminating between two classes of image, 'real' or 'fake', while the generator is responsible for generating 'fake' images that are as close as possible to 'real' images. These two networks are trained to perform their functionalities using a large amount of training data. As a result, the generator can generate fake images that are very similar to real images.

For biometric image generation, there have been studies that generate images for palm-prints [26], irises [27], and fingerprints [28,33]. In [26], the authors used a GAN method to generate realistic palm-print images from a training dataset. In principal, a deep-learning-based system requires a large amount of training data to successfully train a network. However, collecting such data usually requires much effort and expense and sometimes is impossible due to the diversity of input images. To reduce the effects of this problem, several simple techniques, such as cropping-and-scaling, adding noise, and mirroring have been adopted to make the training dataset slightly more generalized. However, these simple methods are not strong enough for full data generalization. As a result [26], the GAN-based method is sufficient for generating realistic palm-print images and consequently helped to reduce the error of the palm-print recognition systems. For other biometric systems, such as the fingerprint and iris, a GAN-based network has also been used to generate images that are close to captured ones [27,28]. In [33], Bontrager et al. showed that fingerprint images generated by a GAN-based network can be used to fool fingerprint recognition systems. This means that the generated images were very similar to actual captured images and also demonstrates that fingerprint recognition systems are vulnerable.

Although GAN-based methods have been widely used for image generation problems, there have been no studies conducted to generate PA images for a face-PAD system. Inspired by the problem of face-PAD systems, we propose a PA image synthesis/generation method for the face-PAD problem based on a GAN. Our study serves two purposes. First, we aim to generate realistic PA images to reduce the effort of PA image acquisition in designing face-PAD systems. Secondly, by training our generation system using not only captured PA face images but also real face images, we tend to generate more trustable PA images, which can fill the gap of missing samples caused by the diversity of attacking methods. Table 1 summarizes the various previous studies that are related to ours.

Table 1. Summary of GAN-based image generation methods for biometrics systems.

Task	Purpose
Fingerprint image generation [28,33]	-Generate realistic fingerprint images that are close to captured fingerprint images [28,33]. -Demonstrate that synthetic fingerprint images are capable of spoofing multiple peoples' fingerprint patterns [33].
Iris image generation [27]	-Generate realistic iris images that are close to captured iris images
Palm-print image generation [26]	-Generate realistic palm-print images and use them as augmented data to train a palm-print recognition system. -Enhance the performance of a palm-print recognition system using generated images.
PA face image generation (Our approach)	-Generate realistic PA face images to reduce the effort required for image acquisition. -Fill the gap of missing samples caused by diversity of attack methods.

3. Proposed Method

In this section, we provide a detailed description of our proposed method for generating PA face images using the deep-learning method based on the CycleGAN network architecture.

3.1. Overview of the Proposed Method

Figure 1 presents an overview of our proposed method for generating PA images.

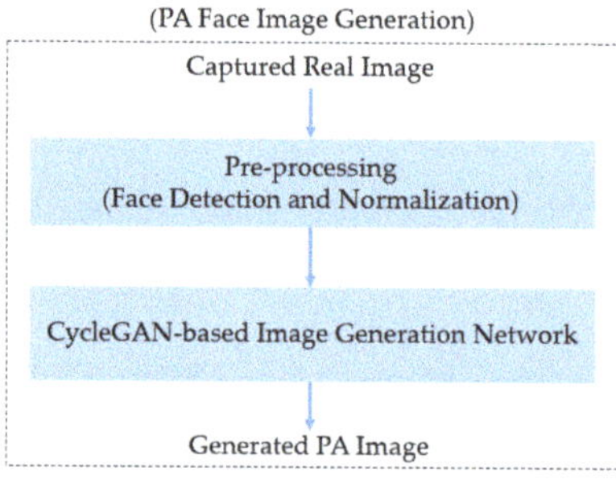

Figure 1. Overview of our proposed method for generating PA face images.

As explained in Section 2, our method aims to generate PA images using a Cycle-GAN network architecture. Therefore, the input of the network is a face image captured in the wild. To learn the characteristics of real and PA images efficiently for the generation problem, the input captured face images are first preprocessed by a face detection and normalization block to detect the face as well as align the face images. This step is explained in more detail in Section 3.2. Using the result of this step, we generate PA images using an image generation network, which is described in Section 3.3.

3.2. Face Detection and Normalization

As shown in Figure 1, our proposed method receives a captured face image as input and generates a PA image at the output. As indicated in previous studies [6–15], the discrimination information between real and PA images mostly appears inside the face region rather than in the background regions. In addition, the purpose of our proposed method is to generate realistic PA images to reduce the effort required in presentation attack image acquisition. Therefore, the input captured face image must be preprocessed to remove the background before using it to generate a PA image.

Generally, an input captured face image contains not only faces but also background regions. Therefore, we perform two preprocessing steps on the input face image: face detection and in-plane rotation compensation [15]. As the first step, we use a face detection method based on the ensemble of regression tree (ERT) proposed by Kazemi et al. [34]. This is an efficient and well-known method for accurate face and face landmark detection. Using this method, we can efficiently locate the face and additional 68 landmark points on the detected face, which can be used to define face shape [15]. Because we are generating PA images by learning the discrimination information between two types of face images, the input face image should be aligned to reduce complexity and misalignment and to let the generator focus on learning the characteristics of the two types of images. Using the detected face and its landmark points, we further perform an in-plane rotation compensation step to compensate the misalignment of the input face image [15]. Figure 2 illustrates the methodology of these steps. In Figure 2a, we illustrate the abstract methodology of the face detection and in-plane rotation compensation steps. A detailed explanation of the mathematical and implementation techniques is provided by Nguyen et al. [15]. In Figure 2b, we show some example results (extracted face images) of the implementation of these steps. As shown in this figure, the final detected face images are aligned to a frontal face without in-plane rotation. This normalization step helps to reduce the effects of non-ideal input images and makes the training procedure more focus on learning the characteristics of images.

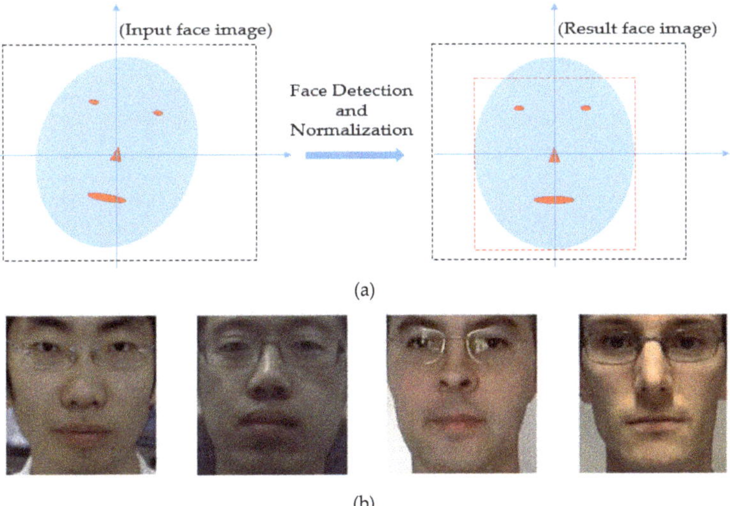

Figure 2. Methodology and an example result of face detection and misalignment compensation steps: (**a**) overview of method for face detection and in-plane rotation compensation with input captured face image (left) and result image (right); (**b**) example results.

3.3. GAN-Based Image Generation Network

As explained in the above sections, our study aims to generate PA images using captured real images as the input of the generation network. Figure 3 illustrates the generation methodology of our study. In Figure 3a, we show a general concept of the distribution of the real and PA images in which the real images are captured by real human faces presenting in front of capturing devices, while the PA images can be obtained by either capturing presentation attack models (photo, video, mask) or generating by an image generation model. In Figure 3b, we show the general concept of our PA image generation framework. As shown in this figure, we can use captured real and PA images to train a generation model to learn the characteristics of these two types of face images and a transformation function from real to PA classes. As a result, we obtain a model to transform a captured real image into a PA image. To construct a generation model, we built a GAN network based on a popular image generation network, namely CycleGAN (as shown in Figure 3b), using two discriminators (D_X and D_Y), which are responsible for distinguishing the real and generated real images (D_X) and distinguishing PA and generated PA images (D_Y), and two generators (G_X and G_Y), which are responsible for generating PA images using real images (G_X) and generating real images using PA images (G_Y). The two discriminator networks share the same discriminator architecture, as described in Section 3.3.1, and the two generator networks share the same generator architecture, as described in Section 3.3.2.

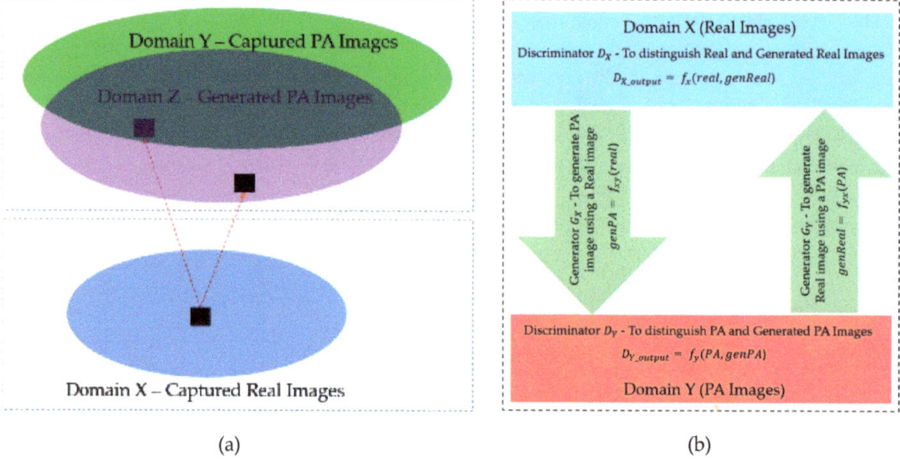

(a)   (b)

Figure 3. Overview of our deep-learning-based image generation network: (**a**) Demonstration of the distribution of real and PA images; (**b**) Overview of our network structure for PA image generation. In this figure, the *real* and *PA* indicate the captured real or presentation attack images; *genReal* and *genPA* indicate the generated real and generated presentation attack images using generator networks and the input *real* and *PA* images; f_x, f_y, f_{xy}, *and* f_{yx} indicates the relation function models by discriminator (D_X, D_Y) and generators (G_X, G_Y), respectively.

To best of our knowledge, our study is the first attempt to use CycleGAN-based network to generated PA face images for face-PAD system. The reason for the use of a CycleGAN-based network to generate PA images in our study is that we tend not only to generate trustable PA images that are close to captured real images as much as possible but also to stimulate new type of presentation attacks that does not exist in training data. For this reason, we should learn the characteristics of both captured real and PA images, and the transformation between the two domains. Because of this reason, we think the CycleGAN-based network is most suitable for our goal.

3.3.1. Discriminator Architecture

The discriminator in the GAN-based network has the responsibility of distinguishing images in one class from images in another. Therefore, a discriminator is essentially a classification network. Inspired by the success of recent deep-learning-based image classification works [35–42], we use a conventional CNN to construct discriminators in our study. In previous studies on the image classification problem [35–42], a deep-learning-based network has been constructed using two separated parts, including convolution layers and fully-connected (dense) layers. Among these parts, the convolution layers are used to extract image features using the convolution operation. Based on the extracted image features, the fully connected layers are used to learn a classifier to classify input features into several predesigned groups. The use of these two parts leads to a high performance in classification systems. However, it has a weakness that it requires a huge number of parameters, which make the classifier complex and difficult to train. To reduce the effects of this problem, we design the discriminators in our GAN-based network by simply using convolution layers to extract image features. As a result, the image classification step is directly executed by comparing the extracted image features with the ground-truth image features of the desired class. In Figure 4, we illustrate the structure of the discriminator used in our study. As shown in this figure, the discriminator contains five convolution layers with a stride value of 2, followed by a leaky rectified linear unit (Leaky ReLU) as the activation function. This network accepts an input color image of 256-by-256 pixels to produce a feature map of 32-by-32 pixels as the output. Table 2 gives a detailed description of the layers and their parameters in the discriminator network. Although pooling layers, such as max or average pooling, are frequently used after convolution layers for feature selection to make the CNN network less invariant to image translation, this is not suitable in our case, which uses only the convolution operation in the discriminator. The reason for this is that pooling layers select a dominant feature for each image patch. As a result, the extracted feature maps are misaligned and give poor classification accuracy. To solve this problem, the discriminator in our study only uses convolution layers with a stride of 2 without a pooling layer. This implementation helps to extract image features in patches (blocks of image) and removes the effects of misalignment on the extracted feature maps.

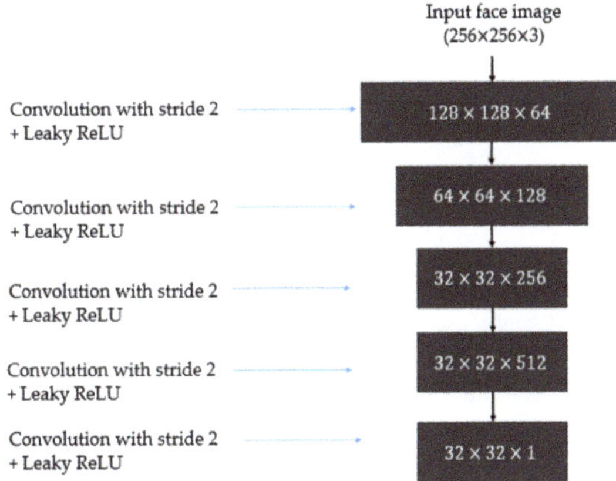

Figure 4. Illustration of the discriminator network used in our study.

Table 2. Detailed description of the discriminator network used in our study (input and output shape is in format: height × width × depth).

Type of Layer	Kernel Size	Stride	Number of Filters	Alpha of Leaky ReLU	Input Shape	Output Shape
Convolution	4 × 4	2	64	-	256 × 256 × 3	128 × 128 × 64
Leaky ReLU	-	-	-	0.2	128 × 128 × 64	128 × 128 × 64
Convolution	4 × 4	2	128	-	128 × 128 × 64	64 × 64× 128
Instance Normalization	-	-	-	-	64 × 64 × 128	64 × 64 × 128
Leaky ReLU	-	-	-	0.2	64 × 64 × 128	64 × 64 × 128
Convolution	4 × 4	2	256	-	64 × 64 × 128	32 × 32 × 256
Instance Normalization	-	-	-	-	32 × 32 × 256	32 × 32 × 256
Leaky ReLU	-	-	-	0.2	32 × 32 × 256	32 × 32 × 256
Convolution	4 × 4	1	512	-	32 × 32 × 256	32 × 32 × 512
Instance Normalization	-	-	-	-	32 × 32 × 512	32 × 32 × 512
Leaky ReLU	-	-	-	0.2	32 × 32 × 512	32 × 32 × 512
Convolution	4 × 4	1	1	-	32 × 32 × 512	32 × 32 × 1

In our study, we only use the convolution layers to extract image features (feature maps) and matching the output of the network with the corresponding label feature maps where feature map of ones represents the ground-truth label of real images, and feature map of zeros represents the ground-truth label of PA images. Therefore, Figure 4 does not contain some layers such as fully connected layer (FC), softmax, or classification. Although we can add these layers (FC, softmax, etc.) to the end of this figure to construct a discriminator as what has done with a normal convolutional neural network, the use of only convolutional layers helps to reduce the number of network parameter and make it not depend on the shape of input images. Consequently, it helps to reduce the overfitting problem that normally occurs in training the deep-learning-based networks.

As shown in Table 2, our discriminator network uses an even kernel in convolution layers. Although odd kernels have been normally used in CNN networks, the even kernel has been used in previous studies for GAN models such as conditional GAN [20], de-blurred GAN [25], pix2pix [31], or CycleGAN-based network [29]. Therefore, we selected to use even kernels in our study.

3.3.2. Generator Architecture

The generator, which is responsible for image generation, is the heart of a GAN-based network. In our study, we use a deep CNN to construct the generator. In detail, the generator is constructed as an auto encoder-decoder network, as shown in Figure 5. At the input, the generator accepts an input image and then performs initial pre-encoding steps, which are composed of three convolution layers to encode the input image. As indicated in previous studies [37,41–43], the performance of deep-learning-based systems can be much enhanced by making them deeper. Using this characteristic, we continue processing the feature maps by applying a sequence of nine residual blocks to further manipulate the image data.

The residual connection was first introduced in the work by He et al. [42]. In their work, they showed that it is difficult to train a very deep neural network that is made by linearly stacking weight layers. The problem is caused by the vanishing gradient problem that normally occurs when the depth of the network increases. Consequently, this problem causes the network's performance saturated and degrading rapidly. To solve this problem, He et al. proposed a new network structure, namely residual block, as shown in Figure 5a, where the residual block uses an identity shortcut connection to skip one or more layers during training the network. As a result, training of some layers becomes easier if they simply are identity mappings.

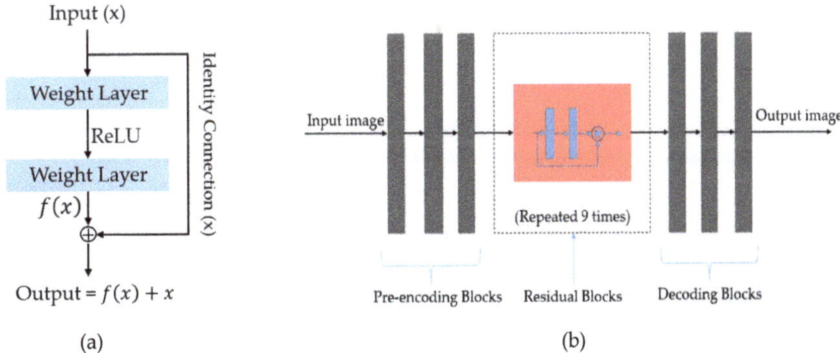

Figure 5. Illustration of the generator network used in our study: (**a**) residual block architecture that is used to increase depth of network; (**b**) draw sketch structure of generator network.

Based on this architecture, we use the residual block in our study to increase the depth of the generator network that is efficient for enhancing the performance of a neural network. The reason is that we not only want to make the network deeper but also make the network easy to train using a residual architecture, which was successfully designed to reduce the effects of the exploring/vanishing gradient problem [42]. To obtain the output image, we use several deconvolution layers on the output of the residual blocks, as shown in Figure 5b. In Table 3, we give a detailed description of the generator network used in our study.

Table 3. Detailed description of the generator network used in our study (input and output shape is in format: height × width × depth).

Type of Layer		Block	Kernel Size	Stride	Number of Filters	Input Shape	Output Shape
Convolution			7×7	1	64	$256 \times 256 \times 3$	$256 \times 256 \times 64$
Instance Normalization			-	-	-	$256 \times 256 \times 64$	$256 \times 256 \times 64$
ReLU			-	-	-	$256 \times 256 \times 64$	$256 \times 256 \times 64$
Convolution			3×3	2	128	$256 \times 256 \times 64$	$128 \times 128 \times 128$
Instance Normalization		Encoding	-	-	-	$128 \times 128 \times 128$	$128 \times 128 \times 128$
ReLU			-	-	-	$128 \times 128 \times 128$	$128 \times 128 \times 128$
Convolution			3×3	2	256	$128 \times 128 \times 128$	$64 \times 64 \times 256$
Instance Normalization			-	-	-	$64 \times 64 \times 256$	$64 \times 64 \times 256$
ReLU			-	-	-	$64 \times 64 \times 256$	$64 \times 64 \times 256$
Residual (Repeated 9 times)	Convolution		3×3	1	256	$64 \times 64 \times 256$	$64 \times 64 \times 256$
	Convolution		3×3	1	256	$64 \times 64 \times 256$	$64 \times 64 \times 256$
Deconvolution			3×3	2	128	$64 \times 64 \times 256$	$128 \times 128 \times 128$
Instance Normalization			-	-	-	$128 \times 128 \times 128$	$128 \times 128 \times 128$
ReLU			-	-	-	$128 \times 128 \times 128$	$128 \times 128 \times 128$
Deconvolution		Decoding	3×3	2	64	$128 \times 128 \times 128$	$256 \times 256 \times 64$
Instance Normalization			-	-	-	$256 \times 256 \times 64$	$256 \times 256 \times 64$
ReLU			-	-	-	$256 \times 256 \times 64$	$256 \times 256 \times 64$
Convolution			7×7	1	3	$256 \times 256 \times 64$	$256 \times 256 \times 3$
Tanh Normalization			-	-	-	$256 \times 256 \times 3$	$256 \times 256 \times 3$

In Tables 2 and 3, the "Normalization" implies the instance-normalization that normalizes the feature maps to a normal distribution with zero-mean and unit variance [25]. The normalization technique is normally used in the neural network to normalize feature maps and make them in the same range and comparable. For the generator network, we tend to generate a real image that normally has pixel values in the range of [0,255] (color image) or [−1, 1] (normalized image) at the output of the network. Therefore, we used the Tanh normalization function at the last layer of the generator to scale the output of the network in the range from −1 to +1.

3.3.3. Calculation of Loss

As discussed in the above sections, our proposed method is based on the CycleGAN network architecture for training an image generation model. To train our network, we must calculate the values of the loss function during the training procedure. For this purpose, let X and Y be two image domains corresponding to the "real" and "PA" classes, respectively. As shown in Figure 3b, we use two discriminator networks (D_X and D_Y) and two generator networks (G_X and G_Y) to construct the GAN network in our study. In each domain, we have one generator and one discriminator, as shown in Figure 3b. In detail, the discriminator D_X is the discriminator in the X domain, which is responsible for discriminating samples in the X domain from those generated by G_Y using samples in the Y domain; the generator G_X is the generator in the X domain, which is responsible for generating samples in the Y domain using input samples in the X domain. Similarly, we have the discriminator D_Y and generator G_Y in the Y domain. G_X is used to generate fake samples of the Y domain using samples in the X domain, and D_Y is used to discriminate the ground-truth samples of the Y domain from those generated by G_X. Therefore, we define the adversarial loss function $L_{GAN}(G_X, D_Y, X, Y)$ as follows:

$$L_{GAN}(G_X, D_Y, X, Y) = E_{y \sim p(y)}[log(D_Y(y)] + E_{x \sim p(x)}[log(1 - D_Y(G_X(x)))] \tag{1}$$

The first term $E_{y \sim p(y)}[f(D_Y(y)]$ is the mean of the loss of the discriminator D_Y using ground-truth samples in the Y domain (PA images), and the second term $E_{x \sim p(x)}[log(1 - D_Y(G_X(x)))]$ is the mean of the loss of D_Y using the PA images generated by G_X because D_Y is responsible for discriminating its ground-truth sample (PA image) from the samples generated by G_X. Similarly, we have the adversarial loss $L_{GAN}(G_Y, D_X, Y, X)$ for G_Y and D_X as follows:

$$L_{GAN}(G_Y, D_X, Y, X) = E_{x \sim p(x)}[log(D_X(x)] + E_{y \sim p(y)}[log(1 - D_X(G_Y(y)))] \tag{2}$$

Equations (1) and (2) describe the loss function using the conventional cross-entropy loss function. However, as indicated by previous research [44], the use of the standard cross-entropy loss function in a deep convolutional GAN (DCGAN) can cause the vanishing gradient problem, and this problem makes the network difficult to train. To overcome this problem and make network training easier, we use the least-squared error instead of conventional cross-entropy for loss calculation in our experiments. As a result, the adversarial loss is as described by Equations (3) and (4) as follows:

$$L_{GAN}(G_X, D_Y, X, Y) = E_{y \sim p(y)}\left[(D_Y(y) - 1)^2\right] + E_{x \sim p(x)}\left[(D_Y(G_X(x)))^2\right] \tag{3}$$

$$L_{GAN}(G_Y, D_X, Y, X) = E_{x \sim p(x)}\left[(D_X(x) - 1)^2\right] + E_{y \sim p(y)}\left[(D_X(G_Y(y)))^2\right] \tag{4}$$

In addition to the adversarial losses, we also use cycle-consistent-loss (cycle-loss) in the reconstructed path to ensure the quality of the reconstruction of the input images using the two generator networks G_X and G_Y. Cycle-loss is defined by Equation (5):

$$L_{cycle}(G_x, G_y) = E_{x \sim p(x)}\left[\left\|G_y(G_x(x)) - x\right\|\right] + E_{y \sim p(y)}\left[\left\|G_x(G_y(y)) - y\right\|\right] \tag{5}$$

As a result, the final loss function used in our study is given by Equation (6), which takes a weighted sum of the adversarial loss and cycleloss. In this equation, λ is used to indicate the weight (importance) of cycleloss over adversarial loss. In our experiments, we used the lamda value of 10 as suggested by Zhu et al. [29].

$$L_{GAN}(G_X, G_Y, D_X, D_Y) = L_{GAN}(G_X, D_Y, X, Y) + L_{GAN}(G_Y, D_X, Y, X) + \lambda L_{cycle}(G_x, G_y) \qquad (6)$$

4. Experimental Results

4.1. Experimental Setups

As explained in our above sections, our study purpose is to generate PA images those are close to the captured PA images, not detecting PA images. Similarity measurements are usually used to evaluate the performance of such systems. As indicated by their meaning, a high performance image generation system has the ability to generate images which are as similar as the ground-truth images. To measure the performance of the image generation model, we followed a well-known quality measurement, namely Frechet Inception Distance (FID) [27,28,45–48]. In addition, we proposed a new quality measurement, namely presentation attack detection distance (padD) as shown in Equation (8). This newly proposed measurement has nice graphical visualization characteristics and it is customized for face-PAD problem. Along with padD measurement, we additionally measured the distribution and the error of face-PAD method with newly generated PA images using attack presentation classification error rate (APCER) measurement, which is followed the ISO/IEC JTC1 SC37-ISO/IEC WD 30107-3 standard for presentation attack detection [49]. The APCER which is along with bona-fide classification error rate (BPCER) and average classification error rate (ACER) are the three popular performance measurements for a presentation attack detection system defined in ISO/IEC JTC1 SC37 standard [49]. By measuring the APCER value, we can evaluate the probability of a generated PA image successfully circumvent a face-PAD system.

For the first measurement method, the FID score is used to measure the quality of the generated images based on the features extracted by a pretrained inception model [43]. This method was proved to work better than the traditional inception score (IS) method [45]. In detail, this method compares the similarity of the two distributions of the extracted features of captured PA images and those of the generated PAD images. For this purpose, a pretrained inception model, which was successfully trained using the ImageNet dataset, is used to extract a 2048-dimensional feature vector for each input image. Suppose that we have N captured PA images and M generated PA images. Using this method, we extract N and M feature vectors for captured PA and generated PA images, respectively. Because the N captured PA images are from the same class (the ground-truth PA images), they form a distribution in a 2048-dimensional space. A similar situation occurs with the M generated images. As a result, we have two distributions for the two classes of images. Suppose these distributions are normal distributions with mean μ and covariance matrix Σ. Then, the FID is given by Equation (7):

$$\text{FID} = \left\| \mu_r - \mu_g \right\|^2 + Tr\left(\Sigma_r + \Sigma_g - 2(\Sigma_r \Sigma_g)^{\frac{1}{2}} \right) \qquad (7)$$

In this equation, the subscripts r and g represent real and generated images, respectively. As shown in Equation (7), the FID measures the dissimilarity between the two distributions in 2048-dimensional space. As a result, a small value of FID indicates a high level of similarity between the two distributions. The FID measurement is based on the texture features extracted from images using a general deep-learning-based feature extraction model (a pre-trained inception model). Therefore, it can be used to assess the quality of the generated image in general. For our specific case of generating PA images for a face-PAD system, we suggest the use of an additional measurement to assess the quality of the generated PA images based on the use of an actual pretrained face-PAD system instead of the inception model, i.e., padD. The concept of the padD measurement is similar to that of the FID, but it uses a different feature extraction method. Because a face-PAD system is designed to discriminate real and PA images, well-generated PA images should have similar characteristics to captured PA images using a face-PAD system. Usually, a face-PAD system receives an image (or sequence of images) to produce a detection score, which stands for probability of the input image belonging to the real or PA class. If the output score is greater than a predefined threshold, the input image is regarded as a real image. Otherwise, the input image is regarded as a PA image. Using the N captured and M generated PA images, we can obtain N and M detection scores for the captured and generated PA image classes, respectively. Finally, we measure the dprime value of the two distributions, as given by in Equation (8), and use this value as the quality measurement of the generated PA images:

$$\text{padD} = \frac{\left| mean_r - mean_g \right|}{\sqrt{\frac{(\sigma_r^2 + \sigma_g^2)}{2}}} \tag{8}$$

As shown in Equation (8), padD measures the distance between the two distributions in one-dimensional space using the dprime method based on two means and standard deviations. As a result, padD is large if the two distributions are very different and becomes smaller as the two distributions become more similar. We can see that padD is a custom FID measurement that is specialized for face-PAD systems. By using both the FID and padD values, we can assess the quality of generated images in more detail. In our experiments, we use two face-PAD systems for measuring the padD value: a deep-learning-based and a handcrafted-based face-PAD system. The deep-learning-based face-PAD system uses a combination of a CNN, recurrent neural network (RNN), and a multi-level local binary pattern (MLBP) for feature extraction and support vector machine (SVM) for classification. The handcrafted-based face-PAD system uses only the MLBP for feature extraction and SVM for classification [50].

To evaluate the performance of our proposed method, we perform experiments using two public datasets: CASIA [7] and Replay-mobile [9]. A detailed description of each dataset is given in Tables 4 and 5, respectively. Originally, these datasets were widely used for training face-PAD systems [7,9,13,15]. The difference between the two datasets is that the CASIA dataset was created for the face-PAD problem in general using a normal camera, while the Replay-mobile dataset is specialized for the mobile environment. As shown in Table 4, the CASIA dataset contains captured real and PA images of 50 persons stored in video format. In total, the CASIA dataset contains 600 video clips (12 video clips (3 real attacks and 9 PAs) per person). By using the face detection method in Section 3.2, we extracted a total of 110,811 face images for the CASIA dataset. The advantage of the CASIA dataset is that it simulates rich attacking methods, including three levels of image resolution (low, normal, and high) and three methods for making PA samples (cut-photo, wrap photo, and video display). As shown in Table 4, the CASIA dataset is pre-divided into training and testing sub-datasets by the dataset's owner for training and testing purposes.

Table 4. Description of the original CASIA dataset used in our experiments.

CASIA Dataset	Training Dataset (20 Persons)		Testing Dataset (30 Persons)		Total
	Real Access	PA Access	Real Access	PA Access	
Number of Videos	60	180	90	270	600
Number of Images	10,940	34,148	16,029	49,694	110,811

Table 5. Description of the original Replay-mobile dataset used in our experiments.

Replay-Mobile Dataset	Training Dataset (12 Persons)		Testing Dataset (12 Persons)		Total
	Real Access	PA Access	Real Access	PA Access	
Number of Videos	120	192	110	192	614
Number of Images	35,087	56,875	32,169	56,612	180,743

The Replay-mobile dataset contains real and PA images of 40 persons from a mobile camera [9]. This dataset is also pre-divided into three sub-datasets for training, testing, and validation. However, we only use the training and testing datasets in our experiments because we do not need to validate the generation model. For both datasets, we use the training dataset to train the generation model and the testing dataset to measure the quality of the generated images.

4.2. Results

As explained in Section 1, the goal of our study is to construct a method for efficiently generating PA images to save efforts in collecting PA images in training a face-PAD system. For this purpose, in this section, we perform various experiments using two public datasets, i.e. CASIA and Replay-mobile, to evaluate the performance of our proposed method in comparison with previous studies. In summary, we first train our proposed image generation models mentioned in Section 3 using these two datasets and the results are presented in this section. Using these trained models, we further evaluate the quality of generated images using two quality measurements, i.e. FID and padD. Finally, we measure the processing time of the image generation model in two hardware systems, including a desktop computer and an embedded system based on an NVIDIA Jetson TX2 board to demonstrate the ability of our proposed method in a real application.

4.2.1. Quality Assessment of Generated Images Using FID Measurement

We show some example result images in Figure 6. In this figure, the left images are captured real images (the input of generation model), the middle images are the corresponding generated PA images, and the right images are reference captured PA images of the same person. As shown in this figure, the generation model can efficiently generate PA images using the captured real images by adding additional effects on the face, such as noise, blurring, color change, and textures. Although these effects can be added to images using conventional methods (adding noise, performing blurring, etc.), they are not manually added but learnt from the captured images in-the-wild. Therefore, we believe that the generated images are more appropriate than the ones using conventional methods.

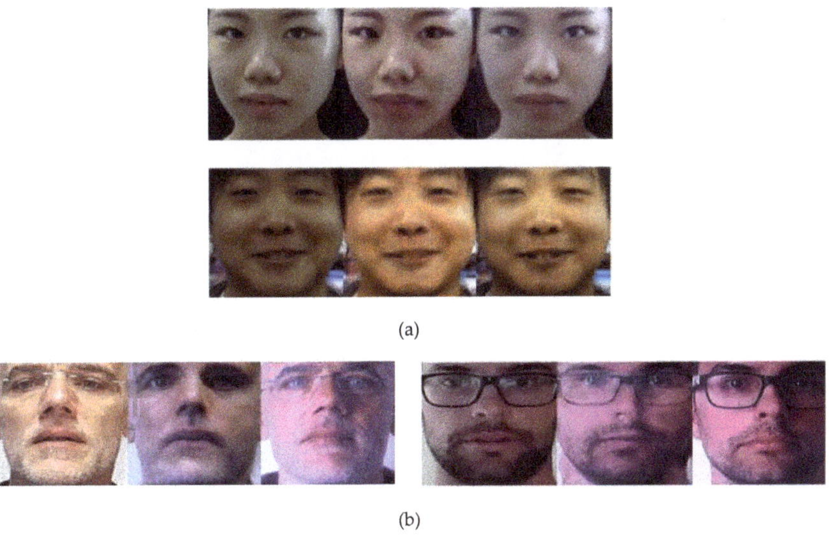

(a)

(b)

Figure 6. Examples of generated images using the trained generation model: (**a**) Images in the CASIA dataset. (**b**) Images in the Replay-mobile dataset. In each trio of images, the left, middle, and right images represent the real image, generated PA image, and reference PA image, respectively.

In the next experiment, we evaluated the quality of generated images using the FID measurement mentioned in Section 4.1. For this purpose, we applied the trained generation model to the CASIA and Replay-mobile testing datasets. Because there is no previous research on this problem, we do not know whether the measured FID in our experiment is good or not. To solve this problem, we additionally measured the FID values between the captured PA images. Because the captured PA images are images captured in-the-wild by simulating attacking methods, they are correct PA images, and measuring the FID between two sets of captured PA images gives us a criterion for evaluating the performance of the generation model. We refer to the FID between the two subsets of captured PA images as the intra-class FID and to the FID between the captured PA images and generated PA images as the inter-class FID in this study.

To measure the intra-class FIDs, we used two different sets of captured PA images: one from the captured PA images in the training dataset and the other from the captured PA images in the testing dataset. This selection ensures two things. First, the images of the two sets are different but cover similar characteristics of PA images as they are from training and testing datasets. Secondly, the size of each set is as large as possible. Even if we divided the captured PA images of either the training or testing dataset into two subsets and measure the FID between these two sets, the number of images in each set would be reduced. As a result, the population of PA images would smaller than it is using our method. For the inter-class FID, we first generated PA images using the captured real images from the testing dataset. With the generated images, we performed the FID measurement using the captured PA images in the testing dataset. The detailed experimental results from the CASIA and Replay-mobile datasets are given in Table 6. As shown in this table, the intra-class FID of the CASIA dataset is approximately 24.614, while the intra-class FID of the Replay-mobile dataset is approximately 37.943. These two FID values are relatively different because the PA images from the two datasets are different. While the CASIA dataset was collected using a commercial camera in good illumination, the Replay-mobile dataset was collected using a mobile camera with uncontrolled light conditions. As a result, the variation of PA face images in the Replay-mobile dataset is large, which resulted in the high intra-FID value. Using the generated images, we obtained an inter-class FID for the CASIA dataset of approximately 28.300, and that of the Replay-mobile dataset was approximately 42.066. Because the intra-class FID was obtained from the ground-truth captured PA images, we can estimate that the intra-class FID should be lower than the inter-class FID because the inter-class FID was obtained using generated PA images. From Table 6, it can be seen that the differences between the intra-class FID and inter-class FID for the CASIA and Replay-mobile datasets are not too high (24.614 vs. 28.3 for the CASIA dataset and 37.943 vs. 42.066 for the Replay-mobile dataset).

Table 6. FID measurements for the captured images versus generated images using our proposed method in comparison with intra-class FID and the model based on conventional cross-entropy loss function.

FID Measurement	Using CASIA Dataset	Using Replay-Mobile Dataset
Intra-class FID	24.614	37.943
Inter-class FID using CycleGAN-based model with conventional cross-entropy loss function	30.968	51.207
Inter-class FID by our proposed method (using least-squared loss function)	28.300	42.066

In addition, we performed experiments using the conventional cross-entropy loss function for a CycleGAN-based image generation model and compared its performance with the least-squared loss function. For this purpose, we measured the FID value between the captured PAD and generated PAD images obtained by a cross-entropy-based CycleGAN model. As explained in Equation (7) of Section 4.1, smaller FID means the higher performance of image generation model. The detail experimental results are given in Table 6. As shown in Table 6, we obtained an FID of 30.968 using a cross-entropy-based

CycleGAN model which is larger than the 28.300 using the least-squared-based CycleGAN model with CASIA dataset. Similarly, we obtained an FID of 51.207 using the cross-entropy-based CycleGAN model which is larger than the 42.066 using the least-squared-based CycleGAN model with Replay-mobile dataset. These results confirmed that the least-squared loss function is better than the conventional cross-entropy loss function in our experiments.

As explained in Section 2, there have been previous studies that generated images between two different domains. Popular methods are DCGAN [27,28], the pix2pix [31], CycleGAN [29], and DualGAN [51] networks. To the best of our knowledge, the pix2pix [31] GAN network requires pairwise images (one for input image, and the other one for ground-truth label image) for learning the relation between the two domains. Therefore, it is not suitable for applying to our study because we are transforming the images between two domains (real vs. PA) without information of pairwise images. The DualGAN [51] is another option (beside CycleGAN) that could be suite for our problem. However, the methodology and structure of DualGAN and CycleGAN is very similar. Therefore, we compared the performance of image generation using DCGAN-based network with our proposed CycleGAN-based method. The experimental results are given in Table 7.

Table 7. Comparison of the measured FIDs in our study with those achieved in previous studies.

Method	DCGAN for Generation				CycleGAN for Generation	
	Iris Images [27]	Fingerprint Images [28]	Replay-Mobile Dataset	CASIA Dataset	Replay-Mobile Dataset	CASIA Dataset
FID	41.08	70.5	65.049	82.400	42.066	28.300

In Table 7, we give a comparison between the FIDs measured in our study and those from previous studies which use DCGAN for image generation problem. Minaee et al. [27] used a GAN to generate iris images. In their study, they showed that the FIDs between the ground-truth and generated images were approximately 41.08 on the IIT-Delhi dataset and 42.1 on the CASIA-1000 dataset. Similarly, the authors of [28] showed that the FID between the ground-truth and generated fingerprint images was approximately 70.5 using a GAN–based method. We can see that the FIDs obtained by our study are much smaller than those obtained by previous studies. Although it is unbalanced to compare the FIDs among different biometrics models because of the difference of image characteristics, we can roughly conclude that our results are comparable or better than those of previous studies.

For ensure a fair comparison, we additionally performed experiments for PA image generation using a DCGAN model. For this purpose, we trained a DCGAN model [27,28] using the CASIA and Replay-mobile datasets and measured the FID between the captured PA and DCGAN-based generated PA images as shown in Table 7, where we obtained an FID of 65.049 in the case of using the captured and generated PA images using DCGAN and the Replay-mobile dataset. This value is much bigger than that of 42.066 using the proposed method. Similarly, we obtained an FID of 82.400 for the case of DCGAN trained on the CASIA dataset. This FID measurement is also much bigger than 28.300 using our proposed method.

Based on these experimental results, we conclude that our proposed method can generate realistic PA images. In addition, the Cycle-GAN-based method is more sufficient than DCGAN-based method, and the Cycle-GAN-based network is a sufficient choice to solve our problem.

4.2.2. Quality Assessment of Generated Images Using padD Measurement on CASIA Dataset

FID measurements have been widely used to evaluate the quality of generated images in general using deep features extracted by a pretrained inception model, which was successfully trained for the general image classification problem. Therefore, FID measurements seem to be too general for our problem. As explained in Section 4.1, our study proposes a new criterion for assessing the quality of generated PA face images called padD. The purpose of this new measurement is to evaluate the quality of generated images for the specific problem of PA image generation. For this purpose, we used an up-to-date face-PAD system [15] to generate decision scores of captured and generated PA images and measure the distance between the two score distributions of these two classes. As a result, if the two distributions are close each other, the generated images have similar characteristics to the captured images. Otherwise, the generated images are different from the captured PA images. One important characteristic of the padD measurement is that it allows a graphical visualization of the distributions of the ground-truth and generated images, which that is not available with the FID. This is because we are working with a one-dimensional feature space instead of a 2048-dimensional feature space. Therefore, the padD measurement gives us a more intuitive measurement than the FID.

As the first experiment in this section, we measured the distributions and padD values for the case of using captured and generated PA images using both face-PAD systems (deep-learning-based and handcrafted-based method). The experimental results are given in Figure 7a,b for the handcrafted-based and the deep-learning-based face-PAD systems, respectively. The specific padD values are listed in Table 8. As shown in Figure 7, the distribution of captured PA images is relatively similar to that of the generated PA images. Numerically, Table 8 shows that the distance (padD) between the two distributions in Figure 7a is approximately 0.610 and that in Figure 7b is approximately 0.711.

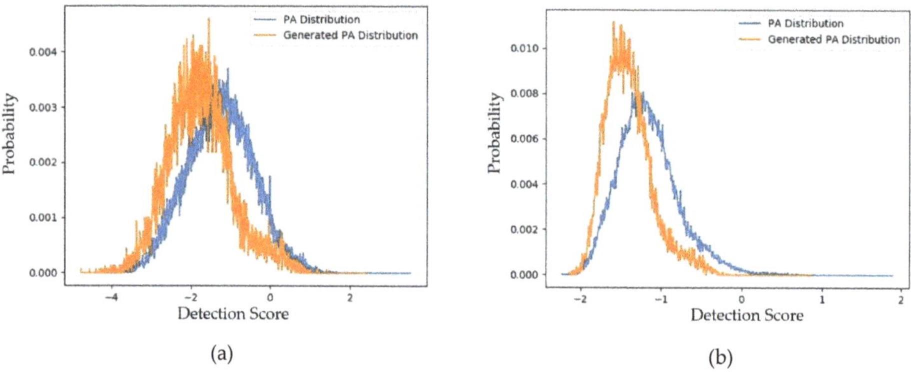

(a)   (b)

Figure 7. Distribution of detection scores for captured PA images versus the generated PA images with the CASIA dataset (**a**) using the handcrafted-based face-PAD system [50]; (**b**) using the deep-learning-based face-PAD system [15].

Table 8. PadD measurements of generated PA images using the CASIA dataset.

Handcrafted-Based PAD Method [50]		Deep-Learning-Based PAD Method [15]	
Captured PA versus Generated PA Images	Captured Real versus Captured PA Images	Captured PA versus Generated PA Images	Captured Real versus Captured PA Images
0.610	2.474	0.711	5.463

To evaluate these above padD measurements, we additionally measured the distributions and padD values for the original (captured real and PA images) CASIA dataset. Figure 8a,b show the distributions of the captured real and captured PA images using the CASIA testing dataset for the handcrafted-based and the deep-learning-based face-PAD systems, respectively. From this figure, it can be observed that the distributions of captured real and PA images were relatively separated. As a classification problem, the errors of this face-PAD system were approximately 0.910% and 9.488% for the deep-learning-based and handcrafted-based method, respectively. As indicated in [15], the error produced by the deep-learning-based method is the smallest compared to other previously proposed face-PAD systems using the CASIA dataset.

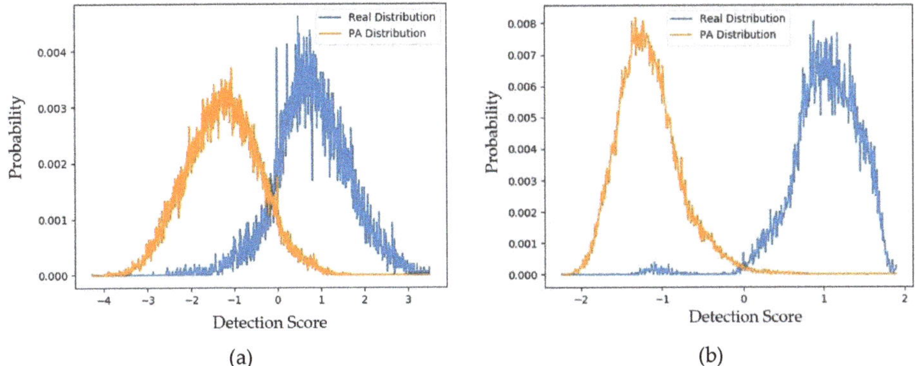

Figure 8. Distribution of detection scores for the original captured data (captured real and captured PA images): (**a**) results obtained using handcrafted-based face-PAD system [50]; (**b**) results obtained using deep-learning-based face-PAD system [15].

Supposing that the two distributions are Gaussian-like, the distance between the two distributions (padD) was measured as 5.463 for the deep-learning-based face-PAD system and 2.474 for the handcrafted-based face-PAD system. This result indicates that the deep-learning-based face-PAD method works well in detecting PA samples in the CASIA dataset. Because we are measuring the padD value for two different types of images, i.e., real and PA images, the measured padD indicates the distance between two different image domains. We see that the padD values in this experiment are much larger than those obtained using the captured and generated PA images in the above experiments (0.610 for the handcrafted-based and 0.711 for the deep-learning-based face-PAD system). This result indicates that the generated PA images have similar characteristics to the captured PA images in the CASIA dataset. We summarize our experimental results in Table 8. As the final experiment in this section, we measured the attack presentation classification error rate (APCER) of the generated PA images using the face-PAD system. By definition, the APCER indicates the proportion of PA images that were incorrectly classified as real images by a face-PAD system. In other words, the APCER represents the possibility of an attack successfully circumventing a face-PAD system. As a result, by measuring the APCER value, we can estimate the quality of generated PA images. The experimental results are shown in Figure 9 and Table 9.

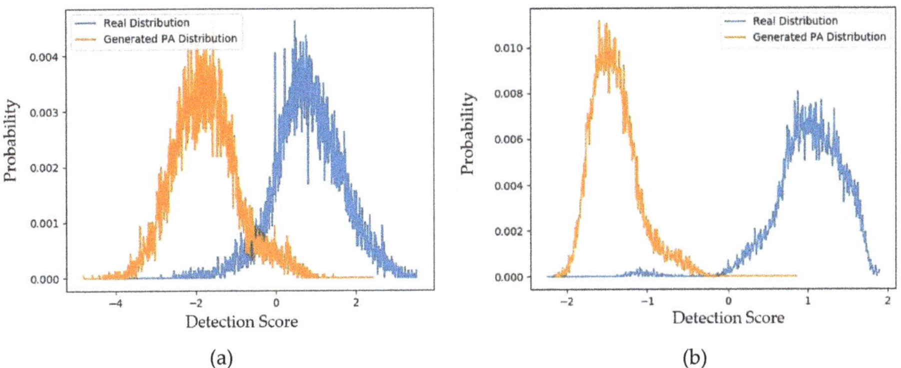

Figure 9. Distribution of detection score of captured real and generated PA images for the CASIA dataset (**a**) using hand-crafted-based face-PAD system [50] and (**b**) using deep-learning-based face-PAD system [15].

Table 9. APCERs of PA images using different face-PAD methods on CASIA dataset (unit: %).

Handcrafted-Based Face-PAD Method [50]		Deep-Learning-Based Face-PAD Method [15]	
Captured PA Images	**Generated PA Images**	**Captured PA Images**	**Generated PA Images**
9.488	4.292	0.784	0.000

As shown in Figure 9, the distributions of captured real and generated PA images are quite far from each other and similar to those in Figure 8. In detail, the padD value for the deep-learning-based face-PAD system is approximately 6.745 and that for the handcrafted-based face-PAD system is approximately 3.128. These values are similar to those using the captured PA images (5.463 and 2.474, respectively). As shown in Table 9, we obtained APCERs of 9.488% and 4.292% for the captured PA and generated PA images, respectively, using the handcrafted-based face-PAD system.

Using the deep-learning-based face-PAD system, we obtained APCER values of 0.784% and 0.000% using the captured PA and generated PA images, respectively. The APCER values produced by the handcrafted-based face-PAD system are much larger than those produced by the deep-learning-based system, which is caused by the fact that the deep-learning-based feature extraction method works much better than the handcrafted-based feature extraction method. By comparing the experimental results for the captured and generated PA images, we see that our approach generates PA images that contain the characteristics of PA images.

4.2.3. Quality Assessment of Generated Images Using padD Measurement on Replay-Mobile Dataset

Similar to the experiments on the CASIA dataset, we performed experiments for the Replay-mobile dataset using the face-PAD systems. First, we measured the distributions and padD values for the use of captured PA versus generated PA images and the use of captured real and PA images. The experimental results of these experiments are given in Figures 10 and 11 and Table 10. Figure 10 shows the distributions of the captured PA and generated PA images of the testing dataset. Similar to the experiments on the CASIA dataset described above, the two distributions (captured and generated PA images) are close to each other. In detail, the padD value for the deep-learning-based face-PAD system is approximately 0.836, and that for the handcrafted-based face-PAD system is approximately 1.214.

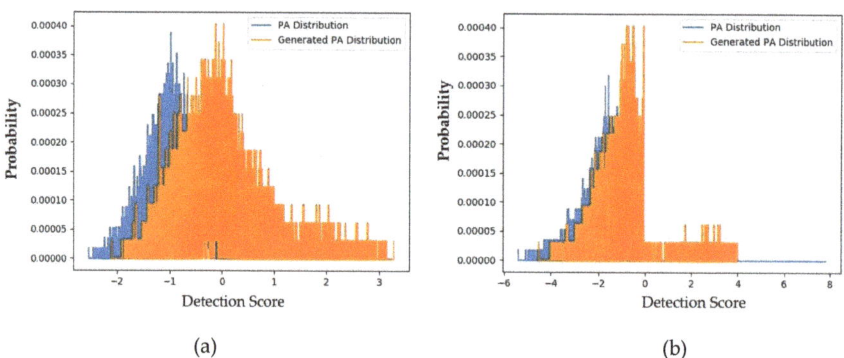

(a) (b)

Figure 10. Distribution of detection scores of captured PA images versus the generated PA images with the Replay-mobile dataset (**a**) using the handcrafted-based face-PAD method [50]; (**b**) using the deep-learning-based face-PAD method [15].

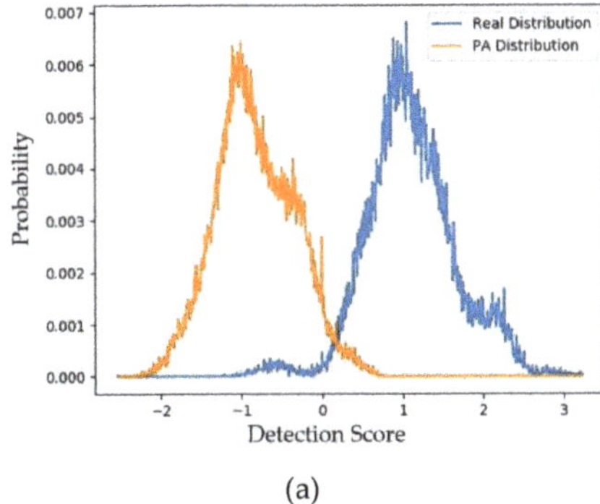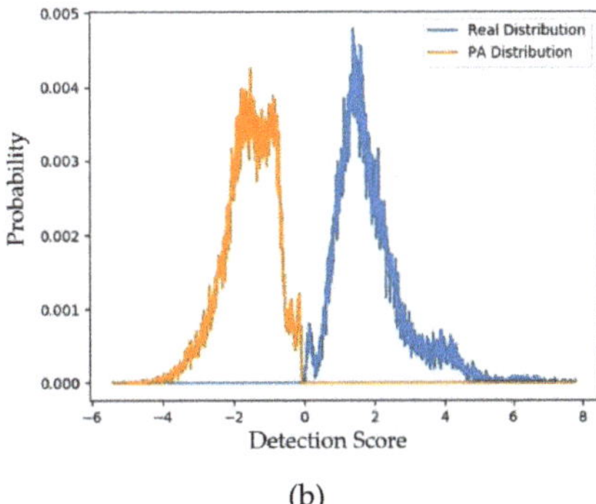

(a) (b)

Figure 11. Distribution of detection scores of captured real versus captured PA images (**a**) using the handcrafted-based face-PAD method [50] and (**b**) using the deep-learning-based face-PAD method [15].

Table 10. padD measurements of generated PA images using Replay-mobile dataset.

Handcrafted-Based Face-PAD Method [50]		Deep-Learning-Based Face-PAD Method [15]	
Captured PA versus Generated PA Images	Captured Real versus Captured PA Images	Captured PA versus Generated PA Images	Captured Real versus Captured PA Images
1.214	3.649	0.836	3.928

Figure 11 shows the distribution of the scores of the captured real and captured PA images. For the deep-learning-based face-PAD system, we obtained a padD value of 3.928, and for the handcrafted-based face-PAD system, we obtained a padD value of 3.649. It is clear that these padD values are much larger than those produced by the captured and generated PA images. Through these results, we can conclude that the generated PA images are close to the captured PA images, while they are far from the captured real face images. In addition, we can see that the distributions of these two types of images do not overlap. This means that although the generated images have similar characteristics to the captured PA images, they are not identical, and the generated images can complement the captured PA images to fill the gap of missing PA samples.

In a subsequent experiment, we measured the APCER of the face-PAD systems using generated PA images. Figure 12 shows the distribution of detection scores of captured real and generated PA images for the deep-learning-based and handcrafted-based face-PAD systems. Similar to Figure 11, the distributions of the real and generated images are relatively separate.

In detail, the two distributions obtained using the handcrafted-based face-PAD system have a padD value of 1.949, and those obtained using the deep-learning-based face-PAD system have a padD value of 3.211. This high padD value indicates that the generated PA images are different from the captured real face images.

Table 11 lists the APCERs obtained in this experiment. Originally, the APCERs were 5.684% and 0.000% for the handcrafted-based and deep-learning-based face-PAD systems, respectively, using the captured data. Using the generated data, these APCER values increased to 41.294% and 1.551%. Although the error caused by the generated PA images in the handcrafted-based face-PAD system is much increased, the error caused by the generated PA images in the deep-learning-based face-PAD system is small. This is caused by the fact that the deep-learning-based method uses a deep CNN-RNN method for feature extraction, which results in higher performance than the handcrafted method. As shown in Figure 12b, the generated PA images have different characteristics to the real images. From this result and the results obtained using the CASIA dataset, we can conclude that the generated images efficiently captured PA features.

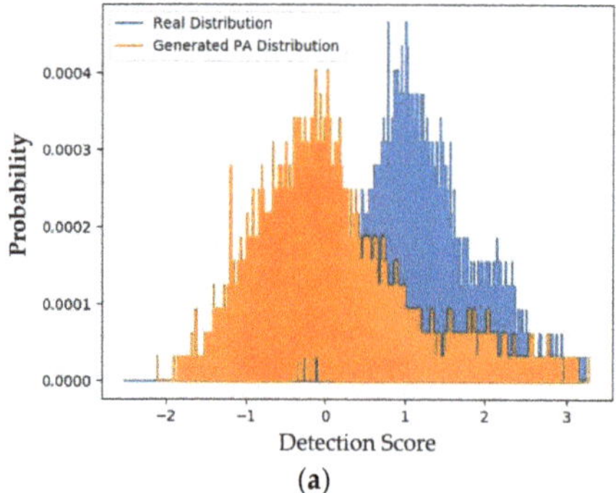 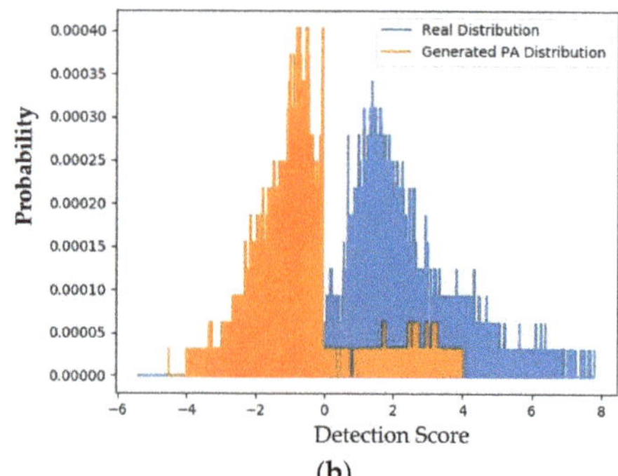

Figure 12. Distribution of detection scores of captured real and generated PA images for the Replay-mobile dataset (**a**) using the handcrafted-based face-PAD method [50] and (**b**) using the deep-learning-based face-PAD method [15].

Table 11. APCERs of PA images using different face-PAD methods (unit: %).

Handcrafted-Based Face-PAD Method [50]		Deep-Learning-Based Face-PAD Method [15]	
Captured PA Images	Generated PA Images	Captured PA Images	Generated PA Images
5.684	41.294	0.00	1.551

We presented our results using the CASIA dataset. Similarly, we presented our results using Replay-mobile dataset. As indicated in these experimental results, the APCER scores of generated PA images are lower than captured PA images, but APCER scores of generated PA images are higher than captured PA ones. The reason for this result is that we trained our PA image generation model using two different datasets which have slightly different characteristics and the amount of PA images. As explained at the beginning of Section 4, the CASIA dataset contains real and PA images of 50 people using various attack methods, including three levels of image quality (low, normal, and high), and three methods for making PA samples (using cut-photo, wrap-photo, and video). Compared to the CASIA dataset, the Replay-mobile dataset only contains PA images for the photo and video attack using a mobile camera. As indicated in the previous study [15], the CASIA dataset has higher complexity of PA images than Replay-mobile dataset, which is indicated by the fact that ACER of an up-to-date face-PAD system [15] is approximately 1.286% and 0.0015% for the CASIA and Replay-mobile dataset, respectively. Because of this reason, we obtained a face-PAD system which covers various kinds of PA images using CASIA dataset (the effects of a new type of PA images on the face-PAD system is small). However, the face-PAD system is more affected by noise and new kind of PA images when it is trained by Replay-mobile dataset because this dataset has limited types of PA images (the effects of a new type of PA images is large). As a result, the APCER of generated PA images is small in the experiment with CASIA dataset, and high in the experiment with Replay-mobile dataset.

4.2.4. Processing Time of the Proposed Approach

As a final experiment, we measured the processing time of our proposed method for generating PA images using the pretrained model to investigate the running speed of our approach. In our experiments, we ran our generation model in two different hardware systems: a general-purpose computer and an embedded system based on the NVIDIA Jetson TX2 board [52]. First, we used a general-purpose computer with an Intel Core i7 central processing unit (CPU) (Intel Corporation, Santa Clara, CA, USA) and 64 GB of RAM. For the deep-learning-based image generation model, we used a TitanX graphics processing unit (GPU) card [53] and the Tensorflow library [54] as the running environment. As the second option, we ran our image generation model on an NVIDIA

Jetson TX2 embedded board, as shown in Figure 13. This is a popular deep-learning-based embedded system developed by NVIDA Corporation, which integrates both the CPU and GPU for deep-learning purposes and has been used for on-board deep-learning processing in self-driving cars. For running a deep-learning-based model, the Jetson TX2 board has an NVIDIA PascalTM-family GPU (256 CUDA cores) with 8 GB of memory shared between the CPU and GPU and 59.7 GB/s of memory bandwidth. Because this board is designed for an embedded system, it uses less than 7.5 W of power. The experimental results are given in Table 12. As shown in this table, it took approximately 29.920 ms to generate a PA image using the general-purpose computer. This means that our generation model can run at a speed of 33.4 frames per second (fps). Using the Jetson TX2 embedded system board, it took approximately 62.423 ms to generate a PA image, which corresponds to 16.02 fps. Compared to the processing time offered by the desktop computer, the Jetson TX2 embedded systems required longer processing time due to its limited computation resources compared to a general-purpose computer. However, with a speed of 16.02 fps with the embedded system and 33.4 fps with the general-purpose computer, we can conclude that our approach is relatively fast and sufficient to run both in general and in embedded environments.

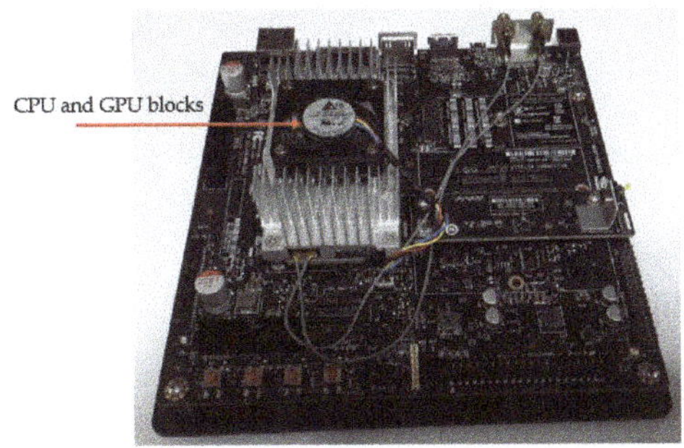

Figure 13. NVIDIA Jetson TX2 on-board embedded system.

Table 12. Processing time of proposed method on a desktop general purpose computer and a Jetson TX2 embedded system (unit: ms).

Desktop Computer	Jetson TX2 Embedded System
29.920	62.423

5. Conclusions

In this paper, we proposed a method for generating PA face images for face-PAD systems. We trained a generation model based on the CycleGAN method using images from two domains, i.e., captured real face images and captured PA images, to learn the characteristics of images in each class and the relations between these two classes. As a result, we showed that the generated PA images are quite similar to but do not overlap with captured PA images, which were collected using a conventional camera and attacking methods. Because the generated images are not identical to the captured PA images, we hope that they can fill the gap of missing samples caused by the lack of PA images because of the diversity of attack methods.

In this study, we aimed to generate PA images to reduce the efforts required for simulating presentation attack methods and PA image acquisition procedure. Therefore, we used a fusion of all kinds of PA images in our experiments without considering every single attack method. Even we can train image generation model using PA images of single available presentation attack method (print attack, display attack etc.), this scheme has some limitations that make it not suitable for our research purpose. First, training an image generation model for every single attack method results in multiple generation models for a single problem. As a result, it wastes processing time, storage, and makes the system complex. Second, as we have explained, the presentation attack detection problem has a special property that we cannot simulate all possible attack methods because of various types of presentation attack instruments and attacking procedures. Therefore, the use of a fusion of existing PA images helps to learn the characteristics of PA images in general to simulate an unknown attack method. In our future work, we plan to use generated images along with captured images to train a face-PAD system to validate the efficiency of the generated images and also to reduce the error of the face-PAD system.

Author Contributions: D.T.N. and G.B. implemented the overall GAN system, and wrote this paper. T.D.P., K.J.N., and K.R.P. helped the experiments and analyzed results. All authors have read and agreed to the published version of the manuscript.

Acknowledgments: This work was supported in part by the National Research Foundation of Korea (NRF) funded by the Korean Government, Ministry of Science and ICT (MSIT), under Grant NRF-2017R1C1B5074062, in part by the NRF funded by the MSIT through the Basic Science Research Program under Grant NRF-2020R1A2C1006179, and in part by the NRF funded by the MSIT, through the Bio and Medical Technology Development Program under Grant NRF-2016M3A9E1915855.

Conflicts of Interest: The authors declare no conflict of interest.

References

1. Jain, A.K.; Ross, A.; Prabhakar, S. An Introduction to Biometric Recognition. *IEEE Trans. Circuits Syst. Video Technol.* **2004**, *14*, 4–20. [CrossRef]
2. Nguyen, D.T.; Park, Y.H.; Lee, H.C.; Shin, K.Y.; Kang, B.J.; Park, K.R. Combining Touched Fingerprint and Finger-Vein of a Finger, and Its Usability Evaluation. *Adv. Sci. Lett.* **2012**, *5*, 85–95. [CrossRef]
3. Taigman, Y.; Yang, M.; Ranzato, M.A.; Wolf, L. DeepFace: Closing the Gap to Human-Level Performance in Face Verification. In Proceedings of the IEEE Conference on Computer Vision and Pattern Recognition, Columbus, OH, USA, 23–28 June 2014; pp. 1701–1708.
4. Lee, W.O.; Kim, Y.G.; Hong, H.G.; Park, K.R. Face Recognition System for Set-Top Box-Based Intelligent TV. *Sensors* **2014**, *14*, 21726–21749. [CrossRef] [PubMed]
5. Zhao, J.; Han, J.; Shao, L. Unconstrained Face Recognition Using a Set-To-Set Distance Measure on Deep Learned Features. *IEEE Trans. Circuits Syst. Video Technol.* **2017**, *28*, 2679–2689. [CrossRef]
6. Maatta, J.; Hadid, A.; Pietikainen, M. Face Spoofing Detection from Single Image Using Micro-Texture Analysis. In Proceedings of the International Joint Conference on Biometric, Washington, DC, USA, 11–13 October 2011; pp. 1–7.
7. Zhang, Z.; Yan, J.; Liu, S.; Lei, Z.; Yi, D.; Li, S.Z. A Face Anti-Spoofing Database with Diverse Attack. In Proceedings of the 5th International Conference on Biometric, New Delhi, India, 29 March–1 April 2012; pp. 26–31.
8. Kim, S.; Ban, Y.; Lee, S. Face Liveness Detection Using Defocus. *Sensors* **2015**, *15*, 1537–1563. [CrossRef]
9. Costa-Pazo, A.; Bhattacharjee, S.; Vazquez-Fernandez, E.; Marcel, S. The Replay-Mobile Face Presentation Attack Database. In Proceedings of the International Conference on the Biometrics Special Interest Group, Darmstadt, Germany, 21–23 September 2016; pp. 1–7.
10. Boulkenafet, Z.; Komulainen, J.; Hadid, A. Face Anti-Spoofing Based on Color Texture Analysis. In Proceedings of the IEEE International Conference on Image Processing, Quebec City, QC, Canada, 27–30 September 2015; pp. 2536–2640.
11. Parveen, S.; Ahmad, S.M.S.; Abbas, N.H.; Adnan, W.A.W.; Hanafi, M.; Naeem, N. Face Liveness Detection Using Dynamic Local Ternary Pattern (DLTP). *Computers* **2016**, *5*, 10. [CrossRef]
12. Menotti, D.; Chiachia, G.; Pinto, A.; Schwartz, W.R.; Pedrini, H.; Falcao, A.X.; Rocha, A. Deep Representation for Iris, Face and Fingerprint Spoofing Detection. *IEEE Trans. Inf. Forensic Secur.* **2015**, *10*, 864–879. [CrossRef]
13. Nguyen, D.T.; Pham, D.T.; Baek, N.R.; Park, K.R. Combining Deep and Handcrafted Image Features for Presentation Attack Detection in Face Recognition Systems Using Visible-Light Camera Sensors. *Sensors* **2018**, *18*, 699. [CrossRef] [PubMed]
14. De Souza, G.B.; Da Silva Santos, D.F.; Pires, R.G.; Marana, A.N.; Papa, J.P. Deep Texture Features for Robust Face Spoofing Detection. *IEEE Trans. Circuits Syst. II-Express* **2017**, *64*, 1397–1401. [CrossRef]

15. Nguyen, D.T.; Pham, D.T.; Lee, M.B.; Park, K.R. Visible-light Camera Sensor-Based Presentation Attack Detection for Face Recognition by Combining Spatial and Temporal Information. *Sensors* **2019**, *19*, 410. [CrossRef] [PubMed]

16. Dongguk Generation Model of Presentation Attack Face Image (DG_FACE_PAD_GEN). Available online: http://dm.dongguk.edu/link.html (accessed on 3 January 2020).

17. Benlamoudi, A.; Zighem, M.E.; Bougourzi, F.; Bekhouche, S.E.; Ouafi, A.; Taleb-Ahmed, A. Face Anti-Spoofing Combining MLLBP and MLBSIF. In Proceedings of the CGE10SPOOFING, Alger, Algérie, 17–18 April 2017.

18. Liu, Y.; Stehouwer, J.; Jourabloo, A.; Liu, X. Deep Tree Learning for Zero-Shot Face Anti-Spoofing. Available online: https://arxiv.org/abs/1904.02860v2 (accessed on 3 January 2020).

19. Goodfellow, I.J.; Pouget-Abadie, J.; Mirza, M.; Xu, B.; Warde-Farley, D.; Ozair, S.; Courville, A.; Bengio, Y. Generative Adversarial Nets. Available online: https://arxiv.org/abs/1406.2661 (accessed on 3 January 2020).

20. Perarnau, G.; Weijer, J.; Raducanu, B.; Alvarez, J.M. Invertible Conditional GAN for Image Editing. Available online: https://arxiv.org/abs/1611.06355 (accessed on 3 January 2020).

21. Zhang, H.; Sindagi, V.; Patel, V.M. Image De-Raining Using a Conditional Generative Adversarial Network. Available online: https://arxiv.org/abs/1701.05957v4 (accessed on 3 January 2020).

22. Ledig, C.; Theis, L.; Huszar, F.; Caballero, J.; Cunningham, A.; Acosta, A.; Aitken, A.; Tejani, A.; Totz, J.; Wang, Z.; et al. Photo-Realistic Single Image Super Resolution Using a Generative Adversarial Network. Available online: https://arxiv.org/abs/1609.04802v5 (accessed on 3 January 2020).

23. Chen, J.; Tai, Y.; Liu, X.; Shen, C.; Yang, J. FSRNet: End-to-End Learning Face Super-Resolution with Facial Priors. Available online: https://arxiv.org/abs/1711.10703v1 (accessed on 3 January 2020).

24. Tan, D.S.; Lin, J.-M.; Lai, Y.-C.; Ilao, J.; Hua, K.-L. Depth Map Up-sampling Via Multi-Modal Generative Adversarial Network. *Sensors* **2019**, *19*, 1587. [CrossRef] [PubMed]

25. Kupyn, O.; Budzan, V.; Mykhailych, M.; Mishkin, D.; Matas, J. DeblurGAN: Blind Motion De-Blurring Using Conditional Adversarial Networks. Available online: https://arxiv.org/abs/1711.07064v4 (accessed on 3 January 2020).

26. Wang, G.; Kang, W.; Wu, Q.; Wang, Z.; Gao, J. Generative Adversarial Network (GAN) Based Data Augmentation for Palm-print Recognition. In Proceedings of the Digital Image Computing: Techniques and Applications (DICTA), Canberra, Australia, 10–13 December 2018; pp. 1–7.

27. Minaee, S.; Abdolrashidi, A. Iris-GAN: Learning to Generate Realistic Iris Images Using Convolutional GAN. Available online: https://arxiv.org/abs/1812.04822v3 (accessed on 3 January 2020).

28. Minaee, S.; Abdolrashidi, A. Finger-GAN: Generating Realistic Fingerprint Images Using Connectivity Imposed GAN. Available online: https://arxiv.org/abs/1812.10482v1 (accessed on 3 January 2020).

29. Zhu, J.-Y.; Park, T.; Isola, P.; Efros, A.A. Unpaired Image-to-Image Translation Using Cycle-Consistent Adversarial Networks. Available online: https://arxiv.org/abs/1703.10593v6 (accessed on 3 January 2020).

30. Chu, M.P.; Sung, Y.; Cho, K. Generative Adversarial Network-based Method for Transforming Single RGB Image into 3D Point Cloud. *IEEE Access* **2019**, *7*, 1021–1029. [CrossRef]

31. Isola, P.; Zhu, J.-Y.; Zhou, T.; Efros, A.A. Image-to-Image Translation with Conditional Adversarial Networks. Available online: https://arxiv.org/abs/1611.07004v3 (accessed on 3 January 2020).

32. Pan, J.; Canton-Ferrer, C.; McGuinnes, K.; O'Connor, N.E.; Torres, J.; Sayol, E.; Giro-i-Nieto, X. SalGAN: Visual Saliency Prediction with Adversarial Networks. Available online: https://arxiv.org/abs/1701.01081v3 (accessed on 3 January 2020).

33. Bontrager, P.; Roy, A.; Togelius, J.; Memon, N. Deepmasterprint: Fingerprint Spoofing via Latent Variable Evolution. Available online: https://arxiv.org/abs/1705.07386v4 (accessed on 3 January 2020).

34. Kazemi, V.; Sullivan, J. One Millisecond Face Alignment with an Ensemble of Regression Trees. In Proceedings of the IEEE Conference on Computer Vision and Pattern Recognition, Columbus, OH, USA, 23–28 June 2014; pp. 1867–1874.

35. Lecun, Y.; Bottou, L.; Bengio, Y.; Haffner, P. Gradient-based Learning Applied to Document Recognition. *Proc. IEEE* **1998**, *86*, 2278–2324. [CrossRef]

36. Krizhevsky, A.; Sutskever, I.; Hinton, G.E. Imagenet Classification with Deep Convolutional Neural Networks. In Proceedings of the Advances in Neural Information Processing Systems, Lake Tahoe, NV, USA, 3–8 December 2012; pp. 1097–1105.

37. Simonyan, K.; Zisserman, A. Very Deep Convolutional Neural Networks for Large-Scale Image Recognition. Available online: https://arxiv.org/abs/1409.1556v6 (accessed on 3 January 2020).

38. Srivastava, N.; Hinton, G.; Krizhevsky, A.; Sutskever, I.; Salakhutdinov, R. Dropout: A Simple Way to Prevent Neural Networks from Overfitting. *J. Mach. Learn. Res.* **2014**, *15*, 1929–1958.

39. Nguyen, D.T.; Kim, K.W.; Hong, H.G.; Koo, J.H.; Kim, M.C.; Park, K.R. Gender Recognition from Human-Body Images Using Visible-Light and Thermal Camera Videos Based on a Convolutional Neural Network for Image Feature Extraction. *Sensors* **2017**, *17*, 637. [CrossRef] [PubMed]

40. Muhammad, K.; Ahmad, J.; Mehmood, I.; Rho, S.; Baik, S.W. Convolutional Neural Networks Based Fire Detection in Surveillance Videos. *IEEE Access* **2018**, *6*, 18174–18183. [CrossRef]

41. Huang, G.; Liu, Z.; Maatern, L.; Weinberger, K.Q. Densely Connected Convolutional Network. Available online: https://arxiv.org/abs/1608.06993v5 (accessed on 3 January 2020).

42. He, K.; Zhang, X.; Ren, S.; Sun, J. Deep Residual Learning for Image Recognition. Available online: https://arxiv.org/abs/1512.03385v1 (accessed on 3 January 2020).

43. Szegedy, C.; Liu, W.; Jia, Y.; Sermanet, P.; Reed, S.; Anguelov, D.; Erhan, D.; Vanhoucke, V.; Rabinovich, A. Going Deeper with Convolutions. Available online: https://arxiv.org/abs/1409.4842v1 (accessed on 3 January 2020).

44. Mao, X.; Li, Q.; Xie, H.; Lau, R.Y.K.; Wang, Z.; Smolley, S.P. Least Squares Generative Adversarial Networks. Available online: https://arxiv.org/abs/1611.04076v3 (accessed on 3 January 2020).

45. Heusel, M.; Ramsauer, H.; Unterthiner, T.; Nessler, B.; Hochreiter, S. GANs Trained by a Two Time-Scale Update Rule Converge to a Local Nash Equilibrium. Available online: https://arxiv.org/abs/1706.08500v6 (accessed on 3 January 2020).

46. Zhang, H.; Goodfellow, I.; Metaxas, D.; Odena, A. Self-Attention Generative Adversarial Networks. Available online: https://arxiv.org/abs/1805.08318v2 (accessed on 3 January 2020).

47. Borji, A. Pros and Cons of GAN Evaluation Measures. *Comput. Vis. Image Underst.* **2019**, *179*, 41–65. [CrossRef]

48. Lucic, M.; Kurach, K.; Michalski, M.; Gelly, S.; Bousquet, O. Are GANs Created Equal? A Large-Scale Study. Available online: https://arxiv.org/abs/1711.10337v4 (accessed on 3 January 2020).

49. ISO Standard. ISO/IEC 30107-3:2017 [ISO/IEC 30107-3:2017] Information Technology—Biometric Presentation Attack Detection—Part 3: Testing and Reporting. Available online: https://www.iso.org/standard/67381.html (accessed on 3 January 2020).

50. Ojala, T.; Pietikainen, M.; Maenpaa, T. Multiresolution Gray-Scale and Rotation Invariant Texture Classification with Local Binary Patterns. *IEEE Trans. Pattern Anal. Mach. Intell.* **2002**, *24*, 971–987. [CrossRef]

51. Yi, Z.; Zhang, H.; Tan, P.; Gong, M. DualGAN: Unsupervised Dual Learning for Image-to-Image Translation. Available online: https://arxiv.org/abs/1704.02510v4 (accessed on 3 January 2020).

52. Jetson TX2 Module. Available online: https://www.nvidia.com/en-us/autonomous-machines/embedded-systems-dev-kits-modules/ (accessed on 3 January 2020).

53. NVIDIA TitanX GPU. Available online: https://www.nvidia.com/en-us/geforce/products/10series/titan-x-pascal/ (accessed on 3 January 2020).

54. Tensorflow Deep-Learning Library. Available online: https://www.tensorflow.org/ (accessed on 3 January 2020).

Article

Deep Active Learning for Surface Defect Detection

Xiaoming Lv, Fajie Duan *, Jia-Jia Jiang, Xiao Fu and Lin Gan

The State Key Lab of Precision Measuring Technology and Instruments, Tianjin University,
Tianjin 300072, China; lvxiaoming1@gmail.com (X.L.); jiajiajiang@tju.edu.cn (J.-J.J.);
fuxiao215@tju.edu.cn (X.F.); ganlin@tju.edu.cn (L.G.)
* Correspondence: fjduan@tju.edu.cn; Tel.: +86-(022)-2789-0261

Received: 14 February 2020; Accepted: 13 March 2020; Published: 16 March 2020

Abstract: Most of the current object detection approaches deliver competitive results with an assumption that a large number of labeled data are generally available and can be fed into a deep network at once. However, due to expensive labeling efforts, it is difficult to deploy the object detection systems into more complex and challenging real-world environments, especially for defect detection in real industries. In order to reduce the labeling efforts, this study proposes an active learning framework for defect detection. First, an Uncertainty Sampling is proposed to produce the candidate list for annotation. Uncertain images can provide more informative knowledge for the learning process. Then, an Average Margin method is designed to set the sampling scale for each defect category. In addition, an iterative pattern of training and selection is adopted to train an effective detection model. Extensive experiments demonstrate that the proposed method can render the required performance with fewer labeled data.

Keywords: surface defect detection; active learning; deep learning

---

1. Introduction

Defects may appear on metallic surfaces due to irresistible influencing factors, such as material characteristics and processing technologies in industrial production. These defects affect not only the qualities, but also the applications of products. Thus, it is of great significance to detect such defects for quality control. As the main solution for defect detection, image processing techniques for detection generally consist of defect localization, recognition, and classification. These methods can be roughly divided into two categories: Traditional image processing and deep learning.

In traditional industries, traditional image processing is usually adopted as a defect detection method combined with hand-crafted features [1], which are exploited to describe the defects. In addition, these approaches require many complex threshold settings, which are sensitive to changes in real-world environments and easily influenced by illumination, background, and so on. Therefore, these approaches are easily affected by environmental noise, which may lead to a poor detection model with weak performance. At the same time, traditional algorithms are weak in processing speeds and cannot meet the requirements of real-time detection. Furthermore, they lack robustness for real-world deployments due to their weak adaption capabilities.

With the rapid development of deep learning, convolutional neural networks (CNN) have been successfully implemented for metallic surface defect detection [2]. For example, Ri-Xian et al. [3] designed a deep confidence network (DCN) for defect detection. Masci et al. [4] proposed a defect detection model for steel classification to reduce the processing time via CNN with a maximum pool. Liu et al. [5] added a feature extraction module based on Faster-RCNN for defect detection. In addition, regional suggestion and regression operation are also widely used in deep learning for defect detection. The former is to generate candidate defect boxes that are classified by CNNs. For example, Girshick developed a defect detection method based on Fast-RCNN [6] to detect five types of defects. The latter can directly perform on the target bounding boxes, such as OverFeat [7,8], SSD [9], YOLO-v2 [10], and YOLO-v3 [11]. OverFeat is a network for integrating recognition, localization, and detection using convolutional networks. The YOLO-v1 method is the base network for YOLO-v3 and YOLO-v3. YOLO-v2 added batch normalization, a high-resolution classifier, multi-scale training, and dimension clusters, while YOLO-v3 used an optimized darknet network and multi-scale prediction. In addition, SSD adopted a softmax classifier, while YOLO methods adopted a logistic classifier. However, all of these approaches assume that a large number of labeled images are available, which require expensive and time-consuming labeling efforts. This problem significantly hinders the deployment of the deep detection models into real-world industries.

As an important branch of machine learning for reducing labeling efforts, active learning has been used to solve data collection problems for many applications. There are situations where unlabeled images are abundant, but manually labeling is expensive. In such a scenario, active learning algorithms can actively query the annotators for labels. Active learning algorithms are mainly aimed at selecting effective data for annotation. Thus, the number of images can often be much lower than the number required in supervised learning. The common query strategies include: (1) Expected model change: Annotate images that would change the current model the most. (2) Uncertainty sampling: Annotate images for which the current model is least certain. (3) Variance reduction: Annotate images that would minimize output variance, which is one of the components of error. Theoretical results indicate that a great active selection strategy can significantly reduce the labeling efforts compared to random selection for obtaining similar accuracy. Although the existing active selection strategies have demonstrated great performance in deep learning, these strategies cannot be directly employed in defect defection. Thus, the active learning for defect detection still faces large challenges. Compared with existing deep detection methods [12,13], active learning methods try to train an effective detection model with the least labeling effort, while deep learning methods annotate all data and train the model with the full training set. Uncertainty is the main problem of performance measurement for deep models, so the active learning methods can select the valuable data for querying annotations.

To effectively reduce the labeling efforts for defect detection, this research introduces active learning into metallic defect detection via uncertainty of data. First, based on the information of the defect images, we propose an Uncertainty Sampling to the product candidate list for annotation. Then, to estimate the sampling scale, an Average Margin method is designed for scale calculation. Finally, a series of experiments is conducted to demonstrate the effectiveness of this method on the NEU-DET [14] dataset.

2. Related Work

2.1. Object Detection for Defect Detection

Over recent years, deep metallic surface defect detection methods have obtained relatively outstanding results in single-image backgrounds [15]. At the same time, deep object detection approaches are the main methods for detecting defects, such as YOLO [8,10,11], Faster-RCNN [6], and SSD [9].

To meet the real-time requirement of object detection, Radmond et al. [8] proposed the YOLO-v1 algorithm to reduce computational complexity and found that end-to-end real-time monitoring is feasible. Meanwhile, combined with machine learning, the extraction of physical information and the construction of structural features show more advantages in the field of defect detection, making a shallow-to-deep transition. Therefore, deep learning plays a crucial role in defect detection at present. Many defect detection methods are proposed based on deep object detection. For example, Feng et al. [16] built an improved version based on the YOLO model, which is widely used in production monitoring, mobile location, and surface defect detection. Ri-Xian et al. [3] proposed to use a convolutional neural network (CNN) to detect defects on the surfaces by trying to overcome the overfitting problem of small data sets that often occur in practical applications. Ren et al. [17]

proposed a surface defect detection model for automatic detection. Azimi et al. [18] used a fully convolutional network for pixel-level segmentation. However, these methods require huge numbers of labeled images, which shall be collected in real-world industries and annotated by great human labors. Thus, it is difficult to deploy them in real-world industries due to the expensive labeling efforts.

2.2. Active Learning

Active learning is a popular approach for reducing the labeling efforts. Its core problem is the sampling strategy, which is used to estimate the labeling value of unlabeled data. Differently from the traditional classifiers that are trained on a large number of data, active learning can speed up the convergence of detection models and reduce the number of labeled data. To effectively reduce the labeling efforts, active learning often selects the samples containing abundant information via specially designed query strategies.

A common active strategy form is the committee [19], which is used to construct a similar classifier for data selection. Then, the voting results are produced by multiple classifiers. In this setting, the samples with the greatest divergence or complexity are selected and labeled. For example, Tuia et al. [20] suggested a simple way to select samples based on the uncertainty: First, all samples are tested by a classifier, and then the most uncertain samples are given to experts for annotations. Xu et al. [21] proposed a density-weighting method to select samples with the richest information based on the potential characteristics of the data. Chakraborty et al. [22] proposed that numerical-optimization-based techniques can also help the selection of useful samples. In addition, Tong et al. [19] proposed a kernel-space-based clustering algorithm for sample selection by using an RBF (radial basis kernel function) to map samples into several clusters in a high-dimensional space, and then selecting the center samples. However, these methods merely consider data characteristics, ignoring the training process of the deep models.

Although defect detection methods based on deep learning have been well studied and greatly improved, there are still expensive and time-consuming labeling efforts for dataset building. Thus, it is important to take the reduction of labeling efforts into consideration for defect detection.

3. Active Learning for Defect Detection

The following sections first introduce an overview of the proposed active learning framework, and then describe the detection model used in the framework in detail. Furthermore, we illustrate the two main components of the framework, i.e., Uncertainty Sampling for Candidates and the Average Margin for Scale.

3.1. Overall Framework

To reduce the labeling efforts in defect detection, we propose an active learning framework that consists of three main modules: The detection model, active strategies, and data. As shown in Figure 1, it utilizes an iterative pattern for selection and annotation. First, a defect detection model is initialized via pre-training weights. Second, the unlabeled images are fed into the detection model for prediction results. Then, combined with the proposed strategies, the uncertain images are selected for querying annotations. After being labeled by annotators, these images are added into the training set, and the detection model is updated from scratch. The above steps are repeated until the required performance is achieved.

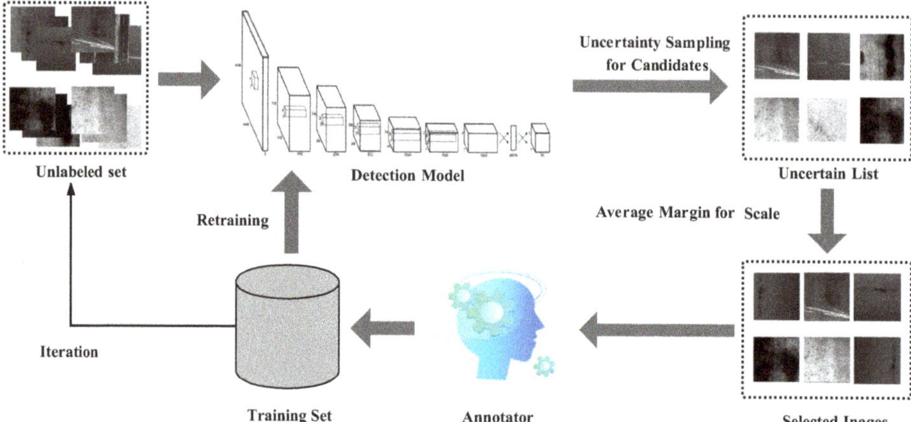

Figure 1. Overview of the proposed active learning framework. First, a defect detection model (neural network) is initialized via pre-training weights. Second, unlabeled images are fed into the detection model for prediction results. Then, combined with the proposed strategies, uncertain images are selected for querying annotations. After being labeled by annotators, the images can be added into the training set, and the detection model is updated from scratch. The above steps are repeated until the required performance is achieved.

3.2. Detection Model

Differently from other object detection methods, YOLO-v2 [23] integrates region prediction and object classification into a single neural network. As illustrated in Figure 2, YOLO-v2 includes a convolutional layer of 3×3 kernel size and a sampling window with a size of 2×2. In addition, the object detection task can be regarded as a regression problem for object region prediction and classification. YOLO-v2 utilizes a single network to directly predict the object boundary and classification probability in an end-to-end way. It adds a batch normalization for each convolution layer and adopts a high-resolution classifier (448×448). In addition, YOLO-v2 introduces anchor boxes to predict bounding boxes.

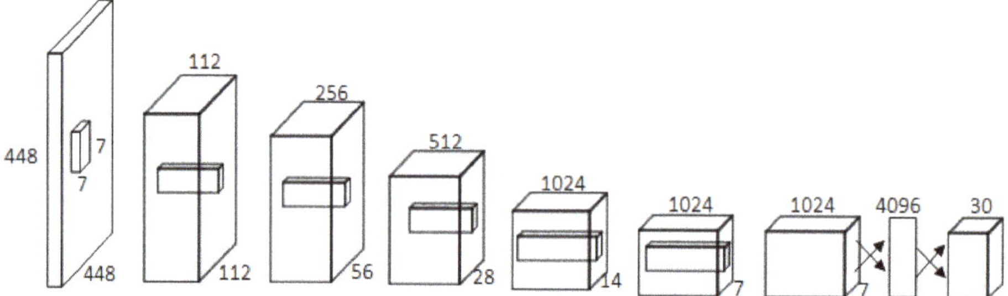

Figure 2. Architecture of the YOLO-v2 model. This network has 24 convolutional layers followed by 2 fully connected layers. Alternating 1×1 convolutional layers reduce the feature space from preceding layers. We pretrain the convolutional layers on the ImageNet classification task at half of the resolution (224×224 input image) and then double the resolution for detection.

3.3. Active Learning for Detection

3.3.1. Uncertainty Sampling for Candidates

In this section, we propose an Uncertainty Sampling for Candidates (USC) algorithm to produce the candidate list for annotation. To estimate the uncertainty of each unlabeled sample, the prediction probabilities for classification are used to calculate the value. Aimed at selecting the most valuable samples for annotation, the outputted probabilities are denoted as follows:

$$P(x_i) = \{ p\,(y_i = 1|x_i; W)\,, p\,(y_i = 2|x_i; W)\,, \ldots, p\,(y_i = n|x_i; W) \},\tag{1}$$

where n is the number of the defect categories.

Note that if the performance of the model is great enough, the model can accurately detect and classify defect images. In this case, the distribution of predicted probabilities has the following characteristics:

- A higher predicted value denotes the higher probability belonging to the corresponding defect.
- There is the maximum value that denotes the probability belonging to the corresponding defect.
- The probability belonging to the corresponding defect is higher than other probabilities of the remaining defects.

To select the most valuable samples, we need to select the sample with inaccurate predictions. According to the above items, we design a predicted margin strategy to estimate the uncertainty of each unlabeled sample, which can be written as:

$$Margin(x_i) = p\,(y_i = j_{max}|x_i; W) - p\,(y_i = j_{smax}|x_i; W)\,,\tag{2}$$

where $p\,(y_i = j_{max}|x_i; W)$ is the maximum probability, $p\,(y_i = j_{smax}|x_i; W)$ is the second maximum probability x_i, and W is the weights of the model.

The predicted margin strategy means that, if a sample is more uncertain, the distribution of predicted probabilities is more average. Thus, the margins among predicted probabilities are small for an uncertain sample. In this case, we select maximum probability and the second one to measure the uncertainty of an unlabeled sample. The smaller value denotes the larger uncertainty of the sample.

3.3.2. Average Margin for Scale

In this section, we design an Average Margin for Scale (AMS) method to measure the sampling scale of the different defects. In fact, a more uncertain defect category needs more unlabeled samples for annotation. Thus, to measure the uncertainty of each defect category, we calculate the average predicted margin for each category. Denote $p(y_i = j|x_i; W)$ as the probability of x_i belonging to the j-th defect category. The average margin for each category is calculated as follows:

$$Magin_c^{avg} = \frac{1}{N} * \sum_{i=1}^{N} Margin_{x_i},\tag{3}$$

where N is the total sample number of the c-th category in the testing set.

The final sample scale can be decided by:

$$Scale = Magin_1^{avg} : Magin_2^{avg} : \ldots : Magin_C^{avg},\tag{4}$$

where $Magin_c^{avg}$ is the mean uncertainty of the c-th category in the testing set, and C is the number of defect categories.

In addition, when selecting uncertain samples as in the above sampling scale, the samples are ranked in the ascending order.

3.3.3. Overview Sampling Algorithm

First, the detection model is initialized by some randomly labeled samples. Then, we calculate the $Margin_x$ value for each unlabeled sample. In addition, the average margin for each category $Magin^{avg}$ in the testing set is calculated. Then, unlabeled samples of each category are ranked in the ascending order. Finally, we select samples according to the scale of $Magin^{avg}$. The algorithm can be referred to Algorithm 1.

Algorithm 1: Active Learning for Defect Detection.

Input: Unlabeled set U, Training set Tr, Testing set Te

1 **Repeat**
2 **for** *each* $x \in U$ **do**
3 $\quad$ calculate $P(x_i)$ according to Formulation 1;
4 $\quad$ calculate $Margin(x_i)$ according to Formulation 2
5 **end**
6 Sort x according to $Margin$ in the ascending order;
7 **for** *each* $x \in Te$ **do**
8 $\quad$ calculate $P(x_i)$ according to Formulation 1;
9 $\quad$ calculate $Margin(x_i)$ according to Formulation 2
10 **end**
11 calculate average $Margin$ for each defect class according to Formulation 3
12 select $x \in U$ according to Formulation 4
13 annotation and update model by Tr
14 **Until**
15 $\quad$ The required performance is met or query budge.

---

4. Experiments

To train an effective detection model, we adopted a small-batch gradient descent method to update the learning rate. The momentum can speed up the convergence of the model, and it is set as 0.9. At the same time, the pre-trained weights are used for the training, and the initialized learning rate was set at 0.005.

4.1. Dataset Introduction

A surface defect dataset called NEU-DET [14] was published by Northeastern University, which collected six typical surface defects on metal surfaces, as shown in Figure 3:

The NEU-DET dataset includes six types of surface defects, i.e., crazing (Cr), rolled-in scale (Rs), patches (Pa), pitted surface (Ps), inclusion (In), and scratches (Sc). The dataset contains 1,800 grayscale images, and each category contains 300 samples. The dataset also provides complete annotations of defect positions and types.

Due to image quality, we merely utilized a subset of NEU-DET, which includes: Patches (Pa), pitted surface (Ps), inclusion (In), and scratches (Sc). This subset has 1200 samples.

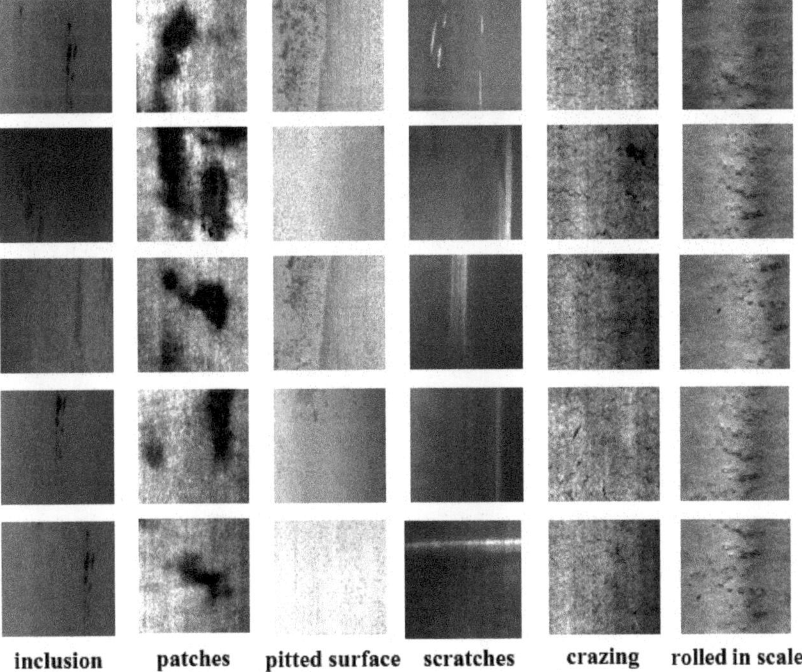

Figure 3. Typical surface defect samples in the NEU-DET dataset.

4.2. Comparisons

Random selection: The random selection (RS) strategy is usually used as a benchmark for the comparison of active learning. In this method, the unlabeled samples are randomly selected in each iteration.

Entropy selection: In this method, entropy value (EN_i) is used to measure the uncertainty of samples. A higher EN_i value represents a greater uncertainty for sample i, which can be defined as follows:

$$EN_i = -\sum_{j=1}^{m} p\left(y_i = j | x_i; W\right) logp\left(y_i = j | x_i; W\right),$$ (5)

where m is the class number and p is the prediction probability for class j.

MS: In this method, all of the unlabeled samples are ranked in an ascending order according to the MS_i value, which can be defined as follows:

$$MS_i = p\left(y_i = j_1 | x_i; W\right) - p\left(y_i = j_2 | x_2; W\right),$$ (6)

where j_1 and j_2 represent the first and second most probable class labels predicted by the model, respectively. The MS value denotes the margin. The smaller value means a higher uncertainty of the sample.

4.3. Evaluation

We adopt Recall, Average Precision (AP), and mean Average Precision (mAP) for performance evaluation. Recall represents the ratio of correctly detected images and all testing images for each defect category. AP represents the average detected precision for each defect category. mAP is the mean of the average detected precision for all defect categories. In addition, we use the full dataset to train the model to obtain the basic required performance, denoted by "Full". This means that all unlabeled images are annotated to train the model.

4.4. Comparison Results

To demonstrate the effectiveness of the proposed method, a series of comparison experiments with Full, MS, EN, and random selection were conducted. From the above experiments, we observed that the performance of the proposed method is better than that of other methods, which indicates the superiority of USC and AMS. Figure 4 exhibits some defect images selected by the active learning strategies of the proposed method.

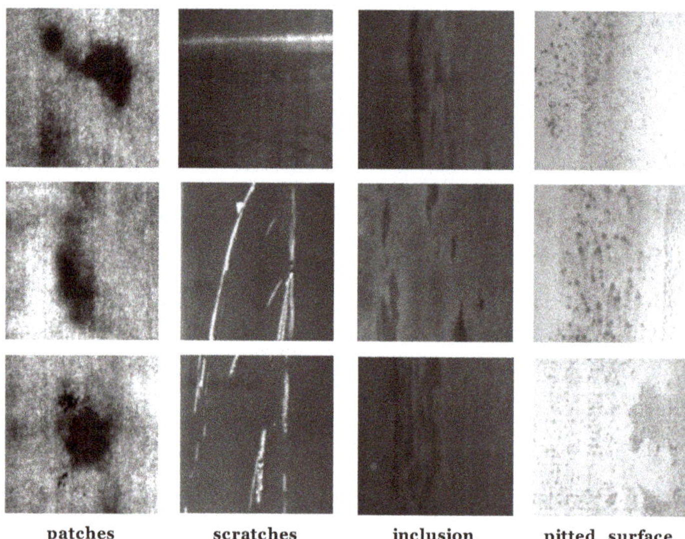

Figure 4. Some images selected by our active learning strategies.

As illustrated in Table 1, the proposed algorithm can achieve the required Recall accuracy with only 21.7% of the annotations of all datasets. Compared with commonly used active strategies, the proposed algorithm can produce competitive results, i.e., 33.3% for MS, 31.6% for EN, and 50.0% for random. It is noticed that, for defect patches (Pa), pitted surfaces (Ps), and scratches (Sc), the proposed algorithm can outperform the Recall results obtained by the full data. In addition, the MS method performed too badly for the pitted surface (Ps), while performing the best for scratches (Sc).

As shown in Table 2, the proposed algorithm can achieve the required Precision accuracy with only 21.7% of the annotations. In detail, for defect inclusion (In), patches (Pa), pitted surfaces (Ps), and scratches (Sc), our algorithm can outperform the Precision results obtained by both the full data and other active learning methods.

As illustrated in Table 3, the proposed algorithm can achieve the required AP accuracy with only 21.7% of the annotations. In detail, for defect patches (Pa) and pitted surfaces (Ps), the proposed algorithm can get the best AP and mAP results. Although other methods may exceed ours in AP accuracy for several defects, they performed with lower mAP results.

The above experimental results indicate that the proposed method obviously performs better than the commonly used active strategies in terms of both Recall Accuracy and Precision. This is because the proposed method can not only select uncertain samples, but also provide a selection scale for each defect category to train the model. Hence, the proposed method demonstrates a competitive advantage in deep defect detection tasks. To clearly exhibit its effectiveness, we also conducted the performance improvement comparison experiments, and the results are discussed below.

Table 1. Comparison of Recall on the NEU-DET dataset. The best results are shown in boldface.

Recall	Type				
	Inclusion (In)	Patches (Pa)	Pitted Surface (Ps)	Scratches (Sc)	Data (%)
Random	0.8333	0.9322	0.6774	0.9394	50.0%
EN	0.8636	0.9322	0.6774	0.9394	31.6%
MS	0.9242	0.9322	0.4839	1.000	33.3%
Full	0.8788	0.8983	0.6129	0.9091	100.0%
Ours	0.8485	0.9153	0.7742	0.9091	21.7%

Table 2. Comparison of Precision on the NEU-DET dataset. The best results are shown in boldface.

Precision	Defect				
	Inclusion (In)	Patches (Pa)	Pitted Surface (Ps)	Scratches (Sc)	Data (%)
Random	0.1291	0.1672	0.0323	0.2583	50.0%
EN	0.1839	0.2696	0.0669	0.2925	31.6%
MS	0.1017	0.2183	0.0498	0.2409	33.3%
Full	0.1213	0.2180	0.0617	0.1899	100%
Ours	**0.1965**	**0.3396**	**0.0774**	**0.3614**	21.7%

Table 3. Comparison of Average Precision (AP) on the NEU-DET dataset. The best results are shown in boldface.

AP	Defect					
	Inclusion (In)	Patches (Pa)	Pitted Surface (Ps)	Scratches (Sc)	mAP	Data (%)
Random	0.5498	0.7498	0.3082	0.5658	0.542	50.0%
EN	0.6359	0.8314	0.1959	**0.8546**	0.629	31.6%
MS	**0.6874**	0.7944	0.1763	0.9104	0.642	33.3%
Full	0.6183	0.7284	0.2103	0.7720	0.582	100%
Ours	0.6390	**0.8269**	**0.3277**	0.7874	**0.645**	21.7%

4.4.1. Performance Improvement Comparison

Figure 5 clearly illustrates the results of the performance improvement comparisons, while Table 4 shows the performance obtained by training on the full data. The horizontal axis represents the annotation percentage, while the vertical axis represents the mAP values. It can be observed that the proposed method can reach the maximum mAP when the percentage is 21.6%, while MS is 33.3% and EN is 31.6%. Therefore, the proposed method performs better than others. This demonstrates that uncertainty sample selection plays an important role in improving the performance.

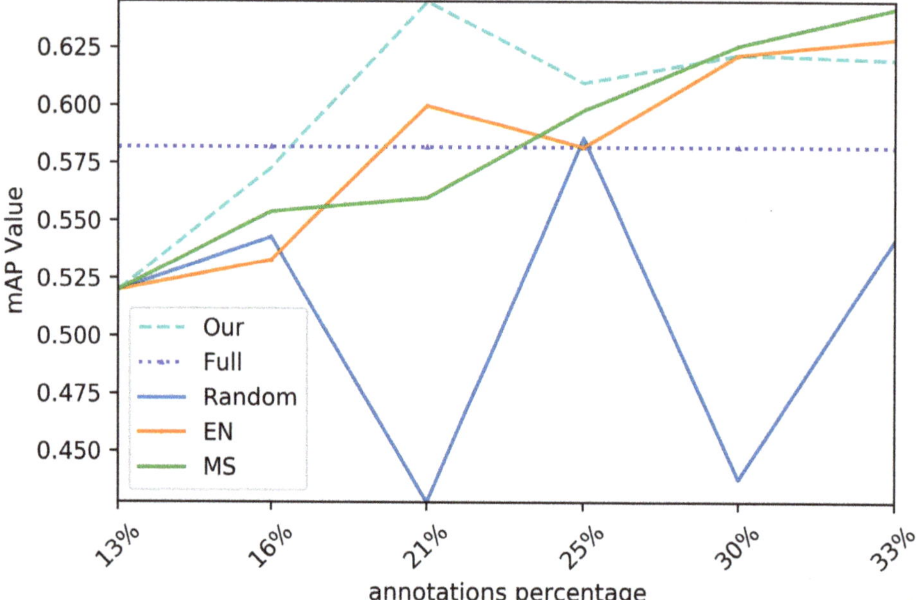

Figure 5. Comparison of mean Average Precision (mAP) for the different strategies.

Table 4. The performance obtained via the full data.

Types	Inclusion (In)	Patches (Pa)	Pitted Surface (Ps)	Scratches (Sc)
Recall	0.8788	0.8983	0.6129	0.9091
Precision	0.1213	0.2108	0.0617	0.1899
AP	0.6183	0.7284	0.2103	0.7720

4.4.2. Query Strategy Analysis

In fact, uncertainty measures the confidence of the current model for a sample, aiming to find the samples with more useful information for model training. The essence of uncertainty is information entropy, which is used to measure the amount of information. A greater information entropy denotes richer information. Thus, many active learning methods [24,25] based on uncertainty are designed to improve the performance of the model and exhibit competitive results.

5. Conclusions

In this paper, we propose an active learning framework for a deep defect detection task, which utilizes the uncertainty of the samples for selection. The main contributions of this work are the following three points: (1) We propose an active learning framework to reduce the labeling efforts for defect detection. The proposed framework adopts an iteration pattern to train the detection model. (2) To select effective data for annotations, we design an uncertainty sampling method to select images according their uncertainty values. (3) To confirm the sampling number for annotations, we design an average margin method to calculate the sampling ratios among defect categories.

Extensive experimental results from a challenging public benchmark demonstrate the effectiveness of the proposed active learning method. In the future, we will explore more effective query strategies to reduce the labeling efforts; the adaptation of active learning is also under consideration.

Author Contributions: Methodology, X.L.; supervision, F.D.; visualization, X.F. and L.G.; writing—original draft, X.L.; writing—review and editing, J.-J.J. All authors have read and agreed to the published version of the manuscript.

Funding: This work was supported in part by the National Key Research and Development Plan (2017YFF0204800), National Natural Science Foundations of China under Grant No. 61971307, 61905175, 51775377, National Science and Technology Major Project (2017-V-0009), the Tianjin Natural Science Foundations of China under Grant No. 17JCQNJC01100, Equipment Pre-Research Field Fund (61405180505,61400040303), Science and Technology on Underwater Information and Control Laboratory (6142218081811), and the Fok Ying Tung Education Foundation (171055).

Conflicts of Interest: The authors declare no conflict of interest.

References

1. Xie, X. A review of recent advances in surface defect detection using texture analysis techniques. *Electron. Lett. Comput. Vision Image Anal.* **2008**, *7*, 1–22. [CrossRef]
2. Wu, X.; Cao, K.; Gu, X. A surface defect detection based on convolutional neural network. In Proceedings of the International Conference on Computer Vision Systems, Shenzhen, China, 10–13 July 2017; pp. 185–194.
3. Liu, R.-X.; Yao, M.-H.; Wang, X.-B. Defects Detection Based on Deep Learning and Transfer Learning. *Metal. Min. Ind.* **2015**. Available online: https://www.researchgate.net/publication/285367015_Defects_detection_based_on_deep_learning_and_transfer_learning (accessed on 12 March 2020).
4. Masci, J.; Meier, U.; Ciresan, D.; Schmidhuber, J.; Fricout, G. Steel defect classification with max-pooling convolutional neural networks. In Proceedings of the 2012 International Joint Conference on Neural Networks (IJCNN), Brisbane, Australia, 10–15 June 2012; pp. 1–6.
5. Liu, Z.; Wang, L.; Li, C.; Han, Z. A high-precision loose strands diagnosis approach for isoelectric line in high-speed railway. *IEEE Trans. Ind. Inf.* **2017**, *14*, 1067–1077. [CrossRef]
6. Ren, S.; He, K.; Girshick, R.; Sun, J. Faster r-cnn: Towards real-time object detection with region proposal networks. In *Advances in Neural Information Processing Systems*; 2015; pp. 91–99. Available online: http://papers.nips.cc/paper/5638-faster-r-cnn-towards-real-time-object-detection-with-region-proposal-networks (accessed on 12 March 2020).
7. Sermanet, P.; Eigen, D.; Zhang, X.; Mathieu, M.; Fergus, R.; LeCun, Y. Overfeat: Integrated recognition, localization and detection using convolutional networks. *arXiv* **2013**, arXiv:1312.6229.
8. Redmon, J.; Divvala, S.; Girshick, R.; Farhadi, A. You only look once: Unified, real-time object detection. In Proceedings of the IEEE conference on computer vision and pattern recognition, Las Vegas, NV, USA, 27–30 June 2016; pp. 779–788.
9. Liu, W.; Anguelov, D.; Erhan, D.; Szegedy, C.; Reed, S.; Fu, C.Y.; Berg, A.C. Ssd: Single shot multibox detector. In Proceedings of the European conference on computer vision, Munich, Germany, 8–14 September 2016; pp. 21–37.
10. Redmon, J.; Farhadi, A. YOLO9000: better, faster, stronger. In Proceedings of the IEEE conference on computer vision and pattern recognition, Honolulu, HI, USA, 21–26 July 2017; pp. 7263–7271.
11. Redmon, J.; Farhadi, A. Yolov3: An incremental improvement. *arXiv* **2018**, arXiv:1804.02767.
12. Hu, J.; Xu, W.; Gao, B.; Tian, G.Y.; Wang, Y.; Wu, Y.; Yin, Y.; Chen, J. Pattern deep region learning for crack detection in thermography diagnosis system. *Metals* **2018**, *8*, 612. [CrossRef]
13. Fan, M.; Wu, G.; Cao, B.; Sarkodie-Gyan, T.; Li, Z.; Tian, G. Uncertainty metric in model-based eddy current inversion using the adaptive Monte Carlo method. *Measurement* **2019**, *137*, 323–331. [CrossRef]
14. Song, K.; Yan, Y. A noise robust method based on completed local binary patterns for hot-rolled steel strip surface defects. *Appl. Surf. Sci.* **2013**, *285*, 858–864. [CrossRef]
15. Kleiner, M.; Geiger, M.; Klaus, A. Manufacturing of lightweight components by metal forming. *Magnesium* **2003**, *530*, 26–66. [CrossRef]

16. Feng, C.; Liu, M.Y.; Kao, C.C.; Lee, T.Y. Deep active learning for civil infrastructure defect detection and classification. In Proceedings of the ASCE International Workshop on Computing in Civil Engineering 2017, Seattle, DC, USA, 25–27, June 2017; pp. 298–306.
17. Ren, R.; Hung, T.; Tan, K.C. A generic deep-learning-based approach for automated surface inspection. *IEEE Trans. Cybern.* **2017**, *48*, 929–940. [CrossRef] [PubMed]
18. Azimi, S.M.; Britz, D.; Engstler, M.; Fritz, M.; Mücklich, F. Advanced steel microstructural classification by deep learning methods. *Sci. Rep.* **2018**, *8*, 1–14. [CrossRef] [PubMed]
19. Tong, S.; Koller, D. Support vector machine active learning with applications to text classification. *J. Mach. Learn. Res.* **2001**, *2*, 45–66.
20. Tuia, D.; Volpi, M.; Copa, L.; Kanevski, M.; Munoz-Mari, J. A survey of active learning algorithms for supervised remote sensing image classification. *IEEE J. Sel. Top. Sign. Proces.* **2011**, *5*, 606–617. [CrossRef]
21. Xu, Z.; Akella, R.; Zhang, Y. Incorporating diversity and density in active learning for relevance feedback. In Proceedings of the European Conference on Information Retrieval, Rome, Italy, 2–5 April 2007; pp. 246–257.
22. Chakraborty, S.; Balasubramanian, V.; Panchanathan, S. Dynamic batch mode active learning. In Proceedings of the Computer Vision and Pattern Recognition (CVPR), Providence, RI, USA, 20–25 June 2011; pp. 2649–2656.
23. Yuan, A.; Bai, G.; Yang, P.; Guo, Y.; Zhao, X. Handwritten English word recognition based on convolutional neural networks. In Proceedings of the 2012 International Conference on Frontiers in Handwriting Recognition, Bari, Italy, 18–20 September 2012; pp. 207–212.
24. Zhu, J.; Wang, H.; Yao, T.; Tsou, B.K. Active learning with sampling by uncertainty and density for word sense disambiguation and text classification. In Proceedings of the 22nd International Conference on Computational Linguistics (Coling 2008), Manchester, UK, 18–22 August 2008; pp. 1137–1144.
25. Yang, Y.; Ma, Z.; Nie, F.; Chang, X.; Hauptmann, A.G. Multi-class active learning by uncertainty sampling with diversity maximization. *Int. J. Comput. Vision* **2015**, *113*, 113–127. [CrossRef]

Article

Multi-Person Pose Estimation using an Orientation and Occlusion Aware Deep Learning Network

Yanlei Gu [1,*], Huiyang Zhang [2] and Shunsuke Kamijo [2]

[1]   College of Information Science and Engineering, Ritsumeikan University, Kusatsu, Shiga 525-8577, Japan
[2]   Institute of Industrial Science, The University of Tokyo, Tokyo 153-8505, Japan;
      zhanghuiyang@kmj.iis.u-tokyo.ac.jp (H.Z.); kamijo@iis.u-tokyo.ac.jp (S.K.)
*   Correspondence: guyanlei@fc.ritsumei.ac.jp; Tel.: +81-77-599-3238

Received: 7 February 2020; Accepted: 10 March 2020; Published: 12 March 2020

Abstract: Image based human behavior and activity understanding has been a hot topic in the field of computer vision and multimedia. As an important part, skeleton estimation, which is also called pose estimation, has attracted lots of interests. For pose estimation, most of the deep learning approaches mainly focus on the joint feature. However, the joint feature is not sufficient, especially when the image includes multi-person and the pose is occluded or not fully visible. This paper proposes a novel multi-task framework for the multi-person pose estimation. The proposed framework is developed based on Mask Region-based Convolutional Neural Networks (R-CNN) and extended to integrate the joint feature, body boundary, body orientation and occlusion condition together. In order to further improve the performance of the multi-person pose estimation, this paper proposes to organize the different information in serial multi-task models instead of the widely used parallel multi-task network. The proposed models are trained on the public dataset Common Objects in Context (COCO), which is further augmented by ground truths of body orientation and mutual-occlusion mask. Experiments demonstrate the performance of the proposed method for multi-person pose estimation and body orientation estimation. The proposed method can detect 84.6% of the Percentage of Correct Keypoints (PCK) and has an 83.7% Correct Detection Rate (CDR). Comparisons further illustrate the proposed model can reduce the over-detection compared with other methods.

Keywords: pose estimation; body orientation; multi-person; multi-task

1. Introduction

Human pose estimation is defined as the problem of the localization of human joints (also known as key points—elbows, wrists, etc.) in images or videos. It has become a highly concerned topic in computer vision. In addition, pose estimation also recently received significant attention from other research fields because of the valuable information contained in data of the human pose.

1.1. Non-Deep Neural Network Approach

In the early stage of the pose estimation research, researchers focus on the simple scenario, and apply the background subtraction to extract the human silhouette from an image sequence. The model-based method enforces a pre-defined human model to be consistent with the extracted silhouette to estimate the human pose [1]. When the model-based method is performed in the multiple views, the silhouette and human pose in 3D can be extracted and estimated [2].

In addition to silhouette information, the temporal information in image sequence is often used to improve the performance of the pose estimation. One kind of these methods is based on detection and tracking, which detects a rough pose in the initial frame and tracks the pose in every continuous frame [3]. After that, the sophisticated model, such as the spatial–temporal Markov Random Field

(MRF) model, is applied for pose estimation in image sequence [4]. A spatial–temporal constraint is also used for optimizing body-part configurations in videos [5]. These models are benefited from the position constraints of both intra-frame joints and inter-frame joints.

Compared to image sequence based pose estimation, extracting a human pose from a single image is a more challenging topic. It requires to detect a human from variant backgrounds and estimate the pose in a large number of degrees of freedom using the limited information in the single frame only. For the single image based pose estimation, Pictorial Structures [6], Deformable Part Models (DPM) [7] and the integration between the two methods [8] are the famous techniques prior to deep neural networks.

1.2. Deep Neural Network for Single Person Pose Estimation

In recent years, Deep Neural Networks (DNN) have achieved high performance and outperformed the conventional methods in many different tasks of computer vision. For example, the proposed Alexnet [9], Visual Geometry Group (VGG) Net [10], GoogLeNet [11] and ResNet [12] improve the accuracy of the object classification step by step. Fast Region-based Convolutional Neural Networks (R-CNN) [13], You Only Look Once (YOLO) [14] and Mask R-CNN [15] also become the new milestones of object detection and segmentation tasks, respectively. In addition, DNNs have been widely used for human pose estimation.

Pose estimation can be classified into single person pose estimation and multi-person pose estimation based on the number of humans that appear in the image. Toshev et al. [16] propose the DeepPose to estimate the pose of a single person from coarse to fine in a cascade DNN structure. However, two limitations exist in their method. Firstly, if the result of the initial pose estimation at the beginning of the cascade DNN is far from the true position, the system will not be able to correct the estimation because the cascade structure is a one-way flow. Secondly, this method only provides one prediction per image, which means there is no possibility to improve the prediction when additional information becomes available for the pose estimation. To address the first issue, Haque et al. [17] and Carreira et al. [18] propose to apply feedback structures to iteratively optimize the pose estimation. Their methods feed the estimated pose from the early iterations to the input end again and gradually refine the pose in the next iterations. As for the second problem, methods using probabilistic heatmaps are proposed as a solution. These heatmaps generated by Convolutional Neural Networks (CNN) turn the joint estimation problem into a pixel-wise classification in pyramid feature maps [19].

In addition, Wei et al. [20] build a convolutional pose machine network for single person pose estimation. Their method learns implicit spatial models via a sequential composition of convolutional architectures, because the easier-to-detect joint can provide strong cues for localizing the difficult-to-detect joint. Similarly, Newell et al. [21] present an hourglass module to capture information at every scale in order to combine features from different stages better. Wang et al. [22] propose a novel densely connected convolutional module-based convolutional neural network to estimate the pose of a single person. Wang et al. [23] apply the multi-scale feature pyramid module to further improve the performance of the deeply learned compositional model of human pose estimation. Chen et al. [24] propose to use generative adversarial networks to exploit the constrained human-pose distribution for improving single-person pose estimation. Szczuko proposes to localize single-person body joints in 3D space based on a single low resolution depth image [25].

1.3. Deep Neural Network for Multi-Person Pose Estimation

However, the pose estimation task becomes more complex in the case of multi-person. Firstly, assembling many homogeneous joints to different persons is a challenging issue. Secondly, occlusion caused by overlapping between multi-persons makes the detection of joint much more difficult. In order to deal with the multi-person pose estimation, the solutions are divided into two types: top-down approaches and bottom-up approaches.

In bottom-up approaches, firstly body part or joint detection is conducted, and then the accurate prediction of the number of people and their poses is performed by person clustering and joint labeling. Pishchulin et al. [26] propose a DeepCut method where Fast R-CNN [13] is used as a body part detector. Then the goal of the task becomes subset partitioning from a graph of all connections among each joint detected from the image, which turns the task into an Integer Linear Programming (ILP) problem. After that, they improve DeepCut by using deeper ResNet architectures [12] to enhance body part detectors and propose image-conditioned pairwise terms that assemble the proposals into a variable number of consistent body part configurations [27]. Kocabas et al. propose to estimate the multi-person pose using a pose residual network [28]. The proposed system is named as the MultiPoseNet and has real time performance. Furthermore, in order to achieve the goal of real-time pose estimation, Cao et al. present a new method using Part Affinity Fields (PAFs) [29], which is also known as OpenPose later. The proposed method firstly generates the joint positions of all of the people in the image by finding the local maximums of joint heatmaps. Then the method connects these joints gradually to construct the person pose structures. This method performs well for visible joints. However, there are many false positives for the occluded joints, because of its aggressive searching of joint positions.

Although bottom-up approaches show a general advantage in runtime analysis, these approaches still suffer from problems that people in the image are in small scale or collusion, because detecting body parts before detecting the person itself becomes even difficult. In this case, top-down approaches could have better results by firstly detecting the person and then proceeding single-person pose estimation in each bounding box. Papandreou et al. [30] propose a method using a two-stage network that uses a person box detection system [31] as a bounding box detector, then predicts the pose of every single person in each bounding box. He et al. [15] also develop a multi-task method based on their former contribution [32], which can be implemented for pose estimation tasks.

1.4. Information Used for Human Pose Estimation

Joint feature is the most direct information for human pose estimation. In addition, human detection [30] and human segmentation [15] have been used for improving the accuracy of pose estimation. Furthermore, in order to handle the self-occlusion in pose estimation, Azizpour et al. [33] and Ghiasi et al. [34] learn templates for occluded versions of each body part. Rafi et al. [35] incorporate the context information of occluding objects to predict the locations of occluded joints. Haque et al. [17] implicitly learn the occlusion through a deep neural network and gave the output of a visibility mask of joints. In our previous work [36], orientation prediction is integrated with single person pose estimation.

Table 1 shows a comparative overview of the different previous frameworks. Non-deep neural network approaches require image sequence [1–5], or hand-crafted features and models [1,6–8] for pose estimation. In addition, the accuracy of non-deep neural network approaches is lower than the following deep neural network approaches [16,18]. The single person pose estimation works with the guarantee of only one person present in the image [16–25,35,36]. However, the scenes with multi-person are exceedingly common in our daily life. Thus, multi-person pose estimation frameworks are more universal for real-world applications. Current multi-person pose estimation approaches are the lack of effectively using occlusion and orientation information, and suffer from the false positive detection problem. By considering the above mentioned points, this paper proposes to integrate the body orientation and occlusion information into the multi-task learning framework for the multi-person pose estimation. In addition, this paper not only simply implements the widely used parallel multi-task network, but also develops the serial multi-task networks to reduce the false positive detection (over-detection) in the multi-person pose estimation.

Table 1. A comparative overview between the different previous frameworks.

Category		Method	Limitation
Non-Deep Neural Network Approach		• Silhouette for pose [1,2] • Spatial + temporal information for pose [3–5] • Pictorial and part structure model for pose [6–8,33,34]	• Requirement of image sequence, hand-crafted features and models • Lower accuracy compared to deep neural network approaches
Deep Neural Network Approach	Single Person Pose Estimation	• DeepPose [16], • Feedback for pose [17,18] • CNN + graphical model for pose [19] • Convolutional pose machines [20] • Stacked hourglass networks for pose [21] • Densely connected convolutional module for pose [22] • Multi-scale compositional models for pose [23] • Generative adversarial network for pose [24] • Pose in depth image [25] • Semantic occlusion model for pose [35] • Orientation for pose [36]	• Limitation of only single person pose in one image
	Multi-Person Pose Estimation	• DeepCut [26] • DeeperCut [27] • MultiPoseNet [28] • Openpose [29] • Two stages (Faster RCNN detector + fully convolutional ResNet) for pose [30] • Mask-RCNN [15]	• Lack of effectively using occlusion and orientation information • False positive detection (over-detection)

This paper has two contributions. First one is to propose a novel multi-task learning framework for the multi-person pose estimation. In the related works, the body segmentation, joint position estimation and joint visibility have been used for the human pose estimation. This paper proposes to add body orientation and mutual-occlusion information into the multi-task learning framework for the multi-person pose estimation. The second contribution of this research to propose to use the serial multi-task networks instead of the widely used parallel multi-task network for the multi-person pose estimation. To prove the advantage of serial multi-task networks, this paper compares the performance of the multi-task learning frameworks with different configurations for the multi-person pose estimation. In addition, for the purpose to train orientation and occlusion recognition tasks, our research team builds a dataset based on images from the Common Objects in Context (COCO) keypoints dataset by adding an orientation and occlusion mask as new annotations. The initial ideas of this paper have been published in our previous conference papers [37]. Compared to our previous conference publication, this paper gives more surveys about the related works. In addition, this paper compares the performance of the different versions of the developed pose estimation networks.

The rest paper is organized as follows: Section 2 describes the proposed pose estimation networks. Section 3 evaluates the performance of the proposed networks. Finally, Section 4 concludes this paper.

2. Deep Neural Network for Multi-Person Pose Estimation

Human pose estimation is a challenging task in computer vision. One of the difficulties is the occlusion problem, including both self-occlusion and mutual-occlusion by other objects. The main reason of self-occlusion is body orientation. For example, if a standing person is facing the right of the camera, his or her left body is probably occluded. On the other hand, the mutual-occlusions may happen due to arbitrary objects. Another difficulty of pose estimation is the left-right similarity problem because of the symmetry of the human body. For example, the left shoulder of a person in the back view is very similar to the right shoulder in the front view. Therefore, to solve these problems, this research proposed to incorporate the body orientation and occlusion condition into the pose estimation.

In this paper, Mask R-CNN [15] was chosen as a basic model to build our multi-task model, because the top-down region proposal network of Mask R-CNN makes it feasible to implement new tasks into the multi-task and the pose estimation task of Mask R-CNN also has a certain space to improve. In the proposed models, body boundary, body orientation and occlusion condition and joint position estimation tasks were integrated in order to improve the accuracy of multi-person pose estimation. Body boundary can be represented as the silhouette. Body orientation is the human body direction relative in the camera space. The occlusion condition in the multi-person case can be separated into two situations: self-occlusion and mutual-occlusion. The more detailed information of these tasks will be described in the following subsections.

The system framework of the proposed method is illustrated in Figure 1. The input image would firstly be resized to 1024 × 1024 in the preprocessing step, then input into a Mask R-CNN layer heads for feature extraction, which consists of the feature pyramid network, region proposal network and a Region of Interest (RoI) align layer at last to generate RoI features. In addition, the two-stage object detection and classification branch in the original Mask R-CNN was also conducted in this layer heads. In fact, this paper also adopted the detection and classification branch of Mask R-CNN to obtain the RoI feature of each human. RoI features were then fed into the multi-task network. The multi-task learning part was the focus of this paper, and the details of the multi-task learning part were explained in the next subsections.

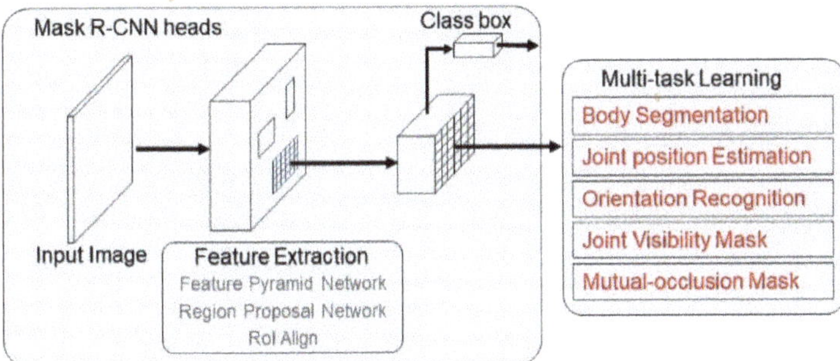

Figure 1. The framework of the proposed multi-task based pose estimation model.

2.1. Parallel Multi-Task Network for Pose Estimation

In this paper, the widely used parallel multi-task network was firstly chosen to integrate the multi-information. The architecture of the parallel multi-task network is illustrated in Figure 2. The network has three branches: body segmentation branch, joint position estimation branch and occlusion-orientation branch. The three branches output five multi-task results: segmentation, joint heatmap, orientation, joint visibility and mutual occlusion. In the training step, the five multi-task results were used for calculating the value of loss function. In this paper, the multi-task loss L was defined as the sum of the loss from each task:

$$L = L_{segm} + L_{joint} + L_{ori} + L_{vis} + L_{occ}, \tag{1}$$

where, L_{segm}, L_{joint}, L_{ori}, L_{vis} and L_{occ} denote the loss function for the task segmentation, joint heatmap estimation, orientation estimation, joint visibility and mutual occlusion estimation, individually. The details of each loss function will be explained in the next subsections.

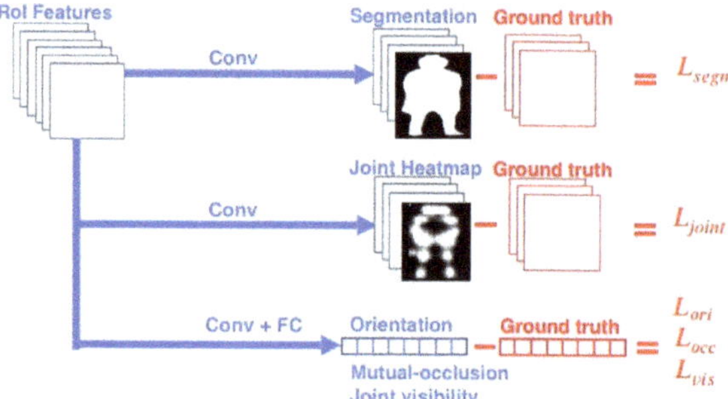

Figure 2. Architecture of the proposed parallel multi-task training network.

2.1.1. Body Segmentation Branch

To build the body segmentation branch, this paper borrowed the segmentation part of Mask R-CNN and changed the number of object classes into 2 (person and background). The architecture of this branch is illustrated in Figure 3. The body segmentation branch predicts masks for each RoI using a Fully Convolutional Networks (FCN) [38]. This branch consists of a stack of four convolutional layers with kernel size 3 and depth 256, followed by a deconvolution layer and a final convolutional layer with kernel size 3. In this branch, the 14 × 14 size RoI features are firstly passed into the segmentation head, and finally converted to an output with the resolution 28 × 28 and depth 2. This structure allows each layer in the branch to maintain the explicit object spatial layout without collapsing it into a vector representation that lacks spatial dimensions.

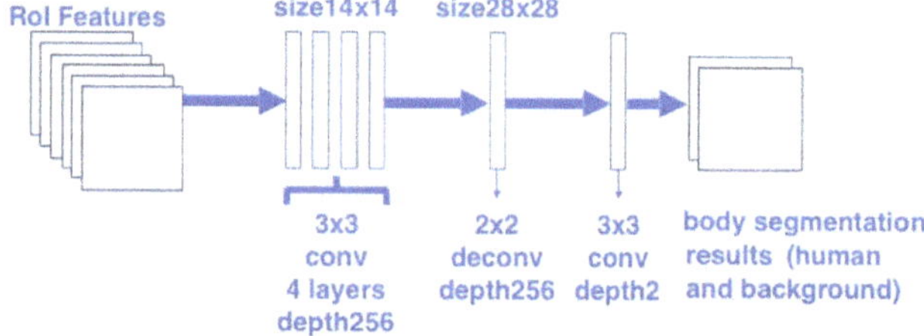

Figure 3. Architecture of the body segmentation branch.

The segmentation loss L_{segm} is defined as a cross entropy between ground truth and segmentation result with considering the class label of the segmentation result:

$$L_{segm} = -\frac{1}{N_{RoI}} \sum_{i=1}^{N_{RoI}} (Segm_i * class_i) * \log Segm_i', \tag{2}$$

where i is the index of RoI. $class_i$ is the ground truth class label. $Segm_i'$ is the segmentation ground truth and $Segm_i$ denotes the segmentation result.

2.1.2. Joint Position Estimation Branch

In the joint position estimation branch, the joint position is considered as a "one-hot" mask. The joint position estimation branch uses the architecture similar to the body segmentation branch for predicting N_{joint} masks. Each mask is corresponding to one of N_{joint} joint types (e.g., left shoulder, right elbow). The architecture of the joint position estimation branch is shown in Figure 4. Similar to the body segmentation task, this paper used the FCN structure to output these N_{joint} masks. For the reason that the joint is much smaller in each RoI, the output joint mask size is set twice as the segmentation mask in order to give accurate results. In this branch, the 14 × 14 RoI features are firstly input into a stack of eight convolutional layers with kernel size 3 and depth 256, then passed through a deconvolution layer with kernel size 2 and a bilinear interpolation up-scaling layer, finally converted to an output with the resolution 56 × 56 and depth N_{joint}.

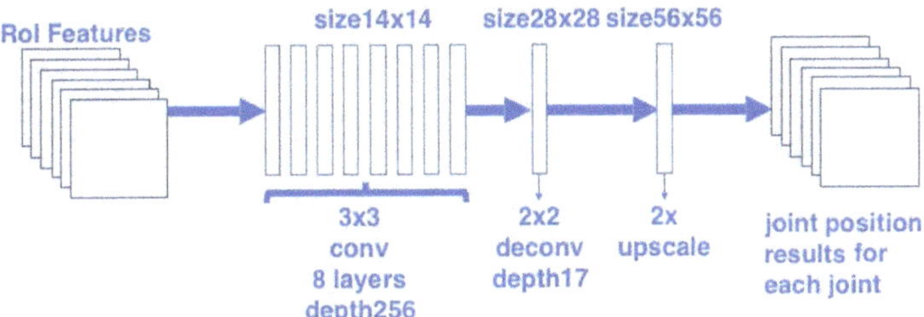

Figure 4. Architecture of the joint position estimation branch.

The joint position loss L_{joint} is defined as Equation (3).

$$L_{joint} = -\frac{1}{17N_{ROI}} \sum_{i=1}^{N_{ROI}} \sum_{j=1}^{N_{joint}=17} \text{Softmax}(K_{ij}) * \log K'_{ij} \tag{3}$$

where the number of joint N_{joint} is set to 17 in this research. K'_{ij} is the ground truth, K_{ij} is the joint position estimation result. Instead of the pixel-wise cross entropy used in body segmentation branch, this paper used one-hot label ground truth where only one correct joint position is true in the joint position loss function. This loss function can force the network to output a probability contribution where only one pixel in the mask gives the peak value after softmax operation.

2.1.3. Orientation-Occlusion Branch

The architecture of the orientation-occlusion branch is shown in Figure 5. This branch consists of a convolutional layer with kernel size 14 × 14 and depth 1024, followed by a fully connected layer. The body orientation is a kind of global information of pose configuration. The body orientation is useful for solving both the self-occlusion problem and left-right similarity problem. For example, if we know a person is facing to right, the body orientation indicates the occlusion of his or her left body. Similarly, if a person is facing the camera, his or her right shoulder is probably on the left side of the image. Our previous research [36] suggests that body orientation could be defined as 8 directions, as shown in Figure 6. When the body is defined as 8 directions, the prediction result of the body orientation can be a vector format with 8 elements.

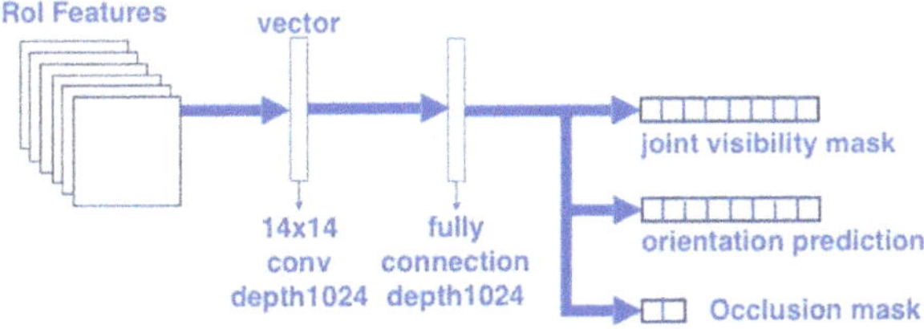

Figure 5. Architecture of the orientation-occlusion branch.

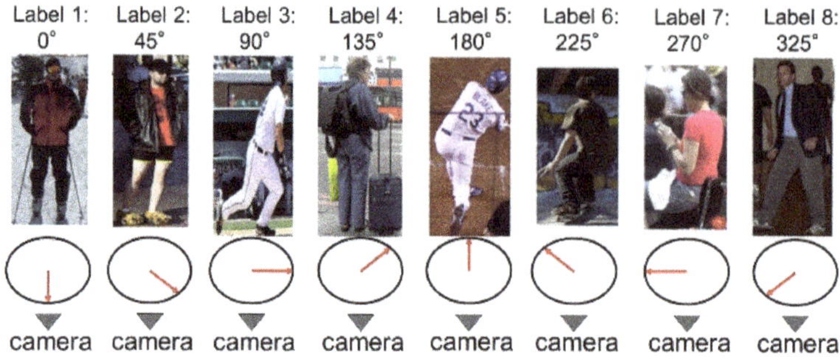

Figure 6. The representations of body orientations.

The term visibility mask is introduced by Haque et al. [17]. However, they only applied the visibility mask in the top view images for self-occlusions. This research extends the application of the visibility mask for more flexible view angles and for both self-occlusion and mutual-occlusion by objects. In this research, the joint visibility mask is a 17-dimensional vector to indicate the visibility of each joint. In addition, the occlusion mask is defined as a 2-dimensional vector. The combination of occlusion mask and joint visibility mask can explain the reason for the invisible joints: only self-occlusion or self-occlusion plus mutual-occlusion. The loss of these 3 outputs are defined as the cross entropy with softmax:

$$L_{ori} = -\frac{1}{N_{ROI}} \sum_{i=1}^{N_{ROI}} \text{Softmax}(Ori_i) * \log Ori_i', \tag{4}$$

$$L_{vis} = -\frac{1}{N_{ROI}} \sum_{i=1}^{N_{ROI}} \text{Softmax}(Vis_i) * \log Vis_i', \tag{5}$$

$$L_{occ} = -\frac{1}{N_{ROI}} \sum_{i=1}^{N_{ROI}} \text{Softmax}(Occ_i) * \log Occ_i', \tag{6}$$

where the terms Ori_i', Vis_i' and Occ_i' are the ground truth labels of body orientation, joint visibility and mutual-occlusion, respectively. The terms Ori_i, Vis_i and Occ_i are the estimation results of body orientation, joint visibility and mutual-occlusion output from this branch.

2.2. Serial Multi-Task Pose Estimation Network

In the parallel multi-task network, the information that can only be shared between different tasks is the RoI features extracted from the Feature Pyramid Network (FPN) and RoI align layer. Especially for the body segmentation task and joint position estimation task, the RoI features is hardly affected by the activation of ground truth label during backpropagation because of the deep convolutional networks. Actually, the information used in human pose estimation is strongly related to each other. Figure 7 shows an example of the relationship between the different tasks. The body segmentation describes the boundary of the human body, it can be a constraint for the joint position estimation. In addition, joint positions are helpful to model the structure of the human pose, which can be a strong reference for body orientation estimation. This is not the only way that this information is related, there are more combinations that the different tasks can benefit from each other. In order to take advantage of these relationships and enhance the connections between different tasks, this paper proposes to use the serial multi-task networks instead of the widely used parallel multi-task network for the multi-person pose estimation.

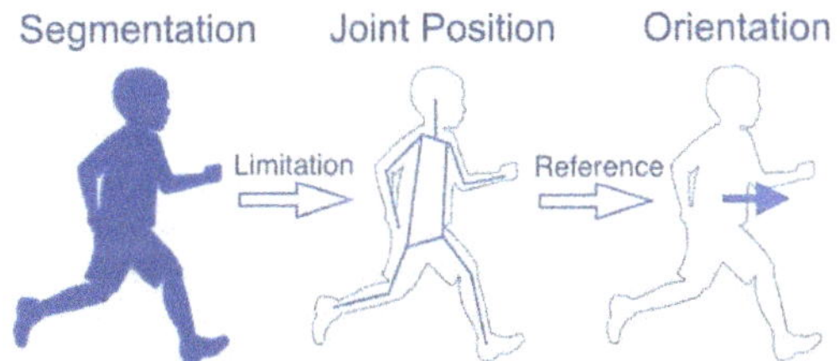

Figure 7. An example of the relationships between different tasks.

Figure 8 shows two forward passing architectures of the proposed serial multi-task networks. Since the orientation and occlusion outputs are vectors that have a smaller amount of data compared to the segmentation mask and joint heatmap, the orientation-occlusion branch is set as the last output layer in this research. For the rest two branches, this paper proposed two models. In the first model, the body segmentation branch was configured at first and the output of segmentation branch was fed into the joint position estimation branch (as shown in Figure 8a). In the other model, the results of the joint position estimation were directly input into other two branches (as shown in Figure 8b). In both two serial multi-task networks, the joint position estimation branch is connected to the orientation-occlusion branch because the joint position is a better reference for the prediction of body orientation than the body segmentation.

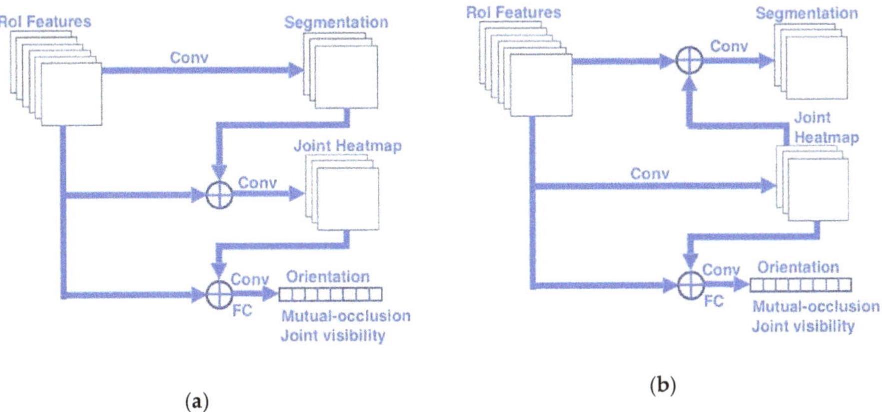

Figure 8. Proposed two serial multi-task networks. (a) Serial multi-task network with "Segmentation→Joint" connection; (b) Serial multi-task network with "Joint→ Segmentation" connection.

As shown in Figure 8a, the output of the body segmentation branch is passed into the joint position estimation branch in the first serial model. Figure 9 illustrates the new architecture of the joint position estimation branch combined with body segmentation. In this new architecture, the body segmentation output (resized to 14 × 14) was added into the RoI features to make the body segmentation be referred into the convolutional process of the joint position estimation.

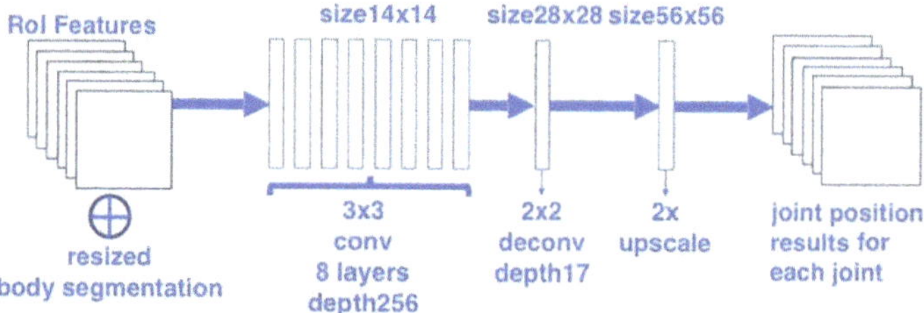

Figure 9. Joint position estimation branch combined with body segmentation (for Figure 8a).

In the second serial model, joint position estimation results were input into the body segmentation branch, as shown in Figure 8b. This model added the intermediate result of the joint position estimation (before the last up-scaling layer (size 28 × 28)) into the body segmentation branch, as shown in Figure 10.

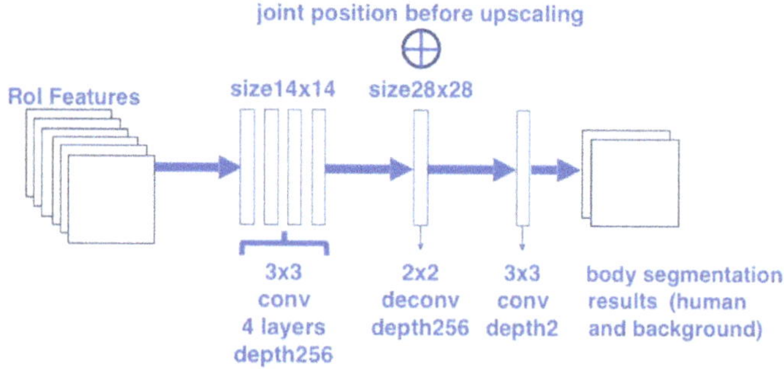

Figure 10. Body segmentation branch combined with joint position (for Figure 8b).

In addition, both two serial models have a connection between the joint position estimation branch and orientation-occlusion branch. Figure 11 shows how to combine the joint position estimation result with the orientation-occlusion branch. This research clips the joint position results to be consistent with the size of RoI features. In this way, the clipped joint position features and RoI features are input into the fully connected network together. The performance of the two serial models will be presented in Section 3.

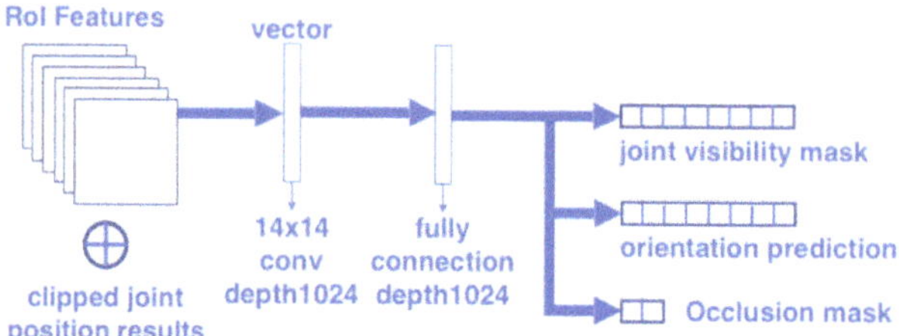

Figure 11. Orientation-occlusion branch combined with joint position (for Figure 8a,b).

3. Results

This section will firstly describe the details of the training dataset, and then demonstrate the evaluations for the proposed models and comparison with other methods.

3.1. COCO Keypoint Dataset

In this paper, the COCO keypoint dataset [39] was used for training and evaluation. COCO (Common Objects in Context) dataset is an open dataset built by Microsoft and Facebook, etc., which has a large volume of images for general object detection and segmentation tasks. Keypoint detection is one of the tasks in the COCO dataset, which requires localization of person keypoints in challenging, uncontrolled conditions. COCO keypoint dataset is the largest 2D pose estimation dataset and has been widely adopted by other multi-person 2D pose estimation algorithms. In addition, the images of the COCO keypoint dataset were captured in human's daily life, such as a party, meeting room and sport field. The images include multi-person with different scales, orientations, occlusions and postures. Figure 12 gives a demonstration of this task, each person in the image has a set of annotations including segmentation, class name (in this task it is "person"), bounding box position, the number of labeled keypoints and a list of body joint information (labeled in the format (x,y,v) with 17 joints, where x,y is the location of joint and v is the visibility of joint that $v = 0$ means not labeled, $v = 1$ means labeled but not visible and $v = 2$ means labeled and visible). The proposed models were trained on the COCO Keypoint Dataset 2017, which includes 56,599 images for training. In the training, the instances with less than three joints were excluded in order to teach the model to learn the whole body structure better and accelerate the converge of the training process.

Figure 12. Demonstrations of Common Objects in Context (COCO) Keypoint Dataset 2017.

3.2. Extended Sub Dataset with Mutual-Occlusion and Body Orientation

Since this paper added the mutual-occlusion estimation and orientation recognition task in the proposed models, which require the annotations that do not exist in the COCO keypoint dataset, our research team built a sub dataset for training these tasks and evaluation of the models. The subdataset included 2306 images from the training dataset of the COCO keypoint dataset and 260 images from the validation dataset in the COCO keypoint dataset. The images include over 8000 human instances in different sizes and situations. The 2306 images are used in training, and the 260 images were used for evaluation.

The label of body orientation follows the definition shown in Figure 6. This paper divided the horizontal space into eight parts with each part of 45 degrees to label different body orientations. However, because some instances are hard to be categorized into certain orientation, for this reason, a label "9" was added to deal with those instance orientations that were hard to distinguish. It means that the prediction result of the body orientation was a vector format with nine elements in this research.

The label system for the mutual-occlusion mask is shown in Figure 13. Unlike the joint visibility mask that is a binary mask for each joint of each instance, the mutual-occlusion mask is defined as a binary mask for the whole instance in one bounding box. It is a classification label that categorizes the person in each bounding box into two classes: occluded and not occluded. To distinguish this mutual-occlusion label from self-occlusion, our research team only labeled the instances that were cropped by the edge of the image or occluded by another person (or object) with Label 1 (occluded). For the instance that has self-occlusion due to its orientation, we labeled it as Label 0 (not occluded).

Figure 13. The label system for the mutual-occlusion mask in our subdataset.

3.3. Dataset for Training and Evaluation

In the research, all the models (will be presented in Tables 2–7) were trained on the training dataset of COCO Keypoint Dataset 2017. More specifically, the proposed models (both the parallel and serial models) were first "pre-trained" on the training dataset of the COCO keypoint dataset for the body segmentation task, joint position estimation task and joint visibility mask task. After that, 2306 images and their annotation (including segmentation, joint position, joint visibility mask, mutual-occlusion and body orientation) in the extended subdataset were reused to retrain all the branches of our proposed models. This retraining processing started from the pretrained networks. For other methods (Openpose [29] and MultiPoseNet [28]) adopted in the comparisons, the models are trained on the same training dataset.

Table 2. Evaluation for joint position estimation of proposed models using the Percentage of Correct Keypoints (PCK; %).

	Nose	Eye	Ear	Shoulder	Elbow	Wrist	Hip	Knee	Ankle	Overall
Joint+Segm	89.3	90.8	90.0	82.6	77.1	73.4	76.7	74.2	72.9	80.8
Joint+Segm With Occ&Ori	90.3	91.6	90.0	84.1	78.4	75.6	75.9	75.8	73.4	82.7
Segm→Joint	89.6	89.9	88.6	84.4	81.8	77.0	79.0	78.2	73.7	82.4
Segm→Joint With Occ&Ori	94.1	94.1	93.3	85.5	82.0	75.2	78.4	77.5	73.9	83.7
Joint→Segm	93.5	93.7	92.8	85.5	81.4	76.1	78.2	76.6	73.6	83.5
Joint→Segm With Occ&Ori	93.7	93.8	93.2	86.9	82.6	77.8	80.5	78.9	74.1	84.6

Table 3. Accuracy of orientation recognition in a strict principle (%).

Label	1	2	3	4	5	6	7	8	9
Degree	0	45	90	135	180	225	270	315	-
Joint+Segm with Occ&Ori	40.2	47.4	58.7	38.6	34.5	38.1	63.1	38.6	48.1
Segm→Joint with Occ&Ori	79.2	50.0	67.7	46.4	80.1	62.3	79.0	42.3	49.4
Joint→Segm with Occ&Ori	88.6	47.2	58.8	50.0	85.9	54.7	75.6	47.1	77.1

Table 4. Accuracy of orientation recognition when a neighbor error allowed (%).

Label	1	2	3	4	5	6	7	8	9
Degree	0	45	90	135	180	225	270	315	-
Joint+Segm with Occ&Ori	91.7	95.8	93.6	86.7	79.0	89.2	93.7	88.7	48.1
Segm→Joint with Occ&Ori	98.2	98.4	100	90.9	83.6	94.4	93.3	94.7	49.4
Joint→Segm with Occ&Ori	98.1	98.5	99.2	90.9	94.7	92.0	93.2	93.8	77.1

Table 5. Comparison between proposed model and other methods using PCK (%).

	Nose	Eye	Ear	Shoulder	Elbow	Wrist	Hip	Knee	Ankle	Overall
OpenPose [29]	92.2	92.1	92.3	89.5	83.9	74.2	83.3	83.5	80.8	85.7
MultiPoseNet [28]	88.9	88.5	88.9	84.5	82.0	79.6	78.6	79.7	78.0	83.2
Joint→Segm WithOcc&Ori	93.7	93.8	93.2	86.9	82.6	77.8	80.5	78.9	74.1	84.6

Table 6. Comparison between proposed models and other methods using Correct Detection Rate (CDR; %).

	Nose	Eye	Ear	Shoulder	Elbow	Wrist	Hip	Knee	Ankle	Overall
OpenPose [29]	72.2	70.5	66.2	74.8	69.3	72.8	72.7	74.8	66.5	71.2
MultiPoseNet [28]	85.6	81.1	77.6	88.3	82.4	80.5	83.7	80.3	71.0	81.2
Joint+Segm with Occ&Ori	82.7	80.4	75.2	86.0	80.0	78.4	88.4	87.7	82.9	82.4
Segm→Joint with Occ&Ori	89.3	86.6	79.2	85.2	77.7	77.8	77.8	85.3	82.7	82.4
Joint→Segm with Occ&Ori	84.7	86.6	80.6	85.2	82.3	81.4	85.8	87.8	79.3	83.7

Table 7. Comparison for processing time of proposed models and other methods.

Method	Processing Time (Second)
OpenPose [29]	0.26
MultiPoseNet [28]	0.05
Joint+Segm with Occ&Ori (Parallel model)	1.16
Segm→Joint with Occ&Ori (Serial model)	1.21
Joint→Segm with Occ&Ori (Serial model)	1.23

In the evaluation for joint position estimation (will be presented in Tables 2, 5 and 6), all the models (including both our proposed models and models in other methods) were evaluated on the validation dataset of the COCO keypoint dataset.

In the evaluation for body orientation recognition (will be presented in Tables 3 and 4), the models (including both the parallel models and serial models) were evaluated on 260 images in the extended subdataset. The 260 images had orientation annotations labeled by our research team.

3.4. Evaluation for Joint Position Estimation

To evaluate the performance for the joint position estimation, this paper used the Percentage of Correct Keypoints (PCK), which is a widely used evaluation metric. The correct keypoint is defined as the predicted joint whose distance from the true joint is less than a given threshold. In this research, the threshold used in the PCK calculation was set as $0.1 \times$ bounding box height. PCK can evaluate how many percentages of keypoints can be correctly detected. Calculation of PCK can be expressed as Equation (7).

$$\text{Percentage of Correct Keypoints} = \frac{\Sigma\left(Joint_{detected} \cap Joint_{groundtruth}\big|_{vis>0}\right)}{\Sigma\left(Joint_{groundtruth}\big|_{vis>0}\right)}, \tag{7}$$

where $Joint_{detected} \cap Joint_{groundtruth}$ is the joints correctly detected by the network and $Joint_{groundtruth}\big|_{vis>0}$ is the visible joints labeled in the ground truth annotations. In the evaluation, the PCK of the joints was calculated by considering visibility > 0.

The evaluation results and comparisons between the different proposed models are shown in Table 2. In Table 2, "Joint+Segm" represents a parallel multi-task model before training on the occlusion and orientation subdataset, "Segm→Joint" represents the serial multi-task model that inputs body segmentation results into the joint position estimation branch, "Joint→Segm" represents the serial multi-task model that inputs joint position estimation results into the body segmentation branch, and "with Occ&Ori" represents models trained on the occlusion and orientation subset.

The evaluation results show that after training on our occlusion and orientation subdataset, the accuracy of each model had an overall increase on each joint. The proposed serial multi-task models obtained higher accuracy than the parallel multi-task model. In addition, the serial model starts from the joint position estimation branch had the best performance. Some results of this model on multi-person images are shown in Figure 14. The examples of incorrect joint position estimation results are shown in Figure 15, where the image on the left top is an example that our model did not separate the two ankles but activated them at the same time, which led to the low accuracy of the ankle. For the human instances with an uncommon pose (image on the right top of Figure 15), our model also could not make a correct estimation. The two images on the bottom show examples where the body segmentation and joint position estimation gave incorrect results together, which is a problem of the multi-task network that we need to solve in the future.

Figure 14. Body segmentation and joint position estimation results on the COCO dataset.

Figure 15. Some incorrect joint position estimation results of the proposed model.

3.5. Evaluation for Orientation Estimation

This paper also evaluated the orientation recognition performance of the proposed models, which was also a widely used benchmark for human pose estimation. Table 3 shows the accuracy of the orientation recognition task of the proposed models. In the first evaluation, the predicted orientation result was correct if the result was exactly the same as the ground truth orientation. We could see the benefit after we input joint position estimation results into the occlusion-orientation branch. The serial model starts from joint position estimation had the best overall accuracy, where we thought body orientation estimation benefitted from the increase of joint position accuracy.

In addition, this paper adopted a compatible principle that allows a neighboring error for the body orientation estimation in the second evaluation. As shown in Table 4, the proposed models show more satisfying performance on the orientation prediction. For the normal eight orientations, the two serial models obtained a similar high accuracy. However, for the other orientation (Label 9), the serial model starts from the joint estimation branch had a better recognition rate than other models. Figure 16 visualizes several results of the whole multi-task outputs of the proposed model, where the red arrow in the center of each bounding box shows the body orientation recognition result, and the number with a blue background on the left top of bounding box represents the mutual-occlusion condition of each person (1 is mutual-occluded and 0 is not mutual-occluded).

Figure 16. Examples of the results for orientation recognition.

3.6. Comparison with Other Methods

In order to demonstrate the effectiveness of the proposed method, this paper compared the proposed multi-task models with other pose estimation methods. The comparison among our models, OpenPose [29] and MultiPoseNet [28], is shown in Table 5. Our serial model starts from joint position estimation obtained higher overall PCK than MultiPoseNet, which benefits from the higher accuracy of the face parts. For the whole-body joints without face parts, our model gave similar PCK as MultiPoseNet. The OpenPose performed higher PCK by taking advantage of its bottom-up architecture and part affinity filed network. However, we found that OpenPose had an obvious over-detection tendency on the "not joint" objects because it did not use the bounding box to limit the area of convolution.

In order to demonstrate the ability to make correct joint detections, this paper defined a Correct Detection Rate (CDR) to evaluate the over-detection tendency, which can be expressed as:

$$\text{Correct Detection Rate} = \frac{\sum\left(Joint_{detected} \cap Joint_{groundtruth}\big|_{vis>0}\right)}{\sum\left(Joint_{detected}\right)}, \tag{8}$$

where $Joint_{detected}$ is the joints detected by the network and $Joint_{groundtruth}$ is the joints labeled in the ground truth annotations. When a joint out of annotation is detected, it will be counted as an over-detection. The comparison between our proposed method and other methods using CDR is shown in Table 6.

This paper evaluated the performance of the human pose estimation algorithms by considering both PCK and CDR. PCK denotes the percentage of correct keypoints and CDR represents the over-detection tendency. Firstly, the proposed model (Joint→Segm with Occ&Ori) outperformed the MultiPoseNet in the comparisons using both PCK and CDR. Secondly, the proposed model (Joint→Segm with Occ&Ori) achieved 84.6% PCK, it was slightly worse than 85.7% of OpenPose in the PCK comparison, as shown in Table 5. However, the proposed model (Joint→Segm with Occ&Ori) had 83.7% CDR, which was much better than the CDR of OpenPose 71.2%, as shown in Table 6. In order to visualize the effect of the over-detection, Figure 17 shows the human pose estimation results generated by OpenPose (left) and the proposed model (Joint→Segm with Occ&Ori; right). It is clearly seen that the result of OpenPose had an over-detection on the right bottom of the left figure. In fact, the low CDR value of the OpenPose method was caused by this kind of over-detection. By analyzing the PCK and CDR, we could understand that OpenPose used an aggressive manner in the detection of the joint position to maintain the higher PCK value, at the same time generated more over-detections. On the contrary, our proposed model dramatically reduced the over-detections and achieved the competitive result on PCK.

Figure 17. Examples of human pose estimation results generated by OpenPose [29] (left) and the proposed model (right).

After the evaluation of the accuracy, the comparison was conducted for the processing speed. The processing speed of each model was calculated by averaging the processing time for each image used in joint position estimation. In the comparison for the processing speed, the processing time of different models were calculated by using the same images, and all experiments were performed on a system with an i7-8700K CPU and Nvidia 1080Ti GPU. The processing time of the parallel multi-task model, the serial multi-task models and other methods are summarized in Table 7. The parallel multi-task model with occlusion and orientation branches spends 1.16 s per image, and both two serial multi-task models need longer processing time than the parallel multi-task model, because later subnetworks need to wait for the earlier sub-networks to finish the process in the serial multi-task models, e.g., as shown in Figure 8b, the joint position estimation branch (label as "Joint Heatmap") should finish the calculation completely, then the segmentation branch and orientation-occlusion branch can use the Joint Heatmap to finish the processing. Thus, this serial framework increased the processing time. In addition, our proposed models had longer processing time than the conventional methods [28,29]. Overall, our proposed model had better accuracy but required more processing time compared to the conventional methods.

This proposed algorithm could be used for some applications that need high accuracy but do not require real-time processing speed, e.g., our previously published works [36,40] propose to analyze the customer pose for marketing. The customer behavior analysis is one of the most concerned topics for retailers because the customer behavior information can indicate the customer interest level to the product in the stores and is helpful to increase the commercial benefit. By checking the body orientation, head orientation and pose interactions between merchandise (such as touching, taking items, returning to shelf or putting into basket), these customer behaviors can reveal their interest level to the merchandise. The proposed methods in this paper can be used to analyze the recorded images by the surveillance camera in the store. High accuracy of the pose estimation is expected for the purpose of the commercial benefit. In addition, the real-time processing speed is not required because the customer behavior analysis could be post-processing. Thus, the customer behavior analysis application is one of the examples that the proposed method can be applied for.

This research has proven that integrating body orientation and occlusion information into the pose estimation multi-task network could improve accuracy. In the future, the temporal information between frames will be considered and utilized for the multi-person pose estimation in video. In addition, the temporal information is expected to be a solution to improve the low accuracy issue for ankle joints.

4. Conclusions

This research presented a multi-person pose estimation method using multi-task deep learning networks. The proposed network model is the extended multi-task network based on a Mask R-CNN layer heads, and it consists of five tasks: (1) joint position estimation, (2) body segmentation, (3) joint visibility mask, (4) body orientation recognition and (5) mutual-occlusion mask, the five tasks are separated into three branches: body segmentation branch, joint position estimation branch and orientation-occlusion branch. This paper first built a parallel multi-task network with each task separately. In order to strengthen the connections between different tasks, this paper further proposed two serial multi-task models. In the evaluation step, this paper first evaluated the accuracy of the joint position estimation and orientation recognition ability of the proposed models. In addition, this paper compared the accuracy of the proposed model with other pose estimation methods by considering two criterions PCK and CDR. The proposed method could detect 84.6% PCK and had 83.7% CDR. Our proposed model could reduce the over-detection compared with other methods.

There are still some problems that need to be solved in the proposed models such as the low accuracy of ankle joints. In the future, we would like to continue improving this model by using the temporal information between frames for multi-person pose estimation in the video.

Author Contributions: Y.G. wrote and revised the manuscript. He conducted the experiments for the revision. He participated in the discussion and development associated with this research. H.Z. developed the algorithm, and conducted the experiments. S.K. is the leader of this research. He proposed the basic idea and participated in the discussion. All authors have read and agreed to the published version of the manuscript.

Funding: This research received no external funding.

Conflicts of Interest: The authors declare no conflict of interest.

References

1.   Sminchisescu, C.; Telea, A. Human Pose Estimation from Silhouettes. A Consistent Approach Using Distance Level Sets. In Proceedings of the 10th international conference in central Europe on computer graphics, visualization and computer vision, Bory, Czech Republic, 4–8 February 2002; pp. 413–420.
2.   Mittal, A.; Zhao, L.; Davis, L.S. Human body pose estimation using silhouette shape analysis. In Proceedings of the IEEE Conference on Advanced Video and Signal Based Surveillance, Miami, FL, USA, 22–22 July 2003; pp. 263–270.
3.   Ramanan, D.; Forsyth, D.A.; Zisserman, A. Tracking people by learning their appearance. *IEEE Trans. Pattern Anal. Mach. Intell.* **2006**, *29*, 65–81. [CrossRef] [PubMed]
4.   Weiss, D.; Sapp, B.; Taskar, B. Sidestepping intractable inference with structured ensemble cascades. In Proceedings of the 24th International Conference on Neural Information Processing Systems, Vancouver, BC, Canada, 6–11 December 2010; pp. 2415–2423.
5.   Li, Q.; He, F.; Wang, T.; Zhou, L.; Xi, S. Human Pose Estimation by Exploiting Spatial and Temporal Constraints in Body-Part Configurations. *IEEE Access* **2016**, *5*, 443–454. [CrossRef]
6.   Felzenszwalb, P.F.; Huttenlocher, D.P. Pictorial structures for object recognition. *Int. J. Comput. Vision* **2005**, *61*, 55–79. [CrossRef]
7.   Yang, Y.; Ramanan, D. Articulated human detection with flexible mixtures of parts. *IEEE Trans. Pattern Anal. Mach. Intell.* **2012**, *35*, 2878–2890. [CrossRef] [PubMed]
8.   Pishchulin, L.; Andriluka, M.; Gehler, P.; Schiele, B. Poselet Conditioned Pictorial Structures. In Proceedings of the IEEE Conference on Computer Vision and Pattern Recognition, Portland, OR, USA, 23–28 June 2013; pp. 588–595.
9.   Krizhevsky, A.; Sutskever, I.; Hinton, G.E. Imagenet classification with deep convolutional neural networks. In Proceedings of the 26th International Conference on Neural Information Processing Systems 2012, Lake Tahoe, CA, USA, 3–8 December 2012; pp. 1097–1105.
10. Simonyan, K.; Zisserman, A. Very deep convolutional networks for large-scale image recognition. Available online: arXivpreprintarXiv:1409.1556 (accessed on 1 December 2019).
11. Szegedy, C.; Liu, W.; Jia, Y.; Sermanet, P.; Reed, S.; Anguelov, D.; Erhan, D.; Vanhoucke, V.; Rabinovich, A. Going deeper with convolutions. In Proceedings of the IEEE Conference on Computer Vision and Pattern Recognition, Boston, MA, USA, 7–12 June 2015; pp. 1–9.
12. He, K.; Zhang, X.; Ren, S.; Sun, J. Deep residual learning for image recognition. In Proceedings of the IEEE Conference on Computer Vision and Pattern Recognition, Las Vegas, NV, USA, 27–30 June 2016; pp. 770–778.
13. Girshick, R. Fast R-CNN. In Proceedings of the IEEE International Conference on Computer Vision, Santiago, Chile, 7–13 December 2015; pp. 1440–1448.
14. Redmon, J.; Farhadi, A. Yolov3: An incremental improvement. Available online: arXivpreprintarXiv: 1804.02767 (accessed on 1 December 2019).
15. He, K.; Gkioxari, G.; Dollár, P.; Girshick, R. Mask R-CNN. In Proceedings of the IEEE International Conference on Computer Vision, Venice, Italy, 22–29 October 2017; pp. 2980–2988.
16. Toshev, A.; Szegedy, C. DeepPose: Human pose estimation via deep neural networks. In Proceedings of the IEEE Conference on Computer Vision and Pattern Recognition, Columbus, OH, USA, 23–28 June 2014; pp. 1653–1660.
17. Haque, A.; Peng, B.; Luo, Z.; Alahi, A.; Yeung, S.; Fei-Fei, L. Towards viewpoint invariant 3D human pose estimation. In Proceedings of the European Conference on Computer Vision, Amsterdam, The Netherlands, 8–16 October 2016; pp. 160–177.

18. Carreira, J.; Agrawal, P.; Fragkiadaki, K.; Malik, J. Human pose estimation with iterative error feedback. In Proceedings of the IEEE Conference on Computer Vision and Pattern Recognition, Las Vegas, NV, USA, 27–30 June 2016; pp. 4733–4742.

19. Tompson, J.J.; Jain, A.; LeCun, Y.; Bregler, C. Joint training of a convolutional network and a graphical model for human pose estimation. In Proceedings of the Neural Information Processing Systems 2014, Montréal, QC, Canada, 8–13 December 2014; pp. 1799–1807.

20. Wei, S.E.; Ramakrishna, V.; Kanade, T.; Sheikh, Y. Convolutional pose machines. In Proceedings of the IEEE Conference on Computer Vision and Pattern Recognition, Las Vegas, NV, USA, 27–30 June 2016; pp. 4724–4732.

21. Newell, A.; Yang, K.; Deng, J. Stacked hourglass networks for human pose estimation. In Proceedings of the European Conference on Computer Vision, Amsterdam, The Netherlands, 8–16 October 2016; pp. 483–499.

22. Wang, Z.; Liu, G.; Tian, G. A parameter efficient human pose estimation method based on densely connected convolutional module. *IEEE Access* **2018**, *6*, 58056–58063. [CrossRef]

23. Wang, R.; Cao, Z.; Wanga, X.; Liu, Z.; Zhu, X. Human pose estimation with deeply learned multi-scale compositional models. *IEEE Access* **2019**, *7*, 71158–71166. [CrossRef]

24. Chen, Y.; Shen, C.; Wei, X.S.; Liu, L.; Yang, J. Adversarial posenet: A structure-aware convolutional network for human pose estimation. In Proceedings of the IEEE International Conference on Computer Vision, Venice, Italy, 22–29 October 2017; pp. 1221–1230.

25. Szczuko, P. Deep neural networks for human pose estimation from a very low resolution depth image. *Multimed. Tools Appl.* **2019**, *78*, 1–21. [CrossRef]

26. Pishchulin, L.; Insafutdinov, E.; Tang, S.; Andres, B.; Andriluka, M.; Gehler, P.V.; Schiele, B. DeepCut: Joint subset partition and labeling for multi person pose estimation. In Proceedings of the IEEE Conference on Computer Vision and Pattern Recognition, Las Vegas, NV, USA, 27–30 June 2016; pp. 4929–4937.

27. Insafutdinov, E.; Pishchulin, L.; Andres, B.; Andriluka, M.; Schiele, B. DeeperCut: A deeper, stronger, and faster multi-person pose estimation model. In Proceedings of the European Conference on Computer Vision, Amsterdam, The Netherlands, 8–16 October 2016; pp. 34–50.

28. Kocabas, M.; Karagoz, S.; Akbas, E. MultiPoseNet: Fast multi-person pose estimation using pose residual network. In Proceedings of the European Conference on Computer Vision, Munich, Germany, 8–14 September 2018; pp. 437–453.

29. Cao, Z.; Simon, T.; Wei, S.E.; Sheikh, Y. Realtime multi-person 2D pose estimation using part affinity fields. In Proceedings of the IEEE Conference on Computer Vision and Pattern Recognition, Honolulu, HI, USA, 21–26 July 2017; pp. 1302–1310.

30. Papandreou, G.; Zhu, T.; Kanazawa, N.; Toshev, A.; Tompson, J.; Bregler, C.; Murphy, K. Towards accurate multi-person pose estimation in the wild. In Proceedings of the IEEE Conference on Computer Vision and Pattern Recognition, Honolulu, HI, USA, 21–26 July 2017; pp. 3711–3719.

31. Huang, J.; Rathod, V.; Sun, C.; Zhu, M.; Korattikara, A.; Fathi, A.; Fischer, I.; Wojna, Z.; Song, Y.; Guadarrama, S.; et al. Speed/Accuracy trade-offs for modern convolutional object detectors. In Proceedings of the IEEE Conference on Computer Vision and Pattern Recognition, Honolulu, HI, USA, 21–26 July 2017; pp. 3296–3297.

32. Ren, S.; He, K.; Girshick, R.; Sun, J. Faster R-CNN: Towards real-time object detection with region proposal networks. In Proceedings of the 28th International Conference on Neural Information Processing Systems, Montreal, QC, Canada, 8–13 December 2014; pp. 91–99.

33. Azizpour, H.; Laptev, I. Object detection using strongly-supervised deformable part models. In Proceedings of the European Conference on Computer Vision, Florence, Italy, 7–13 October 2012; pp. 836–849.

34. Ghiasi, G.; Yang, Y.; Ramanan, D.; Fowlkes, C.C. Parsing occluded people. In Proceedings of the IEEE Conference on Computer Vision and Pattern Recognition, Columbus, OH, USA, 23–28 June 2014; pp. 2401–2408.

35. Rafi, U.; Gall, J.; Leibe, B. A semantic occlusion model for human pose estimation from a single depth image. In Proceedings of the IEEE Conference on Computer Vision and Pattern Recognition Workshops, Boston, MA, USA, 7–12 June 2015; pp. 67–74.

36. Liu, J.; Gu, Y.; Kamijo, S. Integral customer pose estimation using body orientation and visibility mask. *Multimed. Tools Appl.* **2018**, *77*, 26107–26134. [CrossRef]

37. Zhang, H.; Gu, Y.; Kamijo, S. Orientation and occlusion aware multi-person pose estimation using multi-task deep learning network. In Proceedings of the IEEE International Conference on Consumer Electronics, Las Vegas, NV, USA, 11–13 January 2019; pp. 1–5.

38. Long, J.; Shelhamer, E.; Darrell, T. Fully convolutional networks for semantic segmentation. *IEEE Trans. Pattern Anal. Mach. Intell.* **2015**, *39*, 640–651.

39. Lin, T.Y.; Maire, M.; Belongie, S.; Hays, J.; Perona, P.; Ramanan, D.; Dollár, P.; Zitnick, C.L. Microsoft COCO: Common objects in context. In Proceedings of the European Conference on Computer Vision, Zurich, Switzerland, 6–12 September 2014; pp. 740–755.

40. Liu, J.; Gu, Y.; Kamijo, S. Customer behavior classification using surveillance camera for marketing. *Multimed. Tools Appl.* **2017**, *76*, 6595–6622. [CrossRef]

Article

Semi-Supervised Nests of Melanocytes Segmentation Method Using Convolutional Autoencoders

Dariusz Kucharski [1,*], **Pawel Kleczek** [1], **Joanna Jaworek-Korjakowska** [1], **Grzegorz Dyduch** [2] and **Marek Gorgon** [1]

[1] Department of Automatic Control and Robotics, AGH University of Science and Technology, al. A. Mickiewicza 30, 30-059 Krakow, Poland; pkleczek@agh.edu.pl (P.K.); jaworek@agh.edu.pl (J.J.-K.); mago@agh.edu.pl (M.G.)

[2] Chair of Pathomorphology, Jagiellonian University Medical College, ul. Grzegorzecka 16, 31-531 Krakow, Poland; grzegorz.dyduch@cm-uj.krakow.pl

* Correspondence: darekk@agh.edu.pl

Received: 29 December 2019; Accepted: 4 March 2020; Published: 11 March 2020

Abstract: In this research, we present a semi-supervised segmentation solution using convolutional autoencoders to solve the problem of segmentation tasks having a small number of ground-truth images. We evaluate the proposed deep network architecture for the detection of nests of nevus cells in histopathological images of skin specimens is an important step in dermatopathology. The diagnostic criteria based on the degree of uniformity and symmetry of border irregularities are particularly vital in dermatopathology, in order to distinguish between benign and malignant skin lesions. However, to the best of our knowledge, it is the first described method to segment the nests region. The novelty of our approach is not only the area of research, but, furthermore, we address a problem with a small ground-truth dataset. We propose an effective computer-vision based deep learning tool that can perform the nests segmentation based on an autoencoder architecture with two learning steps. Experimental results verified the effectiveness of the proposed approach and its ability to segment nests areas with Dice similarity coefficient 0.81, sensitivity 0.76, and specificity 0.94, which is a state-of-the-art result.

Keywords: deep learning; autoencoders; semi-supervised learning; computer vision; pathology; epidermis; skin

---

1. Introduction

Over the past few decades, both incidence and mortality rates caused by cutaneous melanoma (the most aggressive and lethal skin cancer) among Caucasian populations worldwide has significantly increased [1]. In many countries and regions of the world (such as France, Australia, and Switzerland), the crude incidence rate of cutaneous melanoma increased by 50%–100% between 1993 and 2013 [2]. According to surveys conducted by national health services, melanoma is currently responsible for nearly 70% of all skin cancer-related deaths in the United States and in Australia [3,4]. Although, to date, no effective treatment of melanoma in advanced stages has been developed, an early melanoma (in the in situ stage) is treatable in about 99% of the cases with all but a simple excision [5]. Therefore, the early diagnosis of melanoma, and especially distinguishing melanoma from other types of skin melanocytic lesions, has become an extremely important issue.

The gold standard for diagnosing skin melanocytic lesions (including melanoma) is the histopathological examination—the microscopic examination of tissue in order to study the manifestations of disease [6]. The histopathological criteria for diagnosing skin melanocytic lesions are based on the analysis of such features as lesion's asymmetry, morphometric features of epidermis, proliferation patterns of single melanocytes, and more [7,8]. In particular, the size, shape, position, and distribution of nests of melanocytes (i.e., aggregations of melanin-producing cells, located originally in the bottom layer of the epidermis) are considered. The traditional histopathological examination was carried out manually and with the use of a light microscope. However, such an approach has at least two important drawbacks: quantitative analysis of a large volume of specimens is a laborious activity, and it is a subjectivity-prone task whose results are not reproducible [9,10]. Therefore, there is a need to develop automatic image analysis methods for the diagnostics of skin melanocytic lesions, which will provide accurate, reliable, and reproducible results. The rapid development in the fields of digital pathology and artificial intelligence facilitates research works in this topic.

Artificial intelligence research (AI) is a wide-ranging branch of computer science that was already introduced more than half a century ago [11]. However, the greatest progress has been visible in recent years. Deep learning models, which are the latest but also the most promising methods, have been exploited with impressive results in signal processing and computer-vision tasks. Deep learning is currently one of the most popular advanced neural network models used for image segmentation, classification, and reconstruction for challenges that have not been solved yet, particularly in the field of biological systems and medical diagnostics [12]. Progress in the hardware, software, algorithms, and availability of huge datasets allow employment of high accuracy image recognition systems. A systematic review on the use of deep neural networks including convolutional autoencoders can be found in [13,14].

In this paper, we present a new approach that is one of the first approaches to the segmentation of nests of melanocytes. We examine the possibility to use convolutional autoencoders, which are a specific type of feedforward neural networks where the input is the same as the output. They compress the input into a lower-dimensional code and then reconstruct the output from this representation. We firstly train the network in an unsupervised way to obtain the weights for the encoding part; secondly, we train the network with limited ground-truth images to receive the weights for the decoding part. The network architecture is adapted to the specification of our problem including the densely-connected classifier layer.

To the best of our knowledge, it is the first method to segment nests of melanocytes in histopathological images of H&E-stained skin specimens.

The novelty of this work can be summarized as:

- we present a deep learning based solution for the segmentation of nests on histopathological images,
- we propose a convolutional autoencoder neural network architecture with two semi-supervised training stages for the encoding and decoding parts,
- our method enables the adaptation of the autoencoders for limited ground-truth data amount based on data augmentation and autoencoders specification.

This paper is organized in four sections as follows: Section 1 presents the skin cancer awareness and covers background information on deep learning methods including autoencoders, related works, motivation of the undertaken research, medical devices, and the problem of melanoma misdiagnosis. Section 2 shows in detail the methodology used in this research to segment the nests regions including data specification and preparation, the autoencoder architecture, layers, training, and classification stages. Section 3 presents the training parameters, and conducts tests and results as well as visualization and interpretation of the layers outputs. Section 4 exposes the conclusions and suggests new lines of research.

1.1. Medical Background

Skin melanocytic lesions are neoplasms derived from epidermal melanocytes. The two principal classes of skin melanocytic lesions are "benign" nevi (with no metastatic potential) and "malignant" melanoma (with a metastatic capacity proportional to its thickness).

In the context of melanocytes, the term "nevus" denotes the localized aggregation of nevus cells, i.e., cells derived from melanocytes, as benign neuroectodermal proliferations, neoplasms, or both [15]. Nevus cells, arising as a result of proliferation of melanocytes at the dermal-epidermal junction, are larger than typical melanocytes, do not have dendrites, have more abundant cytoplasm with coarse granules, and are typically grouped in nests [16,17]. Cells within nests are often oval or cuboidal in shape, with clear cytoplasm and variable pigmentation [18]. In the superficial dermis, the cells have an epithelioid cell topography and contain amphophilic cytoplasm with granular melanin. The nuclei have uniform chromatin with a slightly clumped texture. Deeper in the dermis, there is a diminished content of cytoplasm, and the cells resemble lymphocytes and are arranged in linear cords [19]. Examples of nests of nevus cells are shown in Figure 1.

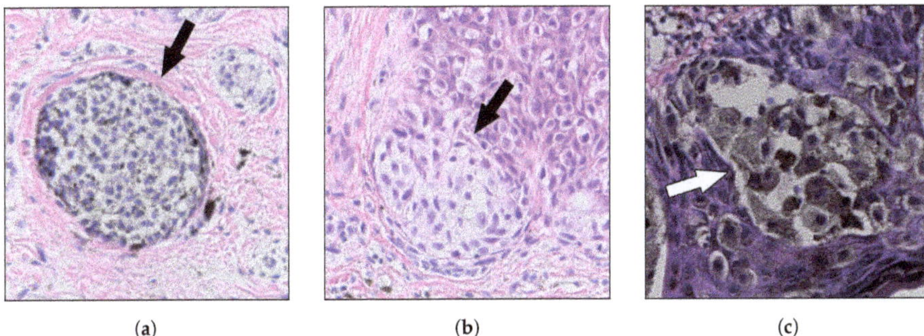

(a)   (b)   (c)

Figure 1. Examples of nests of nevus cells (marked with arrows): (**a**) a dermal nest; (**b**) a nest at the tip of a rete ridge; and (**c**) a nest adjacent to the epidermal plate; note a strong pigmentation of cytoplasm in nevus cells. The structure of nests is highly not-uniform and varies between individual nests.

Melanoma is the generic term for all malignant neoplasms derived from melanocytes, with most melanomas developing through an out-of-control progressive proliferation of melanocytes within the squamous epithelium [8]. Melanomas are the most aggressive and dangerous neoplasms; they grow fast and quickly metastasize to nearby lymph nodes and other organs.

In recent years, the sharp dichotomy between "benign" nevi and "malignant" melanoma started to be considered an oversimplification [7]. No histopathological criterion is entirely specific for nevi (all can be occasionally found in a melanoma) [20] and some criteria are infrequent, or have low specificity, or suffer from high inter-observer variation [21].

Examples of criteria which provide strong evidence in favor of melanoma, but are extremely infrequent, include presence of necroses (sensitivity of 15.3%, specificity of 100%) and melanin in deep cells (sensitivity of 36.1%, specificity of 90.4%). On the other hand, some highly sensitive and specific criteria, such as the presence of mitoses (sensitivity of 81.9%, specificity of 100.0%) might be hard to detect automatically, as demonstrated in the course of the MITOS-ATYPIA-14 challenge and Tumor Proliferation Assessment Challenge 2016—the F_1-score of top-performing methods for mitosis detection was roughly 0.650 [22,23]. The analysis of cases in [21] showed that, in melanomas, except for cytological atypia, not even one of the investigated features was constant.

Moreover, some challenging lesions contain conflicting criteria suggesting opposite diagnoses. Consequently, in many instances, the diagnosis may be subjective because of differences in training and philosophy, and the experience of the observers: accordingly, the same lesion may be classified as dysplastic nevus with severe atypia by one pathologist and as melanoma in situ by another. Therefore, criteria must be used in clusters and ought to be used for specific differential diagnosis (i.e., a specific form of nevus should be differentiated from a type of melanoma morphologically mimicking it) [7]. The above-mentioned issues expose the need for CAD systems which will take into consideration as wide a spectrum of histopathological features as possible.

The histological criteria currently used in the diagnosis of skin melanocytic lesions consist of the analysis of numerous features, such as: lesion's asymmetry, morphometric features of epidermis, proliferation patterns of single melanocytes and nests of nevus cells, cytological atypia, mitoses, and necroses [7,8]. For nests of nevus cells, the following features are typically considered: size, shape, horizontal and vertical proliferation patterns (as well as the regularity for each of those features). These are nests-related diagnostic criteria for some of the most common types of skin melanocytic lesions:

- In common nevi, the nests are roundish or regularly elongated, and typically positioned at the tips of rete ridges [24].
- Melanocytic nevi contain intraepidermal or dermal collections of nevus cells or both. The cells within the junctional nests have round, ovoid, or fusiform shapes and are arranged in cohesive nests [18].
- One of the major histological criteria for the diagnosis of lentigo maligna (a precursor to lentigo maligna melanoma, a potentially serious form of skin cancer) is the absence of intraepidermal nesting [8].
- A junctional dysplastic nevus consists of a proliferation of a variable combination of single melanocytes and nevus cells in nests along the dermal-epidermal junction. If present, nests are often irregular in size and shape and may "bridge" or join together [24].
- In superficial spreading melanoma (SSM) and melanoma in the in situ stage (i.e., entirely restricted to the epidermis, the dermal-epidermal junction, and epithelial appendages), large, irregularly shaped, confluent nests are unevenly distributed along the dermal-epidermal junction, separated one from another by "skip areas" which are either free of melanocytes or with a lesser number of melanocytes arranged in a lentiginous pattern [7,8]. In SSM, nests are also present above the suprapapillary plates at the edges of rete ridges and a discohesive appearance in large nests is often evident. The cellularity, the pigment, and the type of melanocytes vary greatly among nests.
- Melanomas are subdivided in "radial" and "vertical" growth phases: "radial growth phase" includes melanoma in situ and early invasive superficial spreading melanoma and is characterized by small nests in the papillary dermis, whereas nests larger than those at the junction are typical of the "vertical growth phase" [7].

Examples of the above-mentioned melanocytic skin lesions are shown in Figure 2.

1.2. Melanoma Misdiagnosis Problem

The histopathological examination constitutes the gold standard for diagnosing skin melanocytic lesions—other existing forms of examination (such as dermatoscopy) do not yield comparably high diagnostic confidence [25,26]. Nonetheless, even this "gold standard" is far from being perfect and pathologists evaluating the same lesion may not be concordant one with another regarding the diagnosis. Numerous surveys revealed that, even in the case of the histopathological examination, the melanoma misdiagnosis rate may well reach up to 25% [9,27–29]. A diagnostic error with particularly severe consequences is a false-negative diagnosis of melanoma (i.e., a situation when a malignant lesion is misdiagnosed as a benign one)—it delays the start of the treatment, which typically leads to further medical complications resulting even in a patient's death. Three important reasons for the above-mentioned lack of concordance between pathologists and the high melanoma misdiagnosis rate are that histopathological criteria are vaguely defined, in many cases, lesions are evaluated not by dermatopathologists or surgical pathologists but by general histopathologists (who lack profound knowledge of the diagnostic niceties), and the diagnosis is highly subjective as it is based mainly on experience and intuition of a given pathologist.

These obstacles encourage researchers to try to develop new methods for the automatic analysis of histopathological features of skin, such as features related to the nests of melanocytes, which could increase the specificity and sensitivity of the assessment of skin melanocytic lesions.

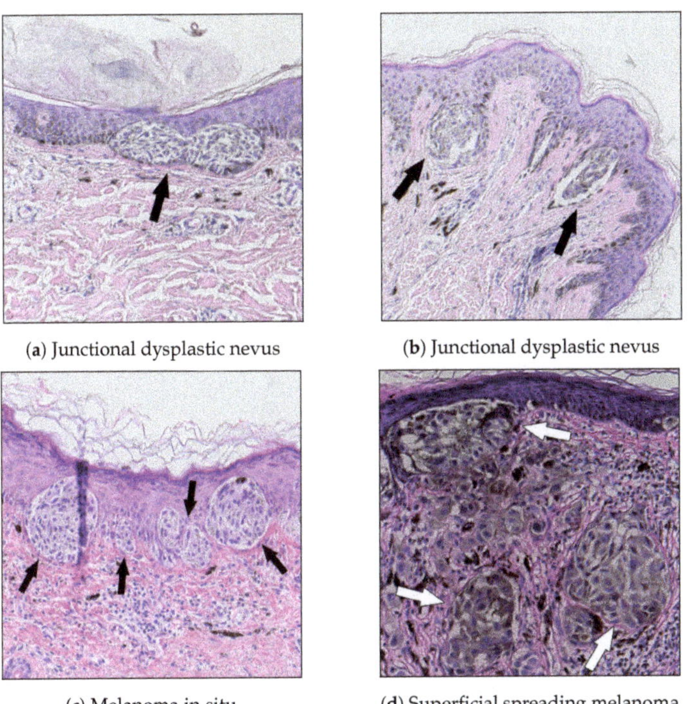

(**a**) Junctional dysplastic nevus

(**b**) Junctional dysplastic nevus

(**c**) Melanoma in situ

(**d**) Superficial spreading melanoma

Figure 2. Examples of melanocytic lesions containing nests of nevus cells: (**a**) in junctional dysplastic nevi, nevus cells are typically arranged in cohesive nests along the dermal-epidermal junction nests and often join together; (**b**) in nevi, nests are often positioned at the tips of rete ridges; (**c**) in melanoma in situ, there are often large, confluent nests, irregular in shape and size, unevenly distributed along the dermal-epidermal junction; and (**d**) in SSM, nests are present above the suprapapillary plate.

1.3. Image Acquisition

The problems plaguing the traditional histopathological examination (mentioned in Section 1) may be addressed by digital pathology—a rapidly growing field primarily driven by developments in technology, which is mainly about analyzing whole slide images (WSIs). WSIs are glass slides of tissue specimens digitized at high magnification and thus able to provide global information for quantitative and qualitative image analysis (Figure 3a) [30,31].

To obtain a WSI image, either an automatic microscope or a WSI scanner is typically used. Modern microscopes have a wide variety of components that can be automated by means of electronic control (mainly shutters, stages, light sources, and focus control) and all these motorized components, sensors, and input devices are typically integrated into a software environment. However, the "traditional" microscope systems still have at least one serious limitation in the context of digital pathology: due to the open construction, they do not provide consistent imaging conditions. This shortcoming was eliminated with the introduction of whole slide scanners—a specially designed microscope under robotic and computer control, which has all components assembled in a special casing (Figure 3b) [32,33].

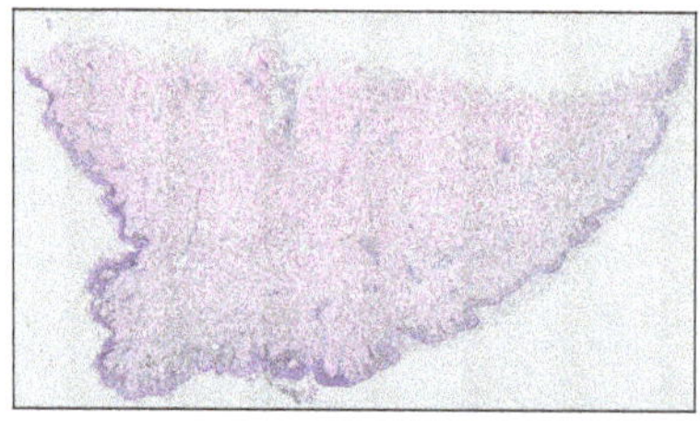

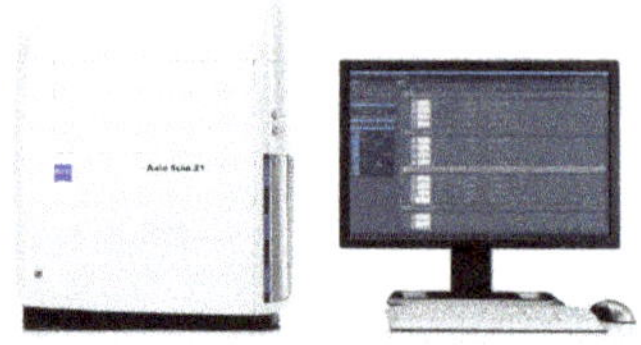

(a) A whole slide image (WSI)   (b) Hardware (left) and software (right)

Figure 3. Whole slide imaging: (**a**) an example of a WSI produced by a scanning system; (**b**) a WSI scanning system consists of a dedicated hardware and software.

After digital data are captured via the camera's charge-coupled device (CCD), the virtual slide is assembled together from large numbers of image frames in one of the following ways, depending on the particular scanner being used: tiling, line scanning, dual sensor scanning, dynamic focusing, or array scanning (the process is performed automatically by a specialized imaging software). However, the stitching process rarely produces a truly seamless image, and thus artifacts (e.g., related to vignetting) are typically observed even in images captured using a whole slide scanner.

For most diagnostic work, the digital slides are routinely scanned at $\times 20$ magnification (at resolution approximately 0.25–0.5 µm/px) [34,35]. The size of a high-resolution whole slide image may well be up to 75,000 $\times$ 50,000 px. When stored in an uncompressed format or using lossless compression, such high-resolution digital slides may result in very large files (on the order of several GB), impacting storage costs and work throughput. Consequently, for routine examination by pathologists, a lossy compression technique (e.g., the JPEG or JPEG 2000 image standard) is typically applied. However, since compressing images in a lossy way renders them virtually useless for various methods of automatic digital image processing and analysis [36], for this sort of research, the TIFF format is usually employed to archive virtual slides.

Nonetheless, in recent years, some context aware image compression methods (i.e., techniques, which compress only irrelevant parts of the image, while leaving regions valuable from the diagnostic point of view intact) were proposed—some notable examples are briefly discussed below. Tellez et al. [37] proposed a method to reduce the size of a gigapixel image while retaining semantic information by shrinking its spatial dimensions and growing along the feature direction—an image is firstly divided into a set of high-resolution patches, then each high-resolution patch is compressed with a neural network mapping every image into a low-dimensional embedding vector, and finally each embedding is placed into an array that keeps the original spatial arrangement intact so that neighbor embeddings in the array represent neighbor patches in the original image. Hernández-Cabronero et al. [38] proposed an optimization method called mosaic optimization for designing irreversible and reversible color transforms simultaneously optimized for any given WSI and the subsequent compression algorithm—the method is designed to enable continuous operation of a WSI scanner. Niazi et al. [39] proposed a pathological image compression framework to address the needs of Big Data image analysis in digital pathology, specifically for breast cancer diagnosis, based on a JPEG2000 image compression standard and the JPEG2000 Interactive Protocol—they suggested to identify "hotspots", i.e., areas in which Ki-67 nuclear protein staining is most prevalent (Ki-67 is an independent breast cancer prognostic marker), and reduce the compression ratio when processing those areas.

1.4. Related Works

There are only a few works in the literature which cover automatic processing of histopathological whole slide images of skin specimens stained with hematoxylin and eosin (H&E), the standard stain in histopathology. Some notable examples include an automated algorithm for the diagnostics of melanocytic tumors by Xu et al. [40] (based on the melanocyte detection technique described in [41] and the epidermis segmentation approach described in [42]), a method capable of differentiating squamous cell carcinoma in situ from actinic keratosis by Noroozi and Zakerolhosseini [43] and a method for classifying histopathological skin images of three common skin lesions: basal cell carcinomas, dermal nevi, and seborrheic keratoses by Olsen et al. [44]. The first two methods are based on classic algorithms for image processing and machine learning, whereas the last one uses deep neural networks. None of the above-mentioned methods considers features related to nests of melanocytes, whereas the size, shape, position, and distribution of nests are among important diagnostic criteria when diagnosing skin melanocytic lesions [7,8].

The characteristics of autoencoders have rendered these models useful in various image processing tasks, such as image denoising and image restoration [45,46]. In particular, they have been successfully used in the field of medical imaging and diagnostics (also for super-resolution images) for image denoising [47], detection of cells [48,49], and the analysis of whole tissue structures [50].

Specifically, two methods for automatic cell segmentation use autoencoders for unsupervised cell detection in histopathological slides: Hou et al. [51] proposed a general-purpose method for nuclei segmentation, whereas Song et al. [52] designed a model to segment erythroid and myeloid cells in bone marrow. These methods are designed to detect individual cells or slightly overlapping cells in cases where the maximum possible size of a cell is precisely determined. However, the nests segmentation task is quite different from the above-mentioned tasks as nests are clusters typically composed of dozens or even hundreds of cells and thus no size- or shape-related criteria can be determined. Moreover, the structure of a nest is highly not-uniform and great inter-nest variability is observed, which further complicate matters.

As histopathological image analysis is the gold standard for diagnosing skin melanocytic lesions and grading skin tissue malignancies, the proposed method will provide a relevant input towards automating this procedure.

2. Convolutional Autoencoder Segmentation Method for Nests of Melanocytes

2.1. Database Specification

To obtain the WSIs, we established scientific cooperation with the University Hospital in Krakow and with the Chair of Pathomorphology of Jagiellonian University Medical College. The dataset included 70. WSIs of selected melanocytic lesions (each image was taken from a separate case): lentigo maligna (22), junctional dysplastic nevus (20), melanoma in situ (13), and superficial spreading melanoma (15). All the histological sections used in the evaluation were prepared from formalin-fixed paraffin-embedded tissue blocks of skin biopsies (each section was about $4\,\mu m$ thick) stained with H&E using an automated stainer. The original images were captured under $10\times$ magnification ($0.44\,\mu m/px$) on Axio Scan.Z1 slide scanner and saved into uncompressed TIFF files whose size varied from 3000×1000 to $20{,}000 \times 30{,}000$ pixels. To verify the results, 39 WSIs (10 of lentigo maligna, 10 of junctional dysplastic nevus, 9 of melanoma in situ, and 10 of superficial spreading melanoma) were paired with the ground truth (binary) segmentation masks of nests of nevus cells prepared manually by an experienced dermatopathologist using GIMP image processing program (i.e., in total, we had 39 manually labeled ground truth images).

Hematoxylin and Eosin (H&E) Staining

Hematoxylin and eosin stain (H&E stain) is the most commonly used stain for light microscopy in histopathology laboratories due to its comparative simplicity and ability to demonstrate a wide range of both normal and abnormal cell and tissue components [53]. The hematoxylin component binds to basophilic structures, such as DNA of cell nuclei, and colors them blue, whereas the eosin colors cell acidophilic structures, such as cytoplasm and most connective tissue fibers, in varying shades and intensities of pink, orange, and red [53,54]. Consequently, all relevant tissue structures in skin specimen are stained and effectively the whole area of the specimen appears in color.

Since the staining protocol cannot be fully standardized (as for instance the staining quality depends on the quality of the dyes used), the variation in staining is an important factor to be considered when designing automatic image processing methods for digital pathology [55].

2.2. Data Preparation

While we were considering a neural network approach to solve the problem of nests segmentation, we were aware of the fact that the input size is strongly related with the final number of parameters that have to be optimized. If the input size is too large, there might be too many model parameters which together with an insufficient number of training examples might lead to overfitting during the training process. For that reason, the input size must be chosen with care. Having a collection of images of different sizes, we decided to split it into small image patches of size 128 × 128 pixels each, which we found balanced between their size and final model's accuracy, and set as a neural network input (Figure 4).

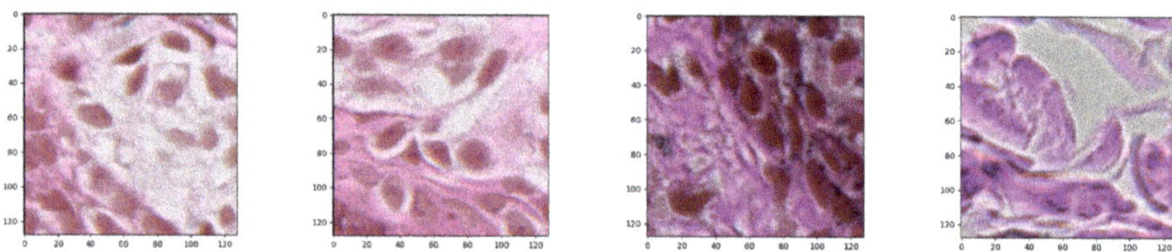

Figure 4. Examples of generated patches of size 128 × 128 pixels each (windows of such size typically include enough context to label the central pixel as either "part of a nest" or "not part of a nest" with high confidence).

Trying to balance classes of the dataset, for those parts of images where nests were not marked, we did not apply an overlapping and simply extracted consecutive examples with 128 px step in width and height. However, while a nest part was found during patches' extraction, an overlapping was applied in order to produce more examples of particular class. Thanks to that trick and an augmentation technique applied later on, we managed to balance classes in the dataset. The augmentation of the nests images was based mainly on the rotation of particular parts by 45 degrees to generate four more examples. Before the data preparation step was applied, the dataset has been divided into three parts—the training, validation, and test part that have been described in detail in Section 3.1. The dataset split step has been applied at the very beginning of the data preparation process on the images, not patches, in order to provide as reliable results as possible.

2.3. The Convolutional Autoencoder

Convolutional autoencoder architecture imposes a bottleneck in the network which forces a compressed knowledge representation of the original input. An autoencoder consists of three components: the encoder, the code, and the decoder as well as a loss function to compare the output with the target (Figure 5).

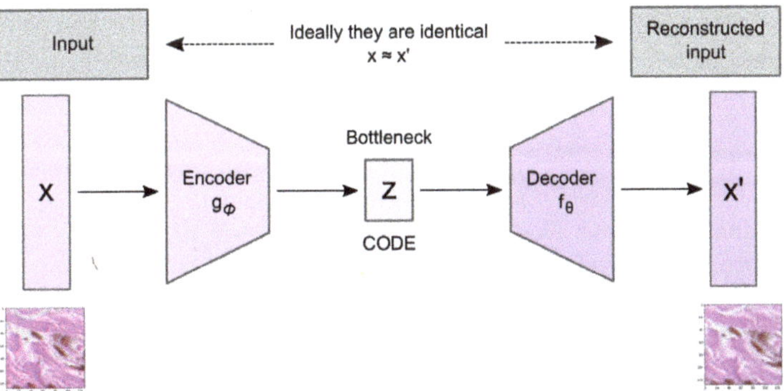

Figure 5. Schema of a basic autoencoder including the encoder, decoder, and code parts. The model contains an encoder function $g(.)$ and a decoder function $f(.)$ parameterized by ϕ and θ, respectively. The low-dimensional code learned for input x in the bottleneck layer is z and the reconstructed input is x'.

Both the encoder and decoder are fully-connected feedforward neural networks where the code is a single layer of a feed forward network (FFN) with specified dimensionality. The encoder function (ϕ) maps the original data X to a latent space F, which is present at the bottleneck. The decoder function, denoted by θ, maps the latent space F at the code to the output, where the output is the same as the input function. Thus, the algorithm is trying to recreate the original image after some generalized nonlinear compression:

$$\phi : \mathcal{X} \to \mathcal{F}$$
$$\psi : \mathcal{F} \to \mathcal{X}$$
$$\phi, \psi = \arg\min_{\phi,\psi} \| X - (\psi \circ \phi) X \|^2 \tag{1}$$

Autoencoders are trained to minimize the loss function which is a reconstruction error such as mean squared error:

$$\mathcal{L}(\mathbf{x}, \mathbf{x'}) = \| \mathbf{x} - \mathbf{x'} \|^2 = \| \mathbf{x} - \sigma'(\mathbf{W'}(\sigma(\mathbf{Wx} + \mathbf{b})) + \mathbf{b'}) \|^2 \tag{2}$$

where $\mathbf{x}$ is usually averaged over some input training set, $\mathbf{W}, \mathbf{W'}$ are weight matrices and $\mathbf{b}, \mathbf{b'}$ are bias vectors for encoder and decoder, respectively.

2.4. Convolutional Autoencoder Architecture

Since our input data consist of images, we use a convolutional autoencoder which is an autoencoder variant where the fully connected layers have been replaced by convolutional layers [56]. The proposed, implemented, and tested convolutional autoencoder architecture has been presented in Figure 6. In our solution, we use the convolution layers with padding along with max-pooling layers in the encoder part. Such combination of layers converts the input image of size $128 \times 128 \times 3$ at the RGB colorspace to an image (code) of size $16 \times 16 \times 64$ at the latent space. In the decoder part, instead of max-pooling, we use the upsampling layer. The decoder converts the code back to an image at the RGB colorspace in its original size (i.e., $128 \times 128 \times 3$).

Convolutional layers apply a convolution operation over the image and perform operation at each point, passing the result to the next layer. Filters belonging to the convolutional layer are trainable feature extractors of size 3×3. As we are solving the segmentation problem, we use padding to avoid spatial dimension decrease. To avoid this, we use 'same' padding of size 2 which is correlated with the convolutional layer filter size 3×3 to preserve as much information about the original input volume as possible to extract those low level features which are necessary for the segmentation approach. Each of the convolutional layers stacks is followed by a rectified linear unit (ReLU) activation function. The purpose of this layer is to introduce nonlinearity to our system that basically has just been computing linear operations during the convolutional layers. The ReLU is currently the most popular nonlinear activation function, defined as the positive part of its argument where x is the input to a neuron:

$$f(x) = \max(0, x) \tag{3}$$

Compared to sigmoid function, the ReLU function is computationally efficient, shows better convergence performance, and alleviates the vanishing gradient problem (the issue where the lower layers of the network train very slowly because the gradient decreases exponentially through the layers).

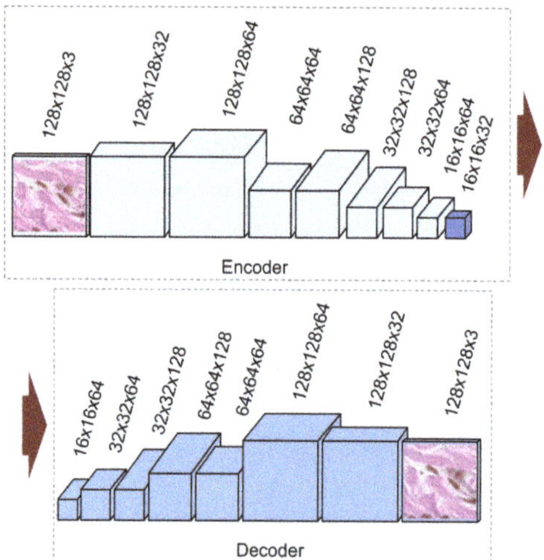

Figure 6. Architecture of the proposed convolutional autoencoder. Each box corresponds to a multichannel feature map. The horizontal arrow denotes transfer between the encoding and decoding parts.

In the encoder part, after the ReLU activation function, we apply a max-pooling layer also referred to as a downsampling layer. The output is the maximum number in every subregion that the 2×2 filter convolves around. The max-pooling layer serves two main purposes: overfitting control and reducing the number of weights (which reduces the computational costs).

The reconstruction process of the autoencoder uses upsampling and convolutions layers. The upsampling layer is a simple layer with no weights, doubling the dimensions of input by repeating the rows and columns of the data. The last convolutional layer is followed by the sigmoid activation function as we are facing the two-class prediction problem while creating the segmentation masks. Sigmoid functions, which are of S-shape, are one of the most widely used activation functions in both machine learning algorithms and deep learning classification layers. A standard choice for a sigmoid function is the logistic function defined as

$$S(x) = \frac{1}{1 + e^{-x}} = \frac{e^x}{e^x + 1}. \tag{4}$$

In Listing 1, we present the proposed autoencoder as well as summarize the layers of the model and their output shapes.

Listing 1. Summary of the implemented convolutional autoencoder model for both training stages. In the first training stage, the last convolutional layer has the output shape of $(128, 128, 3)$ like the input data, while, in the second training stage $128, 128, 1$ as it has been presented in the summary.

Layer (Type)	Output Shape	Nb. of Param.
input_1 (InputLayer)	[(None, 128, 128, 3)]	0
conv2d (Conv2D)	(None, 128, 128, 32)	896
conv2d_1 (Conv2D)	(None, 128, 128, 64)	18,496
max_pooling2d (MaxPooling2D)	(None, 64, 64, 64)	0
conv2d_2 (Conv2D)	(None, 64, 64, 64)	36,928
conv2d_3 (Conv2D)	(None, 64, 64, 128)	73,856
max_pooling2d_1 (MaxPooling2D)	(None, 32, 32, 128)	0
conv2d_4 (Conv2D)	(None, 32, 32, 128)	147,584
conv2d_5 (Conv2D)	(None, 32, 32, 64)	73,792
max_pooling2d_2 (MaxPooling2D)	(None, 16, 16, 64)	0
conv2d_6 (Conv2D)	(None, 16, 16, 64)	36,928
conv2d_7 (Conv2D)	(None, 16, 16, 32)	18,464
conv2d_8 (Conv2D)	(None, 16, 16, 32)	9248
conv2d_9 (Conv2D)	(None, 16, 16, 64)	18,496
up_sampling2d (UpSampling2D)	(None, 32, 32, 64)	0
conv2d_10 (Conv2D)	(None, 32, 32, 64)	36,928
conv2d_11 (Conv2D)	(None, 32, 32, 128)	73,856
up_sampling2d_1 (UpSampling2D)	(None, 64, 64, 128)	0
conv2d_12 (Conv2D)	(None, 64, 64, 128)	147,584
conv2d_13 (Conv2D)	(None, 64, 64, 64)	73,792
up_sampling2d_2 (UpSampling2D)	(None, 128, 128, 64)	0
conv2d_14 (Conv2D)	(None, 128, 128, 64)	36,928
conv2d_15 (Conv2D)	(None, 128, 128, 32)	18,464
conv2d_16 (Conv2D)	(None, 128, 128, 1)	289

Total params: 822,529
Trainable params: 822,529
Non-trainable params: 0

2.5. Semi-Supervised Autoencoder Training Process

Since we had only a few ground-truth examples in comparison with the whole dataset of histopathological images, we took advantage of autoencoders with a semi-supervised learning technique to solve the problem of small and insufficient ground-truth dataset. Semi-supervised, in general, consists of three steps. In the first stage, for the unsupervised part, our autoencoder has been trained to reconstruct the input images (Figure 7).

After the first stage of the semi-supervised training process the encoder's weights have been frozen (fixed) and only the decoder's weights have been reset in order to train the model again, this time to generate masks. This part is a supervised problem where we use both the input images and segmented outcome mask. In the last step, the fine-tuning has been performed—encoder's weights have been unlocked again and the whole autoencoder network has been trained again. For the autoencoder training part, the main assumption is that the network can learn to code patterns and structures found in the image in the latent space. The latent size is usually smaller than the input size which enforces the network to choose only those features which best describe the dataset and skip less relevant ones. Later in the second phase, we utilize the feature extraction part—the encoder along with the latent layer—to use those features in order to distinguish (generate masks) structures in the image. Encoder's weights are blocked because this part of the network is treated as a feature extractor in the second phase of training—while the model for classifying pixels is trained.

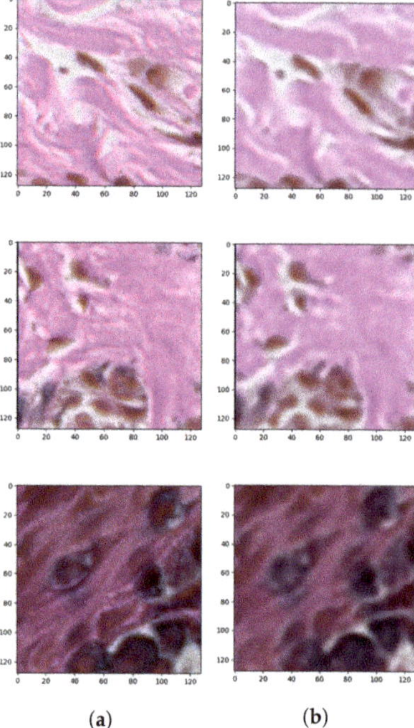

(a)   (b)

Figure 7. Outcomes of the first stage of autoencoder semi-supervised training process showing: (**a**) original patches and (**b**) reconstructed images.

During our research, we have analyzed and tested different deep neural network architectures. However, the presented one has been found to be the most effective. The output layer consisted of three filters of size 3×3 with ReLU activation function which produced an output of size $128 \times 128 \times 3$ (the same as the input). The mean square error has been chosen as the cost function. We applied learning rate decay to ensure that the cost function's minimum is not superpassed due to too large weights update as the optimization process continues.

After we achieved the smallest validation error during the training process of the described autoencoder, the encoder weights were transferred to a new model which had pretty much the same architecture as the one described above. They only differed in the output layer: the autoencoder used for reconstruction produced an RGB image (i.e., an image consisting of three channels), whereas the autoencoder used for segmentation of objects belonging to only one class generates a probability map representing the probability that a given pixel belongs to that class (i.e., an image consisting of only one channel). Therefore, in the autoencoder for the segmentation task which is a binary classification problem, we use the binary cross-entropy loss (BCE) consisting of sigmoid activation function and cross-entropy loss. After applying those changes the training process has been repeated only on supervised examples. This means that the dataset shrunk a lot, but, thanks to initially set weights of the encoder part by the unsupervised training phase, we managed to achieve some decent results. After achieving the minimal training error, in compliance with transfer learning, the pre-trained weights have been unlocked and the fine-tuning has been utilized. Figure 8 presents the generated masks compared to the ground-truth images.

2.6. Convolutional Autoencoder Training Parameters

It has been widely observed that hyperparameters are some of the most critical components of any deep architecture. Hyperparameters are variables that determine the network structure and need to be set before training a deep learning algorithm on a dataset. Hyperparameters presented in Table 1 have been chosen experimentally and set before training our autoencoder.

Table 1. Parameter settings of the autoencoder.

Parameters	Values
Initializer	Glorot uniform (Xavier uniform)
Number of hidden layers	22
Number of hidden conv. layers	16
Learning rate eq. for reconstruction and segmentation	$\log_{10}(0.01 \times \text{epoch} + 0.1) \times 0.001$
Learning rate eq. for fine-tuning	$\log_{10}(0.01 \times \text{epoch} + 0.1) \times 0.0005$
Activation function	ReLU and Sigmoid
Batch size	64
Optimizer	Adam
Loss	Mean Squared Error (reconstruction), Binary crossentropy (segmentation)

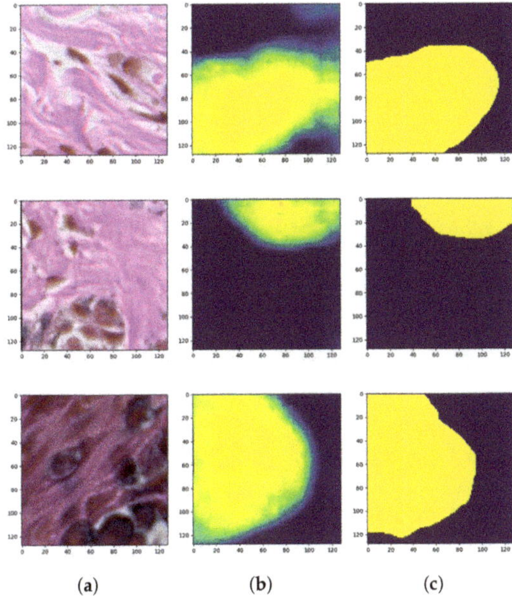

(a)   (b)   (c)

Figure 8. Outcomes of the second training stage of the convolutional autoencoder: (**a**) original images (patches), (**b**) generated masks, and (**c**) ground-truth images.

3. Experimental Results and Evaluation

In order to evaluate the performance of the semi-supervised training strategy of the autoencoder described in the previous section, we conducted a series of experiments. In this section, we present the training parameters, visualization of the convolutional layers as well as the statistical analysis of the obtained results.

3.1. Training, Validation, and Test Sets

The reconstruction stage was performed on all 70 WSIs (both labeled and unlabeled), out of which 49 images (70%) were used for training, 10 (14.3%) for validation, and 11 (15.7%) for testing. The split was performed in a stratification fashion, i.e., in principle, the proportion of the number of labeled to unlabeled images remained constant across all the three sets. The number of patches in each of these sets equaled 223,274, 57,229, and 116,707, respectively.

The segmentation stage was performed on those 39 out of 70 WSIs, which had the ground truth available (i.e., unlabeled images were not used in this stage). Therefore, in this stage, 27 images (69%) were used for training, 6 (15.5%) for validation, and 6 (15.5%) for testing. Each of these sets was created by removing unlabeled images from a corresponding set from the reconstruction stage. The number of patches in training, validation, and test set for the segmentation stage equaled 110,736, 31,566, and 62,718, respectively.

In each stage, the patches were extracted only after splitting the initial dataset (for the given stage) into training, validation, and test sets. Since individual WSIs differed in size, the proportion of patches in those three sets differs between stages.

3.2. Visualization of the Convolutional Autoencoders Layers

Model interpretability of DNN has been an important area of research from the beginning since proposed models achieve high accuracy but at the expense of high abstraction (i.e., accuracy vs. interpretability problem). Visualizing intermediate activations consists of displaying feature maps, filters, and heat maps. The feature maps are the outputs of various convolution and pooling layers in a network. This gives an insight into how an input is decomposed unto the different filters learned by the network (Figure 9). We can observe that the first few layers act as a collection of various edge detectors while the higher-up become increasingly abstract and less visually interpretable.

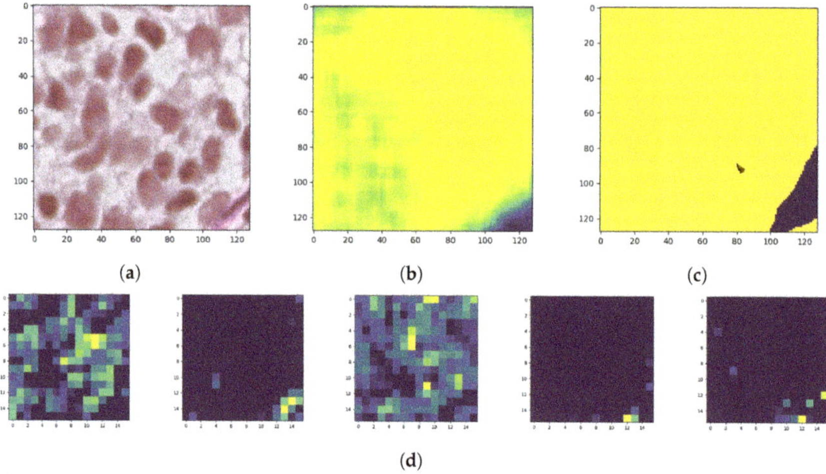

Figure 9. Feature maps of the autoencoder convolutional layers: (**a**) original image, (**b**) generated mask, (**c**) ground-truth image, and (**d**) a partial feature map from latent layers (only a few more interesting activations were included).

The reason for visualizing a feature map for a specific input image is to try to gain some understanding of what features are being detected. In the case of classification errors, it can help us to inspect the results and locate specific objects in an image.

During the first step of the training process, we achieved 0.24 mean square error (MSE) on validation set for reconstruction and 0.50 binary cross-entropy for classifying pixels (with threshold equal to 0.5). The within-sample MSE is computed as

$$\mathrm{MSE} = \frac{1}{n} \sum_{i=1}^{n} (Y_i - \hat{Y}_i)^2. \tag{5}$$

where the vector of n are predictions generated from a sample of "n" data points on all variables, Y is the vector of observed values of the variable being predicted, with $\hat{Y}_i$ being the predicted values.

Figure 10 presents the error rate for reconstruction and segmentation over the training and validation data. Figure 11 presents the learning rate decay for each epoch.

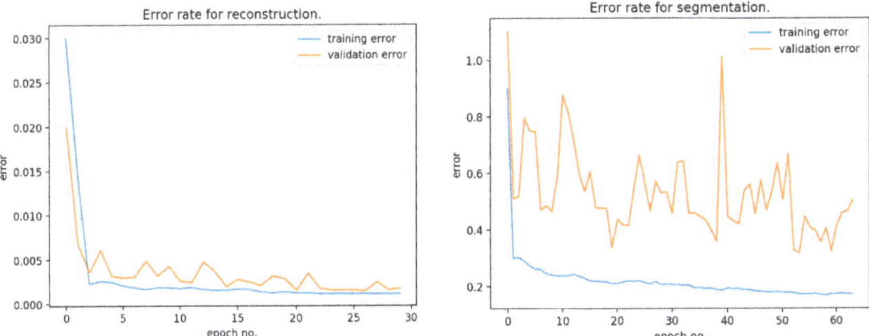

Figure 10. Error rate for reconstruction and segmentation over the training and validation data.

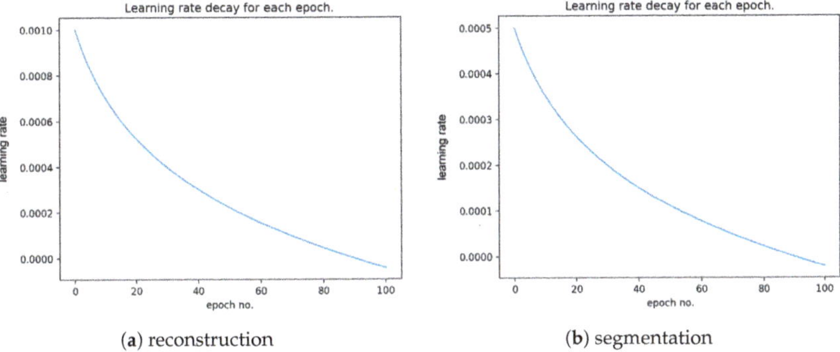

(**a**) reconstruction   (**b**) segmentation

Figure 11. Learning rate decay for each epoch for (**a**) reconstruction and (**b**) segmentation over the training and validation data.

One of the most commonly used performance metrics in segmentation problems is the Sørensen index also known as Dice similarity coefficient (DSC), which compares the similarity of two samples from a statistical population [57].

Given two binary sets, X and Y, the Sørensen's formula is defined as:

$$DSC(X, Y) = \frac{2|X \cap Y|}{|X| + |Y|} \tag{6}$$

Using the definition of true positive (TP), false positive (FP), and false negative (FN), it can be rewritten as:

$$DSC = \frac{2TP}{2TP + FP + FN} \tag{7}$$

where in our case TP denotes correctly detected nests' structure pixels, FP denotes nests structure pixels not detected, and FP denotes background pixels classified as parts of nests. The DSC is a statistical measure that calculates the degree of overlapping between the experimental segmentation and the manual segmentations where possible values of DSC range from 0.0 to 1.0. A perfect classifier or segmentation model achieves DSC of 1.0.

Furthermore, sensitivity and specificity are calculated using the following equation:

$$SEN = \frac{TP}{TP + FN} \tag{8}$$

$$SPE = \frac{TN}{TN + FP} \tag{9}$$

The proposed algorithm achieved an average DSC of 0.81, sensitivity 0.76, and specificity 0.94. Taking into account that the small ground-truth database has been segmented manually, the achieved DSC score is very promising. We observed that the algorithm's misclassified areas that were on the border between the patches as well as the area of the nests was really small containing only few cells.

4. Conclusions

Our results demonstrate the feasibility of segmenting the nests while using a convolutional autoencoder approaches for a small ground-truth database. The achieved results are very promising as there are no other works or solutions conducted so far. Our method allowed for segmenting the nests with DSC 0.81, sensitivity 0.76, and specificity 0.94, respectively. To the best of our knowledge, this is a first-of-its-kind experiment that shows that convolutional autoencoders may be sufficient for histopathological image analysis with small ground-truth databases. Since no histopathological criterion is entirely specific for nevi (all can be occasionally found in a melanoma) and some challenging lesions contain conflicting criteria suggesting opposite diagnoses, criteria must be used in clusters and ought to be used for specific differential diagnosis—hence, the CAD systems should take into consideration as wide a spectrum of histopathological features as possible. Therefore, the proposed solution for nests segmentation is not intended to be used as a standalone tool for melanoma diagnosis but rather in combination with other automatic diagnostic methods and systems—it might help to increase their accuracy and provide additional grounds for a certain diagnosis, especially for pairs of clinical entities lacking strong delimitation (e.g., junctional dysplastic nevus with severe atypia and melanoma in situ).

Future Works

Starting from the described framework, as our results seem very promising, there is still much to improve. For example, by cutting out patches from images, we consciously gave up global spatial dependencies in exchange for usable input size. However, spatial information might be preserved by introducing some recurrent structures inside the neural network architecture which would contain some context from previous examples. On the other hand, such a modification in feeding the model with data which can not be random but in a defined, specified order, changes the current approach for data preparation a lot, which in turn will lead to whole algorithm redesign.

In a follow-up study, we intend to improve the architecture of the proposed network as well as train and validate it on a larger dataset containing images from various laboratories. We then plan to integrate the proposed method with our epidermis segmentation method described in [58] and use the information about the distribution of nests of melanocytes within the epidermis to improve the process of diagnosing skin melanocytic lesions (examples of diagnostic information which could be extracted after such a fusion of methods are provided in Section 1.1 Medical Background). Furthermore, high-resolution histopathological images are very large; therefore, image processing, segmentation, and detection are highly compute intensive tasks, and software implementation requires a significant amount of processing time. To assist the pathologists in real time, special hardware accelerators, which can reduce the processing time, are required.

Author Contributions: Conceptualization, P.K., D.K., and J.J.-K.; methodology, P.K.; software, P.K. and D.K.; validation, P.K., D.K., J.J.-K., and G.D.; formal analysis, P.K., J.J.-K., and M.G.; investigation, P.K. and J.J.-K.; resources, G.D.; data curation, P.K. and G.D.; writing—original draft preparation, P.K. and J.J.-K.; writig—review and editing, M.G. and J.J.-K.; visualization, P.K. and J.J.-K.; supervision, J.J.-K.; project administration, P.K.; funding acquisition, P.K. and D.K. All authors have read and agreed to the published version of the manuscript.

Funding: This work was financed by the National Science Center based on the decision number 2016/23/N/ST7/01361 (to Pawel Kleczek).

Conflicts of Interest: The authors declare no conflict of interest.

References

1. Garbe, C.; Leiter, U. Melanoma epidemiology and trends. *Clin. Dermatol.* **2009**, *27*, 3–9. [CrossRef] [PubMed]
2. Lyon: International Agency for Research on Cancer. Cancer Incidence in Five Continents Time Trends (Electronic Version). Available online: http://ci5.iarc.fr (accessed on 13 February 2020).
3. Cancer Facts & Figures 2016. Available online: http://www.cancer.org/research/cancerfactsstatistics/cancerfactsfigures2016/index (accessed on 13 February 2020).
4. Australian Bureau of Statistics. 3303.0 Causes of Death. Available online: http://www.abs.gov.au/Causes-of-Death (accessed on 13 February 2020).
5. Argenziano, G.; Soyer, P.H.; Giorgio, V.D.; Piccolo, D.; Carli, P.; Delfino, M.; Ferrari, A.; Hofmann-Wellenhof, R.; Massi, D.; Mazzocchetti, G.; et al. *Interactive Atlas of Dermoscopy*; Edra Medical Publishing and New Media: Milan, Italy, 2000.
6. Braun, R.P.; Gutkowicz-Krusin, D.; Rabinovitz, H.; Cognetta, A.; Hofmann-Wellenhof, R.; Ahlgrimm-Siess, V.; Polsky, D.; Oliviero, M.; Kolm, I.; Googe, P.; et al. Agreement of dermatopathologists in the evaluation of clinically difficult melanocytic lesions: How golden is the 'gold standard'? *Dermatology* **2012**, *224*, 51–58. [CrossRef] [PubMed]
7. Massi, G.; LeBoit, P.E. *Histological Diagnosis of Nevi and Melanoma*; Springer: Berlin/Heidelberg, Geramny, 2014.
8. Barnhill, R.L.; Lugassy, C.; Taylor, E.; Zussman, J. Cutaneous Melanoma. In *Pathology of Melanocytic Nevi and Melanoma*; Barnhill, R.L., Piepkorn, M., Busam, K.J., Eds.; Springer: Berlin/Heidelberg, Geramny, 2014; Chapter 10, pp. 331–488.
9. Lodha, S.; Saggar, S.; Celebi, J.; Silvers, D. Discordance in the histopathologic diagnosis of difficult melanocytic neoplasms in the clinical setting. *J. Cutan Pathol.* **2008**, *35*, 349–352. [CrossRef] [PubMed]
10. Lin, M.; Mar, V.; McLean, C.; Wolfe, R.; Kelly, J. Diagnostic accuracy of malignant melanoma according to subtype. *Australas. J. Dermatol.* **2013**, *55*, 35–42. [CrossRef]
11. Ogiela, M.; Tadeusiewicz, R.; Ogiela, L. Graph image language techniques supporting radiological, hand image interpretations. *Comput. Vis. Image Und.* **2006**, *103*, 112–120. [CrossRef]
12. Tadeusiewicz, R. Neural networks as a tool for modeling of biological systems. *BAMS* **2015**, *11*, 135–144. [CrossRef]
13. Bengio, Y.; Courville, A.; Vincent, P. Representation Learning: A Review and New Perspectives. *IEEE Trans. Pattern Anal. Mach. Intell.* **2013**, *35*, 1798–1828. [CrossRef]
14. Shrestha, A.; Mahmood, A. Review of Deep Learning Algorithms and Architectures. *IEEE Access* **2019**, *7*, 53040–53065. [CrossRef]
15. Piepkorn, M.W.; Barnhill, R.L. Common Acquired and Atypical/Dysplastic Melanocytic Nevi. In *Pathology of Melanocytic Nevi and Melanoma*; Barnhill, R.L., Piepkorn, M., Busam, K.J., Eds.; Springer: Berlin/Heidelberg, Geramny, 2014; Chapter 5, pp. 87–154.
16. MacKie, R.M. Disorders of the Cutaneous Melanocyte. In *Rook's Textbook of Dermatology*, 7th ed.; Burns, T., Breathnach, S., Cox, N., Griffiths, C., Eds.; Blackwell Publishing: Hoboken, NJ, USA, 2004; Chapter 38.
17. Reeck, M.C.; Chuang, T.; Eads, T.J.; Faust, H.B.; Farmer, E.R.; Hood, A.F. The diagnostic yield in submitting nevi for histologic examination. *J. Am. Acad. Dermatol.* **1999**, *40*, 567–571. [CrossRef]
18. Barnhill, R.L.; Vernon, S.E.; Rabinovitz, H.S. Benign Melanocytic Neoplasms. In *Color Atlas of Dermatopathology*; Grant-Kels, J.M., Ed.; CRC Press: Boca Raton, FL, USA, 2007; Chapter 18, pp. 247–278.
19. Sardana, K.; Chakravarty, P.; Goel, K. Optimal management of common acquired melanocytic nevi (moles): current perspectives. *Clin. Cosmet. Investig. Dermatol.* **2014**, *7*, 89–103. [CrossRef]
20. Prince, N.M.; Rywlin, A.M.; Ackerman, A.B. Histologic criteria for the diagnosis of superficial spreading malignant melanoma: Formulated on the basis of proven metastatic lesions. *Cancer* **1976**, *38*, 2434–2441. [CrossRef]
21. Urso, C.; Saieva, C.; Borgognoni, L.; Tinacci, G.; Zini, E. Sensitivity and specificity of histological criteria in the diagnosis of conventional cutaneous melanoma. *Melanoma Res.* **2008**, *18*, 253–258. [CrossRef] [PubMed]
22. 22th International Conference on Pattern Recognition (ICPR 2014). MITOS-ATYPIA-14 Challange—Results (Electronic Version). Available online: https://mitos-atypia-14.grand-challenge.org/Results2/ (accessed on 13 February 2020).

23. Veta, M.; Heng, Y.J.; Stathonikos, N.; Bejnordi, B.E.; Beca, F.; Wollmann, T.; Rohr, K.; Shah, M.A.; Wang, D.; Rousson, M.; et al. Predicting breast tumor proliferation from whole-slide images: The TUPAC16 challenge. *Med. Image Anal.* **2019**, *54*, 111–121. [CrossRef] [PubMed]

24. Heenan, P.J.; Elder, D.E.; Sobin, L.H. *Histological Typing of Skin Tumours*, 2nd ed.; Springer: Berlin/Heidelberg, Germany, 1996.

25. Menzies, S.W.; Bischof, L.; Talbot, H.; Gutenev, A.; Avramidis, M.; Wong, L.; Lo, S.K.; Mackellar, G.; Skladnev, V.; McCarthy, W.; et al. The performance of SolarScan: An automated dermoscopy image analysis instrument for the diagnosis of primary melanoma. *Arch. Dermatol.* **2005**, *141*, 1388–1396. [CrossRef]

26. Masood, A.; Ali Al-Jumaily, A. Computer Aided Diagnostic Support System for Skin Cancer: A Review of Techniques and Algorithms. *Int. J. Biomed. Imaging* **2013**, *2013*, 323268. [CrossRef] [PubMed]

27. Farmer, E.R.; Gonin, R.; Hanna, M.P. Discordance in the histopathologic diagnosis of melanoma and melanocytic nevi between expert pathologist. *Hum. Pathol.* **1996**, *27*, 528–531. [CrossRef]

28. Troxel, D. An Insurer's Perspective on Error and Loss in Pathology. *Arch. Pathol. Lab. Med.* **2005**, *129*, 1234–1236.

29. Shoo, B.; Sagebiel, R.; Kashani-Sabet, M. Discordance in the histopathologic diagnosis of melanoma at a melanoma referral center. *J. Am. Acad. Dermatol.* **2010**, *62*, 751–756. [CrossRef]

30. Snead, D.; Tsang, Y.W.; Meskiri, A.; Kimani, P.K.; Crossman, R.; Rajpoot, N.; Blessing, E.; Chen, K.; Gopalakrishnan, K.; Matthews, P. Validation of digital pathology imaging for primary histopathological diagnosis. *Histopathology* **2016**, *68*, 1063–1072. [CrossRef]

31. Pantanowitz, L.; Valenstein, P.; Evans, A.; Kaplan, K.; Pfeifer, J.; Wilbur, D.; Collins, L.; Colgan, T. Review of the current state of whole slide imaging in pathology. *J. Pathol. Informatics* **2011**, *2*, 36. [CrossRef]

32. Nielsen, P.S.; Lindebjerg, J.; Rasmussen, J.; Starklint, H.; Waldstrom, M.; Nielsen, B. Virtual microscopy: An evaluation of its validity and diagnostic performance in routine histologic diagnosis of skin tumors. *Hum. Pathol.* **2010**, *41*, 1770–1776. [CrossRef] [PubMed]

33. Farahani, N.; Parwani, A.V.; Pantanowitz, L. Whole slide imaging in pathology: Advantages, limitations, and emerging perspectives. *Pathol. Lab. Med. Int.* **2015**, *2015*, 23–33. [CrossRef]

34. Thorstenson, S.; Molin, J.; Lundström, C. Implementation of large-scale routine diagnostics using whole slide imaging in Sweden: Digital pathology experiences 2006-2013. *J. Pathol. Informatics* **2014**, *5*, 14. [CrossRef]

35. Stathonikos, N.; Veta, M.; Huisman, A.; van Diest, P. Going fully digital: Perspective of a Dutch academic pathology lab. *J. Pathol. Informatics* **2013**, *4*, 15. [CrossRef]

36. Cree, M.; Jelinek, H. The effect of JPEG compression on automated detection of microaneurysms in retinal images. In Proceedings of the SPIE 6813, Image Processing: Machine Vision Applications, 68130M, San Jose, CA, USA, 27–31 January, 2008; Volume 6813. [CrossRef]

37. Tellez, D.; Litjens, G.; van der Laak, J.; Ciompi, F. Neural Image Compression for Gigapixel Histopathology Image Analysis. *IEEE Trans. Pattern Anal. Machine Intell.* **2019**. [CrossRef]

38. Hernández-Cabronero, M.; Sanchez, V.; Blanes, I.; Aulí-Llinàs, F.; Marcellin, M.W.; Serra-Sagristà, J. Mosaic-Based Color-Transform Optimization for Lossy and Lossy-to-Lossless Compression of Pathology Whole-Slide Images. *IEEE Trans. Med Imaging* **2019**, *38*, 21–32. [CrossRef]

39. Niazi, M.K.K.; Lin, Y.; Liu, F.; Ashok, A.; Marcellin, M.; Tozbikian, G.; Gurcan, M.; Bilgin, A. Pathological image compression for big data image analysis: Application to hotspot detection in breast cancer. *Artif. Intell. Med.* **2019**, *95*, 82–87. [CrossRef]

40. Xu, H.; Lu, C.; Berendt, R.; Jha, N.; Mandal, M.K. Automated analysis and classification of melanocytic tumor on skin whole slide images. *Comput. Med Imaging Graph.* **2018**, *66*, 124–134. [CrossRef]

41. Lu, C.; Mahmood, M.; Jha, N.; Mandal, M. Automated Segmentation of the Melanocytes in Skin Histopathological Images. *IEEE J. Biomed. Health Informatics* **2013**, *17*, 284–296. [CrossRef]

42. Xu, H.; Mandal, M. Epidermis segmentation in skin histopathological images based on thickness measurement and k-means algorithm. *EURASIP J. Image Video Process.* **2015**, *2015*, 1–14. [CrossRef]

43. Noroozi, N.; Zakerolhosseini, A. Differential diagnosis of squamous cell carcinoma in situ using skin histopathological images. *Comput. Biol. Med.* **2016**, *70*, 23–39. [CrossRef] [PubMed]

44. Olsen, T.G.; Jackson, B.; Feeser, T.A.; Kent, M.N.; Moad, J.C.; Krishnamurthy, S.; Lunsford, D.D.; Soans, R.E. Diagnostic performance of deep learning algorithms applied to three common diagnoses in dermatopathology. *J. Pathol. Informatics* **2018**, *9*, 32. [CrossRef]

45. Cho, K. Simple Sparsification Improves Sparse Denoising Autoencoders in Denoising Highly Noisy Images. In Proceedings of the 30th International Conference on International Conference on Machine Learning, Atlanta, GA, USA, 17–19 June 2013; Volume 28, pp. 432–440.

46. Buades, A.; Coll, B.; Morel, J. A review of image denoising algorithms, with a new one. *Multiscale Model. Simul.* **2005**, *4*, 490–530. [CrossRef]

47. Gondara, L. Medical Image Denoising Using Convolutional Denoising Autoencoders. In Proceedings of the 2016 IEEE 16th International Conference on Data Mining Workshops (ICDMW), Barcelona, Spain, 12–15 December 2016; pp. 241–246. [CrossRef]

48. Song, T.H.; Sanchez, V.; EIDaly, H.; Rajpoot, N.M. Hybrid deep autoencoder with Curvature Gaussian for detection of various types of cells in bone marrow trephine biopsy images. In Proceedings of the 2017 IEEE 14th International Symposium on Biomedical Imaging (ISBI 2017), Melbourne, Australia, 18–21 April 2017; pp. 1040–1043. [CrossRef]

49. Xu, J.; Xiang, L.; Liu, Q.; Gilmore, H.; Wu, J.; Tang, J.; Madabhushi, A. Stacked Sparse Autoencoder (SSAE) for Nuclei Detection on Breast Cancer Histopathology Images. *IEEE Trans. Med Imaging* **2016**, *35*, 119–130. [CrossRef] [PubMed]

50. Martinez-Murcia, F.J.; Ortiz, A.; Gorriz, J.; Ramirez, J.; Castillo-Barnes, D. Studying the Manifold Structure of Alzheimer's Disease: A Deep Learning Approach Using Convolutional Autoencoders. *IEEE J. Biomed. Health Informatics* **2019**. [CrossRef]

51. Hou, L.; Nguyen, V.; Kanevsky, A.B.; Samaras, D.; Kurc, T.M.; Zhao, T.; Gupta, R.R.; Gao, Y.; Chen, W.; Foran, D.; et al. Sparse Autoencoder for Unsupervised Nucleus Detection and Representation in Histopathology Images. *Pattern Recognit.* **2019**, *86*, 188–200. [CrossRef]

52. Song, T.; Sanchez, V.; EI Daly, H.; Rajpoot, N.M. Simultaneous Cell Detection and Classification in Bone Marrow Histology Images. *IEEE J. Biomed. Health Informatics* **2019**, *23*, 1469–1476. [CrossRef]

53. Bancroft, J.D.; Layton, C. The hematoxylins and eosin. In *Bancroft's Theory and Practice of Histological Techniques*, 7th ed.; Suvarna, S.K., Layton, C., Bancroft, J.D., Eds.; Churchill Livingstone: London, UK, 2013; Chapter 10, pp. 173–186.

54. Chan, J.K. The wonderful colors of the hematoxylin-eosin stain in diagnostic surgical pathology. *Int. J. Surg. Pathol.* **2014**, *22*. [CrossRef]

55. Piórkowski, A.; Gertych, A. Color Normalization Approach to Adjust Nuclei Segmentation in Images of Hematoxylin and Eosin Stained Tissue. In *Information Technology in Biomedicine*; Pietka, E., Badura, P., Kawa, J., Wieclawek, W., Eds.; Springer International Publishing: Berlin/Heidelberg, Germany, 2019; pp. 393–406.

56. Chollet, F. *Deep Learning with Python*, 1st ed.; Manning Publications Co.: Greenwich, CT, USA, 2017.

57. Sørensen, T.J. A Method of Establishing Groups of Equal Amplitude in Plant Sociology Based on Similarity of Species and Its Application to Analyses of the Vegetation on Danish Commons. *K. Dan. Vidensk. Selsk.* **1948**, *5*, 1–34.

58. Kłeczek, P.; Dyduch, G.; Jaworek-Korjakowska, J.; Tadeusiewicz, R. Automated epidermis segmentation in histopathological images of human skin stained with hematoxylin and eosin. In Proceedings of the SPIE 10140, Medical Imaging 2017: Digital Pathology, 101400M, Orlando, FL, USA, 11–16 February 2017.

Article

Simplified Fréchet Distance for Generative Adversarial Nets

Chung-Il Kim [1], Meejoung Kim [2], Seungwon Jung [1] and Eenjun Hwang [1,*]

[1] School of Electrical Engineering, Korea University, Seoul 02841, Korea; cilkim1@korea.ac.kr (C.-I.K.); jsw161@korea.ac.kr (S.J.)

[2] Research Institute for Information and Communication Technology, Korea University, Seoul 02841, Korea; meejkim@korea.ac.kr

* Correspondence: ehwang04@korea.ac.kr; Tel.: +82-2-3290-3256

Received: 7 January 2020; Accepted: 7 March 2020; Published: 11 March 2020

Abstract: We introduce a distance metric between two distributions and propose a Generative Adversarial Network (GAN) model: the Simplified Fréchet distance (SFD) and the Simplified Fréchet GAN (SFGAN). Although the data generated through GANs are similar to real data, GAN often undergoes unstable training due to its adversarial structure. A possible solution to this problem is considering Fréchet distance (FD). However, FD is unfeasible to realize due to its covariance term. SFD overcomes the complexity so that it enables us to realize in networks. The structure of SFGAN is based on the Boundary Equilibrium GAN (BEGAN) while using SFD in loss functions. Experiments are conducted with several datasets, including CelebA and CIFAR-10. The losses and generated samples of SFGAN and BEGAN are compared with several distance metrics. The evidence of mode collapse and/or mode drop does not occur until 3000k steps for SFGAN, while it occurs between 457k and 968k steps for BEGAN. Experimental results show that SFD makes GANs more stable than other distance metrics used in GANs, and SFD compensates for the weakness of models based on BEGAN-based network structure. Based on the experimental results, we can conclude that SFD is more suitable for GAN than other metrics.

Keywords: image processing; generative models; generative adversarial net

---

1. Introduction

Generative Adversarial Net (GAN) is one of the models drawing attention in the field of machine learning (ML) and computer vision [1]. The model learns the distribution of a given data and generates sample data based on the learning.

Recently, several GAN models have been proposed to deal with different purposes, and the performances of generative models have improved. For instance, Domain Adversarial Neural Network (DANN) and Adversarial Discriminative Domain Adaption (ADDA) considered image-to-image translation [2,3] and GAN with text manifold interpolation and image-text matching discriminator considered text-to-image synthesis [4]. Super-Resolution Generative Adversarial Nets (SRGAN) focused on super-resolution [5], and style transformation [6,7], Context Encoder (CE) and Globally and Locally Consistency Image Completion (GLCIC) considered inpainting [8,9], and Generative Adversarial Nets for Video generation (VGAN) was applied to generate high-dimensional data based on image, video, and audio [10].

The principle of GANs is to set up a game between two players: a generator and a discriminator. The generator generates samples based on the distribution obtained from training data [11]. On the other hand, the discriminator examines whether the input sample is real or fake when a data sample is given. During the training of a model, the generator is trained to deceive the discriminator, while the discriminator is trained to distinguish the generated samples from the real samples correctly.

The optimal generator generates plausible data samples, and this makes the discriminator foolish and unable to work eventually [12]. Owing to many efforts to improve GANs [13–15], the data generated by GANs are so realistic that human beings almost cannot differentiate real data from fake data.

Each loss of generator and discriminator should converge to a constant during the training process for success data generation. If this occurs, then the training process of the GAN is called "stable" [13]. Usually, GANs suffer from unstable training for several reasons. Examples of unstable phenomena include the vanishing the gradient, the gradient becoming too large, or the loss oscillating during training [12,13]. Another problem that GAN is experiencing is the generator collapsing, which produces only a single sample or a small family of very similar samples. These phenomena are called mode collapse and mode drop. Mode collapse is generating the similar or even the same outputs for different random input vectors, while mode drop is concerned with modes being dropped from the output distribution [1]. These phenomena may occur when the distribution of real data cannot be represented correctly because of using inadequate optimizations or insufficient network resources that cause an inability of node counting [1,13,14]. In such conditions, the average of real data distribution is used for mode collapse, while the distribution of real data is ignored for mode drop, during the generation of fake data.

To be an acceptable GAN, the distance between the two distributions of real data and generated data has to be far in the discriminator's viewpoint, while it has to be near the generator's viewpoint. The performance of a GAN, therefore, is closely related to the adopted distance metric in the loss functions, and the instability of networks might be solved by changing the distance metric. One of the studies of this approach was the Wasserstein Generative Adversarial Network (WGAN) [15]. Unlike the original GAN, WGAN applied Wasserstein-1 distance, called Earth Mover's distance (EMD), to measure the distance between two distributions of real data and generated data [16]. The superiority of EMD was considered in terms of stability and quality of a GAN by comparing these values of EMD with those of other distance metrics involved. For instance, WGAN compared EMD with Jensen–Shannon distance (JSD) and Kullback–Leibler (KL) divergence [17,18]. With the advent of WGAN, EMD has been used widely in GANs as a metric in loss functions.

Recently, Fréchet distance (FD), called as Wasserstein-2 distance, has been introduced [19]. FD was initially used as a similarity evaluation index, called Fréchet Inception distance (FID), like the Inception Score (IS) [20]. FID that applied FD was used in the inception v3 model [21]. It used feature values extracted from a pre-trained inception v3 model to evaluate the similarity of real data and generated data [22]. On the other hand, IS was a score correlating human judgment with a pre-trained ImageNet dataset in inception v3 networks [23]. An experiment was conducted to verify the adequateness of FID as an evaluation index, and the superiority of FID over IS demonstrated [20]. Therefore, FID became a primary evaluation index used in GANs [24–26]. However, it is very challenging to apply FD in GANs directly because it requires a longer time and bigger memory compared to other distance metrics [27,28], which is caused by its complexity. To the best of our knowledge, there is no GAN using FD as a metric of training GAN until so far.

In this paper, we introduce a Simplified Fréchet distance (SFD) and propose a GAN model in which SFD is involved. SFD is a simplified and regularized version of FD. That is, SFD reduces the complexity of FD and enables the training process stable representing some of the characteristics of FD. Therefore, a portion of the characteristics of FD could be explored when SFD is used in the training process of GANs. The structure of the proposed GAN model, the Simplified Fréchet Generative Adversarial Networks (SFGAN), is based on the Boundary Equilibrium GAN (BEGAN) [29]. The difference between the two GANs is the distance metric used in the loss functions of the networks. SFGAN uses SFD, while BEGAN uses EMD. In other words, SFGAN is trained by adversarial losses that are defined by SFD among the distributions of input and output. Output distribution is computed through auto-encoder based discriminator by using an adversarial loss.

For demonstrating the superiority and applicability of SFGAN, the experiments are conducted with the CelebA, CIFAR-10, and a 2-D mixture of Gaussians [23,30,31]. Two purposes are considered in the experiment. One is to investigate the stability of training GANs by using SFD and EMD, and the other is to compare JSD, EMD, and SFD between real data and generated data during the training procedure of SFGAN and BEGAN. The trainings are executed up to three million steps to investigate the stability of training and the differences in distance metrics before and after mode collapse and/or mode drop. The same values of hyperparameters are used in the two models. The experiment is conducted five to ten times to see whether the results change with each experiment. It is observed that the differences in results for all experiments are negligible. Experimental results show that the training process of SFGAN seems stable, and neither mode collapse nor mode drop is detected. On the other hand, these phenomena have occurred during the training process of BEGAN, which result in unstable training. Moreover, it is observed that SFD distinguishes the distributions of real data and fake data generated by unstable BEGAN, while EMD sometimes fails it.

The contributions of this study are summarized as follows: (1) A Simplified Fréchet distance is introduced. SFD reduces the complexity of FD, representing some characteristics of FD and enables stable training by compensating the weakness that the models belonging to BEGAN-based network structure have. (2) A new GAN model in which SFD is involved in the loss functions, SFGAN, is proposed. SFGAN is more stable than BEGAN in which EMD applied. (3) SFD is introduced as an evaluation index for detecting mode collapse and/or mode drop during the procedure of training a GAN. It is possible to detect instability of a GAN during the training with SFD alone, without requiring additional models, a balanced dataset or constrained space that other GANs are requiring.

This paper includes the following: The related works are presented in Section 2. In Section 3, we introduce SFD and compare it with the existing distance metrics. The SFGAN model is presented in Section 4, and the stability and effectiveness of the SFGAN model are verified via experiments in Section 5. Lastly, Section 6 concludes the paper.

2. Related Works

There are many studies on GANs [13–15,24,32–47]. In this section, we investigate the studies that consider the stability problem during the training of GANs. The studies dealing with the stability on GANs can be divided into three categories: the GANs that consider stable training by evaluating performances with only IS or FID, the methodologies to analyzing the stability of the GAN model, and the GANs that consider stable training with their own evaluating methods to investigate the stability. Figure 1 presents the known studies for the three categories.

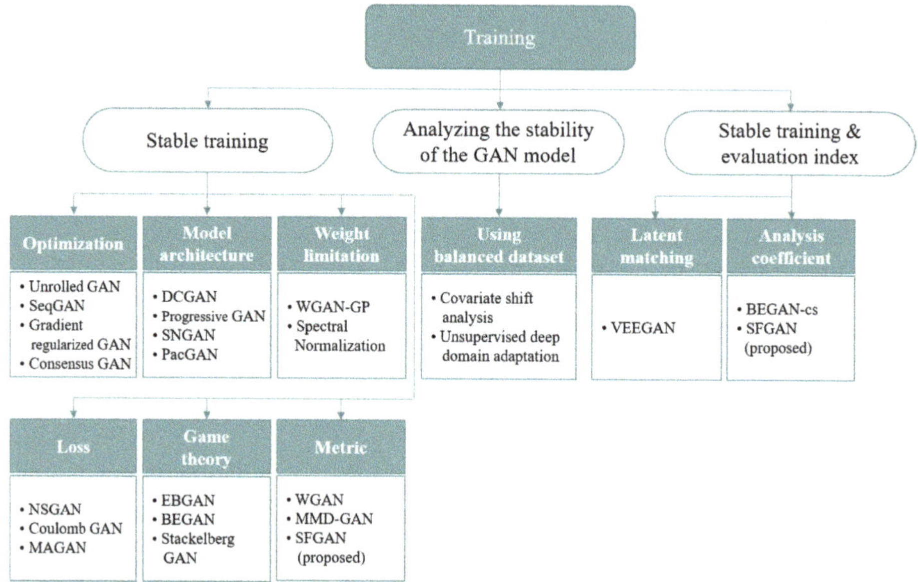

Figure 1. Categorization of studies on Generative Adversarial Networks (GANs) that deal with stability.

2.1. Stable Training

Many studies are dealing with stable training of GANs [13–15,24,32–43]. Notably, several studies [13–15,32,33] also considered on mode collapse and/or mode drop problems. In these studies, IS or FID is used as a performance evaluation index. This category can be classified further according to their purposes or applied methods: optimization, model architecture, weight limitation, loss, and game theory.

2.1.1. Optimization

The studies in the optimization class deal with the optimization of balancing of the generator and discriminator for stable training. Unrolled GAN, SeqGAN, Gradient regularized GAN, and Consensus GAN belong to this class [13,14,32,33].

The network parameters of the generator in the unrolled GAN are updated according to gradient descent (GD) of every step, which reflects the state from the current step to some fixed number of future steps, while those parameters of discriminator are updated according to GD of current step only. Based on the updating rule, the unrolled GAN can mitigate the mode collapse and stabilize the training of GANs. On the other hand, SeqGAN modeled a generator as a stochastic policy in reinforcement learning (RL). SeqGAN performs the gradient policy update directly to avoid the difficulty of differentiation of discrete data such as text generation and music generation. It showed that the performance of SeqGAN outperforms that of the original GAN. A scheme imposing a regularization penalty during the generator update was proposed [32]. This scheme was adopted in the original GAN, and the local stability was confirmed. This local stability is updating the gradient near an equilibrium point. Two major failures caused by GD optimization, vanished real-part and large imaginary-part in the eigenvalues of the Jacobian, were considered in Consensus GAN [33]. By using their Consensus Optimization, Nash equilibrium was found by more powerful optimization than GD based optimization. Nash equilibrium is a state that no player in this state can gain more rewards by changing its own strategy [48]. If the generator and the discriminator reach a Nash equilibrium state, the objective of GANs does not change any more theoretically, so this theory keeps the stability of GANs during the training.

2.1.2. Model Architecture

The GANs in the model architecture class considered network architectures and their parameters for stable training. Deep convolutional GAN (DCGAN), progressive GAN, spectral normalization GAN (SNGAN), and packed GAN (PacGAN) are belonging to this class [24,34–36].

DCGAN is a Convolutional Neural Network (CNN)-based GAN model considering the way of setting parameters and techniques for optimizing GAN [34]. It dealt with the use of batch normalization and activation function. The resolution of images was increased up to 1024 by adding layers, and mini-batch was also considered to improve stability in progressive GAN [24]. SNGAN considered an architecture utilizing the residual block [35,49], and PacGAN considered to use an augmented discriminator. The discriminator in PacGAN maps multiple samples that are jointly coming from either real data or the generator to a single label [36].

2.1.3. Weight Limitation

In the weight limitation class, several techniques are dealing with the discriminator's weights in a network's nodes. The techniques are schemes to control the instability of GANs.

Gradient-penalty and spectral normalization are typical examples in this class [37]. The main idea of WGAN gradient-penalty (WGAN-GP) is considering the Lipschitz-1 constraint in WGAN. The gradient penalty was added to the loss of WGAN, which is directly constraining the gradient norm of the discriminator. By adjusting this method, WGAN-GP outperformed original WGAN in terms of stable training and similarity between real data and generated data. SNGAN belonging to the model architecture class also proposed a spectral normalization technique to increase the stability of training [34]. Unlike gradient-penalty, spectral normalization does not depend much on the current generative distribution but regularizes the weights of nodes in a network. Training with spectral normalization was compared with that of gradient-penalty, and it was concluded that the former does not easily destabilize with a high learning rate while the latter destabilizes.

2.1.4. Loss

Other recent approaches are considering new loss functions. NSGAN considered a risk that the gradient of the generator would vanish when the original GAN loss is used [12]. For preventing the risk, a function was proposed to maximize the loss of generator, and better performance was obtained than that of the original GAN models. Margin adaption for GAN (MAGAN) evaluated the performance improvement of stable training by using an adaptive hinge loss, which estimates the appropriate margin of the loss [38]. MAGAN not only generated diverse datasets but also achieved an improvement in terms of IS compared to energy-based GAN (EBGAN) and boundary equilibrium GAN (BEGAN) [29,39]. Coulomb GAN trained the networks using the Coulomb potential equation that makes samples attracted to the training samples but repulsed to each generated sample [40]. It was shown that Coulomb GAN has only one Nash equilibrium.

2.1.5. Game Theory

In the game theory class, most studies used the Nash equilibrium [48]. For instance, EBGAN and BEGAN considered Nash equilibrium [29,38]. These two models generate realistic data successfully and hardly fail to learn the distribution of data. Stackelberg GAN was inspired by the Stackelberg competition of game theory [41]. It is known that the Stackelberg model can be used to find the perfect Nash equilibrium of sub-games. Experiments verified the effectiveness of the Stackelberg competition by using a multi-generator architecture.

2.1.6. Metric

The other approach of dealing with instability is considering distance metrics used in GANs. As far as we know, Wasserstein GAN (WGAN) is the first study to improve the learning stability of GANs by defining a new distance metric between data distributions [15]. It was shown that traditional distances such as JSD are insufficient for data training in GANs. As an approach to mitigate this problem, EMD was applied to WGAN. MMD-GAN used a maximum mean discrepancy (MMD) as a distance and adopted auto-encoder [42,43]. The definition of MMD can be found in Appendix A. This model's discriminator was trained via MMD with adversarially learned kernels. Although it was obtained that the IS of MMD-GAN was higher than WGAN when the experiment was conducted with CIFAR-10, there are two problems in this model: (i) The performance of MMD-GAN comes only by using either per-pixel reconstruction error term or gradient penalty. (ii) The fine learning from data seems to discourage to contract the discriminator outputs of real data using MMD [50].

Improving distance metrics in GANs has the following advantages: (1) No additional network models may be required for stable training a GAN, and existing losses can be used as they are. (2) The state of mode collapse and/or mode drop during the training process can be identified through the proposed metric. The training curves and sample graphs by WGAN and MMD-GAN showed the relation between the loss and the sample quality. As our study proposes a metric, the proposed model can also take the advantages that these studies grouped by this section have.

2.2. Analyzing the Stability of the GAN Model

Although analyzing the stability of the GAN model did not affect directly stable training of GANs, studies in this category detected and showed these phenomena based on the data generated by the trained model. Covariate shift analysis was a scheme to add a multi-class classifier in a balanced multi-class dataset to investigate whether the data generated by GAN was biased [44]. Unsupervised deep domain adaptation was a scheme to extend covariate shift analysis to an unbalanced dataset with the existence of a balanced dataset [45].

2.3. Stable Training and An Evaluation Index

In this category, there are several studies recently. For instance, an algorithm was proposed in Variational Encoder Enhancement GANs (VEEGAN) to estimate mode collapse [46]. It was conducted by training a multi-layer neural network with sample data and the standard deviation of the data. Boundary Equilibrium Generative Adversarial Nets-Constrained Space (BEGAN-CS) added an embedding space-constrained loss in BEGAN and showed the stability improvement by using the proportional coefficient's variation during training [47]. Although the results of VEEGAN and BEGAN-CS were noticeable, they had limitations that require additional CNN models and a constraint of latent space for a discriminator, respectively.

Our model, SFGAN, detects mode drop and (or) mode collapse during training. The details of SFGAN are presented in Section 3. Experimental results of SFGAN and BEGAN presented in Section 5 show that the output values of the two models are related to the mode drop and (or) mode collapse of BEGAN.

3. Simplified Fréchet Distance

In this section, we introduce SFD and present its advantages. Section 3.1 defines notations to describe image distribution and then introduces the Fréchet distance and SFD based on the defined notations. The advantages of SFD are investigated in Section 3.2 by comparing it with other distance metrics using two different examples.

3.1. Simplified Fréchet Distance

We introduce distance metrics of image distributions. For defining distance metrics, images have to be converted to numerical values. Consider a color image that has h and w pixels for height and width, respectively. As an image usually consisted of three channels, R, G, and B, and its values are numbers, without loss of generality, we assume that each pixel of an image has a number for each channel. Then, the image has $3hw$ pixels in total. Let X be a random vector whose components consist of a random variable $X^c_{i,j}$, where $X^c_{i,j}$ is the value of pixel (i,j) for $c = $ R, G, B. Then X can be written as

$$X = (X^c_{i,j}), i = 1, \cdots, h, \ j = 1, \cdots, w, \ c = \text{R, G, B.} \tag{1}$$

From now on, we call X as an 'image vector' and describe distance metrics in terms of the image vectors. Figure 2 illustrates the way of converting an image to an image vector.

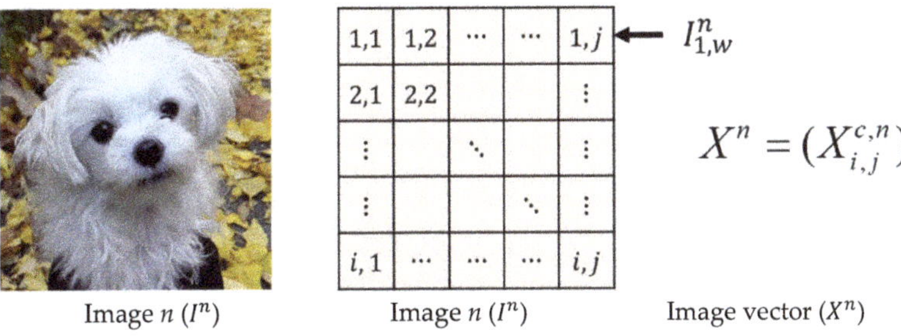

Image n (I^n)   Image n (I^n)   Image vector (X^n)

Figure 2. Conversion of an image to an image vector.

We first describe FD of images and then introduce SFD. Assume that there are k images and denote an image vector corresponding to the n-th image as X^n. Then, X^n is given by $X^n = (X^{c,n}_{i,j})$, $i = 1, \cdots, h, \ j = 1, \cdots, w, \ c = $ R, G, B.

Let X and Y be random vectors given by image vectors and F and G be their distributions, respectively. Let $m_\bullet$ and $C_\bullet$ be the mean vector and covariance matrix of a variable $\bullet$, respectively, where $\bullet = X, Y$. Then, m_X and C_X of X are defined by

$$m_X = (m_{X^c_{i,j}}) = \left(\frac{1}{k} \sum_{n=1}^{k} X^{c,n}_{i,j} \right) \text{ and } C_X = (X^c_{i,j} - m_{X^c_{i,j}})^{\text{T}} (X^c_{i,j} - m_{X^c_{i,j}}), \tag{2}$$
$$i = 1, \cdots, h, \ j = 1, \cdots, w, \ c = \text{R, G, B.}$$

respectively, where T in C_X represents transpose of a matrix.

Definition 1. *Fréchet Distance.*

The Fréchet distance $Fr(F, G)$ between two distributions F and G is defined by

$$Fr^2(F, G) = \min_{X,Y} E|X - Y|^2, \tag{3}$$

where E represents the expectation, and the minimization is taken over all random variable X and Y having distributions F and G, respectively [19].

In particular, if X and Y follow multivariate normal distributions $Fr^2(F, G)$ is given by

$$Fr^2(F, G) = \|m_X - m_Y\|^2 + \text{tr}\left\{ C_X + C_Y - 2(C_X C_Y)^{1/2} \right\}, \tag{4}$$

where 'tr' in Equation (4) represents the trace of a matrix [51].

FID was a metric providing a better result for measuring the similarity between the two distributions of generated data and the real data. For instance, FID was compared with the IS in experiments using various data [20]. However, the covariance term in Equation (4) has drawbacks in adopting FD as a loss to train GAN. These drawbacks were demonstrated in [28] by the empirical results using the MNIST [52], Fashion-MNIST [53], CIFAR-10 [31], and CelebA [30] datasets. In [28], each dataset was divided into two groups, and the FID was used as the similarity index between the groups. It was observed that estimating a total covariance matrix can be unnecessary and unproductive. It was also mentioned that a constrained version of FID might be enough to represent distances between data. Based on this, it seems relevant to delete the covariance term in FID. By considering this aspect, it may not be a problem to apply the distance metric without covariance in FD to the data, not the inceptionv3 feature of data. Furthermore, the larger the dimension size of the datasets is, the higher the computational load on the covariance matrix is. These facts motivate SFD. That is, SFD simplifies and regularizes the covariance term in FD to reduce the complexity of FD and to learn stably, respectively. This distance metric makes applicable FD in the training process of GANs with less computing load. For this purpose, we assume that all components of both random vectors X and Y are independent. There is no guarantee that these components are independent, and they may be dependent on the real world. In the field of deep learning, however, such an assumption was used in several studies, and the better results were obtained under the assumption [54–56]. The independence of the two variables does not imply that they have the same variances. The SFD is introduced with this assumption.

Definition 2. *Simplified Fréchet Distance.*

The Simplified Fréchet distance $SF(F, G, \alpha)$ between two multivariate normal distributions F and G with coefficient α is given by

$$SF^2(F, G, \alpha) = \|m_X - m_Y\|^2 + \frac{1}{\alpha}\|\sigma_X - \sigma_Y\|^2, \tag{5}$$

where $\sigma_\bullet^2$ is the variance of a random variable $\bullet$ and α is a constant for regularization.

In the following, $SF(F, G, \alpha)$ is representing $\sqrt{SF^2(F, G, \alpha)}$.

3.2. Advantages of Simplified Fréchet Distance

To investigate the advantages of SFD, we consider two examples. Although these examples might be extreme cases, they can appear in the training process of GANs. For the two examples, SFD is compared with two distance metrics: JSD and EMD. The definitions of JSD and EMD can be found in Appendix A.

Example 1. *(Learning parallel distribution) The distributions of real data and estimated data are parallel.*

It was shown that JSD is inadequate, while EMD is adequate to measure data distributions that are parallel [15]. We consider the two-dimensional random vectors X and Y whose components are normally distributed with means $(0,0)$ and $(\theta,0)$, and covariance matrices $\begin{pmatrix} 0 & 0 \\ 0 & 1 \end{pmatrix}$ and $\begin{pmatrix} 0 & 0 \\ 0 & 1 \end{pmatrix}$, respectively. Then, the distributions of X and Y are parallel.

Example 2. *The distributions of real data and estimated data are a couple of univariate normal distributions with mean zero.*

Let X and Y be the one-dimensional random variables distributed according to normal probability density function (pdf) with means 0 and 0, variances 1 and δ^2, respectively.

The three distances for the two examples are represented in Table 1. The detailed derivation of the obtained values can be found in Appendix B. According to the table, JSD is a constant or log2, regardless of $\theta, \theta \neq 0$. On the other hand, EMD and SFD are varying according to θ for Example 1. These values imply that JSD cannot distinguish the given distributions, while EMD and SFD can distinguish those. For Example 2, on the other hand, EMD has zero as a lower bound regardless of δ, while JSD and SFD depend on the δ. That is, EMD cannot distinguish the given distributions, while JSD and SFD can distinguish those. Figure 3 illustrates the obtained three distances given in Table 1 for varying θ and δ with $\alpha = 1$ for SFD; (a) Example 1 with $\theta \in [-1, 1]$, (b) Example 2 with $\delta \in [0.04, 2]$. From the two examples, it is noticed that SFD is the only distance metric that can always be expressed in terms of the respective parameters θ and δ.

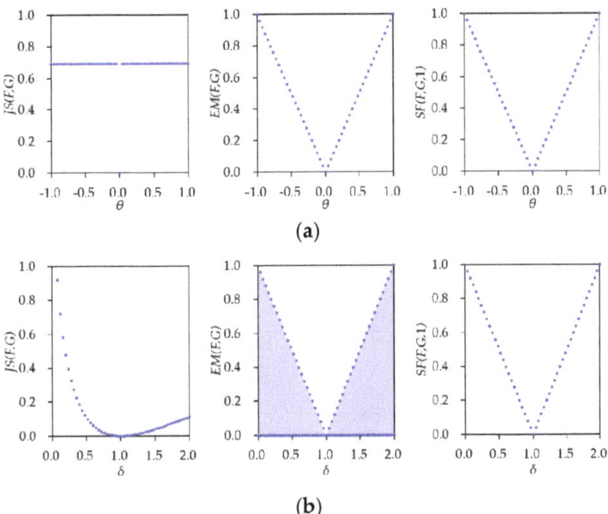

Figure 3. Description of Jensen–Shannon distance (JSD), Earth Mover's distance (EMD), and Simplified Fréchet distance (SFD) with $\alpha = 1$ for the examples: (**a**) Example 1 for $\theta \in [-1, 1]$; (**b**) Example 2 for $\delta \in [0.04, 2]$.

This disadvantage of EMD is not limited to a univariate normal distribution, as shown in Example 2. By Equation (A22), it is noteworthy that the lower bound of EMD is zero for other cases, such as the given two distributions are multivariate normal distributions with the same mean but other than zero. In this case, EMD cannot distinguish the two distributions.

Table 1. Comparison of JSD, EMD, and SFD for the examples.

Distance Metric	Example 1	Example 2
Jensen–Shannon distance (JSD)	$\begin{matrix} \log 2, & \theta \neq 0 \\ 0, & \theta = 0 \end{matrix}$	$\frac{1}{2} \log \frac{\delta^2 + 1}{2\delta}$
Earth Mover's distance (EMD)	$\lvert \theta \rvert$	$[0, \lvert \delta - 1 \rvert]$
Simplified Fréchet distance (SFD)	$\lvert \theta \rvert$	$\frac{1}{\sqrt{\alpha}} \lvert \delta - 1 \rvert$

4. Simplified Fréchet GAN

In this section, we present a GAN model, simplified Fréchet GAN (SFGAN). SFGAN uses the SFD in the calculation of the discriminator loss and the generator loss in a GAN.

Since GANs are trained by an adversarial loss that is based on the discriminator output, the output must have a multivariate normal distribution to apply SFD. As far as we know, however, no studies have considered multivariate normal distributions as outputs. It is widely accepted in the image processing field that the distribution of a lot of images is assumed as multivariate normal [57–59], and the distribution of output data will be the same as that of input data through an auto-encoder. For this reason, we design the discriminator in the form of an auto-encoder.

Three candidates can be considered for the baseline of our model; BEGAN, EBGAN, and MMD-GAN. The architectures of all three models include an auto-encoder. However, EBGAN cannot apply measures based on data distributions because EBGAN uses errors per pixel. Therefore, EBGAN is excluded from the candidates. As MMD-GAN requires additional reconstruction error term, the error has to be defined additionally as the form of per-pixel error. Therefore, this model is not relevant to the baseline model. Since BEGAN is only required to replace distance metric, it seems to be the better model than the other two models.

For this reason, we select BEGAN as the baseline model of SFGAN. The network architecture and used losses in the network of SFGAN are the same as those of BEGAN. The only difference is the used distance metric, SFD for SFGAN, and EMD for BEGAN. Figure 4 illustrates the model architecture and procedure of SFGAN. The procedure, including the data flow of SFGAN, is described in Table 2.

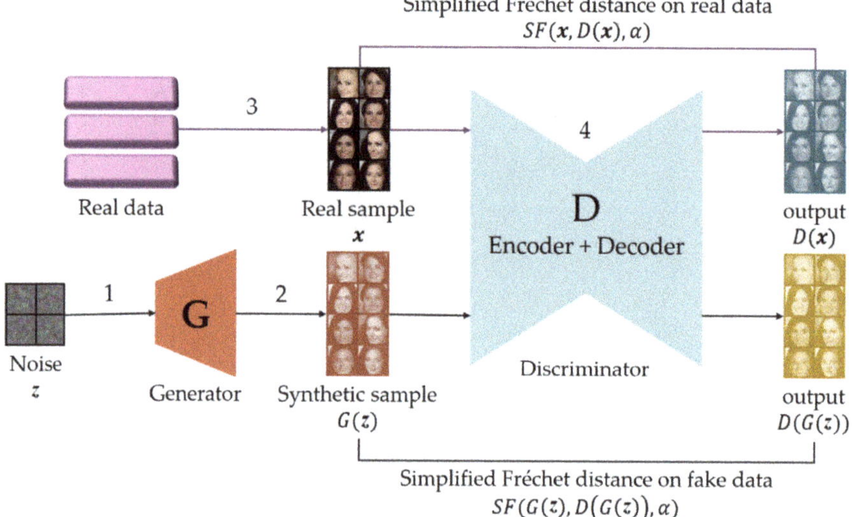

Figure 4. Schematic diagram of the SFGAN algorithm. The numbers on the arrow indicate those steps in the procedure of SFGAN.

Table 2. Procedure of SFGAN.

The Procedure of SFGAN
1. Generate n noise samples $z = \{z^1, \cdots, z^n\}$. Each z^i is vector given by $z^i = (z^i_1, z^i_2, \ldots, z^i_d)$, where d is the number of dimensions, and each component of z^i is randomly generated according to a uniform distribution in the interval $[-1,1]$ with pdf p_z.
2. The generator G generates fake data $G(z)$ based on the generated samples z.
3. n samples $x = \{x^1, \cdots, x^n\}$ are taken from real data with pdf p_{data}.
4. Auto-encoder D receives $G(z)$ and x as inputs and produces $D(G(z))$ and $D(x)$ as outputs.

The detailed structure in D will be explained according to Figure 5. The loss functions in SFGAN are as same as those of BEGAN except involved distance metric in the functions. The losses consist of discriminator loss L_D and generator loss L_G, which is given by

$$L_D = \mathbb{E}_{x \sim p_{data}(x)}[SF(x, D(x), \alpha)] - k_t \mathbb{E}_{z \sim p_Z(z)}[SF(G(z), D(G(z)), \alpha)] \tag{6}$$

and

$$L_G = \mathbb{E}_{z \sim p_Z(z)}[SF(G(z), D(G(z)), \alpha)], \tag{7}$$

where $\mathbb{E}_{r \sim p}$ represents the expectation of variable r with pdf p. k_t is a variable for stable learning at t step which is updated by proportional control given by

$$k_{t+1} = k_t + \lambda_k(\gamma(SF(x, D(x), \alpha)) - SF(G(z), D(G(z)), \alpha)), \tag{8}$$

where $k_0 = 0$, $k_t \in [0, 1]$, λ_k is the proportional gain depending on k_t, and γ is a hyperparameter for controlling image diversity taking values in the interval [0,1]. Note that the small value of γ gives low diversity of generated data. Both of discriminator and generator are trained using the GD method with values of Equations (6) and (7), respectively.

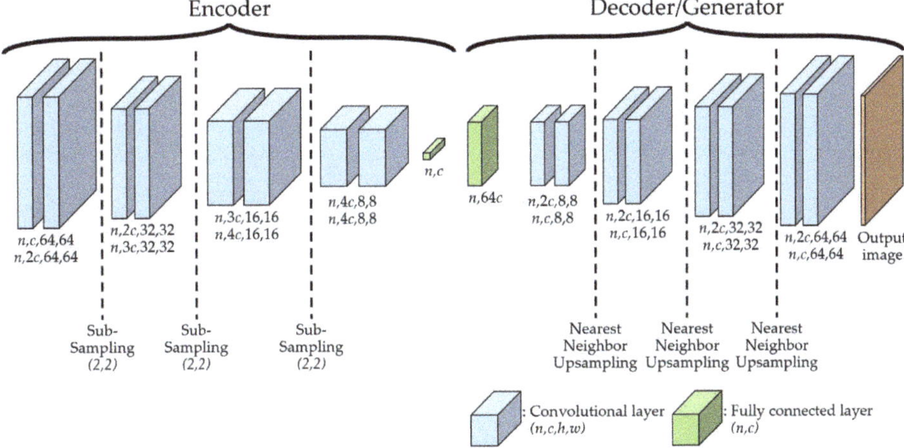

Figure 5. The architecture of SFGAN for discriminator and generator in 64 × 64 resolution with batch size n and filter number c.

Figure 5 illustrates the overall network architecture of SFGAN. This architecture consists of two parts: encoder and decoder/generator. The decoder and the generator have the same structure with different weights. The discriminator consists of an encoder and a decoder. All convolutional layers in the encoder are the same as the layers in the generator except for the number of filters. In the sub-sampling layers of the encoder, the input of a layer is downsampled using stride two, which reduces the input size as it passes through the layers. On the other hand, in the up-sampling layers of the decoder/generator, the input of a layer is up-sampled using the nearest neighbor method, which enlarges the input size as it passes through the layers.

The input of encoder is the real data or the generated data, while the input of decoder is the output of the encoder, which is called as a hidden variable. The input of the generator is the noise sample vectors z and it gets into the fully connected layer. All convolutional layers use 3 × 3 convolutions with exponential linear units [60] and are repeated twice for better output. The batch size of the discriminator is twice of that for generator because discriminator has to afford to deal with both real data and generated data. In the figure, n and c represent the batch size and the number of filters, respectively.

5. Experiments

In this section, we present experimental results. We implemented two experiments. One is for stability comparison of generation models, and the other is for comparison of distance metrics to detect mode collapse and/or mode drop during training the network. For performance comparison, BEGAN is executed in the experiment. SFGAN and BEGAN are trained until mode collapse and/or mode drop occur to compare the stability. In the former experiment, the images generated during the training at every 1000 steps, the measured values of losses, and the k_ts at that time are compared. In the latter experiments, JSD, EMD, and SFD between the generated data and the training data of the two models are measured. As a criterion for the similarity between real data and fake data, FID is considered for the two models.

In the following, the experimental setting is explained in Section 5.1, and the image stability on sequential steps by SFGAN and BEGAN are presented in Section 5.2. The results for verifying the stability of SFGAN and the performance comparison of the two models are presented in Sections 5.3 and 5.4, respectively.

5.1. Experimental Setting

For training, two computers are used. One of them is composed of Intel® CORE™ CPU, NVIDIA GTX 1080ti as GPU with 24 GB RAM, while the other is composed of Intel® Xeon® CPU E5-2680 v4, NVIDIA RTX TITAN as GPU with 128 GB RAM. All experiments are implemented by the TensorFlow library [61].

The CelebA [30], CIFAR-10 [23], and a mixture of Gaussian distributions are used to train the GAN models. These datasets are commonly used in GANs research. The CelebA is a collection of human face images. It is effective at testing qualitative results because human faces are good at recognizing defects [29]. In the experiment, the images of 64×64 and 128×128 resolutions will be generated for CelebA. The CIFAR-10 is a set of widely used images in the image-based machine learning studies. CelebA has only human faces, and the number of training images is 202,599, while CIFAR-10 contains various images such as trucks, frogs, birds, and ships, which are hardly the same objects, and the number of images is 60,000, relatively small. The last dataset consists of 2-D random variables from the mixture of Gaussian distributions. Eight random variables are distributed in a circle, and each random variable consists of x and y coordinates. The expectation of each random variable depends on the position of the random variable, and its standard deviation is fixed as 0.02. The test with this dataset was proposed in unrolled GAN [13] to evaluate the performance of discriminator. In the test, it was assumed as unstable if any one of eight distributions are not learned. With the same stability criterion, the stability of SFGAN is also investigated with this dataset.

Figure 6 shows samples from CelebA, CIFAR-10, and the mixture of Gaussian. Table 3 summarizes parameters and the corresponding values used in the experiments. Note that the parameter values are the same used in BEGAN-cs [50] or BEGAN. It is noteworthy that two regularization constants are used in the experiments. The value one is for detecting mode collapse and/or mode drop while 12,288 is for training the networks, which is obtained for acceptable learning of 64×64 resolution images during the experiments. The experiments are conducted five to ten times to see if the results fluctuate per each experiment. It is observed that the differences among results are negligible. The presented SFD values are one-dimensional values, which are obtained from the experimental results divided by the number of dimensions. Adam is used as an optimizer to both models since this optimizer is invariant to a diagonal rescaling of the gradients [62].

Figure 6. Samples of datasets: (**a**) CelebA; (**b**) CIFAR-10; (**c**) a mixture of Gaussian.

Table 3. Hyperparameters used in the experiments.

Hyperparameter	Value
Batch size (n)	64
CelebA resolutions ($h \times w$)	64×64, 128×128
CIFAR-10 resolution ($h \times w$)	32×32
Channel unit (c)	64
Regularization coefficient for training (α)	12,288
Regularization coefficient for detecting (α)	1
Adam (β_1, β_2)	0.9, 0.999
Proportional gain (λ_k)	0.001
Diversity ratio (γ)	0.5
Learning rate	0.001
Total global steps	3000,000

5.2. Stability of Training

5.2.1. The Generated Images of CelebA: 64×64 Resolution

Figure 7a,b show 16 sample images in 64×64 resolution generated in four steps by BEGAN and SFGAN, respectively. 16 samples (i.e., l is set to 16) are chosen from a random uniform distribution for the generator's input at the beginning, and the images generated from the samples are monitored in every 1000 steps until 3000k steps. It is challenging to say mode collapse or mode drop numerically, but the generated samples allow us to determine if the training was as intended. In the 200k step, the training result of BEGAN is similar to that of SFGAN. However, at 970k step, the generated images by BEGAN became weird while SFGAN still generated acceptable human faces. At 3000k step, BEGAN generated very similar images even for different inputs. On the other hand, SFGAN generated diverse human-like face images steadily until 3000k steps. In other words, BEGAN started to generate weird similar face images after 968k steps on the average in a total of ten experiments. This implies that the mode collapse occurs at 968k step, and BEGAN never optimized or restored to stable status after all. On the other hand, SFGAN keeps on generating face images until 3000k steps without mode collapse. Therefore, it can be said that the training process of SFGAN is stable until 3000k steps. However, the example in Figure 7 appears hard to assume that a mode drop has occurred in both BEGAN and SFGAN.

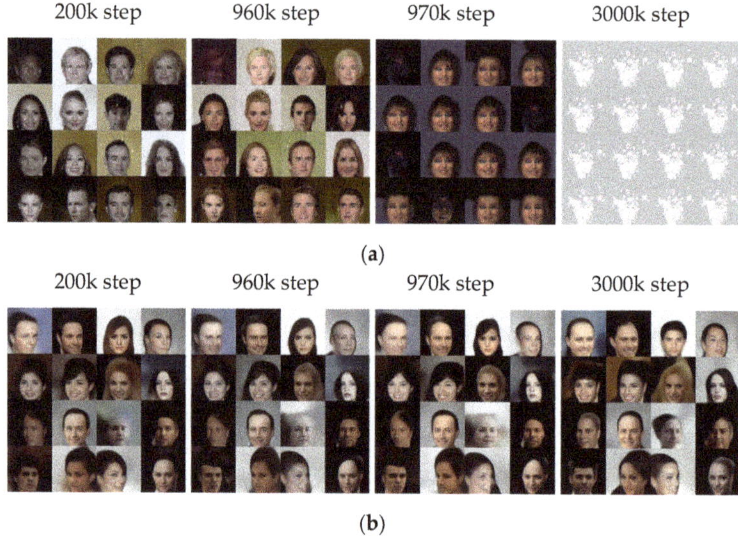

Figure 7. A total of 16 samples generated at 200k, 960k, 970k, and 3000k steps trained with CelebA dataset in 64 × 64 resolution by: (**a**) BEGAN; (**b**) SFGAN with $\alpha = 12,288$.

Figure 8 compares the two losses, k_ts, and the generated images during the training process of the two models. Figure 8a,b shows the losses of discriminator and generator of the models, respectively. The amplitudes of both losses of discriminator and generator of BEGAN increase after 970k steps, while those of SFGAN do not change much and even seem to converge. The loss differences for the two models appeared the difference between generated images, as shown in Figure 8d. That is, the generator of BEGAN fails to generate human-like faces at 970k step, while the generator of SFGAN keeps on generating human-like faces until 3000k steps. Figure 8c shows that the change of k_t is negligible if GANs are in the equilibrium state. That is, the more stable the training gives a smaller variation of k_t [50]. In early training stages, both of BEGAN and SFGAN are tending to generate easy-to-reconstruct data by auto-encoder because the real data distribution has not been learned accurately yet. BEGAN seems to find a stable value as it gradually descends. When BEGAN started to generate images that are not human-like faces, however, k_t decreases rapidly. This means that the discriminator's loss is reduced faster than that of the generator. In other words, the discriminator won the generator. SFGAN, on the other hand, k_t increases and then decreases slowly until 150,000 steps. After 150,000 steps, no abrupt decrements are observed for SFGAN as BEGAN does have. Based on this observation, we may derive the following: (1) If k_t is not zero and has small vibration or converging to a constant that is not zero, then the network can be considered as in a stable state. (2) If k_t converges to zero, the generator's loss is too large, and the network is far from the equilibrium state.

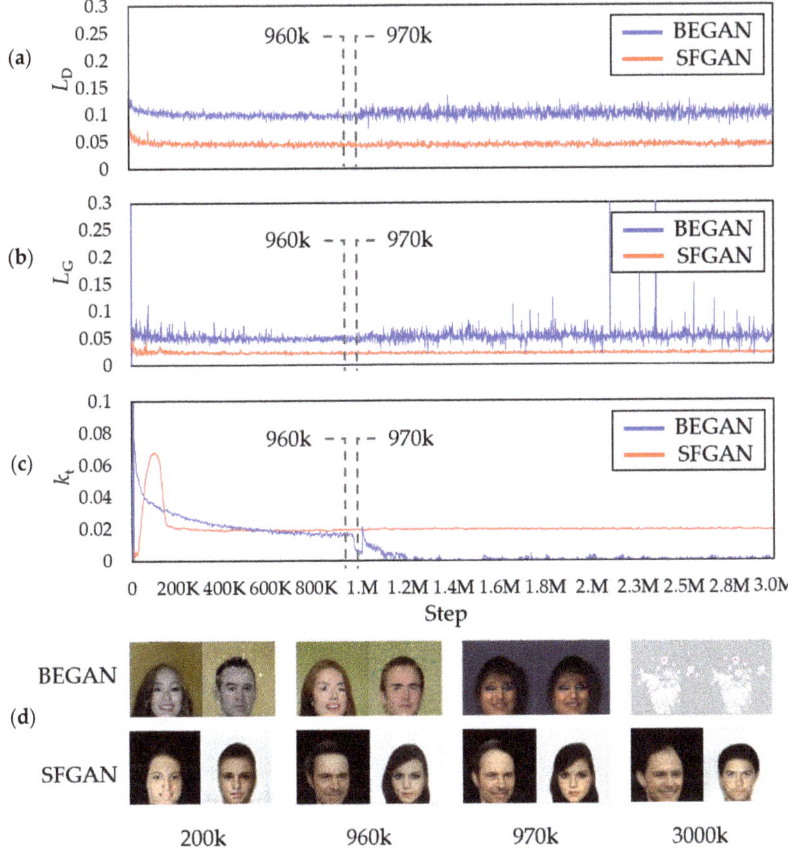

Figure 8. Comparison of BEGAN and SFGAN trained with CelebA dataset in 64 × 64 resolution: (**a**) Discriminator loss L_D; (**b**) generator loss L_G; (**c**) variable k_t; (**d**) samples during the training process of BEGAN and SFGAN with $\alpha = 12,288$.

5.2.2. The Generated Images of CelebA: 128 × 128 Resolution

Figure 9 presents the results for the images with 128 × 128 resolution. In the 200k step, the training result of BEGAN is similar to that of SFGAN. However, it is observed that the mode collapse occurs 520k, which is earlier than that with 64 × 64 resolution in BEGAN. At 3000k step, BEGAN generated the same images even if different inputs are given. Based on the results with two different resolutions, the mode collapse seems to occur faster as the image resolution increases in BEGAN. This phenomenon seems caused by the insufficiency of the weight parameters in the model network because the mode collapse and/or mode drop can easily occur when the number of weight parameters is insufficient [63].

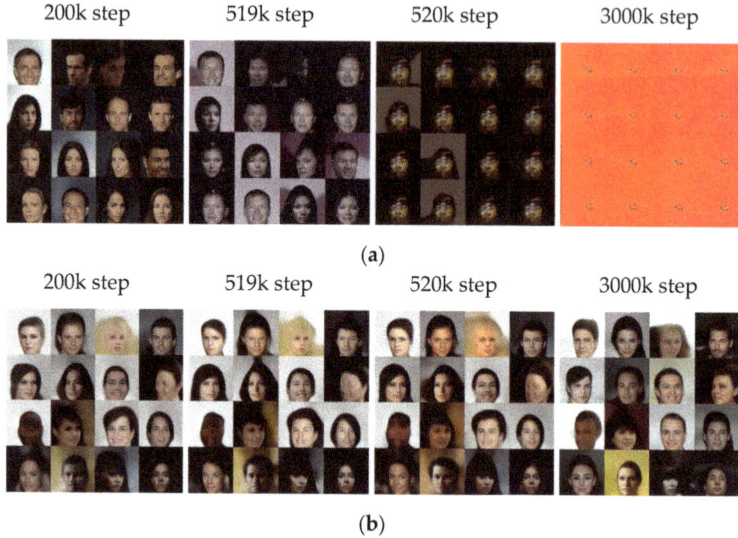

Figure 9. A total of 16 samples generated at 200k, 519k, 520k, and 3000k steps trained with CelebA dataset in 128×128 resolution by: (**a**) BEGAN; (**b**) SFGAN with $\alpha = 12,288$.

Figure 10 compares the two losses, k_ts, and the generated images during the training process of the two models. Figure 10a,b show that the differences in both discriminators' losses are insignificant, while those of both generators' losses seem significant. The loss of generator of BEGAN fluctuates significantly after 700k steps compared to the previous steps, while the loss of generator of SFGAN is almost consistent. The instant of the abrupt changes in loss of generator with this resolution is different from that with a smaller resolution, 520k steps. This can be interpreted in two ways; (i) The exact instant is missing because the losses are measured at every 1000 steps. (ii) The mode collapse and/or mode drop can occur even if the generator learns stably. Figure 10c illustrates k_ts for every 1000 steps of the two models, and Figure 10d presents the created samples corresponding to 200k steps, 519k steps, 520k steps, and 3000k steps. At 520k steps, k_t of BEGAN dropped rapidly, and the similar images are generated. The rapid drop of k_t implies that the loss of discriminator does not change much during its update, and this may result in the model collapse or mode drop. This phenomenon did not occur in SFGAN until 3000k steps, as same as 64×64 resolution.

If we compare the results of two resolutions and two models, the followings are concluded; (i) BEGAN requires one and two more convolutional layers for the generator and discriminator, respectively, if the resolution becomes twice. (ii) The used network architecture is not a good model for large sizes of images. (iii) Even though the network structure of SFGAN is the same as BEGAN, it is less affected by the architecture and, therefore, resolution. This phenomenon seems to owe to using SFD.

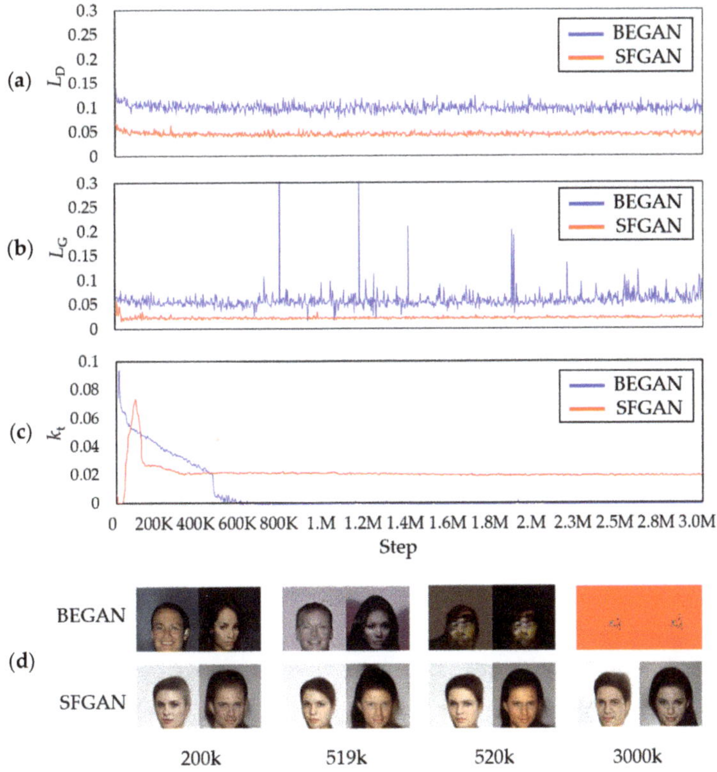

Figure 10. Comparison of BEGAN and SFGAN trained with CelebA dataset in 128×128 resolution: (**a**) Discriminator loss L_D; (**b**) generator loss L_G; (**c**) variable k_t; (**d**) samples during the training process of BEGAN and SFGAN with $\alpha = 12,288$.

5.2.3. The Generated Images of CIFAR-10: 32×32 Resolution

The same experiments are conducted with CIFAR-10. Figures 11 and 12 present the results with 32×32 resolution. Figure 11 compares the images generated by the two models at 397k, 427k, 457k, and 3000k steps. Overall, it is difficult to figure out the images created by both models. According to [63], BEGAN-based models perform slightly better than DCGAN in training CIFAR-10. Therefore, it can be derived that the obtained unclear images may be caused by the network structure of BEGAN. The figure shows that BEGAN seems to produce relatively sharp images initially up to 397k steps compared to SFGAN. However, similar images are generated from 427k, and it seems to fail to generate different images from 457 k steps. The generated images are all the same images at 3000k steps finally. On the other hand, SFGAN generates different images continuously, even though the images are blurry from the beginning. Figure 12 compares the two losses, k_ts, and the generated images during the training process of the two models. The losses of discriminator seem stable in both models, while the losses of the generator are not. That is, the loss of generator in BEGAN increases slightly from about 400k steps and then fluctuates after all. This phenomenon may result in a lack of diversity in the generated images.

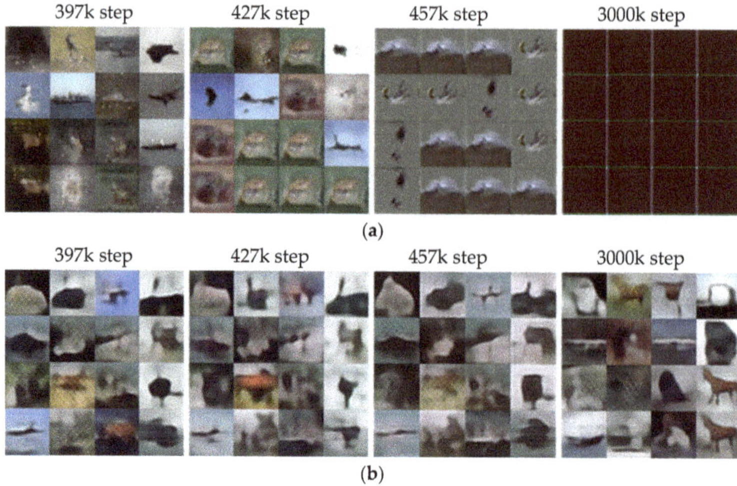

Figure 11. A total of 16 samples generated at 397k, 427k, 457k, and 3000k steps trained with CIFAR-10 dataset by: (**a**) BEGAN; (**b**) SFGAN with $\alpha = 12,288$.

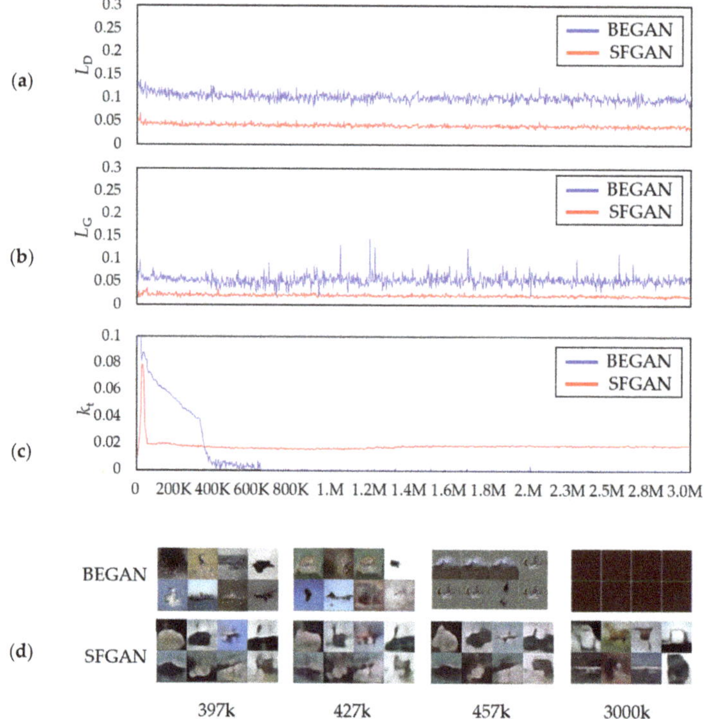

Figure 12. Comparison of BEGAN and SFGAN trained with CIFAR-10 dataset: (**a**) Discriminator loss L_D; (**b**) generator loss L_G; (**c**) variable k_t; (**d**) samples during the training process of BEGAN and SFGAN with $\alpha = 12,288$.

On the other hand, for SFGAN, the amplitude of loss of the generator gradually decreases, staying around small values near zero. As shown in Figure 12c, the k_t values for BEGAN drop rapidly at 397k steps, 427k steps, and 457 steps, which seems to be associated with the reduction of diversity in the generated images, as shown in Figure 12d. The k_ts for SFGAN does not drop abruptly, except in the initial stage of training. The stable k_ts seem to correspond to the generated images of SFGAN.

Comparing the results of two datasets, CelebA and CIFAR-10, the followings are concluded: (i) BEGAN and SFGAN can make human faces up to 500k steps when training a CelebA dataset, but it is not valid for the CIFAR-10. This phenomenon seems to owe to the number of training data because the number of images of CelebA is approximately 3.4 times that of CIFAR-10. (ii) Based on the training results of CIFAR-10, the capacities of SFD and BED as distance metrics are similar in training, while SFD is better than EMD in stability and performance.

5.3. Mixture of Gaussian Dataset

In this section, we compare the qualitative results of a mixture of Gaussian. Figure 13c shows examples of images used as training data. The example images are dawn by randomly generated 100 samples from the mixture of Gaussian distribution. Figure 13a,b presents the generated images by BEGAN and SFGAN, respectively. As shown in the figure, both models learn roughly the circular positions of the random variables. However, the densities of each eight centers of BEGAN are relatively low compared to those of SFGAN. These results imply that SFD enables us to learn the distribution at least as same as or better than EMD.

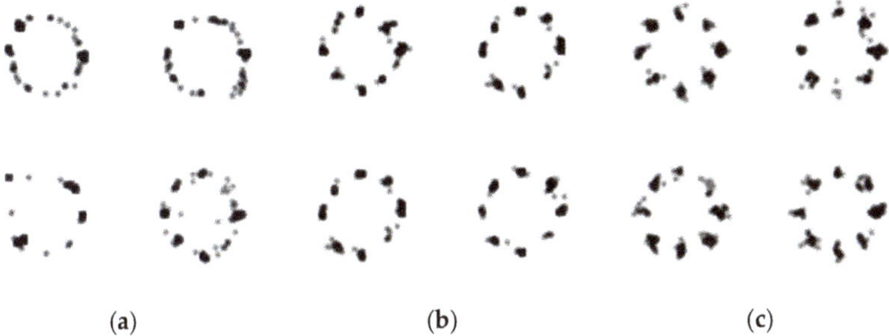

(a)   (b)   (c)

Figure 13. Four batches of 100 sample generation results trained with a 2-D mixture of Gaussian by: (a) BEGAN; (b) SFGAN with $\alpha = 12,288$; (c) training data.

5.4. Detecting Mode Collapse Using Distances

Figure 14 compares (a) JSD, (b) EMD, (c) SFD between the training data and the generated data for every 1000 steps during the training process of BEGAN and SFGAN with $\alpha = 12,288$ for (a) and (b), and $\alpha = 1$ for (c), and (d) the generated samples presented in Figure 14 d. At a glance, each of the three distance metrics for BEGAN is more fluctuate than that of SFGAN, especially after 970k. For BEGAN, the following are observed; For JSD, it is difficult to find the exact spot where the distance is distinguished from detecting the mode collapse even though the amplitude of the values increases between 960k and 970k. For EMD, the range of distances slightly increases from 960k to 970k after mode collapse occurs. However, there are overlapping ranges before and after the mode collapse occurs, which seems to owe the definition of EMD. In the case of SFD, however, the ranges of distance differ significantly before and after the mode collapse, compared to those of JSD and EMD. In other words, mode collapse and/or mode drop detection is detected better by SFD than by JSD and EMD. As already seen in Section 5.2, no collapse occurs in SFGAN, and distance metrics verifies this. As a result, the distance values of SFGAN in (a), (b), and (c) can imply the stable state.

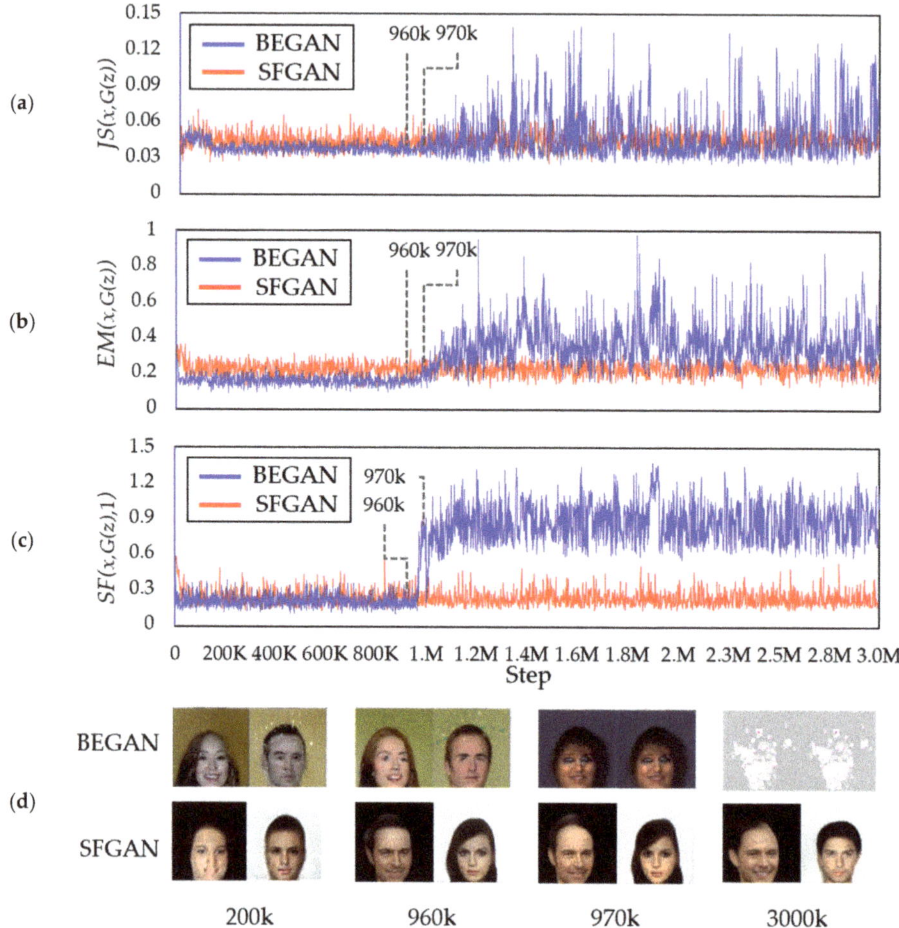

Figure 14. Comparison of distances between the distributions of training data and data generated by BEGAN and SFGAN: (**a**) JSD; (**b**) EMD; (**c**) SFD with $\alpha = 1$; (**d**) generated samples of the two models.

In Table 4, the mean values of the three distance metrics are compared for two groups: a group from 1 step to 968k step and a group from 969k step to three million steps, for the two models. The presented values in the table are the averages of ten experiments. In BEGAN, all values of JSD, EMD, and SFD for the second group are increased compared to the corresponding value for the first group. The biggest increment is observed in SFD, followed by JSD and EMD. This increment shows that mode collapse and/or mode drop signs can be captured without the inception v3 model or balanced dataset, which are regarded requirements for detecting these phenomena. All values of the three metrics for SFGAN are slightly increased in the second group compared to the corresponding value for the first group, even if mode drop and (or) mode collapse did not yet appear. This phenomenon can be interpreted as two situations: (1) SFGAN's learning is almost balanced, which derives similar values of distances. (2) SFGAN remains the possibility to occur mode drop and (or) mode collapse.

Table 4. Mean values of JSD, EMD, and SFD of two groups; Before 968k (from 1 step to 968k) and After 968k (from 969k step to three million steps) for BEGAN and SFGAN.

Model	BEGAN Step Group		SFGAN Step Group	
	Before 968k	After 968k	Before 968k	After 968k
JSD	0.038065	0.048083	0.044120	0.045827
EMD	0.152573	0.365471	0.227430	0.229367
SFD	0.197894	0.867904	0.237440	0.240151

5.5. Quantitative Comparison

The performance of SFGAN is evaluated in two ways; comparison of FID index with BEGAN and comparison with well-known GAN models.

Figure 15 shows the FID [20] between the real and the generated data for BEGAN and SFGAN for every 1000 steps up to 2000k steps. As the green circle indicates, the minimum values of FID for both models are 32.88 and 32.4, respectively, which are almost the same. The FID values of BEGAN increased suddenly around 970k, while those of SFGAN remain steady. However, BEGAN could no longer maintain its quality after 970k, while SFGAN maintained its quality until 2000k steps. This FID value demonstrated that SFGAN is not in mode drop or mode drop phenomenon.

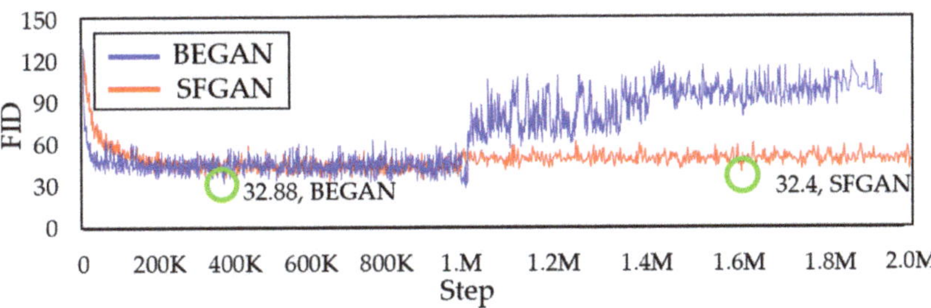

Figure 15. Comparison of FID between the training data and the generated data by BEGAN and SFGAN.

The obtained FID indices of the two models are compared with those of well-known GAN models. Table 5 summarizes the FID indices of those GAN models, including SFGAN. The FID indices of DCGAN, WGAN, and WGAN-GP in the table are from [28], which obtained by using CelebA and CIFAR-10.

Table 5. Comparison of FID indices by several GAN models.

Model	CelebA	CIFAR-10
DCGAN	65.6 ± 4.2	72.7 ± 3.6
WGAN	41.3 ± 2.0	55.2 ± 2.3
WGAN-GP	30.0 ± 1.0	55.8 ± 0.9
BEGAN	38.9 ± 0.9	71.4 ± 1.6
SFGAN	38.1 ± 0.8	68.4 ± 1.3

For the CelebA dataset, WGAN-GP performs best. This superiority comes from the process of calculating the gradient of the discriminator. The calculation of gradient in WGAN-GP is executed the forward and backward propagation as a whole. Even though SFGAN is in the second position, it is the best FID value in the models that do not calculate the gradient. The models with BEGAN-based architecture, such as BEGAN and SFGAN, appear to be better models for learning CelebA data than the models with DCGAN-based architecture. Note that DCGAN, WGAN, and WGAN-GP belong to models with DCGAN-based architecture.

Meanwhile, when training CIFAR-10, BEGAN and SFGAN are worse than WGAN and WGAN-GP. These performances are because the models with BEGAN-based architecture do not train the dataset sufficiently, which seems to owe the number of filters of a layer [63]. In other words, it is assumed that the number of filters in each layer within the BEGAN-based architecture is smaller than that of the DCGAN-based architecture.

Even though the training processes of some experiments turn out to be unstable and the same or blurred images are generated, it is difficult to figure out the reason for the results. Maybe it is because of mode drop or mode collapse or both. At the current level of researches, it is not very easy to find out the relationship between the stability of the training process and the two phenomena, quantitatively measure and distinguish the two phenomena.

6. Conclusions

We have introduced a distance metric SFD and proposed a SFGAN model. SFD has used for two cases: one is in loss functions of SFGAN, and the other is as a measure to detect mode drop and mode collapse during the training process. SFGAN has implemented using several datasets, including CelebA and CIFAR-10, and is compared with BEGAN that is using auto-encoder and EMD. Experimental results show that the training process of SFGAN is more stable than that of BEGAN under the same conditions. Also, it is verified that SFD is an acceptable distance metric presenting better results than the existing distance metrics such as JSD and EMD in detecting mode drop and/or mode collapse. This study will be extended to apply SFD in the field of GAN and apply SFGAN to various datasets such as ImageNet. The relationship between stability and mode collapse and/or mode drop and how to distinguish between mode collapse and mode drop will be studied in the future.

Author Contributions: Conceptualization, C.-I.K. and E.H.; methodology, C.-I.K. and M.K.; software, C.-I.K. and S.J.; validation, M.K. and S.J.; formal analysis, C.-I.K. and M.K.; investigation, C.-I.K. and S.J.; resources, C.-I.K., S.J. and E.H.; data curation, C.-I.K. and M.K.; writing—original draft preparation, C.-I.K.; writing—review and editing, M.K., S.J. and E.H.; visualization, C.-I.K., M.K. and E.H.; supervision, E.H. All authors have read and agreed to the published version of the manuscript.

Funding: This research was supported in part by the Korea Electric Power Corporation (grant number: R18XA05) and in part by the Mid-career Research Program through NRF grant funded by the MEST (NRF-2019R1A2C1002706).

Conflicts of Interest: The authors declare no conflict of interest.

Appendix A

Definition A1. *The Jensen–Shannon distance (JSD).*

$$JS(F,G) = \frac{1}{2}KL\left(F\|\frac{F+G}{2}\right) + \frac{1}{2}KL\left(G\|\frac{F+G}{2}\right), \tag{A1}$$

where $KL(F\|G)$ is the Kullback–Leibler (KL) divergence defined by

$$KL(F\|G) = \int f(x) \log\left(\frac{f(x)}{g(x)}\right)dx, \tag{A2}$$

where both $f(x)$ and $g(x)$ are absolutely continuous with $F(A) = \int_A f(x)dx$ and $G(A) = \int_A g(x)dx$ [18].

JSD is based on the KL divergence considering symmetric property which KL divergence does not have. Since JSD is bounded and symmetric [64], it is the first distance metric applied in GANs. However, it has the disadvantage of not being able to learn the data distribution in low dimension [15].

Definition A2. *The Earth-Mover distance (EMD or Wasserstein-1).*

$$EM(F,G) = \inf_{\gamma \in \Theta} \mathbb{E}_{(X,Y)\sim\gamma}[|X - Y|], \tag{A3}$$

where Θ is the set of all joint distributions whose marginal distributions of X and Y are F and G, respectively. The EMD considers the minimal cost for transformation of distributions.

Definition A3. *Maximum Mean Discrepancy (MMD)*

$$MMD(\psi,F,G) = \sup_{f\in\psi}\left(\mathbb{E}_{x\sim F}[f(x)] - \mathbb{E}_{y\sim G}[f(y)]\right), \tag{A4}$$

where $\psi = \{h : \chi \to \mathbb{R} : \chi$ is a metric space$\}$ and F and G are the distribution functions of x and y, $x, y \in \chi$, respectively [44].

Appendix B

For Example 1, the JSD, EMD, and SED are computed as follows:

Lemma A1.

$$(a)\,JS(F,G) = \begin{cases} \log 2, & \theta \neq 0, \\ 0, & \theta = 0. \end{cases} \tag{A5}$$

$$(b)\,EM(F,G) = |\theta| \tag{A6}$$

$$(c)\,SF(F,G,\alpha) = |\theta| \tag{A7}$$

Proof. (a) If $\theta = 0$, then F and G are the same distribution. Therefore, $\frac{F+G}{2} = F = G$ and this gives

$$JS(F,G) = \frac{1}{2}KL(F\|F) + \frac{1}{2}KL(F\|F) = KL(F\|F) = \int f(x) \log\left(\frac{f(x)}{f(x)}\right)dx = \int f(x)\cdot 0\,dx = 0 \tag{A8}$$

If $\theta \neq 0$, then

$$\begin{aligned}
JS(F,G) &= \tfrac{1}{2}KL\left(F\|\tfrac{F+G}{2}\right) + \tfrac{1}{2}KL\left(G\|\tfrac{F+G}{2}\right) \\
&= \tfrac{1}{2}\int f(x)\log\left(\tfrac{2f(x)}{f(x)+g(x)}\right)dx + \tfrac{1}{2}\int g(x)\log\left(\tfrac{2g(x)}{f(x)+g(x)}\right)dx \\
&= \tfrac{1}{2}\int f(x)\log\left(\tfrac{2f(x)}{f(x)}\right)dx + \tfrac{1}{2}\int g(x)\log\left(\tfrac{2g(x)}{g(x)}\right)dx \\
&= \tfrac{\log 2}{2}\int f(x)dx + \tfrac{\log 2}{2}\int g(x)dx \\
&= \log 2.
\end{aligned} \tag{A9}$$

(b) By Jensen's inequality, the following relation holds.

$$\inf_{\gamma\in\Theta}\left|\mathbb{E}_{(X,Y)\sim\gamma}[X - Y]\right| \leq \inf_{\gamma\in\Theta}\mathbb{E}_{(X,Y)\sim\gamma}[|X - Y|] \leq \inf_{\gamma\in\Theta}\sqrt{\mathbb{E}_{(X,Y)\sim\gamma}\|X - Y\|^2}. \tag{A10}$$

That is,

$$\inf_{\gamma\in\Theta}\left|\mathbb{E}_{(X,Y)\sim\gamma}[X] - \mathbb{E}_{(X,Y)\sim\gamma}[Y]\right| \leq EM(F,G) \leq \sqrt{Fr^2(F,G)} \tag{A11}$$

and this gives

$$\inf_{\gamma\in\Theta}|m_X - m_Y| = |\theta| \leq EM(F,G) \leq \sqrt{Fr^2(F,G)}. \tag{A12}$$

FD, the upper bound of Equation (A10), for this example is as follows:

$$Fr^2(F,G) = \theta^2 + \text{tr}\left[\left(\begin{array}{cc} 0 & 0 \\ 0 & 1 \end{array}\right) + \left(\begin{array}{cc} 0 & 0 \\ 0 & 1 \end{array}\right) - 2\left(\left(\begin{array}{cc} 0 & 0 \\ 0 & 1 \end{array}\right)\left(\begin{array}{cc} 0 & 0 \\ 0 & 1 \end{array}\right)\right)^{\frac{1}{2}}\right] = \theta^2. \tag{A13}$$

Therefore, $EM(F,G)$ is given by

$$EM(F,G) = |\theta|. \tag{A14}$$

$$(c)SF^2(F,G,\alpha) = \theta^2 + \frac{1}{\alpha}\cdot 0^2 = \theta^2. \tag{A15}$$

□

The JSD, EMD, and SFD for Example 2 are computed as follows.

Lemma A2.

$$(a)JS(F,G) = \frac{1}{2}(\log\frac{\delta^2+1}{2\delta}), \tag{A16}$$

$$(b)\ 0 \leq EM(F,G) \leq |\delta - 1|, \tag{A17}$$

$$(c)SF(F,G,\alpha) = \frac{1}{\sqrt{\alpha}}|\delta - 1|. \tag{A18}$$

Proof. (a) Since F and G are belonging to the same mixture family, $JS(F,G)$ can be expressed as a Jensen–Bregman divergence [65]. Therefore, it can be written as:

$$JS(F,G) = \frac{1}{2}\left(\frac{1}{2}m_X^T C_X^{-1} m_X + \frac{1}{2}m_Y^T C_Y^{-1} m_Y - m_{1/2}^T C_{1/2}^{-1} m_{1/2} + \log\frac{|C_X|^{1/2}|C_Y|^{1/2}}{|C_{1/2}|}\right), \tag{A19}$$

where $C_{1/2} = \left(\frac{1}{2}C_X^{-1} + \frac{1}{2}C_Y^{-1}\right)^{-1}$ and $m_{1/2} = C_{1/2}\left(\frac{1}{2}C_X^{-1}m_X + \frac{1}{2}C_Y^{-1}m_Y\right)$. Substituting the mean and covariance of X and Y respectively, $C_{1/2}$ and $m_{1/2}$ are simply written as

$$C_{1/2} = \left(\frac{1}{2}1 + \frac{1}{2}\delta^{-2}\right)^{-1} = \frac{2\delta^2}{\delta^2+1} \text{ and } m_{1/2} = C_{1/2}\left(\frac{1}{2}\cdot 1 \cdot 0 + \frac{1}{2}\cdot\delta^{-2}\cdot 0\right) = \frac{2\delta^2}{\delta^2+1}\cdot 0 = 0, \tag{A20}$$

respectively. Therefore, the JSD between F and G is computed as

$$JS(F,G) = \frac{1}{2}\left(\frac{1}{2}0\cdot 1\cdot 0 + \frac{1}{2}0\cdot\delta^{-2}\cdot 0 - 0\cdot\frac{\delta^2+1}{2\delta^2}\cdot 0 + \log\frac{1^{1/2}|\delta^2|^{1/2}}{\left|\frac{2\delta^2}{\delta^2+1}\right|}\right) == \frac{1}{2}\left(\log\frac{\delta^2+1}{2\delta}\right) \tag{A21}$$

(b) By Equation (A10), we obtain

$$\inf_{\gamma\in\Theta}|m_X - m_Y| = 0 \leq EM(F,G) \leq \sqrt{Fr^2(F,G)}. \tag{A22}$$

In this example, $m_X = m_Y = 0$, $C_X = 1^2$, and $C_Y = \delta^2$. Therefore, FD, the upper bound of Equation (A22), is as follows:

$$Fr^2(F,G) = 1^2 + \delta^2 - 2(1^2\delta^2)^{\frac{1}{2}} = (\delta - 1)^2. \tag{A23}$$

Finally, $EM(F,G)$ is given by

$$\inf_{\gamma\in\Theta}\|0 - 0\| = 0 \leq EM(F,G) \leq |\delta - 1|. \tag{A24}$$

(c) In this example, $m_X = m_Y = 0$, $\sigma_X = 1^2$, and $\sigma_Y = \delta^2$, therefore, $SF^2(F, G, \alpha)$ is computed as

$$SF^2(F, G, \alpha) = \|m_X - m_Y\|^2 + \frac{1}{\alpha}\|\sigma_X - \sigma_Y\|^2 = \|0 - 0\|^2 + \frac{1}{\alpha}\|1 - \delta\|^2 \tag{A25}$$

and this gives the result. □

References

1. Borji, A. Pros and Cons of GAN evaluation measures. *arXiv* **2018**, arXiv:1802.03446. [CrossRef]
2. Ganin, Y.; Ustinova, E.; Ajakan, H.; Germain, P.; Larochelle, H.; Laviolette, F.; Marchand, M.; Lempitsky, V. Domain-adversarial training of neural networks. *J. Mach. Learn. Res.* **2016**, *17*, 1–35.
3. Tzeng, E.; Hoffman, J.; Saenko, K.; Darrell, T. Adversarial Discriminative Domain Adaptation. In Proceedings of the IEEE Conference on Computer Vision and Pattern Recognition, Honolulu, HI, USA, 21–26 July 2017.
4. Reed, S.; Akata, Z.; Yan, X.; Logeswaran, L.; Schiele, B.; Lee, H. Generative adversarial text to image synthesis. *arXiv* **2016**, arXiv:1605.05396.
5. Ledig, C.; Theis, L.; Huszár, F.; Caballero, J.; Cunningham, A.; Acosta, A.; Aitken, A.; Tejani, A.; Totz, J.; Wang, Z. Photo-Realistic Single Image Super-Resolution Using a Generative Adversarial Network. In Proceedings of the IEEE Conference on Computer Vision and Pattern Recognition, Honolulu, HI, USA, 21–26 July 2017.
6. Gatys, L.A.; Ecker, A.S.; Bethge, M. Image Style Transfer Using Convolutional Neural Networks. In Proceedings of the IEEE Conference on Computer Vision and Pattern Recognition, Las Vegas, NV, USA, 27–30 June 2016; pp. 2414–2423.
7. Johnson, J.; Alahi, A.; Fei-Fei, L. Perceptual Losses for Real-Time Style Transfer and Super-Resolution. In Proceedings of the European Conference on Computer Vision, Amsterdam, The Netherlands, 8–16 October 2016; pp. 694–711.
8. Pathak, D.; Krahenbuhl, P.; Donahue, J.; Darrell, T.; Efros, A.A. Context Encoders: Feature Learning by Inpainting. In Proceedings of the IEEE Conference on Computer Vision and Pattern Recognition, Las Vegas, NV, USA, 27–30 June 2016.
9. Iizuka, S.; Simo-Serra, E.; Ishikawa, H. Globally and locally consistent image completion. *ACM Trans. Graph.* **2017**, *36*, 107. [CrossRef]
10. Vondrick, C.; Pirsiavash, H.; Torralba, A. Generating Videos with Scene Dynamics. In Proceedings of the Advances in Neural Information Processing Systems, Barcelona, Spain, 5–10 December 2016; pp. 613–621.
11. Goodfellow, I. NIPS 2016 tutorial: Generative adversarial networks. *arXiv* **2016**, arXiv:1701.00160.
12. Goodfellow, I.; Pouget-Abadie, J.; Mirza, M.; Xu, B.; Warde-Farley, D.; Ozair, S.; Courville, A.; Bengio, Y. Generative Adversarial Nets. In Proceedings of the Advances in Neural Information Processing Systems, Montreal, QC, Canada, 8–13 December 2014; pp. 2672–2680.
13. Metz, L.; Poole, B.; Pfau, D.; Sohl-Dickstein, J. Unrolled generative adversarial networks. *arXiv* **2016**, arXiv:1611.02163.
14. Yu, L.; Zhang, W.; Wang, J.; Yu, Y. SeqGAN: Sequence Generative Adversarial Nets with Policy Gradient. In Proceedings of the Thirty-First AAAI Conference on Artificial Intelligence, San Francisco, CA, USA, 4–9 February 2017; pp. 2852–2858.
15. Arjovsky, M.; Chintala, S.; Bottou, L. Wasserstein GAN. *arXiv* **2017**, arXiv:1701.07875.
16. Rubner, Y.; Tomasi, C.; Guibas, L.J. The earth mover's distance as a metric for image retrieval. *Int. J. Comput.* **2000**, *40*, 99–121.
17. Lin, J. Divergence measures based on the Shannon entropy. *IEEE Trans. Inf. Theory* **1991**, *37*, 145–151. [CrossRef]
18. Kullback, S.; Leibler, R.A. On information and sufficiency. *Ann. Math. Stat.* **1951**, *22*, 79–86. [CrossRef]
19. Fréchet, M. Sur la distance de deux lois de probabilité. *C. R. Hebd. S©Ances Acad. Sci.* **1957**, *244*, 689–692.
20. Heusel, M.; Ramsauer, H.; Unterthiner, T.; Nessler, B.; Klambauer, G.; Hochreiter, S. GANs trained by a two time-scale update rule converge to a Nash equilibrium. *arXiv* **2017**, arXiv:1706.08500.
21. Szegedy, C.; Vanhoucke, V.; Ioffe, S.; Shlens, J.; Wojna, Z. Rethinking the Inception Architecture for Computer Vision. In Proceedings of the IEEE Conference on Computer Vision and Pattern Recognition, Las Vegas, NV, USA, 27–30 June 2016; pp. 2818–2826.

22. Salimans, T.; Goodfellow, I.; Zaremba, W.; Cheung, V.; Radford, A.; Chen, X. Improved Techniques for Training GANs. In Proceedings of the Advances in Neural Information Processing Systems, Barcelona, Spain, 5–10 December 2016; pp. 2234–2242.

23. Fei-Fei, L. ImageNet: Crowdsourcing, benchmarking & other cool things. In Proceedings of the CMU VASC Seminar, PA, USA, 7 June 2010; pp. 18–25.

24. Karras, T.; Aila, T.; Laine, S.; Lehtinen, J. Progressive growing of GANs for improved quality, stability, and variation. *arXiv* **2017**, arXiv:1710.10196.

25. Karras, T.; Laine, S.; Aila, T. A Style-Based Generator Architecture for Generative Adversarial Networks. In Proceedings of the IEEE Conference on Computer Vision and Pattern Recognition, Long Beach, CA, USA, 16–20 June 2019; pp. 4401–4410.

26. Karras, T.; Laine, S.; Aittala, M.; Hellsten, J.; Lehtinen, J.; Aila, T. Analyzing and improving the image quality of style GAN. *arXiv* **2019**, arXiv:1912.04958.

27. Kim, C.; Jung, S.; Moon, J.; Hwang, E. Detecting mode drop and collapse in GANs using simplified frèchet distance. *J. KIISE* **2019**, *46*, 1012–1019. [CrossRef]

28. Lucic, M.; Kurach, K.; Michalski, M.; Gelly, S.; Bousquet, O. Are Gans Created Equal? A Large-Scale Study. In Proceedings of the Advances in Neural Information Processing Systems, Montréal, CA, USA, 3–8 December 2018; pp. 700–709.

29. Berthelot, D.; Schumm, T.; Metz, L. BEGAN: Boundary equilibrium generative adversarial networks. *arXiv* **2017**, arXiv:1703.10717.

30. Liu, Z.; Luo, P.; Wang, X.; Tang, X. Deep Learning Face Attributes in the Wild. In Proceedings of the IEEE International Conference on Computer Vision, Santiago, Chile, 7–13 December 2015; pp. 3730–3738.

31. Krizhevsky, A.; Hinton, G. Learning Multiple Layers of Features from Tiny Images. Available online: https://www.cs.toronto.edu/~{}kriz/learning-features-2009-TR.pdf (accessed on 8 April 2009).

32. Nagarajan, V.; Kolter, J.Z. Gradient Descent GAN Optimization is Locally Stable. In Proceedings of the Advances in Neural Information Processing Systems, Long Beach, CA, USA, 4–9 December 2017; pp. 5585–5595.

33. Mescheder, L.; Nowozin, S.; Geiger, A. The Numerics of GANs. In Proceedings of the Advances in Neural Information Processing Systems, Long Beach, CA, USA, 4–9 December 2017; pp. 1825–1835.

34. Radford, A.; Metz, L.; Chintala, S. Unsupervised representation learning with deep convolutional generative adversarial networks. *arXiv* **2015**, arXiv:1511.06434.

35. Miyato, T.; Kataoka, T.; Koyama, M.; Yoshida, Y. Spectral normalization for generative adversarial networks. *arXiv* **2018**, arXiv:1802.05957.

36. Lin, Z.; Khetan, A.; Fanti, G.; Oh, S. PacGAN: The power of two samples in generative adversarial networks. *arXiv* **2017**, arXiv:1712.04086.

37. Gulrajani, I.; Ahmed, F.; Arjovsky, M.; Dumoulin, V.; Courville, A.C. Improved Training of Wasserstein GANs. In Proceedings of the Advances in Neural Information Processing Systems, Long Beach, CA, USA, 4–9 December 2017; pp. 5767–5777.

38. Wang, R.; Cully, A.; Chang, H.J.; Demiris, Y. Magan: Margin adaptation for generative adversarial networks. *arXiv* **2017**, arXiv:1704.03817.

39. Zhao, J.; Mathieu, M.; LeCun, Y. Energy-based generative adversarial network. *arXiv* **2016**, arXiv:1609.03126.

40. Unterthiner, T.; Nessler, B.; Seward, C.; Klambauer, G.; Heusel, M.; Ramsauer, H.; Hochreiter, S. Coulomb GANs: Provably optimal nash equilibria via potential fields. *arXiv* **2017**, arXiv:1708.08819.

41. Zhang, H.; Xu, S.; Jiao, J.; Xie, P.; Salakhutdinov, R.; Xing, E.P. Stackelberg GAN: Towards Provable Minimax Equilibrium via Multi-Generator Architectures. *arXiv* **2018**, arXiv:1811.08010.

42. Gretton, A.; Borgwardt, K.M.; Rasch, M.J.; Schölkopf, B.; Smola, A. A kernel two-sample test. *J. Mach. Learn. Res.* **2012**, *13*, 723–773.

43. Li, C.-L.; Chang, W.-C.; Cheng, Y.; Yang, Y.; Póczos, B. Mmd Gan: Towards Deeper Understanding of Moment Matching Network. In Proceedings of the Advances in Neural Information Processing Systems, Long Beach, CA, USA, 4–9 December 2017; pp. 2203–2213.

44. Santurkar, S.; Schmidt, L.; Madry, A. A Classification-Based Study of Covariate Shift in GAN Distributions. In Proceedings of the International Conference on Machine Learning, Jinan, China, 26–28 May 2018; pp. 4487–4496.

45. Wilson, G.; Cook, D.J. A survey of unsupervised deep domain adaptation. *arXiv* **2019**, arXiv:1812.02849.

46. Srivastava, A.; Valkoz, L.; Russell, C.; Gutmann, M.U.; Sutton, C. VEEGAN: Reducing Mode Collapse in GANs using Implicit Variational Learning. In Proceedings of the Advances in Neural Information Processing Systems, Long Beach, CA, USA, 4–9 December 2017; pp. 3308–3318.
47. Chang, C.-C.; Hubert Lin, C.; Lee, C.-R.; Juan, D.-C.; Wei, W.; Chen, H.-T. Escaping From Collapsing Modes in a Constrained Space. In Proceedings of the European Conference on Computer Vision (ECCV), Munich, Germany, 8–14 September 2018; pp. 204–219.
48. Schelling, T.C. *The Strategy of Conflict*; Harvard University Press: Cambridge, London, UK, 1980.
49. He, K.; Zhang, X.; Ren, S.; Sun, J. Deep Residual Learning for Image Recognition. In Proceedings of the IEEE Conference on Computer Vision and Pattern Recognition, Las Vegas, NV, USA, 27–30 June 2016; pp. 770–778.
50. Wang, W.; Sun, Y.; Halgamuge, S. Improving MMD-GAN Training with Repulsive Loss Function. *arXiv* **2018**, arXiv:1812.09916.
51. Dowson, D.; Landau, B. The Fréchet distance between multivariate normal distributions. *J. Multivar. Anal.* **1982**, *12*, 450–455. [CrossRef]
52. LeCun, Y.; Bottou, L.; Bengio, Y.; Haffner, P. Gradient-based learning applied to document recognition. *Proc. IEEE* **1998**, *86*, 2278–2324. [CrossRef]
53. Xiao, H.; Rasul, K.; Vollgraf, R. Fashion-mnist: A novel image dataset for benchmarking machine learning algorithms. *arXiv* **2017**, arXiv:1708.07747.
54. Huang, X.; Belongie, S. Arbitrary Style Transfer in Real-Time with Adaptive Instance Normalization. In Proceedings of the IEEE International Conference on Computer Vision, Venice, Italy, 22–29 October 2017; pp. 1501–1510.
55. Ulyanov, D.; Vedaldi, A.; Lempitsky, V. Improved Texture Networks: Maximizing Quality and Diversity in Feed-Forward Stylization and Texture Synthesis. In Proceedings of the IEEE Conference on Computer Vision and Pattern Recognition, Honolulu, HI, USA, 21–26 July 2017; pp. 6924–6932.
56. Ghiasi, G.; Lee, H.; Kudlur, M.; Dumoulin, V.; Shlens, J. Exploring the structure of a real-time, arbitrary neural artistic stylization network. *arXiv* **2017**, arXiv:1705.06830.
57. Mittal, A.; Soundararajan, R.; Bovik, A.C. Making a "completely blind" image quality analyzer. *IEEE Signal Process. Lett.* **2012**, *20*, 209–212. [CrossRef]
58. Sharifi, K.; Leon-Garcia, A. Estimation of shape parameter for generalized Gaussian distributions in subband decompositions of video. *IEEE Trans. Circuits Syst. Video Technol.* **1995**, *5*, 52–56. [CrossRef]
59. Hardie, R.C.; Barnard, K.J.; Armstrong, E.E. Joint MAP registration and high resolution image estimation using a sequence of undersampled images. *IEEE Trans. Image Process.* **1997**, *6*, 1621–1633. [CrossRef]
60. Clevert, D.-A.; Unterthiner, T.; Hochreiter, S. Fast and Accurate Deep Network Learning by Exponential Linear Units (ELUs). *arXiv* **2015**, arXiv:1511.07289.
61. Abadi, M.; Barham, P.; Chen, J.; Chen, Z.; Davis, A.; Dean, J.; Devin, M.; Ghemawat, S.; Irving, G.; Isard, M. Tensorflow: A System for Large-Scale Machine Learning. In Proceedings of the 12th USENIX Symposium on Operating Systems Design and Implementation, Savannah, GA, USA, 2–4 November 2016; pp. 265–283.
62. Kingma, D.P.; Ba, J. Adam: A method for stochastic optimization. *arXiv* **2014**, arXiv:1412.6980.
63. Arora, S.; Zhang, Y. Do GANs actually learn the distribution? An empirical study. *arXiv* **2017**, arXiv:1706.08224.
64. Endres, D.M.; Schindelin, J.E. A new metric for probability distributions. *IEEE Trans. Inf. Theory* **2003**, *49*, 1858–1860. [CrossRef]
65. Nielsen, F. On a generalization of the Jensen-Shannon divergence and the JS-symmetrization of distances relying on abstract means. *arXiv* **2019**, arXiv:1904.04017.

Article

Color-Guided Depth Map Super-Resolution Using a Dual-Branch Multi-Scale Residual Network with Channel Interaction

Ruijin Chen [1,2] **and Wei Gao** [1,2,*]

[1] National Laboratory of Pattern Recognition (NLPR), Institute of Automation, Chinese Academy of Sciences, Beijing 100190, China; ruijin.chen@nlpr.ia.ac.cn

[2] School of Artificial Intelligence, University of Chinese Academy of Sciences, Beijing 100049, China

[*] Correspondence: wgao@nlpr.ia.ac.cn; Tel.: +86-10-8254-4618

Received: 24 February 2020; Accepted: 7 March 2020; Published: 11 March 2020

Abstract: We designed an end-to-end dual-branch residual network architecture that inputs a low-resolution (LR) depth map and a corresponding high-resolution (HR) color image separately into the two branches, and outputs an HR depth map through a multi-scale, channel-wise feature extraction, interaction, and upsampling. Each branch of this network contains several residual levels at different scales, and each level comprises multiple residual groups composed of several residual blocks. A short-skip connection in every residual block and a long-skip connection in each residual group or level allow for low-frequency information to be bypassed while the main network focuses on learning high-frequency information. High-frequency information learned by each residual block in the color image branch is input into the corresponding residual block in the depth map branch, and this kind of channel-wise feature supplement and fusion can not only help the depth map branch to alleviate blur in details like edges, but also introduce some depth artifacts to feature maps. To avoid the above introduced artifacts, the channel interaction fuses the feature maps using weights referring to the channel attention mechanism. The parallel multi-scale network architecture with channel interaction for feature guidance is the main contribution of our work and experiments show that our proposed method had a better performance in terms of accuracy compared with other methods.

Keywords: depth map; super-resolution; guidance; residual network; channel interaction

---

1. Introduction

With the development of 3D technologies, such as 3D reconstruction, robot interaction, and virtual reality, the acquisition of precise depth information as the basis of 3D technology has become very important. At present, depth maps can be obtained conveniently using low-cost depth cameras. However, depth maps obtained under such hardware constraints are usually of low resolution. To use low-cost depth maps in 3D tasks, we need to perform super-resolution (SR) processing on low-resolution (LR) depth maps to obtain high-resolution (HR) depth maps.

The main difficulty of depth map SR tasks is that the spatial downsampling of HR images to LR images will result in the loss and distortion of details, and this phenomenon will become more serious as the downscaling factor increases. When we want to recover HR images from LR images using simple upsampling, an edge blur and other detail distortion problems will appear. To cope with these problems, methods of using HR intensity images to guide the upsampling process of LR images have been proposed. The realization of these methods is based on the corresponding association relationship between HR intensity images and LR depth maps in the same scene. If the resolution of intensity image and target HR depth map are the same, edges of the intensity image and the target HR depth map can be regarded as basically corresponding, and therefore discontinuities in the intensity image can help to locate discontinuities in the target HR depth map during upsampling on the LR depth map. Although the introduction of intensity image guidance during the upsampling process will alleviate the blur of details like edges, extra textures may be introduced into the generated HR depth map owing to the inconsistency of the structure between the depth map and the intensity image.

We proposed an end-to-end, multi-scale deep map SR network, which consists of two branches, namely the RGB image branch (Y-branch) and the depth map branch (D-branch). Each branch is mainly composed of residual levels at multiple scales, and each residual level has two functional structures of feature extraction and upsampling. Among them, feature extraction is achieved by connecting several residual groups, each of which contains several residual blocks. As the key to residual structure, the internal short-skip connections of residual blocks and the long-skip connections in residual groups and levels enable the main road of branch network to learn the high-frequency information of the RGB image or depth map at different scales. Feature extraction parts in every residual level correspond one-to-one, which means that channel-wise, high-frequency features learned by each residual block of the Y-branch can be input into the corresponding residual block of the D-branch. On this foundation, we utilized a channel attention mechanism to rescale the channel-wise feature maps and fuse these features from two branches to implement guidance from the RGB image to the depth map. Under this kind of guidance, features in the HR depth map are supplemented, meanwhile weights in the aforementioned channel-wise feature rescaling limits the addition of artifacts from the RGB image. Compared with many existing methods, we input the LR depth map and HR RGB image directly into the network instead of inputting a bicubic interpolation of the LR depth map. Experiments indicate that our proposed method achieved great performances when recovering an HR depth map from an LR depth map with different upscaling factors.

The main contributions of our work are:

1. We designed a multi-scale residual network with two branches to realize an end-to-end LR depth map super-resolution under the guidance from an HR color image.
2. We applied a channel attention mechanism [1] to learn the features of a depth map and RGB image and fuse them via weights; furthermore, we tried to avoid copying artifacts to the depth map while ensuring the guidance from RGB image worked.
3. We discuss the detailed steps toward realizing image-wise upsampling and end-to-end training of this dual-branch, multi-scale residual network.

2. Related Works

There have been many methods proposed to complete the task of depth map SR reconstruction. Based on whether the method uses the guidance of an intensity image, the methods for depth map super-resolution can be divided into two categories, namely methods only based on depth maps and methods based on depth maps and intensity images.

Regarding methods based on depth maps, some methods are based on filters. The filter-based methods calculate the depth value of a pixel using its local information. Narayanan et al. [2] proposed a modified adaptive Wiener filter and a spatially adaptive signal-to-noise ratio estimate for reconstructing HR JPEG2000-compressed images. Lu et al. [3] used image segmentation and proposed a smoothing method to reconstruct the depth structure of each segmentation. Some methods are based on a dictionary that employs the relationship between each patch pair of LR and HR depth maps through sparse coding. Kwon et al. [4] defined an upscaling problem and introduced a scale-dependent dictionary. Xie et al. [5] proposed a framework that reconstructs a depth map's edge firstly and then reconstructs the HR depth map. These methods based on a dictionary usually require image block extraction and pre-processing operations that are difficult to implement for an end-to-end image super-resolution. In addition, it is hard to establish correct mapping between LR and HR image blocks in the dictionary. Some methods are based on a convolution neural network (CNN) and differ from dictionary-based methods by not explicitly learning a mapping dictionary. Dong et al. [6] proposed an SR reconstruction method called a super-resolution convolutional neural network (SRCNN) based on a

CNN, which uses three convolution layers to non-linearly map a LR feature space to a HR feature space. This network has a relatively simple structure and small receptive fields such that it can only learn a few features. Kim et al. [7] proposed a VDSR (Very Deep Super Resolution) network that has 20 layers and learns more features. VDSR pre-processes the input depth map using bicubic interpolation that affects the network's learning of the LR depth map's original information and introduces artifacts to the reconstructed HR depth map. Lai et al. [8] proposed a Laplacian pyramid SR network called LapSRN that gradually reconstructs the sub-band residuals of HR images and uses transposition convolution to generate HR images. The input of LapSRN is an LR image without bicubic interpolation such that artifacts can be avoided. However, checkerboard artifacts [9] will occur if network parameters, such as the kernel size, are set improperly.

Regarding methods based on depth maps and intensity images, some methods are based on filters. He et al. [10] enhanced an LR depth map by assuming a linear relationship between the patches of the image for guidance and the output depth map. Barron and Poole [11] proposed a fast bilateral solver that can be used for enhancing the depth map under the guidance from a color image. Some methods are based on optimization. In these methods, depth upsampling is defined as an optimization problem in which if a pixel's neighboring pixels have similar colors in the intensity image but different values in the depth map, then this pixel will be given a large loss value and the total loss of all pixels needs to be minimized. Diebel et al. [12] proposed a MRF (Markov Random Fields) formula containing a data term from an LR depth map and a smooth term from an HR intensity image. Park et al. [13] integrated edge, gradient, and segmentation from an HR color image to design the anisotropic affinities of the regularization terms. Ferstl et al. [14] used a secondary generalized variable guided by an anisotropic diffusion tensor extracted from an HR color image to limit a regularized HR depth map. Zuo et al. [15,16] measured the discontinuities of edges between a color image and a depth map in an MRF, and these discontinuities can be reflected in the edge weight of the minimum spanning tree. Yang et al. [17] proposed a novel depth map SR method guided by a color image by using an auto-regression model. All these optimization-based methods are based on the assumption that the edges of a color image and a depth map have consistency. However, textures in a color image may not have corresponding regions in a depth map, which will override the assumption of consistency and introduce artifacts to the reconstructed HR depth map. Some methods are based on a dictionary. Kiechle et al. [18] proposed a dual-mode co-sparse analysis model that reconstructs a depth map by capturing the interdependence between the intensity of a color image and the depth of a depth map. Some methods are based on a CNN. Riegler et al. [19] designed a kind of special end-to-end deep convolution neural network (DCNN) to learn data terms and regulation terms in an MRF that reconstructs an HR depth map. Zhou et al. [20] developed a new DCNN to jointly learn nonlinear projection equations when noise occurs. Yang et al. [21] learned joint features to obtain an HR depth map guided by the edge attention map extracted from an HR color image. Ye et al. [22] designed a kind of DCNN to learn the binary map of depth edge positions from an LR depth map under the guidance of a corresponding HR color image. These DCNNs introduce noise to the output HR depth map by inputting the interpolated LR depth map, which is ineffective for processing features in the high-frequency domain. Hui et al. [23] proposed a DCNN that accepts multi-scale guidance from an HR intensity image and mainly learns features in the high-frequency domain. Zuo et al. [24] proposed a data-driven approach based on a CNN with local residual learning introduced in each scale-dependent reconstruction sub-network and global residual learning is utilized to learn the difference between the upsampled depth map and the ground truth. Zuo et al. [25] proposed a DCNN to reconstruct the HR depth map guided by the intensity image, where dense connections and sub-networks recover the high-frequency details from coarse to fine. These DCNNs adopt a residual network or multi-scale upsampling mechanism like our proposed network but the ways in which the intensity image guides the process are different, which determines a difference in the severity of artifacts. Voynov et al. [26] tried to avoid artifacts for virtual reality applications and they measured the quality of a depth map upsampling using renderings of the resulting 3D surfaces.

In recent years, there have been a lot of remarkable works in single-image super-resolution (SISR) tasks, which have common ground with our depth map reconstruction task. Lim et al. [27] developed a multi-scale deep SR system that can reconstruct HR images of different upscaling factors in a single model. Zhang et al. [28] proposed a residual dense network that uses a residual dense block to extract local features with a contiguous memory mechanism and then learned global hierarchical features by fusing dense local features jointly and adaptively. Zhang et al. [1] proposed the very deep residual channel attention networks formed by residuals in a residual structure and a channel attention mechanism such that channel-wise features are treated differently. Liu et al. [29] proposed a kind of non-local module to capture deep feature correlations between each location and its neighborhood and employed the recurrent neural network structure for deep feature propagation. Qiu et al. [30] proposed an embedded block residual network where different modules restore the information of different frequencies for a texture SR. Hu et al. [31] proposed a channel-wise and spatial feature modulation network where LR features can be transformed to high informative features using feature-modulation memory modules. Jing et al. [32] took the LR image and its downsampled resolution (DR) and upsampled resolution (UR) versions as inputs and learned the internal structure coherence with the pairs of UR-LR and LR-DR to generate a hierarchical dictionary. In addition to SISR, multi-image super-resolution (MISR) has gained attention and there have already been some deep learning methods focusing on it. Haris el at. [33] proposed a recurrent backprojection network (RBPN) that integrates spatial and temporal contexts from continuous video frames using a recurrent encoder–decoder module that fuses multi-frame information with a single-frame SR method for the target frame. Molini et al. [34] proposed a CNN-based technique called DeepSUM to exploit spatial and temporal correlations for the SR of a remote sensing scene from multiple unregistered LR images. DeepSUM has three stages including shared 2D convolutions to extract high-dimensional features from the inputs, a subnetwork proposing registration filters, and 3D convolutions for the slow fusion of the features. DeepSUM++ [35] evolved from DeepSUM and shows that non-local information in a CNN can exploit self-similar patterns to provide the enhanced regularization of SR.

3. Proposed Dual-Branch Multi-Scale Residual Network with Channel Interaction

In this study, we supposed that an LR depth map D_l is obtained by downsampling its corresponding target HR depth map D_h and an HR RGB image Y_h of the same scene is available. Y_h and D_l of the same scene are the inputs of our network, and the goal is to reconstruct and output D_h end to end at an upscaling factor s.

In the following, we take $s = 8$ as an example to show our network structure (see Figure 1).

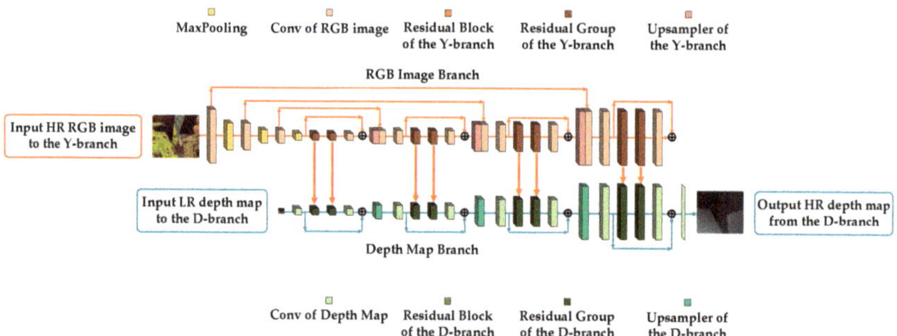

Figure 1. The architecture of our network for 8× upsampling. HR: High-resolution, LR: Low-resolution.

3.1. RGB Image Network Branch

The main role of the RGB image network branch is to provide guidance for the feature map extraction of the deep map network branch. In general, the structure of the Y-branch can be divided into three functional parts. The first part is to downscale the input RGB image by a factor of 2 through a convolution layer and a maxpooling layer for $m = \log(s)$ times until the resolution of the feature maps is the same as the input depth map (see Figure 1). Since the sample network has an upscaling factor of 8, such a downsampling operation is executed three times in total. The feature maps obtained in the first part can be expressed as follows:

$$F^Y_{DW(1)} = \mathbf{W}^Y_{DW(1)} * Y_h + \mathbf{b}^Y_{DW(1)} \tag{1}$$

$$F^Y_{DW(i)} = \mathbf{W}^Y_{DW(i)} * F^Y_{DW(i-1)} + \mathbf{b}^Y_{DW(i)} \tag{2}$$

$$F^Y_{DW(2i')} = \mathrm{MaxPool}\left(F^Y_{DW(2i'-1)}\right) \tag{3}$$

where $i = \{3, 5, \dots, 2m - 1\}$, $i' = \{1, 2, \dots, m\}$. The operator $*$ represents convolution and $\mathbf{W}^Y_{DW}$ is a kernel of size 3×3 and $\mathbf{b}^Y_{DW}$ is a bias vector. The superscript Y means that features or blobs belong to the Y-branch and subscript DW stands for the whole downscaling part.

The second part is the parallel network structure matching with the D-branch, which includes a nested structure of residual blocks, groups, and levels. As the most basic constituent unit in the network structure, the residual block of the Y-branch matches with the residual block at the same location in the D-branch. Despite this one-to-one relationship, the residual block for feature extraction in the Y-branch consists of two convolution layers and one PReLU (Parametric Rectified Linear Unit) layer, which is simpler relative to that in the D-branch. After the second convolution operation in the block, the generated feature maps are input into the matched residual block in the D-branch and concatenate feature maps of the depth map guided from the RGB image. In addition, the input feature maps of each residual block are added to the feature maps obtained after feature extraction, which is called a short-skip connection inside the block. Based on the residual block, a residual group is composed of several connected residual blocks and one convolution layer. Similar to a short-skip connection, a long-skip connection is implemented by adding the input and output of each residual group. In the same way, several residual groups and one convolution layer are connected to constitute a residual level and a long-skip connection is also realized in each level using the same addition of input and output. Figure 2 shows the structure of a residual block and a residual group in the Y-branch. The feature maps generated by each residual level l can be expressed as follows:

$$F^Y_{DF(1)} = H^Y_{DF(1)}\left(F^Y_{DW(2m)}\right) \tag{4}$$

$$F^Y_{DF(l)} = H^Y_{DF(1)}\left(F^Y_{UP(l-1)}\right) \tag{5}$$

where $l = \{2, 3, \dots, m + 1\}$. $H^Y_{DF}(\cdot)$ donates the deep feature extraction and F^Y_{UP} represents the feature maps from the third part of the Y-branch. In each residual level l, the feature maps generated by each group g can be expressed as follows:

$$F^Y_{l,1} = H^Y_{l,1}\left(F^Y_{l,0}\right) \tag{6}$$

$$F^Y_{l,g} = H^Y_{l,g}\left(F^Y_{l,g-1}\right) \tag{7}$$

$$F^Y_{DF(l)} = F^Y_{l,0} + \mathbf{W}^Y_l F^Y_{l,G} \tag{8}$$

where $g = \{2, 3, \dots, G\}$, and G is the number of residual groups in a level. $F^Y_{l,0}$ is the input of the residual level. $H^Y_{l,g}(\cdot)$ donates the function of the gth residual group. $F^Y_{l,g-1}$ and $F^Y_{l,g}$ are the input and output of gth residual group, respectively. $\mathbf{W}^Y_l$ is the weight set of the tail convolution layer. In each residual group g, the feature maps generated by each residual block b can be expressed as follows:

$$F^Y_{g,1} = H^Y_{g,1}\left(F^Y_{g-1}\right) \tag{9}$$

$$F_{g,b}^Y = H_{g,b}^Y\left(F_{g,b-1}^Y\right) \tag{10}$$

$$F_g^Y = F_{g-1}^Y + \mathbf{W}_g^Y F_{g,B}^Y \tag{11}$$

where $b = \{2, 3, \ldots, B\}$, and B is the number of residual blocks in a group. F_{g-1}^Y and F_g^Y are the input and output of gth group, respectively. $H_{g,b}^Y(\cdot)$ donates the function of the bth residual block. $F_{g,b-1}^Y$ and $F_{g,b}^Y$ are the input and output of the bth residual block, respectively. $\mathbf{W}_g^Y$ is the weight set of the tail convolution layer. In each residual block b, the basic operations can be expressed as follows:

$$h\left(F_b^Y\right) = \mathbf{W}_{b,2}^Y * \left(\sigma\left(\mathbf{W}_{b,1}^Y * F_{b-1}^Y + \mathbf{b}_{b,1}^Y\right)\right) + \mathbf{b}_{b,2}^Y \tag{12}$$

$$F_b^Y = F_{b-1}^Y + h\left(F_b^Y\right) \tag{13}$$

where $h(\cdot)$ denotes the high-frequency feature maps of the input. $\sigma(\cdot)$ donates the activation function PReLU. F_{b-1}^Y and F_b^Y are the input and output of the bth residual block, respectively. $\mathbf{W}_{b,1}^Y$ and $\mathbf{W}_{b,2}^Y$ are kernels of size 3×3, and $\mathbf{b}_{b,1}^Y$ and $\mathbf{b}_{b,2}^Y$ are the bias vectors.

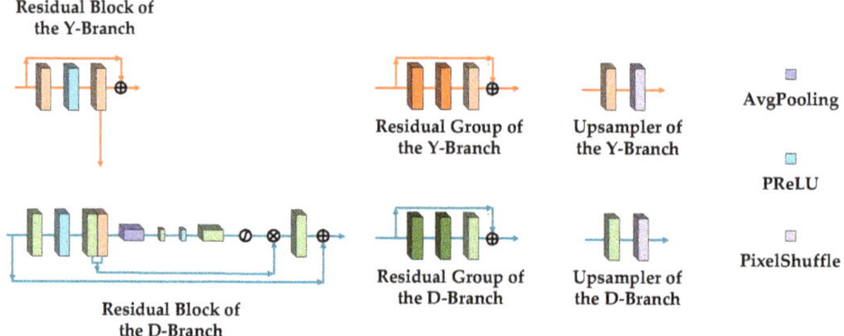

Figure 2. The structure of residual block, residual group and upsampler.

The third part of the Y-branch is the resolution enlarging level. This part consists of an upsampler and a convolution layer, and all these layers are connected after the residual level. The upsampler here is composed of a convolution layer and a pixel-shuffling layer. Corresponding to the initial downscaling steps, feature maps are upscaled by a factor of 2 after each residual level and resolution enlarging level. Furthermore, the feature maps from the first part concatenate the feature maps that have the same resolution after upsampling, and then perform a convolution operation (see Figure 2). This design means the upsampled feature maps become supplemented by feature maps with an original high resolution from the first part such that more structured features at different scales can be retained in the network for the processing that follows, meaning that enough guidance is provided to the D-branch. The feature maps generated by the third part can be expressed as follows:

$$F_{UP(l')}^Y = \mathbf{W}_{l',2}^Y * \left(\text{PixelShuffle}\left(\mathbf{W}_{l',1}^Y * F_{DF(l')}^Y + \mathbf{b}_{l',1}^Y\right), F_{DW(2m-2l'+1)}^Y\right) + \mathbf{b}_{l',2}^Y \tag{14}$$

where $l' = \{1, 2, \ldots, m\}$. $\mathbf{W}_{l',1}^Y$ and $\mathbf{W}_{l',2}^Y$ are kernels of size 3×3, and $\mathbf{b}_{l',1}^Y$ and $\mathbf{b}_{l',2}^Y$ are the bias vectors.

Referring to Shi et al. [36], the pixel-shuffling layer rearranges the elements of a $H \times W \times C \cdot r^2$ blob B to a blob of shape $rH \times rW \times C$, where r is the upscaling factor and $H \times W$ is the size of C feature maps. Mathematically, the pixel-shuffling operation can be described as follows:

$$\text{PixelShuffle}(B)_{x,y,c} = B_{x/r,y/r,C \cdot r \cdot \text{mod}(y,r) + C \cdot \text{mod}(x,r) + c} \tag{15}$$

where x and y are the output pixel coordinates of the cth feature map in HR space. The feature maps from the LR space are built into HR feature maps through the pixel-shuffling layer.

3.2. Depth Map Network Branch

The task of the depth map network branch is to complete the super-resolution of an LR depth map under guidance from the parallel Y-branch. Compared to the Y-branch, due to the low resolution of the input depth map, the D-branch is mainly composed of two parts, the residual levels and the resolution enlarging levels, without the downscaling part. Except for this difference in architecture, the nested structure of the residual blocks, groups, and the short- or long-skip connections in the D-branch still exist as in the Y-branch. However, the composition of the residual block that contains convolution layers, PReLU layers, and average-pooling layer in the D-branch is more complicated than that in the Y-branch. The whole feature extraction procedure of this kind of residual block is explained as follows. The input feature maps are processed using convolution, PReLU, and convolution first, and then the feature maps from the Y-branch are concatenated. After the subsequent average pooling, convolution, PReLU, convolution again, and applying the sigmoid function, the weights are generated and multiplied by the previous concatenated feature maps to generate new feature maps that not only integrate the structure information coming from the RGB image, but also prevent unreasonable textures from appearing. In addition to these internal structures, the short-skip connection still exists and adds the input and the output of each residual block. Figure 2 shows the structure of the residual block and residual group in the D-branch. The feature maps generated by each residual level l can be expressed as follows:

$$F^D_{DF(1)} = H^D_{DF(1)}\left(\mathbf{W}^D_0 * D_l + \mathbf{b}^D_0\right) \tag{16}$$

$$F^D_{DF(l)} = H^D_{DF(l)}\left(F^D_{UP(l-1)}\right) \tag{17}$$

where $l = \{2, 3, \dots, m + 1\}$. The superscript D means that features or blobs belong to the D-branch. $\mathbf{W}^D_0$ and $\mathbf{b}^D_0$ are a kernel of 3×3 and a bias vector to the head convolution layer for initial feature extraction, respectively. $H^D_{DF}(\cdot)$ denotes the deep feature extraction and F^D_{UP} represents the feature maps from the second part of the D-branch. In each residual level l, the feature maps generated by each group g can be expressed as follows:

$$F^D_{l,1} = H^D_{l,1}\left(F^D_{l,0}\right) \tag{18}$$

$$F^D_{l,g} = H^D_{l,g}\left(F^D_{l,g-1}\right) \tag{19}$$

$$F^D_{DF(l)} = F^D_{l,0} + \mathbf{W}^D_l F^D_{l,G} \tag{20}$$

where $g = \{2, 3, \dots, G\}$, and G is the number of residual groups in a level. $F^D_{l,0}$ is the input of the residual level. $H^D_{l,g}(\cdot)$ denotes the function of the gth residual group. $F^D_{l,g-1}$ and $F^D_{l,g}$ are the input and output of the gth residual group, respectively. $\mathbf{W}^D_l$ is the weight set of the tail convolution layer. In each residual group g, the feature maps generated by each residual block b can be expressed as follows:

$$F^D_{g,1} = H^D_{g,1}\left(F^D_{g-1}\right) \tag{21}$$

$$F^D_{g,b} = H^D_{g,b}\left(F^D_{g,b-1}\right) \tag{22}$$

$$F^D_g = F^D_{g-1} + \mathbf{W}^D_g F^D_{g,B} \tag{23}$$

where $b = \{2, 3, \dots, B\}$, and B is the number of residual blocks in a group. F^D_{g-1} and F^D_g are the input and output of the gth group, respectively. $H^D_{g,b}(\cdot)$ denotes the function of the bth residual block. $F^D_{g,b-1}$ and $F^D_{g,b}$ are the input and output of the bth residual block, respectively. $\mathbf{W}^D_g$ is the weight set of the tail convolution layer. In each residual block b, the basic operations can be expressed as follows:

$$h\left(F^D_b\right) = \mathbf{W}^D_{b,2} * \left(\sigma\left(\mathbf{W}^D_{b,1} * F^D_{b-1} + \mathbf{b}^D_{b,1}\right)\right) + \mathbf{b}^D_{b,2} \tag{24}$$

$$F_b^D = F_{b-1}^D + R_b^D\left(h\left(F_b^D\right), h\left(F_b^Y\right)\right) \cdot \left(h\left(F_b^D\right), h\left(F_b^Y\right)\right) \qquad (25)$$

where $h(\cdot)$ denotes the high-frequency feature maps of the input. $\sigma(\cdot)$ denotes the activation function PReLU. F_{b-1}^D and F_b^D are the input and output of the bth residual block, respectively. $\mathbf{W}_{b,1}^D$ and $\mathbf{W}_{b,2}^D$ are kernels of size 3×3, and $\mathbf{b}_{b,1}^D$ and $\mathbf{b}_{b,2}^D$ are the bias vectors. $R_b^D(\cdot)$ denotes the function of the channel interaction.

Except for the difference in the residual block, the D-branch directly employs the upsampler and the convolution layer as a resolution enlarging level to upscale the feature maps without concatenating feature maps from the branch itself due to the lack of a downscaling part. The resolution enlarging level is arranged to be connected after the residual level, which is one of the steps used to gradually achieve super-resolution. Finally, a convolution layer is connected after the last residual layer to convert the feature maps into a depth map to generate a target HR depth map as the whole dual-branch network's output (see Figure 2). The feature maps generated by the second part can be expressed as follows:

$$F_{UP(l')}^D = \mathbf{W}_{l',2}^D * \text{PixelShuffle}\left(\mathbf{W}_{l',1}^D * F_{DF(l')}^D + \mathbf{b}_{l',1}^D\right) + \mathbf{b}_{l',2}^D \qquad (26)$$

where $l' = \{1, 2, \ldots, m\}$. $\mathbf{W}_{l',1}^D$ and $\mathbf{W}_{l',2}^D$ are kernels of size 3×3, and $\mathbf{b}_{l',1}^D$ and $\mathbf{b}_{l',2}^D$ are the bias vectors.

At the end of our network is a convolution layer that reconstructs feature maps into an output HR depth map $\widetilde{D}_h$ as follows:

$$\widetilde{D}_h = \mathbf{W}_{REC}^D * F_{DF(m+1)}^D + \mathbf{b}_{REC}^D \qquad (27)$$

where $\mathbf{W}_{REC}^D$ is a kernel of size 3×3, and $\mathbf{b}_{REC}^D$ is the bias vector.

Our network is optimized with a loss function L_1. Given a training set $\left\{Y_{h'}^i, D_{l'}^i, D_h^i\right\}_{i=1}^N$, which contains N HR RGB images and LR depth maps as inputs, along with their HR depth map counterparts, our network is trained by minimizing the L_1 loss function

$$L(\Theta) = \frac{1}{N} \sum_{i=1}^N \|\widetilde{D}_h^i - D_h^i\|_1 \qquad (28)$$

where Θ denotes the parameter set of our network. This L_1 loss function is optimized using a stochastic gradient descent.

3.3. Channel Interaction

Channel attention is a channel-wise feature interaction and change mechanism proposed by Zhang et al. [1], whose goal is to allow the network to pay more attention to features that contain more information. This mechanism originates from two points. One is that there are abundant low-frequency and valuable high-frequency components in LR space. The low-frequency components are mostly flat, and the high-frequency components are mostly regions full of details, such as edges and textures. Another is that each filter of the convolution layer has a local receptive field such that convolution fails to use contextual information outside the local region. In response to these two points, the channel attention mechanism uses global average pooling to obtain channel-wise global spatial information and employs a gating mechanism to capture the dependencies between channels. This gating mechanism can not only learn nonlinear interactions, but also avoids mutual exclusion between channel-wise features. The coefficient factors learned by the gating mechanism are the weights for rescaling the channels. The channel attention mechanism operates between the channel-wise features learned from the input image. We further extended this mechanism to the guidance from the RGB image to the depth map, which makes the features learned by dual-network branches interact with each other.

There are two types of channel interactions in our network. The first one is the concatenation of the feature maps before downscaling and after upsampling in the Y-branch, and then executing the convolution operation for new channel-wise feature maps. This is a relatively common channel-wise interaction procedure, which guarantees that the feature maps of all the channels affect each other equally. The reason for adopting this kind of equal channel interaction is that due to the beginning downscaling part, the loss of details in the previous residual level needs to be supplemented for feature extraction and network learning of the next residual level at a larger scale. Furthermore, the supplemented feature maps also help the guidance provided for the D-branch. The second way channel interaction occurs is through the weight of each channel, which is calculated through a series of functions and decides the influence of its channel in the process of generating new feature maps after the feature maps of each residual block in the D-branch concatenates the feature maps from the Y-branch. The guidance from the Y-branch to the D-branch is realized in this way for the channels from the Y-branch, which can affect all the channels in the residual block. However, each channel from the Y-branch has an unequal influence and interacts with each other according to different weights such that the structured features that have a corresponding relationship between the RGB image and depth map are emphasized and the inconsistent features without such a relationship suppressed. Small weights limit the appearance of artifacts introduced by the feature maps from the Y-branch.

As $R_b^D(\cdot)$ denotes the entire operation of channel interaction, we suppose that $X = \left[x_1^Y, \ldots, x_c^Y, \ldots, x_C^Y, x_1^D, \ldots, x_c^D, \ldots, x_C^D\right]$ is an input, which has C feature maps with a size of $H \times W$ from the Yth and Dth branches separately. The channel-wise statistic $z \in \mathcal{R}^{2C}$ can be obtained by shrinking X, and the cth element of z is:

$$z_c = \text{AveragePool}(x_c) = \frac{1}{H \times W} \sum_{h=1}^{H} \sum_{w=1}^{W} x_c(h, w) \tag{29}$$

where $x_c(h, w)$ is the value at position (h, w) of the cth feature x_c from either the Yth or Dth branch. Therefore, we obtain the weight coefficient using the function:

$$s = f\left(\mathbf{W}_U^D \sigma\left(\mathbf{W}_D^D z\right)\right) \tag{30}$$

where $f(\cdot)$ and $\sigma(\cdot)$ denote the sigmoid and PReLU functions, respectively. $\mathbf{W}_D^D$ is the weight set of a convolution layer that downscales channels with a reduction ratio r. In our experiments, r was set to 16. $\mathbf{W}_U^D$ is also a weight set of a convolution layer that upscales channels with the same ratio r. Then, we can rescale x_c by:

$$\hat{x}_c = s_c \cdot x_c \tag{31}$$

4. Evaluation

4.1. Network Training

The data set for experiments in this paper was the same as in Hui et al. [23], which consisted of 58 RGBD images from the MPI (Max-Planck Institute) Sintel depth dataset and 34 RGBD images from the Middlebury dataset. Among them, a total of 82 RGBD images made up the training set for our network training, and the other 10 images composed the test set for validation. Our experiments included SR reconstruction of an LR depth map with upscaling factors of 2, 3, 4, 8, and 16 separately. Considering that a factor of 2 was the initial base, we first trained a network with an upscaling factor of 2 whose Y-branch was pre-trained using 1000 images from the NYUv2 (New York University Version 2) dataset [37]; then, the entire network was trained using these 1000 RGB images and depth maps, and finally, the aforementioned training dataset containing 82 RGBD images were used for network fine-tuning. Based on the trained network with an upscaling factor of 2, other networks with upscaling factors of 3, 4, 8, and 16 were further fine-tuned using the same 82 RGBD images.

In terms of the details of training, we gathered LR depth maps to form the training dataset at different upscaling factors by downscaling the corresponding HR depth maps through bicubic interpolation. In the process of training, we did not input large-size images or depth maps into our network directly, but split each one into small overlapping patches and did some common data enhancement before a patch entered the network. The size of these patches was set according to the upscaling factor. The upscaling factors were {2,3,4,8,16}, the corresponding size of the input depth map's patch were $\{48^2, 48^2, 48^2, 24^2, 12^2\}$, and the sizes of the input RGB image's patch were $\{96^2, 144^2, 192^2, 192^2, 192^2\}$. Furthermore, the other settings of the network training included the choice of the loss function, optimizer, learning rate, etc. We chose L_1 as the loss function, used the ADAM optimizer where $P_1 = 0.8$, $P_2 = 0.999$, $\varepsilon = 10^{-8}$ and the initial learning rate was set to 10^{-4}. The learning rate was halved after every 200 epochs. We trained all these network models using PyTorch on a GTX 1080 GPU.

4.2. Evaluation on the Middlebury Dataset

In order to compare our method with the experimental results of other studies, we used the root mean squared error (RMSE) as an evaluation criterion. Referring to Hui et al. [23], we evaluated our algorithm using Middlebury RGBD datasets whose holes were filled. The dataset was divided into three sets, namely *A*, *B*, and *C*. Data in the table came from References [2,3,6,10,12–14,16–18,23–25]. At each upscaling factor, the best RMSE result of all the algorithms listed in the table is in bold and the sub-optimal result is underlined. For dataset *C*, the comparison was only performed until the upscaling factor increased to 8 because the resolution of the input depth map was too low to reconstruct the HR depth map when the upscaling factor was 16. In addition, the experimental results at the upscaling factor of 3 were not put into the three tables because the other algorithms cannot reconstruct depth maps at a factor that is not a power of 2.

Tables 1–3 are records of the evaluation on sets *A*, *B*, and *C* separately, and our algorithm showed an excellent performance compared with the others. When the upscaling factor was small, the gap between the algorithms was not huge, but the advantage of our method was obvious with after increasing the upscaling factor. This phenomenon shows that it is feasible to use an HR RGB image to guide an LR depth map super-resolution in a multi-scaled way if the LR depth map has poor quality and lacks high-frequency information. This condition is a challenge to all the image SR methods. Since References [23,24] adopt a multi-scale mechanism and References [24,25] are built on a residual structure, we focused on the comparison of the experiment results between theirs and ours. According to Table 1, the average RMSE of our network on dataset *A* at the upscaling factors of {2, 4, 8, 16} were {0.37, 0.78, 1.27, 1.89}, which outperformed Hui et al. [23] with gains of {0.09 (+19.6%), 0.15 (+16.1%), 0.23 (+15.3%), 0.71 (+27.3%)}, outperformed Zuo et al. [24] with gains of {0.15 (+28.8%), 0.22 (+22.0%), 0.35 (+21.6%), 0.73 (+27.9%)} and outperformed Zuo et al. [25] with gains of {0.06 (+14.0%), 0.15 (+16.1%), 0.28 (+18.1%), 0.61 (+24.4%)}. On dataset *B*, our network outperformed Hui et al. [23] with gains of {0.07 (18.4%), 0.13 (+15.9%), 0.32 (+22.2%), 0.75 (+31.5%)}, outperformed Zuo et al. [24] with gains of {0.31 (+50%), 0.39 (+36.1%), 0.56 (+33.3%), 1.2 (+42.4%)}, and outperformed Zuo et al. [25] with gains of {0.21 (+40.4%), 0.31 (+31%), 0.51 (+31.3%), 1.09 (+40.1%)}. On dataset *C*, our network outperformed Hui et al. [23] with gains of {0.35 (+38.9%), 0.53 (+24.3%), 0.96 (23.3%)} at the upscaling factors of {2, 4, 8}. Overall, our network substantially reduced the RMSE using these three datasets in the mean sense compared with other methods. Although our network only had sub-optimal results in several cases, such as for Venus in dataset *C*, it is still reasonable to infer that special optimization may be required for some isolated samples.

Table 1. Quantitative comparison (in RMSE) on dataset *A*.

Method Used	Art				Books				Moebius			
	2x	4x	8x	16x	2x	4x	8x	16x	2x	4x	8x	16x
Bilinear	2.83	4.15	6.00	8.93	1.12	1.67	2.39	3.53	1.02	1.50	2.20	3.18
Narayanan [2]	2.76	3.10	3.51	–	1.17	1.24	1.82	–	0.99	1.03	1.76	–
MRFs [12]	3.12	3.79	5.50	8.66	1.21	1.55	2.21	3.40	1.19	1.44	2.05	3.08
Park et al. [13]	2.83	3.50	4.17	6.26	1.09	1.53	1.99	2.76	1.06	1.35	1.80	2.38
Guided [10]	2.93	3.79	4.97	7.88	1.16	1.57	2.10	3.19	1.10	1.43	1.88	2.85
Kiechle et al. [18]	1.25	2.01	3.23	5.77	0.65	0.92	1.27	1.93	0.64	0.89	1.27	2.13
Ferstl et al. [14]	3.03	3.79	4.79	7.10	1.29	1.60	1.99	2.94	1.13	1.46	1.91	2.63
Lu et al. [3]	–	–	5.80	7.65	–	–	2.73	3.55	–	–	2.42	3.12
SRCNN [6]	1.13	2.02	3.83	7.27	0.52	0.94	1.73	3.10	0.54	0.91	1.58	2.69
MSF [16]	3.01	3.70	4.66	6.68	1.25	1.63	2.02	2.84	1.13	1.51	2.06	2.93
Hui et al. [23]	0.66	1.47	2.46	4.57	0.37	0.67	1.03	1.60	0.36	0.66	1.02	1.63
MFR-SR [24]	0.71	1.54	2.71	4.35	0.42	0.63	1.05	1.78	0.42	0.72	1.10	1.73
RDN-GDE [25]	0.56	1.47	2.60	4.16	0.36	0.62	1.00	1.68	0.38	0.69	1.05	1.65
Ours	**0.44**	**1.17**	**1.96**	**3.24**	**0.35**	**0.60**	**0.96**	**1.24**	**0.32**	**0.58**	**0.89**	**1.18**

Table 2. Quantitative comparison (in RMSE) on dataset *B*.

Method Used	Dolls				Laundry				Reindeer			
	2x	4x	8x	16x	2x	4x	8x	16x	2x	4x	8x	16x
Bicubic	0.91	1.31	1.86	2.63	1.61	2.41	3.45	5.10	1.94	2.81	3.99	5.82
Narayanan [2]	0.84	1.25	1.69	–	1.34	1.87	2.65	–	1.79	2.02	2.40	–
Park et al. [13]	0.96	1.30	1.75	2.41	1.55	2.13	2.77	4.16	1.83	2.41	2.99	4.29
Ferstl et al. [14]	1.12	1.36	1.86	3.57	1.99	2.51	3.76	6.41	2.41	2.71	3.79	7.27
Kiechle et al. [18]	0.70	0.92	1.26	1.74	0.75	1.21	2.08	3.62	0.92	1.56	2.58	4.64
AP [17]	1.15	1.35	1.65	2.32	1.72	2.26	2.85	4.66	1.80	2.43	2.95	4.09
SRCNN [6]	0.58	0.95	1.52	2.45	0.64	1.18	2.43	4.58	0.77	1.50	2.86	5.25
MSF [16]	1.15	1.43	1.80	2.49	1.93	2.37	3.18	4.58	2.36	2.76	3.53	4.74
Hui et al. [23]	0.35	0.69	1.05	1.60	0.37	0.79	1.51	2.63	0.42	0.98	1.76	2.92
MFR-SR [24]	0.60	0.89	1.22	1.74	0.61	1.11	1.75	3.01	0.65	1.23	2.06	3.74
RDN-GDE [25]	0.56	0.88	1.21	1.71	0.48	0.96	1.63	2.86	0.51	1.17	2.05	3.58
Ours	**0.27**	**0.64**	**0.99**	**1.34**	**0.34**	**0.64**	**1.06**	**1.50**	**0.33**	**0.78**	**1.31**	**2.04**

Table 3. Quantitative comparison (in RMSE) on dataset *C*.

Method Used	Tsukuba			Venus			Teddy			Cones		
	2x	4x	8x	2x	4x	8x	2x	4x	8x	2x	4x	8x
Kiechle et al. [18]	3.65	6.21	10.08	0.61	0.82	1.17	1.20	1.82	2.37	1.47	2.97	4.52
Ferstl et al. [14]	5.25	7.35	–	1.11	1.74	–	1.69	2.60	–	2.19	3.50	–
Lu et al. [3]	–	10.29	13.77	–	1.73	2.13	–	2.72	3.47	–	3.99	5.34
SRCNN [6]	3.28	7.94	11.28	0.46	0.79	1.71	1.17	1.99	3.25	1.48	3.59	5.18
Hui et al. [23]	1.85	4.29	8.43	**0.14**	**0.35**	1.04	0.71	1.49	2.76	0.91	2.60	4.23
Ours	**0.91**	**2.75**	**6.18**	0.21	0.42	**0.95**	**0.55**	**1.34**	**2.16**	**0.51**	**2.09**	**3.33**

Figure 3 shows the results of our network on dataset *A* with an upscaling factor of 8. To further verify the effectiveness of the network structure we designed, we selected several regions full of details in each HR depth map to observe the differences between our SR results and the ground truths. We examined the effect of our network in terms of two aspects. One aspect was concerned with whether the regions containing edges were blurred after super-resolution. In Figure 3, we marked these regions with blue boxes in (a–c), and give the contrast between the ground truths and our SR results in (d). It is obvious that edges in our SR results were as sharp as those in the ground truths. Generally, deeper networks like ours can learn more complex and finer features, including edges. On the other hand, we examined whether the artifacts existed in the reconstructed HR depth maps. We marked the regions containing textures in the HR RGB image but were complanated in the corresponding HR depth map with red boxes. The contrasts between the reconstructed results and ground truths given in (e) demonstrate that artifacts disappeared after super-resolution. From these results, we can conclude that our proposed method can perform finer depth map SR reconstruction while suppressing the introduction of artifacts.

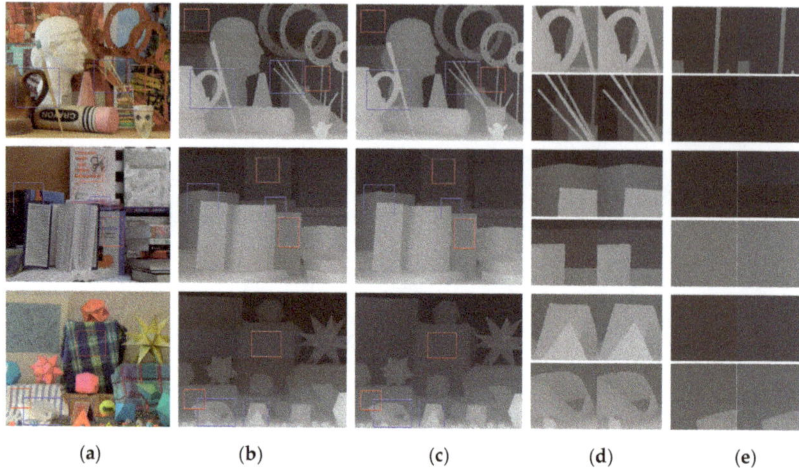

Figure 3. Upsampled depth maps for dataset *A* with an upscaling factor of 8. (**a**) HR RGB images for input, (**b**) ground-truth HR depth maps, (**c**) upsampled results from our network, (**d**) regions inside blue boxes from (**b**) (left) and (**c**), and (**e**) regions inside red boxes from (**b**) (left) and (**c**).

4.3. Evaluation of Generalization

To test the generalization of our proposed network, we selected three RGBD images from different databases to form a new dataset *Mixture* in which image Lucy from the SimGeo dataset [26], image Plant from the ICL-NUIM (Imperial College London- National University of Ireland Maynooth) dataset [38], and image Vintage from Middlebury dataset were considered. The model we used for evaluation was the same as the model tested on datasets *A*, *B*, and *C* without fine-tuning, and the evaluation criterion was still the RMSE. We mainly tested our method at the upscaling factors of 4 and 8, in comparison with methods from References [23,26,39–41]. Our method produced the best performance on the image from the Middlebury dataset and performed nearly 20% better than the sub-optimal result (see Table 4). On the ICL-NUIM dataset, our method's performance was similar to other methods. However, the results on image Lucy indicated that our network was not suitable for this dataset, which means the generalization ability of our network needs to be improved in the future. Figure 4 shows the results of our network on dataset *Mixture* with an upscaling factor of 4. Details in blue boxes were enlarged and shown in columns (d) and (e).

Table 4. Quantitative comparison (in RMSE) on dataset *Mixture*.

Method Used	Lucy	Plant		Vintage	
	4x	4x	8x	4x	8x
Bicubic	0.27	0.25	0.29	0.26	0.30
PDN [39]	0.25	0.27	0.31	0.32	0.35
SRfS [40]	0.37	0.28	0.31	0.35	0.38
DG [41]	0.25	0.27	0.29	0.29	0.30
Hui [23]	0.26	**0.23**	0.29	0.29	0.36
DIP-V [26]	**0.22**	0.26	**0.28**	0.34	0.44
Ours	0.40	0.24	0.31	**0.19**	**0.24**

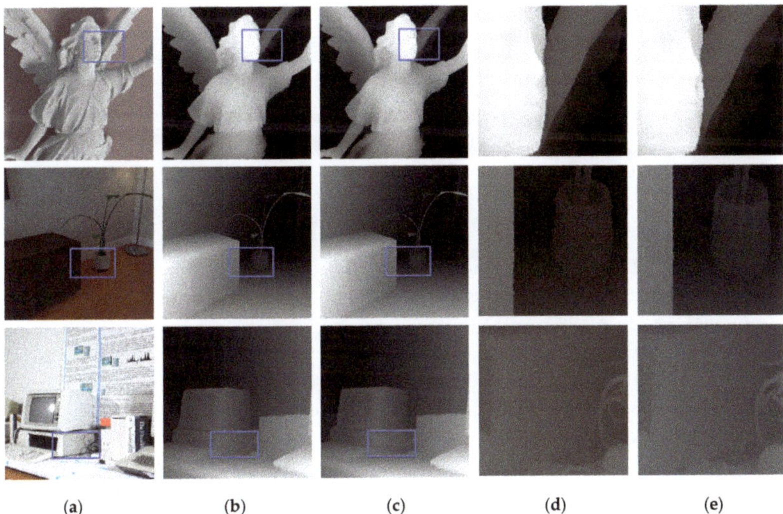

(a)   (b)   (c)   (d)   (e)

Figure 4. Upsampled depth maps for dataset *Mixture* with an upscaling factor of 4. (**a**) HR RGB images for input, (**b**) ground-truth HR depth maps, (**c**) upsampled results from our network, (**d**) regions inside blue boxes from (**b**), and (**e**) regions inside blue boxes from (**c**).

In Table 5, we provide the time taken by our network and other methods [6,7,23] to upscale the depth map from different low resolutions to full resolution. The computation time of Hui et al. [23] was calculated by upsampling image Art using dataset *A*, and we completed the same experiment on a GTX 1080 GPU using Python. Bicubic, SRCNN, and VDSR were written in MATLAB and Guo et al. [42] provides information about the average running time.

Table 5. Computation time (seconds).

Method Used	2x	3x	4x	8x	16x
Bicubic	0.01	0.01	0.01	0.01	0.01
SRCNN [6]	46.63	46.55	46.87	–	–
VDSR [7]	0.44	0.44	0.45	0.44	0.47
Hui [23]	0.247	–	0.296	0.326	0.368
Ours	4.17	3.99	5.21	6.72	39.89

5. Conclusions

We proposed a dual-branch residual network that realizes LR depth map super-resolution with channel interaction and multi-scale residual levels under the guidance of an HR RGB image. In the design of the network structure, we made the residual levels of the RGB image branch and the depth map branch parallel for not only the corresponding feature extraction process, but also the guidance process from the RGB image branch to the depth map branch. Furthermore, the channel interaction via weights avoided introducing artifacts into the upscaled depth map. Using a multi-scale method for upscaling the LR depth map helped to alleviate the blur of the HR depth map that is caused by upsampling to a high resolution in one step. The experiments showed that our method performed excellently compared with other methods, especially when the upscaling factor was large. In the future, we hope to explore other methods for the channel-wise feature fusion and go further in the residual network design. In addition, the RGB image branch, as an auxiliary role in our network, has more layers than the depth map branch, which gives room for improved performance regarding compressing the layers of the whole network.

433

Author Contributions: Conceptualization, R.C. and W.G.; methodology, R.C.; software, R.C.; validation, R.C.; formal analysis, R.C.; investigation, R.C.; resources, R.C.; data curation, R.C.; writing—original draft preparation, R.C.; writing—review and editing, R.C. and W.G.; visualization, W.G.; supervision, W.G.; project administration, W.G.; funding acquisition, W.G. All authors have read and agreed to the published version of the manuscript.

Funding: This research was funded by the National Key R&D Program of China (grant number 2016YFB0502002), and the National Natural Science Foundation of China (NSFC) (grant numbers 61872361, 61991423, and 61421004).

Conflicts of Interest: The authors declare no conflict of interest.

References

1. Zhang, Y.; Li, K.; Li, K.; Wang, L.; Zhong, B.; Fu, Y. Image Super-Resolution Using Very Deep Residual Channel Attention Networks. In Proceedings of the European Conference on Computer Vision, Munich, Germany, 8–14 September 2018.

2. Narayanan, B.N.; Hardie, R.C.; Balster, E. Multiframe Adaptive Wiener Filter Super-Resolution with JPEG2000-Compressed Images. *EURASIP J. Adv. Signal Process.* **2014**, *55*, 1–18. [CrossRef]

3. Lu, J.; Forsyth, D. Sparse Depth Super Resolution. In Proceedings of the IEEE Conference on Computer Vision and Pattern Recognition, Boston, MA, USA, 7–12 June 2015; pp. 2245–2253.

4. Kwon, H.; Tai, Y.W.; Lin, S. Data-Driven Depth Map Refinement via Multi-Scale Sparse Representation. In Proceedings of the IEEE Conference on Computer Vision and Pattern Recognition, Boston, MA, USA, 7–12 June 2015; pp. 159–167.

5. Xie, J.; Feris, R.S.; Sun, M. Edge-Guided Single Depth Image Super Resolution. *IEEE Trans. Image Process.* **2016**, *25*, 428–438. [CrossRef] [PubMed]

6. Dong, C.; Loy, C.; He, K.; Tang, X. Image Super-Resolution Using Deep Convolutional Networks. *PAMI* **2015**, *38*, 295–307. [CrossRef] [PubMed]

7. Kim, J.; Lee, J.K.; Lee, K.M. Accurate Image Super-Resolution Using Very Deep Convolutional Networks. In Proceedings of the IEEE Conference on Computer Vision and Pattern Recognition, Las Vegas, NV, USA, 26 June–1 July 2016; pp. 1646–1654.

8. Lai, W.; Huang, J.; Ahuja, N.; Yang, M. Deep Laplacian Pyramid Networks for Fast and Accurate Super-Resolution. In Proceedings of the IEEE Conference on Computer Vision and Pattern Recognition, Honolulu, HI, USA, 21–26 July 2017; pp. 624–632.

9. Odena, A.; Dumoulin, V.; Olah, C. Deconvolution and Checkerboard Artifacts. *Distill* **2016**, *1*, e3. [CrossRef]

10. He, K.; Sun, J.; Tang, X. Guided Image Filtering. *IEEE Trans. Pattern Anal. Mach. Intel.* **2013**, *6*, 1397–1409. [CrossRef] [PubMed]

11. Barron, J.T.; Poole, B. The Fast Bilateral Solver. In Proceedings of the European Conference on Computer Vision, Amsterdam, The Netherlands, 8–16 October 2016; pp. 617–632.

12. Diebel, J.; Thrun, S. An Application of Markov Random Fields to Range Sensing. In Proceedings of the 19th Annual Conference on Neural Information Processing Systems, Vancouver, BC, Canada, 5–8 December 2005.

13. Park, J.; Kim, H.; Tai, Y.W.; Brown, M.; Kweon, I. High Quality Depth Map Upsampling for 3D-TOF Cameras. In Proceedings of the International Conference on Computer Vision, Barcelona, Spain, 6–13 November 2011; pp. 1623–1630.

14. Ferstl, D.; Reinbacher, C.; Ranftl, R.; Rüther, M.; Bischof, H. Image Guided Depth Upsampling Using Anisotropic Total Generalized Variation. In Proceedings of the International Conference on Computer Vision, Sydney, Australia, 1–8 December 2013; pp. 993–1000.

15. Zuo, Y.; Wu, Q.; Zhang, J.; An, P. Explicit Edge Inconsistency Evaluation Model for Color-Guided Depth Map Enhancement. *IEEE Trans. Circuit Syst. Video Techol.* **2018**, *28*, 439–453. [CrossRef]

16. Zuo, Y.; Wu, Q.; Zhang, J.; An, P. Minimum Spanning Forest with Embedded Edge Inconsistency Measurement Model for Guided Depth Map Enhancement. *IEEE Trans. Image Process.* **2018**, *27*, 4145–4149. [CrossRef] [PubMed]

17. Yang, J.; Ye, X.; Ki, K.; Hou, C.; Wang, Y. Color-Guided Depth Recovery from RGB-D Data Using an Adaptive Autoregressive Model. *TIP* **2014**, *23*, 3962–3969. [CrossRef]

18. Kiechle, M.; Hawe, S.; Kleinsteuber, M. A Joint Intensity and Depth Co-Sparse Analysis Model for Depth Map Super-Resolution. In Proceedings of the International Conference on Computer Vision, Sydney, Australia, 1–8 December 2013; pp. 1545–1552.

19. Riegler, G.; Rüther, M.; Bischof, H. Atgv-Net: Accurate Depth Super-Resolution. In Proceedings of the European Conference on Computer Vision, Amsterdam, The Netherlands, 8–16 October 2016; pp. 268–284.

20. Zhou, W.; Li, X.; Reynolds, D. Guided Deep Network for Depth Map Super-Resolution: How much can color help? In Proceedings of the IEEE International Conference on Acoustics, Speech and Signal Processing, New Orleans, LA, USA, 5–9 March 2017; pp. 1457–1461.

21. Yang, J.; Lan, H.; Song, X.; Li, K. Depth Super-Resolution via Fully Edge-Augmented Guidance. In Proceedings of the IEEE Visual Communications and Image Processing, St. Petersburg, FL, USA, 10–13 December 2017; pp. 1–4.

22. Ye, X.; Duan, X.; Li, H. Depth Super-Resolution with Deep Edge-Inference Network and Edge-Guided Depth Filling. In Proceedings of the IEEE International Conference on Acoustics, Speech and Signal Processing, Seoul, South Korea, 22–27 April 2018; pp. 1398–1402.

23. Hui, T.-W.; Loy, C.C.; Tang, X. Depth Map Super-Resolution by Deep Multi-Scale Guidance. In Proceedings of the European Conference on Computer Vision, Amsterdam, The Netherlands, 8–16 October 2016; pp. 353–369.

24. Zuo, Y.; Wu, Q.; Fang, Y.; An, P.; Huang, L.; Chen, Z. Multi-Scale Frequency Reconstruction for Guided Depth Map Super-Resolution via Deep Residual Network. *IEEE Trans. Circuit Syst. Video Techol.* **2020**, *30*, 297–306. [CrossRef]

25. Zuo, Y.; Fang, Y.; Yang, Y.; Shang, X.; Wang, B. Residual Dense Network for Intensity-Guided Depth Map Enhancement. *Inf. Sci.* **2019**, *495*, 52–64. [CrossRef]

26. Voynov, O.; Artemov, A.; Egiazarian, V.; Notchenko, A.; Bobrovskikh, G.; Burnaev, E. Perceptual Deep Depth Super-Resolution. In Proceedings of the International Conference on Computer Vision, Seoul, Korea, 27 October–2 November 2019; pp. 5653–5663.

27. Lim, B.; Son, S.; Kim, H.; Nah, S.; Lee, K.M. Enhanced Deep Residual Networks for Single Image Super-Resolution. In Proceedings of the IEEE Conference on Computer Vision and Pattern Recognition, Honolulu, HI, USA, 21–26 July 2017; pp. 1132–1140.

28. Zhang, Y.; Tian, Y.; Kong, Y.; Zhong, B.; Fu, Y. Residual Dense Network for Image Super-Resolution. In Proceedings of the IEEE Conference on Computer Vision and Pattern Recognition, Salt Lake City, UT, USA, 18–22 June 2018; pp. 2472–2481.

29. Liu, D.; Wen, B.; Fan, Y.; Loy, C.C.; Huang, T.S. Non-Local Recurrent Network for Image Restoration. In Proceedings of the Neural Information Processing Systems, Montreal, QC, Canada, 2–7 December 2018; pp. 1673–1682.

30. Qiu, Y.; Wang, R.; Tao, D.; Cheng, J. Embedded Block Residual Network: A Recursive Restoration Model for Single-Image Super-Resolution. In Proceedings of the International Conference on Computer Vision, Seoul, Korea, 27 October–2 November 2019; pp. 4180–4189.

31. Hu, Y.; Li, J.; Huang, Y.; Gao, X. Channel-wise and Spatial Feature Modulation Network for Single Image Super-Resolution. *IEEE Trans. Circuit Syst. Video Techol.* **2019**. [CrossRef]

32. Jing, P.; Guan, W.; Bai, X.; Guo, H.; Su, Y. Single Image Super-Resolution via Low-Rank Tensor Representation and Hierarchical Dictionary Learning. *Multimed. Tools Appl.* **2020**. [CrossRef]

33. Haris, M.; Shakhnarovich, G.; Ukita, N. Recurrent Back-Projection Network for Video Super-Resolution. In Proceedings of the IEEE Conference on Computer Vision and Pattern Recognition, Long Beach, CA, USA, 15–21 June 2019; pp. 3897–3906.

34. Molini, A.B.; Valsesia, D.; Fracastoro, G.; Magli, E. DeepSUM: Deep Neural Network for Super-Resolution of Unregistered Multitemporal Images. *IEEE Trans. Geosci. Remote Sens.* **2020**. [CrossRef]

35. Molini, A.B.; Valsesia, D.; Fracastoro, G.; Magli, E. DeepSUM++: Non-local Deep Neural Network for Super-Resolution of Unregistered Multitemporal Images. *arXiv* **2020**, arXiv:2001.06342. [CrossRef]

36. Shi, W.; Caballero, J.; Huszar, F.; Totz, J.; Aitken, A.P.; Bishop, R.; Rueckert, D.; Wang, Z. Real-Time Single Image and Video Super-Resolution Using an Efficient Sub-Pixel Convolutional Neural Network. In Proceedings of the IEEE Conference on Computer Vision and Pattern Recognition, Las Vegas, NV, USA, 26 June–1 July 2016; pp. 1874–1883.

37. Silberman, N.; Kohli, P.; Hoiem, D.; Fergus, R. Indoor Segmentation and Support Inference from RGBD Images. In Proceedings of the European Conference on Computer Vision, Florence, Italy, 7–13 October 2012.

38. Handa, A.; Whelan, T.; McDonald, J.; Davison, A.J. A Benchmark for RGB-D Visual Odometry, 3D reconstruction and SLAM. In Proceedings of the IEEE Conference on Robotics and Automation, Hong Kong, China, 31 May–5 June 2014; pp. 1524–1531.

39. Riegler, G.; Ferstl, D.; Ruther, M.; Bischof, H. A Deep Primal-Dual Network for Guided Depth Super-Resolution. In Proceedings of the British Machine Vision Conference, York, UK, 19–22 September 2016.
40. Haefner, B.; Queau, Y.; Mollenhoff, T.; Cremers, D. Fight Ill-Posedness with Ill-Posedness: Single-shot Variational Depth Super-Resolution from Shading. In Proceedings of the IEEE Conference on Computer Vision and Pattern Recognition, Salt Lake City, UT, USA, 18–22 June 2018; pp. 164–174.
41. Gu, S.; Zuo, W.; Guo, S.; Chen, Y.; Chen, C.; Zhang, L. Learing dynamic guidance for depth image enhancement. In Proceedings of the IEEE Conference on Computer Vision and Pattern Recognition, Honolulu, HI, USA, 21–26 July 2017; pp. 712–721.
42. Guo, C.; Li, C.; Guo, J.; Cong, R.; Fu, H.; Han, P. Hierarchical Features Driven Residual Learning for Depth Map Super-Resolution. *IEEE Trans. Image Process.* **2019**, *28*, 2545–2557. [CrossRef] [PubMed]

Article

An Efficient Building Extraction Method from High Spatial Resolution Remote Sensing Images Based on Improved Mask R-CNN

Lili Zhang [1,*], Jisen Wu [1], Yu Fan [1], Hongmin Gao [1] and Yehong Shao [2]

[1]   College of Computer and Information Engineering, Hohai University, Nanjing 211100, China;
       171307030009@hhu.edu.cn (J.W.); fanyu@hhu.edu.cn (Y.F.); gaohongmin@hhu.edu.cn (H.G.)
[2]   Arts and Science, Ohio University Southern, Ironton, OH 45638, USA; yehongshao@gmail.com
*   Correspondence: lilzhang@hhu.edu.cn

Received: 4 January 2020; Accepted: 28 February 2020; Published: 6 March 2020

Abstract: In this paper, we consider building extraction from high spatial resolution remote sensing images. At present, most building extraction methods are based on artificial features. However, the diversity and complexity of buildings mean that building extraction methods still face great challenges, so methods based on deep learning have recently been proposed. In this paper, a building extraction framework based on a convolution neural network and edge detection algorithm is proposed. The method is called Mask R-CNN Fusion Sobel. Because of the outstanding achievement of Mask R-CNN in the field of image segmentation, this paper improves it and then applies it in remote sensing image building extraction. Our method consists of three parts. First, the convolutional neural network is used for rough location and pixel level classification, and the problem of false and missed extraction is solved by automatically discovering semantic features. Second, Sobel edge detection algorithm is used to segment building edges accurately so as to solve the problem of edge extraction and the integrity of the object of deep convolutional neural networks in semantic segmentation. Third, buildings are extracted by the fusion algorithm. We utilize the proposed framework to extract the building in high-resolution remote sensing images from Chinese satellite GF-2, and the experiments show that the average value of IOU (intersection over union) of the proposed method was 88.7% and the average value of Kappa was 87.8%, respectively. Therefore, our method can be applied to the recognition and segmentation of complex buildings and is superior to the classical method in accuracy.

Keywords: building extraction; convolutional neural networks; mask R-CNN; high-resolution remote sensing image

---

1. Introduction

With the development of remote sensing satellite technology and the demand of urbanization, it has become an important research field to automatically and accurately extract building objects from remote sensing images. There are many studies on building extraction approaches from remote sensing images based on artificial design features, and these approaches can be divided into three categories. The first is the building extraction method based on edge and corner detection and matching. In this method, feature matching is carried out through edge and corner information of the building to complete building extraction [1]. The second is the building extraction method combined with elevation information, and this kind of method uses elevation information to separate out non-ground points, and then detect buildings by combining the common edge features, spectral features and other artificial features [2]. The last is the object-oriented building extraction method. This kind of method uses edge information to segment the remote sensing image initially so that homogeneous pixels make up objects of different sizes, then extract them by using the unique spectral information, shape and texture features of the buildings [3]. Due to different shooting angles, light and other factors, remote sensing images in different periods have considerable internal variability, and buildings have a variety of structure, texture and spectral information, therefore, the methods above cannot perform well in the extraction of complex buildings.

In recent years, deep learning technology has ushered in a new wave of revival. At present, deep learning technology represented by deep convolutional neural networks has achieved excellent results in the field of computer vision [4–6]. Compared with the traditional method of feature extraction in artificial design, deep convolutional neural networks can obtain the structure, texture, semantics and other information of the object through multiple convolutional layers, and their performance is closer to visual interpretation in object recognition. The experiments in [7,8] had the advantages of deep convolutional neural networks in the field of object detection, but also revealed the problems of local absence and edge blur in image segmentation. This phenomenon is especially serious when the hardware equipment is insufficient or the dataset is small. Figure 1 shows this phenomenon using a mask image [9]. The main reasons for poor extraction results are as follows. Firstly, the lack of data sets makes the convolutional neural network unable to effectively learn the hierarchical contextual image features. Secondly, adding more layers to the suitably deep model leads to higher training errors [10]. Lastly, the prediction ability of CNNs (convolutional neural networks) with low computation is at the expense of a decrease in output resolution.

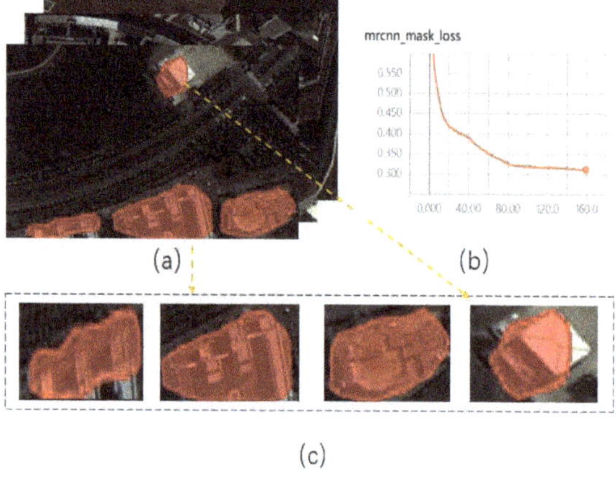

Figure 1. Building extraction based on Mask R-CNN. (**a**) The whole building extraction; (**b**) Loss function curve; (**c**) The single building extraction.

This paper investigates the possibility of artificial features to optimize the building extraction results of convolutional neural networks. A fusion method of artificial edge features and convolutional neural network recognition results is proposed to address the problem of building extraction accurately.

2. Related Work

(A) Artificial design model: this model based on artificial features has been widely used in the detection of remote sensing images. Combining image segmentation technology, spectral constraint, shadow constraint and shape constraint, Ding et al. proposed a new building extraction framework based on the MBI (morphological building index) [11]. Aytekin et al. designed a general algorithm for automatically extracting buildings from multispectral images by using spectral and spatial characteristics [12]. Jimenez et al. proposed an efficient implementation of MBI and MSI algorithms, which were specially developed for GPU [13]. Jinxing et al. introduced a method based on multi-level segmentation and multi-feature fusion for building detection in remote sensing images [14]. Cui et al. proposed a method based on the Hough transform combining edge and region to extract complex buildings [15].

(B) Deep learning model: convolutional neural networks have been successfully applied to natural image categorization recently. Studies have shown that the deep learning method can effectively improve the accuracy of building extraction. Guo et al. proposed a series of convolutional neural networks that can be applied to a pixel level classification framework for township building identification [16]. Kang et al. proposed a general framework for building classification based on convolutional neural networks [17]. Makantasis et al. addressed the problem of man-made object detection from hyperspectral data through a deep learning classification framework [18]. Nogueira et al. analyzed three strategies of applying convolutional neural networks to remote sensing images [19]. The results show that fine tuning is the best training strategy of convolutional neural networks applied to remote sensing images. Yu et al. proposed a convolutional neural network remote sensing classification model based on PTL-CFS (parameter transfer learning and correlation-based feature selection), which can accelerate the convergence speed of CNN loss function [20].

Mask R-CNN was proposed after R-CNN [21], Fast R-CNN [22], and Faster R-CNN [23]. The architecture of R-CNN is divided into three parts: firstly, 2000 candidate regions are extracted by selective search; secondly, the extracted candidate regions are extracted by a multi-layer convolutional neural network; lastly, support vector machine (SVM) and linear regression model are used to classify and regress the object.

Although the R-CNN has high extraction accuracy, it is difficult to train and has lower execution time of inference. To solve this problem, Fast R-CNN and Faster R-CNN are optimized in many ways: 1. The deep convolutional neural network is used to extract the features of the original image directly. 2. The RPN (region proposal network) is used instead of the selective search to extract the candidate regions so as to reduce the training time of the model. 3. The end-to-end training is realized by using fully connected layers instead of SVM. 4. The unified feature map size of the region of interest pooling (ROI pooling) is proposed to meet the input requirements of fully connected layers.

Mask R-CNN adds a fully convolutional network (FCN) [24] branch to Faster R-CNN to segment the object. At the same time, the ROI (region of interest) align method based on the bilinear interpolation algorithm is proposed to solve the problem of the pixel offset between the input image and the feature map caused by ROI pooling.

Mask R-CNN architecture is shown in Figure 2. The original image is processed by the multi-layered convolutional network to obtain hierarchical contextual image features, then the candidate region is extracted by region selection networks, and the ROI pooling of the feature map in Faster R-CNN is solved by using ROI align based on the bilinear interpolation algorithm. In addition, FCN is introduced as a branch of the model to achieve accurate segmentation of objects.

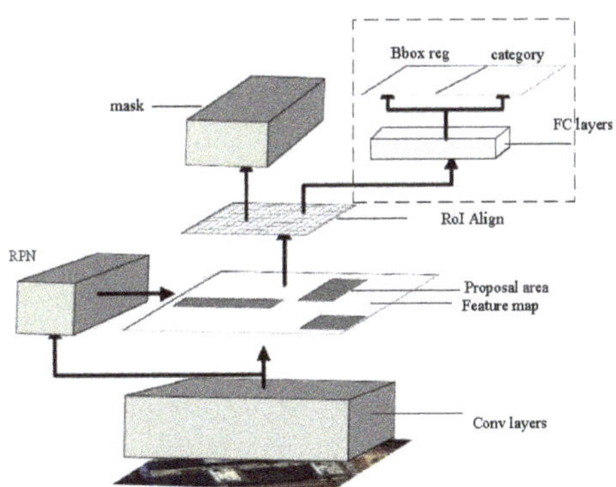

Figure 2. Architecture of the Mask R-CNN.

3. Fast and Effective Building Extraction Method

Due to the high precision of Mask R-CNN in natural image segmentation, this paper will improve the architecture of Mask R-CNN to get our framework, which is applicable to building extraction. For simplicity, we abbreviate this method to MRFS (Mask R-CNN Fusion Sobel). Our method can be summarized as follows:

(1) Image preprocessing: High resolution remote sensing images usually include panchromatic images and multispectral images. Panchromatic images have high resolution and little spectral information. Multispectral images have low resolution and rich spectral information. Both are not conducive to dense pixelwise labeling. Therefore, we enhance image information through fusing the panchromatic images and multispectral images.

(2) Constructing a network model: Our network architecture is improved based on Mask R-CNN. The feature pyramid network is utilized for feature fusion in order to improve the final detection accuracy. Compared with the Mask R-CNN model, RestNet50 (residual network) is used as the pre-training model in this paper to extract building features, and to remove the branch of boundary fitting and category judgment.

(3) Training of the network model: We adopt the cross-validation method to train our model.

(4) Detection: The trained model is used for building extraction, and the results are used as the input data of the proposed method.

(5) Combining edge features: The remote sensing image is segmented by edge features, and the building extraction results in the previous step are optimized by the results.

3.1. Image Preprocessing

A GF-2 remote sensing image is selected as the raw data, including the panchromatic image and multispectral image. We fuse the panchromatic images and multispectral images to get high-resolution multispectral images.

In order to ensure that the high-resolution multispectral images can preserve the color, texture and spectral information of remote sensing images effectively, the nearest-neighbor diffusion pan-shaping algorithm is used for image fusion, and the fusion maps are shown in Figure 3.

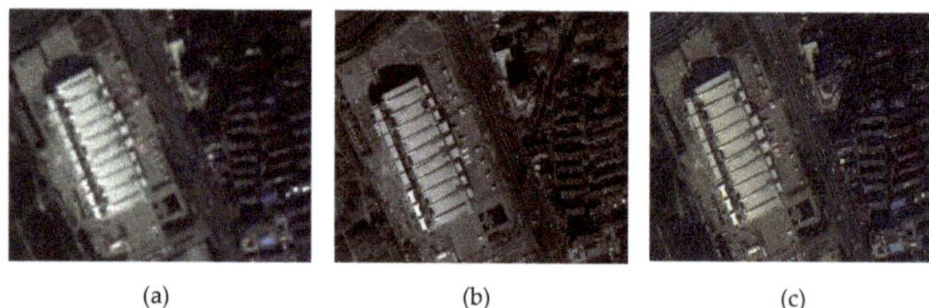

(a)   (b)   (c)

Figure 3. Remote sensing image. (**a**) Multispectral image; (**b**) Panchromatic image; (**c**) Fusion image based on nearest-neighbor diffusion pan-shaping algorithm.

3.2. Network Model

Mask R-CNN models have made remarkable achievements in the field of image segmentation. Compared with the single-stage convolutional neural networks, Mask R-CNN has higher precision in image segmentation. The architecture of the model is shown in Figure 4. In the original image, the multi-layer convolutional neural network is used to obtain the high-dimensional feature map, and candidate regions are extracted by RPN. The corresponding position offset between the feature maps in Faster R-CNN and the original map is solved by using the pooling layer ROI align based on the bilinear interpolation algorithm. In addition, the model achieves object segmentation by fully convolutional networks as a branch.

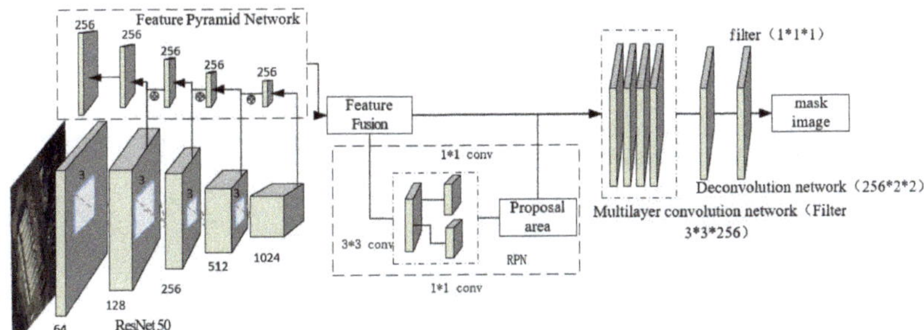

Figure 4. Single class extraction model of the convolution network based on Mask R-CNN.

Compared with multi-object segmentation, building extraction in remote sensing images is essentially a binary classification task. Therefore, the branch for category detection in the model is removed to optimize the training time of the model. For our improved Mask R-CNN in this paper, the cross-entropy loss function is used as follows.

3.2.1. Loss function

In our network architecture, the output of the RPN recommendation area is used as the input of the fully convolutional layer, and FCN is used to complete a per-pixel classification. The cross-entropy loss function is defined as:

$$\text{Lx} = L_{mask} = -\frac{1}{m}\sum_{i=1}^{m} L(x_i) \tag{1}$$

$$\text{L}(x_i) = -\frac{1}{n}\sum_{j=1}^{n}\left[y_j lna_j + \left(1 - y_j\right)\ln\left(1 - a_j\right)\right] \tag{2}$$

where L(x) denotes the loss function of the total training samples; L_{mask} is the loss function of the mask branches; m is the total number of samples; Lx_i is the loss value of a single sample; n is the number of pixels of a single sample; y_j is the expected output of a single pixel; and a_j is the output of the neural network.

The output layer of Mask R-CNN uses the sigmoid function as the activation function, and uses the average binary value of each sample as a loss function. Cross entropy is used to train the back propagation of convolutional neural networks. Building extraction is a binary classification for a single pixel. The expected output of a single pixel can be expressed as 0 or 1. From the cross entropy function, when the expected value of a pixel is 0 or 1, and the predicted value a_j approaches the expected value y_j, the cross entropy loss function $\text{L}(x_i)$ approaches 0; otherwise, $\text{L}(x_i)$ is close to infinity.

3.2.2. Dataset Construction

We annotate different remote sensing images and use them as the experimental data of our method. The dataset includes the visual interpretation of buildings from the GF-2 remote sensing image and a building dataset from (https://www.cs.toronto.edu/~vmnih/data/). In summary, the training datasets contain 3231 images. In order to increase the sample size, the data sets are rotated at 90°, 180°, and 270°, and flipped vertically and horizontally. In order to ensure the uniform distribution of data, the data sets are randomly divided into training sets and test sets with the proportion of 7:3.

In order to verify the performance of MRFS on different kinds of buildings, this paper selects three kinds of buildings according to their structure and distribution characteristics. This is shown in Figure 5.

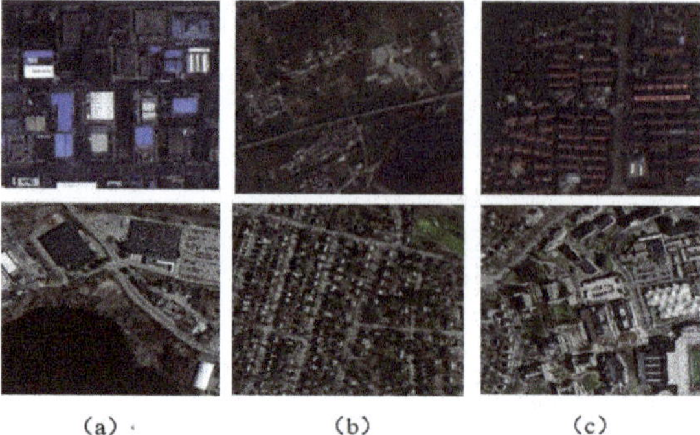

(a)   (b)   (c)

Figure 5. Validate dataset. (**a**) Large regular building group. (**b**) Village buildings. (**c**) Medium sized regular buildings.

3.3. Combining Artificial Edge Features

Deep convolutional networks have a high accuracy of the recognition and localization of the object, but there are certain shortcomings in the segmentation of the object, which are mainly manifested on the edge extraction and the integrity of the object. In order to solve these problems, we propose an optimization method of object extraction based on deep convolutional neural networks combined with artificial edge features.

The edge detection points of the Sobel operator are accurately located. There are fewer edge detection errors, and the operator detection edge points correspond to the actual edge points one by one. Therefore, we use the Sobel operator to obtain the gradient amplitude and gradient direction.

In order to reduce the influence of image noise on the edge detector, a Gaussian filter is used for the smoothing operation. The Gaussian filtering results are shown in Figure 6.

Figure 6. Gaussian filtering result.

For the Gaussian smoothed image, the Sobel operator is used to obtain the gradient amplitude and gradient direction. The specific formula is as follows:

$$G_X = \begin{bmatrix} -1 & 0 & -1 \\ -2 & 0 & -2 \\ -1 & 0 & -1 \end{bmatrix} * A \tag{3}$$

$$G_y = \begin{bmatrix} +1 & +2 & -1 \\ 0 & 0 & 0 \\ -1 & -2 & -1 \end{bmatrix} * A \tag{4}$$

where A denotes the original image, and G_X and G_y are the first derivative values in the horizontal and vertical directions. Thus, the gradient G and direction θ of the pixel can be determined.

$$M = \sqrt{G_x{}^2 + G_y{}^2} \tag{5}$$

$$\theta = \arctan\left(\frac{G_y}{G_x}\right) \tag{6}$$

where M denotes the edge strength of the image; and θ the edge direction.

In order to solve the shortcomings of convolutional neural networks on building extraction, we propose optimizing the extraction results by an artificial design of edge features.

The process in detail is as follows:

a. Use the Sobel operator to detect edges of remote sensing images and apply the watershed algorithm to perform label segmentation on gradient images, which is shown in Figure 7.
b. The trained convolutional neural network is used to build the extraction model and get the map of the building extraction.
c. Get the area of the building object in step (b) and the area of the object in the corresponding position in step (a). Establish a judgment function including the threshold value λ, as shown in formula 7. When the pixel value of the object occupied by the mask is greater than a certain threshold, the object is marked as a building object.

$$\Upsilon(x_i, y_i) = sign(x_i - \lambda y_i) \tag{7}$$

where γ is 0, 1 or -1 for the building mark. If γ is 1 or 0, then object i is marked as building, and if γ is -1, object i is marked as non-building. Here x_i denotes the number of pixels of the mask occupying object i, y_i is the number of pixels of object i, and λ is the threshold.

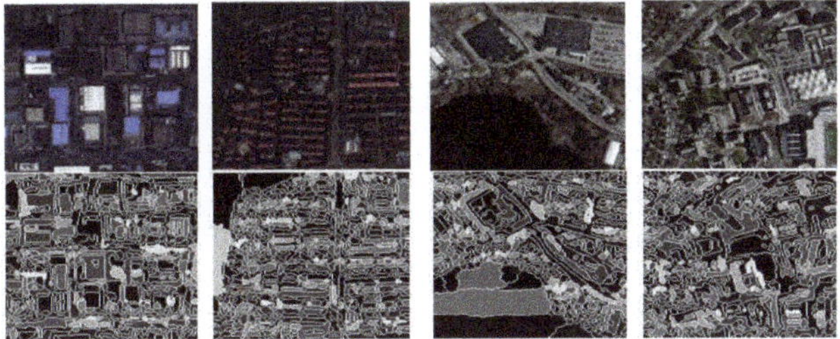

Figure 7. Remote sensing image after edge segmentation.

4. Experiments

4.1. Setting

All of the experiments are implemented on a single NVIDIA GeForce GTX1070 with 8 GB memory. The training time of the u-net model is 5 h. Mask R-CNN uses ResNet50 as the backbone, with a batch size of 2. We use a weight decay of 0.001 and momentum of 0.9. The training time of Mask R-CNN and the MRFS model is 2 days, and 120 K iterations are completed. We use a threshold $\lambda = 0.6$.

4.2. Evaluation Criteria

In this paper, the three measures, IoU, detection accuracy (pixel accuracy), and Kappa coefficients, are selected to evaluate our method. They are commonly used in remote sensing classifications. The IoU describes the degree of overlap between the predicted value and the authenticity value; the detection accuracy is used to measure the proportion of correct prediction results; the Kappa coefficient is usually used to measure the accuracy of remote sensing image classifications. Their formulae are shown in Equation (8) to Equation (10):

$$IoU = \frac{Area(P) \cap Area(T)}{Area(P) \cup Area(T)} \tag{8}$$

$$P = \frac{TP}{TP + FP} \tag{9}$$

$$K = \frac{p_0 - p_e}{1 - p_e} \tag{10}$$

In Equation (8), $Area(P)$ is the prediction area and $Area(T)$ is the true value area. In Equation (9), P is the detection accuracy rate, TP is the correct detection, and FP is the error detection. In Equation (10), K is the kappa coefficient, p_0 explains the proportion of correct cells, and p_e is the proportion of misinterpretations caused by chance.

4.3. Analysis

4.3.1. Comparison of loss function curves

In order to analyze the effect of three convolution neural network models, we compared the loss function curves of u-net [7], Mask R-CNN and MRFS. As shown in Figure 8, we record the training error every five epochs and plot the loss function curve.

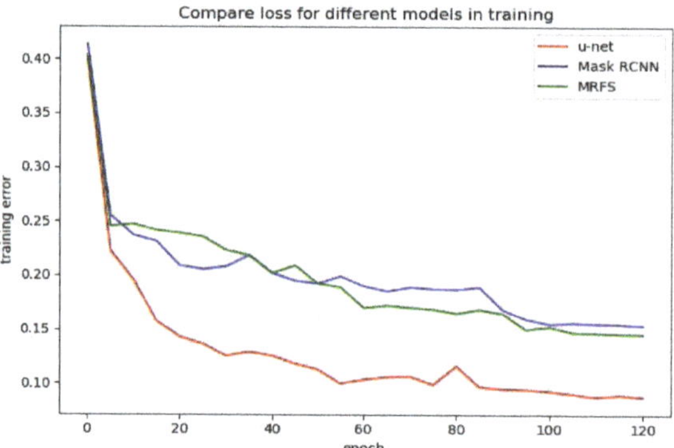

Figure 8. Comparison of u-net, mask R-CNN and MRFS on training error.

Compared with u-net, Mask R-CNN and the MRFS model have higher training error because the deeper network has a higher training error [10]. It can be seen from Table 1 and Figure 8 that when the training error converges to a bad local minimum, the method proposed in this paper has a better performance.

4.3.2. Evaluation of different types of buildings

We use the same dataset to train the convolutional neural networks and divide it into three parts to test according to the building characteristics. The experiments are shown in Figures 9–11. This paper compares the performance of three convolution neural networks in three different regions, analyzes the shortcomings of convolution neural networks in building extraction of high-resolution remote sensing images, and verifies the effectiveness of the proposed method in this paper.

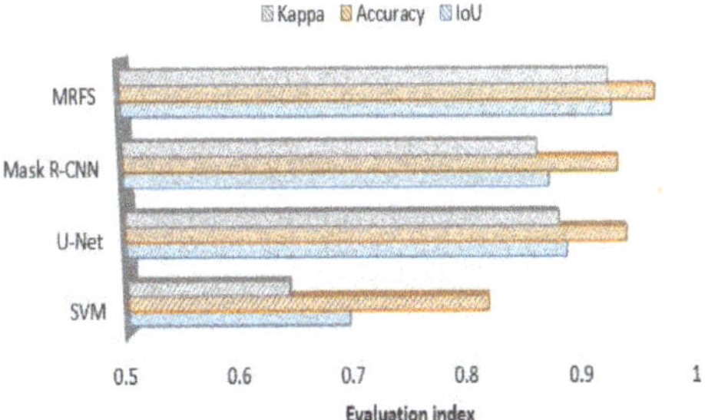

Figure 9. Comparison of MRFS, Mask R-CNN, SVM and KNN(k-Nearest Neighbor) on large building extractions.

(1) In this paper, three evaluation indexes are used to measure the extraction results of SVM, u-net, Mask R-CNN and MRFS. Figure 9 shows the high recognition ability of the above method for large regular buildings. As shown in Figure 12, MRFS has obvious advantages in object integrity and edge extraction compared with the classical convolutional neural network algorithm.

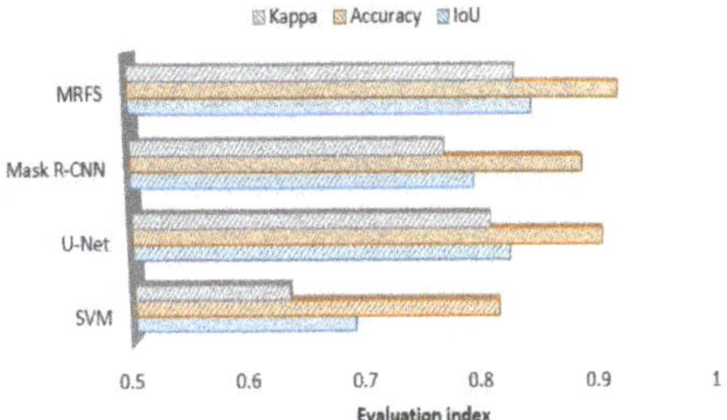

Figure 10. Comparison of MRFS, Mask R-CNN, u-net, SVM and KNN on the village buildings extraction.

(2) For the village buildings, Figure 10 shows that compared with the SVM, the convolutional networks model has better extraction results, which proves that the deep convolutional neural network has advantages on the extraction of complex buildings. Due to the poor edge of rural buildings, the evaluation result of MRFS is lower than that of area a.

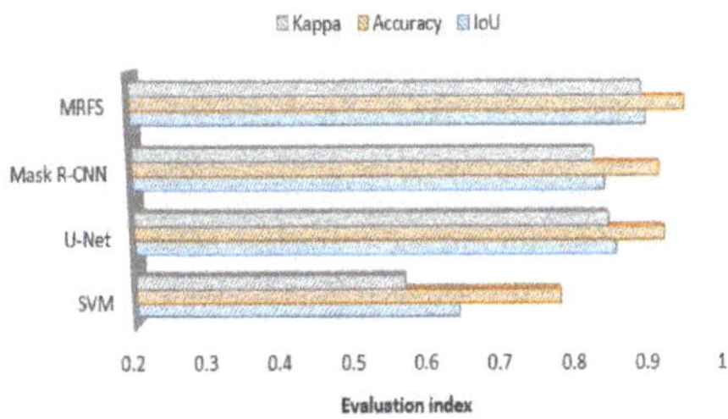

Figure 11. Comparison of MRFS, Mask R-CNN, SVM and KNN on medium sized area c.

(3) Comparing the performance of the four methods on medium sized buildings (as shown in Figure 5c), Figure 11 shows that the convolutional neural network has a high consistency with the real value of the building label. At the same time, it shows that the artificial feature can optimize the result of the convolution neural network.

4.3.3. Comparison of Single building extraction

In order to further analyze the efficiency of these four methods in remote sensing image building extraction, Figure 12 shows the building extraction map of the four methods, where the black and the white represent background and building respectively. SVM is conducive to the overall recognition of buildings, but it cannot effectively distinguish between the cement floors and buildings, and there are a lot of false extractions. Mask R-CNN and u-net have achieved high accuracy in building recognition and building locating, However, compared with visual interpretation, they are still insufficient on edge extraction and object integrity.

The main error sources of classical convolution neural networks are as follows: firstly, a convolutional neural network has a high requirement on hardware equipment. The batch size in this paper is 2. In convolutional neural networks, large batches usually make the network converge faster, but due to the limitation of memory resources, large batches may lead to insufficient memory or program kernel crash. Secondly, in the convolution neural network, the pooling layer is used to reduce the model parameters, and the deconvolution operation will also affect the integrity of object extraction to a certain extent. In summary, the difference between the proposed method and the classical convolutional neural network in building recognition is mainly focused on the optimization of building edges, which can solve the incompleteness of the building extraction.

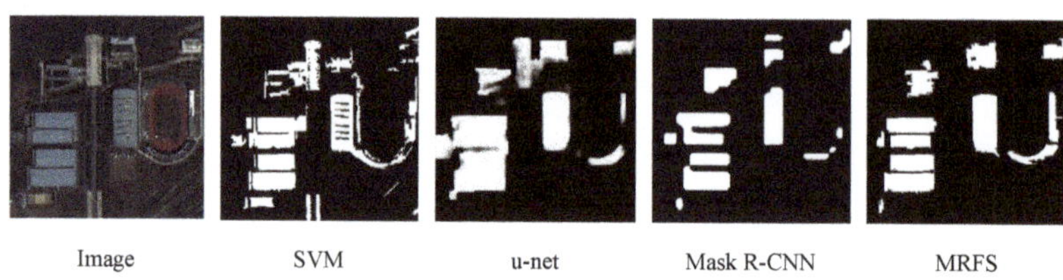

Figure 12. Building extraction result maps of different methods.

Figure 12 shows that the convolutional neural network performs poorly in edge extraction and the integrity of the object but the MRFS can deal with such problems more effectively. Figure 13 shows the extracted results of the single building objects after enlargement. Compared with convolution neural networks, the support vector machine has better edge segmentation, but misses some areas of complex buildings. For Mask R-CNN, u-net and MRFS proposed in this paper, the convolutional neural network has better efficiency to locate buildings in complex areas.

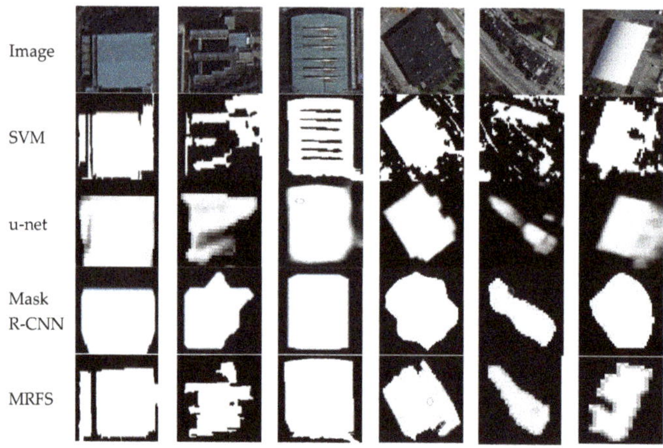

Figure 13. Comparison of single building detection maps.

Table 1 shows that due to the complex structure and various materials of remote sensing image buildings, object-oriented building extraction methods are prone to large-scale error recognition, and convolutional neural networks solve this problem well. The MRFS method we proposed in this paper solves the problems of edge extraction and the integrity of the object. Therefore, compared with the Mask R-CNN and other methods, our method has better efficiency on different building extractions.

Table 1. Quantitative evaluation of different methods on building extraction.

Test Area	IoU				Pixel Accuracy				Kappa			
	SVM	u-net	Mask R-CNN	MRFS	SVM	u-net	Mask R-CNN	MRFS	SVM	u-net	Mask R-CNN	MRFS
Area a	0.697	0.890	0.871	0.925	0.822	0.941	0.931	0.961	0.643	0.881	0.861	0.921
Area b	0.691	0.826	0.794	0.841	0.818	0.905	0.885	0.914	0.634	0.808	0.768	0.827
Area c	0.647	0.861	0.841	0.894	0.786	0.925	0.914	0.943	0.571	0.848	0.825	0.886
mean	0.679	0.859	0.836	0.887	0.808	0.924	0.910	0.940	0.616	0.846	0.818	0.878

4.3.4. Edge feature fusion parameter λ

In order to analyze the influence of the parameters of the MRFS, this paper carries out a number of experiments by changing the object selection threshold parameter λ. The values of λ are set as 0.3, 0.4, 0.5, 0.6, 0.7 and 0.8, respectively. Figure 14 shows the changes of IoU, detection accuracy and kappa with the threshold parameter λ. In this experiment, λ is in the range of [0.4,0.6], and the change of each index is relatively small.

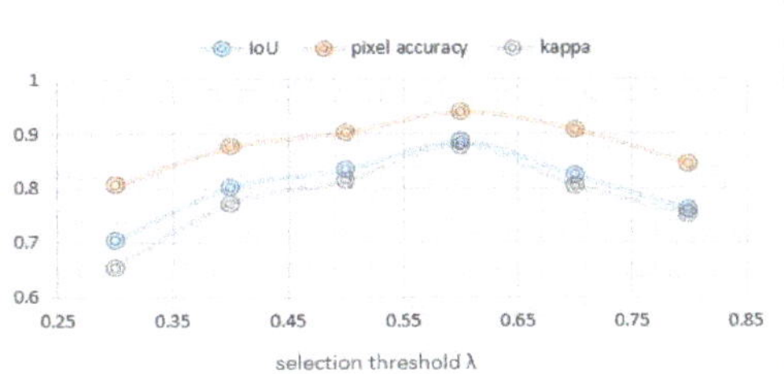

Figure 14. Effect of threshold parameter λ on accuracy.

5. Conclusions

In this paper, the design of deep convolutional neural networks in semantic segmentation is applied to extract building from high-resolution remote sensing images. Based on the characteristics of the deep convolutional network, building recognition and high-precision extraction in high-resolution remote sensing images were realized. Concerning the problems of poor edge recognition and incomplete extraction of the convolutional networks on the building extraction from remote sensing images, an optimization method combining edge features was proposed to improve the efficiency of the network model on building extraction. Experiments on 3231 images and 20,000 building objects were carried out to verify the effectiveness of the method we proposed in this paper. Compared with the classical convolutional network models, the accuracy rate and integrity of building extraction were improved. In the future, the relationship between the selection of the threshold parameter λ and the Mask R-CNN training results will be analyzed in detail, and the method will be improved by automatically getting the model parameters.

Author Contributions: Conceptualization, L.Z. and J.W.; Methodology, L.Z. and J.W.; Validation, H.G. and Y.F.; Resources, Y.F and Y.S.; Data Curation, H.G and Y.S.; Writing-Original Draft Preparation, L.Z and J.W.; Writing-Review & Editing, L.Z and J.W.; Supervision, L.Z.; Funding Acquisition, L.Z. and H.G. All authors have read and agreed to the published version of the manuscript.

Funding: This research was funded by National Natural Science Foundation of China (No. 61671201, No. 61701166), PAPD and the Natural Science Foundation of JiangSu Province (No. BK20170891).

Conflicts of Interest: The authors declare no conflict of interest.

References

1. Jung, C.R.; Schramm, R. Rectangle detection based on a windowed Hough transform. In Proceedings of the 17th Brazilian Symposium on Computer Graphics and Image Processing, Foz do Iguaçu, Brazil, 17–20 October 2004; pp. 113–120.
2. Ahmadi, S.; Zoej, M.J.V.; Ebadi, H.; Moghaddam, H.A.; Mohammadzadeh, A. Automatic urban building boundary extraction from high resolution aerial images using an innovative model of active contours. *Int. J. Appl. Earth Obs. Geoinf.* **2010**, *12*, 150–157. [CrossRef]
3. Myint, S.W.; Gober, P.; Brazel, A.; Grossman-Clarke, S.; Weng, Q. Per-pixel vs. object-based classification of urban land cover extraction using high spatial resolution imagery. *Remote Sens. Environ.* **2011**, *115*, 1145–1161. [CrossRef]
4. Krizhevsky, A.; Sutskever, I.; Hinton, G.E. Imagenet classification with deep convolutional neural networks. In Proceedings of the Advances in Neural Information Processing Systems 25, Lake Tahoe, NV, USA, 3–6 December 2012; pp. 1097–1105.
5. Li, Z.; Peng, C.; Yu, G.; Zhang, X.; Deng, Y.; Sun, J. Light-head R-CNN: In defense of two-stage object detector. *arXiv* **2017**, arXiv:1711.07264. Available online: http://dwz.date/CyZ (accessed on 22 November 2017).
6. He, K.; Zhang, X.; Ren, S.; Sun, J. Spatial Pyramid Pooling in Deep Convolutional Networks for Visual Recognition. *IEEE Trans. Pattern Anal. Mach. Intell.* **2014**, *37*, 1904–1916. [CrossRef] [PubMed]
7. Ronneberger, O.; Fischer, P.; Brox, T. U-Net: Convolutional Networks for Biomedical Image Segmentation. In *Medical Image Computing and Computer-Assisted Intervention—MICCAI 2015, Proceedings of the International Conference on Medical Image Computing and Computer-Assisted Intervention, 5–9 October, Munich, Germany*; Springer: Cham, Switzerland, 2015; pp. 234–241.
8. Chen, L.C.; Papandreou, G.; Schroff, F.; Adam, H. Rethinking atrous convolution for semantic image segmentation. *arXiv* **2017**, arXiv:1706.05587. Available online: http://dwz.date/Czd (accessed on 5 December 2017).
9. He, K.; Gkioxari, G.; Dollár, P.; Girshick, R. Mask R-CNN. In Proceedings of the IEEE International Conference on Computer Vision, Venice, Italy, 22–29 October 2017; pp. 2961–2969.
10. He, K.; Zhang, X.; Ren, S.; Sun, J. Deep residual learning for image recognition. In Proceedings of the IEEE Conference on Computer Vision and Pattern Recognition, Las Vegas, NV, USA, 10 April 2016; pp. 770–778.
11. Ding, Z.; Wang, X.Q.; Li, Y.L.; Zhang, S.S. Study on Building Extraction from High-Resolution Images Using Mbi. *Int. Arch. Photogramm. Remote Sens. Spat. Inf. Sci.* **2018**, *42*, 3. [CrossRef]
12. Aytekin, Ö.; Erener, A.; Ulusoy, İ.; Düzgün, Ş. Unsupervised building detection in complex urban environments from multispectral satellite imagery. *Int. J. Remote Sens.* **2012**, *33*, 2152–2177. [CrossRef]
13. Jiménez, L.I.; Plaza, J.; Plaza, A. Efficient implementation of morphological index for building/shadow extraction from remotely sensed images. *J. Supercomput.* **2017**, *73*, 482–494. [CrossRef]
14. Chen, J.; Wang, C.; Zhang, H.; Wu, F.; Zhang, B.; Lei, W. Automatic detection of low-rise gable-roof building from single submeter SAR images based on local multilevel segmentation. *Remote Sens.* **2017**, *9*, 263. [CrossRef]
15. Cui, S.; Yan, Q.; Reinartz, P. Complex building description and extraction based on Hough transformation and cycle detection. *Remote Sens. Lett.* **2012**, *3*, 151–159. [CrossRef]
16. Guo, Z.; Chen, Q.; Wu, G.; Xu, Y.; Shibasaki, R.; Shao, X. Village Building Identification Based on Ensemble Convolutional Neural Networks. *Sensors* **2017**, *17*, 2487. [CrossRef] [PubMed]
17. Kang, J.; Körner, M.; Wang, Y.; Taubenböck, H.; Zhu, X.X. Building instance classification using street view images. *ISPRS J. Photogramm.* **2018**, *145*, 44–59. [CrossRef]
18. Makantasis, K.; Karantzalos, K.; Doulamis, A.; Loupos, K. Deep learning-based man-made object detection from hyperspectral data. In Proceedings of the 11th International Symposium, ISVC 2015, Las Vegas, NV, USA, 14–16 December 2015; pp. 717–727.
19. Nogueira, K.; Penatti, O.A.B.; Santos, J.A.D. Towards Better Exploiting Convolutional Neural Networks for Remote Sensing Scene Classification. *Pattern Recognit.* **2016**, *61*, 539–556. [CrossRef]

20. Yu, X.; Dong, H. PTL-CFS based deep convolutional neural network model for remote sensing classification. *Computing* **2018**, *100*, 773–785. [CrossRef]
21. Girshick, R.; Donahue, J.; Darrell, T.; Malik, J. Rich feature hierarchies for accurate object detection and semantic segmentation. In Proceedings of the IEEE Conference on Computer Vision and Pattern Recognition, Columbus, OH, USA, 24–27 June 2014; pp. 580–587.
22. Girshick, R. Fast R-CNN. In Proceedings of the IEEE International Conference on Computer Vision, Santiago, Chile, 13–16 December 2015; pp. 1440–1448.
23. Ren, S.; He, K.; Girshick, R.; Sun, J. Faster R-CNN: Towards real-time object detection with region proposal networks. In Proceedings of the Neural Information Processing Systems, Montreal, QC, Canada, 7–12 December 2015.
24. Long, J.; Shelhamer, E.; Darrell, T. Fully Convolutional Networks for Semantic Segmentation. In Proceedings of the IEEE Conference on Computer Vision and Pattern Recognition, Boston, MA, USA, 8–10 June 2015; pp. 3431–3440.

Article

A Multi-Task Framework for Facial Attributes Classification through End-to-End Face Parsing and Deep Convolutional Neural Networks

Khalil Khan [1,6,*], **Muhammad Attique** [2,*], **Rehan Ullah Khan** [3,6], **Ikram Syed** [4] and **Tae-Sun Chung** [5]

[1] Department of Electrical Engineering, University of Azad Jammu and Kashmir, Muzaffarabad 13100, Pakistan
[2] Department of Software, Sejong University, Seoul 05006, Korea
[3] Department of Information Technology, College of Computer, Qassim University, Al-Mulida 51431, Saudi Arabia; Re.khan@qu.edu.sa
[4] Department of Computer Science, The Superior College, Lahore 54000, Pakistan; ikram.syed@superior.edu.pk
[5] Department of Computer Engineering, Ajou University, Ajou 16499, Korea; tschung@ajou.ac.kr
[6] Intelligent Analytics Group (IAG), College of Computer, Qassim University, Al-Mulida 51431, Saudi Arabia
*   Correspondence: khalil.khan@ajku.edu.pk (K.K.); attique@sejong.ac.kr (M.A.)

Received: 13 November 2019; Accepted: 30 December 2019; Published: 7 January 2020

Abstract: Human face image analysis is an active research area within computer vision. In this paper we propose a framework for face image analysis, addressing three challenging problems of race, age, and gender recognition through face parsing. We manually labeled face images for training an end-to-end face parsing model through Deep Convolutional Neural Networks. The deep learning-based segmentation model parses a face image into seven dense classes. We use the probabilistic classification method and created probability maps for each face class. The probability maps are used as feature descriptors. We trained another Convolutional Neural Network model by extracting features from probability maps of the corresponding class for each demographic task (race, age, and gender). We perform extensive experiments on state-of-the-art datasets and obtained much better results as compared to previous results.

Keywords: face image analysis; deep learning; face parsing; facial attributes classification

---

1. Introduction

Face image analysis describes several face perception tasks, including face recognition, race classification, face detection, age classification, gender recognition, etc. These demographic attributes have been given immense attention in recent computer vision research due to large scale applications. Face analysis plays a crucial role in different real-world applications, including image augmentation, animations, biometrics, visual surveillance, human-computer interaction, and many other commercial applications. Despite significant research developments, face analysis is still challenging due to various reasons such as complex facial expressions, poor imagery conditions, and complex background. Face analysis has more complications in particular in the unconstrained and 'in the wild' conditions. Motivated by all the above reasons, we propose a multi-task framework that is targeting jointly three facial attributes, including race, age, and gender classification.

Besides a large number of benefits of an autonomous classification of gender, race, and age, there are certain social and ethical issues related to such classification. Some clinical practices believe that race and ethnic classification provides the crucial genetic surrogates that might be helpful in regimens treatment predictions. Due to such factors, and the increasing influence and attention given to the racial disparities in social and health, the definition of race have undergone scientific scrutiny [1]. Such practices have been seen from poor to modern societies. Moreover, the fields of cancer research, treatment, and prevention are facing the complexities of exploiting race and ethnic features for predicting outcomes medical decisions [2]. With reference to the gender classification, the authors in [3] report that female ratings of ethical judgment are consistently higher than that of males across two out of three moral issues examined (i.e., sales and retails) and ethics theories. The analysis of gender-based discrimination in [4] shows that worker characteristics and job search methods do account, although little of the gender gap in earnings.

There are multiple benefits to age, race and gender classification. With the increased use of smart devices, the autonomous recognition of age and gender can provide a large number of application-oriented benefits. One of the most beneficial is the recommendation systems. When the age and gender of a child are recognized, it should be helpful for many applications such as YouTube. YouTube can then use this information to recommend autonomously the age-based filtered videos. This can help in presenting related information to the user. Such recognition is also useful for autonomous parental controls of the websites and video services. The applications should thus provide a better experience, control, and security if the age of a particular user is correctly recognized. Similar other benefits can be exhibited by the computer-based applications if gender and age are recognized. Many shopping recommendation systems can present customized items to users just by recognizing their gender and age.

Compared to age and gender, we find that the justification and uses of race classification are expressively limited, and thus race classification is not only a sensitive and challenging matter, but many societies consider it an unethical process. Because it is believed that it could create and motivate social problems among masses. However, we believe that race classification can also be useful to several applications and scenarios. For example, the advanced countries experience an influx of illegal immigrants seeping into the country through several un-explored channels by the security agencies. The autonomous recognition and classification of the race at a number of locations inside the country can be very useful in this regard. Moreover, the border control can use such classification for better understanding and blocking of forged identities. Facebook and other social applications can use race classification for recommending related information, including but not limited to friends and products.

Typically, each of these facial attributes classification (race, age, and gender) are addressed individually through different set of methods [5–12]. We argue, all these tasks can be addressed in a single framework if sufficient information about different face parts is provided. In the proposed framework, we provide various face parts information through a prior segmentation model, which we develop through Deep Convolutional Networks (DCNNs). The psychology literature also confirmed the fact that different face parts help the human visual system to recognize face identity, and all face parts information is mutually related [13,14]. Therefore, the performance of all face related applications can be improved if a well-segmented face image having sufficient face parts information is given as input to the model.

The literature reports various methods to address human face analysis. Among all reported methods, face analysis through landmarks information is frequently used by researchers [8,15]. However, the performance in such cases is highly dependent on accurate facial landmarks information, which in real-world scenarios is again challenging [9,11,16]. These landmarks location identification is greatly effected with image rotation, occlusions, or if images are with poor quality. Similarly, landmarks extraction is again difficult if the images are collected in far-field imagery conditions. Due to all the problems mentioned above, we approach the face image analysis differently, i.e., providing prior face parts information through face image parsing.

We introduce a new framework in which face parts information is provided through a prior face segmentation model, which we develop through DCNNs. We address the three demographic tasks (race, age, and gender classification) through the face parts information provided previously.

The proposed model is a joint estimation probability task that tackles it through DCNNs. The multi-task model can be formulated as;

$$(r, a, g) = \arg\max_{r,a,g} p(r, a, g | \mathbf{I}, \mathbf{B}) \tag{1}$$

where race, age, and gender are represented by r, a and g respectively. The input face image is represented by I and the bounding box by B in Equation (1).

Multi-class face segmentation (MCFS) is already addressed by researchers [17–19]. Previously, face parsing was considered as three or sometimes four-class classification problem. In MCFS [17], face parsing was extended to six classes, including skin, hair, back, nose, mouth, and eyes. The MCFS [17] was developed through traditional machine learning methods (TMLMs). We addressed face parsing through DCNNs instead of TMLMs, obtained much better results as compared to previous results. Moreover, we extended our current research work to seven classes by adding eyebrow class. Additionally, MCFS [17] was evaluated on a minimal set of images, which we extended into three large datasets. We also extended our work to a joint task of race, age, and gender recognition. To summarize, the contributions of this paper are:

- We propose a new face parsing method through DCNNs, known as MCFP-DCNNs. We develop a unified human face analysis framework using the face parts information provided by a prior MCFP-DCNNs model. The multi-task framework is addressing the three demographic tasks (race, age, and gender) in a single architecture, which we named RAG-MCFP-DCNNs.
- We conduct detailed experiments on state-of-the-art (SOA) databases for face parsing, race, age, and gender classification. We obtained significant improvement in performance on both controlled and unconstrained databases for all four tasks.

The structure of the remaining paper is as follows: Section 2 describes related work for all the four cases, i.e., face parsing, race, age, and gender recognition. The databases used in the proposed work are discussed in Section 3. The proposed face parsing model is presented in Section 4. The multi-task face analysis framework is discussed in Section 5. All obtained results are discussed and compared with SOA in Section 6. The paper is summarized with some future directions in Section 7.

2. Related Work

Human face analysis is a well explored research area in computer vision. In this Section of the paper we review SOA methods used to address face parsing and remaining three demographic tasks.

2.1. Face Parsing

Face parsing methods can be categorized into two groups: local and global based methods. Local face parsing methods trained separate models for different face components such as eyes, nose, mouth, etc. For example, Luo et al. [20] proposed a method segmenting each face part separately. An interlinked DCNNs based method was proposed by Zhou et al. [21]. The approach proposed in [21] is benefiting from the complex sort of designing. The interlinked DCNNs can pass specific information between fine and coarse levels bidirectionally, consequently getting better performance at the expense of large computational cost and memory. A shallow DCNNs method having better computational cost as compared to the last mentioned method is proposed in [22]. SOA accuracy is obtained with [22] having a very fast running speed.

In global face parsing methods, a semantic label is predicted for each pixel over the whole image. Correlation between different face parts through different modeling methods is performed in some cases, as Epitome Model [23] and exemplar modeling method [24]. The underlying layout of the whole face image is performed through DCNNs. For example Aaron et al. [25] used facial landmarks information combined with DCNNs to address face parsing. Saito et al. [26] proposed that the computational cost of the face parsing can be much reduced with DCNNs, which makes a network fit for real-time applications.

Most of the methods mentioned above (algorithms with satisfactory performance) treated facial parts globally and inherently integrated them prior to the face image layout. Pixel labeling accuracy of all these methods was less because individual face parts were not focused upon. Moreover, most of these methods were evaluated with limited databases or images in the databases were collected in very constrained imaging conditions. Additionally, none of these methods addressed maximum face classes, but in most cases, only three or four classes were considered. We evaluated our framework on three large databases, namely, LFW-LP [27], HELEN [28], and FASSEG [29]. These databases include both low and high-resolution images. Images collected in very unconstrained conditions are also included. Moreover, unlike the previous methods considering a few semantic classes, we extend our face parsing work to seven semantic labels.

2.2. Race Classification

Race classification is a well-explored research area, but still, it is challenging due to certain reasons mentioned in the introduction portion of the paper. Recently, a method is proposed by Saliha et al. [30] for race classification. The proposed method combined local binary pattern information and logistic regression on a framework called Spark. Local binary patterns were used for feature extraction, and Spark's regression for classification. The method was evaluated on two databases, namely FERET [31] and CAS-PEAL [32]. Two major races, Asian and Non-Asian, were included in the experimentation.

In holistic race classification methods, the face image is considered as one-dimensional feature vector, and some features are extracted. For example, Gutta et al. [33–35] used the RBF neural network and decision tree for race classification. The work was validated on FERET [31] dataset. Another race classification system was developed by Lu and Jain [36] through discriminant analysis. The system was tested on Asian and Non-Asian races. A support vector machine (SVM) classifier was used as a classification tool in another method proposed in [37]. The framework proposed in [37] was evaluated on a subset of face images from the FERET [31] database.

Manesh et al. [38] extracted face features from images through Gabor filter and used SVM for classification. The method proposed in [38] was evaluated on CAS-PEAL [32] and FERET [31]. Another approach [39] addressed the race classification through skin information and some secondary features such as lips and forehead information. For experiments, Yale [40] and FERET [31] databases were used. The framework classified five race classification, including Asian, American, Caucasian, African, and American. A comprehensive algorithm classifying three races Oriental, European, and African were classified by Salah et al. [41]. Face features were extracted through uniform local binary patterns combined with Haar Wavelet transform. For classification, K-nearest neighbors (KNN) was used. Some more methods addressing race classification through holistic methods can be explored in [42–44].

All the methods mentioned above are performed on a smaller or subset of a larger database. One method which was evaluated on comparatively larger dataset is reported by Xi et al. [45]. In this method, face information was extracted through color features. The performance of the framework was evaluated on the MBGC, having three classes of images. Han et al. [46] proposed another approach which was using biologically inspired features and hierarchical classifiers. Two large scale databases were used by the authors for experimentation, including MORPH2 and PCSO.

DCNNs are extensively used in different computer vision applications due to their excellent performance. A method proposed by Zhang et al. [47] used stacked spare auto encoding for features extraction. The classification was performed with regression Soft-Max method. Another flexible DCNNs method was proposed by Wei et al. [48]. Due to several different object segment hypotheses, this method was also called Hypothesis-CNNs-Pooling. The method proposed by Anwar and Nadeem [42] used DCNNs for feature extraction but performed classification through SVM.

2.3. Age Classification

Age classification can be studied both as regression and classification problem. Age is associated with a certain group in age classification, while the exact age of a person is estimated in the regression case. Our current research study regarding age is limited to classification only. Two survey papers are reported in literature [49,50], which addressed both age classification and estimation. The survey papers reported all the databases to date and also presented an overview of various age estimation methods as well.

Kwon et al. [51] proposed an age classification algorithm by extracting face information and training a classification tool. Face wrinkles information was used as features by the authors. Extension of the age classification using wrinkles information was done in another paper [52]. First facial features localization was performed, and then proper modeling strategy was adapted. Craniofacial growth information was extracted through anthropometric and psychophysical evidences, and modeling was performed. Accurate face features localization is necessary for the last mentioned approach. In some examples, when face features were not localized, the performance of the framework was drastically effected.

A new class of gender recognition methods was proposed known as AGing PatErn subspace methods [53,54]. Regression models were trained in these methods. For training regression models, features from face images that are related to aging were extracted. Both of these methods reported some excellent results as compared to SOA. Two main weaknesses faced by these methods, firstly, it was mandatory for face images to be frontal and well-aligned. Secondly, these methods are well suited for images collected in very controlled environmental conditions. The performance of these methods decreased as exposed to an unconstrained outdoor environment.

Another algorithm that used cost-sensitive hyperplanes information ranking way was introduced by Chang et al. [55]. It was a multi-stage learning algorithm which they named 'a grouping estimation fusion' (DEF). Another method that used features selection procedure was proposed in [56]. All the above-mentioned methods have shown good results in images collected in indoor conditions; however, when exposed to the real-world scenario, a drastic drop in performance was noted.

2.4. Gender Classification

Gender recognition received immense attention for many years due to its large scale applications in face analysis, particularly face recognition [57], soft-biometrics [58], and human-computer interaction [59].

Makinen and Raisamo [60] investigated gender recognition thoroughly in their work. Neural networks was used by early researcher to address gender classification [61]. However, very few (only 90) face images were used by Golom et al. [61]. Jia et al. [62] trained a gender classifier using four million weakly marked images. Similarly, Moghaddam and Yang [63] used SVM with some dimensionality reduction features for gender classification. Another paper [64] used Adaboost classifier for gender classification.

Antipov et al. [65] used deep learning architecture for gender recognition. The authors claimed that much improved performance can be achieved with less training data. The model was validated with CASIA [66] dataset, having 494,414 face images. Jia et al. [62], in another paper, collected a large dataset of five million weakly labeled images. Gender recognition through face segmentation is already explored in another work [67,68]. However, the work proposed in [67,68] has been validated on very limited data and through traditional machine learning methods.

2.5. Multi Tasks Framework

A framework addressing gender and age was prosed by Toews and Arbel [69]. The proposed model is a view-point invariant appearance model that is robust to rotations at the local scale level. Another algorithm proposed by Yu et al. [70] was based on gait and linear discriminant algorithms.

A benchmark for both age and gender was proposed in [71]. Khan et al. [72] suggested another algorithm, also called semantic pyramid gender and action recognition method, which addressed both gender and action recognition. Chen et al. [43] proposed a multi modeling mechanism, which combined both the text and image information. Higher accuracy was reported as compared to SOA with the proposed model. Another generic algorithm proposed in [46] estimated gender, race, and age in a single framework.

The performance of different visual recognition tasks was much improved with recently introduced deep learning architectures. The three demographic attributes (race, age, and gender) were also explored in a single model through these deep learning architectures. For example, a hybrid approach for age and gender was introduced in [73]. DCNNs were used for features extraction and for classification extreme machine learning (EML) strategy was adapted. The proposed method was named CNNs-ELMs due to the joint venture of DCNNs and EML. The CNNs-ELMs was evaluated on two challenging databases MORPH-II [59] and Adience [71].

All the methods mentioned above made lots of progress towards mature face image analysis systems. However, these methods were designed either for non-automated estimation algorithms or worked well in constrained and controlled imaging conditions. Both appearance and geometric based methods were facing some serious problems, we approached face image analysis through a different idea. Our face image analysis and attributes classification idea is novel; in a sense, we approach the face analysis task through a prior face segmentation method. Initially, we segment a face image into seven parts, including mouth, hair, back, skin, nose, eyes, and eyebrow. We used a probabilistic classification strategy and modeled a DCNNs based framework for each demographic task, i.e., race, age, and gender recognition. We test our framework on SOA databases, obtaining superior results as compared to previous results.

3. Used Datasets

In this Section of the paper, we discuss different face image databases we used to evaluate our framework.

3.1. Face Parsing

To the best of our knowledge, three authentic databases are publically available for different face parts labeling. Details of these datasets are as follow;

- **LFW-PL**: We evaluate our face parsing part with LFW-PL [27]. Some recent methods [26,74] already use the LFW-PL [27] for face parsing. We use a subset of training and testing images. For fair and more exact comparisons, we conduct experiments on the same set of images as in [75]. The LFW-PL contains 2927 images with size 250×250, which are all collected in the wild conditions. The ground truth data are created manually through commercial editing software. All face images are labeled to three classes, including back, skin and hair.
- **HELEN**: The HELEN [28] database contains class labels for 11 categories. This database contains 2330 images, each with size 400×400. These images are also manually labeled. The database is divided into a training set (2000) and a validation set (330). We keep the experimental setup as in [75]. Although the HELEN database is a comparatively large database having 11 dense classes, the ground truth labeling is not very precise. Especially the hair class is mostly mislabeled with skin in most of the cases.
- **FASSEG**: The FASSEG [29] consists of both frontal and profile face images. Frontal01, frontal02, and frontal03 contain frontal images of 220 faces along with ground truth data. The subset multipose01 contains profile face images of more than 200 faces. The FASSEG images are taken from other publically available datasets, and ground truth data is created through manual editing tool. The images contain both high and low-resolution data. The illumination conditions and facial expressions are also changing in some cases. The dataset is very precise as ground truth

data is created with extreme care. Figures 1 and 2 show some images from the FASSEG [29] database. Original images are shown in row 1, ground truth in row 2 and the segmentation results in row 3.

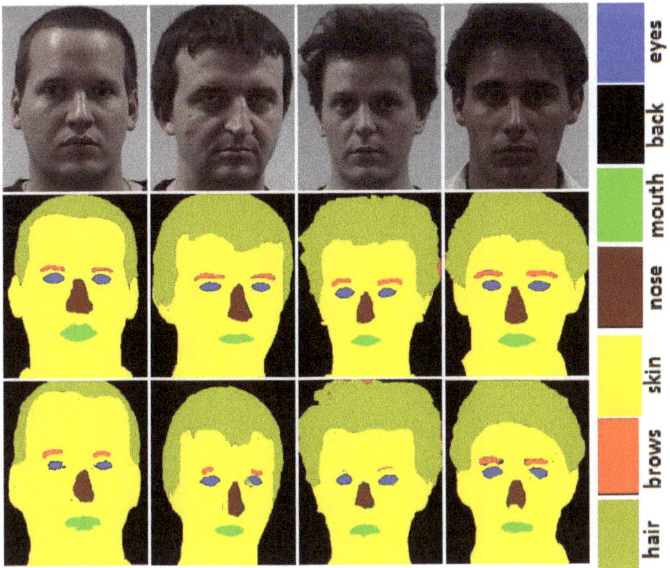

Figure 1. Face parsing results for FASSEG [29] frontal images. First row show: original images, second row: ground truth, and third row: face parsing results.

Figure 2. Face segmentation results for profile images for FASSEG [29] database. Order of the images, row 1 shows: original images, row 2: ground truth, and row 3: face parsing results.

3.2. Race, Age, and Gender

- **CAS-PEAL** The CAS-PEAL [32] is a face database used for various tasks such as head pose estimation, gender recognition, race classification, etc. It is a larger dataset with 99,594 face images. The CAS-PEAL [32] is collected with a large number of face images having 1040 subjects. The dataset is sufficiently large, but the complexity level of the images is not higher, making the dataset a bit simple. We used CAS-PEAL [32] for race classification in the proposed work. Figure 3, row 1 shows some images from the CAS-PEAL [32] database. The ground truth images manually labeled to build DCNNs model are shown in row 2, whereas the segmentation results with proposed DCNNs model are shown in row 3.

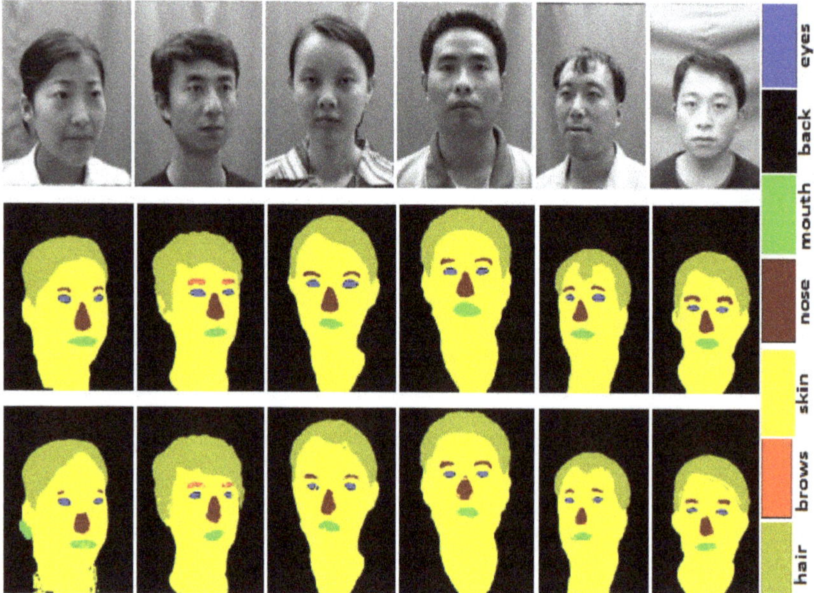

Figure 3. Face images from CAS-PEAL [32] dataset in row 1, ground truth in row 2, and face segmentation results in row 3.

- **FERET:** This is an old dataset which is used for various face analysis tasks such as face recognition, head pose estimation, gender recognition, etc. The FERET [31] dataset is collected in very constrained lab conditions, and gender information is also provided for each participant. It is a medium-sized dataset with 14,126 face images. However a sufficient number of participants are included in the database collection with 1199 subjects. We use the colored version of the FERET [31]. The participants include variations in facial expressions, changing in lighting conditions, which make the database a bit challenging. we evaluate our race and gender recognition part with FERET [31] database. Figure 4 , row 1 shows some face images from the FERET [31] database. The ground truth images manually annotated are shown in row 2, whereas the segmentation results in row 3.
- **LFW:** The LFW [76] database consists of 13,233 face image collected from 5749 participants. The dataset is collected in very unconstrained environmental conditions. All the face images are collected from the internet, with very poor resolution. The LFW [76] is a very unbalanced database, as the number of female participants are 2977, whereas male candidates are 102,566. We use this database for evaluating our gender recognition part.

- **Adience:** The Adience [71] is a new database that was released in 2018. We evaluate our age and gender classification part with Adience [71]. The database is collected in the wild and real-world conditions. The images are collected through smart phones Much complexities are added to the images to make the database rather challenging; such as pose variation, changing lighting conditions, noise, etc. are present in the images. It is a comparatively larger dataset with more than 25,580 face images. Sufficient number of candidates are included (2284) in the dataset collection. Information about the exact age of each participant is not provided, instead each participant is assigned to eight age groups, i.e., [0,2], [4,6], [8,13], [15,20], [25,32], [38,43], [48,53], [60,+]. The database is freely available for downloading from the Open University of Israel.

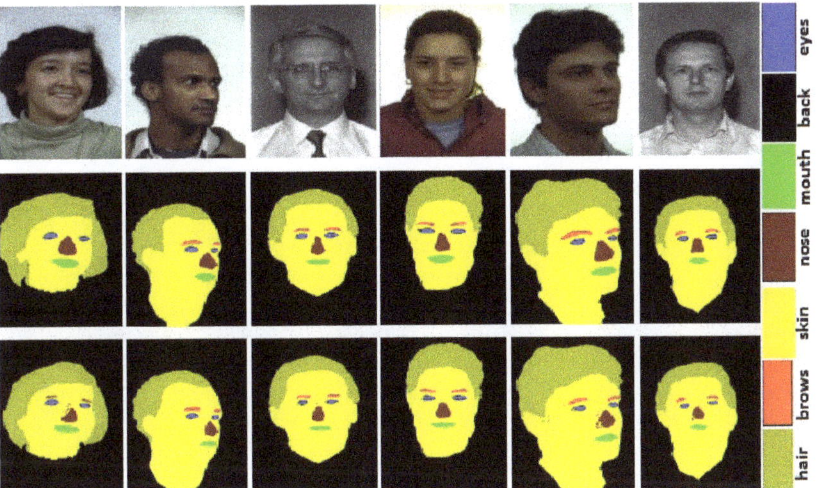

Figure 4. Original face images from FERET [31] database in row 1, ground truth in row 2, and face parsing results in row 3.

4. Proposed Face Parsing Framework (MCFP-DCNNs)

In this Section of the paper, we present the DCNNs we used to build our face parsing model. We make this model for each demographic task, i.e., race, age, and gender.

Face parts are not localized in face images with some datasets. We apply a face detection algorithm in the start if needed. Many excellent face detection algorithms are reported in the literature; we use a deep learning-based face detector reported in [77]. After face detection, we re-scaled all face images to a fixed size 227×227. Details of the proposed DCNNs architecture is in the following paragraphs;

Architecture

The performance of the DCNNs based model depends on several parameters; for example, the size of the kernels used, the convolutional layer numbers, and filters in every layer. In our DCNNs model, we used three convolutional (C1–C3) layers, each followed by a max-pooling layer (P1–P3). The size of the kernel in the convolutional layer was set as 5×5. The down sampling stride for both convolutional and max-pooling layer was fixed at two. We kept the kernel size 3×3 in the max-pooling layer. Table 1 shows details about each convolution layer, kernel size, stride, and feature maps. Various parameters setting of the proposed CNNs is presented in Table 2.

Table 1. Information about each convolutional neural networks (CNNs) layer.

Layer	kernel Size	Stride	Feature Maps	Output Size
Input image	–	–	–	227×227
C-1	5×5	2	96	112×112
P-1	3×3	2	96	56×56
C-2	5×5	2	256	27×27
P-2	3×3	2	256	12×12
C-3	5×5	2	512	5×5
P-3	3×3	2	512	2×2

Table 2. Parameters setting for CNNs training.

Parameters	Vales
Epochs	30
Batch size	125
Momentum	0.9
Base learning rate	10^{-4}

For activation function we used rectified linear unit (ReLu). After each convolutional layer we embedded pooling layer. For pooling layer we used max-pooling.

A complete DCNNs model has three main parts, i.e., convolutional layers, pooling layer, and fully connected layers. We represented the kernels as $N \times M \times C$ where N and M represent the height and width of the filter and C represents the channel. The pooling layers filters are represented by $P \times Q$, where P represents height and Q width of the filter. The fully connected layer is the final layer which performs the task of classification. For better optimization of the DCNNs and more exploring deep learning architecture, readers are recommended to read Goodfellow et al.'s book [78].

5. Proposed RAG-MCFP-DCNNs

Initially, we develop a segmentation model MCFP-DCNNs for each demographic task, i.e., race, age, and gender. The MCFP-DCNNs assigns a semantic class label to each pixel of a face image. We use a probabilistic classification strategy and generated probability maps (PMAPs) for each face class. The PMAPs are computed by converting the probability of each pixel to a gray-scale image. In PMAPs, higher intensity represents a higher value of probability for the most likely class on their respective position and vice versa.

Figure 5 shows some images of FASSEG [29] database and their respective probability maps for seven classes. The better segmentation for a specific class, the higher will be the predicted probability value and vice versa. As a result, a brighter PMAP on the respective position will be obtained. From Figure 5, it is clear that some good results are produced by the segmentation model for skin, back, and hair. These regions can be easily differentiated from the others in a respective PMAP image, as can be seen from column 4, 6, and 8 (PMAPs for hair, back, and skin) in Figure 5. The segmentation results also confirm this fact because much better results are obtained for these classes as compared to minor classes. On the other hand, PMAPs for minor classes can not be differentiated from the remaining parts in a respective PMAPs, leading to a confusing situation. PMAPs for minor classes are shown in column 2, 3, 5, and 7 (brows, eyes, mouth, and nose).

We investigated thoroughly which PMAPs are more helpful in age, race, and gender classification. The PMAPs which are helpful for the respective task are then used for feature extraction.

We presents summary of the proposed RAG-MCFP-DCNNs in Algorithm 1. Initially a segmentation model is developed through CNNs. For the classification of race, age, and gender we use PMAPs created during segmentation. We use these probability maps as features descriptors. We extracted features from these PMAPs through deep convolutional neural networks. After extracting features from PMAPs, feature vectors for the corresponding classes are concatenated to a single unique feature vector which is given to Soft-Max classifier. We use 10-fold cross validation experiments in our work. We represent PMAPs generated for each semantic class as $PMAP_{nose}$, $PMAP_{eyebrow}$, $PMAP_{back}$, $PMAP_{mouth}$, $PMAP_{eyes}$, $PMAP_{skin}$, and $PMAP_{hair}$.

Figure 5. Example face images from FASSEG [29] database, probability maps in the order such that: column 1 shows: original images, 2: eyebrow, 3: eyes, 4: hair, 5: mouth, 6: back, 7: nose, and 8: skin class.

Algorithm 1 proposed RAG-MCFP-DCNNs algorithm

Input: $M_{train} = \{(I_i, T_i)\}_{i=1}^{j}$, M_{test}.

where the DCNNs model is trained through training data represented as M_{train} and tested through M_{test}. The input training image is represented as I and the ground truth data is $T(i,j) \in \{1,2,3,4,5,6,7\}$.

a: Face parsing part:

Step a.1: Training a face parsing model DCNNs through training images and class labels.

Step a.2: Using the probabilistic classification strategy and producing PMAPs for each semantic class, represented as:

$PMAP_{skin}, PMAP_{mouth}, PMAP_{eyes}, PMAP_{nose}, PMAP_{hair}, PMAP_{back}$, and $PMAP_{eyebrow}$

b. race, age and gender classification part:

Training a second DCNNs for each demographic class (race, age, and gender) by extracting infomration from PMAPs of the corresponding classes such that;

if race classification:

$$f = PMAP_{skin} + PMAP_{mouth} + PMAP_{eyes} + PMAP_{nose} + PMAP_{hair} + PMAP_{brows}$$

Else if age classification:
$$f = PMAP_{skin} + PMAP_{mouth} + PMAP_{eyes} + PMAP_{nose} + PMAP_{brows}$$

Else if gender recognition:
$$f = PMAP_{skin} + PMAP_{eyes} + PMAP_{brows} + PMAP_{nose} + PMAP_{mouth}$$

where f is the feature vector.

Output: estimated race, age and gender.

For each face analysis task (race, age, and gender), we train a second DCNNs using the corresponding PMAPs as features descriptors. The DCNNs extract features from the corresponding classes which are used to train and test race, age, and gender classification module.

5.1. Race Classification

We classify face images into two races, i.e., Asian and Non-Asian. For race classification, we used two datasets, namely CAS-PEAL [32] and colored version of FERET [31]. The CAS-PEAL [32] is a Chinese database containing images collected in different poses. We named images of CAS-PEAL [32] as Asian class. The colored version of FERET [31] contains 12,332 face images. All images in FERET [31] are Non-Asian; hence, we named these images as Non-Asian class. Sample face images from both CAS-PEAL [32] and FERET [31] are shown in Figures 3 and 4.

We manually labeled 200 face images from each race class of each database. We used the manually labeled images to build an MCFP-DCNNs model, as discussed in Section 4. For all face images of each database, we generated PMAPs. When a test face image was provided as input to the RAG-MCFP-DCNNs, the model predicted PMAPs from the segmentation part for all seven face classes.

To know which face parts help in race classification, we conducted a set of qualitative and quantitative experiments. A graph in Figure 6 shows which face part contributes towards race classification. From the Figure 6 it is clear that six face classes contribute towards age classification. We utilized PMAPs for eyes, nose, mouth, skin, eyebrows, and hair. We extracted features from PMAPs of the above mentioned six classes through DCNNs. For classification we used Soft-Max classifier as in the first case. We kept 10-fold cross validation experimental setup for race classification. Images that were used to build the MCFP-DCNNs model were excluded from the testing phase.

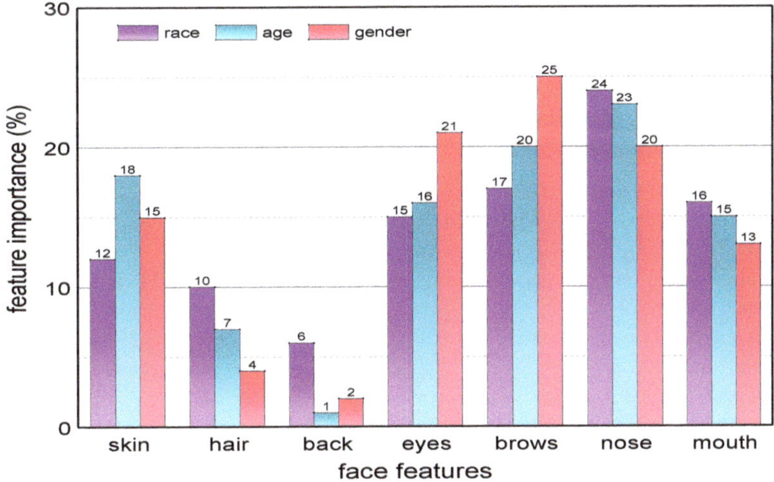

Figure 6. Feature importances of 'skin', 'hair', 'back', 'mouth', 'eyes', 'brows', and 'nose' for three facial attributes (race, age, and gender) classification.

5.2. Age Classification

In our age classification module, each face image is given a specific age category within certain categories. We manually labeled 20 images from each age group. We used the manually labeled images to build an MCFP-DCNNs model for age classification.

We investigated during experiments that each face part has certain contribution towards age classification. Figure 6 shows how different features contributes to the age classification. From Figure 6, it is clear that skin, nose, eyes, mouth, and eyebrows contributes significantly towards age classification. We also notice that using all class information makes the algorithm computationally quite expensive. Therefore, we used a subset of all the seven classes for age classification.

All the testing images were passed to the MCFP-DCNNs to get a probability value for each class. We generated PMAPs for each age category image and each class. We extracted features from the PMAPs through DCNNs. For all face images of each database, we generated PMAPs. When a test face image is provided as input to the MCFP-DCNNs, the model predicts PMAPs from the segmentation part for all seven face classes.

We used Soft-Max for classification, as previously. We kept 10-fold cross validation experimental setup during our experiments. We excluded face images that were previously used to build the MCFP-DCNNs model for age classification.

5.3. Gender Recognition

We manually labeled 50 images from each gender for gender recognition. We built a DCNNs based model for gender classification. We performed intensive experiments to know which face parts help in gender classification. Human face anatomy also helps in gender classification. These fundamental differences also help us to develop a gender classification module; the information is summarized as follows:

- Face anatomy reports that the male forehead is more significant than the female forehead. In most of the cases, male hairline lags behind as compared to female. Is the case of baldness (males only) hairline is missing entirely. All this results in a more massive forehead in males compared to females. We assigned to all forehead a skin label. Hence our MCFP-DCNNs model creates a probability map for skin, which is on the larger brighter area in males compared to females.

- Visually, female eyelashes are curly and comparatively larger. These eyelashes are misclassified with hair and in some cases with eyebrows. Although, labeling accuracy is reduced for segmentation part with this misclassification, however, this misclassification helps in gender classification. The MCFP-DCNNs model generated a brighter PMAPS for males as compared to females.

- Qualitative results reveal that the male nose is larger than the female nose. The male body is larger, which needs a sufficient supply of air towards the lungs. This results in larger nostrils and a giant nose for males compared to females. We also encode this information in the form of PMAPs through MCFP-DCNNs.

- Literature reports very complex geometry for hair. For humans, it is easy to identify the region between hair and face parts, but for the computer, it is not an easy task. Our MCFP-DCNNs model reports excellent labeling accuracy for hair class. From segmented images it can be seen how efficiently hairline is detected by MCFP-DCNNs. We also encoded this information in PMAPs for hair class and used it in gender classification.

- Eyebrows is another class that helps immensely in gender classification. It is generally noticed that female eyebrows are thinner, well managed, and curly at the ends. On the other hand, male eyebrows are thicker, mismanaged. We obtained better labeling accuracy for eyebrow from our face parsing model.

- Mouth is another class that also helps in gender recognition. Female lips are visible and very clear; in the male in some cases (in images), the upper lip is even missing. We encoded this information as well and used it in our modeling process.

Due to the reasons mentioned above, we use PMAPs of five classes including skin, nose, eyes, brows, and mouth to build the second stage of DCNNs model. We perform 10-fold cross-validation experiments to evaluate our model more precisely. We excluded all those images from the testing phase, which were previously used to build MCFP-DCNNs.

6. Results and Discussion

6.1. Experimental Setup

Hardware Platform: For experiments we used intel i7 CPU. RAM of the system was 16 G while graphical processing unit was NVIDIA 840 M. We used TensorFlow and Keras for experiments. We trained the model for 30 Epochs and the batch size was 125.

From Figure 7 it is clear that the mis classification rate reduces as the number of Epochs are increased. At 25 Epochs the miss classification rate for training data almost reaches to 0. This graph is for training data of the face parsing part for HELEN [28] database only.

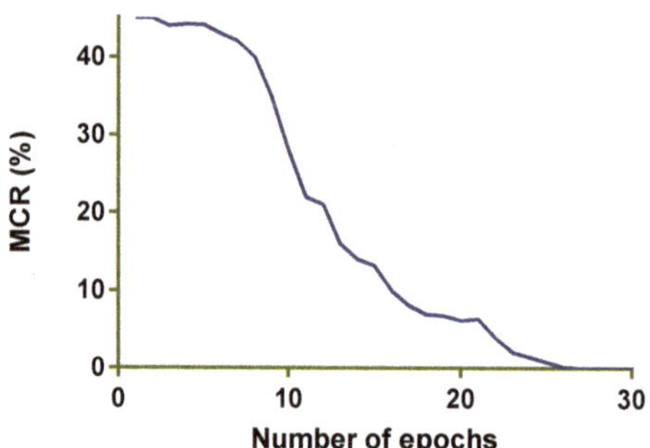

Figure 7. The mis-classification rate (%) for training data (for HELEN [28] database only).

6.2. Face Parsing Results

Previously, FASSEG [29] was evaluated with pixel labeling accuracy (PLA). The PLA compares the estimated segmentation with manually annotated labels. For fair comparison with SOA We also report the face parsing results in the form of confusion matrices.

F-measure is a common metric used in the literature for evaluating a face parsing framework. We used F-measure for our evaluation of our work fot two dataset LFW-PL [27] and HELEN [28]. We used the same setting as in [48,75,79]. The total face images in HELEN are 2330. We used 2000 for training, 230 for the validation, and remaning 100 images for the testing phase.

The LFW-PL [27] is a database with 2972 face images. All these images were manually annotated with three labels: hair, back, and skin. As in [21,27], we also used 1500 images for training, 500 for validation, and 927 for the testing phase. However, we re-annotate the remaining four parts, namely, nose, eyebrow, mouth, and eyes. We reprocess the already annotated parts if needed, as the manual annotation is not precise for LFW-PL [27]. For more accurate comparison, we kept the testing part images unchanged except adding additional four parts labels for our own experiments.

We summarize the key points of the face segmentation results as following:

FASSEG: Some face Images from FASSEG are shown in Figures 1 and 2. Images in Figure 1 are showing some segmentation results; we categorize these as good results. Original face images are placed in the first row, ground truth in the second row, and segmented images in row three. It is also clear from these images that better segmentation is produced for frontal images as compared to profile images, as expected. It can also be observe from these images that better segmentation results are produced for larger classes (skin, hair, and back) while comparatively poor segmentation results are noticed for smaller classes (eyes, nose, mouth, and eyebrows).

There is very limited research work on FASSEG [29] database. Results reported for FASSEG till date are shown in Table 3. From the Table 3 it is clear that we have much better results as compared to previous results. The results reported in [80] do not consider an eyebrows class; we also added eyebrow parts in our current research work.

HELEN: Previously, background class was not considered in face parsing. We believe sometime background class also helps in some real-world application scenarios. We included background class also and reported results in our paper. Results reported with our proposed face parsing model and its comparison with SOA are shown in Table 4. From Table 4, it is clear that most of the previous work does not consider hair class in segmentation due to its complex geometry. We included hair class as well in our face parsing model.

It is clear from Table 4 that we have better results as compared to previous results. However, the performance of the proposed MCFP-DCNNs is poor with minor classes (eyes, brow, and mouth). We obtained better results on three major classes (skin, hair, and back) and one minor class (nose). However, the overall results for the face parsing is improved as the contribution of the major classes in the face is more as compared to minor classes.

LFW-PL: Results for the LFW-PL [27] database is reported in Table 5. From the Table 5, it is clear that we have improvements in results for two major classes, hair and background. However, previous results surpassed us in one case [23], as can be seen. The overall performance is also comparatively poor, as compared to previous results [23]. The possible reason: we included four smaller classes which are comparatively difficult and lower PLA as well. The lower PLA values for smaller classes decreased the overall performance of the framework as well. We also noted the poor performance of the prosed MCFP-DCNNs for background class for LFW-PL [27] database. This confirms the fact that the proposed algorithm is not working good with complicated background scenarios, as LFW-LP [27] images have complex background.

We also noticed that segmentation is highly depended on the quality of face images. For example, we noted poor segmentation results for LFW-PL [27] and much better results on FASSEG [29] and HELEN [28].

Table 3. Face parsing: comparison of the MCFP-DCNN with SOA on FASSEG (frontal) DB. The reported results are based on pixel labeling accuracy.

Method	Eyes	Brows	Mouth	Nose	Skin	Hair	Back	Overall
Khan et al. [80]	60.75	–	84.2	61.25	94.66	95.81	91.50	–
MCFP-DCNNs	84.30	86.25	89.30	87.8	96.58	98.2	94.54	95.12

Table 4. Face parsing: comparison of the MCFP-DCNN with SOA on HELEN DB. The reported results are for F1 measure.

Method	Eyes	Brows	Mouth	Nose	Skin	Hair	Back	Overall
Smith et al. [24]	78.5	92.2	85.7	92.2	88.2	–	–	80.4
Zhou et al. [81]	87.4	81.3	92.6	95.0	–	–	–	87.3
Liu et al. [75]	76.8	71.3	84.1	90.9	91.0	–	–	84.7
Liu et al. [22]	86.8	77.0	89.1	93.0	92.1	–	–	88.6
Wei et al. [48]	84.7	78.6	89.1	93.0	91.5	–	–	90.2
Jonathan et al. [23]	89.7	85.9	95.2	95.6	95.3	88.7	–	93.1
MCFP-DCNN	78.6	83.2	88.5	97.2	96.2	98.4	86.2	95.2

Table 5. Face parsing: comparison of the MCFP-DCNN with SOA on LFW-PL DB. The reported results are for F1 measure.

Method	Eyes	Brows	Mouth	Nose	Skin	Hair	Back	Overall
Liu et al. [75]	–	–	–	–	93.93	80.70	97.10	95.12
Long et al. [82]	–	–	–	–	92.91	82.69	96.32	94.13
Chen et al. [83]	–	–	–	–	92.54	80.14	95.65	93.44
Chen et al. [84]	–	–	–	–	91.17	78.85	94.95	92.49
Zhou et al. [81]	–	–	–	–	94.10	85.16	96.46	95.28
Liu et al. [22]	–	–	–	–	97.55	83.43	94.37	95.46
Jonathan et al. [23]	–	–	–	–	98.77	88.31	98.26	96.71
MCFP-DCNN	78.2	68.3	72.5	85.7	96.8	94.2	97.2	93.25

Race, Age, and Gender Classification Experimental Setup

In this Subsection, the experimental setup for training and testing data of race, age, and gender classification is presented.

Race classification: For race classification, we used two datasets, including FERET [31] and CAS-PEAL [32]. We selected 100 images randomly from each dataset. The 200 chosen images were excluded from the testing phase of race classification.

Gender classification: We used three datasets for gender recognition, including LFW [76], Adience [71], and FERET [31]. We selected 50 images each from these datasets, constituting a total of 150 images. As in race classification, the training phase images were excluded from the testing phase.

Race Classification: We used the Adience [71] dataset for age classification. The Adience dataset contains eight different age categories. We manually labeled 20 images from each group. In this way, the total number of training images we selected were 160. In all the three cases above, the selection of the images for the training phase was random. The training phase images were not included in the testing phase. Moreover, to validate the model and results more precisely, we conducted 10-fold cross validation experiments for all the three cases (race, age, and gender classification).

6.3. Race Classification

To know how much each facial part contributes to specific demographic task, we exploited the feature importance measure, which is returned by a Random Forest implementation as in [85]. Figure 6 shows the feature importance of all the facial parts for all three tasks. From Figure 6, it is clear that the nose has a maximum and background minimum contribution towards race classification.

We report race classification results with classification accuracy. For race classification, we used two datasets, namely CAS-PEAL [32] and FERET [31]. The first database represents Asian and the later Non-Asian class. We manually annotated 100 images from each of these databases for training a DCNNs based model.

The MCFP-DCNNs built for the race was used to create PMAPs for each image in the testing phase. We created PMAPs for all images of both classes. We built another DCNNs using the PMAPs as descriptors and extracting features from the corresponding PMAPs. We performed 10-fold cross-validation experiments in our work. We excluded 200 images that were previously used to train MCFP-DCNNs.

For race classification, we investigated the possible combination of facial features. We noticed during these experiments the contribution of each face part towards race classification. We utilized six face features, excluding background to train a second stage of DCNNs.

We reported results and comparison with SOA in Table 6. From Table 6, it is clear that we have perfect results for Asians and better results as compared to previous results for the Non-Asian class.

From Table 6, it is clear that we used only two classes (Asian and Non-Asian) for experimentation. Although we evaluated our work on two large databases, but the number of races to be classify were limited.

Table 6. Comparison between proposed method and related works on race classification.

Database	Method	Asian (Accuracy%)	Non-Asian (Accuracy%)
RAG-MCFP-DCNNs	–	100	96.4
Manesh et al. [38]	FERET [31] and CAS-PEAL [32]	98	96
Muhammed et al. [86]	FERET [31]	99.4	–
Chen and Ross [43]	CAS-PEAL [32]	98.7	–
Anwar and Naeem [42]	FERET	98.28	–

The computational cost is another factor that we did not consider in our work. One main limitation of the deep learning architectures is a substantial computational cost, which we also faced in our work. Our approach may lag as compared to previous methods if compared computationally, as we built two DCNNs models for complete face analysis.

6.4. Age Classification

We reported our age classification results with classification rate, as in race classification. We used Adience [71] for age classification. This database has eight different age categories. We labeled 20 face images from each category. We built our age classification model with 160 manually labeled images.

The MCFP-DCNNs model was used to create PMAPs for each testing image. After creating PMAPs for all images and all eight classes, we performed 10-fold cross-validation experiments. We excluded all 160 images from the testing phase, which were previously used to build an MCFP-DCNNs model.

We investigated all possible combinations of facial features for age classification. Figure 6 shows which face part has major contribution toward age classification. Again nose has most contribution and back least contribution towards age classification. We used all six face classes, excluding the background class to built a DCNNs model.

We show our reported results and its comparison with SOA in Table 7. It can be seen from Table 7 that we have much better results as compared to SOA on age classification for Adienece [71] database.

We manually labeled ground truth data using an image editing software. We did not use any automatic manually labeling tool. This ground truth labeling has two significant drawbacks. First, such sort of labeling highly depends on the subjective perception of a subject who is involved in all labeling process. Accurate label providing in such cases is tough, specifically differentiating the boundary region between two or more regions is highly challenging. For example, it is very difficult to distinguish the skin region from the nose and vice versa. Second, this ground truth labeling is a very time consuming process. One main drawback of our proposed method is; this research work is limited to age classification only, which is due to tedious labeling process. We do not consider age estimation, because, in that case, a large number of images needed to be labeled. Moreover, the computational cost of the framework will also be sufficiently large.

Table 7. Comparative experiments on age classification using Adience database.

Database	Method	Classification Aaccuracy (%)
Adience	RAG-MCFP-DCNNs	69.4
	Dehghan et al. [87]	61.3
	Hou et al. [88]	61.1
	Hassner et al. [89]	50.7
	Hernandez et al. [90]	51.6
	CNN-ELM [73]	52.3

6.5. Gender Recognition

As in the other two cases, we try all possible combinations of facial features for gender classification. After experimentation, we conclude to use five parts, i.e., nose, mouth, eyebrows, eyes, and skin. Figure 6 shows contributions of each face part in gender classification. We created manually labeled images from each male and female gender. We randomly took 30 images from each gender and each database to train an MCFP-DCNNs model. The total training images were 180. As in the case of age classification, we excluded 180 images that were previously used to build an MCFP-DCNNs model.

We perform gender classification with three datasets, including Adience [71], LFW [76], and FERET [31]. In Table 8, we show classification accuracy for all the three datasets. Table 8 also compares our reported results with SOA. For gender classification, we perform 10-fold cross-validation experiments, as in the previous two cases. We obtained better results as compared to previous results, as can be seen from Table 8.

Table 8. Comparative experiments on gender recognition using Adience, LFW and FERET data-sets.

Database	Method	Classification Accuracy (%)
Adience	RAG-MCFP-DCNNs	93.6
	Levi et al. [91]	86.8
	Lapuschkin et al. [92]	85.9
	CNNs-EML [73]	**77.8**
	Hassner et al. [89]	79.3
LFW	Van et al. [93]	94.4
	RAG-MCFP-DCNNs	94.1
	HyperFace [94]	94.0
	LNets+ANet [95]	94.0
	Moeini et al. [96]	93.6
	PANDA-1 [47]	92.0
	ANet [56]	91.0
	Rai and Khanna [97]	89.1
FERET	RAG-MCFP-DCNNs	100
	Moeini et al. [96]	99.5
	Tapia and Perez [98]	99.1
	Rai and Khanna [97]	98.4
	Afifi and Abdelrahman [99]	99.4
	A priori-driven PCA [100]	84.0

As a whole, we noticed the performance of the proposed RAG-MCFP-DCNNs very interesting. We introduced an idea of human face image analysis, which is using different face parts information provided by a segmentation model. We reached an important observation stating: *"face parts parsing and different visual recognition tasks are closely related, the better segmentation, better results for the three tasks will be observed"*.

7. Conclusions

We proposed an end-to-end face parsing method which tries to address three face image analysis tasks, including race, age, and gender classification. We trained the MCFS-DCNNs model through a DCNNs model by extracting information from various face parts. The MCFS-DCNNs classified every pixel to one of the seven categories (hair, eyebrows, eyes, skin, nose, back, and mouth). We used probabilistic classification method to generate PMAPS for seven face classes. We built another DCNNs model by extracting features from the corresponding PMAPs for each of the three demographic tasks (race, age, and gender). We performed a series of experiments to investigate which face parts help in the race, age, and gender classification. We validate our experiments on seven face databases, obtaining much better results as compared to SOA.

We argue that sufficient information is provided by the face parsing model for different visual recognition tasks. We provide a route towards other complicated face image analysis problems. For example, we intend to add complicated facial expressions, head pose estimation, and many other applications to the framework. We are also planning to optimize the segmentation part to improve the performance of the face parsing part of the framework.

Author Contributions: Conceptualization, K.K. and M.A.; methodology, K.K.; software, R.U.K.; validation, K.K. and T.-S.C.; formal analysis, K.K.; investigation, R.U.K.; resources, T.-S.C.; data curation, K.K.; writing—original draft preparation, K.K.; writing—review and editing, K.K., T.-S.C., I.S. and M.A.; visualization, K.K.; supervision, K.K.; project administration, K.K. All authors have read and agreed to the published version of the manuscript.

Acknowledgments: This research was partially supported by Basic Science Research Program through the NRF, Korea funded by the Ministry of Education (2019R1F1A1058548) and the Ajou university research fund.

Conflicts of Interest: The authors declare no conflict of interest.

References

1. Bonham, V.L.; Knerr, S. Social and ethical implications of genomics, race, ethnicity, and health inequities. *Semin. Oncol. Nurs.* **2008**, *24*, 254–261. [CrossRef] [PubMed]
2. Rebbeck, T.R.; Halbert, C.H.; Sankar, P. Genetics, epidemiology and cancer disparities: Is it black and white? *J. Clin. Oncol.* **2006**, *24*, 2164–2169. [CrossRef] [PubMed]
3. Nguyen, N.T.; Basuray, M.T.; Smith, W.P.; Kopka, D.; McCulloh, D. Moral Issues and Gender Differences in Ethical Judgement Using Reidenbach and Robin's (1990) Multidimensional Ethics Scale: Implications in Teaching of Business Ethics. *J. Bus. Ethics* **2008**, *77*, 417–430. [CrossRef]
4. Huffman, M.L.; Torres, L. Job Search Methods: Consequences for Gender-Based Earnings Inequality. *J. Vocat. Behav.* **2001**, *58*, 127–141. [CrossRef]
5. Asthana, A.; Zafeiriou, S.; Cheng, S.; Pantic, M. Robust discriminative response map fitting with constrained local models. In Proceedings of the 2013 IEEE Conference on Computer Vision and Pattern Recognition, Portland, OR, USA, 23–28 June 2013.
6. Belhumeur, P.N.; Jacobs, D.W.; Kriegman, D.J.; Kumar, N. Localizing parts of faces using a consensus of exemplars. In Proceedings of the CVPR 2011, Providence, RI, USA, 20–25 June 2011.
7. Cao, X.; Wei, Y.; Wen, F.; Sun, J. Face alignment by explicit shape regression. In Proceedings of the CVPR 2012, Providence, RI, USA, 16–21 June 2012.
8. Dantone, M.; Gall, J.; Fanelli, G.; van Gool, L. Real-time facial feature detection using conditional regression forests. In Proceedings of the CVPR 2012, Providence, RI, USA, 16–21 June 2012.
9. Murphy-Chutorian, E.; Trivedi, M.M. Head pose estimation in computer vision: A survey. *IEEE TPAMI* **2009**, *31*, 607–626. [CrossRef]

10. Saragih, J.M.; Lucey, S.; Cohn, J.F. Deformable model fitting by regularized landmark mean-shift. *IJCV* **2011**, *91*, 200–215. [CrossRef]

11. Shan, C.; Gong, S.; McOwan, P.W. Facial expression recognition based on local binary patterns: A comprehensive study. *IVC* **2009**, *27*, 803–816. [CrossRef]

12. Xiong, X.; de la Torre, F. Supervised descent method and its applications to face alignment. In Proceedings of the CVPR 2013, Portland, OR, USA, 23–28 June 2013.

13. Davies, G.; Ellis, H.; Shepherd, J. *Perceiving and Remembering Faces*; Academic Press: Cambridge, MA, USA, 1981.

14. Sinha, P.; Balas, B.; Ostrovsky, Y.; Russell, R. Face recognition by humans: Nineteen results all computer vision researchers should know about. *Proc. IEEE* **2006**, *94*, 1948–1962. [CrossRef]

15. Gross, R.; Baker, S. Generic vs. person specific active appearance models. *IVC* **2005**, *23*, 1080–1093. [CrossRef]

16. Haj, M.A.; Gonzalez, J.; Davis, L.S. On partial least squares in head pose estimation: How to simultaneously deal with misalignment. In Proceedings of the CVPR 2012, Providence, RI, USA, 16–21 June 2012.

17. Khan, K.; Mauro, M.; Leonardi, R. Multi-class semantic segmentation of faces. In Proceedings of the 2015 IEEE International Conference on Image Processing (ICIP), Quebec City, QC, Canada, 27–30 September 2015; pp. 827–831.

18. Khan, K.; Ahmad, N.; Ullah, K.; Din, I. Multiclass semantic segmentation of faces using CRFs. *Turk. J. Electr. Eng. Comput. Sci.* **2017**, *25*, 3164–3174. [CrossRef]

19. Khan, K.; Mauro, M.; Migliorati, P.; Leonardi, R. Head pose estimation through multi-class face segmentation. In Proceedings of the 2017 IEEE International Conference on Multimedia and Expo (ICME), Hong Kong, China, 10–14 July 2017; pp. 253–258.

20. Luo, P.; Wang, X.; Tang, X. Hierarchical face parsing via deep learning. In Proceedings of the IEEE Conference on Computer Vision and Pattern Recognition, Providence, RI, USA, 16–21 June 2012; pp. 2480–2487.

21. Zhou, Y.; Hu, X.; Zhang, B. Interlinked convolutional neural networks for face parsing. In Proceedings of the International Symposium on Neural Networks, Jeju, Korea, 15–18 October 2015; pp. 222–231.

22. Liu, S.; Shi, J.; Liang, J.; Yang, M.-H. Face parsing via recurrent propagation. *arXiv* **2017**, arXiv:1708.01936.

23. Warrell, J.; Prince, S.J.D. Labelfaces: Parsing facial features by multiclass labeling with an epitome prior. In Proceedings of the 2009 16th IEEE International Conference on Image Processing (ICIP), Cairo, Egypt, 7–10 November 2009; pp. 2481–2484.

24. Smith, B.M.; Zhang, L.; Brandt, J.; Lin, Z.; Yang, J. Exemplar-based face parsing. In Proceedings of the IEEE Conference on Computer Vision and Pattern Recognition, Portland, OR, USA, 23–28 June 2013; pp. 3484–3491.

25. Jackson, A.S.; Valstar, M.; Tzimiropoulos, G. A CNN cascade for landmark guided semantic part segmentation. In Proceedings of the European Conference on Computer Vision (ECCV), Amsterdam, The Netherlands, 11–14 October 2016; pp. 143–155.

26. Saito, S.; Li, T.; Li, H. Real-time facial segmentation and performance capture from rgb input. In Proceedings of the European Conference on Computer Vision, Amsterdam, The Netherlands, 11–14 October 2016; pp. 244–261.

27. Kae, A.; Sohn, K.; Lee, H.; Learned-Miller, E. Augmenting CRFs with Boltzmann machine shape priors for image labeling. In Proceedings of the IEEE Conference on Computer Vision and Pattern Recognition, Portland, OR, USA, 23–28 June 2013; pp. 2019–2026.

28. Smith, B.; Zhang, L.; Brandt, J.; Lin, Z.; Yang, J. Exemplarbased face parsing. In Proceedings of the CVPR 2013, Portland, OR, USA, 23–28 June 2013.

29. Benini, S.; Khan, K.; Leonardi, R.; Mauro, M.; Migliorati, P. FASSEG: A FAce semantic SEGmentation repository for face image analysis. *Data Brief* **2019**, *24*, 103881. [CrossRef] [PubMed]

30. Mezzoudj, S.; Ali, B.; Rachid, S. Towards large-scale face-based race classification on spark framework. *Multimed. Tools Appl.* **2019**. [CrossRef]

31. Phillips, P.J.; Wechsler, H.; Huang, J.; Rauss, P.J. The FERET database and evaluation procedure for face-recognition algorithms. *Image Vis. Comput.* **1998**, *16*, 295–306. [CrossRef]

32. Gao, W.; Cao, B.; Shan, S.; Chen, X.; Zhou, D.; Zhang, X.; Zhao, D. The cas-peal large-scale chinese face database and baseline evaluations. *IEEE Trans. Syst. Man Cybern. Part A Syst. Hum.* **2008**, *38*, 149–161.

33. Gutta, S.; Huang, J.; Jonathon, P.; Wechsler, H. Mixture of experts for classification of gender, ethnic origin, and pose of human faces. *IEEE Trans. Neural Netw.* **2000**, *11*, 948–960. [CrossRef]

34. Gutta, S.; Wechsler, H. Gender and ethnic classification of human faces using hybrid classifiers. In Proceedings of the IJCNN'99 International Joint Conference on Neural Networks, Washington, DC, USA, 10–16 July 1999; Volume 6, pp. 4084–4089.

35. Gutta, S.; Wechsler, H.; Phillips, P.J. Gender and ethnic classification of face images. In Proceedings of the Third IEEE International Conference on Automatic Face and Gesture Recognition, Nara, Japan, 14–16 April 1998; pp. 194–199.

36. Lu, X.; Jain, A.K. Ethnicity identification from face images. *Proc. SPIE* **2004**, *5404*, 114–123.

37. Ou, Y.; Wu, X.; Qian, H.; Xu, Y. A real time race classification system. In Proceedings of the 2005 IEEE International Conference on Information Acquisition, Hong Kong, China, 27 June–3 July 2005; p. 6.

38. Manesh, F.S.; Ghahramani, M.; Tan, Y.-P. Facial part displacement effect on template-based gender and ethnicity classification. In Proceedings of the 2010 11th International Conference on Control Automation Robotics & Vision, Singapore, 7–10 December 2010; pp. 1644–1649.

39. Roomi, S.M.M.; Virasundarii, S.; Selvamegala, S.; Jeevanandham, S.; Hariharasudhan, D. Race classification based on facial features. In Proceedings of the 2011 Third National Conference on Computer Vision, Pattern Recognition, Image Processing and Graphics, Hubli, India, 15–17 December 2011; pp. 54–57.

40. Minear, M.; Park, D.C. A lifespan database of adult facial stimuli. *Behav. Res. Methods Instrum. Comput.* **2004**, *36*, 630–633. [CrossRef]

41. Salah, S.H.; Du, H.; Al-Jawad, N. Fusing local binary patterns with wavelet features for ethnicity identification. *World Acad. Sci. Eng. Technol. Int. J. Comput. Inf. Syst. Control. Eng.* **2013**, *7*, 471.

42. Anwar, I.; Islam, N.U. Learned features are better for ethnicity classification. *Cybern. Inf. Technol.* **2017**, *17*, 152–164. [CrossRef]

43. Chen, H.; Gallagher, A.; Girod, B. Face modeling with first name attributes. *IEEE Trans. Pattern Anal. Mach. Intell.* **2014**, *36*, 1860–1873. [CrossRef] [PubMed]

44. Lee, S.; Huh, J.-H. An effective security measures for nuclear power plant using big data analysis approach. *J. Supercomput.* **2019**, *75*, 4267–4294. [CrossRef]

45. Xie, Y.; Luu, K.; Savvides, M. A robust approach to facial ethnicity classification on large scale face databases. In Proceedings of the 2012 IEEE Fifth International Conference on Biometrics: Theory, Applications and Systems (BTAS), Arlington, VA, USA, 23–27 September 2012; pp. 143–149.

46. Han, H.; Otto, C.; Liu, X.; Jain, A.K. Demographic estimation from face images: human vs. machine performance. *IEEE Trans. Pattern Anal. Mach. Intell.* **2015**, *37*, 1148–1161. [CrossRef] [PubMed]

47. Zhang, N.; Paluri, M.; Ranzato, M.; Darrell, T.; Bourdev, L. Panda: Pose aligned networks for deep attribute modeling. In Proceedings of the IEEE Conference on Computer Vision and Pattern Recognition, Columbus, OH, USA, 24–27 June 2014; pp. 1637–1644.

48. Wei, Y.; Xia, W.; Lin, M.; Huang, J.; Ni, B.; Dong, J.; Zhao, Y.; Yan, S. HCP: A flexible CNN framework for multi-label image classification. *IEEE Trans. Pattern Anal. Mach. Intell.* **2016**, *38*, 1901–1907. [CrossRef] [PubMed]

49. Ragheb, A.R.; Kamsin, A.; Ismail, M.A.; Abdelrahman, S.A.; Zerdoumi, S. Face recognition and age estimation implications of changes in facial features: A critical review study. *IEEE Access* **2018**, *6*, 28290–28304.

50. Fu, Y.; Guo, G.; Huang, T.S. Age synthesis and estimation via faces: A survey. *IEEE Trans. Pattern Anal. Mach. Intell.* **2010**, *32*, 1955–1976.

51. Kwon, Y.H.; da Vitoria Lobo, N. Age classification from facial images. In Proceedings of the CVPR 1994, Seattle, WA, USA, 21–23 June 1994; pp. 762–767.

52. Ramanathan, N.; Chellappa, R. Modeling age progression in young faces. In Proceedings of the CVPR 2006, New York, NY, USA, 17–22 June 2006; pp. 387–394.

53. Geng, X.; Zhou, Z.H.; Smith-Miles, K. Automatic age estimation based on facial aging patterns. *IEEE Trans. Pattern Anal. Mach. Intell.* **2007**, *29*, 2234–2240. [CrossRef]

54. Guo, G.; Fu, Y.; Dyer, C.R.; Huang, T.S. Image-based human age estimation by manifold learning and locally adjusted robust regression. *IEEE Trans. Image Process.* **2008**, *17*, 1178–1188.

55. Chang, K.Y.; Chen, C.S. A learning framework for age rank estimation based on face images with scattering transform. *IEEE Trans. Image Process.* **2015**, *24*, 785–798. [CrossRef]

56. Li, C.; Liu, Q.; Dong, W.; Zhu, X.; Liu, J.; Lu, H. Human age estimation based on locality and ordinal information. *IEEE Trans. Cybern.* **2015**, *45*, 2522–2534. [CrossRef]

57. Kumar, N.; Berg, A.; Belhumeur, P.; Nayar, S. Attribute and smile classifiers for face verification. In Proceedings of the ICCV 2009, Kyoto, Japan, 29 September–2 October 2009.

58. Demirkus, M.; Precup, D.; Clark, J.J.; Arbel, T. Soft Biometric trait classification from real worl face videos conditioned on head pose estimation. In Proceedings of the 2012 IEEE Computer Society Conference on Computer Vision and Pattern Recognition Workshops, Providence, RI, USA, 16–21 June 2012.

59. Ricanek, K.; Tesafaye, T. Morph: A longitudinal image database of normal adult age-progression. In Proceedings of the Seventh International Conference on Automatic Face and Gesture Recognition, Southampton, UK, 10–12 April 2006; pp. 341–345.

60. Makinen, E.; Raisamo, R. Evaluation of gender classification methods with automatically detected and aligned faces. *IEEE Trans. Pattern Anal. Mach. Intell.* **2008**, *30*, 541–547. [CrossRef]

61. Golomb, B.A.; Lawrence, D.T.; Sejnowski, T.J. Sexnet: A neural network identifies sex from human faces. In Proceedings of the 1990 Conference on Advances in Neural Information Processing Systems 3, Denver, CO, USA, 26–29 November 1990; pp. 572–577.

62. Jia, S.; Lansdall-Welfare, T.; Cristianini, N. Gender classification by deep learning on millions of weakly labelled images. In Proceedings of the IEEE International Conference on Data Mining Workshops (ICDMW), Barcelona, Spain, 12–15 December 2016; pp. 462–467.

63. Moghaddam, B.; Yang, M.-H. Learning gender with support faces. *IEEE Trans. Pattern Anal. Mach. Intell.* **2002**, *24*, 707–711. [CrossRef]

64. Baluja, S.; Rowley, H.A. Boosting sex identification performance. *Int. J. Comput. Vis.* **2006**, *71*, 111–119. [CrossRef]

65. Antipov, G.; Berrani, S.A.; Dugelay, J.L. Minimalistic cnnbased ensemble model for gender prediction from face images. *Pattern Recognit. Lett.* **2016**, *70*, 59–65. [CrossRef]

66. Yi, D.; Lei, Z.; Liao, S.; Li, S.Z. Learning face representation from scratch. *arXiv* **2014**, arXiv:1411.7923.

67. Khan, K.; Attique, M.; Syed, I.; Gul, A. Automatic Gender Classification through Face Segmentation. *Symmetry* **2019**, *11*, 770. [CrossRef]

68. Khan, K.; Attique, M.; Syed, I.; Sarwar, G.; Irfan, M.A.; Khan, R.U. A Unified Framework for Head Pose, Age and Gender Classification through End-to-End Face Segmentation. *Entropy* **2019**, *21*, 647. [CrossRef]

69. Toews, M.; Arbel, T. Detection, localization, and sex classification of faces from arbitrary viewpoints and under occlusion. *IEEE Trans. Pattern Anal. Mach. Intell.* **2009**, *31*, 1567–1581. [CrossRef]

70. Yu, S.; Tan, T.; Huang, K.; Jia, K.; Wu, X. A study on gait-based gender classification. *IEEE Trans. Image Process.* **2009**, *18*, 1905–1910. [PubMed]

71. Eidinger, E.; Enbar, R.; Hassner, T. Age and gender estimation of unfiltered faces. *IEEE Trans. Inf. Forensics Secur.* **2014**, *9*, 2170–2179. [CrossRef]

72. Khan, F.S.; van de Weijer, J.; Anwer, R.M.; Felsberg, M.; Gatta, C. Semantic pyramids for gender and action recognition. *IEEE Trans. Image Process.* **2014**, *23*, 3633–3645. [CrossRef] [PubMed]

73. Duan, M.; Li, K.; Yang, C.; Li, K. A hybrid deep learning CNN–ELM for age and gender classification. *Neurocomputing* **2018**, *275*, 448–461. [CrossRef]

74. Lin, J.; Yang, H.; Chen, D.; Zeng, M.; Wen, F.; Yuan, L. Face Parsing with RoI Tanh-Warping. In Proceedings of the IEEE Conference on Computer Vision and Pattern Recognition, Long Beach, CA, USA, 16–20 June 2019; pp. 5654–5663.

75. Liu, S.; Yang, J.; Huang, C.; Yang, M.-H. Multi-objective convolutional learning for face labeling. In Proceedings of the IEEE Conference on Computer Vision and Pattern Recognition, Boston, MA, USA, 7–12 June 2015; pp. 3451–3459.

76. Huang, G.; Mattar, M.; Berg, T.; Learned-Miller, E. Labeled faces in the wild: A database for studying face recognition in unconstrained environments. In Proceedings of the Workshop on Faces in'RealLife'Images: Detection, Alignment, and Recognition, Marseille-France, 12–18 October 2008.

77. Liu, W.; Anguelov, D.; Erhan, D.; Szegedy, C.; Reed, S.; Fu, C.-Y.; Berg, A.C. SSD: Single shot multibox detector. In Proceedings of the Computer Vision—ECCV 2016, Amsterdam, The Netherlands, 11–14 October 2016; Leibe, B., Matas, J., Sebe, N., Welling, M., Eds.; Springer International Publishing: Cham, Switzerland, 2016; pp. 21–37.

78. Goodfellow, I.; Bengio, Y.; Courville, A. *Deep Learning*; MIT Press: Cambridge, MA, USA, 2016.

79. Yamashita, T.; Nakamura, T.; Fukui, H.; Yamauchi, Y.; Fujiyoshi, H. Cost-alleviative learning for deep convolutional neural network-based facial part labeling. *IPSJ Trans. Comput. Vis. Appl.* **2015**, *7*, 99–103. [CrossRef]

80. Khan, K.; Ahmad, N.; Khan, F.; Syed, I. A framework for head pose estimation and face segmentation through conditional random fields. *Signal, Image Video Process.* **2019**, 1–8. [CrossRef]

81. Zhou, L.; Liu, Z.; He, X. Face parsing via a fully-convolutional continuous CRF neural network. *arXiv* **2017**, arXiv:1708.03736.

82. Long, J.; Shelhamer, E.; Darrell, T. Fully convolutional networks for semantic segmentation. In Proceedings of the IEEE Conference on Computer Vision and Pattern Recognition, Boston, MA, USA, 7–12 June 2015; pp. 3431–3440.

83. Chen, L.-C.; Barron, J.T.; Papandreou, G.; Murphy, K.; Yuille, A.L. Semantic image segmentation with task-specific edge detection using CNNs and a discriminatively trained domain transform. In Proceedings of the IEEE Conference on Computer Vision and Pattern Recognition, Las Vegas, NV, USA, 26 June–1 July 2016; pp. 4545–4554.

84. Chen, L.-C.; Papandreou, G.; Kokkinos, I.; Murphy, K.; Yuille, A.L. Deeplab: Semantic image segmentation with deep convolutional nets, atrous convolution, and fully connected CRFs. *arXiv* **2016**, arXiv:1606.00915.

85. Pedregosa, F.; Varoquaux, G.; Gramfort, A.; Michel, V.; Thirion, B.; Grisel, O.; Blondel, M.; Prettenhofer, P.; Weiss, R.; Dubourg, V.; et al. Scikit-learn: Machine learning in Python. *J. Mach. Learn. Res.* **2011**, *12*, 2825–2830.

86. Muhammad, G.; Hussain, M.; Alenezy, F.; Bebis, G.; Mirza, A.M.; Aboalsamh, H. Race classification from face images using local descriptors. *Int. J. Artif. Intell. Tools* **2012**, *21*, 1250019. [CrossRef]

87. Dehghan, A.; Ortiz, E.G.; Shu, G.; Masood, S.Z. Dager: Deep age, gender and emotion recognition using convolutional neural network. *arXiv* **2017**, arXiv:1702.04280.

88. Hou, L.; Yu, C.P.; Samaras, D. Squared earth mover's distance-based loss for training deep neural networks. *arXiv* **2016**, arXiv:1611.05916.

89. Hassner, T.; Harel, S.; Paz, E.; Enbar, R. Effective face frontalization in unconstrained images. In Proceedings of the CVPR, Boston, MA, USA, 7–12 June 2015.

90. Hernández, S.; Vergara, D.; Valdenegro-Toro, M.; Jorquera, F. Improving predictive uncertainty estimation using Dropout–Hamiltonian Monte Carlo. *Soft Comput.* **2018**, 1–16. [CrossRef]

91. Levi, G.; Hassner, T. Age and gender classification using convolutional neural networks. In Proceedings of the IEEE Conference on Computer Vision and Pattern Recognition Workshops, Boston, MA, USA, 7–12 June 2015; pp. 34–42.

92. Lapuschkin, S.; Binder, A.; Muller, K.-R.; Samek, W. Understanding and comparing deep neural networks for age and gender classification. In Proceedings of the IEEE International Conference on Computer Vision, Venice, Italy, 22–29 October 2017; pp. 1629–1638.

93. van de Wolfshaar, J.; Karaaba, M.F.; Wiering, M.A. Deep convolutional neural networks and support vector machines for gender recognition. In Proceedings of the IEEE Symposium Series on Computational Intelligence, Cape Town, South Africa, 7–10 December 2015; pp. 188–195.

94. Ranjan, R.; Patel, V.M.; Chellappa, R. Hyperface: A deep multi-task learning framework for face detection, landmark localization, pose estimation, and gender recognition. *IEEE Trans. Pattern Anal. Mach. Intell.* **2019**, *41*, 121–135. [CrossRef] [PubMed]

95. Kumar, N.; Belhumeur, P.N.; Nayar, S.K. FaceTracer: A Search Engine for Large Collections of Images with Faces. In Proceedings of the European Conference on Computer Vision (ECCV), Marseille, France, 12–18 October 2008; pp. 340–353.

96. Moeini, H.; Mozaffari, S. Gender dictionary learning for gender classification. *J. Vis. Commun. Image Represent.* **2017**, *42*, 1–13. [CrossRef]

97. Rai, P.; Khanna, P. An illumination, expression, and noise invariant gender classifier using two-directional 2DPCA on real Gabor space. *J. Vis. Lang. Comput.* **2015**, *26* (Suppl. C), 15–28. [CrossRef]

98. Tapia, J.E.; Perez, C.A. Gender classification based on fusion of different spatial scale features selected by mutual information from histogram of lbp, intensity, and shape. *IEEE Trans. Inf. Forensics Secur.* **2013**, *8*, 488–499. [CrossRef]

99. Afifi, M.; Abdelhamed, A. AFIF4: Deep gender classification based on AdaBoost-based fusion of isolated facial features and foggy faces. *arXiv* **2017**, arXiv:1706.04277.

100. Thomaz, C.; Giraldi, G.; Costa, J.; Gillies, D. A priori-driven PCA. In Proceedings of the Computer Vision ACCV 2012 Workshops, Lecture Notes in Computer Science, Daejeon, Korea, 5–9 November 2013; pp. 236–247.

Article

EEG-Based Multi-Modal Emotion Recognition using Bag of Deep Features: An Optimal Feature Selection Approach

Muhammad Adeel Asghar [1], Muhammad Jamil Khan [1], Fawad [1], Yasar Amin [1], Muhammad Rizwan [2], MuhibUr Rahman [3,*], Salman Badnava [4,*] and Seyed Sajad Mirjavadi [5]

[1] Department of Telecommunication Engineering, University of Engineering and Technology, Taxila 47050, Pakistan; adeel.asghar@students.uettaxila.edu.pk (M.A.A.); muhammad.jamil@uettaxila.edu.pk (M.J.K.); engr.fawad@students.uettaxila.edu.pk (F.); yasar.amin@uettaxila.edu.pk (Y.A.)

[2] Department of Computer Engineering, University of Engineering and Technology, Taxila 47050, Pakistan; muhammad.rizwan@uettaxila.edu.pk

[3] Department of Electrical Engineering, Polytechnique Montreal, Montreal, QC H3T 1J4, Canada

[4] Department of Computer Science and Engineering, College of Engineering, Qatar University, P.O. Box 2713 Doha, Qatar

[5] Department of Mechanical and Industrial Engineering, College of Engineering, Qatar University, P.O. Box 2713 Doha, Qatar; seyedsajadmirjavadi@gmail.com

* Correspondence: muhibur.rahman@polymtl.ca (M.R.); sb1107439@qu.edu.qa (S.B.)

Received: 30 October 2019; Accepted: 26 November 2019; Published: 28 November 2019

Abstract: Much attention has been paid to the recognition of human emotions with the help of electroencephalogram (EEG) signals based on machine learning technology. Recognizing emotions is a challenging task due to the non-linear property of the EEG signal. This paper presents an advanced signal processing method using the deep neural network (DNN) for emotion recognition based on EEG signals. The spectral and temporal components of the raw EEG signal are first retained in the 2D Spectrogram before the extraction of features. The pre-trained AlexNet model is used to extract the raw features from the 2D Spectrogram for each channel. To reduce the feature dimensionality, spatial, and temporal based, bag of deep features (BoDF) model is proposed. A series of vocabularies consisting of 10 cluster centers of each class is calculated using the k-means cluster algorithm. Lastly, the emotion of each subject is represented using the histogram of the vocabulary set collected from the raw-feature of a single channel. Features extracted from the proposed BoDF model have considerably smaller dimensions. The proposed model achieves better classification accuracy compared to the recently reported work when validated on SJTU SEED and DEAP data sets. For optimal classification performance, we use a support vector machine (SVM) and k-nearest neighbor (k-NN) to classify the extracted features for the different emotional states of the two data sets. The BoDF model achieves 93.8% accuracy in the SEED data set and 77.4% accuracy in the DEAP data set, which is more accurate compared to other state-of-the-art methods of human emotion recognition.

Keywords: emotion recognition; brain computer interface; bag of deep features; continuous wavelet transform

---

1. Introduction

Brain–computer interface has been used for decades in the biomedical engineering field to control devices using brain signals [1]. The electroencephalogram (EEG) signals captured from the electrodes placed on the human skull [2] are used to classify and detect human emotions. Many researchers have conducted a lot of studies about the recognition of emotions through EEG signals. However, emotion recognition is still a challenging task for machines to recognize. With the advancements of machine learning tools, there is a growing need for automatic human emotion recognition [3]. Human emotional states are associated with the participant's perception and apprehension. Emotional awareness is of great importance in other areas such as cognitive sciences, computer science, psychology, life sciences, and artificial intelligence [4]. Due to the growing demands of mobile applications, emotion recognition is also becoming an essential part of providing emotional care to people. Human emotions can be recognized from speech, image, or video graphics, but the system for these types of recognition systems are much expensive. The task of recognizing emotions from brain signals is yet challenging due to the lack of temporal boundaries; also, different participants perceive the unusual amount of emotions in different ways [5]. Previously researchers have found new ways to understand and discover emotions through speech, images, videos, or BCI technology. In connection with non-invasive techniques, the brain–computer interface (BCI) provides a gateway for obtaining EEG signals related to emotional stimuli. The signals collected by the BCI help to better understand the emotional response, but it is still unclear how we can accurately and extensively decipher emotions [6,7]. BCI and biosignal acquisition techniques have grown considerably allowing real-time analysis of biosignals to quantify relevant insights such as the mental and emotional state of the user [8]. EEG signals are collected using 10–20 international systems for electrodes placement used to decode the information [9,10].

Despite many of BCI's techniques for capturing EEG signals for emotion recognition, however, there is still room for improvement in extracting spatial functions, including accuracy, interpretability, and utility of online applications. The most consistent approach proposed by [11] is to normalize the common spatial pattern (CSP). Normalized CSP is used to extract features to achieve good decoding accuracy. The purpose of normalizing the CSP is to reduce the effects of noise and artifacts that occur in the raw EEG signal. Ref. [12] also used CSP features for motor rehabilitation in virtual reality (VR) control.

Ref. [13] designed a filter for selecting features using CSP. As experienced by previous BCI researchers [13,14], the manual selection of the best filter for each subject is still tricky. Ref. [13] suggested optimal filtering to select filters for all subjects automatically. Choosing the best filter removes the access noise from the signal, resulting in a superior accuracy of the classification performance.

In the past, researchers used time-frequency distribution and spectral analysis methods, such as the discrete wavelet transformation method (DWT) [15] and the Fourier transformation method (FT) [5]. However, given the complex and subjective nature of the emotional state, it is challenging to introduce a general method for analyzing different emotional states. For non-stationary EEG signals, the frequency components change with time and frequency component information is not enough for the classification of human emotions. Therefore, to acquire the full knowledge of signal frequency in the spatial and temporal domain, another technique used is a continuous wavelet transform (CWT). Several techniques for automated classification of Human emotions from EEG signals are proposed using different machine learning techniques. Due to the non-linear behavior of EEG signals recognizing emotions for a different subject is a challenging task. Therefore, the selection of channels and features is crucial in recognizing human emotions accurately. Large feature dimensions have high calculation costs and a broad set of training data. Various techniques proposed by the researchers [14,16] decompose the signal into a series of features to deal with extensive data. Wavelet-based techniques such as empirical mode decomposition (EMD), discrete wavelet transform (DWT), and wavelet packet decomposition (WPD) are used to decompose signals. Ref. [16] used multi-scale PCA along with WPD to decompose and remove noise from the signal. When classifying EEG signals for motor rehabilitation, they achieve a classification accuracy of 92.8%. The wavelets based feature extraction technique is proposed in [17] for emotion classification on the SEED dataset. They used flexible analytical wavelet transform (FAWT) for channel decomposition. FAWT decomposition is a channel-specific technique that selects specific channels based on machine learning. The accuracy of 83.3% is achieved using SVM on the SEED dataset. Ref. [18] proposed a method of evolutionary feature selection in which frontal ad occipital channels were selected for classification and achieved an accuracy of 90%. Feature selection methods are effective in eliminating irrelevant features and maximizing the performance of the classifier to reduce high dimensions automatically. Of the many techniques that can be applied to prevent selection problems, the simplest is a filtering method based on ranking techniques. The filter method selects functions by scoring and ordering tasks based on their relevance and defining thresholds for filtering out irrelevant features. This method is intended to filter less relevant and noisy features from the set of features to improve classification performance. Filtering methods applied to the emotion classification system include Pearson correlation [19], correlation-based feature reduction [20,21], and canonical correlation analysis (CCA). However, the filter method has two possible disadvantages, assuming that all features are independent of each other [22]. The first disadvantage is that there is a risk that features are thrown away that are not relevant when viewed separately, but that may be relevant in combination with other features. The second disadvantage is the ability to choose individual related functions that can cause duplication. To overcome this issue, the evolutionary-based feature selection method proposed by [23] evaluated on DEAP and MAHNOB dataset. Differential evolutionary (DE) based features selection method was classified using a probabilistic neural network (PNN) and achieved a classification accuracy of 77.8% and 79.3% on MAHNOB and DEAP datasets, respectively. Text-based and speech-based emotions are also proposed by [3], which used the Mel frequency cepstral coefficients (MFCC) and reported the overall accuracy of 71.04% on IEMOCAP dataset. Multivariate empirical mode decomposition (MEMD) in [15] also used to decompose channels up to 18 out of 32. It decomposes the signal into amplitude and frequency modulated (AM–FM) oscillations known as intrinsic mode features (IMFs). Two-dimensional emotional states in arousal and valence dimensions are classified in [15] using SVM and ANN classifiers. In other studies, differential entropy is calculated on different frequency bands associated with EEG rhythms. Beta and gamma rhythms are the most effective for emotion recognition [24,25]. The authors discovered that 18 different linear and non-linear features were time-frequency domain features using spatial-temporal recurrent neural networks (STRNN). Spatial and temporal dependency model is designed to select features. Ref. [26] investigated the dynamic system features of EEG measurements and other aspects that are important for cross-target emotion recognition (e.g., databases for different EEG channels and emotion analysis). The recursive emotion feature elimination (RFE) method proposed by authors to eliminate repetitive features ad reduce the feature dimension. They achieve an average accuracy of the rating of 59.06% and 83.33% Using physiological signals (DEAP) and SJTU sentiment EEG data set (SEED) databases, respectively. Recently, many articles have been published in the field of emotion recognition [27–29] with the help of EEG signals. In comparison with traditional methods that use deep learning, there is a possibility of recognizing emotions in multi-channel EEG signals. However, two challenges remain. First, is deciding how to obtain relevant information from the time domain, the frequency domain, and the time–frequency characteristics to the emotional state EEG signal.

Ref. [30] states that the selection of specific EEG channels is essential for multi-channel EEG-based emotion recognition. Ref. [30] shows 32 channels and ten specific channels (F3, F4, Fp1, Fp2, P3, P4, T7, T8, O1, and O2 for emotion recognition). Experiments show better results when 10 channels are used compared to all 32 channels. Ref. [31] suggested the emotion recognition method based on the entropy of samples. Their experimental results corresponding to channels related to the emotional state are primarily from the frontal lobe areas, namely F3, CP5, FP2, FZ, and FC2. Few studies have analyzed the spatial domain characteristics of multi-channel EEG, and it may also contain important information. There were also some spatial features. The study is limited to the asymmetry between electrode pairs [32].

In the past, researches used to recognize human emotion from the selective number of channels, which may increase the computational speed but decreases the accuracy rate of understanding emotions. In this paper, we presented an accurate multi-modal EEG-based human recognition using a bag of deep features (BoDF), which ultimately reduces the size of features from all the channels which are used for detecting brain signals. First, a feature vector is obtained from a time-frequency representation of a preprocessed EEG emotions dataset using continuous wavelet transform (CWT). The features of all subjects from the AlexNet model are collected, which uses 2D images as input. BoDF method is proposed to arrange features as a single matrix and reduce the feature vector using k-means clustering. The reduced feature vector for all channels is then classified. Three states of participants as positive, negative, and neutral for SEED dataset and two states as arousal and valence are classified using all kernels of support vector machine (SVM) and k-nearest neighbor (k-NN) classifiers.

The rest of this article is organized as follows: Section 2 describes the dataset and electrode channel mapping. Section 3 introduces the emotion recognition framework and the proposed deep feature model using SVM and k-NN classifiers. In Section 4, experimental results are discussed by comparing other emotion recognition models. Section 5 concludes this work.

2. Materials

2.1. SEED Dataset

The emotion recognition task is carried on SEED [33] dataset developed by SJTU. The EEG dataset was collected by Prof. Bao Liang Lu at brain-like computing and machine Intelligence (BCMI) laboratory. This publicly available dataset contains multiple physiological signals with emotion evaluation, which makes it a well-formed multi-modal dataset for emotion recognition. In this data set, 15 participants were subjected to watch four minutes of six video clips. Each clip is well-edited that can be understood without explanation and exhibits the maximum emotional meanings. The detail of each clip used in acquiring EEG data are listed in Table 1.

Table 1. SEED dataset overview.

No.	Emotion Label	Film Clip Source
1	Negative	Tangshan Earthquake
2	Negative	1942
3	Positive	Lost in Thailand
4	Positive	Flirting scholar
5	Positive	Just another Pandora's Box
6	Neutral	World Heritage in Chine

The data was collected from 15 Chinese participants (7 males, 8 females), who were aged between 22–24. Each participant's data includes 15 trials, and, in each trial, the experiment performed twice. The RAW EEG signals were first segmented and then downsampled to 200 Hz. A bandpass frequency filter of 0–75 Hz was also applied to remove noise from the signal and EMG signals. The data was collected from 62 channels of each participant using 10–20 International standard [33].

2.2. DEAP Dataset

DEAP dataset is also publicly available as an on-line dataset for emotion classification [34]. The dataset was generated by a team of researchers at the Queen Mary University of London. The DEAP dataset contains multiple physiological signals for the evaluation of emotions. 32 channel EEG data were collected from 32 subjects. EEG signals were recorded by showing 40 preselected music videos, each with a duration of 60 seconds. The signals were then downsampled to 128 Hz and de-noised using bandpass and lowpass frequency filters. DEAP dataset can be classified using Russell's circumplex model [35] as shown in Table 2.

Table 2. DEAP dataset classes.

No.	Emotion Label	States
1	LAHV (Low Arousal High Valence)	Alert
2	HALV (High Arousal Low Valence)	Calm
3	HAHV (High Arousal High Valence)	Happy
4	LALV (Low Arousal Low Valence)	Sad

To visualize the scale by Russell's model DEAP dataset uses self-assessment manikins (SAMs) [36] with the real numbers 1–9. To make SAM as a continuous scale, each subject was asked to tick any number in between 0–10 from the list provided [34]. The scale based on self assessment rating were selected as 1–5 and 5–9 [37–41]. If the rating was greater than or equal to 5, the label was set to "high", and if it was less than 5, the label was set to "low". Thus, to create a total of four labels: high arousal low valence (HALV), low arousal high valence (LAHV), high arousal high valence (HAHV), and low arousal low valence (LALV). The emotional states describe by the given labels are shown in Table 2.

In this work, both the dataset were analyzed separately. In the SEED dataset, three classes are used as positive/negative/neutral. Positive class indicates the subject is in a happy mood; the negative class indicates the sad emotion of the subject while the neutral class tells the normal behavior of the subject. In the DEAP dataset, there are two types of labels named valance and arousal with a scale of 1–9. Labels 1–4 represent the valance, and 5–9 represents the arousal scale. Detailed information about the datasets can be found in [34].

2.3. Electrode to Channel Mapping

Both the datasets acquired EEG signals using the 10–20 International System [42]. The 10–20 international system describes the position of the electrodes to be placed on the human scalp for detecting EEG signals. The "10" and "20" indicate that the actual distance between adjacent electrodes is 10% or 20% of the distance between the front and back or right and left sides of the skull. Figure 1 shows the mapping of electrodes to channel number describes by both the datasets. The first 32 channels are used by the DEAP dataset, while all 62 channels are used by the SEED dataset to acquire EEG signals. In the figure, the first number in Figure 1 represents the channel number of the DEAP data set, and the second number is the channel number of the SEED data set (DEAP channel number, SEED channel number). Nasion in the figure represents the frontal part of the head, and channels are named using the 10–20 international system.

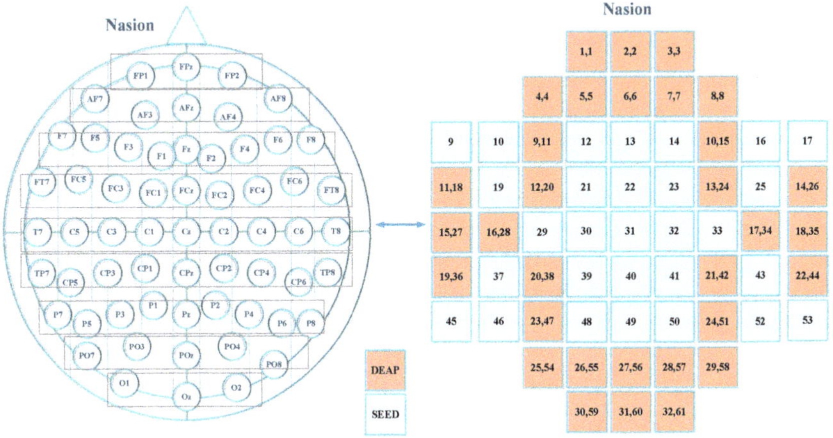

Figure 1. Electrode to channel mapping.

3. Methodology

This section explores step by step working of the proposed model and overall architecture. Figure 2 presents the framework of our work. First, the preprocessed EEG signals of both the datasets are used for time-frequency representation using the continuous wavelet transform filterbank. The feature extraction stage extracts the feature from all the channels of both the datasets.

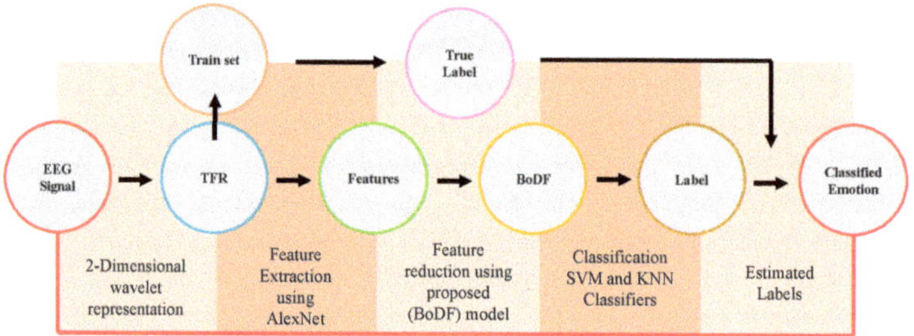

Figure 2. Framework.

3.1. Time Frequency Representation

The preprocessed EEG signal is used for time-frequency representation (TFR) of signals. Traditional emotion recognition models directly extract features from the EEG signals results in low accuracy and abortive results. In this paper, the one-dimensional EEG signal is presented as a two-dimensional model of EEG signal as time–frequency representation to analyze the signal in a better way and extract the desired features. This is achieved by using a continuous wavelet transform.

Contineous Wavelet Transform

Continuous wavelet transform (CWT) expresses the signals in terms of wavelet functions [43], which are localized in both time and frequency domain. CWT provides the complete representation of the 1D signal by letting the translation and scale parameters of the wavelets, varying continuously. Continuous EEG signal $x(t)$ for TFR $\chi_\omega(a, b)$ can be expressed as:

$$\chi_\omega(a,b) = \frac{1}{\sqrt{|a|}} \int_{-\infty}^{\infty} x(t)\tilde{\psi}\left(\frac{t-b}{a}\right) dt, \tag{1}$$

where, a is the scaling parameter that should be greater than 0 and $a \in R$; b is translation and $b \in R$; t is the time instant; $\tilde{\psi}(t)$ is called the mother wavelet which is also continuous for both time and frequency domain

Mother wavelet provides the scaling and translation of the original wavelet $x(t)$. The original signal is then formulated again using Equation (2).

$$x(t) = C_\psi^{-1} \int_{-\infty}^{\infty} \int_{-\infty}^{\infty} \chi_\omega(a,b)\frac{1}{\sqrt{|a|}}\tilde{\psi}\left(\frac{t-b}{a}\right) ab\frac{da}{a^2} \tag{2}$$

where C_ψ is called the wavelet admissible constant whose value satisfies between

$$0 < C_\psi < \infty$$

And expressed as:

$$C_\psi = \int_{-\infty}^{\infty} \frac{\tilde{\hat{\psi}}(\omega)\hat{\psi}(\omega)}{|\omega|}d\omega \tag{3}$$

An admissible wavelet must integrate to zero. For this $x(t)$ is recovered by using the second inverse wavelet shown in Equation (2).

$$x(t) = \frac{1}{2\pi \tilde{\psi}(1)} \int_{-\infty}^{\infty} \int_{-\infty}^{\infty} \frac{1}{a^2} \chi_\omega(a,b) exp\left(\frac{i(t-b)}{a}\right) dbda \qquad (4)$$

The wavelet for each time is defined as:

$$\psi(t) = \omega(t)exp(it) \qquad (5)$$

where, $\omega(t)$ is the window. Using the filterbank in CWT, we gather TFR of each channel. Figure 3 represents the TFR of different classes of SEED and DEAP. The TFR of different classes can be differentiated. Before extracting features, TFR images are first resized to 227×227 and separated for testing and training the model using 20x fold validation.

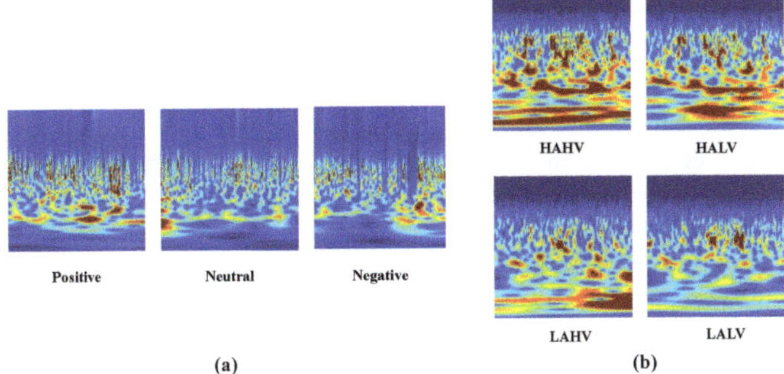

Figure 3. Continuous wavelet transform. (**a**) is the 3 classes TFR of SEED dataset. Positve, negative and netral classes of SEED dataset is clearly differentiable. (**b**) shows TFR of 4 classes in DEAP dataset.

3.2. Feature Extraction

In this paper, the time and frequency domain features were extracted using AlexNet. AlexNet performance was superior to earlier methods. AlexNet is a multilayer deep neural network (DNN) model consisting of 23 layers followed by the classification output layer [44]. AlexNet has an extensive network structure of 60 million parameters and 650,000 neurons [45].

AlexNet is a network of vast deep neural networks (DNN) with the combination of three layers: convolutional, max pooling, and fully connected layer. The TFR images were forwarded through the pre-trained AlexNet model. Figure 4 shows the overall architecture of AlexNet used in this article. AlexNet has five convolutional layers, followed by three fully connected layers. We did not use the last segment in our work, which is softmax and performs the classification. The main advantage of AlexNet is its rapid down sampling through stride convolutions and max-pooling layers. The AlexNet model has low computational complexity due to its lower layer count in comparison to other models. The two-dimensional TFR images are first resized to $227 \times 227 \times 3$ before extracting features. The images used are the 2D representation of an EEG signal. Therefore, different emotional states have different corresponding regions. Hence, it is more sensitive to differences in spatial information between different states. Each feature obtained from the last fully connected layer has 4096 attributes. The 'fc7' layer outputs the feature vector of each channel.

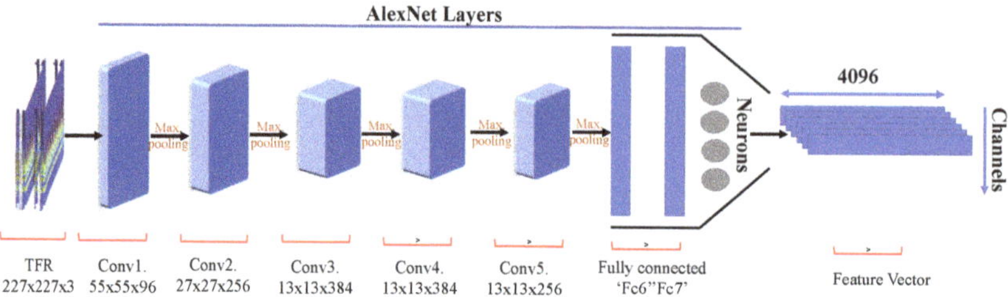

Figure 4. AlexNet layer architecture

In Figure 4 the TFR images are convolved using all conv layers and max-pooling is used to reduce the dimension of an image after each convolution. Rectified linear unit (ReLU) is used as the activation function and is expressed as

$$ReLU = max(0, x) \tag{6}$$

It selects the maximum value between 0 and the input x. The full connection layer 'fc7' gives the output with 4096 attributes. Figure 5 represents the tree of the output feature dimension vector of both the datasets. SEED dataset uses three classes; the experiment was performed on 15 subjects with five trials of an experiment for each category. The number of channels used for each participant was 62 [33]. After two-dimensional representation, the signal across each channel is represented as an image. Therefore, 62 images are created for each participant using TFR with the 4096 dimension vector of the 'fc7' layer. Hence, the total output feature vector obtained from the SEED dataset is 13,950×4096. Similarly, for the DEAP dataset, the nine class data were performed on 32 subjects twice generates a feature vector of 18,432×4096.

Figure 5. Feature tree.

3.3. Bag of Deep Features (BoDF)

Bag of features has been used to order less collection of features for image classification in computer vision [46]. We proposed the bag of deep features (BoDF) model for the reduction of features up to suitable value. Using this model, the feature size is significantly reduced to save time for training the dataset. A large number of features usually take long processing time and poor classification performance. In our article, we have been using all the channels to recognize emotions; therefore, feature with high dimension results in high computational cost for training the dataset. Previously researchers used only eight or 12 channels, which also results in lower accuracy by ignoring the rest of the channels. Hence, there is a trade-off between accuracy and the total number of channels used for classification [47].

In this paper, we utilize all the channels of SEED and DEAP dataset and achieve better results. Figure 6 illustrates the overall structure of the proposed method in which the total number of features with dimension 13,950×4096 for SEED dataset and 18,432 × 4096 for DEAP dataset are fed to BoDF model for feature reduction. The BoDF model consists of two steps in which features extracted from AlexNet are reduced using k-means clustering. k-mean clustering groups similar features and represent one feature vector. The first stage reduces the feature vector based on mean clustering. Features are further reduced in the second stage by calculating the histogram features. This model selects the number of features that are closer to the centroid for each class while ignoring others, which are not very useful features for emotions.

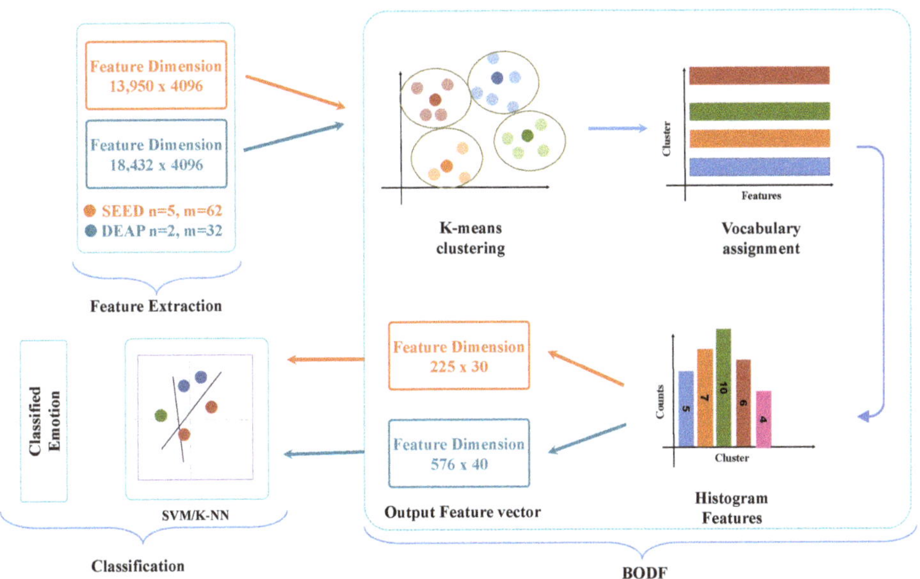

Figure 6. Bag of deep features (BoDF).

3.3.1. Stage 1: k-Mean Clustering

In step 1 of BoDF, the features are extracted from the AlexNet model of each class clustered using k-means clustering. Clustering is used to group similar features that belong to the same class. The k-means algorithm suits well for large datasets while other available clustering techniques suffer overfitting when dealing with large dataset [48]. The k-means algorithm is used to group redundant features by comparing the distance Initially, k is chosen randomly, and k is the number of clusters to group the features. At k = 10, all features are clustered correctly. That is, the features of the two data sets are clustered in 10 comparable groups. First, the distance of each characteristic value is calculated using Euclidean distance. The distance formula is used to compare each feature value and goes to the specific cluster with the shortest distance. For each cluster, the mean value is calculated by taking the average of all attribute values in a specific cluster. The average feature values are then re-evaluated until the average of the centers converges. For SEED, feature vectors of 13,950 × 4096 and 18,432 × 4096 are clustered in 30 × 4096 and 40 × 4096 in the SEED and DEAP datasets, respectively. The 10 × 4096 for each class feature vector is called a vocabulary.

Choosing 'k' clusters (centroids) is also a difficult task for achieving the best possible results. The universal answer for determining the value of k does not exist and starts with an arbitrary value instead. Hit and trial methods were used to select the value of k. This commonly used method is to measure the difference between the sum of the results of squared errors at different values of k. Sum of the squared errors for all values of k is calculated using Equation (7)

$$\text{Sum of the squared errors} = \sum_{i=1}^{k}(x_i - c_i)^2, \tag{7}$$

where, x_i is the input feature vector; c_i centroid of the ith cluster; k is the cluster number.

In Figure 7 graph shows that the value of the sum of the squared errors is quite small at k = 10.

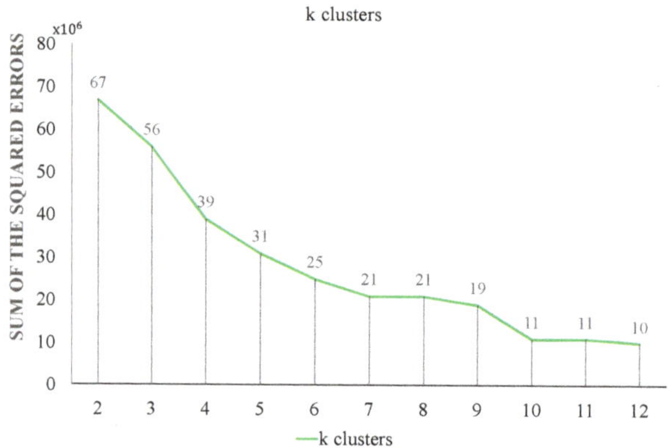

Figure 7. Selecting value of k.

3.3.2. Stage 2: Histogram Features

The histogram of vocabulary is calculated in the second step. The vocabulary compares each feature vector of 30 × 4096 and 40 × 4096 with the EEG data set of all channels. For the SEED dataset, we used features one by one at 30 × 4096 to compare with 62 channel features and calculate the frequency of occurrence. So we get a histogram feature of 225 × 30 with each attribute value being a histogram between 0 and 30. Similarly, we get a histogram feature vector of 576 × 40 for a DEAP dataset with four classes and 32 channels. This step has considerably reduced the feature size. The histogram features indicate the frequency of the best feature for each class.

3.4. Classification

As stated earlier, we use the AlexNet model only for extracting features; we did not use the last three layers of AlexNet models, which performs classification of the provided data. AlexNet is a pre-trained model and has the capability of recognizing up to 10,000 objects [44]. In our case, we have three and four classes for SEED and DEAP datasets, respectively. Therefore, our classifier selection is based on the classification output. For this work, we select a support vector machine (SVM) and k-nearest neighbor (k-NN) classifiers that outperform the classification performance based on four classes.

In this study, emotions of both datasets from all the kernels of SVM [49] and k-NN [50] are used for classification purposes. The principle of each classification is based on aggregation. In the SVM classifier [51], data is assigned to higher dimensions the optimum hyperspace for separation of space and data is built in this space. This classifier is a secondary programming problem [51]. In the training phase, SVM creates a model, maps the decision boundaries of each class, and specifies the hyperplane that separates the other classes. Increasing the hyperspace margin increases the distance between classes for better accuracy of the classification. SVM is used because it can effectively perform for non-linear classification [52]. In this study, BoDF feature vector is fed to SVM classifier to distinguish between happy, sad, and healthy emotional states for SEED dataset and two-dimensional Valence and Arousal states for the DEAP dataset. The SVM classifier is the kernel-based classifier. The kernel function is the mapping procedure performed on a training set to improve the correspondence with a linearly separable dataset. The purpose of mapping is to increase the dimensions of the dataset and execute it efficiently using kernel functions. Some of the most commonly used kernel functions are linear, quadratic, and polynomial Gaussian (RBF) kernels.

The second classifier we used for classification is the k-NN classifier. It is the instance-based classifier that classifies objects based on their closest training examples in feature space [52]. The object is classified as a majority of neighbors; that is, the object is assigned to the most common class of k-nearest neighbors, where k is a positive integer [53]. In the k-NN algorithm, the classification of the new test feature vector is determined by the class of the k-nearest neighbors [54]. Here in our work, we implement a k-NN algorithm using a Euclidean distance metric to find the nearest neighbor. The Euclidean distance between two points x and y is calculated using Equation (8). Where N is the number of features; 45 and 288 for SEED and DEAP dataset, respectively. The classification performance is determined on different values of k ranging from 1–10, and results showed that the best accuracy achieved at k = 2.

$$d(x,y) = \sum_{i=1}^{N} \sqrt{x_i^2 - y_i^2} \tag{8}$$

For accessing the classification performance using SVM and k-NN classifiers, we used $20\times$ cross-validation, which is useful for classification [55]. Classification performance achieved from both the classifiers is discussed with the results in the next section.

4. Results and Discussion

The accuracy performance of the proposed BoDF-method is evaluated on both the SEED and DEAP datasets. The features extracted from the proposed method are then classified using all the kernels of SVM and k-NN classifier. The results achieved from both the classifiers are also compared with the earlier studies and benchmark results of SEED and DEAP dataset. Feature vector extracted from AlexNet are reduced to 225×30 and 576×40 dimension vector for SEED and DEAP dataset, respectively in BODF-model. After classification, three emotional states positive/negative/neutral of all 15 participants of SEED dataset are classified.

Figure 8 shows the variation in classification accuracy for different kernels of the SVM classifier for two datasets. Linear and cubic classifiers show minimum variation in the classification accuracy as compared to other kernels. The maximum classification accuracy of 93.8% is achieved using the cubic kernel, as evident from the figure. Figure 8b shows the classification accuracy with the k-NN classifier. The result shows that the fine kernel of k-NN works better for CWT based features. Figure 8b shows the variation in classification accuracy for different kernels of the k-NN classifier. Cosine and coarse classifiers show minimum variation but minimum classification accuracy as compared to other kernels. The maximum classification accuracy of 91.4% is achieved using the fine and medium kernel, as evident from the figure.

The average classification accuracy on both datasets using SVM and K-NN classifiers are shown in Table 3.

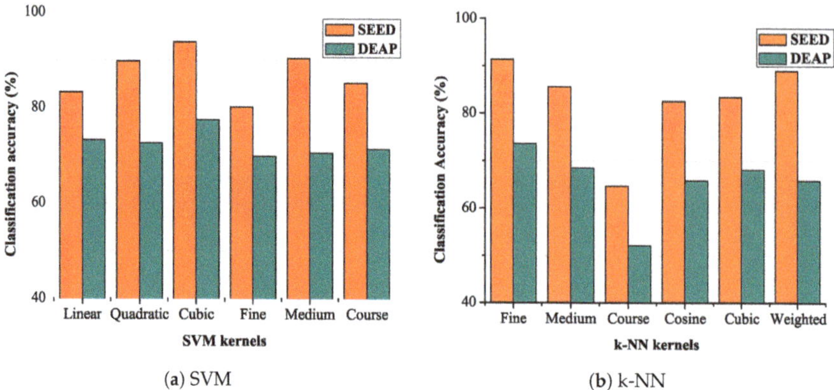

(a) SVM   (b) k-NN

Figure 8. Classification accuracy.

Table 3. Average classification accuracy (%).

Classifier	SEED		DEAP	
	k Value	Accuracy	k Value	Accuracy
SVM	10	93.8	10	77.4
	8	92.6	8	76.3
	6	92.4	6	76.1
	4	91.8	4	75.3
	2	90.9	2	75.1
k-NN	10	91.4	10	73.6
	8	90.2	8	71.1
	6	87.4	6	69.8
	4	87.1	4	68.5
	2	86.6	2	67.3

In reference to the comparison table, our proposed model has a higher classification accuracy than earlier studies. It can be seen that the classification accuracy achieved from the SVM classifier is more as compared emotion recognition using the k-NN classifier (see Table 4). BoDF model proposed in this paper recognize emotions without channel decomposition.

Table 4. Comparison on publicly available dataset with previous studies.

Ref.	Features	Dataset	Number of Channels	Classifier	Accuracy (%)
[3]	MOCAP	IMOCAP	62	CNN	71.04
[4]	MFM	DEAP	18	CapsNet	68.2
[17]	MFCC	SEED	12	SVM	83.5
				Random Forest	72.07
		DEAP	6	Random Forest	72.07
[15]	MEMD	DEAP	12	ANN	75
				k-NN	67
[24]	STRNN	SEED	62	CNN	89.5
[26]	RFE	SEED	18	SVM	90.4
		DEAP	12	SVM	60.5
[23]	DE	MAHNOB	18	PNN	77.8
		DEAP	32	PNN	79.3
Our work	DWT-BODF	SEED	62	SVM	93.8
				k-NN	91.4
		DEAP	32	SVM	77.4
				k-NN	73.6

The proposed BoDF model employs the vocabulary set of size 30 features, which produces a small feature dimension. However, the lowest computation complexity is possible by using the AlexNet model with the lowest number of layers. Feature reduction based on BoDF model also shows that using clustering and selecting mean features results in higher classification accuracy. Figure 9 clearly indicates that higher efficiency is achieved when trials or channels are reduced to its maximum value.

The lower feature dimensionality provides the superiority of the proposed architecture on other models in terms of its memory requirement. Secondly, the proposed model can more accurately classify the emotion based on the EEG signal in comparison to other models. However, the proposed model requires the vocabulary set before the feature extraction process. This will increase the computational complexity of the model. In the future, more robustness in recognizing emotion with low computational complexity can be achieved using different pooling algorithms.

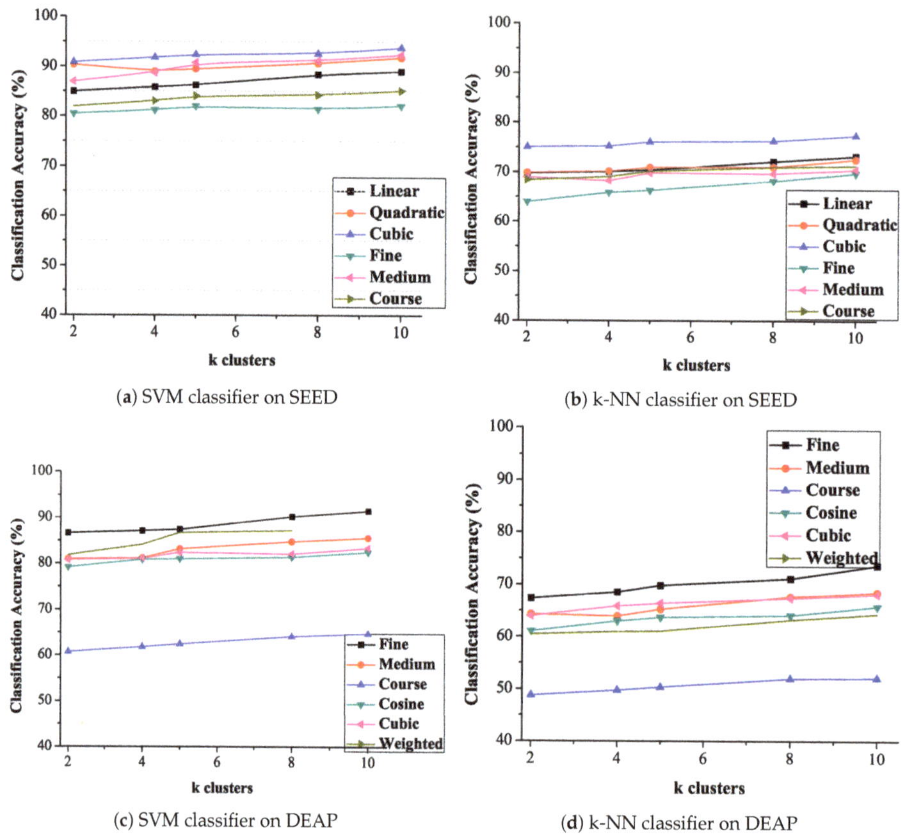

(**a**) SVM classifier on SEED

(**b**) k-NN classifier on SEED

(**c**) SVM classifier on DEAP

(**d**) k-NN classifier on DEAP

Figure 9. Classification accuracy at different value of k-clusters.

5. Conclusions

This emotion recognition model, based on bag of deep features (BoDF), is intended to achieve a higher classification accuracy for SEED and DEAP data sets. The results obtained with the proposed method without channel decomposition are higher than the benchmark results of both datasets. We have also found that the proposed method has better classification performance when the feature size is reduced to the optimum level. Reducing the number of channels also degrades the classification performance. All EEG signals are analyzed based on temporal and spatial characteristics. In the BoDF model, the k-means clustering algorithm is used to reduce the size of the feature without disturbing the overall accuracy of the proposed model. Three states for SEED and four emotional states for the DEAP dataset of each participant are recognized using multiple support vector machines (M-SVM) and k-NN classifiers. The results show that our method is superior to previous studies with the same data set. In the BoDF model without channel degradation, it has been demonstrated that the properties obtained from all channels are useful for classifying human emotions using EEG signals. The rating accuracy obtained with multiple SVMs is higher than the rating accuracy of the k-NN classifier. Our study has shown that it achieves higher classification accuracy compared to the emotion recognition method of the conventional channel.

Author Contributions: Formal analysis, F.; Funding acquisition, M.R. (MuhibUr Rahman) and S.B.; Investigation, S.S.M.; Methodology, F.; Project administration, Y.A.; Supervision, M.J.K. and M.R. (Muhammad Rizwan); Writing—original draft, M.A.A.

Funding: This research was funded by Higher Education Commission (HEC): Tdf/67/2017.

Acknowledgments: The publication of this article was funded by the Qatar National Library. Seyed Sajad Mirjavadi also appreciates the help from Fidar Project Qaem Company (FPQ).

Conflicts of Interest: The authors declare no conflict of interest.

References

1. Tarnowski, P.; Kołodziej, M.; Majkowski, A.; Rak, R.J. Combined analysis of GSR and EEG signals for emotion recognition. In Proceedings of the 2018 International Interdisciplinary PhD Workshop (IIPhDW), Swinoujście, Poland, 9–12 May 2018; pp. 137–141.
2. Faust, O.; Hagiwara, Y.; Hong, T.J.; Lih, O.S.; Acharya, U.R. Deep learning for healthcare applications based on physiological signals: A review. *Comput. Methods Progr. Biomed.* **2018**, *161*, 1–13. [CrossRef]
3. Tripathi, S.; Beigi, H. Multi-Modal Emotion recognition on IEMOCAP Dataset using Deep Learning. *arXiv* **2018**, arXiv:1804.05788.
4. Hao, C.; Liang, D.; Yongli, L.; Baoyun, L. Emotion Recognition from Multiband EEG Signals Using CapsNet. *Sensors* **2019**, *19*, 2212. [CrossRef]
5. Anagnostopoulos, C.-N.; Iliou, T.; Giannoukos, I. Features and classifiers for emotion recognition from speech: A survey from 2000 to 2011. *Artif. Intell. Rev.* **2015**, *43*, 155–177. [CrossRef]
6. Tzirakis, P.; Trigeorgis, G.; Nicolaou, M.A.; Schuller, B.W.; Zafeiriou, S. End-to-End Multimodal Emotion Recognition Using Deep Neural Networks. *IEEE J. Sel. Top. Signal Process.* **2017**, *11*, 1301–1309. [CrossRef]
7. Aloise, F.; Aricò, P.; Schettini, F.; Salinari, S.; Mattia, D.; Cincotti, F. Asynchronous gaze-independent event-related potential-based brain-computer interface. *Artif. Intell. Med.* **2013**, *59*, 61–69. [CrossRef] [PubMed]
8. Aric, P.; Borghini, G.; di Flumeri, G.; Sciaraffa, N.; Babiloni, F. Passive BCI beyond the lab: Current trends and future directions. *Physiol. Meas.* **2018**, *39*, 08TR02. [CrossRef]
9. Di Flumeri, G.; Aricò, P.; Borghini, G.; Sciaraffa, N.; Maglione, A.G.; Rossi, D.; Modica, E.; Trettel, A.; Babiloni, F.; Colosimo, A.; et al. EEG-based Approach-Withdrawal index for the pleasantness evaluation during taste experience in realistic settings. In Proceedings of the 2017 39th Annual International Conference of the IEEE Engineering in Medicine and Biology Society (EMBC), Seogwipo, Korea, 11–15 July 2017; pp. 3228–3231.
10. Aricò, P.; Borghini, G.; di Flumeri, G.; Bonelli, S.; Golfetti, A.; Graziani, I.; Pozzi, S.; Imbert, J.; Granger, G.; Benhacene, R.; et al. Human Factors and Neurophysiological Metrics in Air Traffic Control: A Critical Review. *IEEE Rev. Biomed. Eng.* **2017**, *10*, 250–263. [CrossRef]
11. Lotte, F.; Guan, C. Regularizing common spatial patterns to improve BCI designs: Unified theory and new algorithms. *IEEE Trans. Biomed. Eng.* **2010**, *58*, 355–362. [CrossRef]
12. Coogan, G.; He, B. Brain-computer interface control in a virtual reality environment and applications for the Internet of things. *IEEE Access* **2018**, *6*, 840–849. [CrossRef]
13. Song, L.; Epps, J. Classifying EEG for brain-computer interface: Learning optimal filters for dynamical system features. *Comput. Intell. Neurosci.* **2007**, *2007*, 57180. [CrossRef] [PubMed]
14. Sadiq, T.; Yu, X.; Yuan, Z.; Fan, Z.; Rehman, A.U.; Li, G.; Xiao, G. Motor imagery EEG signals classification based on mode amplitude and frequency components using empirical wavelet transform. *IEEE Access* **2019**, *7*, 678–692. [CrossRef]
15. Mert, A.; Akan, A. Emotion recognition from EEG signals by using multivariate empirical mode decomposition. *Pattern Anal. Appl.* **2018**, *21*, 81–89. [CrossRef]
16. Kevric, J.; Subasi, A. Comparison of signal decomposition methods in classification of EEG signals for motor-imagery BCI system. *Biomed. Signal Process. Control* **2017**, *31*, 398–406. [CrossRef]
17. Gupta, V.; Chopda, M.D.; Pachori, R.B. Cross-Subject Emotion Recognition Using Flexible Analytic Wavelet Transform From EEG Signals. *IEEE Sens. J.* **2019**, *19*, 2266–2274. [CrossRef]
18. Zangeneh Soroush, M.; Maghooli, K.; Setarehdan, S.K.; Motie Nasrabadi, A. Emotion classification through nonlinear EEG analysis using machine learning methods. *Int. Clin. Neurosci. J.* **2018**, *5*, 135–149. [CrossRef]
19. Kroupi, E.; Yazdani, A.; Ebrahimi, T. EEG correlates of different emotional states elicited during watching music videos. In *Affective Computing and Intelligent Interaction*; Springer: Berlin/Heidelberg, Germany, 2011; pp. 457–466.

20. Nie, D.; Wang, X.-W.; Shi, L.-C.; Lu, B.-L. EEG-based emotion recognition during watching movies. In Proceedings of the 2011 5th international IEEE/EMBS Conference on Neural Engineering (NER), Cancun, Mexico, 27 April–1 May 2011; pp. 667–670.

21. Schaaff, K.; Schultz, T. Towards emotion recognition from electroencephalographic signals. In Proceedings of the 2009 3rd International Conference on Affective Computing and Intelligent Interaction and Workshops, Amsterdam, The Netherlands, 10–12 September 2009; pp. 1–6.

22. Zhang, S.; Zhao, Z. Feature selection filtering methods for emotion recognition in Chinese speech signal. In Proceedings of the 2008 9th International Conference on Signal Processing, Beijing, China, 26–29 October 2008; pp. 1699–1702.

23. Nakisa, B.; Rastgoo, M.N.; Tjondronegoro, D.; Chandran, V. Evolutionary computation algorithms for feature selection of EEG-based emotion recognition using mobile sensors. *Expert Syst. Appl.* **2018**, *93*, 143–155. [CrossRef]

24. Zhang, T.; Zheng, W.; Cui, Z.; Zong, Y.; Li, Y. Spatial-Temporal Recurrent Neural Network for Emotion Recognition . *IEEE Trans. Cybern.* **2019**, *49*, 839–847. [CrossRef]

25. Duan, R.; Zhu, J.; Lu, B. Differential Entropy Feature for EEG-based Emotion Classification. In Proceedings of the 6th International IEEE EMBS Conference on Neural Engineering (NER), San Diego, CA, USA, 6–8 November 2013; pp. 81–84.

26. Li, X.; Song, D.; Zhang, P.; Zhang, Y.; Hou, Y.; Hu, B. Exploring EEG features in cross-subject emotion recognition . *Front. Neurosci.* **2018**, *12*, 162. [CrossRef]

27. Jirayucharoensak, S.; Panngum, S.; Israsena, P. EEG-based emotion recognition using deep learning network with principal component based covariate shift adaptation. *Sci. World J.* **2014**, *2014*, 627892. [CrossRef]

28. Liu, W.; Zheng, W.L.; Lu, B.L. Emotion Recognition Using Multimodal Deep Learning. In Proceedings of the 23rd International Conference onNeural Information Processing, Kyoto, Japan, 16–21 October 2016; pp. 521–529.

29. Yang, B.; Han, X.; Tang, J. Three class emotions recognition based on deep learning using staked autoencoder. In Proceedings of the 2017 10th International Congress on Image and Signal Processing, BioMedical Engineering and Informatics (CISP-BMEI), Shanghai, China, 14–16 October 2017.

30. Thammasan, N.; Moriyama, K.; Fukui, K.; Numao, M. Continuous music-emotion recognition based on electroencephalogram. *IEICE Trans. Inf. Syst.* **2016**, *99*, 1234–1241. [CrossRef]

31. Jie, X.; Cao, R.; Li, L. Emotion recognition based on the sample entropy of EEG. *Bio-Med. Mater. Eng.* **2014**, *24*, 1185.

32. Liu, Y.J.; Yu, M.; Zhao, G.; Song, J.; Ge, Y.; Shi, Y. Real-time movie-induced discrete emotion recognition from EEG signals. *IEEE Trans. Affect. Comput.* **2017**, *9*, 550–562. [CrossRef]

33. Zheng, W.-L.; Lu, B.-L. Investigating Critical Frequency Bands and Channels for EEG-based Emotion Recognition with Deep Neural Networks. *IEEE Trans. Auton. Ment. Dev. (IEEE TAMD)* **2015**, *7*, 162–175. [CrossRef]

34. Koelstra, S.; Muehl, C.; Soleymani, M.; Lee, J.-S.; Yazdani, A.; Ebrahimi, T.; Pun, T.; Nijholt, A.; Patras, I. DEAP: A Database for Emotion Analysis using Physiological Signals. *IEEE Trans. Affect. Comput.* **2012**, *3*, 18–31. [CrossRef]

35. García-Martínez, B.; Martinez-Rodrigo, A.; Alcaraz, R.; Fernández-Caballero, A. A Review on Nonlinear Methods Using Electroencephalographic Recordings for Emotion Recognition. *IEEE Trans. Affect. Comput.* **2019**, doi:10.1109/TAFFC.2018.2890636. [CrossRef]

36. Morris, J.D. SAM. The Self-Assessment Manikin an Efficient Cross-Cultural Measurement of Emotional Response. *Advert. Res.* **1995**, *35*, 63–68.

37. Li, X.; Zhang, P.; Song, D.W.; Yu, G.L.; Hou, Y.X.; Hu, B. EEG Based Emotion Identification Using Unsupervised Deep Feature Learning. In Proceedings of the SIGIR2015Workshop on Neuro-Physiological Methods in IR Research, Santiago, Chile, 13 August 2015.

38. Naser, D.S.; Saha, G. Classification of emotions induced by music videos and correlation with participants' rating. *Expert Syst. Appl.* **2014**, *41*, 6057–6065.

39. Naser, D.S.; Saha, G. Recognition of emotions induced by music videos using DT-CWPT. In Proceedings of the IEEE Indian Conference on Medical Informatics and Telemedicine (ICMIT), Kharagpur, India, 28–30 March 2013; pp. 53–57.

40. Chung, S.Y.; Yoon, H.J. An effective classification using Bayesian classifier and supervised learning. In Proceedings of the 2012 12th International Conference on Control, Automation and Systems, Je Ju Island, Korea, 17–21 October 2012; pp. 1768–1771.

41. Wang, D.; Shang, Y. Modeling Physiological Data with Deep Belief Networks. *Int. J. Inf. Educ. Technol.* **2013**, *3*, 505–511.

42. Homan, R.; Herman, J.P.; Purdy, P. Cerebral location of international 10–20 system electrode placement. *Electroencephalogr. Clin. Neurophysiol.* **1987**, *664*, 376–382. [CrossRef]

43. Vuong, A.K.; Zhang, J.; Gibson, R.L.; Sager, W.W. Application of the two-dimensional continuous wavelet transforms to imaging of the Shatsky Rise plateau using marine seismic data. *Geol. Soc. Am. Spec. Pap.* **2015**, *511*, 127–146.

44. Krizhevsky, A.; Sutskever, I.; Hinton, G.E. ImageNet Classification with Deep Convolutional Neural Networks. *Adv. Neural Inf. Process. Syst.* **2012**, *25*, 1097–1105. [CrossRef]

45. Xie, C.; Shao, Y.; Li, X.; He, Y. Detection of early blight and late blight diseases on tomato leaves using hyperspectral imaging. *Sci. Rep.* **2015**, *5*, 16564. [CrossRef] [PubMed]

46. O'Hara, S.; Draper, B.A. Introduction to the Bag of Features Paradigm for Image Classification and Retrieval. *arXiv* **2011** ;arXiv:1101.3354

47. Elazary, L.; Itti, L. Interesting objects are visually salient. *J. Vis.* **2008**, *8*, 1–15. [CrossRef] [PubMed]

48. Ester, M.; Kriegel, H.P.; Sander, J.; Xu, X. *A Density-Based Algorithm for Discovering Clusters in Large Spatial Databases with Noise*; AAAI Press: Menlo Park, CA, USA, 1996; pp. 226–231.

49. Tsai, C.-F.; Hsu, Y.-F.; Lin, C.-Y.; Lin, W.-Y. Intrusion detection by machine learning: A review. *Expert Syst. Appl.* **2009**, *36*, 11994–12000. [CrossRef]

50. Li, M.; Xu, H.P.; Liu, X.W.; Lu, S.F. Emotion recognition from multichannel EEG signals using K-nearest neighbor classification. *Technol. Health Care* **2018**, *29*, 509–519. [CrossRef]

51. Wichakam, I.; Vateekul, P. An evaluation of feature extraction in EEG-based emotion prediction with support vector machines. In Proceedings of the 11th international Joint Conference on Computer Science and Software Engineering, Chon Buri, Thailand, 14–16 May 2014.

52. Palaniappan, R.; Sundaraj, K.; Sundaraj, S. A comparative study of the SVM and K-nn machine learning algorithms for the diagnosis of respiratory pathologies using pulmonary acoustic signals. *BMC Bioinform.* **2014**, *27*, 223. [CrossRef]

53. Hmeidi, I.; Hawashin, B.; El-Qawasmeh, E. Performance of KNN and SVM classifiers on full word Arabic articles. *Adv. Eng. Inf.* **2008**, *22*, 106–111. [CrossRef]

54. Pan, F.; Wang, B.; Hu, X.; Perrizo, W. Comprehensive vertical sample-based KNN/LSVM classification for gene expression analysis. *J. Biomed. Inform.* **2004**, *37*, 240–248. [CrossRef]

55. Khandoker, A.H.; Lai DT, H.; Begg, R.K.; Palaniswami, M. Wavelet-based feature extraction for support vector machines for screening balance impairments in the elderly. *IEEE Trans. Neural Syst. Rehabil. Eng.* **2007**, *15*, 587–597. [CrossRef] [PubMed]

MDPI

St. Alban-Anlage 66

4052 Basel

Switzerland

Tel. +41 61 683 77 34

Fax +41 61 302 89 18

www.mdpi.com

Sensors Editorial Office

E-mail: sensors@mdpi.com

www.mdpi.com/journal/sensors